Local Approximation Techniques in Signal and Image Processing

Also Available from SPIE Press Exclusively Published by New Age International Pvt. Ltd.

Fourier-Transform Spectroscopy Instrumentation Engineering
Vidi Saptari
ISBN: 978-81-224-2823-0, Original Edition Price: $ 48.00, Indian Edition Price: Rs. 395.00

Tissue Optics : Light Scattering Methods and *Instruments for Medical Diagnosis,* 2nd Ed.
Valery Tuchin
ISBN: 978-81-224-2829-2, Original Edition Price: $ 99.00, Indian Edition Price: Rs. 995.00

Local Approximation Techniques in Signal and Image Processing
Vladimir Katkovnik, Karen Egiazarian, Jaakko Astola
ISBN: 978-81-224-2830-8, Original Edition Price: $ 99.00, Indian Edition Price: Rs. 695.00

Logic-based Nonlinear Image Processing
Stephen Marshall
ISBN: 978-81-224-2825-4, Original Edition Price: $ 51.00, Indian Edition Price: Rs. 395.00

Introduction to the Optical Transfer Function
Charles S. Williams, Orville A. Becklund
ISBN: 978-81-224-2831-5, Original Edition Price: $ 78.00, Indian Edition Price: Rs. 695.00

Digital and Analog Fiber Optic Communications for CATV and FTTx Applications
Avigdor Brillant
ISBN: 978-81-224-2832-2, Original Edition Price: $ 115.00, Indian Edition Price: Rs. 999.00

Infrared Fibers and Their Applications
James A. Harrington
ISBN: 978-81-224-2827-8, Original Edition Price: $ 78.00, Indian Edition Price: Rs. 495.00

Optoelectronics of Solar Cells
Greg P. Smestad
ISBN: 978-81-224-2826-1, Original Edition Price: $ 46.00, Indian Edition Price: Rs. 395.00

MOEMS: Micro-Opto-Electro-Mechanical Systems
Manouchehr E. Motamedi
ISBN: 978-81-224-2833-9, Original Edition Price: $ 113.00, Indian Edition Price: Rs. 795.00

Introduction to Image Stabilization
Scott W. Teare, Sergio R. Restaino
ISBN: 978-81-224-2822-3, Original Edition Price: $ 51.00, Indian Edition Price: Rs. 395.00

Fundamentals of Polarimetric Remote Sensing
John R. Schott
ISBN: 978-81-224-2834-6, Original Edition Price: $ 68.00, Indian Edition Price: Rs. 395.00

Fundamentals of Infrared Detector Materials
Michael A. Kinch
ISBN: 978-81-224-2824-7, Original Edition Price: $ 61.00, Indian Edition Price: Rs. 395.00

Local Approximation Techniques in Signal and Image Processing

Vladimir Katkovnik
Karen Egiazarian
Jaakko Astola

SPIE
PRESS
Bellingham, Washington USA

NEW AGE

First Indian Sub-Continent Edition : 2010
by New Age International (P) Ltd., Publishers

Branches:

- 36, Malikarjuna Temple Street, Opp. ICWA, Basavanagudi, **Bangalore**-560 004. Tel.: (080) 26677815, Telefax: 26615247, E-mail: bangalore@newagepublishers.com
- 26, Damodaran Street, T. Nagar, **Chennai**-600 017. Tel.: (044) 24353401, Telefax: 24351463, E-mail: chennai@newagepublishers.com
- CC-39/1016, Carrier Station Road, Ernakulam South, **Cochin**-682 016. Tel.: (0484) 2377004, Telefax: 4051303, E-mail: cochin@newagepublishers.com
- Hemsen Complex, Mohd. Shah Road, Paltan Bazar, Near Starline Hotel, **Guwahati**-781 008. Tel.: (0361) 2513881, Telefax: 2543669, E-mail: guwahati@newagepublishers.com
- No. 105, 1st Floor, Madhiray Kaveri Tower, 3-2-19, Azam Jahi Road, Nimboliadda, **Hyderabad**-500 027. Tel.: (040) 24652456, Telefax: 24652457, E-mail: hyderabad@newagepublishers.com
- RDB Chambers (Formerly Lotus Cinema)106A, 1st Floor, S.N. Banerjee Road, **Kolkata**-700 014. Tel.: (033) 22273773, Telefax: 22275247, E-mail: kolkata@newagepublishers.com
- 16-A, Jopling Road, **Lucknow**-226 001. Tel.: (0522) 2209578, 4045297, Telefax: 2204098, E-mail: lucknow@newagepublishers.com
- 142C, Victor House, Ground Floor, N.M. Joshi Marg, Lower Parel, **Mumbai**-400 013. Tel.: (022) 24927869, Telefax: 24915415, E-mail: mumbai@newagepublishers.com
- 22, Golden House, Daryaganj, **New Delhi**-110 002. Tel.: (011) 23262370, 23262368, Telefax: 43551305, E-mail: sales@newagepublishers.com

Printed in India at Anand Sons, Delhi.
C-09-11-4214

ISBN: 978-81-224-2830-8
Original Edition Price: $ 99.00
Indian Edition Price: Rs. 695.00

PUBLISHING FOR ONE WORLD
NEW AGE INTERNATIONAL (P) LIMITED, PUBLISHERS
4835/24, Ansari Road, Daryaganj, New Delhi-110002
Visit us at **www.newagepublishers.com**

Contents

Preface

This book deals with a wide class of novel and efficient adaptive signal processing techniques developed to restore signals from noisy and degraded observations. Imaging provides a clear example of this type of problem. Digital images produced by a variety of physical units, including still/video cameras, electron microscopes, radar, x-rays, ultrasound devices, etc., are used for various purposes, including entertainment, medical, business, industrial, military, civil, security, and scientific applications. In today's computer networks, the number of circulating digital images is enormous and continues to increase rapidly. Digital images can become distorted for many reasons, such as blurring and changes in illumination, motion, and processing, including transmission, quantization, smoothing, compression, and geometric transformations.

In many cases useful information and high quality must be extracted from the imaging. However, often raw signals are not directly suitable for this purpose and must be processed in some way. Such processing is called *signal reconstruction.*

This book is devoted to a recent and original approach to signal reconstruction based on combining two independent ideas: local polynomial approximation and the intersection of confidence interval rule. Local polynomial approximation is applied for linear and nonlinear estimation using a polynomial data fit in a sliding window. The window size, which is also interpreted as a scale, is one of the key parameters of this technique. The terms "window size," "bandwidth," and "scale" are synonymous here.

The local polynomial approximation is combined with an adaptive window size selection procedure called the intersection of confidence interval rule. The idea of this adaptation is as follows. The algorithm searches for the largest local area of the point of estimation where the local polynomial approximation assumptions fit well to the data. The estimates are calculated for a grid of window sizes and compared. The adaptive window size is defined as the largest of those windows in the set for which the estimate does not differ significantly from the estimators corresponding to the smaller window sizes.

In the approach considered in this book, the intersection of confidence interval rule defines the best scale for each point/pixel of the signal. In this way, we arrive at a varying adaptive-scale signal processing (imaging) with a pointwise adaptive-scale selection. The adaptive estimator is always nonlinear, even for a linear local polynomial approximation, because the nonlinearity of the method is incorporated in the intersection of the confidence interval rule itself. It is proved that these adaptive estimators allow one to get a near-optimal quality of the signal recovery.

The local polynomial approximation (LPA) plus the intersection of confidence interval (ICI) rule is an efficient tool for signal (image) processing, especially for denoising, interpolation, differentiation, and inverse problems.

The local approximation (nonparametric regression) techniques considered here signify that no prior limit is placed on the number of unknown parameters used to model the signal. Such theories of estimation are necessarily quite different from traditional (parametric regression) statistical models with a small number of parameters specified in advance.

Experiments demonstrate the state-of-art performance of the new algorithms, which on many occasions visually and quantitatively outperform the best existing methods.

Historically, the nonparametric regression technique is a predecessor of wavelets. It demonstrates a tremendous breakthrough in adaptive methods overall. This current development is almost unknown to the signal processing community, which is mainly dominated by the wavelet paradigm.

In this book we present basic concepts, methodology, and theory of this modern, spatially adaptive (nonparametric-regression-based) signal and image processing. The advanced performance of the algorithms is illustrated by applications to various image processing problems.

We present and interpret the nonparametric regression LPA as a filter design technique.

The ICI adaptive-scale selection concerns adaptive neighborhood filtering. The ICI as well as the nonparametric regression approach overall admit the derivation of the adaptive algorithms from the optimization of the estimation accuracy. In this way we develop a regular approach to the adaptive neighborhood filtering design that is universally applicable to various problems.

We focus on material that we believe is fundamental and has a scope of applications that is not limited to solutions of specialized problems. Therefore, the book is prepared for a broad readership. To avoid mathematical technicalities, we prefer to postpone the presentation of proofs until the Appendix.

In addition to explanations of the formal mathematics required for background, all methods are discussed in light of their implementations and applications.

Local polynomial approximation is an old idea. However, when combined with the new adaptation technique, it becomes a novel and powerful tool. The new approach and new algorithms are mainly illustrated for imaging applications. However, they are quite general in nature and can be applied to any scalar or multidimensional data. These new methods can be exploited as independent tools as well as jointly with conventional techniques.

The emphasis of the book is on multivariate problems. However, the introduction guides the reader through the most important ideas and techniques relating to univariate signals.

The mathematics used in the book is motivated by and restricted to the necessity of making the ideas and algorithms clear. Its complexity remains at a level well within the grasp of college seniors and first-year graduate students who have had courses in mathematical analysis, vectors, matrices, probability, and statistics. All statements are explained in the text, and the proofs presented in the Appendix are not necessary for understanding the algorithms. The more formal mathematical sections marked by asterisks may be omitted on a first reading.

The book is written as a self-contained text and includes review comments on the basic literature in each of the subjects treated.

Numerical experiments are necessary to fully understand the algorithms and methods developed in this book. We provide MATLAB[1] implementation of the basic algorithms presented as well as programs used in the illustrating examples. These programs are available at http://www.cs.tut.fi/~lasip. We support this software by updating the programs and including those programs that implement new algorithms and methods.

We hope that readers will find this book highly valuable in their research, data processing and analysis, and data interpretation.

The primary target readers are researchers, engineers, and graduate and post-graduate students working in signal and image processing fields who develop and study methods and algorithms. We trust that the book will be interesting to those coming to the area for the first time as well as to readers who are more familiar with the field. The book can be useful for teaching in electrical engineering departments. The secondary beneficiaries of the book are readers who develop and use algorithms for specific applications, in particular, specialists working in the areas of signal and image processing, communications, geoscience and remote sensing, medicine, and biology imaging.

The aim of this book is to advocate the modern adaptive nonparametric techniques that have proven efficient for image and signal processing and to provoke further research in both theory and applications.

Many people have been of great help to our work on this book. In particular, we thank Professors Alexandre Tsybakov (University 6, Paris) and Vladimir Spokoiny (Weierstrass-Institute, Berlin) for fruitful discussions.

We especially thank our students, Alessandro Foi and Dmitry Paliy, for dedicated participation in the book project. We note the significant contribution of Alessandro Foi for developing ideas and algorithms for directional anisotropic and recursive image processing.

Tampere, Finland
April 2006

V. Katkovnik
K. Egiazarian
J. Astola

---

[1]MATLAB® is a registered trademark of The Mathwork, Inc.

Notation and Abbreviations

Signals, variables, parameters

y	true signal
z	observed signal
h	scalar/vector nonnegative scale parameter of estimate
$\hat{y}_h$	estimate of y of scale h
$\hat{y}_h^{(r)}$	estimate of the derivative $y^{(r)}$ of scale h
x	argument variable (integer or continuous)
X_s	coordinates of observations
m	order (power) of LPA
r	order of the derivative
Δ	sampling interval
ϕ	vector of basis function in LPA
w	window function
v	kernel of blurring (convolution) operator
Γ	threshold in the ICI rule
d	dimensionality of signals
$r = (r_1 \ldots, r_d)$	multi-index order of the derivative
$m = (m_1, \ldots, m_d)$	multi-index order of LPA
$h = (h_1, \ldots, h_d)$	multi-index scale parameter
$\Delta = (\Delta_1, \ldots, \Delta_d)$	multi-index sampling interval

Symbols

g	kernel of smoothing operator (filter)
$g^{(r)}$	kernel of rth differentiating operator (filter)
∂_x^r	derivative operator: derivative on x of order r
$(g \circledast y)(x)$	convolution (discrete/continuous) of g and y
$arg\ \max_x f(x)$	the value of x that maximizes $f(x)$
$arg\ \min_x f(x)$	the value of x that minimizes $f(x)$
σ_z	standard deviation of random z
$var\{z\}$	variance of random z
$E\{z\}$	expected value of random z
$cov\{x, y\}$	covariance of random x and y

$(\cdot)^T$	transpose of vector or matrix
$(\cdot)^*$	complex conjugate of vector or matrix
$(\cdot)^H$	Hermitian transpose of vector or matrix
$\delta_{k,l}$	Kronecker delta
$\delta(x)$	Dirac delta
$O(x)$	a small value of order x
$o(x)$	a small value of the order smaller than x
$\triangleq$	equal by definition
$\simeq$	equal approximately
$\sim$	proportional
$\|\|x\|\|$	Euclidian norm of a vector x
$\mathbb{C}^\alpha$	continuous differentiable functions with bounded pth-order derivatives
$\mathbf{H}^{\alpha,\gamma}$	Hölder class of functions
$[z]_+$	$[z]_+ = z$ if $z \geq 0$ and $[z]_+ = 0$ if $z < 0$
$\langle a, b\rangle$	inner product of vectors a and b
$\lfloor\cdot\rfloor$	integer part of number
$\int_{R^d} = \int$	d-dimensional integration
$1(x) = 1$	indicator function
B^{-1}	inverse matrix
$B^\#$	pseudoinverse matrix
$\det B$	determinant of matrix

Sets

$\mathbb{R}$	scalar real numbers
$\mathbb{R}_+$	nonnegative scalar real numbers
$\mathbb{Z}$	integer numbers
$\mathbb{Z}_+$	nonnegative integer numbers
$\mathbb{R}^d$	space of real d-dimensional vectors
$\mathbb{R}_+^d$	space of non-negative real d-dimensional vectors
$\mathbb{Z}^d$	space of d-dimensional vectors with integer elements
$\mathbb{Z}_+^d$	space of d-dimensional vectors with integer nonnegative elements
L_2	space of squared integrable functions
$l_2(\mathbb{Z}^d)$	space of squared summable sequences defined in $\mathbb{Z}^d$-space

Abbreviations

IFT	integral Fourier transform
DFT	discrete Fourier transform, discrete in both spatial and frequency variables
IDFT	inverse discrete Fourier transform

DTFT	discrete time Fourier transform, discrete in time/spacial variables and continuous in frequency variables
LSM	least-square method
ML	maximum likelihood
MSE	mean-square error
LPA	local polynomial approximation
ICI	intersection of confidence intervals
LPA-ICI	adaptive local polynomial approximation algorithm equipped with ICI rule
LPA-ICI(AV)	LPA-ICI algorithm with the variance-adaptive ICI
LPA(AV)-ICI(AV)	LPA-ICI algorithm with the variance-adaptive ICI and with variance-adaptive LPA estimation
$\mathcal{N}(m, \sigma^2)$	Gaussian distribution with expectation m and variance σ^2
i.i.d.	independent identically distributed
p.d.f.	probability density function

Chapter 1

Introduction

The book is devoted to signal and image reconstruction based on two independent ideas: *local approximation* for the design of linear and nonlinear filters (estimators) and *adaptation* of these filters to unknown smoothness of the signal of interest.

As flexible universal tools we use *local polynomial approximation* (LPA) for *approximation* and *intersection of confidence intervals* (ICI) for *adaptation*.

The LPA is applied for linear filter design using a polynomial fit in a sliding window. The window as well as the order of the polynomial defines a desirable filter.

The *window size* is considered as a *varying adaptation parameter* of the filter.

The ICI is an *adaptation algorithm*. It searches for a largest local window size where LPA assumptions fit well to the observations. It is shown that the ICI adaptive LPA is efficient and allows for a nearly optimal quality of estimation.

To deal with anisotropic signals, in particular those that are typical in imaging, narrowed nonsymmetric directional windows are used. The corresponding directional (multidirectional) LPA filters equipped with the ICI adaptive window sizes demonstrate an advanced performance.

The LPA combined with the *maximum likelihood* and *quasi-likelihood* are used for the design of *nonlinear filters* (estimators). The ICI rule used for the window size selection of these filters defines nonlinear filters as adaptive to the unknown smoothness of the reconstructed signal.

LPA-ICI algorithms are powerful recent tools in signal and image processing. These algorithms demonstrate the state-of-art performance and can visually and quantitatively outperform some of the best existing methods.

The ICI rule is universal and can be applied with many existing linear and nonlinear filters (not only with the LPA-based filters) where the bias-variance trade-off is valid for selecting tunable parameters.

The LPA-ICI techniques define a wide class of pointwise spatially/time adaptive filters applicable in many scientific and engineering fields.

In this introduction we start with motivation, ideas, and constructive comments to review basic elements of the techniques presented in the book. These comments are presented for univariate signals while the book is addressed mainly to multidimensional problems.

1.1 Linear Local Approximation

1.1.1 Windowing

Let y be a scalar function of a scalar argument x. We wish to reconstruct this function using the observations

$$z_s = y(X_s) + \varepsilon_s, \qquad s = 1, \dots, n, \tag{1.1}$$

where the observation coordinates X_s are known and the ε_s are zero-mean random errors.

We make no global assumptions about y but assume that locally it can be well approximated with members of a simple class of parametric functions. Taylor's series can serve as a good approximation of a smooth function.

In a neighborhood of x this series gives for $y(X_s)$:

$$y(X_s) \simeq y(x) - y^{(1)}(x)(x - X_s) + y^{(2)}(x)(x - X_s)^2/2 + \cdots . \tag{1.2}$$

Since the function y and the derivatives $y^{(1)}$ and $y^{(2)}$ are unknown in Eq. (1.2), we may look to fit data $y(X_s)$ in the form

$$y(X_s) \simeq C_0 - C_1(x - X_s) + C_2(x - X_s)^2/2 + \cdots , \tag{1.3}$$

where the coefficients C_0, C_1, and C_2 in Eq. (1.3) are used as estimates for $y(x)$, $y^{(1)}(x)$, and $y^{(2)}(x)$, respectively.

Modeling in the form (1.3) allows this fruitful interpretation of the coefficients C_k as the derivatives of y.

Note that we use the first-order polynomial x in Eq. (1.3) with the minus in order to have C_1 as the estimator of $y^{(1)}(x)$. Of course, Eq. (1.3) can be written with positive signs for all polynomials,

$$y(X_s) \simeq C_0 + C_1(x - X_s) + C_2(x - X_s)^2/2 + \cdots , \tag{1.4}$$

but then the estimate of $y^{(1)}(x)$ is $-C_1$.

In order to preserve the meaning of the coefficients C_k as the estimators of the derivatives we always use the odd-order polynomials in the model (1.3) with the negative signs.

How do we formalize a local fit? The weighted least-square method gives the following criterion:

$$J(x, C) = \sum_s w(x - X_s)e_s^2, \; e_s = z_s - y(x, X_s),$$
$$y(x, X_s) = C_0 - C_1(x - X_s) + C_2(x - X_s)^2/2, \tag{1.5}$$

where w is a window function; $y(x, X_s)$ denotes a polynomial approximation; e_s is a residual between the observation z_s and the approximation $y(x, X_s)$; and C is a vector with components C_0, C_1, and C_2.

The window function w is "centered" on the point of interest x. The coordinate x positions the window, and $w(x - X_s)$ in Eq. (1.5) defines a set of neighborhood observations X_s and their weights in estimation for x.

Figure 1.1 illustrates this sliding window concept showing $w(x - X_s)$ for a few x and corresponding sets of observations in Eq. (1.5).

Minimizing Eq. (1.5) on C, one may find an estimate of the parameter C:

$$\widehat{C}(x) = \arg\min_{C} J(x, C). \tag{1.6}$$

It is clear that, in general, the coefficients C are different for each x. The notation $\widehat{C}(x)$ emphasizes that the estimate of C depends on x. The sliding window estimation makes the coefficients C in the model (1.3) varying on x.

The model (1.3) can be zero, first, second, or higher orders depending on the power of used polynomials. The parameter $\widehat{C}(x)$ immediately gives estimates of the function y and the derivatives $y^{(r)}$:

$$\hat{y}(x) = \widehat{C}_0, \quad \hat{y}^{(1)}(x) = \widehat{C}_1, \quad \hat{y}^{(2)}(x) = \widehat{C}_2. \tag{1.7}$$

1.1.2 Nonparametric estimation

The sliding window estimates (1.5)–(1.6) belong to the class of so-called nonparametric estimates. Let us clarify what we mean here by parametric and nonparametric estimation.

For a parametric estimation, the criterion (1.5) and the estimates (1.6) are transformed to

$$\begin{aligned} J(C) &= \sum_s w_s e_s^2, \qquad e_s = z_s - y(X_s), \\ y(X_s) &= C_0 - C_1 X_s + C_2 X_s^2/2, \end{aligned} \tag{1.8}$$

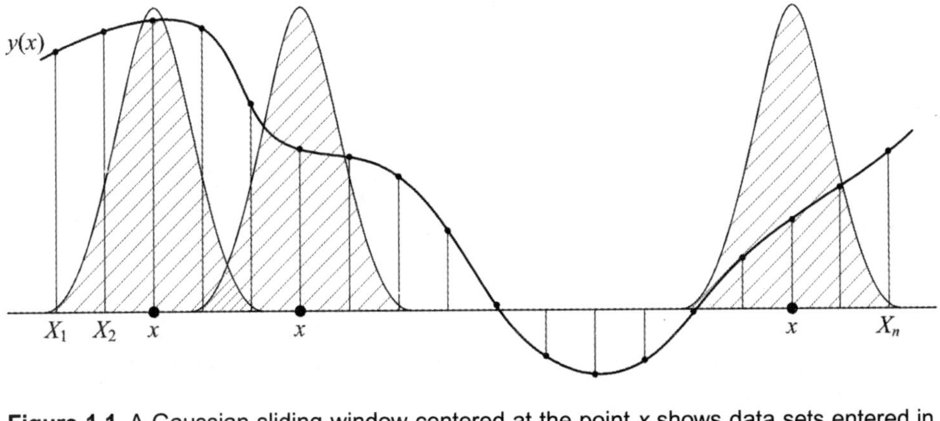

Figure 1.1 A Gaussian sliding window centered at the point x shows data sets entered in the estimate for the point x and the weights of the included observations.

where the weights w_s are fixed and

$$\widehat{C} = \arg\min_{C} J(C). \tag{1.9}$$

A zero-order parametric model with $y(x) = C_0$ gives an estimate $\hat{y}(x) = \widehat{C}_0$ constant for all x. The first-order parametric model $y(x) = C_0 - C_1 x$ gives the estimate $\hat{y}(x) = \widehat{C}_0 - \widehat{C}_1 x$ as linear on x, and the second-order parametric model $y(x) = C_0 - C_1 x + C_2 x^2/2$ obviously gives the estimate as quadratic on x.

The order of the polynomial model completely defines the type of estimation curve (constant, linear, or quadratic).

It is quite different for the sliding window estimates (1.5)–(1.6), where the order of the polynomial does not specify explicitly the type of estimation curve, which essentially depends on observed data.

The windowed estimate (1.6) cannot be represented as a polynomial function of the order given in the local model. In slightly different terms, the estimate curve (1.5)–(1.6) cannot be parameterized as a polynomial function of the given order, thus explaining the terms parametric and nonparametric estimation.

As an important technical point defining the difference between the parametric and nonparametric estimates, note that the weights w_s are constant in Eq. (1.8) and vary with x in Eq. (1.5).

The models of the signal are used in very different ways in the nonparametric and parametric estimations. In the parametric approach, the model $\hat{y}(x) = \widehat{C}_0 - \widehat{C}_1 x + \widehat{C}_2 x^2/2$ is global and used in this form for all x. Thus, if we need to calculate the derivatives, they are $\hat{y}^{(1)}(x) = -\widehat{C}_1 + \widehat{C}_2 x$ and $\hat{y}^{(2)}(x) = \widehat{C}_2$.

In the nonparametric approach we use the model $\hat{y}(x, X_s) = \widehat{C}_0(x) - \widehat{C}_1(x)(x - X_s) + \widehat{C}_2(x)(x - X_s)^2/2$ in Eq. (1.5) for $X_s = x$ only, giving

$$\hat{y}(x) = \hat{y}(x, X_s)|_{x=X_s} = \widehat{C}_0(x).$$

For the estimation of the derivatives we have

$$\hat{y}^{(1)}(x) = \partial_{X_s} \hat{y}(x, X_s)|_{x=X_s} = \widehat{C}_1(x)$$

and

$$\hat{y}^{(2)}(x) = \partial^2_{X_s} \hat{y}(x, X_s)|_{x=X_s} = \widehat{C}_2(x).$$

Thus, in the nonparametric estimation, the model is explicitly local and used in a local pointwise manner. The coefficients in Eqs. (1.5–1.6) are calculated for a fixed x and the model is used for this x only.

The dependence on x in the parametric case is defined by the basis functions (polynomials) of the model and the coefficients are fixed. In the nonparametric case, everything is completely different: The dependence on x is defined by the varying coefficients $\widehat{C}_0(x)$, $\widehat{C}_1(x)$, $\widehat{C}_2(x)$ while the basis functions are used with a fixed zero value of the argument.

This nonparametric nature of the windowed estimation follows from the technique used for each x in Eq. (1.5) when fit with different weights for different data.

The localization relaxes the sensitivity of the standard parametric models to the order of the model and transforms the polynomial fit into a much more flexible tool than can be expected from the standard parametric approach.

The estimate $\hat{y}(x) = \widehat{C}_0(x)$ (as well as the estimates of the derivatives) can be given in the form of the kernel operator (filter) linear on the following observations:

$$\hat{y}(x) = \sum_s g(x, X_s) z_s. \tag{1.10}$$

The following notation is used here:

$$\begin{aligned} g(x, X_s) &= w(x - X_s)\phi^T(x - X_s)\Phi^{-1}\phi(0), \\ \Phi &= \sum_s w(x - X_s)\phi(x - X_s)\phi^T(x - X_s), \end{aligned} \tag{1.11}$$

where $\phi(x) = (1, -x, x^2/2, \ldots)^T$ is a vector of polynomials of the LPA. This is a general formula where the length of the vector ϕ is equal to the order of the polynomials plus one.

1.1.3 Scale

The size of the window function w in Eq. (1.5) is an important and natural parameter of the local approximation. If the window is large and prescribes equal weights to all residuals, then the nonparametric estimate coincides with the parametric one. The difference between the parametric and nonparametric approaches disappears. In this case, the estimation curve is constant, linear, or quadratic, depending on the polynomial degree used in the model, and enables a maximal smoothing of random noise in observations.

If the window is of a minimal size, then the estimation curve goes exactly through the nearby observation points and there is no smoothing of the data. In this way the window size controls the level of smoothing enabled by the estimator.

The curves in Figs. 2.6 and 2.9 in Chapter 2 illustrate the role of window size (with notation h for this parameter). The zero-order LPA is used for filtering a finite number of observations in Fig. 2.6. For a large window size (large h) the estimate is a constant equal to the mean value of the observations. For a very small window size, the estimate is a piecewise constant going exactly through the observed values of the signal. For intermediate values of the window size, we obtain curves demonstrating different smoothness and different closeness to the observations.

The first-order LPA is used for filtering the same observations in Fig. 2.9. For the large window size, the estimate is a linear function corresponding to the linear least square fit of the signal. For a very small window size, the estimate is nearly piecewise linear going through the observation points. For intermediate values of the window size, we have curves demonstrating different smoothness and different closeness to the observations.

Mathematically, the window size is introduced by a special *scale parameter* h with the window function w in Eq. (1.5) replaced by

$$w_h(x) = w(x/h)/h, \qquad h > 0. \tag{1.12}$$

A smaller or larger h narrows or widens the window $w_h(x)$, respectively. The scalar factor $1/h$ in Eq. (1.12) serves for normalization of the window. If $\int w(x)dx = 1$, then $\int w(x/h)/h dx = 1$ for all $h > 0$.

With this scale parameter, the estimate (1.10) is transformed to

$$\hat{y}_h(x) = \sum_s g_h(x, X_s) z_s \tag{1.13}$$

with the notation

$$\begin{aligned} g_h(x, X_s) &= w_h(x - X_s)\phi_h^T(x - X_s)\Phi_h^{-1}\phi_h(0), \\ \Phi_h &= \sum_s w_h(x - X_s)\phi_h(x - X_s)\phi_h^T(x - X_s). \end{aligned} \tag{1.14}$$

We use the scale parameter h in the window function w_h and for the polynomials. In the above formulas,

$$\phi_h(x) = \phi(x/h).$$

Thus, the polynomials in $\phi_h(x)$ are also scaled by h.

The index h in $\hat{y}_h$ and g_h emphasizes the special role of the scale parameter h for the estimate as well as for the kernel of the estimator.

With this notation, the local polynomial model in Eq. (1.5) has the form

$$y_h(x, X_s) = \phi_h^T(x - X_s)C. \tag{1.15}$$

Under some appropriate assumptions, the discrete local estimator (1.13) becomes shift-invariant, and for the continuous data it can be represented in the form of the integral kernel transformation:

$$\hat{y}_h(x) = \frac{1}{h}\int g\left(\frac{x-u}{h}\right)z(u)du = \int g(u)z(x - hu)du, \tag{1.16}$$

where g is a kernel depending on the window w and the order of the polynomials,

$$\begin{aligned} g(x) &= w_h(x)\phi^T(x)\Phi^{-1}\phi(0), \\ \Phi &= \int w(x)\phi(x)\phi^T(x)dx. \end{aligned} \tag{1.17}$$

The integral form (1.16) makes obvious an interpretation of the window size h as the *scale parameter* of the estimate and of the LPA.

In the first formula in Eq. (1.16), h defines the scale of the kernel $g((x - u)/h)/h$ convolved with the signal $z(u)$ independent on h, while in the second formula h defines a scale of the signal $z(x - hu)$ convolved with the kernel $g(u)$ independent on h.

The following moment conditions hold for the LPA kernels of the order m:

$$\int g(x)dx = 1, \qquad \text{for } m = 0,$$

$$\int g(x)dx = 1, \quad \int xg(x)dx = 0, \qquad \text{for } m = 1, \tag{1.18}$$

$$\int g(x)dx = 1, \quad \int xg(x)dx = 0, \quad \int x^2 g(x)dx = 0, \qquad \text{for } m = 2.$$

Using the Taylor series,

$$y(x - hu) = y(x) - hy^{(1)}(x)u + h^2 y^{(2)}(x)u^2/2 - h^3 y^{(3)}(x)u^3/6,$$

and inserting it into the second formula in Eq. (1.16) instead of $z(x - hu)$, we find

$$\hat{y}_h(x) = y(x) - hy^{(1)}(x) \int g(u)udu + 0(h^2), \qquad \text{for } m = 0,$$

$$\hat{y}_h(x) = y(x) + h^2 y^{(2)}(x) \int g(u)u^2 du/2 + 0(h^3), \qquad \text{for } m = 1, \tag{1.19}$$

$$\hat{y}_h(x) = y(x) - h^3 y^{(3)}(x) \int g(u)u^3 du/6 + 0(h^4), \qquad \text{for } m = 2.$$

All of these estimates give the true value of $y(x)$ as the main term of the series. The error of the estimation is proportional to h for $m = 0$, to h^2 for $m = 1$, and to h^3 for $m = 2$. For small h this shows the order of the error, which depends on the order of the polynomial approximation. In any case, if $h \to 0$, the estimate becomes accurate as $\hat{y}_h(x) \to y(x)$.

The parameter h controls the smoothing of y. Larger and smaller h result in stronger and weaker smoothing of y, respectively.

In the frequency domain, the Eq. (1.16) becomes

$$\widehat{Y}_h(\lambda) = G(\lambda h)Y(\lambda), \tag{1.20}$$

where the capital letters denote the integral Fourier transform of the corresponding variable; in particular, we have for $y(x)$

$$Y(\lambda) = \mathcal{F}\{y(x)\} = \int y(x) \exp(-i\lambda x)dx.$$

The filter $G(\lambda) = \mathcal{F}\{g(x)\}$ obtained by the local polynomial approximation is a typical low-pass filter such that $G(0) = 1$ and $|G(\lambda)| \to 0$ as $|\lambda| \to \infty$.

The parameter h scales/dilates the frequency characteristic of the filter (Fig. 1.2), and defines a bandpass width of the filter, which is narrowed with $h \to \infty$ and widened with $h \to 0$.

Various terms are used in publications for the parameter h, such as window size, scale, bandwidth, or dilation. In this book the term *scale* is mainly used.

The integral form of the local approximation operator clearly reveals the scaling role of the parameter h.

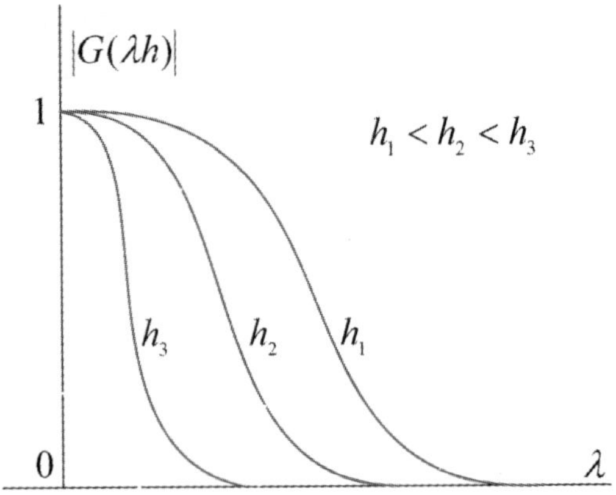

Figure 1.2 Amplitude frequency characteristics of the continuous kernels for different scales $h_1 < h_2 < h_3$.

1.1.4 Ideal scale

If the observation model includes the random noise in Eq. (1.1), the estimation error

$$e_y(x, h) = y(x) - \hat{y}_h(x) = m_{\hat{y}_h}(x, h) + e_y^0(x, h)$$

consists of the systematic $m_{\hat{y}_h}(x, h)$ and random $e_y^0(x, h)$ components

$$m_{\hat{y}_h}(x, h) = y(x) - E\{\hat{y}_h(x)\},$$
$$e_y^0(x, h) = e_y(x, h) - m_{\hat{y}_h}(x, h).$$

In the asymptotic analysis the systematic error is evaluated according to Eq. (1.16) as

$$m_{\hat{y}_h}(x, h) = y(x) - \int g(u)y(x - hu)du.$$

Assuming that the noise is independent with an identical distribution for all observations, the variance of the random components is defined as

$$\sigma^2_{\hat{y}_h}(x, h) = \sigma^2 \frac{\Delta}{h} B_g, \qquad \text{where } B_g = \int g^2(u)du, \tag{1.21}$$

Δ is a small sampling period, and σ^2 is the variance of the noise in Eq. (1.1).

According to Eq. (1.19), we may conclude that the main term of the systematic error is

$$|m_{\hat{y}_h}(x, h)| = h^{m+1}|y^{(m+1)}(x)|A_g, \tag{1.22}$$

where m is the order of the polynomial approximation and $A_g = |\int g(u)u^{m+1}du|/(m+1)!$.

The estimation variance (1.21) is a monotonically decreasing function of h while the bias (1.22) is the monotonically increasing function of h. This means that there is a bias-variance balance giving the optimal value h^* to the scale h.

The deviation from this optimal value means that the bias is increasing and the variance is decreasing if $h > h^*$, and the bias is decreasing and the variance is increasing if $h < h^*$.

This optimal h can be found by minimizing the mean-square error:

$$l_{\hat{y}_h}(x, h) \triangleq E\{e_y^2(x, h)\} = m_{\hat{y}_h}^2(x, h) + \sigma_{\hat{y}_h}^2(x, h).$$

Inserting Eqs. (1.21) and (1.22) into this formula, we have the mean-square risk function,

$$l_{\hat{y}_h}(x, h) = \left[h^{m+1} A_g y^{(m+1)}(x)\right]^2 + \sigma^2 \frac{\Delta}{h} B_g,$$

convex on h.

The equation $\partial l_{\hat{y}_h}(x, h)/\partial h = 0$ gives, for the optimal scale,

$$h^* = \left(\frac{\sigma^2 B_g \Delta}{2(m+1)[A_g y^{(m+1)}(x)]^2}\right)^{1/(2m+3)}. \tag{1.23}$$

This optimal solution is called the *ideal* or *oracle* scale and depends on the derivative $y^{(m+1)}(x)$, which is unknown. The ideal scale cannot be used in practice without estimation of this derivative. Nevertheless, it gives important information about the behavior of the optimal scales and estimates. In particular, it shows that the optimal scale at point x should be large or small for small or large $|y^{(m+1)}(x)|$, respectively.

Another conclusion is that the varying scale is crucial in the areas where the behavior of y is essentially varying on x because it is identified by the derivative $y^{(m+1)}(x)$. In the areas where this derivative is more or less invariant, the invariant scale can be exploited.

Figure 1.3 illustrates the idea of the bias-variance trade-off. Note that the ideal scale h^* is always smaller than the h corresponding to the equality

$$|m_{\hat{y}_h}(x, h)| = \sigma_{\hat{y}_h}(h).$$

Thus, in the ideal balance of the bias versus the variance, the bias is smaller than the standard deviation of error.

1.1.5 Adaptive varying scale

Theoretical analysis and experiments show that the local approximation estimates can be efficient only with a correct selection of the scale h. It can be varying or invariant, but it should be properly selected. The scale selection is a necessary element of a successful local approximation estimation. Without automatic scale selection, the local approximation, even though it is a reasonable idea, does not work.

In signal processing and statistics, scale selection as it is defined above is the subject of many publications exploiting different ideas and techniques.

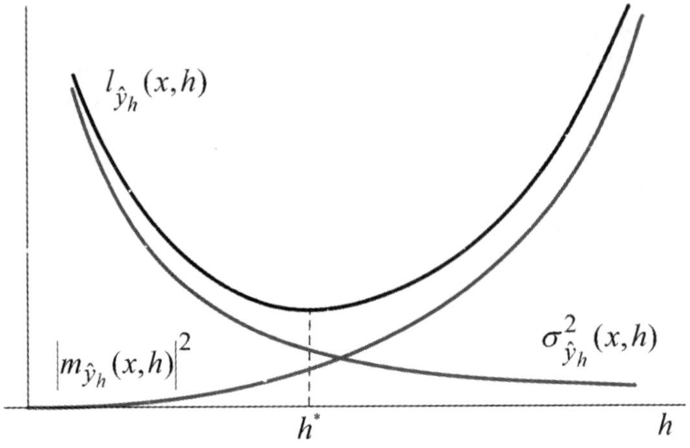

Figure 1.3 The squared bias $|m_{\hat{y}_h}(x, h)|^2$ and the variance $\sigma^2_{\hat{y}_h}(x, h)$ illustrating the bias-variance trade-off. The "ideal" balance is achieved at $h = h^*$ minimizing the mean-square loss function $l_{\hat{y}_h}(x, h)$.

Recently, a novel class of algorithms known under the generic name *Lepski's approach* has been introduced in statistics and shown to be efficient in theory and practice. These algorithms are proposed for the pointwise varying-scale adaptive nonparametric estimation. In this book we develop as the basic adaptation algorithm one version of this approach, called the intersection of confidence interval (ICI) rule.

The algorithm is simple in implementation. Let H be a set of the ordered scale values $H = \{h_1 < h_2 < \cdots < h_J\}$. The estimates $\hat{y}_h(x)$ are calculated for $h \in H$ and compared. A special statistic is exploited to identify a scale close to the ideal one. This statistic needs only the estimates $\hat{y}_h(x)$ and the variances of these estimates, $\sigma^2_{\hat{y}_h}(x, h)$, both calculated for $h \in H$.

A layout of the LPA-ICI adaptive algorithm is shown in Fig. 1.4. The LPA designed filters with the kernels $g_{h_1}, \ldots, g_{h_J}$ are applied to the input signal z. The outputs of these filters $\hat{y}_{h_1}, \ldots, \hat{y}_{h_J}$ are inputs of the ICI algorithm, which defines the adaptive-scale estimates $\hat{y}_{h^+(x)}$ in a pointwise manner for each x.

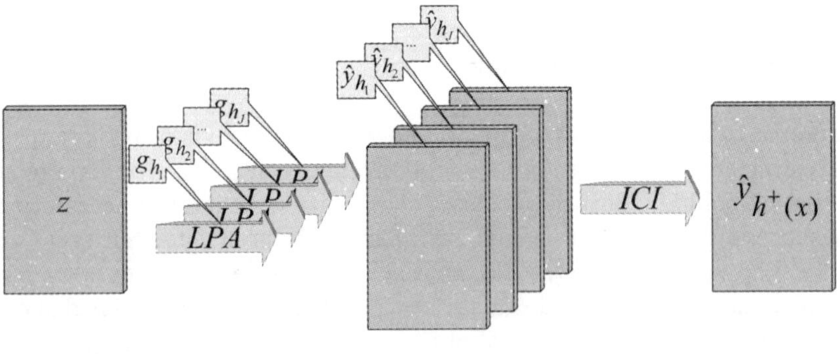

Figure 1.4 A layout of the adaptive-scale LPA-ICI algorithm.

1.2 Anisotropy

1.2.1 Univariate case

The local polynomial approximation motivated by the Taylor series in Eq. (1.2) is based on the assumption that the function y is smooth and differentiable. This means that (at least locally) the behavior of the function is more or less the same in all directions around the point of interest x. This sort of isotropy assumes, in particular, that the following three definitions of the derivative are equivalent:

$$\begin{aligned}\partial y/\partial x &= \lim_{h\to 0}\frac{y(x+h)-y(x)}{h}\\ &= \lim_{h\to 0}\frac{y(x)-y(x-h)}{h}\\ &= \lim_{h\to 0}\frac{y(x+h/2)-y(x-h/2)}{2h}.\end{aligned}$$

Thus, for derivative estimation we can use the right, left, or symmetric finite-differences and the results are approximately equal. This simple hypothesis, which is used by default in most signal processing algorithms, does not hold at singular points or so-called change points, where the behavior of y changes rapidly.

Directional local approximation is developed to deal with the anisotropy of y. For univariate y, the directionality of estimation assumes the use of right and left windows. We say that the window is right or left if $w(x) = 0$ for $x < 0$ or $x > 0$, respectively. This allows us to get good approximations near the jump points of curves.

Let us look at a simple example. The signal is piecewise constant having zero value for $x < a$ and a value of 1 for $x \geq a$ (Fig. 1.5a). The observation noise is an independent standard Gaussian. Let the estimator be a sample mean (symmetric window estimate) with the estimate for the point x^0 in the form

$$\hat{y}_h(x^0) = \frac{1}{\#}\sum_{|X_s - x^0|\leq h} z_s,$$

where # is a number of observations used in the estimate.

Figure 1.5(a) shows the estimation intervals for different h. It is obvious that for $x^0 < a$ the best ideal estimator should have $h = a - x^0$, because in this case the sample mean estimate is unbiased (the signal has a constant value on this interval) with the minimum variance while the estimation interval is maximal. Figure 1.5(b) illustrates the nonsymmetric sample means when the right/left means (right/left windows) are used with varying window sizes:

$$\hat{y}_h^{right}(x^0+0) = \frac{1}{\#}\sum_{x^0\leq X_s\leq x^0+h} z_s,$$

$$\hat{y}_h^{left}(x^0-0) = \frac{1}{\#}\sum_{x^0-h\leq X_s< x^0} z_s.$$

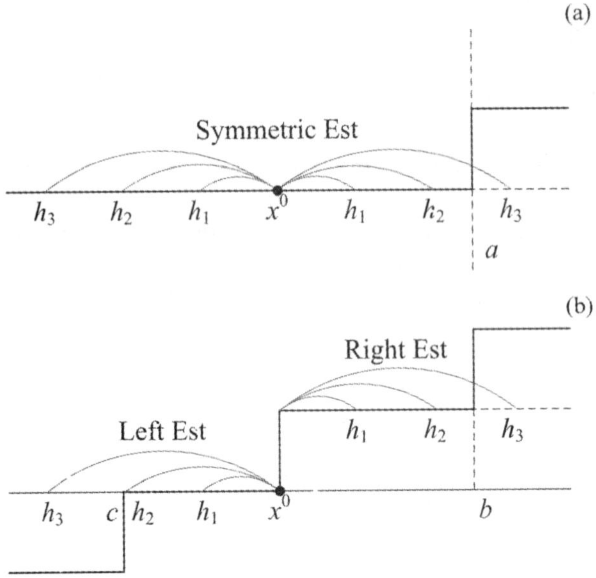

Figure 1.5 Symmetric, right/left estimation, and scale selection.

In order to emphasize their differences with respect to the symmetric window estimate, we consider the estimation at the jump point $x = x^0$. The plus and minus zero in the argument of the estimate mean the plus and minus of some small values, respectively. This is used to make the function well-defined at the point of estimation. The argument $x^0 + 0$ says that the estimation point is from the right-hand side of the jump point, while the argument value $x^0 - 0$ corresponds to estimation for the point from the left-hand side of the jump.

It is clear that the right window estimator is well-suited for estimation at $x = x^0 + 0$, because it can use a large window size estimate with a maximum size $b - x^0$.

Both the symmetric and left window estimates are not able to work well in this case because they can use only small windows (which can be very noisy) or the larger window estimates (with a definitely large bias error).

In a similar way the left window estimate works well for $x = x^0 - 0$ with the maximum window size $x^0 - c$, while the right and symmetric estimators are not able to give good results because they can use only very small windows.

1.2.2 Multivariate case

Points, lines, edges, and textures are specific elements in images. They are locally defined by position, orientation, and size. Often being of small size, these specific features encode a great proportion of information contained in images. The corresponding image intensity is a typical example of an anisotropic function y and points, lines, and edges are singular change points and change surfaces in images.

What makes the problem more specific is that reconstruction of y near the change points is of prime interest because these points present the most valuable imaging information.

For a multivariate y for directional estimation we consider a ball of radius h defining a spherical neighborhood of x,

$$\mathcal{B}_h = \{u : ||x - u|| \leq h\}. \tag{1.24}$$

Next we introduce a sectorial partition of this ball with K nonoverlapping sectors S_θ^h having x as a common vertex. Here θ gives a direction of the sector and h is a length of the sector.

One can exploit the LPA to design the estimate and the estimate kernel for x using the sectors as the estimate support. In this way, we obtain K estimates $\hat{y}_{h,\theta_i}(x)$, $i = 1, \ldots, K$, with K nonoverlapping supports $S_{\theta_i}^h$ covering the ball $\mathcal{B}_h$.

The ICI is used to find the adaptive varying scale for each sector independently. Thus, we arrive at K adaptive-scale estimators $\hat{y}_{h,\theta_i}(x)$, which have different adaptive scales/sizes for each sectorial support $S_{\theta_i}^{h^+(\theta_i)}$ (Fig. 1.6a). The union of the supports of these sectors is not a ball $\mathcal{B}_h$ of radius h, but a star-shaped set that can have a sophisticated configuration (Fig. 1.6b). Being defined by the ICI rule, this star-shaped set is a shape-adaptive neighborhood of the x used for estimation. In this way we obtain an adaptive pointwise segmentation of the image.

Figure 1.7 illustrates the implementation of the concept of the adaptive star-shaped neighborhoods for *Lena* test images for a few pixels in the image. These adaptive sets cover nearly perfect neighborhoods with more or less homogeneous behavior of the intensity function where the smooth polynomial approximation works well.

The adaptive scaling and segmentation can be used for many purposes—in particular, in edge detection, motion estimation, pattern recognition, and so on.

In denoising, the final estimate is calculated as a weighted mean of the sectorial estimates. Thus, we obtain a highly efficient anisotropic adaptive-scale estimator with the angular resolution depending on a number of sectors K.

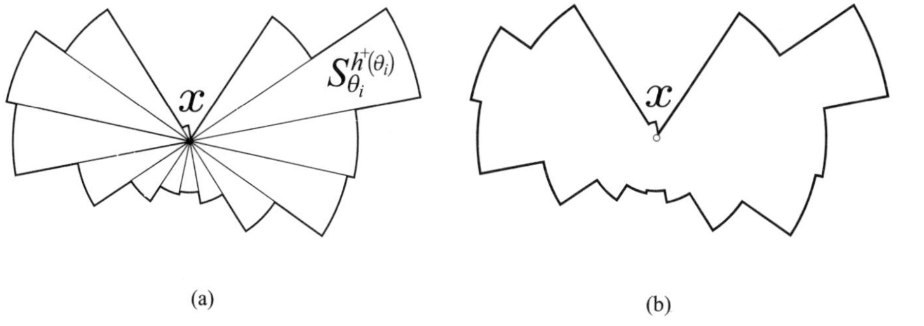

Figure 1.6 A star-shaped adaptive neighborhood used in anisotropic directional estimation.

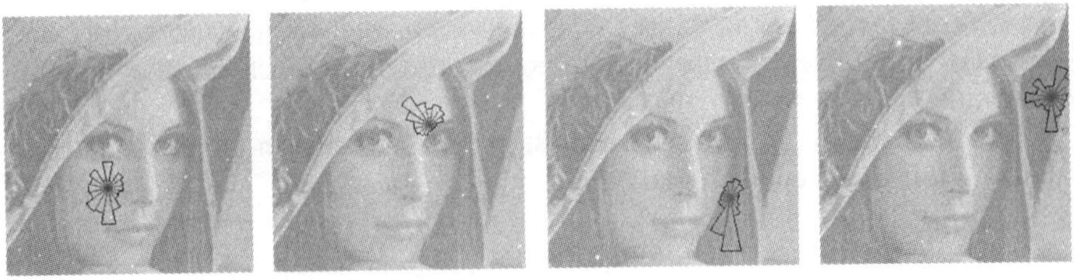

Figure 1.7 An LPA-ICI pointwise adaptive segmentation of the *Lena* test image.

1.3 Nonlinear Local Approximation

1.3.1 Likelihood and quasi-likelihood

Our previous discussion of the LPA concentrated on the object function y, mainly ignoring the noise in the observation model (1.1). However, the noise in Eq. (1.1) is an equally important element of the observation model that becomes an extremely sensitive issue when noise characteristics (e.g., the variance) are signal-dependent.

To deal with this sort of complex scenario, local versions of the likelihood and quasi-likelihood methods are developed. These methods allow us to exploit explicit probabilistic characteristics of observations to obtain optimal noise reduction procedures.

In statistics, the observation model usually means a joint probability density $f(z, y)$ of the observation pairs $\{z_s, X_s\}_{s=1,\dots,n}$. In this probability density the argument z corresponds to observations and y is a parameter of interest, usually $y = E\{z\}$. This modeling is quite different from the more restrictive standard additive model (1.1).

The maximum likelihood (ML) is a universal statistical technique that is used if the distribution of observations is known. The optimality of the ML estimate means that this method is able to extract full information about the estimated parameters available in the given observations.

The local ML is a nonparametric counterpart of the widely used parametric ML. It extends the scope of the parametric ML to a broader class of functions y. The local likelihood does not assume a global parametric modeling of y but fits this model locally within the sliding window.

The local likelihood and local log-likelihood use the window function w for localization in a neighborhood of the point of interest x and replace the global parametric model with a local one.

Using the polynomial linear model (1.15), the local log-likelihood can be presented in the form

$$
\begin{aligned}
L_h(x, C) &= \sum_s \ln f(z_s, y_h(x, X_s)) w_h(x - X_s), \\
y_h(x, v) &= C^T \phi_h(x - v), \;\; \phi_h(x) = \phi(x/h), \\
w_h(x) &= w(x/h)/h,
\end{aligned}
\tag{1.25}
$$

where h stays for the scale parameter used both in the window function w and for the scaling of the polynomials ϕ.

For y we use the model $y_h(x, v)$ linear on C. This type of model is known in statistics as a *generalized linear model*. For the local log-likelihood it becomes a *local (nonparametric) generalized linear model.*

The standard nonlocal likelihood uses the global parametric model on C, which is assumed to be valid for all x. The local likelihood relaxes this assumption and assumes that the used linear model is valid only locally in some neighborhood of the estimation point x.

Following the idea of the local nonparametric estimation, we use the criterion $L_h(x, C)$ for computing C. The model $y_h(x, v)$ is exploited in the pointwise manner only. In this way we arrive at the following local ML version of the LPA:

$$\begin{aligned} \widehat{C} &= \arg\max_{C} L_h(x, C), \\ \hat{y}_h(x) &= \widehat{C}^T \phi_h(0). \end{aligned} \tag{1.26}$$

For Gaussian noise with a constant variance, this approach gives the linear LPA estimates in Eq. (1.10).

In general, the estimates in Eq. (1.26) are nonlinear. This nonlinearity concerns two different aspects of the problem. First, the equation

$$\partial L_h(x, C)/\partial C = 0$$

is nonlinear with respect to C, which means that there is no analytical solution for $\widehat{C}$ and numerical methods should be used for maximization in Eq. (1.26). Second, the estimate $\hat{y}_h(x)$ is nonlinear with respect to the observations z_s.

There exist many circumstances in which, even though the full likelihood is unknown, one can specify the relationship between the mean and the variance. In this situation, estimation of the mean (regression, local regression) can be achieved by replacing the log-likelihood by a quasi-likelihood $Q(z, y)$. The quasi-likelihood has been proposed as a distribution-free method (note that the ML is a distribution-based method).

One assumes that we have independent observations with the expectations y and the variance of these observations is modeled as

$$\sigma^2(y) = \sigma^2\rho(y), \tag{1.27}$$

where σ^2 is unknown and ρ is a given positive function of y called a *variance function*. Thus, the variance is considered as a product of some given function of the estimated y and the unknown invariant scalar σ^2.

In many applications the noise that corrupts the signal is signal-dependent, the most widely encountered types being multiplicative, Poisson, film-grain, and speckle noise. Their common feature is that the variance of the noise is related to the magnitude of the signal.

There are many adaptive-neighborhood approaches to filtering signals and images corrupted by signal-dependent noise. Instead of using fixed-sized, fixed-shaped neighborhoods, statistics of noise and signals are computed within variable-sized, variable-shaped neighborhoods that are selected for every point of estimation.

In the developed quasi-likelihood approach we are concerned with this class of problems. The quasi-likelihood is used to derive the estimate relevant to the observation model and then the ICI enables the neighborhood adaptivity close to the ideal one.

1.3.2 Robust *M*-estimation

The term "M-estimate" is used for estimates obtained by minimizing a sample mean of a nonquadratic residual loss function. The M-estimation nonlinear version of the linear LPA estimate (1.5) is different by a nonquadratic residual loss function $\psi(\cdot)$:

$$\widehat{C}(x,h) = \arg\min_{C \in \mathbb{R}^M} J_h(x,C),$$

$$J_h(x,C) = \sum_s w_h(x - X_s)\psi(e_s), \tag{1.28}$$

$$e_s = z_s - y(x, X_s),\ y_h(x, X_s) = C^T \phi_h(x - v),$$

with the function estimate in the usual form,

$$\hat{y}_h(x) = \widehat{C}^T \phi_h(0). \tag{1.29}$$

Here $y_h(x, X_s)$ is the LPA of the signal y as it is exploited in Eq. (1.5). The vector C is an unknown parameter of this model found by minimization of the nonquadratic criterion $J_h(x, C)$ in Eq. (1.28). Equations (1.28)–(1.29) define a LPA nonparametric M-estimate of $y(x)$.

The median estimates corresponding to the standard median with $y(x, X_s) = C_0$ in Eq. (1.28) and the local polynomial median with the power $m > 0$ for $y(x, X_s)$ are particular cases of M-estimates with

$$\psi(e) = |e|. \tag{1.30}$$

There are various ideas behind the selection of the loss function ψ for M-estimation. One of them is to design a robust estimate. The concept of robustness has numerous and different interpretations in literature. *Huber's* minimax estimation theory proposes a regular approach for designing the robust estimates in the sense that these estimates are insensitive (to some extent) with respect to variations of the unknown noise distribution.

1.3.3 Adaptive varying scale

Under unrestrictive assumptions the ICI adaptive-scale selection rule is applicable for the nonlinear nonparametric estimates (1.26) and (1.28). The estimates $\hat{y}_h$ and their variances should be calculated for a set scale, $h \in H$, as the input of the ICI rule.

In general, the scale-adaptive nonlinear estimates are much more demanding than the linear ones in terms of the required computational load, in that they concern calculation of both the estimates and the variances. For one particular case, when the LPA of the zero power is used, the quasi-likelihood and the maximum likelihood estimates (for the exponential class of distributions) become linear estimates. The ICI yields neighborhoods where the unknown signal is nearly constant. Thus, by using the linear estimates in conjunction with the ICI rule, the pointwise segmentation of the data is produced into neighborhoods with nearly constant values of the underlining signal.

Fast algorithms are developed for nonlinear estimates of Eqs. (1.26) and (1.28) that exploit this idea. The linear estimates are calculated for all scales, and the ICI algorithm selects the maximum scales where the signal continues to be nearly constant. In this way the difficulties of nonlinear estimations are overcome and algorithms that are incomparably faster than any standard nonlinear adaptive-scale estimators are developed.

1.4 Multiresolution Analysis

Based on the LPA linear filters we introduce a novel multiresolution (MR) nonparametric regression analysis. This MR approach gives varying-scale adaptive estimators that are quite different from those defined by the ICI algorithm. Instead of selecting the estimate with a single best scale h, we build a nonlinear estimator using the estimate $\hat{y}_h$ with all available scales $h \in H$.

The denoising is produced in two steps. The first step transforms the data into noisy spectrum coefficients (MR analysis). In the second step, these noisy estimates of the spectrum are filtered by thresholding procedures and then exploited for estimation (MR synthesis). In this way an extension of the conventional nonparametric regression approach is achieved and a wider class of adaptive-scale estimators with potentially better performance is obtained.

1.5 Imaging Applications

One fundamental branch in signal processing concerns image reconstruction problems—the restoration of an optically and electronically corrupted digital image or a sequence of digital images. Corruption can be the result of some indirect observations (experiments) when what we observe is treated as a linear or nonlinear transformation of the object of interest. The problem is to reconstruct the image relevant to the observed phenomenon.

We illustrate performance of the developed techniques considering the following imaging problems:

- Denoising with invariant variance Gaussian noise,
- Denoising with signal-dependent noise,
- Differentiation of noisy images and edge detection,

- Motion estimation from an image sequence,
- Deblurring of noisy images,
- Inverse halftoning,
- Super-resolution imaging from a sequence of images,
- 3D image deblurring (3D microscopic imaging),
- Photon-limited imaging with direct and indirect (blurred) observations subject to a Poisson distribution,
- Computerized tomography.

The developed imaging techniques are universal and can also be applied in the following areas where our current interests are concentrated:

- Blind deblurring with parametric and nonparametric uncertainty of the point spread function;
- Focus-from-defocus, depth-from-defocus, and 3D imaging from defocus;
- Imaging for PET and SPET tomography;
- SAR and ISAR image processing;
- Time-frequency analysis;
- Digital holography.

1.6 Overview of the Book

In Chapters 2 and 3, basic local approximation ideas and techniques are presented, and the roles of the window function and scale are discussed. The polynomial smoothness of LPA kernels is defined in terms of the vanishing moment equations and in the frequency domain. It is shown that the designed filters are reproducing (accurately) for the polynomial signals. The nonparametric approach is exploited for filtering, interpolation, and differentiation.

Shift-invariant kernel estimates are derived for regular grid signals. All results are given in multivariate notation equally applicable for uni- and multivariate signals.

In Chapter 4, the integral local polynomial approximation is studied. These continuous variable filters and transformations are important on their own as well as limit representations for the discrete ones. The limit integral forms are used for accuracy analyses of discrete filters and for accuracy optimization. The integral LPA estimates equipped with scale parameters are treated as delta-sequences for estimation and differentiation. Corresponding links with a theory of generalized functions are also discussed.

The accuracy analysis of the discrete LPA estimates is the subject of Chapter 5. Formulas for the bias are derived for some classes of signals, and the ideal scale balancing the bias and variance of the estimates is studied.

Adaptive-scale selection is the subject of Chapter 6. The varying adaptive-scale selection is considered to be an efficient tool to deal with nonstationary signals having spatially varying properties. *Lepski's* general approach and the intersection of confidence (ICI) rule are introduced as basic adaptation tools for spatially varying adaptation. The ICI rule is derived from optimization of the mean-square error. The idea as well as the implementation and properties of the ICI rule are presented. Simulation experiments demonstrate a high sensitivity of the ICI rule to singularities and change points in signals and its efficiency for adaptive filtering.

The directional LPA is presented in Chapter 7 as a further development of the local approximation idea. It is proposed to deal with the anisotropy of signals having different behaviors in different directions. The directionality of the approximation is enabled by narrowed sectorial windows. The directional kernels use polynomials of different orders and different scales for different directions. The designing of this sort of filter for estimation (smoothing) and differentiation is given.

The directional ICI adaptive-scale estimation is the subject of Chapter 8. The main goal is to obtain a shape-adaptive neighborhood for the best approximation of anisotropic signals. The approach is based on a sectorial partition of the local neighborhood with adaptive sizes of the sectors defined by the ICI rule. A bank of scale-adaptive directional filters is used with the final estimate found by aggregation of the directional estimates.

Recursive LPA-ICI adaptive algorithms are proposed as a special class of adaptive filters with a cyclic repetition of the multidirectional LPA-ICI filtering. The advance performance of the LPA-ICI directional algorithms is demonstrated for image denoising problems.

For differentiation, a novel concept of the anisotropic gradient is developed. The algorithm looks for the adaptive neighborhood where the signal allows a linear local approximation and the gradient is calculated using observations only from this neighborhood. Specific features of this concept are illustrated for *shading-from-depth* problems.

Scale-adaptive image *deblurring techniques* are considered in Chapter 9. A basic topic concerns the deblurring of a noisy image provided that the point-spread function of the blur is shift-invariant. Developed algorithms combine the regularized inverse and regularized Wiener inverse filtering procedures, both using the LPA-ICI scale-adaptive technique. Fast implementation of the algorithms is based on using convolution operations for computing.

The ICI scale adaptivity is proved to be very efficient. It is shown that the newly developed algorithms demonstrate state-of-the-art performance. This technique is generalized for 3D deblurring problems.

Nonlinear nonparametric estimation (filtering) is the subject of Chapter 10. Observation modeling and performance criterion are considered as the main reasons to replace linear estimates by more demanding nonlinear ones. At first, a class of nonparametric robust estimates based on LPA and the M-estimation is considered. Links between the robust and minimax estimations are discussed. Accuracy analyses and scale optimizations show that the ICI rule is applicable for the adaptive varying-scale selection. Adaptive-scale median filters are considered as a special class of these adaptive M-estimates.

The nonlinear *Box-Cox and Anscombe* transformations allow the stabilization of the variance in observations. In particular, this approach is efficient to deal with the Poisson distribution data. Using this concept we develop adaptive algorithms where the nonlinear estimation is reduced to a linear one with nonlinear transforms before and after filtering.

A general approach to nonlinear estimation problems based on the local likelihood and quasi-likelihood is considered in Chapter 11. The LPA of the signal of interest is used as the basic design tool. The accuracy results for the bias and variance of the estimates are obtained. It is shown that the only difference versus the corresponding results for the linear estimate concerns the formulas for the variance. It follows that the ICI rule can be modified and used for the nonlinear adaptive varying-scale filtering.

The quasi-likelihood is a distribution-free version of the likelihood that does not require the accurate probabilistic observation model. The reweighted least-square algorithms are used for the quasi-likelihood estimate calculation. The quasi-likelihood-based methods are relevant to a wide class of applications where the variance of observations is signal-dependant.

Fast algorithms implementing the quasi-likelihood scale-adaptive estimates are developed based on linear convolution estimates completed by the ICI rule. As a particular application of the developed adaptive nonlinear estimates, we consider the counting (Poisson distribution) and binary (Bernoulli distribution) observations.

Image reconstruction problems concerning Poisson distribution data are the subject of Chapter 12. Important areas of application include digital photography, biology, and medical imaging problems. In many cases crucial high-quality imaging requirements are in clear contradiction to very noisy raw data, with the noise variance depending on the signal. Nonlinear methods based on likelihood and quasi-likelihood are standard tools for this sort of problem.

A number of novel algorithms are proposed using the nonlinear LPA joint with the ICI adaptive-scale selection. Special attention is paid to the development of more computationally productive methods based on fast linear estimates completed by the ICI adaptive-scale selection. It is shown that these methods are able to give high quality with a reasonable complexity of computing.

A novel multiresolution (MR) nonparametric regression analysis is considered in Chapter 13 as a further development and an alternative to the LPA-ICI adaptive technique. This processing includes the following three successive steps: MR analysis, thresholding (filtering in the transform domain), and synthesis. The final estimate is a composition of the estimates of different scales. In contrast to it, the ICI rule defines a unique best scale in a pointwise manner for each estimation point.

Mathematical results concerning proofs of propositions as well as derivations of some formulas are collected in the Appendix.

Chapter 2

Discrete LPA

The idea of local smoothing and local approximation is so natural that it is not surprising it has appeared in many branches of science. Citing [145] we mention early works in statistics using local polynomials by the Italian meteorologist Schiaparelli (1866) and the Danish actuary Gram (1879) (famous for developing the Gram-Schmidt procedure for the orthogonalization of vectors).

In the 1960s and 1970s the idea became the subject of intensive theoretical study and applications, in statistics by Nadaraya [158], Watson [230], Parzen [175], and Stone [215], and in engineering sciences by Brown [19], Savitzky and Golay [196], Petersen [179], Katkovnik [98], [99], [101], and Cleveland [29].

The local polynomial approximation as a tool appears in different modifications and under different names, such as moving (sliding, windowed) least square, Savitzky-Golay filter, reproducing kernels, and moment filters. We prefer the term LPA with a reference to publications on nonparametric estimation in mathematical statistics where the advanced development of this technique can be seen.

In this chapter the discrete local approximation is presented in a general multivariate form. In the introductory section (Sec. 2.1), we discuss an observation model and multi-index notation for multivariate data, signals, and estimators.

Section 2.2 starts with the basic ideas of the LPA presented initially for 2D signals typical for image processing. The window function, the order of the LPA model, and the scaling of the estimates are considered in detail. Further, the LPA is presented for the general multivariate case with estimates for smoothing and differentiation. Estimators for the derivatives of arbitrary orders are derived. Examples of 1D smoothers demonstrate that the scale (window size) parameter selection is of importance.

The LPA estimates are given in kernel form in Sec. 2.3. The polynomial smoothness of the kernels and their properties are characterized by the vanishing moment conditions.

The LPA can be treated as a design method for linear smoothing and differentiating filters. Links of the LPA with the nonparametric regression concepts are clarified in Sec. 2.4, and the LPA for interpolation is discussed briefly in Sec. 2.5.

2.1 Introduction

2.1.1 Observation modeling

Since the LPA is a method that is valid for signals of any dimensionality, we mainly use a multidimensional notation because it is universal and convenient. For instance, we say that the signal $y(x)$ depends on $x \in \mathbb{R}^d$, i.e., the argument x is a d-dimensional vector, and we use it with $d = 1$ for scalar signals in time and with $d = 2$ for 2D images. The multidimensional notation allows us to present results in a general form valid for both scalar time and 2D image and signal processing.

The symbol d is reserved for the dimensionality of the argument-variable x. Thus, x is a vector with elements $x_1, \ldots, x_d$.

Suppose that we are given observations of y in the form $y_s = y(X_s)$, $s = 1, \ldots, n$. Here X_s stays for a location of the sth observation, and the coordinates of this location are

$$X_s = [x_1(s), \ldots, x_d(s)].$$

The integers s number observations and their locations.

Noisy observations with additive noise can be given in the following standard model commonly used for image and signal processing:

$$z_s = y(X_s) + \varepsilon_s, s = 1, \ldots, n, \tag{2.1}$$

where the additive noise ε_s is an error of the sth experiment usually assumed to be random, zero-mean independent for different s with $E\{\varepsilon_s\} = 0$, $E\{\varepsilon_s^2\} = \sigma^2$.

Equation (2.1) is a *direct observation model* where a signal y is measured directly and the additive noise only defines a data degradation.

Suppose we are able to observe $(y \circledast v)(x)$, where v is a *blurring kernel* or *point spread function* (PSF) of a linear convolution. The blurring phenomenon modeled by the PSF v (continuous or discrete) is very evident in many signal and image applications. Moreover, we assume that the data are noisy, so that we observe z_s given by

$$z_s = (y \circledast v)(X_s) + \varepsilon_s. \tag{2.2}$$

This equation is an *indirect observation model* with a signal y measured not directly but after some transformation.

A data degradation is defined by this transformation as well as by the additive noise. It is assumed in image processing that all signals in Eqs. (2.1) and (2.2) are defined on a 2D rectangular *regular grid*:

$$X = \{X_s : s = 1, \ldots, n\}$$

with pixel (grid node) coordinates

$$X_s = [x_1(s), x_2(s)]. \tag{2.3}$$

In this notation the image pixels are numbered by s taking values from 1 through n. The coordinates of these pixels can be given explicitly as

$$X = \{k_1\Delta_1, k_2\Delta_2 : k_1 = 1, \ldots, n_1, k_2 = 1, \ldots, n_2\}, \qquad n = n_1 n_2, \tag{2.4}$$

where Δ_1 and Δ_2 are the sampling intervals for x_1 and x_2, respectively.

In signal processing literature, square brackets are sometimes used for discrete arguments to differentiate them from continuous arguments. For example, the notation $y(k_1\Delta_1, k_2\Delta_2)$ is for a continuous argument function and $y[k_1, k_2]$ is for a discrete one with

$$y(k_1\Delta_1, k_2\Delta_2) = y[k_1, k_2]. \tag{2.5}$$

For this discrete argument y we also can use the general notation $y(X_s)$ provided that $X_s = (k_1\Delta_1, k_2\Delta_2)$.

The LPA considered in the book is applicable to observations given on irregular grids, when the coordinates X_s and the distances between X_s can be arbitrary. In image processing this sort of situation appears when data are destroyed or lost. Figure 2.1 illustrates the cases of regular and irregular grids for 2D data.

2.1.2 Classes of signals

A basic objective of data processing is to reconstruct (estimate) y from the noisy observations $\{z_s\}$ with pointwise errors that are as small as possible. It is assumed that y is an unknown deterministic. In comparison with stochastic models of y, this means that the main intention is to obtain the best results for every realization of y, even if they are generated by a probabilistic phenomenon.

The first category is that of *parametric models* $y(x, \theta)$, where θ is a vector of unknown parameters. These models seek a signal y with parameter θ that gives an approximation of the relation between x and y. We predict y based on the argument variable x assuming that θ is given. The estimation is concentrated around methods for finding θ from observations

$$z_s = y(X_s, \theta) + \varepsilon_s, \qquad s = 1, \ldots, n.$$

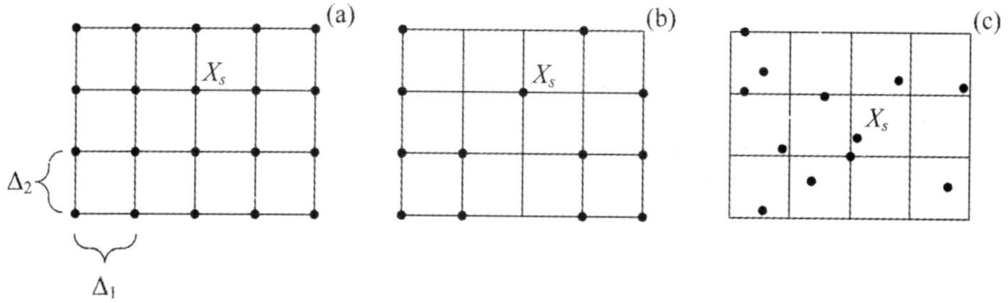

Figure 2.1 Types of data grids: (a) regular with sampling intervals Δ_1 and Δ_2 on the arguments x_1 and x_2; (b) regular with missing data; and (c) irregular with varying intervals between observations X_s.

The second category is that of *nonparametric models*, which assume that a parametric representation of y with a reasonably small number of parameters θ is unknown or does not exist. This y can be quite arbitrary, say, with discontinuities and varying curvature. Examples of nonparametric and parametric curves (linear $y(x,\theta) = \theta_1 + \theta_2 x$ and harmonic $y(x,\theta) = \theta_1 \cos(\theta_2 x + \theta_3) + \theta_4$) are shown in Fig. 2.2.

In this book we deal with the nonparametric category of signals y. Usually it is assumed that there is some continuous argument signal $y(x)$, and the observed $y(X_s)$ are sampled values of this continuous signal. Different classes of continuous argument signals can be exploited. We restrict our analysis to the class $\mathbb{C}^\alpha$ of continuous differentiable signals y of d variables with bounded αth-order derivatives, which are defined mathematically as follows:

$$\mathbb{C}^\alpha = \left\{ y : \max_{r_1+\cdots+r_d=\alpha} \left| \partial_{x_1}^{r_1} \ldots \partial_{x_d}^{r_d} y(x) \right| = L_\alpha(x) \leq \overline{L}_\alpha, \forall x \in \mathbb{R}^d \right\}, \tag{2.6}$$

where $L_\alpha(x)$ and $\overline{L}_\alpha$ are finite and $\alpha = r_1 + \cdots + r_d$ is the order of the derivative used in this definition.

The class $\mathbb{C}^\alpha$ is composed of signals having bounded all possible derivatives of the order α.

Let $d = 1$. Then the class $\mathbb{C}^\alpha$ is defined as

$$\mathbb{C}^\alpha = \{ y : |\partial_{x_1}^\alpha y(x_1)| = L_\alpha(x_1) \leq \overline{L}_\alpha, \forall x_1 \in \mathbb{R} \}.$$

The class includes all signals with the bounded αth-order derivative.

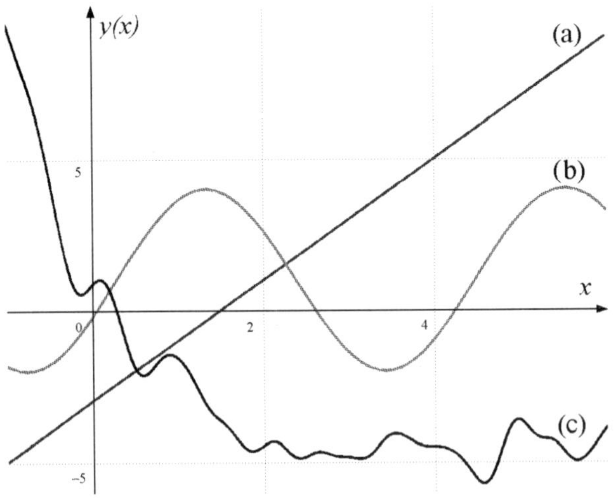

Figure 2.2 Parametric and nonparametric models: (a) linear parametric model, (b) cosine parametric model, and (c) nonparametric function with unknown parametric representation.

Let $d = 2$. Then

$$\mathbb{C}^{\alpha} = \left\{ y : \max_{r_1+r_2=\alpha} \left| \partial_{x_1}^{r_1} \partial_{x_2}^{r_2} y(x_1, x_2) \right| = L_{\alpha}(x_1, x_2) \leq \overline{L}_{\alpha}, \forall (x_1, x_2) \in \mathbb{R}^2 \right\}.$$

In this definition, the maximum for each $x = (x_1, x_2)$ is calculated over all derivatives of the order $\alpha = r_1 + r_2$. For $\alpha = 1$, it assumes that

$$\max(|\partial_{x_1} y(x)|, |\partial_{x_2} y(x)|) = L_1(x),$$

while for $\alpha = 2$ it means

$$\max(|\partial_{x_1}^2 y(x)|, |\partial_{x_2}^2 y(x)|, |\partial_{x_1} \partial_{x_2} y(x)|) = L_2(x).$$

Definition (2.6) is global because the conditions hold for all x. For images where singularities and discontinuities of y carry the most important image features, a local smoothness of y is more realistic. Therefore, the class $\mathbb{C}^{\alpha}$ can be replaced by

$$\mathbb{C}_A^{\alpha} = \left\{ y : \max_{r_1+\cdots+r_d=\alpha} \left| \partial_{x_1}^{r_1} \ldots \partial_{x_d}^{r_d} y(x) \right| = L_{\alpha}(x) \leq \overline{L}_{\alpha}, \forall x \in A \subset \mathbb{R}^d \right\}, \qquad (2.7)$$

where the argument x is restricted to a subset A.

The following piecewise smooth models are useful for many applications.

Let an area where y is defined be separated into q regions $A_i, i = 1, \ldots, q$, and each of these regions be a connected set with edges (boundaries G_i). The signal y is assumed to be smooth differentiable within A_i:

$$y(x) = \sum_{i=1}^{q} y_i(x) \mathbf{1}[x \in A_i], \qquad (2.8)$$

where $\mathbf{1}[x \in A_i]$ is an indicator of the region A_i, $\mathbf{1}[x \in A_i] = 1$ if $x \in A_i$ and zero otherwise. More specifically, y_i belongs to a class of continuous differentiable signals $\mathbb{C}_{A_i}^{\alpha}$. The parameters α, $L_{\alpha}(x)$, and $\overline{L}_{\alpha}$ can be different for the different y_i in Eq. (2.8).

The piecewise constant y,

$$y(x) = \sum_{i=1}^{q} a_i \mathbf{1}[x \in A_i], \qquad (2.9)$$

is a particular case of Eq. (2.8) with constant values within each region A_i. Figure 2.3 shows an example of a piecewise constant y defined according to Eq. (2.9) as well as the corresponding regions A_i.

In the models (2.8) and (2.9) $y_i(x)$, and a_i, as well as the regions A_i are usually unknown. The boundaries G_i define change points of the piecewise smooth y in Eq. (2.8). The estimation of y can be produced in different ways. One possible approach deals with a two-stage procedure that includes estimating the boundaries G_i at the first stage, which defines the regions A_i. The second stage is a parametric or nonparametric fitting y_i in A_i.

Another approach is connected with the concept of spatially adaptive estimation. In this context, change points can be viewed as a sort of anisotropic behavior

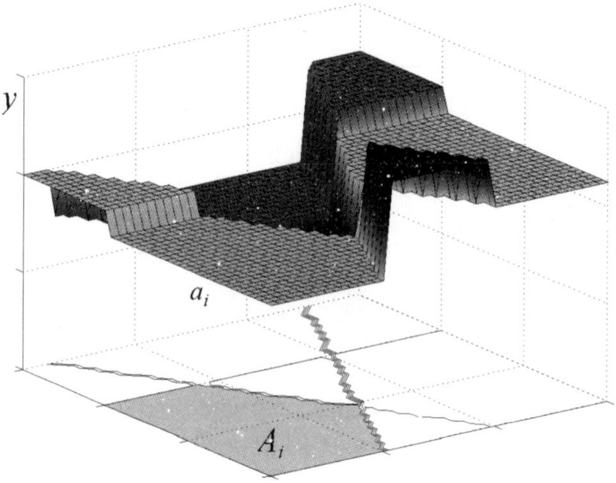

Figure 2.3 A piecewise constant 2D function.

of the estimated signal. One may therefore apply the same procedure for all x, for instance, nonlinear wavelet, ridgelet, or curvelet estimators, and the analysis focuses on the quality estimation when change-points are incorporated into the model. With this approach, the main intention is to estimate the signal but not the locations of the change-points, which are treated as elements of the signal surface.

In this book we will follow the second approach. The objective is to develop a method that simultaneously adapts to the anisotropic smoothness of the estimated curve and is also sensitive to the discontinuities in the curves and their derivatives.

In general, conditions (2.6) and (2.7) define nonparametric signals that do not have parametric representations. The following important class of parametric y can be interpreted as a special case of the nonparametric class $\mathbb{C}^{\alpha}$.

Let $L_{m+1}(x) \equiv 0$ for $x \in \mathbb{R}^d$. Then y is polynomial of the degree m. Indeed, the condition $L_{m+1}(x) \equiv 0$ means that y is a solution of the set of differential equations

$$\partial_{x_1}^{r_1} \cdot \ldots \cdot \partial_{x_d}^{r_d} y(x) = 0$$

for all r_i such that $r_1 + \cdots + r_d = m + 1$, and this solution is a polynomial. For this class of polynomials we use the notation

$$\mathcal{P}^m = \{y : \partial_{x_1}^{r_1} \cdot \ldots \cdot \partial_{x_d}^{r_d} y(x) = 0, r_1 + \cdots + r_d = m + 1, \forall x \in \mathbb{R}^d\}, \qquad (2.10)$$

where m stays for the power of the polynomial. It is obvious that $\mathcal{P}^m \subset \mathbb{C}^{m+1}$ with $\overline{L}_{m+1} = 0$.

2.1.3 Multi-index notation

Multi-index notation is used to shorten expressions that contain many indices. Let x be a d-dimensional vector, $x \in \mathbb{R}^d$, and y be a scalar function of x,

$y(x) = y(x_1, \dots, x_d)$. A multi-index r is a d-tuple of nonnegative integers r_j, $j = 1, \dots, d$,

$$r = (r_1, r_2, \dots, r_d),$$

where $r_s \geq 0$. The notation $|r|$ is used to denote the length $\sum_{s=1}^{d} r_s$.

For polynomial expressions in coordinates x_j, the multi-index notation provides a convenient shorthand:

$$x^k \triangleq x_1^{k_1} x_2^{k_2} \cdots x_d^{k_d},$$

where x and k are d-tuples. Further,

$$\frac{x^k}{k!} \triangleq \frac{x_1^{k_1}}{k_1!} \frac{x_2^{k_2}}{k_2!} \cdots \frac{x_d^{k_d}}{k_d!}, \tag{2.11}$$

where $k!$ means a product of $k_j!$, $j = 1, \dots, d$, and $x^k/k!$ is a product of the ratios $x_j^{k_j}/k_j!$, $j = 1, \dots, d$.

The differentiation operator of the order r can be written in the following equivalent forms:

$$\partial_x^r y(x) = \partial_{x_1}^{r_1} \partial_{x_2}^{r_2} \dots \partial_{x_d}^{r_d} y(x) = \partial_{(x_1, x_2, \dots, x_d)}^{(r_1, r_2, \dots, r_d)} y(x), \tag{2.12}$$

where the elements of the tuple r in the superscript of ∂_x^r are the orders of the partial derivatives calculated on the variables given by the d-tuple x shown as the subscript. This subscript x can be omitted when the meaning of the differentiation variables is obvious. Then the derivative is written as $\partial^r y(x)$.

The differentiation of the polynomial x^k is defined by

$$\partial_x^r x^k = k! \frac{x^{k-r}}{(k-r)!} \text{ if } k \geq r, \tag{2.13}$$

where $k \geq r$ means that $k_1 \geq r_1, \dots, k_d \geq r_d$. In the standard scalar notation, Eq. (2.13) means

$$\partial_x^r x^k = \partial_{x_1}^{r_1} x_1^{k_1} \cdot \dots \cdot \partial_{x_d}^{r_d} x_d^{k_d} = \frac{k_1!}{(k_1 - r_1)!} x_1^{k_1 - r_1} \cdot \dots \cdot \frac{k_d!}{(k_d - r_d)!} x_d^{k_d - r_d}. \tag{2.14}$$

For the polynomial (2.11), this differentiation gives the following formula from Eq. (2.13):

$$\partial_x^r \frac{x^k}{k!} = \frac{1}{k!} \partial^r (x^k) = \frac{x^{k-r}}{(k-r)!} \text{ if } k \geq r. \tag{2.15}$$

Consider the Taylor series of the d variables in the scalar notation,

$$y(x_1 + u_1, \dots, x_d + u_d) = \sum_{r_1=0}^{\infty} \cdots \sum_{r_d=0}^{\infty} \frac{u_1^{r_1} \cdots u_d^{r_2}}{r_1! \cdots r_d!} \partial_{(x_1, \dots, x_d)}^{(r_1, \dots, r_d)} y(x_1, \dots, x_d). \tag{2.16}$$

Using the multi-index notation and tuples, this can be given in the compact form

$$y(x+u) = \sum_{r=0}^{\infty} \frac{u^r}{r!} \partial_x^r y(x), \tag{2.17}$$

where $\sum_{r=0}^{\infty} = \sum_{r_1=0}^{\infty} \cdots \sum_{r_d=0}^{\infty}$.

Expressions (2.11)–(2.17) demonstrate that, in multi-index notation, multivariate formulas look like usual scalar ones. However, the multivariate nature of these formulas is intact. In fact, the multi-index notation corresponds exactly to the so-called *array arithmetic operations* as they are defined in MATLAB. These operations are carried out element-by-element and can be used with multidimensional data arrays. In MATLAB the period character (.) distinguishes the array operations from the usual ones.

2.2 Basis of LPA

The idea of the LPA is simple and natural. It is assumed that the signal y is well approximated by a polynomial in some neighborhood of the point of interest x. We find the coefficients of the polynomial fit by the weighted least-square method (LSM) and use this approximation to calculate the estimate for the point of interest x, called "center" of the LPA. Note that the term "center" does not assume a central position of x with respect to neighboring observations. This term aims to stress that the LPA is applied to obtain the estimate for this particular point x. The local expansion is applied to calculate the estimate for the given x only. The calculations are repeated for each point. This pointwise procedure determines the nonparametric character of the LPA estimation.

The LPA technique is valid for any dimensionality d of x. Starting from $d = 2$, we deal first with the 2D images. These provide a good opportunity for clear illustrations. At the same time, $d = 2$ is an obvious multidimensional case that is not so trivial as the scalar one, $d = 1$.

2.2.1 Idea of LPA

Let $x \in \mathbb{R}^2$ be a "*center*" (desired point) of the LPA. The estimate for the point $v \in \mathbb{R}^2$ in a neighborhood of the center x is of the form

$$y_h(x, v) = C^T \phi_h(x - v), \ \phi_h(x) = \phi(x/h), \tag{2.18}$$

$$\begin{aligned} \phi(x) &= [\phi_1(x), \phi_2(x), \ldots, \phi_M(x)]^T, \\ C &= (C_1, C_2, \ldots, C_M)^T, x = (x_1, x_2), v = (v_1, v_2), \end{aligned} \tag{2.19}$$

where $\phi \in \mathbb{R}^M$ is a vector of linearly independent 2D polynomials

$$(-1)^{k_1+k_2} \frac{x_1^{k_1}}{k_1!} \frac{x_2^{k_2}}{k_2!} \tag{2.20}$$

with the polynomial degrees $k_1 + k_2$ from 0 up to m, and $C \in \mathbb{R}^M$ is a vector of the parameters. The minus sign in this equation enables proper signs to estimates of the derivative, which are obtained using this model. Details concerning the sign selection in this equation will be discussed later.

Through this book, the character h, $h > 0$, is reserved to denote a *scale*. It can be a scale of the estimate or a scale of variables, but it is always a scale. In $\phi_h(x) = \phi(x/h)$, it is a scale of the polynomials. Used as a subscript, h should not be confused with integer subscripts denoting the items of the vector ϕ in Eq. (2.19). The elements of the vector $\phi_h(x)$ are

$$\phi_{i,h}(x) = \phi_i(x/h), \qquad i = 1, \ldots, M.$$

The total number of polynomials in $\phi = (\phi_1, \ldots, \phi_M)^T$ and the unknown parameters in the vector $C = (C_1, \ldots, C_M)^T$ are equal to

$$M = \frac{(m+2)!}{2 \cdot m!} = (m+2)(m+1)/2.$$

The corresponding ϕ_i, $i = 1, \ldots, M$, form a complete set of the polynomials of the power m. By definition, the *completeness* for the polynomials means that for any x and v

$$\phi(x - v) = A(v)\phi(x), \tag{2.21}$$

where $A(v)$ is an $M \times M$ matrix depending on v only. Therefore, Eq. (2.21) means that any polynomial of degree equal to or less than m can be written as a linear combination of $\{\phi_i\}$. For $m = 0, 1, 2,$ and 3, complete sets of the 2D linearly independent polynomials (2.20) are of the form

$$\begin{aligned}
&\phi_1 = 1, \quad \text{for } m = 0; \\
&\phi_2 = -x_1, \quad \phi_3 = -x_2, \quad \text{for } m = 1; \\
&\phi_4 = x_1^2/2, \quad \phi_5 = x_2^2/2, \quad \phi_6 = x_1x_2, \quad \text{for } m = 2; \\
&\phi_7 = -x_1^3/6, \quad \phi_8 = -x_2^3/6, \quad \phi_9 = -x_1^2x_2/2, \quad \phi_{10} = -x_1x_2^2/2, \quad \text{for } m = 3;
\end{aligned} \tag{2.22}$$

with $M = 1$ for $m = 0$, $M = 3$ for $m = 1$, $M = 6$ for $m = 2$, and $M = 10$ for $m = 3$. For each m the complete set of the polynomials in Eq. (2.22) includes the row corresponding to this m as well as all previous rows with smaller values of m.

It is a good exercise to check the completeness of Eq. (2.21) for the sets of polynomials in Eq. (2.22) corresponding to different m and to obtain the corresponding matrices $A(v)$.

Formula (2.18) can be interpreted as a truncated Taylor series that looks like the following in the scalar form:

$$y_h(x, v) = \sum_{i=1}^{M} C_i \phi_{i,h}(x - v). \tag{2.23}$$

Model (2.18) is exploited to fit the discrete observations z_s as they are introduced in Eq. (2.1).

The following weighted least-square criterion $J_h(x, C)$ is used in the LPA to find the coefficient C in Eq. (2.18) (e.g., [31], [46], [77], [98], [99], [100], [102], [145]):

$$J_h(x, C) = \sum_s w_h(x - X_s)[z_s - y_h(x, X_s)]^2, \tag{2.24}$$

where the window function w formalizes the localization of fitting with respect to the center x and

$$w_h(x) = \frac{1}{h^2} w\left(\frac{x}{h}\right). \tag{2.25}$$

The scale parameter $h > 0$ determines the size of the window used in $J_h(x, C)$.

Let us clarify the meaning of the local model used in Eq. (2.24). Consider the residual $z_s - y_h(x, X_s)$ from the criterion (2.24) assuming that the observations are noiseless and the model $y_h(x, X_s)$ fits $z_s = y_s = y(X_s)$. According to the general formula (2.16), the Taylor series for $y(X_s)$ with the reference point x is

$$\begin{aligned} z_s &= y(X_s) \\ &= y(x) - \partial_{x_1} y(x) \cdot [x_1 - x_1(s)] - \partial_{x_2} y(x) \cdot [x_2 - x_2(s)] \\ &\quad + \partial_{x_1}^2 y(x) \cdot [x_1 - x_1(s)]^2 \tfrac{1}{2} + \partial_{x_2}^2 y(x) \cdot [x_2 - x_2(s)]^2 \tfrac{1}{2} \\ &\quad + \partial_{x_1} \partial_{x_2} y(x) \cdot [x_1 - x_1(s)][x_2 - x_2(s)] + \cdots . \end{aligned} \tag{2.26}$$

In parallel, consider the model $y_h(x, X_s)$ in Eq. (2.23) with the polynomials of the corresponding powers:

$$\begin{aligned} y_h(x, X_s) &= C_1 - C_2 \cdot [x_1 - x_1(s)]/h - C_3 \cdot [x_2 - x_2(s)]/h \\ &\quad + C_4 \cdot [x_1 - x_1(s)]^2 \frac{1}{2h^2} + C_5 \cdot [x_2 - x_2(s)]^2 \frac{1}{2h^2} \\ &\quad + C_6 \cdot [x_1 - x_1(s)][x_2 - x_2(s)] \frac{1}{h^2} + \cdots . \end{aligned}$$

Using Eq. (2.26), the residuals $z_s - y_h(x, X_s)$ in Eq. (2.24) can be given in the form

$$\begin{aligned} z_s - y_h(x, X_s) &= [y(x) - C_1] + \left[C_2 \frac{1}{h} - \partial_{x_1} y(x)\right][x_1 - x_1(s)] \\ &\quad + \left[C_3 \frac{1}{h} - \partial_{x_2} y(x)\right][x_2 - x_2(s)] + \left[\partial_{x_1}^2 y(x) - C_4 \frac{1}{h^2}\right] \\ &\quad \cdot [x_1 - x_1(s)]^2/2 + \left[\partial_{x_2}^2 y(x) - C_5 \frac{1}{h^2}\right] \cdot [x_2 - x_2(s)]^2/2 \\ &\quad + \left[\partial_{x_1} \partial_{x_2} y(x) - C_6 \frac{1}{h^2}\right] \cdot [x_1 - x_1(s)][x_2 - x_2(s)]. \end{aligned} \tag{2.27}$$

Because the polynomials in Eq. (2.27) are linearly independent, minimizing the weighted residuals in Eq. (2.24) is equivalent to minimizing the squared $[y(x) -$

$C_1]^2$, $[C_2/h - \partial_{x_1} y(x)]^2$, $[C_3/h - \partial_{x_2} y(x)]^2$, and so on. In this way C_1 gives the estimate of $y(x)$, C_2/h gives the estimate of the derivative $\partial_{x_1} y(x)$, C_3/h gives the estimate of the derivative $\partial_{x_2} y(x)$, and so on. The formulas (2.27) show that the use of the polynomials of Eq. (2.20) in Eq. (2.23) results in the coefficients C of this expansion giving the estimates of the derivatives of y.

The negative sign for the odd powers $k_1 + k_2$ in Eq. (2.20) enables the correct signs for the derivative estimates. The parameter h is an important element of the estimates scaling the elements of the vector C. The expansions (2.26)–(2.27) provide an initial important flavor of the LPA approach as well as the accurate meaning of its elements.

Let us go further to some general aspects of this technique. In the model $y_h(x, v)$ used in Eq. (2.24), the argument x is considered as a fixed invariant parameter defining the center of the approximation, and the approximations for different X_s are obtained by using this model with the varying second argument v assuming $v = X_s$. Figure 2.4 illustrates the idea and the meaning of the LPA center.

The data are given on the grid $X = \{X_s, s = 1, 2, \ldots, n\}$ for Eq. (2.4) and a location of the center (LPA reference point) x defines the type of estimation problem considered. If $x = X_k \in X$, then the LPA estimate approximates the values y_k or smooths the noisy data z_k. If $x \notin X$ and x is inside of image boundaries, then the LPA estimate interpolates the function $y(x)$ for the argument value x using the observations given on the grid X. If $x \notin X$ and x is outside of the image boundaries, the LPA estimate extrapolates the signal $y(x)$ outside the area covered by X.

The size of the neighborhood used in the estimation depends on the scale h. Larger values of h result in a larger neighborhood.

Depending on the x value in Eq. (2.24) we can treat the LPA estimates as approximating/smoothing, interpolating, or extrapolating the data defined on the data grid X.

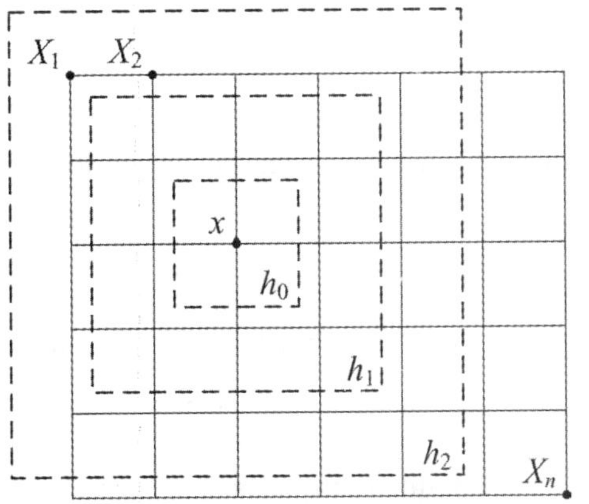

Figure 2.4 The LPA center x and its neighborhood. The scale parameter h taking values $h_0 < h_1 < h_2$ defines the size of the square neighborhood of the center x.

2.2.1.1 *Signal estimation*

Let $\hat{y}_h(x)$ denote the LPA estimate (approximate) of $y(x)$, where the subscript h shows a dependence of the estimate on this scale parameter. This estimate is defined according to Eq. (2.18) as

$$\hat{y}_h(x) = y_h(x, v)|_{v=x} = y_h(x, x),$$

where the expansion (2.18) is used to calculate the estimate for $v = x$ only.

This is one of the key ideas of the pointwise nonparametric estimate design. We introduce the estimate $y_h(x, v)$ as an expansion in a neighborhood of x, we estimate the coefficient C of this expansion, and, finally, we use this expansion only to estimate the signal at the argument $v = x$.

It follows from Eq. (2.18) that

$$\hat{y}_h(x) = y_h(x, v)|_{v=x} = \widehat{C}^T \phi(0). \tag{2.28}$$

Note that $\phi(0) = \phi_h(0)$. For the polynomials (2.22) it yields

$$\hat{y}_h(x) = \widehat{C}^T \phi(0) = \widehat{C}_1, \tag{2.29}$$

because $\phi(0)$ is a vector-column $M \times 1$ with the single nonzero first element equal to 1,

$$\phi(0) = (1, 0, \ldots, 0)^T.$$

2.2.1.2 *Derivative estimation*

Note that, in the model (2.18), the argument x of $y_h(x, v)$ is considered as the fixed invariant parameter, and values of $y_h(x, v)$ for different X_s in Eq. (2.24) are obtained by using $y_h(x, v)$ with the varying second argument v assuming $v = X_s$. Thus, in a neighborhood of x, the model $y_h(x, v)$ is a function of the argument v, provided there is a fixed x. The derivative of $y_h(x, v)$ on v is used as an estimate of the derivative of $y(x)$.

To obtain an estimate of the derivative $\partial^{(r_1, r_2)} y(x)$, we calculate the derivative $\partial_{v_1}^{r_1} \partial_{v_2}^{r_2} y_h(x, v)$, and following the very idea of the pointwise estimation, use it for $v = x$:

$$\begin{aligned}
\hat{y}_h^{(r_1, r_2)}(x) &= \partial_{v_1}^{r_1} \partial_{v_2}^{r_2} y_h(x, v)|_{v=x} \\
&= C^T \partial_{v_1}^{r_1} \partial_{v_2}^{r_2} \phi((x - v)/h)|_{v=x} \\
&= (-1/h)^{r_1 + r_2} C^T \partial_{x_1}^{r_1} \partial_{x_2}^{r_2} \phi(0).
\end{aligned}$$

This derivative estimate is obtained in the form

$$\hat{y}_h^{(r_1, r_2)}(x) = (-1/h)^{r_1 + r_2} \widehat{C}^T \partial_{(x_1, x_2)}^{(r_1, r_2)} \phi(0). \tag{2.30}$$

For the polynomial $\phi(x)$ in (2.18) composed from (2.20),

$$\partial^{(r_1,r_2)}_{(x_1,x_2)}\phi(0) = (0,\ldots,0,\underbrace{(-1)^{r_1+r_2}}_{r_1\cdot r_2-th},0,\ldots,0)^T \tag{2.31}$$

is a vector-row $1\times M$ with the only nonzero element corresponding to the polynomial $(-1)^{r_1+r_2}x_1^{r_1}x_2^{r_2}/r_1!r_2!$ in $\phi(x)$. This nonzero derivative is equal to $(-1)^{r_1+r_2}$.

Assume that the polynomials in ϕ are ordered according to the example in Eq. (2.22). Inserting Eq. (2.31) in Eq. (2.30) gives for the derivatives

$$\begin{aligned}
&\hat{y}_h^{(1,0)}(x) = \widehat{C}_2/h, && \hat{y}_h^{(0,1)}(x) = \widehat{C}_3/h, && \hat{y}_h^{(2,0)}(x) = \widehat{C}_4/h^2,\\
&\hat{y}_h^{(0,2)}(x) = \widehat{C}_5/h^2, && \hat{y}_h^{(1,1)}(x) = \widehat{C}_6/h^2, && \hat{y}_h^{(3,0)}(x,h) = \widehat{C}_7/h^3,\\
&\hat{y}_h^{(0,3)}(x,h) = \widehat{C}_8/h^3, && \hat{y}_h^{(2,1)}(x,h) = \widehat{C}_9/h^3, && \hat{y}_h^{(1,2)}(x,h) = \widehat{C}_{10}/h^3.
\end{aligned} \tag{2.32}$$

Thus, the vector C in the LPA model (2.18) gives estimates of both the signal and the corresponding derivatives. This clear link of C with the signal and derivative estimates is important for the understanding of the nature of nonparametric estimation.

The LPA model (2.18) of the degree m can be used to estimate the derivative $\partial_{x_1}^{r_1}\partial_{x_2}^{r_2}y(x)$ of the orders $r=(r_1,r_2)$, $|r_1+r_2|\leq m$.

Concerning the meaning of the LPA center x we may conclude that, as we already said for the signal estimation, the LPA derivative estimates can be treated as approximation, interpolation, or extrapolation of the derivative $\partial_{x_1}^{r_1}\partial_{x_2}^{r_2}$, depending on $x\in X$ or $x\notin X$ and on the x location inside or outside of the data set X, respectively.

The local approximation in the form of Eqs. (2.18) and (2.24) is applicable not only for the polynomials in the form of Eq. (2.20) but also for other polynomials as well as for any basis functions $\phi(x)$ in Eq. (2.18) that are reasonable for the local fit. The estimates of the signals and the derivatives are always defined by the general formulas (2.28) and (2.30). However, in general, these estimates (signal and derivative) can depend on all items of the vector C.

The simple coefficient-wise estimates shown in Eqs. (2.29) and (2.32) are valid only for the polynomials (2.20).

It deserves to be mentioned that Eq. (2.30) is not a unique definition of the derivative estimate. The estimate of the derivative can be defined as the corresponding derivative of the signal-estimate $\hat{y}_h(x)$ (see Sec. 4.4):

$$\hat{y}_h^{(r_1,r_2)}(x) = \partial_{x_1}^{r_1}\partial_{x_2}^{r_2}\hat{y}_h(x). \tag{2.33}$$

The estimates (2.30) and (2.33) can be very different at least because the derivatives of the window function $w(x)$ appear in Eq. (2.33) while they are not used in Eq. (2.30).

2.2.2 Windowing and scale

The window w is a function conventionally satisfying the following properties:

$$w(x) \geq 0, \qquad w(0) = \max_x w(x), \qquad \int_{\mathbb{R}^2} w(x)dx = 1. \tag{2.34}$$

The values of w are nonnegative and define the comparative weights of the squared residuals in Eq. (2.24). Roughly speaking, these weights depend on the distances between the observations X_s and the reference point x. The condition $w(0) = \max_x w(x)$ means that the maximum weight is given to the observation with $X_s = x$. This maximum could be achieved for a set of x close to $x = 0$. Then, the same weights are shared by a number of observations. The normalization $\int_{\mathbb{R}^2} w(x)dx = 1$ is convenient but not compulsory.

If the window w is a rectangular function, all observations enter into the LPA criterion with equal weights. Nonrectangular windows such as triangular, quadratic, Gaussian, and so on, prescribe higher weights to observations z_s, with X_s closer to the reference point x.

Table 2.1 shows some typical 1D window functions used in nonparametric estimations. Let us also mention windows conventional in signal processing and associated with the names Kaiser, Hamming, Bartlett, Blackman, Chebyshev, and so on.

Note that the *B-splines* can be used as the windows for the LPA. Let us call $I(t)$ the indicator function of the interval $[-1/2, 1/2]$ and $I^{(r)}(x)$ be the r-time convolution of $I(x)$ with itself, that is,

$$\begin{aligned} I^{(r)}(x) &= \int_{-\infty}^{\infty} I(u)I^{(r-1)}(x-u)du, \qquad r = 1, \ldots, q, \\ I^{(0)}(x) &= I(x). \end{aligned} \tag{2.35}$$

By definition, $I^{(q)}(x)$ is the *B-spline* of order q. This function is a nonnegative polynomial of degree q on the interval $[0, q+1]$, with the maximum value at the middle of the interval. It is also known that $I^{(q)}(x)$ approaches the Gaussian distribution as q is increasing.

Table 2.1 1D window functions

Name	$w(x)$
Rectangular symmetric	$1,\ \lvert x\rvert \leq 1/2$
Rectangular nonsymmetric	$1,\ 0 \leq x \leq 1$
Exponential	$\frac{1}{2}\exp(-\lvert x\rvert)$
Gaussian	$\exp(-x^2/2)/\sqrt{2\pi}$
Epanechnikov	$\frac{3}{4}(1-x^2),\ \lvert x\rvert \leq 1$
Bisquare window	$(1-x^2)^2,\ \lvert x\rvert \leq 1$
Tricube	$(1-\lvert x\rvert^3)^3,\ \lvert x\rvert \leq 1$
Triweight window	$(1-x^2)^3,\ \lvert x\rvert \leq 1$

A multiplicative window

$$w(x) = w_1(x_1)w_2(x_2), \tag{2.36}$$

where $w_1(x_1)$ and $w_2(x_2)$ are functions of scalar arguments, is commonly used for multivariate data.

There is a simple way to create nontrivial 2D windows that are different from the simple multiplicative ones in Eq. (2.36). Let us replace the argument x_1 in $w_1(x_1)$ by the norm $||x||$, where x is a 2D vector and the norm is not exclusively Euclidian. Then, after normalization we obtain 2D window functions satisfying the conditions in Eq. (2.34).

Examples of the 1D windows are shown in Fig. 2.5. They are given in two versions: symmetric and nonsymmetric left/right windows. The nonsymmetric windows are obtained from the symmetric ones by using nonpositive and nonnegative values of the argument.

Often the symmetry of the window is assumed by default. It is a standard in nonparametric estimation and in signal processing to use symmetric support windows. We will show that the symmetry is not always a good choice and the nonsymmetry combined with the adaptive selection of the scale h can enable a serious improvement in results.

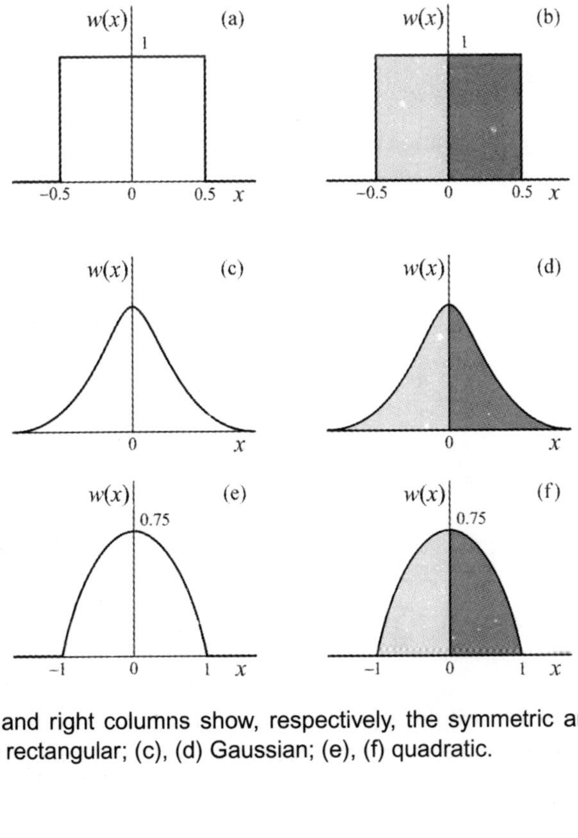

Figure 2.5 Left and right columns show, respectively, the symmetric and nonsymmetric windows: (a), (b) rectangular; (c), (d) Gaussian; (e), (f) quadratic.

By definition, the *support* of a function is a set of argument values where the function takes nonzero values. For the nonnegative window w, this means that

$$supp(w) = \{x : w(x) > 0\}.$$

For example, let $w(x) = 0$ if $||x|| = \sqrt{x_1^2 + x_2^2} > 1$. Then $w_h(x) = 0$ if $||x|| > h$ and

$$supp(w_h) = \{x : ||x|| \leq h\}. \tag{2.37}$$

Thus, the parameter h defines the size of the support of the estimate.

Further, we are mainly concerned with the case when the window $w(x)$ is finite, that is, it has a finite support. Note that the finite support windows in Table 2.1 can be continuous or discontinuous at the boundary of the support. For example, the symmetric rectangular window is discontinuous at $|x| = 1/2$ and the rectangular nonsymmetric one is discontinuous at $x = 0$ and $x = 1$. The Epanechnikov, bisquare, treecube, and treeweight windows are continuous at the boundary $|x| = 1$. The bisquare, treecube, and treeweight windows are even smoother (flatter) at this boundary than the Epanechnikov window, because they have derivatives equal to zero at $|x| = 1$. These are the first-order derivative for the bisquare window and the first and second derivatives for the treecube and treeweight windows. The smoothness of the window enables the smoothness of the LPA estimates.

2.2.3 Estimate calculation

2.2.3.1 Estimate of C

Let us produce a minimization of $J_h(x, C)$ in Eq. (2.24) with respect to C. The criterion $J_h(x, C)$ is quadratic and the corresponding minimum condition has the form

$$\partial_C J_h(x, C) = 0,$$

giving a linear, so-called normal equation:

$$\begin{aligned} \Phi_h \widehat{C} &= \sum_s w_h(x - X_s)\phi_h(x - X_s)z_s, \\ \Phi_h &= \sum_s w_h(x - X_s)\phi_h(x - X_s)\phi_h^T(x - X_s). \end{aligned} \tag{2.38}$$

Provided that the matrix Φ_h is nonsingular, $\det \Phi_h \neq 0$, the solution can be written as

$$\widehat{C} = \Phi_h^{-1} \sum_s w_h(x - X_s)\phi_h(x - X_s)z_s. \tag{2.39}$$

In what follows we use the arguments x and h for $\widehat{C}$ to emphasize the dependence of this solution on the scale parameter h as well as the fact that the estimate is calculated for each reference point x. We may finalize the result in the following form:

$$\hat{C}(x,h) = \arg\min_{C \in \mathbb{R}^M} J_h(x,C),$$
$$\hat{C}(x,h) = \Phi_h^{-1} \sum_s w_h(x - X_s)\phi_h(x - X_s)z_s, \tag{2.40}$$
$$\hat{C}(x,h) = [\hat{C}_1(x,h), \ldots, \hat{C}_M(x,h)]^T.$$

2.2.3.2 Estimate of signal

Let us use $\hat{C}(x,h)$ in Eq. (2.28) instead of C. Then

$$\hat{y}_h(x) = y_h(x,x) = \hat{C}^T(x,h)\phi(0). \tag{2.41}$$

2.2.3.3 Estimate of derivative

Let us use $\hat{C}(x,h)$ in Eq. (2.30) instead of C. We then obtain the derivative estimates in the form

$$\hat{y}_h^{(r_1,r_2)}(x) = (-1/h)^{r_1+r_2} \cdot \hat{C}^T(x,h)\phi^{(r_1,r_2)}(0), \tag{2.42}$$

where

$$\phi^{(r_1,r_2)}(0) = \partial_{x_1}^{r_1}\partial_{x_2}^{r_2}\phi(x)|_{x=0}.$$

For $m = 3$ and the polynomials of the form (2.22), we find that $\hat{y}_h(x) \triangleq \hat{C}_1(x,h)$ is the estimate of the signal, and $\hat{C}_l(x,h)$, $l = 2, \ldots, 10$, define the estimates of the derivatives $y^{(1,0)}(x)$, $y^{(0,1)}(x)$, $y^{(2,0)}(x)$, $y^{(0,2)}(x)$, $y^{(1,1)}(x)$, and so on, respectively, as is in Eq. (2.32). The estimate of the signal is given only by the first item of the vector $\hat{C}(x,h)$; other items of $\hat{C}(x,h)$ give the estimates of the derivatives ordered according to the order of the 2D polynomials in the vector $\phi(x)$.

Formulas (2.41) and (2.42) as well as (2.32) give universal recipes for how the LPA estimates can be calculated. In fact, we use the moving weighted least-square method to find coefficients $\hat{C}$ of the polynomial fit that are exploited as the pointwise estimates of the signal and the derivatives. In this form, the approach is a universal calculation scheme that can be applied to regular and irregular data grids as well as for arbitrary windows w.

2.2.4 Multivariate estimates

Using the multi-index notations, the estimates (2.41)–(2.42) can be written in a compact form that is valid for the general multivariate case $x \in \mathbb{R}^d$. Let $x, X_s \in \mathbb{R}^d$.

Then the estimate of the derivative $\partial_x^r y(x)$ can be written as

$$\begin{aligned}\hat{y}_h^{(r)}(x) &= (-1/h)^{|r|} \cdot \widehat{C}^T(x,h)\phi^{(r)}(0),\\ \widehat{C}(x,h) &= \Phi_h^{-1} \sum_s w_h(x - X_s)\phi_h(x - X_s)z_s,\end{aligned} \tag{2.43}$$

where

$$\begin{aligned}\Phi_h &= \sum_s w_h(x - X_s)\phi_h(x - X_s)\phi_h^T(x - X_s),\\ \phi^{(r)}(0) &= \partial_x^r \phi(x)|_{x=0}.\end{aligned} \tag{2.44}$$

Note that the matrix Φ_h is symmetric, $\Phi_h = \Phi_h^T$.

Here, $r = (r_1, \ldots, r_d)$ is the multi-index showing the estimated derivative. For the polynomial model, ϕ is a vector of the polynomials of the powers $k = (k_1, \ldots, k_d)$:

$$(-1)^{|k|} x^k / k!. \tag{2.45}$$

The multi-index notation in Eq. (2.11) is applied here and values of k are defined by the set

$$\{k \in \mathbb{Z}_+^d : \forall |k| \leq m\},$$

where $\mathbb{Z}_+^d$ is a d-dimensional set of nonnegative integers. We use in Eq. (2.45) all polynomials of the powers from 0 to m. In this way we can derive the estimates of any derivative of the order r such that

$$|r| \leq m.$$

For $r = 0$, the formula (2.43) defines the estimate of $y(x)$:

$$\hat{y}_h(x) = \widehat{C}^T(x,h)\phi(0), \tag{2.46}$$

where $\widehat{C}(x,h)$ and Φ are defined in Eqs. (2.43) and (2.44), respectively.

We remind the reader that in this general form the estimates are applicable not only for the polynomials in Eq. (2.45) but for any linearly independent functions reasonable for the local fit of y and its derivatives. Even if the matrix Φ_h is singular, the solution of the normal equations (2.38) exists and can be written using the pseudoinverse matrix $\Phi_h^{\#}$ instead of the inverse Φ_h^{-1}.

2.2.5 Examples

Let us now consider a few 1D and 2D numerical examples that highlight some of the principal properties of the introduced LPA estimators. These examples demonstrate the nonparametric character of the estimates as well as their asymptotic behavior for small- and large-scale values as $h \to 0$ and $h \to \infty$.

2.2.5.1 *Zero-order 1D estimator*

Let us assume that the zero-order LPA is used, $m = 0$, and observations are given without noise, $z_s = y_s = y(X_s)$. It follows from Eq. (2.46) that the LPA model has the form of the moving weighted average:

$$\hat{y}_h(x) = \sum_s g_h(x, X_s) y_s,$$
$$g_h(x, X_s) = \frac{w_h(x - X_s)}{\sum_s w_h(x - X_s)}, \quad w_h(x) = \frac{1}{h} w\left(\frac{x}{h}\right), \tag{2.47}$$

where the X_s are arbitrary.

(a) Effects of scaling: nonparametric estimation

A sequence of data, $n = 7$, is given on an irregular grid covering the interval [0, 1]. The estimates are calculated for all $x \in (0, 1)$ and for different values of h: small ($h = 1/120$), large ($h = 4$), and two intermediate ($h = 1/24$ and $h = 1/6$). We use the Gaussian window $w(x) = \exp(-x^2/2)/\sqrt{2\pi}$. The corresponding estimates $\hat{y}_h(x)$ are shown in Fig. 2.6.

For large h ($h = 4$, asymptotically $h \to \infty$) the weights $g_h(x, X_s)$ are the same for all observations and the estimate becomes the sample mean:

$$\hat{y}_h(x) \simeq \hat{y}_\infty(x) = \frac{1}{n} \sum_{s=1}^{n} y(X_s).$$

In the alternative limit case of small h ($h = 1/120$, asymptotically $h \to 0$),

$$\hat{y}_h(x) \simeq \hat{y}_0(x) = \lim_{h \to 0} \hat{y}_h(x).$$

It can be seen from Fig. 2.6 that this limit estimate is a piecewise constant signal going exactly through the observations $y(X_s)$ at $x = X_s$. This estimate has jumps at $x = (X_s + X_{s+1})/2$, where it is equal to the mean of two neighboring observations,

$$\hat{y}_0\left(\frac{X_s + X_{s+1}}{2}\right) = \frac{y(X_s) + y(X_{s+1})}{2}.$$

These jumps, marked by black circles in Fig. 2.6, occur exactly at the middle points between the observations.

Let $x_{\min}$ and $x_{\max}$ be the minimum and maximum values of X_s, respectively. For $x < x_{\min}$ and $x > x_{\max}$, the limit estimate $\hat{y}_0(x)$ takes the values of the corresponding nearest first and last observations in the grid. For $h = 1/24$, the estimation curve is smooth and close to the observations,

$$\hat{y}_h(X_s) \simeq y(X_s).$$

It also shows a reasonable interpolation for x values between the observations X_s. For the larger $h = 1/6$, the observations are definitely oversmoothed and the estimates are far from the observed $y(X_s)$.

Figure 2.6 demonstrates two important facts, the first of which is the nonparametric character of the estimate. Remember that here we use the zero-order polynomial fit. However, because of the local nature of this fit, the estimate $\hat{y}_h(x)$ is constant only for $h \to \infty$ and far from being constant for small h. Secondly, the curves show how important the selection of the scale h is. We can obtain a large variety of curves from the constant one obtained for large h to a piecewise constant going through the observation points obtained for small h. Depending on h, the estimates can be very different and for the selection of h, in particular, an adaptive data-driven selection is a key question of nonparametric estimation overall.

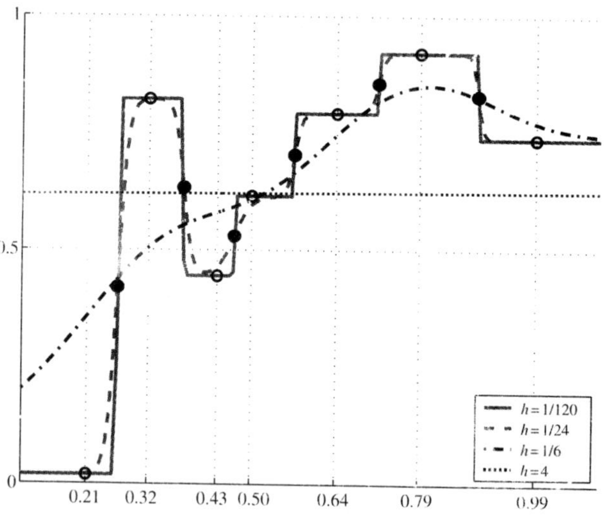

Figure 2.6 Reconstruction of *y* given by a few irregular grid observations marked by circles (o). The LPA of the degree $m = 0$ is used with the Gaussian window. The estimates of different scales h are shown. For large $h = 4$, the estimate is a constant equal to the sample mean of the observations. For small $h = 1/120$, the estimate is a piecewise constant function. The change-point values of this estimate are marked by black circles. For $h = 1/24$, the smooth estimate gives values both close to the observed values and reasonable interpolation for x between the observation. For $h = 1/6$, the estimation curve oversmooths the observations.

(b) "Oracle" scale selection

The scale selection is a much more complex problem than the calculation of the estimate $\hat{y}_h(x)$. This problem is addressed in detail in Chapter 6. Here we provide some illustrations concerning the so-called oracle scales. The "oracle" or "ideal" scale assumes that the true signal is available and its value can be used for evaluation of the accuracy of estimation and for optimization of the accuracy by selection of the proper scales. The oracle scale enables one to analyze the potential of the estimates as well as to study the accuracy of the estimates depending on the parameters of the estimator.

Let us illustrate the selection of the scale h by two different modes. First, we select the best scale h assuming that this h is constant and independent on x. In other words, the scale is the same for all estimates $\hat{y}_h(x)$. This scale is called *global* or *invariant*. Second, let the scale be local or varying. The best scale is selected for each x, and, as a result, the scale depends on x and the estimate is of the form $\hat{y}_{h(x)}(x)$. Compare the estimate $\hat{y}_h(x)$ with the signal's true value $y(x)$ at $x = x_t$. The estimation error is

$$e_h(x_t) = y(x_t) - \hat{y}_h(x_t).$$

Let us use two performance criteria to average these errors. The first global criterion averages the errors over all x_t. It has the form of the sample mean of the squared errors,

$$R^2(h) = \frac{1}{n}\sum_{t=1}^{n} e_h^2(x_t), \tag{2.48}$$

considered as a function of h. This accuracy criterion is used for the selection of the global invariant scale according to the rule

$$h^* = \arg\min_h R^2(h). \tag{2.49}$$

The second local criterion is calculated as a sliding average with the window length $2k + 1$:

$$R^2(x, h) = \frac{1}{2k+1}\sum_{|x-x_t|\leq k} e_h^2(x_t). \tag{2.50}$$

The parameter k is used to smooth observation errors and yield more stable results. The varying oracle scale is defined by the rule

$$h^*(x) = \arg\min_h R^2(x, h). \tag{2.51}$$

The observations are given on the regular grid as $X_s = (s-1)/n, s = 1, \ldots, n$, $n = 100$. The corresponding noisy observations z_s with the additive Gaussian noise having standard deviation $\sigma = 0.2$ are shown in Fig. 2.7(a). A piecewise constant signal y can be seen in Fig. 2.8(a).

The estimates are defined by the formula (2.47) where the noisy z_s are substituted instead of y_s. We use three different windows w in this formula: the symmetric Gaussian, the nonsymmetric left Gaussian, and nonsymmetric right Gaussian. Thus, we study symmetric and nonsymmetric estimates.

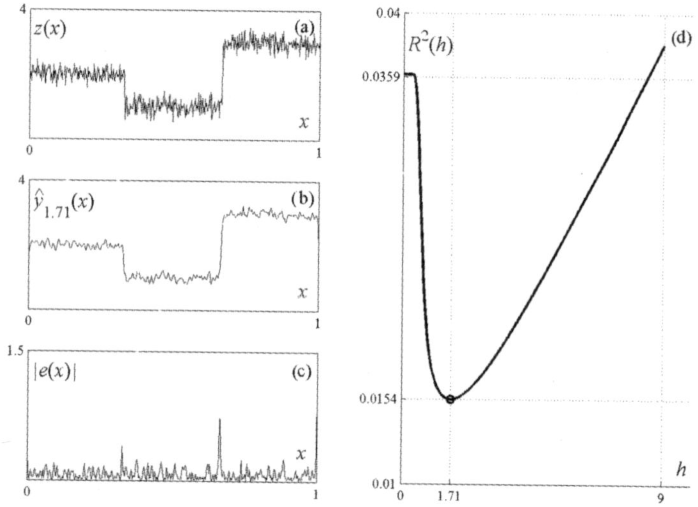

Figure 2.7 The estimator with the ideal invariant scale: (a) the noisy signal; (b) the estimate; (c) the estimation error; (d) the mean-square estimation error as a function of the scale h with the minimum value at $h^* = 1.71$.

For the invariant scale selection the symmetric and nonsymmetric estimators yield more or less equivalent results, which are shown in Fig. 2.7 for the symmetric Gaussian window only. The curve in Fig. 2.7(d) shows the criterion (2.48) as a function of h with a clear optimum value $R^2(h^*) = 0.0154$ at $h^* = 1.71$. The curve in Fig. 2.7(b) presents the estimates with this ideal global scale of $h = h^*$.

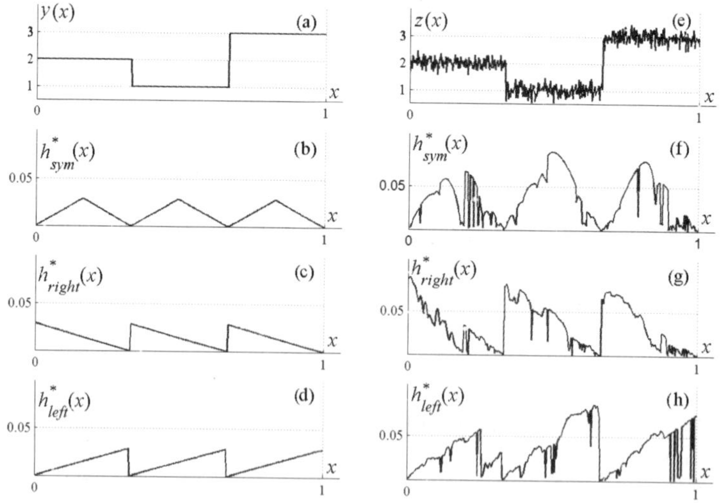

Figure 2.8 The estimator with the ideal varying scale: (a) true signal; (b)–(d) the ideal scales for the symmetric, left, and right estimators (nearly zero noise level); (e) noisy signal, $\sigma = 0.2$; and (f)–(h) the ideal scales for the symmetric, left, and right estimators for noisy observations.

The estimation errors are given by curve (c) with the maximum errors occurring at the change-points of the signal. The maximum errors are naturally always at the change-points because the estimator with a comparatively large window/scale is not able to accurately catch the signal with large variations in a small area.

The varying-scale selection is shown in Fig. 2.8. We consider two cases that are different by the level of the noise. Let us start with the case when the level of noise is very small. Curves (b)–(d) show the ideal scales (window sizes) as defined by Eq. (2.51) for the symmetric, right, and left estimators, respectively. It is interesting to analyze the character of these ideal $h^*(x)$. The ideal scale $h^*(x)$ is large for x that are far from the jumps of the signal y and from the boundaries of the interval $[0, 1]$ and becomes smaller when x is approaching these discontinuity points. This behavior of $h^*(x)$ is explained as follows. If there is noise in the observations (even a small one), the ideal estimator always tries to use a scale h as large as possible. However, near discontinuities, the large changes in the signal result in large errors in estimation. This prevents the ideal estimator from using a large h.

These ideal scales are essentially different for the symmetric, right, and left estimators. The symmetric window estimator takes a maximum value of the ideal scale at the middle points of segments where the signal is invariant. This ideal value decreases to zero while the argument approaches the left and right change-points. This happens because at the change-point a symmetric neighborhood of x becomes infinitely small where the signal has a correct constant value.

The right-hand estimator has a maximum ideal scale at the left-hand point of the segments where the signal is invariant. At this point the right-hand estimator is able to use all data located on the right-hand side of x. The ideal scale decreases while x increases and approaches the change-point where the ideal scale takes the zero value. The left-hand estimator demonstrates the opposite behavior with the maximum ideal scale on the right-hand side of the segments with the invariant signal values.

Curves (f)–(h) show the ideal scales obtained for the noisy observations shown in Fig. 2.8(e) obtained with a larger value of $\sigma = 0.2$. Overall, the character of these noisy ideal scales is similar to the behavior observed in the ideal scales (b)–(d) for a small level of noise. In particular, for a symmetric estimator they have maximum values at the middle points of the segments with the invariant signal values and for the nonsymmetric estimators at the boundaries of these segments.

Note that the scales obtained for the noisy data (curves (f)–(h)) are essentially larger than the scales (curves (b)–(d)) obtained for nearly noiseless observations. This happens because a large level of noise requires larger optimal window sizes for the estimates.

The global accuracy criterion (2.48) with the varying ideal scales

$$R^2(h^*) = \frac{1}{n}\sum_{t=1}^{n}[e(x_t)]^2,$$

$$e(x_t) = y(x_t) - \hat{y}_{h^*(x_t)}(x_t)$$

gives the following figures:

$$R^2(h^*_{sym}(x)) = 0.0016,\ R^2(h^*_{right}(x)) = 0.0011,\ R^2(h^*_{left}(x)) = 0.0018. \quad (2.52)$$

We remind the reader that the ideal invariant scale estimator gives $R^2 = 0.0154$, which is much worse than the mean-square values in Eq. (2.52). Thus, the varying ideal scale estimators are able to demonstrate a much better performance than the estimators with the ideal but invariant scale.

2.2.5.2 First-order 1D estimator

Let us go to the first-order LPA, $m = 1$, when the LPA model includes the polynomials of the zero and first degrees and the two coefficients C_1 and C_2 can be found from the normal equations (2.38).

If the moment matrix Φ_h is denoted as

$$\Phi_h = \begin{pmatrix} S_0 & S_1 \\ S_1 & S_2 \end{pmatrix},$$

$$S_l(x) = (-h)^{-l} \sum_s w_h(x - X_s)(x - X_s)^l, \qquad l = 0, 1, 2,$$

then the solution for $\widehat{C}(x, h)$ can be presented in the form

$$\begin{aligned} \hat{y}_h(x) &= \widehat{C}_1(x, h) = \sum_s g_h(x, X_s) z_s, \\ g_h(x, X_s) &= \frac{w_h(x - X_s)[S_2(x) + S_1(x)(x - X_s)/h]}{S_2(x)S_0(x) - S_1^2(x)}, \end{aligned} \qquad (2.53)$$

for the signal estimate, and in the form

$$\begin{aligned} \hat{y}_h^{(1)}(x) &= \widehat{C}_2(x, h)/h = \sum_s g_h^{(1)}(x, X_s) z_s, \\ g_h^{(1)}(x, X_s) &= -\frac{w_h(x - X_s)[S_1(x) + S_0(x)(x - X_s)/h]}{(S_2(x)S_0(x) - S_1^2(x))h}, \end{aligned} \qquad (2.54)$$

for the derivative estimate.

We consider the estimates $\hat{y}_h(x)$ for the scenario given in Fig. 2.6. Thus, $z_s = y_s$ and the reconstruction of y is based on a few irregular grid observations marked by open circles. The LPA of the degree $m = 1$ is used with the Gaussian symmetric window. The estimates of the different scales h are shown in Fig. 2.9. For the largest $h = 4$, the estimate is a linear function coinciding with the standard parametric LSM fit by the linear polynomial model. For the smallest $h = 1/24$, the estimate is a nearly piecewise linear function going through the two nearest observations. For $h = 1/12$, the smooth estimate gives both values close to the observed values and a reasonable interpolation for x between the observations. For $h = 1/3$, the estimation curve oversmooths the observed values. Comparing these curves with

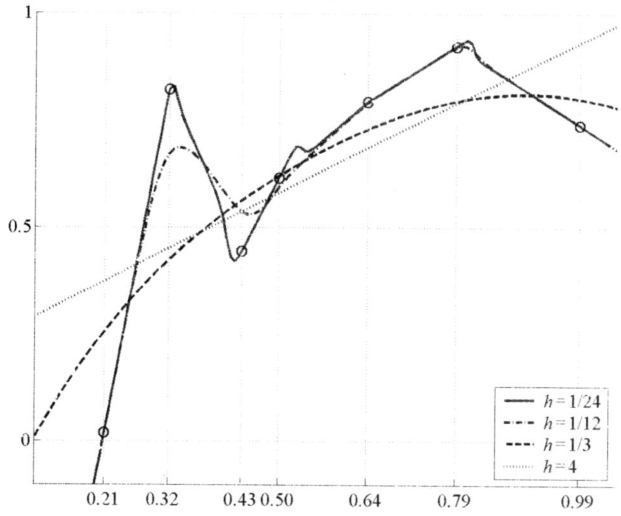

Figure 2.9 Reconstruction of y given by a few irregular grid observations marked by open circles (o). The LPA of the degree $m = 1$ is used with the Gaussian window. Estimates with different scales h are shown. For the large $h = 4$, the estimate is a linear function coinciding with the parametric least-square mean error fit by the linear polynomial model. For smallest $h = 1/24$, the estimate is a nearly piecewise linear function going through the observations. For $h = 1/12$, the smooth estimate gives both values close to the observed values and reasonable interpolation for x between the observation. For $h = 1/3$, the estimation curve oversmooths the observations.

the ones shown in Fig. 2.6, we may conclude that the estimates are essentially different depending on the degree of the LPA.

2.2.5.3 Zero-order 2D estimator

The zero-order formula (2.47) is valid for multidimensional signals without any changes. The behavior of these 2D estimates for different h is qualitatively similar to the one discussed above for the 1D case. Namely, for large h, the estimates coincide with the corresponding parametric LSM fits. For the zero order it means that this estimate is a constant and equal to the sample mean of the 2D observation. For small h, this estimate is a piecewise constant function going through the nearest observations. Figure 2.10 shows this limit estimate as h is very small. We use the 2D standard Gaussian probability density for the window function w. The scale h is the standard deviation of this density. Naturally, the obtained 2D piecewise constant curve has a much more complex structure than the 1D limit curve in Fig. 2.6. The 1D change-points in Fig. 2.6 are replaced by change-lines in Fig. 2.10. The limit estimate (for $h \to 0$) can be represented in the form of Eq. (2.8) as

$$\hat{y}_0(x) = \frac{1}{\sum_s \mathbf{1}[X_s \in A(x)]} \sum_s y_s \mathbf{1}[X_s \in A(x)],$$

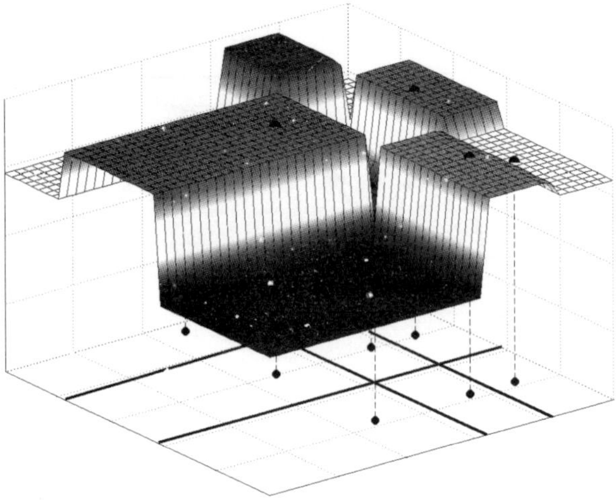

Figure 2.10 The LPA estimate with $m = 0$ and small h. This limit estimate is a piecewise constant surface going through the nearest observations.

where $A(x)$ is a set of the observations closest to x and $\sum_s \mathbf{1}[X_s \in A(x)]$ is the number of observations in the set $A(x)$.

This limit estimate produces a segmentation of the 2D space with constant values of the estimate in the corresponding subareas. The Gaussian window w works as a distance measure between the estimation point x and the observations X_s and defines the limit sets $A(x)$.

2.2.5.4 First-order 2D estimator

Let us consider the 2D estimate defined by the formulas (2.46) for $m = 1$ with $\phi(x) = (1, -x_1, -x_2)^T$ and the vector $C = (C_1, C_2, C_3)$. As above, the 2D standard Gaussian probability density with h for the standard deviation is used for the window function. It can be seen that, for large h, this estimate is equivalent to the linear parametric 2D plane obtained by the LSM.

The behavior of the estimate for small h is more interesting and nontrivial because it is a nearly piecewise linear 2D function similar to its 1D counterpart shown in Fig. 2.9. This limit estimate is shown in Fig. 2.11 for the same data as in Fig. 2.10. For each x the estimate $\hat{y}_h(x)$ is a linear 2D function going through three observations of $y(X_s)$ with coordinates X_s closest to the current estimation point x.

For values of h that are not so small and not too large we obtain intermediate results that are somewhere in between the two limit estimates corresponding to $h \to 0$ and $h \to \infty$. It is clear that, depending on h, we may have a huge variety of the estimates. This demonstrates once more that the selection of h is an important point of the considered class of nonparametric estimates.

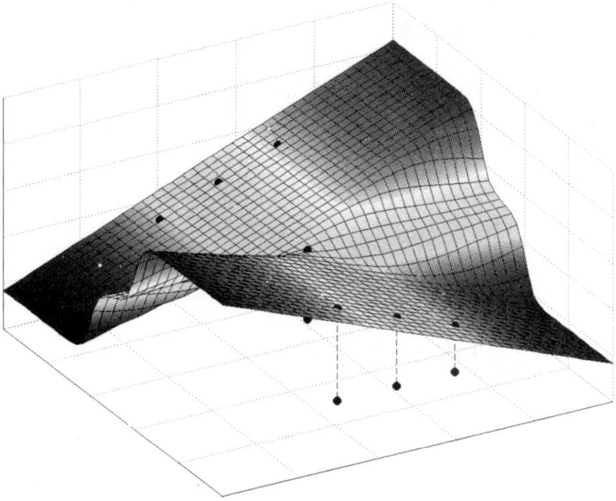

Figure 2.11 The LPA estimate with $m = 1$ and small h. This limit estimate is a piecewise linear surface going through the nearest observation points.

2.3 Kernel LPA estimates

2.3.1 Estimate of signal

Consider the general multivariate case $x \in R^d$. Inserting $\widehat{C}(x, h)$ from of Eq. (2.43) into the estimate of Eq. (2.46) we find the signal estimate as an output of the linear filter:

$$\hat{y}_h(x) = \sum_s g_h(x, X_s) z_s, \tag{2.55}$$

$$\begin{aligned} g_h(x, X_s) &= w_h(x - X_s)\phi_h^T(x - X_s)\Phi_h^{-1}\phi(0), \\ \Phi_h &= \sum_s w_h(x - X_s)\phi_h(x - X_s)\phi_h^T(x - X_s), \end{aligned} \tag{2.56}$$

where $g_h(x, X_s)$ is the impulse response of this filter for a fixed x. Note that the summation $\sum_s$ in Eqs. (2.55–2.56) and in what follows is produced over all observation points X_s. For the image defined on the regular grid X in Eq. (2.4), this means a summation over all pixels in the image.

The matrices Φ_h and Φ_h^{-1} are symmetric and $g_h(x, X_s)$ can be given in the equivalent form

$$g_h(x, X_s) = w_h(x - X_s)\phi^T(0)\Phi_h^{-1}\phi_h(x - X_s).$$

The formulas in Eq. (2.56) define a *smoothing filter* (*smoothing operator*) of the polynomial order (of the order) m.

2.3.2 Estimate of derivative

Substituting $\widehat{C}(x, h)$ in Eq. (2.43) into $\hat{y}_h^{(r)}(x)$ we obtain an estimate of the derivative $\partial^r y(x)$ in the form

$$\hat{y}_h^{(r)}(x) = \sum_s g_h^{(r)}(x, X_s) z_s, \tag{2.57}$$

$$\begin{aligned} g_h^{(r)}(x, X_s) &= (-1/h)^{|r|} w_h(x - X_s) \phi_h^T(x - X_s) \Phi_h^{-1} \phi^{(r)}(0), \\ \Phi_h &= \sum_s w_h(x - X_s) \phi_h(x - X_s) \phi_h^T(x - X_s), \end{aligned} \tag{2.58}$$

with $g_h^{(r)}(x, X_s)$ as the impulse response of this differentiating filter for a fixed x. The formulas in Eq. (2.58) define a *differentiation filter* (*differentiation operator*) of the polynomial order (of the order) m.

Concerning the terminology, we note that in statistics the weights $g_h(x, X_s)$ and $g_h^{(r)}(x, X_s)$ are named *kernels* and then the estimates (Eqs. (2.55) and (2.57)) are called *kernel estimates*. In image processing, the term "mask" is standard for the weights $g_h(x, X_s)$ and $g_h^{(r)}(x, X_s)$. The parameter h simultaneously defines the scale as well as the size of the support of the kernels $g_h^{(r)}(x, X_s)$ and $g_h(x, X_s)$. A kernel with a finite support is called *finite* and the corresponding filter is a *finite impulse response* (*FIR*) filter. The observations z_s and the estimates $\hat{y}_h^{(r)}$ are, respectively, the input and output signals of these filters (operators). The kernels g_h and $g_h^{(r)}$ are called *polynomials* of the order m if the vectors $\phi_h(x)$ in Eqs. (2.56) and (2.58) are composed from the polynomials in Eq. (2.45) of the power m,

$$(-1)^{|k|} x^k / k!, \ k \in \mathbb{Z}_+^d : \forall |k| \leq m. \tag{2.59}$$

Mainly we use the estimates with the polynomial approximation, and even the term LPA assumes this sort of basis function to be used.

2.3.3 Reproducing polynomial kernels

The introduced LPA estimates are accurate for polynomials, meaning that the LPA gives accurate values of the signal if the observations are samples of polynomials. For derivative estimations, the differentiating LPA estimate is accurate if the observations are given as samples of a polynomial signal. In both cases the estimates are also polynomials. Thus, the LPA defines the estimates that are reproducing for polynomials. We call the corresponding kernels g_h and $g_h^{(r)}$ *reproducing polynomial kernels* and the corresponding estimates $\hat{y}_h(x)$ and $\hat{y}_h^{(r)}(x)$ *reproducing polynomial estimates*. The accurate meaning of this concept is given in the following proposition.

Proposition 1 *Let the estimates $\hat{y}_h(x)$ in Eq. (2.55) and $\hat{y}_h^{(r)}(x)$ in Eq. (2.57), $x \in \mathbb{R}^d$, be the LPA estimates of the order m and y be a polynomial, $y \in \mathcal{P}^m \subset \mathbb{R}^d$,*

$z_s = y(X_s)$. *Then, for any* $h > 0$ *such that* $\det \Phi_h \neq 0$,

$$\hat{y}_h(x) = y(x), x \in \mathbb{R}^d, \tag{2.60}$$

$$\hat{y}_h^{(r)}(x) = y^{(r)}(x), x \in \mathbb{R}^d. \tag{2.61}$$

Proof. According to Eq. (2.21), any polynomial y of the power m at $x = X_s$ can be represented in the form

$$y_s = y(X_s) = y((X_s - x) + x) = \phi_h^T(x - X_s)a(x), \tag{2.62}$$

where $a(x) \in \mathbb{R}^M$ is a vector-coefficient independent of X_s.

Inserting Eq. (2.62) in Eq. (2.55) instead of z_s we obtain

$$\begin{aligned}\hat{y}_h(x) &= \sum_s g_h(x, X_s)y(X_s)\\ &= \sum_s g_h(x, X_s)\phi_h^T(x - X_s)a(x)\\ &= \phi^T(0)\Phi_h^{-1}\sum_s w_h(x - X_s)\phi_h(x - X_s)\phi_h^T(x - X_s)a(x)\\ &= \phi^T(0)\Phi_h^{-1}\Phi_h a(x) = \phi^T(0)a(x).\end{aligned} \tag{2.63}$$

For $X_s = x$, Eq. (2.62) gives $y(x) = \phi^T(0)a(x)$. Then, it follows from Eq. (2.63) that

$$\hat{y}_h(x) = y(x), \tag{2.64}$$

which proves Eq. (2.60).

Substitute Eq. (2.62) in Eq. (2.57) instead of z_s. Transformations show that

$$\begin{aligned}\hat{y}_h^{(r)}(x) &= \sum_s g_h^{(r)}(x, X_s)y(X_s)\\ &= (-1/h)^{|r|}[\phi^{(r)}(0)]^T\Phi_h^{-1}\sum_s w_h(x - X_s)\phi_h(x - X_s)\phi_h^T(x - X_s)a(x)\\ &= (-1/h)^{|r|}[\phi^{(r)}(0)]^T a(x).\end{aligned} \tag{2.65}$$

Let us differentiate $y(X_s)$ in Eq. (2.62) on X_s. Then

$$\partial_{X_s}^r y(X_s) = (-1/h)^{|r|}[\partial_{(x-X_s)/h}^r \phi_h^T(x - X_s)]a(x).$$

and for $x = X_s$ we yield

$$\partial_x^r y(x) = (-1/h)^{|r|}[\phi^{(r)}(0)]^T a(x). \tag{2.66}$$

Comparing the right-hand sides of Eqs. (2.65) and (2.66) we conclude that

$$\hat{y}_h^{(r)}(x) = y^{(r)}(x), \tag{2.67}$$

which proves Eq. (2.61). □

The results in Eqs. (2.60) and (2.61) are valid for any scale parameter h, and for any x and X_s with the only requirement that the matrix Φ_h is nonsingular, i.e., $\det \Phi_h \neq 0$. Thus, the filters designed according to the LPA method give accurate values of the signal and its derivatives (of the orders $(r_1 + \cdots + r_d) \leq m$) for any polynomial input signal of degrees up to m. The designed linear filters can be applied to data given on any regular or irregular grids—in particular, to data with lost observations and for data interpolation problems when the center x of the LPA does not belong to the regular grid.

2.4 Nonparametric regression

This section presents the local approximation approach from a statistical perspective.

2.4.1 Regression function

Suppose that we are given random pairs of observations,

$$\{z_s, X_s\}, s = 1, \ldots, n,$$

where $z_s \in \mathbb{R}$ is viewed as a random observation at $X_s \in \mathbb{R}$ and X_s also can be random. Let $f(z, x)$ be a joint probability density of the random $\{z_s, X_s\}$. The conditional probability density $f(z|x)$ of z provided that the random $X = x$ is defined as $f(z|x) = f(z, x) / \int_{\mathbb{R}} f(z, x)dz$ and the conditional expectation of z is of the form (e.g. [174]):

$$E\{z|X = x\} = \int_{\mathbb{R}} zf(z|x)dz = \frac{\int_{\mathbb{R}} zf(z, x)dz}{\int_{\mathbb{R}} f(z, x)dz}, \tag{2.68}$$

where $E\{z|X = x\}$ denotes the conditional expectation of the random z provided that a fixed $X = x$.

This conditional expectation, which is considered to be a function of x, is called a *regression*. Let us use the symbol y for this regression:

$$y(x) = \frac{\int_{\mathbb{R}} zf(z, x)dz}{\int_{\mathbb{R}} f(z, x)dz}. \tag{2.69}$$

It is well-known that this regression is the best least-squares error estimate of z provided that X is fixed, which explains the interest in regressions of random data to explain and predict data.

It is clear for the model (2.1) that $E\{z|X = x\} = y(x)$. Thus, $y(x)$ in Eq. (2.1) is a regression of the random z on x. This is a special case of the general regression definition (Eq. (2.68)) specified by the assumptions that the noise is additive, the variance of the noise is invariant, and the X_s are deterministic.

If the joint density $f(z, x)$ is given, the regression can be calculated according to the definition in Eq. (2.69). However, if the density $f(z, x)$ is unknown, the direct use of Eq. (2.69) is not possible. If the density $f(x, y)$ can be specified by a finite number of parameters, then the regression $y(x)$ is defined by these parameters and we arrive at the parametric estimation problem, where regression estimation is reduced to the estimation of unknown parameters. If the density $f(x, y)$ does not allow this sort of parameterization, the density as well as the regression are nonparametric, and nonparametric methods should be used for estimation.

2.4.2 Nadaraya-Watson estimate

Historically, the first nonparametric estimates of the regression y were proposed independently in 1964 based on two different concepts. Both works were concerned with the case of a scalar regression with $d = 1$.

2.4.2.1 *Nadaraya estimate*

The Nadaraya estimate [158] has been introduced as a heuristic idea in the form of the weighted sum of the observations:

$$\hat{y}_h(x) = \frac{1}{\sum_s w_h(x - X_s)} \sum_s w_h(x - X_s) z_s, \tag{2.70}$$

where a nonnegative window $w_h(x - X_s)$ defines a measure of a distance between the observation X_s and the target point x. These normalized distances define the estimate in Eq. (2.70) in the form of Eq. (2.55) as

$$\hat{y}_h(x) = \sum_s g_h(x, X_s) z_s, \; g_h(x, X_s) = \frac{w_h(x - X_s)}{\sum_s w_h(x - X_s)}, \tag{2.71}$$

with the kernel $g_h(x, X_s)$.

2.4.2.2 *Watson estimate*

The Watson estimator [230], also in the form of Eq. (2.70), has been derived (not heuristically proposed as in [158]) using the definition of the regression as the conditional expectation in Eq. (2.69). Let $\hat{f}_h(z, x)$ be the Parzen kernel estimator of the probability density $f(z, x)$ [175],

$$\hat{f}_h(z, x) = \frac{1}{nh^2} \sum_s \rho\left(\frac{z - z_s}{h}, \frac{x - X_s}{h}\right), \tag{2.72}$$

$$\int_{\mathbb{R}} \rho(z, x) dx dz = 1, \text{ and } \int_{\mathbb{R}} z\rho(z, x) dz = 0. \tag{2.73}$$

The kernel $\rho \geq 0$ is assumed to be even on both arguments, $\rho(z, x) = \rho(-z, x)$ and $\rho(z, x) = \rho(z, -x)$, and n is the total number of observations. Substitute $\hat{f}_h(z, x)$ in Eq. (2.72) in Eq. (2.69) instead of $f(z, x)$ and calculate the corresponding integrals.

We have for $\int_{\mathbb{R}} f(z, x)dz$:

$$\int_{\mathbb{R}} \hat{f}_h(z, x)dz = \frac{1}{nh^2}\sum_s \int_{\mathbb{R}} \rho\left(\frac{z - z_s}{h}, \frac{x - X_s}{h}\right)dz$$
$$= \frac{1}{nh}\sum_s \int_{\mathbb{R}} \rho\left(v, \frac{x - X_s}{h}\right)dv$$
$$= \frac{1}{nh}\sum_s w\left(\frac{x - X_s}{h}\right) = \frac{1}{n}\sum_s w_h(x - X_s), \qquad (2.74)$$

where it is denoted as

$$w(x) = \int_{\mathbb{R}} \rho(v, x)dv,\ w_h(x) = \frac{1}{h}w\left(\frac{x}{h}\right).$$

In a similar way we obtain for $\int_{\mathbb{R}} zf(z, x)dz$:

$$\int_{\mathbb{R}} z\hat{f}_h(z, x)dz = \frac{1}{nh^2}\sum_s \int_{\mathbb{R}} z\rho\left(\frac{z - z_s}{h}, \frac{x - X_s}{h}\right)dz$$
$$= \frac{1}{nh^2}\sum_s \left[\int_{\mathbb{R}} (z - z_s)\rho\left(\frac{z - z_s}{h}, \frac{x - X_s}{h}\right)dz\right.$$
$$\left. + z_s \int_{\mathbb{R}} \rho\left(\frac{z - z_s}{h}, \frac{x - X_s}{h}\right)dz\right].$$

Here the first integral is equal to zero because of Eq. (2.73). Then

$$\int_{\mathbb{R}} z\hat{f}_h(z, x)dz = \frac{1}{nh^2}\sum_s z_s \int_{\mathbb{R}} \rho\left(\frac{z - z_s}{h}, \frac{x - X_s}{h}\right)dz = \frac{1}{n}\sum_s z_s w_h(x - X_s). \qquad (2.75)$$

Inserting Eqs. (2.74) and (2.75) into Eq. (2.69), we calculate the estimate of the conditional expectation $y(x)$ as

$$y(x) = \frac{\int_{\mathbb{R}} zf(z, x)dz}{\int_{\mathbb{R}} f(z, x)dz} \simeq \frac{\frac{1}{n}\sum_s z_s w_h(x - X_s)}{\frac{1}{n}\sum_s w_h(x - X_s)}$$
$$= \frac{1}{\sum_s w_h(x - X_s)}\sum_s w_h(x - X_s)z_s.$$

This proves the estimate in Eq. (2.70). Thus, the same estimator is obtained from a different starting point using the statistical definition of regression.

2.4.2.3 LPA estimate

We already met the Nadaraya-Watson estimate in Eq. (2.47), which was derived from the LPA. However, let us once more demonstrate how the Nadaraya-Watson estimator appears using the LPA approach.

Let us apply in Eq. (2.24) the LPA model of the zero degree, $m = 0$. Then, the vector C is a scalar and the estimate is a solution of the problem

$$\hat{y}_h(x) = \widehat{C}(x,h) = \arg\min_C \Big(\sum_s w_h(x - X_s)[z_s - C]^2\Big). \tag{2.76}$$

A simple calculation shows that

$$\partial_C \sum_s w_h(x - X_s)[z_s - C]^2 = 0 \Longrightarrow C \sum_s w_h(x - X_s) = \sum_s w_h(x - X_s) z_s,$$

which immediately gives $C = \sum_s w_h(x - X_s)z_s / \sum_s w_h(x - X_s)$ and the formula (2.70) for the estimate $\hat{y}_h(x)$.

We can arrive at the same estimator (Eq. (2.70)) starting from three different concepts: the heuristic idea of Nadaraya, the statistically optimal estimator of Watson, and the LPA. It follows from the last approach that the Nadaraya-Watson estimator is a restrictive zero-order model where the estimated function is assumed to be locally constant.

2.5 Nonparametric interpolation

Let us consider the interpolation problem in the following classical setting. The signal of continuous argument y should be reconstructed for $x \in X \subset \mathbb{R}^d$ from discrete observations $y_s = y(X_s)$. It is assumed that the observations are accurate,

$$z_s = y_s, \qquad s = 1, \ldots, n,$$

and a reconstruction $\hat{y}_h(x)$ is a subject of requirements,

$$\hat{y}_h(X_s) = y_s, \qquad s = 1, \ldots, n, \tag{2.77}$$

that is, the curve $\hat{y}_h(x)$ goes through the observation points.

The standard LPA nonparametric technique can be used and can be quite efficient for interpolation. However, the ordinary windows discussed in Sec. 2.2.2 give the estimates satisfying the conditions in Eq. (2.77) only for small h as $h \to 0$.

These limit estimates are piecewise constant for $m = 0$ or piecewise linear for $m = 1$. In general, they do not give a smooth interpolation for all x.

Special windows w can be used in the LPA to satisfy two seemingly contradictory requirements: polynomial smoothness of the estimates for intermediate values of x, and $\hat{y}_h(X_s) = y_s$ for the nodes $x = X_s$. These special windows are introduced as singular functions taking infinite values at $x = 0$. They give infinite weights to the residuals in Eq. (2.24) for $x = X_s$. As a result, the estimate $\hat{y}_h(x)$ enables zero residuals for these points and the conditions in Eq. (2.77) are automatically satisfied for any h.

Following [130], we present the following singular interpolation windows:

$$w(x) = ||x||^{-\alpha},\ \alpha > 0, \tag{2.78}$$

$$w(x) = \begin{cases} \cos^2(||x||\pi/2)/||x||^2, & \text{if } ||x|| \leq 1, \\ 0, & \text{otherwise}, \end{cases} \tag{2.79}$$

$$w(x) = ||x||^{-2}[1 - ||x||]_+^2, \tag{2.80}$$

where

$$[z]_+ = z \quad \text{if } z \geq 0 \qquad \text{and} \qquad [z]_+ = 0 \quad \text{if } z < 0.$$

The plots of these windows for $x \in \mathbb{R}$ are shown in Fig. 2.12.

Let us consider the Nadaraya-Watson estimate in Eq. (2.70) with the window in Eq. (2.78). The kernel of this estimate is

$$g_h(x, X_s) = \frac{w[(x - X_s)/h]}{\sum_s w[(x - X_s)/h]} = \frac{||x - X_s||^{-\alpha}}{\sum_s ||x - X_s||^{-\alpha}}, \tag{2.81}$$

with the corresponding estimate as

$$\hat{y}_h(x) = \sum_s g_h(x, X_s) y_s = \frac{\sum_s ||x - X_s||^{-\alpha} y_s}{\sum_s ||x - X_s||^{-\alpha}}. \tag{2.82}$$

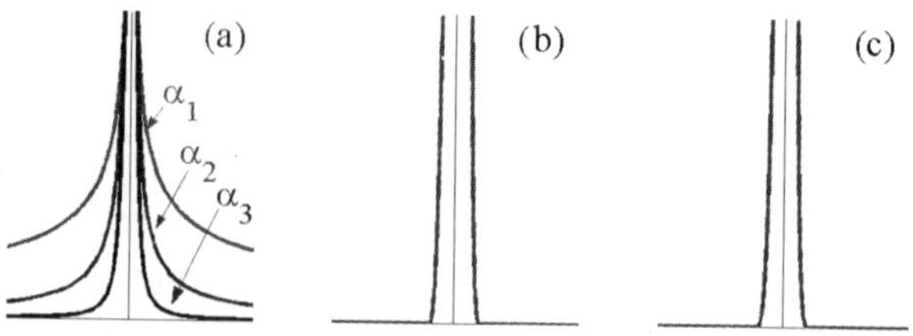

Figure 2.12 Interpolation windows given by Eq. (2.78), $\alpha_3 < \alpha_2 < \alpha_1$, (2.79), and (2.80), respectively.

The following interesting properties can be verified for this estimate provided that $\alpha > 1$ (the 1D case is considered in [130]):

1. The estimate $\hat{y}_h(x)$ has an infinite number of continuous derivatives for $x \neq X_s$;
2. For large $||x - X_s|| \to \infty$,

$$\hat{y}_h(x) \to \frac{1}{n} \sum_s y_s; \tag{2.83}$$

3. At the observation points, $x \to X_k$, the following limit passages hold for the kernel and signal estimates:

$$\lim_{x \to X_k} g_h(x, X_s) = \delta_{s,k},$$

and

$$\lim_{x \to X_s} \hat{y}_h(x) = y(X_s). \tag{2.84}$$

4. At the observation points, $x \to X_s$, the following limit passage holds for the derivative of the signal estimate:

$$\lim_{x \to X_s} \partial_x \hat{y}_h(x) = 0. \tag{2.85}$$

Thus, the requirement in Eq. (2.77) holds. The last property, Eq. (2.85), shows that there is an effect of a "flat area" at the observation points because the derivative of the estimate is equal to zero [130].

For LPA with $m > 0$ the following can be stated for the estimates with the windows (2.78)–(2.80):

1. The interpolation requirements in Eq. (2.77) hold;
2. For larger $||x - X_s|| \to \infty$, the estimate $\hat{y}_h(x)$ is a polynomial of degree m;
3. The effect of the flat area in Eq. (2.85), in general, has no place,

$$\lim_{x \to X_s} \partial_x \hat{y}_h(x) \neq 0. \tag{2.86}$$

The singularity of the window function w_h means that the elements of the matrix Φ_h in Eq. (2.56) have infinite values at $x = X_s$. This matrix can be ill-conditioned or even singular, which means that the calculation of the estimates for x close to the grid points can cause computing problems.

Figure 2.13 illustrates how the interpolation LPA works. Consider the interpolation produced according to the formulas in Eq. (2.47) ($m = 0$) and Eq. (2.53) ($m = 1$) with the interpolation window in Eq. (2.78), where $\alpha = 2$. Curves (a) and (b) in Fig. 2.13 are shown for the LPA of orders $m = 0$ and $m = 1$, respectively.

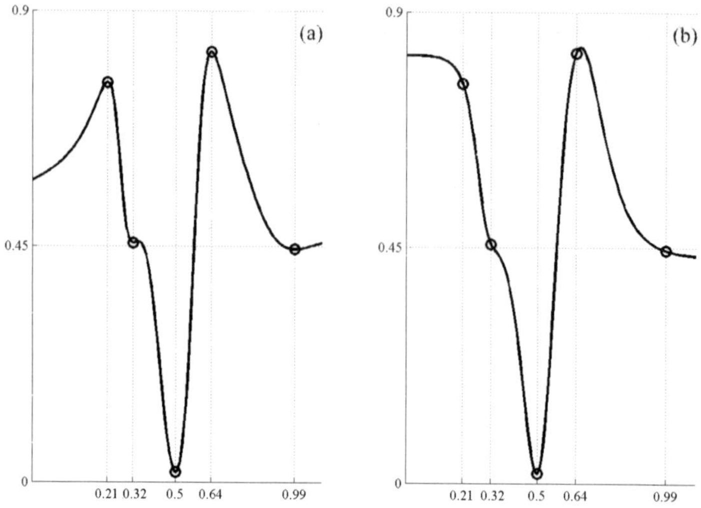

Figure 2.13 Interpolation produced using the LPA of the orders $m = 0$ [curve (a)] and $m = 1$ [curve (b)] with the interpolation window of (2.78), $\alpha = 2$. Observations are shown by circles.

Both interpolation curves go through the observations and demonstrate a reasonable interpolation of a few given observations. The flat area effect is clearly seen at the observation points for $m = 0$. The behavior of these curves between the observations for $x \in [0.21, 0.99]$ is nearly identical. The essential difference can be noted outside of the observation interval for $x < 0.21$ or $x > 0.99$.

Chapter 3

Shift-Invariant LPA Kernels

Shift-invariant kernels and estimates are introduced in Sec. 3.1, where the formulas for the kernels are derived. The smoothness of the shift-invariant kernels and the reproducing properties of the estimates are specified. The corresponding vanishing moment conditions are presented in Sec. 3.2.

The frequency domain characterization of the LPA kernels and estimates are presented in Sec. 3.3. Using orthonormal polynomials allows us to obtain analytical formulas for smoothing and differentiating LPA kernels.

Numerical 2D examples of LPA filters are discussed in Sec. 3.4 for uniform and Gaussian windows. In Sec. 3.5 the conventional differentiators are considered with connections to the general LPA approach.

3.1 Regular Grid Kernels

The estimators considered in the previous chapter are quite general. They are applicable for regular/irregular and finite/infinite data grids as well as for finite and infinite window supports.

Here we concentrate on *shift-invariant* (*homogeneous*) kernels. These are a special class of the LPA estimates that naturally appears when the grid is assumed to be regular. These kernels give estimates that are simple to implement and the implementation can be done by fast convolution algorithms.

Let the LPA be applied to 2D image data defined on a rectangular finite grid in Eq. (2.4). The summation on X_s in the estimation formulas in Eqs. (2.55) and (2.57) is produced over image pixels only. Figure 2.4 shows the pixels used in the square support LPA depending on h. If the estimation point x is far from the image boundary, then the entire square window w can be inside the image and the total window support is exploited in estimation.

However, if x is close to the image boundaries or h is too large, then part of the support is outside the image area and only a part of this square support works. We may conclude that in the estimates in Eqs. (2.55) and (2.57) the active support (really used in estimation) depends on x and h. Even more, the estimate kernel is data-dependent and data-adaptive.

For all x and h, the LPA enables an accurate polynomial fit of the data provided that the number of pixels in the active support area is sufficient. This estimator reproduces polynomial components of the image intensity regardless of the location of the support with respect to the boundaries. As a result, these LPA estimators do not have specific boundary effects typical for the standard filtering based on convolution operation. However, this essential advantage is at the price of a higher computational load. It can be seen that the symmetric matrix

$$\Phi_h = \sum_s w_h(x - X_s)\phi_h(x - X_s)\phi_h^T(x - X_s),$$

used in the estimates depends on x in the areas close to the boundaries because a number of pixels in the sum depends on the location of the window support.

As a result, the kernels $g_h(x, X_s)$ and $g_h^{(r)}(x, X_s)$ are *shift-variant* or *shift-dependent* (*heterogeneous*) and depend on both arguments x (reference or estimation point) and X_s (observation location).

In this chapter we introduce *shift-invariant* or *shift-independent* kernel estimates. Let us assume that the observation points X_s and the estimation points x belong to the same regular grid and this grid is infinite. Then, the summation $\sum_s$ in the formulas for the kernels as well as in the formulas for the estimates is data-invariant and produced for all points of this grid. Thus, the matrix Φ_h is invariant. The weights $g_h(x, X_s)$, and $g_h^{(r)}(x, X_s)$ become shift-invariant and depend on the difference of the arguments x and X_s only:

$$\begin{aligned} g_h(x, X_s) &= g_h(x - X_s), \\ g_h^{(r)}(x, X_s) &= g_h^{(r)}(x - X_s). \end{aligned}$$

This conclusion immediately follows from the general formulas in Eqs. (2.56) and (2.58).

The estimates in Eqs. (2.55) and (2.57) can be represented in the convolution form with the shift-invariant on x (independent on x) kernels:

$$\begin{aligned} \hat{y}_h(x) &= \sum_s g_h(u_s)z(x - u_s) = \sum_s g_h(x - X_s)z(X_s), \\ g_h(u_s) &= w_h(u_s)\phi_h^T(u_s)\Phi_h^{-1}\phi(0), \end{aligned} \tag{3.1}$$

and

$$\begin{aligned} \hat{y}_h^{(r)}(x) &= \sum_s g_h^{(r)}(u_s)z(x - u_s) = \sum_s g_h^{(r)}(x - X_s)z(X_s), \\ g_h^{(r)}(u_s) &= (-1/h)^{|r|} w_h(u_s)\phi_h^T(u_s)\Phi_h^{-1}\phi^{(r)}(0), \\ \Phi_h &= \sum_s w_h(u_s)\phi_h(u_s)\phi_h^T(u_s). \end{aligned} \tag{3.2}$$

In this representation the differences

$$u_s = x - X_s$$

are relative coordinates of the image pixels X_s with respect to the estimation point x.

The considered passage to the shift-invariant estimates is valid for the general multidimensional signals, $x \in \mathbb{R}^d$. Let X_{Zd} be the infinite regular d-dimensional grid:

$$X_{\mathbb{Z}^d} = \{l_1\Delta_1, \ldots, l_d\Delta_d: l_j \in \mathbb{Z}, j = 1, \ldots, d\}, \tag{3.3}$$

where Δ_j are the sampling intervals on the corresponding arguments. Assuming that $x \in X_{\mathbb{Z}^d}$ and $X_s \in X_{\mathbb{Z}^d}$, the above formulas for the estimates and the kernels can be applied for the general multivariate case.

The convolutions in Eqs. (3.1) and (3.2) can be rewritten in the following forms:

$$\hat{y}_h(x) = \sum_{l\in\mathbb{Z}^d} g_h(l\Delta)z(x - l\Delta), \tag{3.4}$$

$$g_h(l\Delta) = w_h(l\Delta)\phi_h^T(l\Delta)\Phi_h^{-1}\phi(0), \tag{3.5}$$

$$\Phi_h = \sum_{l\in\mathbb{Z}^d} w_h(l\Delta)\phi_h(l\Delta)\phi_h^T(l\Delta), \tag{3.6}$$

and

$$\hat{y}_h^{(r)}(x) = \sum_{l\in\mathbb{Z}^d} g_h^{(r)}(l\Delta)z(x - i\Delta), \tag{3.7}$$

$$g_h^{(r)}(l\Delta) = (-1/h)^{|r|}w_h(l\Delta)\phi_h^T(l\Delta)\Phi_h^{-1}\phi^{(r)}(0), \tag{3.8}$$

where ϕ in Eq. (2.19) is a vector of polynomials in Eq. (2.45):

$$\phi(l\Delta) = [\phi_1(l\Delta), \phi_2(l\Delta), \ldots, \phi_M(l\Delta)]^T.$$

The multi-index notation is exploited in these formulas, in particular,

$$l\Delta = (l_1\Delta_1, \ldots, l_d\Delta_d) \quad \text{and} \quad \sum_{l\in\mathbb{Z}^d} = \sum_{l_1=-\infty}^{\infty} \cdots \sum_{l_d=-\infty}^{\infty}.$$

Using the standard notation for multidimensional convolution, the estimates in Eqs. (3.4) and (3.7) can be represented in the following compact form:

$$\begin{aligned} \hat{y}_h(x) &= (g_h \circledast z)(x), \\ \hat{y}_h^{(r)}(x) &= (g_h^{(r)} \circledast z)(x), x \in X_{\mathbb{Z}^d}. \end{aligned} \tag{3.9}$$

If the parentheses and square brackets in Eq. (2.5) are used for the continuous and discrete arguments, such that $y(k\Delta) = y[k]$, the above formulas can be given in the form omitting the sampling period Δ, which gives for Eqs. (3.4–3.9)

$$\begin{aligned}\hat{y}_h[k] &= \sum_{l\in\mathbb{Z}^d} g_h[l]z[k-l] = \sum_{l\in\mathbb{Z}^d} g_h[k-l]z[l],\\ g_h[l] &= w_h[l]\phi_h^T[l]\Phi_h^{-1}\phi[0],\ \Phi_h = \sum_{l\in\mathbb{Z}^d} w_h[l]\phi_h[l]\phi_h^T[l],\end{aligned} \tag{3.10}$$

and

$$\begin{aligned}\hat{y}_h^{(r)}[k] &= \sum_{l\in\mathbb{Z}^d} g_h^{(r)}[l]z[k-l] = \sum_{l\in\mathbb{Z}^d} g_h^{(r)}[k-l]z[l],\\ g_h^{(r)}[l] &= (-1/h)^{|r|}w_h[l]\phi_h^T[l]\Phi_h^{-1}\phi^{(r)}[0],\end{aligned} \tag{3.11}$$

where

$$\phi[l] = (\phi_1[l], \phi_2[l], \ldots, \phi_M[l])^T,$$

and the standard notation for the d-dimensional discrete convolutions are

$$\begin{aligned}\hat{y}_h[k] &= (g_h \circledast z)[k],\\ \hat{y}_h^{(r)}[k] &= (g_h^{(r)} \circledast z)[k],\ k\in\mathbb{Z}^d.\end{aligned} \tag{3.12}$$

3.2 Vanishing Moments

The reproducing polynomial properties given for the polynomial kernels and estimates in Proposition 1 are valid for the shift-invariant kernels and estimates. We formulate these results in the form of special propositions.

Proposition 2 *Let $\hat{y}_h[k]$ and $\hat{y}_h^{(r)}[k]$ in Eq. (3.12), $k\in\mathbb{Z}^d$, be the LPA estimates of the order m; the signal y be a polynomial, $y\in\mathcal{P}^m\subset\mathbb{R}^d$; and $z(k\Delta) = y(k\Delta) = y[k]$. Then, for any h, $\Delta > 0$ such that* $\det\Phi_h \neq 0$,

$$\hat{y}_h[k] = y[k],\ k\in\mathbb{Z}^d, \tag{3.13}$$

$$\hat{y}_h^{(r)}[k] = y^{(r)}[k],\ k\in\mathbb{Z}^d. \tag{3.14}$$

This is a special case of Proposition 1, and the proof is identical to that given for Proposition 1.

Proposition 3 *Let the kernels $g_h[k]$ and $g_h^{(r)}[k]$, $k\in\mathbb{Z}^d$, be polynomials of the order m. Then the following vanishing moment conditions hold for these kernels*:

$$\sum_{l\in\mathbb{Z}^d} g_h[l]l^i = \delta_{i,0},\ 0\le|i|\le m, \tag{3.15}$$

$$\frac{1}{i!}\sum_{l\in\mathbb{Z}^d} g_h^{(r)}[l]l^i = \frac{(-1)^{|r|}}{\Delta^r}\delta_{i,r},\ 0\le|i|\le m,\ 0<|r|\le m, \tag{3.16}$$

where

$$l^i = l_1^{i_1} \cdot \ldots \cdot l_d^{i_d} \quad \text{and} \quad \frac{l^i}{i!} = \frac{l_1^{i_1}}{i_1!} \cdot \ldots \cdot \frac{l_d^{i_d}}{i_d!}.$$

Proof. The formulas (3.15)–(3.16) follow from Proposition 2 if we assume in the formulas (3.13)–(3.14) that $y[k] = (\Delta k)^i / i!$, $i \in \mathbb{Z}_+^d$, $|i| \leq m$.

Substituting this polynomial into $\sum_{l \in \mathbb{Z}^d} g_h[l] y[k-l] = y[k]$, we obtain

$$\sum_{l \in \mathbb{Z}^d} g_h[l] \frac{[\Delta(k-l)]^i}{i!} = \frac{(\Delta k)^i}{i!}. \tag{3.17}$$

For $k = 0$, this gives

$$\sum_{l \in \mathbb{Z}^d} g_h[l] \frac{[\Delta(-l)]^i}{i!} = \frac{0^i}{i!}, \qquad |i| \leq m, \tag{3.18}$$

and proves Eq. (3.15).

Substituting this polynomial into $\sum_{l \in \mathbb{Z}^d} g_h^{(r)}[l] y[k-l] = y^{(r)}[k]$, we obtain for $r = i$:

$$\sum_{l \in \mathbb{Z}^d} g_h^{(r)}[l] \frac{[\Delta(k-l)]^r}{r!} = 1.$$

It is used here that $\partial_x^r (x^r / r!) = 1$; see formulas (2.13).

Then, for $k = 0$, we have

$$\sum_{l \in \mathbb{Z}^d} g_h^{(r)}[l] \frac{[\Delta(-l)]^r}{r!} = 1 \Longrightarrow \sum_{l \in \mathbb{Z}^d} g_h^{(r)}[l] \frac{l^r}{r!} = \frac{(-1)^{|r|}}{\Delta^r}, \tag{3.19}$$

which proves in Eq. (3.16) for $i = r$.

Consider the derivatives $\partial_x^r (x^i / i!)$ where $i \neq r$; see formula (2.13).

In the multi-index notation, $i \neq r$ means that there exists at least one q, $q = 1, \ldots, d$, such that $i_q \neq r_q$. We may check that $\partial_x^r (x^i / i!)|_{x=0} = 0$ in both cases $i > r$ and $i < r$. It then follows that

$$\sum_{l \in \mathbb{Z}^d} g_h^{(r)}[l] \frac{[\Delta(k-l)]^i}{i!} = 0 \Longrightarrow \tag{3.20}$$

$$\sum_{l \in \mathbb{Z}^d} g_h^{(r)}[l] \frac{[\Delta(k-l)]^i}{i!}|_{k=0} = 0 \Longrightarrow \sum_{l \in \mathbb{Z}^d} g_h^{(r)}[l] \frac{l^i}{i!} = 0,$$

which proves in Eq. (3.16) for $i \neq r$. □

For $d = 1$, the vanishing moment conditions in Eq. (3.15) for the smoothing kernel are

$$\sum_{l=-\infty}^{\infty} g_h[l]l^i = \delta_{i,0},\ 0 \le i \le m, \tag{3.21}$$

that is, the kernel is normalized, $\sum_{l=-\infty}^{\infty} g_h[l] = 1$, and has zero higher moments up to order m.

For the differentiation kernel,

$$\frac{1}{i!}\sum_{l=-\infty}^{\infty} g_h^{(r)}[l]l^i = \begin{cases} (-1/\Delta)^r, & \text{if } r = i, \\ 0, & \text{if } r \ne i, \end{cases} \tag{3.22}$$
$$0 < i \le m, \qquad 1 \le r \le m.$$

For $i = r$, the moment of this kernel is equal to $(-1/\Delta)^r$ and all other moments are zero.

3.3 Frequency Domain

Classical discrete signal processing algorithms are based mostly on the shift-invariant (time-invariant) linear operators in Eqs. (3.4) and (3.7). In an explicit form these operators are given by formulas (3.10) and (3.11) defining the discrete convolutions in Eq. (3.12). The frequency characteristics and the Fourier transform are efficient methods for characterizing this type of operator as well as for implementing fast algorithms for calculations.

We start this section with a brief reminder of some basic concepts to introduce the necessary notations and multivariate formulas. The material in this section can be found in many sources, in particular, details and proofs can be found, for example, in [18], [173], and [229].

3.3.1 Frequency response

Let $z[k]$, $k \in \mathbb{Z}^d$, be a discrete multidimensional signal (sequence) defined on a regular d-dimensional grid and let this signal be a complex-valued harmonic:

$$z[k] = z(k\Delta) = e^{jk\Delta\omega}, \tag{3.23}$$

where $j^2 = -1$ with real multivariate frequencies ω, $\omega \in \mathbb{R}^d$. In scalar variables this formula means

$$z[k_1, \ldots, k_d] = z(k_1\Delta_1, \ldots, k_d\Delta_d) = e^{jk_1\Delta_1\omega_1} \cdot \ldots \cdot e^{jk_d\Delta_d\omega_d} = e^{j\sum_{s=1}^{d} k_s\Delta_s\omega_s},$$

as $\Delta\omega$ is the multi-index $\Delta\omega = (\Delta_1\omega_1, \ldots, \Delta_d\omega_d)$. Inserting Eq. (3.23) into Eq. (3.10), we immediately obtain

$$\begin{aligned}\hat{y}_h[k] &= \sum_{l\in\mathbb{Z}^d} g_h[l]z[k-l] = \sum_{l\in\mathbb{Z}^d} g_h[l]e^{j(k-l)\Delta\omega} \\ &= e^{jk\Delta\omega}\sum_{l\in\mathbb{Z}^d} g_h[l]e^{-jl\Delta\omega} = G_h(e^{-j\Delta\omega})\cdot z[k],\end{aligned} \tag{3.24}$$

where the *transfer function or frequency response* of the operator g_h is defined as

$$G_h(e^{-j\Delta\omega}) = \sum_{l\in\mathbb{Z}^d} g_h[l]e^{-jl\Delta\omega}. \tag{3.25}$$

The input-output link of the operator in Eq. (3.24) is presented as a product of the transfer function and the harmonic input signal,

$$\hat{y}_h[k] = G_h(e^{-j\Delta\omega})\cdot z[k]. \tag{3.26}$$

As a standard in this analysis, the normalized frequency is used instead of ω:

$$\bar{\omega} = \Delta\omega\,. \tag{3.27}$$

Then

$$\begin{aligned}\hat{y}_h[k] &= G_h(e^{-j\bar{\omega}})\cdot z[k], \\ G_h(e^{-j\bar{\omega}}) &= \sum_{l\in\mathbb{Z}^d} g_h[l]e^{-jl\bar{\omega}}.\end{aligned} \tag{3.28}$$

If $\Delta = 1$, then $\bar{\omega} = \omega$ and the representation of the frequency characteristic in the normalized frequency $\bar{\omega}$ coincides with its representation in the original frequency ω.

In notation based on the concept of the inner product of vectors, the above formulas can be given as

$$z[k] = z(k\Delta) = e^{-j\langle k,\Delta\omega\rangle}, \tag{3.29}$$

where $k = (k_1, \ldots, k_d)^T$ and $\Delta\omega = (\Delta_1\omega_1, \ldots, \Delta_d\omega_d)^T$ are d vectors and the inner product of these two vectors is

$$\langle k, \Delta\omega\rangle = \sum_{s=1}^{d} k_s\Delta_s\omega_s.$$

It is obvious that formulas (3.23) and (3.29) are equivalent. In this notation, the transfer function $G_h(e^{-j\bar{\omega}})$ has the form

$$G_h(e^{-j\bar{\omega}}) = \sum_{l\in\mathbb{Z}^d} g_h[l]e^{-j\langle l,\bar{\omega}\rangle}, \tag{3.30}$$

where the inner product of the vectors l and $\bar{\omega}$ means $\langle l, \bar{\omega}\rangle = \sum_{s=1}^{d} l_s\bar{\omega}_s$.

The transfer function G_h is complex-valued and has the amplitude $|G_h|$ and the phase φ which are accordingly defined through real and imaginary parts of G_h:

$$\begin{aligned} G_h(e^{-j\bar{\omega}}) &= |G_h(e^{-j\bar{\omega}})|e^{j\varphi(\bar{\omega})}, \\ |G_h(e^{-j\bar{\omega}})| &= \sqrt{\mathrm{Re}^2 G_h(e^{-j\bar{\omega}}) + \mathrm{Im}^2 G_h(e^{-j\bar{\omega}})}, \\ \varphi(\bar{\omega}) &= \arg G_h(e^{-j\bar{\omega}}), \\ \cos\varphi(\bar{\omega}) &= \frac{\mathrm{Re} G_h(e^{-j\bar{\omega}})}{|G_h(e^{-j\bar{\omega}})|},\ \sin\varphi(\bar{\omega}) = \frac{\mathrm{Im} G_h(e^{-j\bar{\omega}})}{|G_h(e^{-j\bar{\omega}})|}. \end{aligned} \tag{3.31}$$

The amplitude and phase considered as functions of d-dimensional frequency $\bar{\omega}$ or ω define the amplitude and phase *characteristics* of the operator. The amplitude characteristic $|G_h(e^{-j\bar{\omega}})|$ gives the amplitude ratio of the input and output harmonic signals while $\varphi(\bar{\omega})$ is a phase difference between these signals:

$$\hat{y}_h[k] = |G_h(e^{-j\bar{\omega}})|e^{j(k\bar{\omega}+\varphi(\bar{\omega}))}.$$

For differentiating operators the frequency characteristic is defined similar to Eq. (3.28) as

$$G_h^{(r)}(e^{-j\bar{\omega}}) = \sum_{l\in\mathbb{Z}^d} g_h^{(r)}[l]e^{-jl\bar{\omega}}, \tag{3.32}$$

with the amplitude and phase characteristics calculated according to Eq. (3.31) with G_h replaced by $G_h^{(r)}$.

We conclude this section with some notes concerning the eigenfunctions and eigenvalues of the considered shift-invariant linear operators. The kernel g_h defines the smoothing linear discrete operator written as the convolution in Eq. (3.12). The input signal z is called the *eigenfunction* of this operator if the output is equal to the input within the invariant factor such that

$$(g_h \circledast z)[k] = \lambda \cdot z[k].$$

This invariant factor λ is the *eigenvalue* of the operator.

Formula (3.26) says that for the harmonic input, $z[k]$ in Eq. (3.23) is

$$(g_h \circledast z)[k] = G_h(e^{-j\bar{\omega}}) \cdot z[k].$$

This proves that the smoothing operator with the kernel g_h has the multivariate harmonic function (Eq. (3.23)) as the eigenfunction and the frequency response $\lambda = G_h(e^{-j\bar{\omega}})$ as the eigenvalue. Of course, this statement is true for any shift-invariant operator, including the differentiating linear operators, defined by Eq. (3.11). The harmonic signal is the eigenfunction of this operator and the frequency response calculated for the frequency of the input harmonic is the corresponding eigenvalue of the operator. This is true for any frequency of the input signal. Thus, there exists a continuum of eigenfunctions and eigenvalues for shift-invariant operators.

3.3.2 Discrete-time Fourier transform

The *discrete-time/space Fourier transform* (*DTFT*) gives a different interpretation of the frequency characteristics of the shift-invariant linear system.

By definition, the DTFT of the signal $z[k]$ is

$$\mathcal{F}\{z[k]\} = Z(e^{-j\bar{\omega}}) = \sum_{k\in\mathbb{Z}^d} e^{-jk\bar{\omega}} z[k]. \tag{3.33}$$

Here we use the normalized frequency $\bar{\omega}$ of Eq. (3.27).

Formula (3.33) defines a multidimensional version of the discrete-time Fourier transform that is used for time sequences. Let us apply the DTFT to the convolution in Eq. (3.10) or Eq. (3.12). Routine transformations show that

$$\begin{aligned}\mathcal{F}\{\hat{y}_h[k]\} &= \sum_{k\in\mathbb{Z}^d} e^{-jk\bar{\omega}} \left(\sum_{l\in\mathbb{Z}^d} g_h[l] z[k-l] \right) \\ &= \sum_{(k-l)\in\mathbb{Z}^d} e^{-j(k-l)\bar{\omega}} z[k-l] \times \sum_{l\in\mathbb{Z}^d} e^{-jl\bar{\omega}} g_h[l] = Z(e^{-j\bar{\omega}}) G_h(e^{-j\bar{\omega}}),\end{aligned} \tag{3.34}$$

where

$$G_h(e^{-j\bar{\omega}}) = \mathcal{F}\{g_h[k]\} = \sum_{l\in\mathbb{Z}^d} e^{-jl\bar{\omega}} g_h[l] \tag{3.35}$$

is the DTFT of the kernel g_h. With the notation

$$Y_h(e^{-j\bar{\omega}}) = \mathcal{F}\{\hat{y}_h[k]\}$$

formula (3.34) is

$$Y_h(e^{-j\bar{\omega}}) = G_h(e^{-j\bar{\omega}}) Z(e^{-j\bar{\omega}}), \tag{3.36}$$

which shows that the DTFT of the output signal is a product of the DTFT of the input signal $Z(e^{-j\bar{\omega}})$ and the transfer function $G_h(e^{-j\bar{\omega}})$. Thus the convolution in the space domain in Eq. (3.12) is replaced by the product of Eq. (3.36) in the frequency domain. This demonstrates that the transfer function can be defined as a ratio of the DTFT of the output and input signals as

$$G_h(e^{-j\bar{\omega}}) = \frac{Y_h(e^{-j\bar{\omega}})}{Z(e^{-j\bar{\omega}})}. \tag{3.37}$$

Formula (3.37) gives the transfer function for an arbitrary input signal provided that a DTFT of the signal exists. This definition is different from the one given in Eq. (3.24) for the harmonic input signal. However, of course, the transfer function is completely the same.

The transfer function G_h and the DTFT of the signals Y_h and Z in Eq. (3.36) are functions of a continuous normalized frequency $\bar{\omega} \in \mathbb{R}^d$. These functions are periodical with the period 2π on each element of $\bar{\omega} = (\bar{\omega}_1, \ldots, \bar{\omega}_d)$. Usually, the basic interval for these frequencies is defined as a rectangle:

$$-\pi < \bar{\omega} \leq \pi, \tag{3.38}$$

that is,

$$-\pi < \bar{\omega}_l \leq \pi, \qquad -\pi/\Delta_l < \omega_l \leq \pi/\Delta_l, \qquad l = 1, \ldots, d.$$

The inverse discrete-time Fourier transform (IDTFT) allows us to reconstruct the discrete sequence $z[k]$ from its DTFT:

$$z[k] = \mathcal{F}^{-1}\{Z(e^{-j\bar{\omega}})\} = \frac{1}{(2\pi)^d}\int_{-\pi}^{\pi} Z(e^{-j\bar{\omega}})e^{jk\bar{\omega}}d\bar{\omega}. \tag{3.39}$$

This last formula is very interesting. It states that the sequence $z[k]$ defined in the spatial domain (with k integers) can be accurately reconstructed from its spectrum $Z(e^{-j\bar{\omega}})$. In the integral in Eq. (3.39), the multivariate signal $z[k]$ is represented as the integral (the sum of infinite combinations) of harmonics $e^{jk\bar{\omega}}$ with complex-valued amplitudes $Z(e^{-j\bar{\omega}})$. These harmonics work as elementary functions ("atoms") of a signal decomposition while the spectrum $Z(e^{-j\bar{\omega}})$ gives sufficient analysis information for a perfect reconstruction of the signal.

In a similar way, the *transfer function* of the *differentiation operator* in Eq. (3.11) is of the form

$$G_h^{(r)}(e^{-j\bar{\omega}}) = \mathcal{F}\{g_h^{(r)}[k]\} = \sum_{k\in\mathbb{Z}^d} g_h^{(r)}[k]e^{-jk\bar{\omega}}, \tag{3.40}$$

and the input-output link for the differentiating operator is

$$Y_h^{(r)}(e^{-j\bar{\omega}}) = G_h^{(r)}(e^{-j\bar{\omega}})Z(e^{-j\bar{\omega}}). \tag{3.41}$$

Let $X(e^{-j\bar{\omega}})$ and $Y(e^{-j\bar{\omega}})$ be the DTFT of the sequences $x[k]$ and $y[k]$. Then, the following important *Parseval formula* is valid:

$$\sum_{k\in\mathbb{Z}^d} x[k]y[k] = \frac{1}{(2\pi)^d}\int_{-\pi}^{\pi} X(e^{-j\bar{\omega}})Y^*(e^{-j\bar{\omega}})d\bar{\omega}, \tag{3.42}$$

where the star (*) stands for complex conjugate variables.

For $x[k] = y[k]$, it follows from Eq. (3.42) that

$$\sum_{k\in\mathbb{Z}^d} x^2[k] = \frac{1}{(2\pi)^d}\int_{-\pi}^{\pi} |X(e^{-j\bar{\omega}})|^2 d\bar{\omega}. \tag{3.43}$$

The sum $\sum_{k\in\mathbb{Z}^d} x^2[k]$ is the total energy of the signal sequence $x[k]$, and the right-hand side of Eq. (3.43) gives the same energy by computing the integral in the frequency domain. Eq. (3.43) shows that the amplitude spectrum $|X(e^{-j\bar{\omega}})|^2$ can be viewed as the frequency distribution of the signal energy, and is a standard useful interpretation of the Parseval formula.

3.3.3 Vanishing moments

There are simple links between the frequency characteristics and the moments of the kernels that allow us to transfer the vanishing moment condition of Proposition 3 into the frequency domain. We can derive from Eq. (3.35) that

$$G_h(e^{-j\bar{\omega}})|_{\bar{\omega}=0} = \sum_{l \in \mathbb{Z}^d} g_h[l], \tag{3.44}$$

$$\partial^i_{j\bar{\omega}} G_h(e^{-j\bar{\omega}})|_{\bar{\omega}=0} = (-1)^{|i|} \sum_{l \in \mathbb{Z}^d} l^i \cdot g_h[l]. \tag{3.45}$$

Remember, that $\partial^i_{j\bar{\omega}} = \partial^{(i_1,\ldots,i_d)}_{(j\bar{\omega}_1,\ldots,j\omega_d)}$. From here on, $|_{\bar{\omega}=0}$ means that $\bar{\omega} = 0$ in the corresponding expression.

Using formulas (3.44)–(3.45), we may conclude that the moment conditions of Eq. (3.15) in the frequency domain are equivalent to

$$G_h(e^{-j\bar{\omega}})|_{\bar{\omega}=0} = 1, \tag{3.46}$$

$$\partial^i_{j\bar{\omega}} G_h(e^{-j\bar{\omega}})|_{\bar{\omega}=0} = 0, \qquad 0 < |i| \leq m. \tag{3.47}$$

For the differentiating operator, using Eqs. (3.44–3.45) we arrive at the following frequency domain equivalent of the conditions of Eq. (3.16):

$$G_h^{(r)}(e^{-j\bar{\omega}})|_{\bar{\omega}=0} = 0, \tag{3.48}$$

$$\frac{1}{i!}\partial^i_{j\bar{\omega}} G_h^{(r)}(e^{-j\bar{\omega}})|_{\bar{\omega}=0} = \frac{1}{\Delta^r}\delta_{i,r}, \qquad 0 \leq |i| \leq m, \quad 0 < |r| \leq m. \tag{3.49}$$

The low-pass, high-pass, and bandpass filters have the typical amplitude frequency characteristics $|G_L|$, $|G_H|$, and $|G_B|$, respectively, depicted in Fig. 3.1 for the 1D case. The low-pass filter attenuates a high-frequency signal and lets low-frequency signal components pass through with small deterioration. With the low-pass filter, $|G_L|$ has values close to 1 for low frequencies and values close to zero for larger frequencies. Conversely, the high-pass filter has the amplitude characteristic $|G_H|$ with values close to zero for low frequencies and the large values of $|G_H|$ for high-frequency signals. The amplitude frequency characteristic $|G_B|$ of the *bandpass* filter has low values at the zero and high frequencies and lets pass through a signal of some intermediate frequencies.

It follows from Eq. (3.46) that

$$\left|G_h(e^{-j\bar{\omega}})\right|\Big|_{\bar{\omega}=0} = \left|G_h(1)\right| = 1.$$

The amplitude frequency response of this filter is equal to 1 at zero frequency. The further conditions in Eq. (3.47) say that the amplitude frequency response is flat at zero frequency, and this flatness (a number of zero derivatives) of $G_h(e^{-j\bar{\omega}})$ is defined by the order m of the kernel. Overall, this means that the LPA smoother is a low-pass filter with the typical frequency characteristic shown in Fig. 3.1.

For the differentiating kernel we have for $\bar{\omega} = 0$ that $G_h^{(r)}(e^{-j\bar{\omega}})|_{\bar{\omega}=0} = 0$ from Eq. (3.48). The zero value of the derivatives in Eq. (3.49) show that the frequency

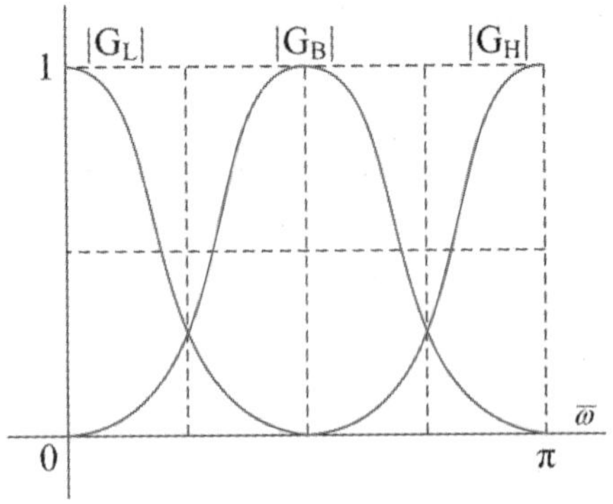

Figure 3.1 Amplitude frequency responses of low-pass $|G_L|$, high-pass $|G_H|$, and bandpass $|G_B|$ filters.

characteristic $|G_h^{(r)}(e^{-j\bar{\omega}})|$ takes zero values and is flat at $\bar{\omega} = 0$. Thus, for frequencies near $\bar{\omega} = 0$, the frequency response of the differentiating filter corresponds to the behavior of the bandpass or high-pass filters, as is illustrated in Fig. 3.1.

3.3.4 Discrete Fourier transform

In Sec. 3.3.2 we considered the frequency characteristics of linear operators defined for continuous frequency ω. In this section we go to the discrete Fourier transform, where the variables are discrete in both spatial (time) and frequency domains. This is presented in the form of a *discrete Fourier transform* (DFT). For the discrete frequency of DFT we use the notation f. To give the formulas in the familiar standard notation, we consider here the case $d = 2$, while generalization for $d > 2$ as well as the formulas for $d = 1$ are similar and can be easily derived.

If a discrete sequence $x[k_1, k_2]$ is periodic with periods n_1 and n_2, that is,

$$x[k_1, k_2] = x[k_1 + n_1 l_1, k_2 + n_2 l_2], \quad l_1, l_2 \in \mathbb{Z},$$

then its discrete Fourier series representation is given by

$$X[f_1, f_2] = \sum_{k_1=0}^{n_1-1} \sum_{k_2=0}^{n_2-1} x[k_1, k_2] W_{n_1}^{f_1 k_1} W_{n_2}^{f_2 k_2},$$

$$0 \le f_1 \le n_1 - 1, \quad 0 \le f_2 \le n_2 - 1, \tag{3.50}$$

$$x[k_1, k_2] = \frac{1}{n_1 n_2} \sum_{f_1=0}^{n_1-1} \sum_{f_2=0}^{n_2-1} X[f_1, f_2] W_{n_1}^{-f_1 k_1} W_{n_2}^{-f_2 k_2},$$

$$0 \le k_1 \le n_1 - 1, \quad 0 \le k_2 \le n_2 - 1, \tag{3.51}$$

where W_N is the Nth root of unity,

$$W_N = e^{-j2\pi/N}. \tag{3.52}$$

Equation (3.50) defines the DFT of the sequence $x[k_1, k_2]$ with the frequencies f_1, f_2 normalized to be integers from the intervals $[0, n_1 - 1]$ and $[0, n_2 - 1]$, respectively. Equation (3.51) defines the *inverse discrete Fourier transform* (IDFT) and restores the sequence $x[k_1, k_2]$ from its discrete spectrum $X[f_1, f_2]$ as a linear combination of the complex-valued harmonics $W_{n_1}^{-f_1 k_1} W_{n_2}^{-f_2 k_2}$.

Let us verify Eqs. (3.50–3.51). First note that

$$\sum_{f_1=0}^{n_1-1} W_{n_1}^{f_1} = 0 \quad \text{and} \quad \sum_{f_1=0}^{n_1-1} W_{n_1}^{f_1 r_1} = n_1 \delta_{r_1} \qquad \text{for integer } r_1, \tag{3.53}$$

$$\sum_{f_2=0}^{n_2-1} W_{n_2}^{f_2} = 0 \quad \text{and} \quad \sum_{f_2=0}^{n_2-1} W_{n_2}^{f_2 r_2} = n_2 \delta_{r_2} \qquad \text{for integer } r_2.$$

Substituting Eq. (3.50) into Eq. (3.51) and using Eq. (3.53), we obtain

$$\begin{aligned}
x[k_1, k_2] &= \frac{1}{n_1 n_2} \sum_{f_1=0}^{n_1-1} \sum_{f_2=0}^{n_2-1} W_{n_1}^{-f_1 k_1} W_{n_2}^{-f_2 k_2} \sum_{l_1=0}^{n_1-1} \sum_{l_2=0}^{n_2-1} x[l_1, l_2] W_{n_1}^{f_1 l_1} W_{n_2}^{f_2 l_2} \\
&= \frac{1}{n_1 n_2} \sum_{l_1=0}^{n_1-1} \sum_{l_2=0}^{n_2-1} x[l_1, l_2] \sum_{f_1=0}^{n_1-1} \sum_{f_2=0}^{n_2-1} W_{n_1}^{-f_1 k_1} W_{n_2}^{-f_2 k_2} W_{n_1}^{f_1 l_1} W_{n_2}^{f_2 l_2} \\
&= \frac{1}{n_1 n_2} \sum_{l_1=0}^{n_1-1} \sum_{l_2=0}^{n_2-1} x[l_1, l_2] \sum_{f_1=0}^{n_1-1} \sum_{f_2=0}^{n_2-1} W_{n_1}^{-f_1 (k_1 - l_1)} W_{n_2}^{\ f_2 (k_2 - l_2)} \\
&= \sum_{l_1=0}^{n_1-1} \sum_{l_2=0}^{n_2-1} x[l_1, l_2] \delta_{k_1 - l_1} \delta_{k_2 - l_2} = x[k_1, k_2].
\end{aligned}$$

Thus, we arrive at the identity $x[k_1, k_2] = x[k_1, k_2]$ and prove that the signal can be reconstructed from its spectrum. In a similar way, substituting Eq. (3.51) into Eq. (3.50), we arrive at the identity $X[f_1, f_2] = X[f_1, f_2]$. This proves Eq. (3.50).

All the familiar properties of Fourier transform hold, taking periodicity into account. In particular, a convolution is now a periodic convolution. For Eq. (3.10) this means that $g_h[k_1, k_2]$ and $z[k_1, k_2]$ are periodic with periods n_1 and n_2 on the variables k_1 and k_2. Then, the periodic convolution is calculated for one period only:

$$\begin{aligned}
\hat{y}_h[k_1, k_2] &= \sum_{l_1=0}^{n_1-1} \sum_{l_2=0}^{n_2-1} g_h[l_1, l_2] \cdot z[(k_1 - l_1)\ mod\ n_1, (k_2 - l_2)\ mod\ n_2] \\
&= \sum_{l_1=0}^{n_1-1} \sum_{l_2=0}^{n_2-1} g_h[(k_1 - l_1)\ mod\ n_1, (k_2 - l_2)\ mod\ n_2] \cdot z[l_1, l_2].
\end{aligned} \tag{3.54}$$

In this formula, $z[(k_1 - l_1)\ mod\ n_1,\ (k_2 - l_2)\ mod\ n_2]$ means that the signal z is defined on the segment $[0, n_1 - 1] \centerdot [0, n_2 - 1]$ with the periodical extension out of this segment. The formula $g_h[(k_1 - l_1)\ mod\ n_1, (k_2 - l_2)\ mod\ n_2]$ has the same meaning for g_h.

For the periodic convolution in Eq. (3.54), the DFT gives

$$\hat{Y}_h[f_1, f_2] = G_h[f_1, f_2]Z(f_1, f_2), \tag{3.55}$$

where

$$\hat{Y}_h[f_1, f_2] = \sum_{k_1=0}^{n_1-1}\sum_{k_2=0}^{n_2-1} \hat{y}_h[k_1, k_2]W_{n_1}^{f_1k_1}W_{n_2}^{f_2k_2},$$

$$G_h[f_1, f_2] = \sum_{k_1=0}^{n_1-1}\sum_{k_2=0}^{n_2-1} g_h[k_1, k_2]W_{n_1}^{f_1k_1}W_{n_2}^{f_2k_2},$$

$$Z(f_1, f_2) = \sum_{k_1=0}^{n_1-1}\sum_{k_2=0}^{n_2-1} z[k_1, k_2]W_{n_1}^{f_1k_1}W_{n_2}^{f_2k_2},$$

$$0 \le f_1 \le n_1 - 1, \qquad 0 \le f_2 \le n_2 - 1.$$

The Parseval formula for the periodic signals $x[k_1, k_2]$ and $y[k_1, k_2]$ is of the form

$$\sum_{k_1=0}^{n_1-1}\sum_{k_2=0}^{n_2-1} x[k_1, k_2]y[k_1, k_2] = \frac{1}{n_1n_2}\sum_{f_1=0}^{n_1-1}\sum_{f_2=0}^{n_2-1} X[f_1, f_2]Y^*[f_1, f_2]. \tag{3.56}$$

When $x[k_1, k_2] = y[k_1, k_2]$,

$$\sum_{k_1=0}^{n_1-1}\sum_{k_2=0}^{n_2-1} x^2[k_1, k_2] = \frac{1}{n_1n_2}\sum_{f_1=0}^{n_1-1}\sum_{f_2=0}^{n_2-1} |X[f_1, f_2]|^2. \tag{3.57}$$

The DFT is very important for computational reasons, because it can be implemented using the fast algorithms for n being an integer degree of 2, $n = 2^L$.

The DFT routinely is applied to finite-length sequences. We wish to emphasize once more that this implies that these sequences are periodically extended beyond the intervals where these sequences are defined. This point requires us to be accurate in our application of the DFT.

In the multi-index notation, we have for formulas (3.50)–(3.51)

$$X[f] = \sum_{k=0}^{n-1} x[k]W_n^{fk}, \qquad 0 \le f \le n - 1, \tag{3.58}$$

$$x[k] = \frac{1}{n}\sum_{f=0}^{n-1} X[f]W_n^{-fk}, \qquad 0 \le k \le n - 1, \tag{3.59}$$

where $f = (f_1, f_2)$, $n = (n_1, n_2)$, $W_n^{fk} = W_{n_1}^{f_1 k_1} W_{n_2}^{f_2 k_2}$, and $1/n = 1/(n_1 n_2)$. For the Parseval formulas (3.56)–(3.57), this gives

$$\sum_{k=0}^{n-1} x[k]y[k] = \frac{1}{n}\sum_{f=0}^{n} X[f]Y^*[f] \tag{3.60}$$

when $x[k] = y[k]$ and

$$\sum_{k=0}^{n-1} x^2[k] = \frac{1}{n}\sum_{f=0}^{n-1} |X[f]|^2. \tag{3.61}$$

In this multi-index notation, the Parseval formulas (3.60) and (3.61) are valid for signals of any dimensionality d.

3.3.5 Examples

3.3.5.1 *Smoothing filters*

Figures 3.2 and 3.3 show the kernels g_h and the amplitude frequency characteristics $|G_h(e^{-j\bar{\omega}})|$ for 1D smoothing filters with the symmetric rectangular and Gaussian windows of the orders $m = 0$ and $m = 2$. These are low-pass filters.

We show the kernels and frequency characteristics for small and large values of h, $h_2 > h_1$, to demonstrate the influence of the scale h on the characteristics of

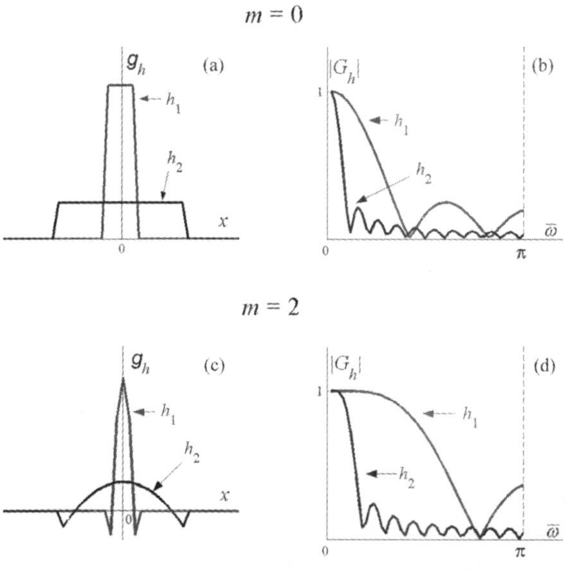

Figure 3.2 The kernels g_h and frequency characteristics $|G_h|$ of the discrete smoothing filters. The LPA design with the rectangular window and the orders $m = 0$ and $m = 2$.

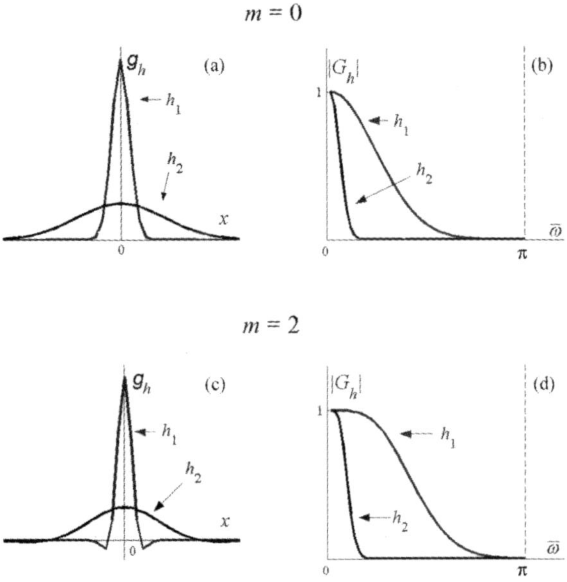

Figure 3.3 The kernels g_h and frequency characteristics $|G_h|$ of the discrete smoothing filters. The LPA design with the Gaussian window and the orders $m = 0$ and $m = 2$.

the operators. Comparison of the curves in the figures show:

(a) Increasing h narrows the low-pass band of the filter;
(b) Increasing m widens the low-pass band of the filter and makes flatter the maximum of $|G_h(e^{-j\bar{\omega}})|$ at the zero frequency;
(c) The kernel and amplitude frequency responses of the filters based on the Gaussian window are smoother than the corresponding characteristics of the filters based on the rectangular window.

3.3.5.2 Differentiating filters

Let us consider the differentiating kernels. It follows from Eq. (3.48) that amplitude frequency response at zero frequency is equal to 0. The differentiation filter is a typical bandpass filter.

Figures 3.4 and 3.5 show the kernels and the amplitude frequency characteristics of the 1D differentiating filters for the rectangular and Gaussian windows for the orders $m = 1$ and $m = 3$. The curves show:

(a) Increasing h narrows the bandpass of the filter;
(b) Increasing m widens the bandpass band of the filter and makes the minimum at the zero frequency flatter;
(c) The filters based on the Gaussian window have smoother kernels and frequency characteristics than those for the filters based on the rectangular window.

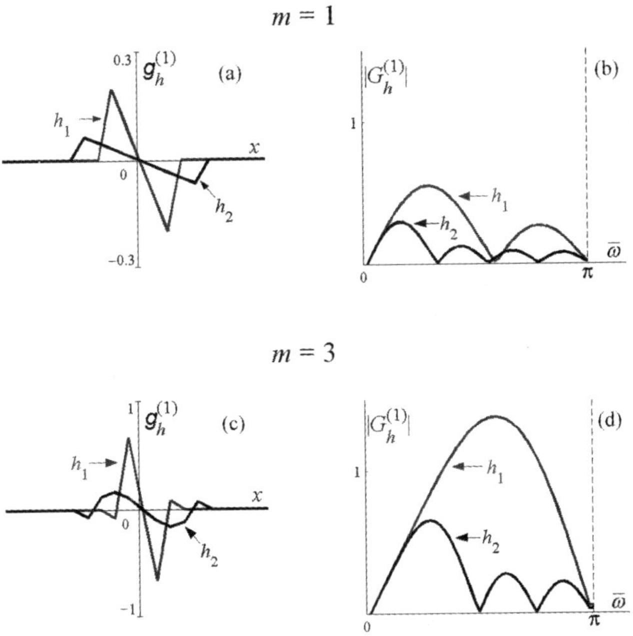

Figure 3.4 The kernels $g_h^{(1)}$ and frequency characteristics $|G_h^{(1)}|$ of the discrete differentiation filters. The LPA design with the rectangular window and the orders $m = 1$ and $m = 3$.

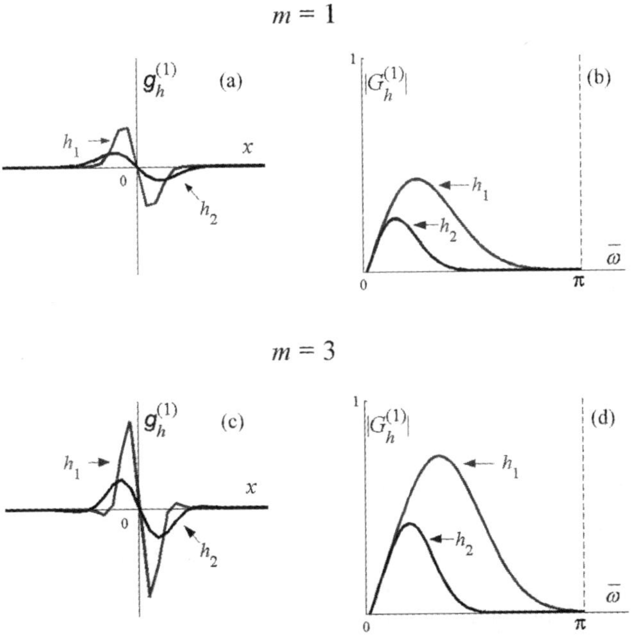

Figure 3.5 The kernels $g_h^{(1)}$ and frequency characteristics $|G_h^{(1)}|$ of the discrete differentiation filters. The LPA design with the Gaussian window and the orders $m = 1$ and $m = 3$.

3.3.5.3 Analytical regular grid kernels

Formulas (3.10) and (3.11) define the smoothing and differentiating kernels on the regular grid in the form

$$g_h[l] = w_h[l]\phi_h^T[l]\Phi_h^{-1}\phi[0] \tag{3.62}$$

and

$$g_h^{(r)}[l] = (-1)^{|r|}w_h[l]\phi_h^T[l]\Phi_h^{-1}\phi^{(r)}[0], \qquad l \in \mathbb{Z}^d, \tag{3.63}$$

with

$$\Phi_h = \sum_{l\in\mathbb{Z}^d} w_h[l]\phi_h[l]\phi_h^T[l]. \tag{3.64}$$

In this section we consider some particular cases when these formulas give analytical results. As a general idea we use the polynomials $\phi_h[l]$, which are orthonormal with the weights $w_h[l]$. Then, Φ_h becomes the identical matrix and formulas (3.62)–(3.63) are simplified to

$$\begin{aligned} g_h[l] &= w_h[l]\phi_h^T[l]\phi[0], \\ g_h^{(r)}[l] &= (-1)^{|r|}w_h[l]\phi_h^T[l]\phi^{(r)}[0]. \end{aligned} \tag{3.65}$$

The orthonormal polynomials are quite different from the ones given in the standard form $(-1)^{|k|}x^k/k!$ in Eq. (2.45). They are not monomials of a single power k but are composed of the monomials of different powers. Then, a few elements of the vectors $\phi(0)$ and $\phi^{(r)}(0)$ can be nonzero. This makes a difference with the monomials in Eq. (2.22) or Eq. (2.45) having single nonzero elements in $\phi(0)$ and $\phi^{(r)}(0)$. Examples of the analytical 1D and 2D kernels for different windows w are derived in Sec. 14.1 of the Appendix.

3.3.5.4 Shift-invariant and shift-variant kernels

A difference between the shift-varying Nadaraya-Watson estimate in Eq. (2.70) with the window in Eq. (14.10) and the corresponding shift-invariant estimate is that the former gives an accurate approximation of the zero order for any finite grid of data while the latter assumes that the data grid is regular and infinite.

Note once more that the Nadaraya-Watson estimate as well as all LPA estimates in the form of Eq. (2.55), in general, are shift-variant, that is, their kernels depend on the data grid.

When we apply these shift-variant and shift-invariant estimates to a finite sequence given on a regular grid, the only difference appears near the boundaries of data sets. Let us illustrate this typical boundary problem. Assume that the window w is rectangular symmetric, $w_h[l_1] = (1/2h + 1)$, $|l_1| \leq h$. Then the corresponding shift-invariant kernel (see formula (14.12)) has the form

$$g_h[l_1] = w_h[l_1].$$

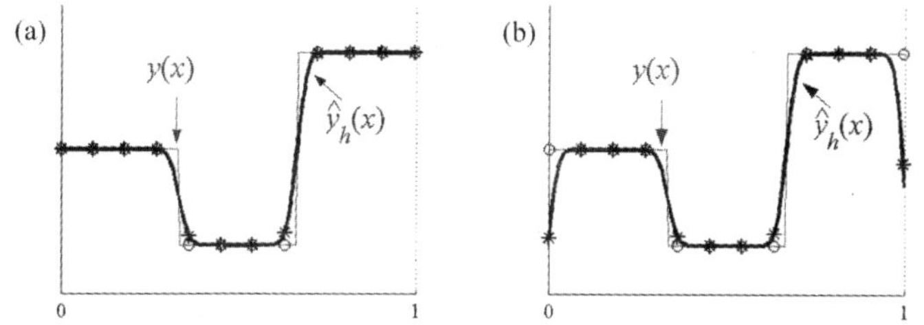

Figure 3.6 Shift-variant (a) and shift-invariant (b) estimates. The shift-invariant estimate gives large errors at the boundaries $x = 0$ and $x = 1$, while the shift-variant estimate is able to obtain accurate results for these points.

Figure 3.6 shows the estimates obtained for a piecewise constant signal. The difference between the estimates is clearly seen at the boundary points $x = 0$ and $x = 1$.

The shift-invariant estimator with the symmetric kernel $g_h[l_1]$ completes the data defined only on the interval [0, 1] by observations outside this interval, which by default are equal to zero. This results in large estimation errors at $x = 0$ and $x = 1$ for this shift-invariant estimator [Fig. 3.6(b)]. The shift-variant estimator in Eq. (2.70) enables an accurate LPA fit for any finite data grid and in this way gives accurate results near the boundaries, as can be seen in Fig. 3.6(a).

3.4 Numerical Shift-Invariant Kernels

Here we present examples of 2D shift-invariant kernels for smoothing and differentiating discrete filters. The general formulas (3.10) and (3.11) with $\Delta_1 = \Delta_2 = 1$ are applied for calculations. The used polynomials as well as the windows can be quite arbitrary, provided that the support of the window has a size sufficient for fitting the local polynomials of the corresponding powers. To obtain a clear visualization, the kernels are shown in two graphical forms using 2D and 3D imaging.

It is assumed that all kernels are finite and defined on square supports. The support can be symmetric or nonsymmetric with respect to the center of the LPA. All supports have $k_1 = 0$ and $k_2 = 0$ as the central points. These symmetric and nonsymmetric quadrant supports are shown in Fig. 3.7. The origin (0, 0) is the reference point of the LPA. The support of the symmetric kernel is denoted as Q_0 with notation Q_i, $i = 1, 2, 3, 4$, for the quadrant supports. If the window w in Eqs. (3.10) and (3.11) is a finite support function, then the kernel is finite with the same support. It is convenient to assume that the window w is always finite. For the infinite window, say Gaussian, this means that the truncated Gaussian window is used.

In this section we show results only for the square symmetric supports, postponing the discussion of the nonsymmetric case until Chapter 7. In what follows, we consider square uniform and Gaussian window filters of the order $m = 2$ and of support size 13×13.

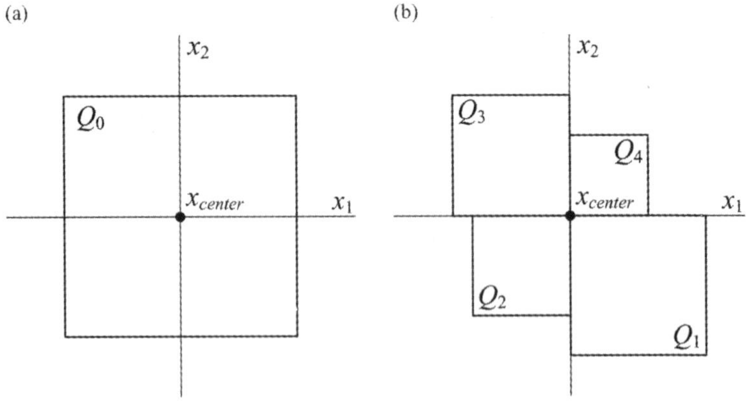

Figure 3.7 Symmetric Q_0 and nonsymmetric quadrant Q_i, $i = 1, 2, 3, 4$, supports/masks.

3.4.1 Square uniform window

The uniform square window means that w has equal values inside the square support. The LPA with degree $m = 0$ results in a trivial kernel with constant values of g_h in the square window/support. Then, the estimate of the signal is the sample mean of the data in the window.

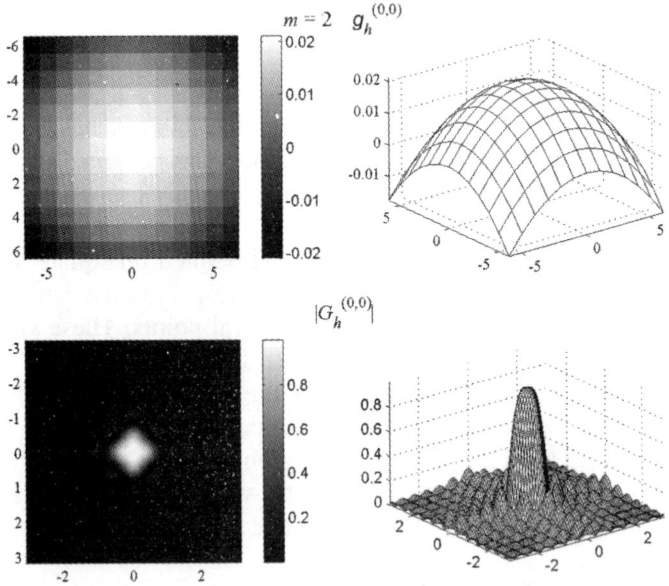

Figure 3.8 The smoothing kernel g_h and its amplitude frequency characteristic $|G_h|$: square window, $m = 2$. The low-pass filter with a peak of the frequency characteristic at $\bar{\omega} = 0$.

The degree $m = 1$ gives the same smoothing kernel g_h, which follows from the symmetry of the window (see the discussion on the role of the window symmetry in Sec. 14.1). The degree $m = 1$ enables one to obtain the differentiation kernels $g_h^{(1,0)}$ and $g_h^{(0,1)}$. These differentiating kernels have a clear linear structure with the vertical and horizontal bands corresponding to equal values of the weights in $g_h^{(1,0)}$ and $g_h^{(0,1)}$. We do not show these trivial kernels.

The degree $m = 2$ results in more interesting, nontrivial kernels. We show a full set of these kernels for $m = 2$: g_h, $g_h^{(1,0)}$, $g_h^{(0,1)}$, $g_h^{(2,0)}$, $g_h^{(0,2)}$, and $g_h^{(1,1)}$. The smoothing kernel g_h in Fig. 3.8 is a symmetric quadratic function with a peak at $k_1 = 0$, $k_2 = 0$. Its frequency characteristic is typical for the low-pass filter with a peak at $\bar{\omega}_1 = 0$, $\bar{\omega}_2 = 0$.

The differentiation kernels $g_h^{(1,0)}$, $g_h^{(0,1)}$ in Figs. 3.9 and 3.10 are clearly directional, forming weighted finite differences in the directions of the estimated derivatives $\partial^{(1,0)}$, $\partial^{(0,1)}$for the arguments x_1 and x_2, respectively. The frequency characteristics have zero values at $\bar{\omega}_1 = 0$, $\bar{\omega}_2 = 0$, and show that the differentiation kernel defines bandpass filters. These frequency characteristics are also directional with two distinctive peaks.

The differentiation kernels $g_h^{(2,0)}$, $g_h^{(0,2)}$ in Figs. 3.11 and 3.12 are also directional, forming second-order weighted finite differences in the directions of the estimated derivatives $\partial^{(2,0)}$, $\partial^{(0,2)}$for the arguments x_1 and x_2, respectively. The two-peak frequency characteristics have zero values at $\bar{\omega}_1 = 0$, $\bar{\omega}_2 = 0$.

The differentiation kernel $g_h^{(1,1)}$ for estimation of the cross-derivative $\partial^{(1,1)}$ is shown in Fig. 3.13. Its frequency characteristic has zero value at $\bar{\omega}_1 = 0$, $\bar{\omega}_2 = 0$, and four distinctive peaks.

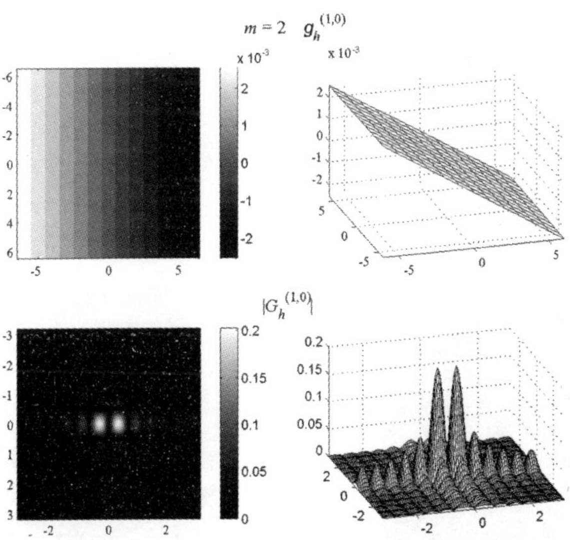

Figure 3.9 The differentiation kernel $g_h^{(1,0)}$ and its amplitude frequency characteristic $|G_h^{(1,0)}|$: square window, $m = 2$. The bandpass filter, $|G_h^{(1,0)}| = 0$ at $\bar{\omega} = 0$.

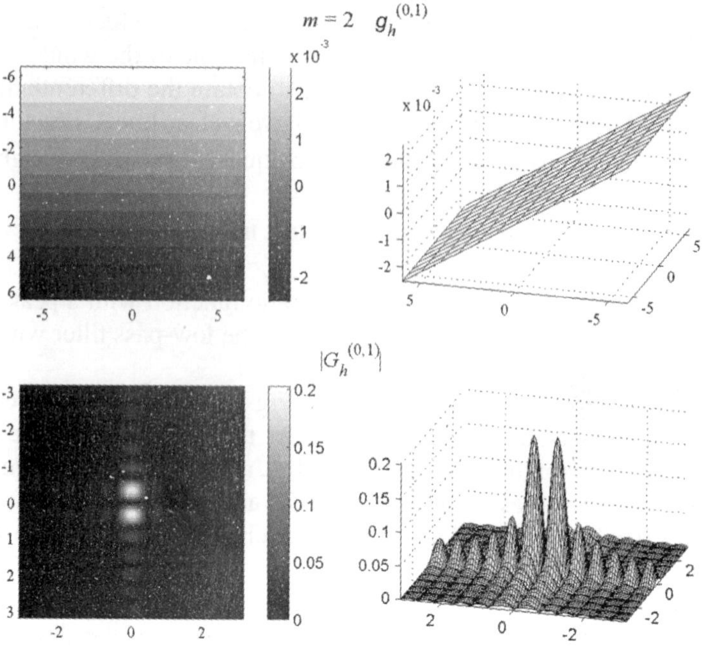

Figure 3.10 The differentiation kernel $g_h^{(0,1)}$ and its amplitude frequency characteristic $|G_h^{(0,1)}|$: square window, $m = 2$. The bandpass filter, $|G_h^{(0,1)}| = 0$ at $\bar{\omega} = 0$.

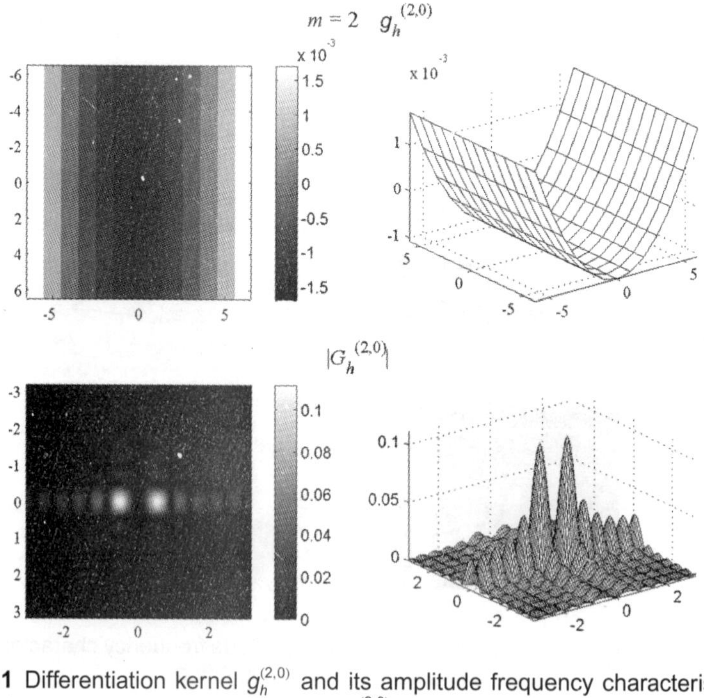

Figure 3.11 Differentiation kernel $g_h^{(2,0)}$ and its amplitude frequency characteristic $|G_h^{(2,0)}|$: square window, $m = 2$. The bandpass filter, $|G_h^{(2,0)}| = 0$ at $\bar{\omega} = 0$.

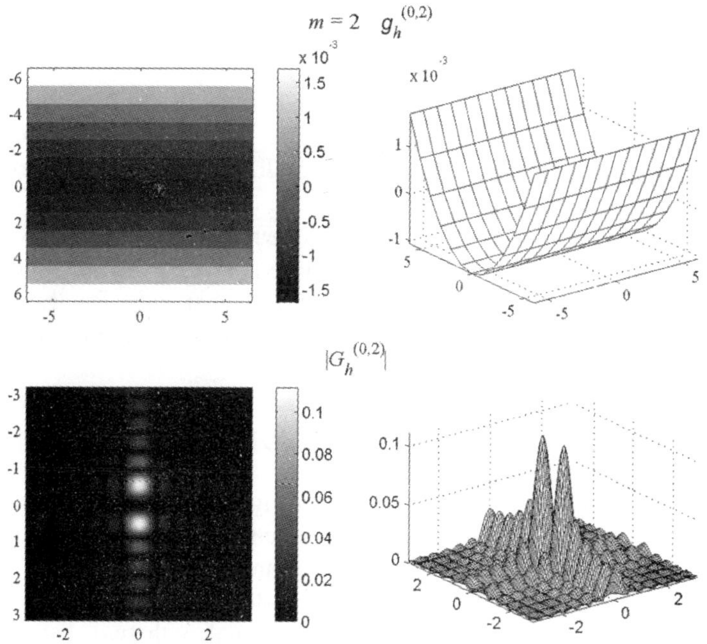

Figure 3.12 The differentiation kernel $g_h^{(0,2)}$ and its amplitude frequency characteristic $|G_h^{(0,2)}|$: square window, $m = 2$. The bandpass filter, $|G_h^{(0,2)}| = 0$ at $\bar{\omega} = 0$.

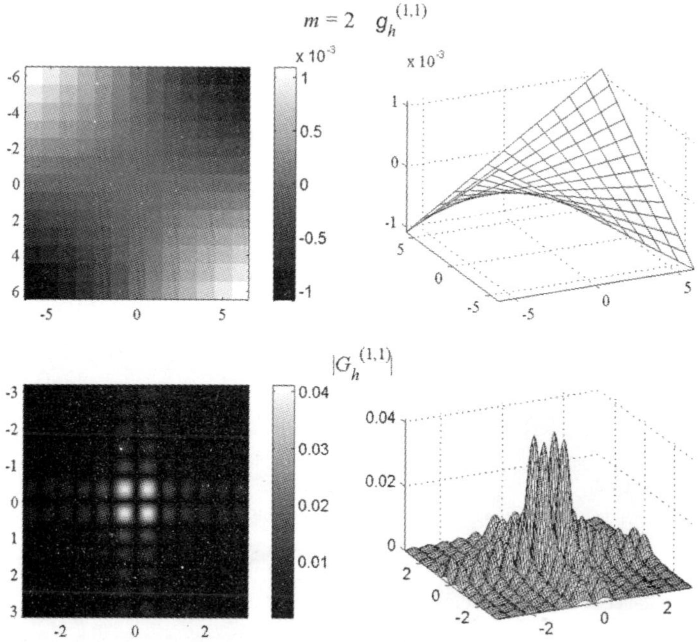

Figure 3.13 The differentiation kernel $g_h^{(1,1)}$ and its amplitude frequency characteristic $|G_h^{(1,1)}|$: square window, $m = 2$. The bandpass filter, $|G_h^{(1,1)}| = 0$ at $\bar{\omega} = 0$.

3.4.2 Gaussian window

The symmetric smoothing and differentiating kernels of the order $m = 2$ for the Gaussian symmetric window are shown in Figs. 3.14–3.19. The window w used here is a 2D standard Gaussian density. Overall, the Gaussian kernels and their frequency characteristics are different from the ones considered for the uniform window, mainly by their smoothness. One of the reasons for us to show the results for both the square and Gaussian windows here is that the linear finite-difference structures of the derivative estimates are more transparent for the square uniform window while the smoothness of the Gaussian kernels and their frequency characteristics is more appealing.

The smoothing kernel g_h in Fig. 3.14 is a symmetric function with a peak at $k_1 = 0$, $k_2 = 0$. Its frequency characteristic is typical for a low-pass filter with a peak at $\bar{\omega}_1 = 0$, $\bar{\omega}_2 = 0$.

The differentiation kernels $g_h^{(1,0)}$, $g_h^{(0,1)}$ in Figs. 3.15 and 3.16 are directional, forming weighted finite differences in the directions of the estimated derivatives $\partial^{(1,0)}$, $\partial^{(0,1)}$, respectively, for the arguments x_1 and x_2. The frequency characteristics have zero values at $\bar{\omega}_1 = 0$, $\bar{\omega}_2 = 0$. These differentiation filters are bandpass filters. The frequency characteristics are directional with two distinctive peaks. The differentiation kernels $g_h^{(2,0)}$, $g_h^{(0,2)}$ in Figs. 3.17 and 3.18 are also directional forming finite differences in the directions of the estimated derivatives $\partial^{(2,0)}$, $\partial^{(0,2)}$. The two-peak frequency characteristics have zero values at $\bar{\omega}_1 = 0$, $\bar{\omega}_2 = 0$. The differentiation kernel $g_h^{(1,1)}$ for estimation of the cross-derivative $\partial^{(1,1)}$ is shown in

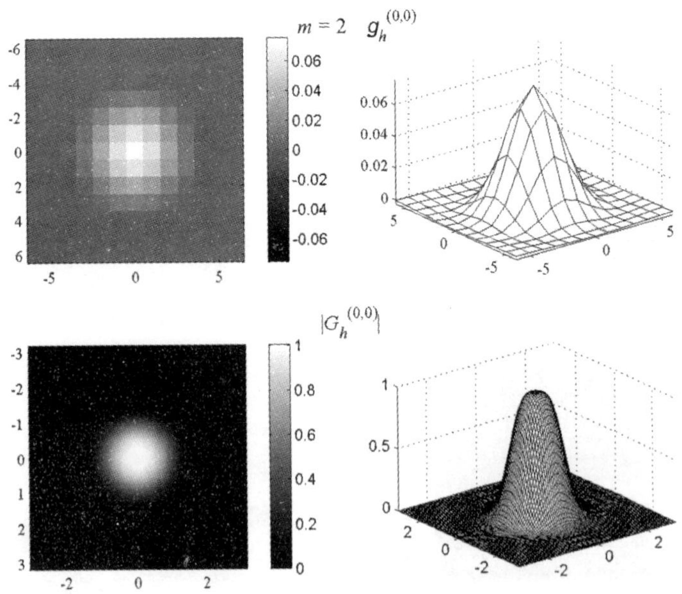

Figure 3.14 The smoothing kernel g_h and its amplitude frequency characteristic $|G_h|$: Gaussian window, $m = 2$. The low-pass filter with a peak of the frequency characteristic at $\bar{\omega} = 0$.

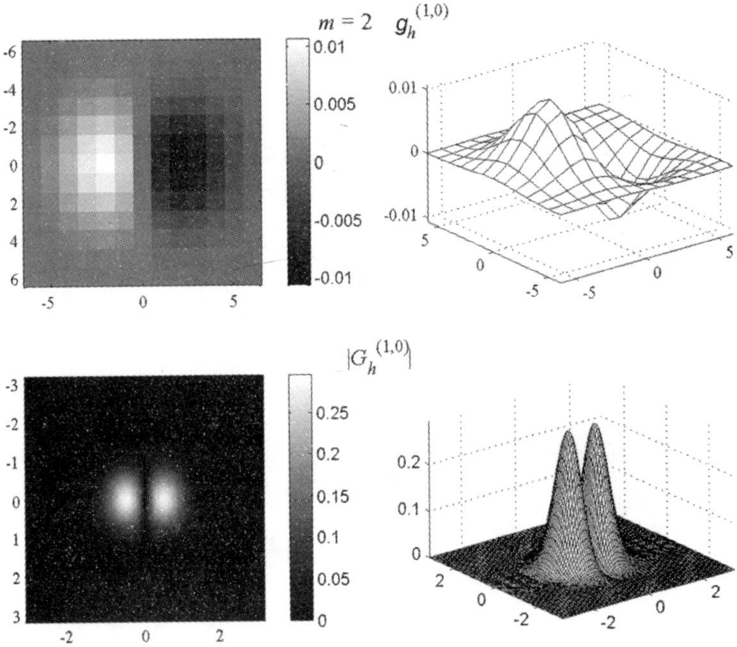

Figure 3.15 The differentiation kernel $g_h^{(1,0)}$ and its amplitude frequency characteristic $|G_h^{(1,0)}|$: Gaussian window, $m = 2$. The bandpass filter, $|G_h^{(1,0)}| = 0$ at $\bar{\omega} = 0$.

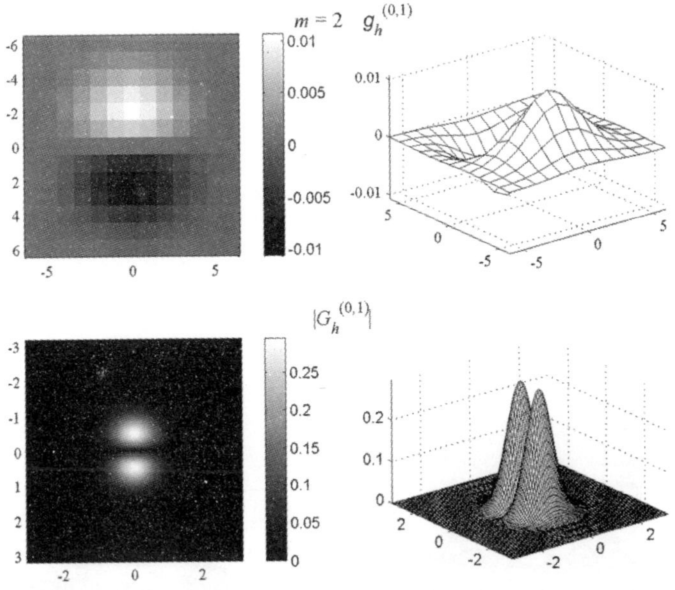

Figure 3.16 The differentiation kernel $g_h^{(0,1)}$ and its amplitude frequency characteristic $|G_h^{(0,1)}|$: Gaussian window, $m = 2$. The bandpass filter, $|G_h^{(0,1)}| = 0$ at $\bar{\omega} = 0$.

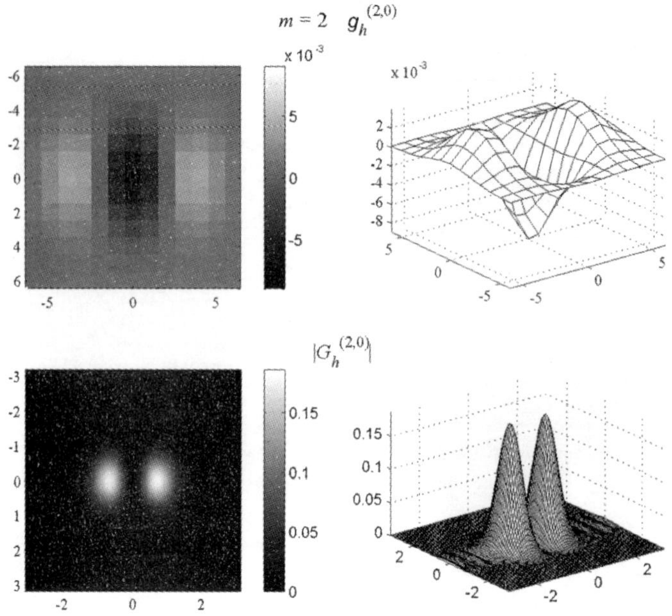

Figure 3.17 The differentiation kernel $g_h^{(2,0)}$ and its amplitude frequency characteristic $|G_h^{(2,0)}|$: Gaussian window, $m = 2$. The bandpass filter, $|G_h^{(2,0)}| = 0$ at $\bar{\omega} = 0$.

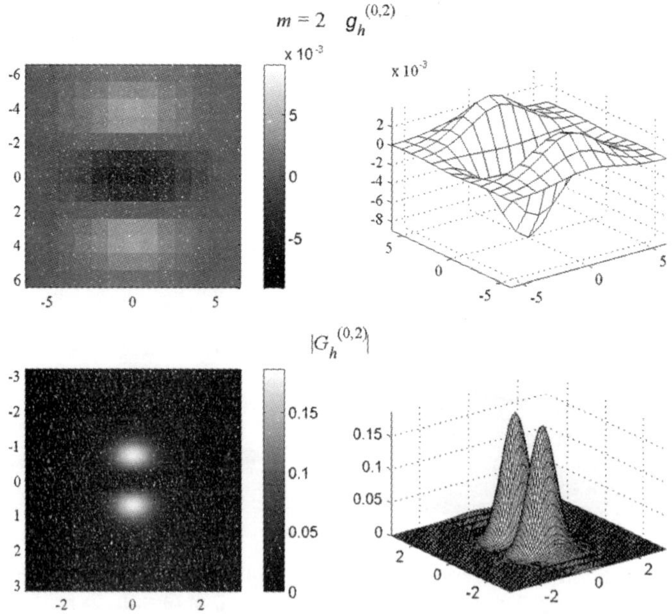

Figure 3.18 The differentiation kernel $g_h^{(0,2)}$ and its amplitude frequency characteristic $|G_h^{(0,2)}|$: Gaussian window, $m = 2$. The bandpass filter, $|G_h^{(0,2)}| = 0$ at $\bar{\omega} = 0$.

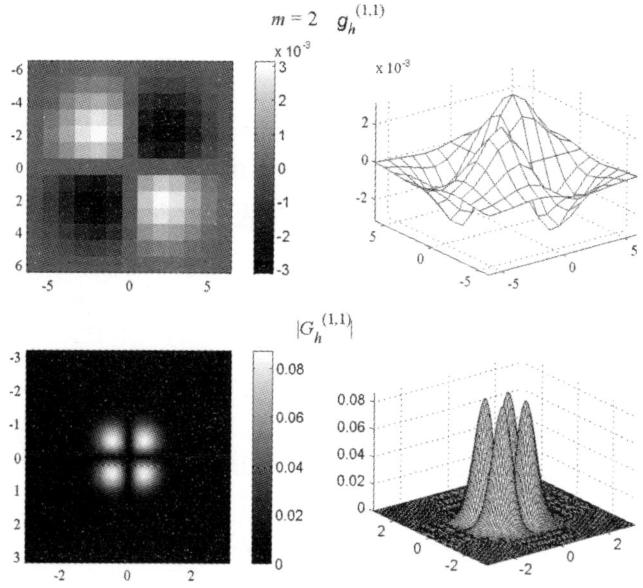

Figure 3.19 The differentiation kernel $g_h^{(1,1)}$ and its amplitude frequency characteristic $|G_h^{(1,1)}|$: Gaussian window, $m = 2$. The bandpass filter, $|G_h^{(1,1)}| = 0$ at $\bar{\omega} = 0$.

Fig. 3.19. Its frequency characteristic has zero value at $\bar{\omega}_1 = 0$, $\bar{\omega}_2 = 0$ and four distinctive peaks.

3.5 Numerical Differentiation

Derivatives are frequently required by numerical procedures across many disciplines. Image processing methods, iterative nonlinear equation solvers, optimization algorithms, and integration algorithms for differential equations require derivative information about the functions upon which they work.

A number of techniques have been developed for numerical differentiation. Roughly speaking, they fall in two categories: difference methods and statistical methods. The first category, especially the central finite-difference formula, is well known. These methods have the advantage of simplicity and are considered by many authors to yield satisfactory results when the function is given very precisely.

The problem of numerical differentiation is known to be sensitive with respect to small perturbations that lead to large errors in the computed derivative. In many applications it is necessary to estimate the derivative of a function given by noisy observations. Statistical methods are used in this situation to attenuate noise effects. The LPA derivative estimates belong to the class of statistical methods using the polynomial approximation to smooth the noise. In this section we interpret the well-known difference methods in terms of the developed technique and discuss how they differ from the LPA differentiators.

3.5.1 Univariate differentiation

3.5.1.1 *Two-point estimates*

From calculus, the derivative of the function y on x is defined as

$$y^{(1)}(x) = \lim_{\Delta \to 0} \frac{y(x+\Delta) - y(x)}{\Delta}. \tag{3.66}$$

This equation motivates a simple approximation for the first derivative called a forward difference:

$$y^{(1)}(x) \simeq \hat{y}_{\Delta}^{(1)}(x) = \frac{y(x+\Delta) - y(x)}{\Delta}. \tag{3.67}$$

This estimate is from a class of two-point estimates using two values of y. Other simple two-point formulas,

$$\hat{y}_{\Delta}^{(1)}(x) = \frac{y(x) - y(x-\Delta)}{\Delta}, \tag{3.68}$$

$$\hat{y}_{\Delta}^{(1)}(x) = \frac{y(x+\Delta) - y(x-\Delta)}{2\Delta}, \tag{3.69}$$

are known as backward and symmetric differences, respectively.

These estimates are motivated by the derivative definition in Eq. (3.66) and assume that Δ is small. To evaluate the accuracy of these formulas, we can use the Taylor series for

$$y(x+\Delta) = y(x) + \Delta y^{(1)}(x) + \Delta^2 y^{(2)}(x)/2 + \Delta^3 y^{(3)}(x)/6 + \cdots .$$

Inserting this series in the first estimate of Eq. (3.67), we can see that

$$\begin{aligned}\hat{y}^{(1)}(x) &= \frac{1}{\Delta}(-y(x) + y(x) + \Delta y^{(1)}(x) + \Delta^2 y^{(2)}(x)/2 + \cdots) \\ &\approx y^{(1)} + \Delta y^{(2)}(x)/2 + \cdots .\end{aligned}$$

Thus, the forward difference estimate gives the derivative within the error of the order $\Delta y^{(2)}(x)/2$. The error of the derivative estimate is a linear function of Δ and proportional to the second derivative $y^{(2)}(x)$.

The calculations for the backward difference in Eq. (3.68) give the equivalent result

$$\begin{aligned}\hat{y}^{(1)}(x) &= \frac{1}{\Delta}(y(x) - y(x) + \Delta y^{(1)}(x) - \Delta^2 y^{(2)}(x)/2 + \cdots) \\ &\approx y^{(1)}(x) - \Delta y^{(2)}(x)/2 + \cdots .\end{aligned}$$

with the only difference being that the main error term has an opposite sign.

Let us look at the symmetric estimate. It gives

$$\hat{y}^{(1)}(x) = \frac{1}{2\Delta}[y(x+\Delta) - y(x-\Delta)] = y^{(1)}(x) + \Delta^2 y^{(3)}(x)/6 + \cdots .$$

This formula shows that the symmetric difference is a much more accurate result for small Δ because the error is proportional to Δ^2.

Another principal difference between the symmetric and nonsymmetric forward and backward estimates is that the symmetric estimate is accurate for any quadratic y (when $y^{(3)} = 0$), while the nonsymmetric estimates are accurate only for any linear y (when $y^{(2)} = 0$). Thus, the symmetry makes a big difference in the performance of the derivative estimate.

Now, let us go to parallels between the difference estimates motivated by the derivative definition and the estimates derived from the LPA. To make this comparison explicitly clear, rewrite the estimates in Eqs. (3.67–3.69) as the convolution

$$\hat{y}^{(1)}(x) = \sum_{l=-\infty}^{\infty} y(x - \Delta l) g^{(1)}[l] = y(x - \Delta) g^{(1)}[1] + y(x) g^{(1)}[0] + y(x + \Delta) g^{(1)}[-1], \tag{3.70}$$

where $g^{(1)}[l]$ is the kernel of the estimate. Comparing Eq. (3.70) with Eqs. (3.67–3.69) we may conclude that:

for the forward difference

$$g^{(1)}[l] = \frac{1}{\Delta}(-2l - 1), \qquad l = 0, -1; \tag{3.71}$$

for the backward difference

$$g^{(1)}[l] = \frac{1}{\Delta}(1 - 2l), \qquad l = 0, 1; \tag{3.72}$$

and for the symmetric difference

$$g^{(1)}[l] = \frac{1}{2\Delta}(-l), \qquad l = -1, 1. \tag{3.73}$$

Naturally, for the two-point difference methods only two values of $g^{(1)}[l]$ shown in Eqs. (3.71–3.73) are nonzero. For $\Delta = 1$ and $h = 1$, the kernels $g^{(1)}[l]$ in Eqs. (14.26) and (3.72) are identical. In a similar way, we may see that the symmetric difference kernel in Eq. (3.73) is identical to $g^{(1)}[l]$ in Eq. (14.14) again for $\Delta = 1$ and $h = 1$.

The standard finite-difference derivative estimates in Eqs. (3.67–3.69) can be treated as a special case of the LPA differentiating operators corresponding to a rectangular window, $m = 1$, and, what is more important, to the minimum number of points (observations) used in the estimate.

3.5.1.2 Multiple-point estimates

Formal techniques exploited for the multiple-point estimate are based on polynomial interpolation formulas (e.g., [20], [34]). In particular, the approximation of y can be given in the form

$$\hat{y}(x) = \sum_{l} y(l)\varphi_l(x),$$
$$\varphi_l(x) = \frac{\Pi_{k\neq l}(x-k)}{\Pi_{k\neq l}(l-k)}, \tag{3.74}$$

where $\varphi_l(x)$ is the Kronecker function for integer x, as $\varphi_l(l) = 1$ and $\varphi_l(k) = 0$ if $k \neq l$. Thus, $\hat{y}(l) = y(l)$ and $\hat{y}(x)$ is the polynomial of the power defined by the number of points in Eq. (3.74). The approximation $\hat{y}$ is differentiated on x. Assuming $x = k$ in this derivative, we obtain a multipoint differentiation formula in which the number of nodes defines the order of the polynomial approximation. Increasing the number of points automatically increases the order of the approximating polynomial.

For instance, the following three-point derivative estimates can be obtained from Eq. (3.74) using the second-order polynomial and the corresponding three points, symmetric and nonsymmetric, with respect to the estimation point x:

$$\hat{y}^{(1)}(x) = \frac{1}{2\Delta}[-3y(x) + 4y(x+\Delta) - y(x+2\Delta)],$$
$$\hat{y}^{(1)}(x) = \frac{1}{2\Delta}[-y(x-\Delta) + y(x+\Delta)],$$
$$\hat{y}^{(1)}(x) = \frac{1}{2\Delta}[y(x-2\Delta) - 4y(x-\Delta) + 3y(x)].$$

These estimates are forward, symmetric, and backward differences. Presented in the convolution form [Eq. (3.70)], these estimates have the following kernels, respectively,

$$g^{(1)}[0] = \frac{-3}{2\Delta}, \qquad g^{(1)}[-1] = \frac{2}{\Delta}, \qquad g^{(1)}[-2] = \frac{-1}{2\Delta}, \tag{3.75}$$
$$g^{(1)}[0] = 0, \qquad g^{(1)}[-1] = \frac{1}{2\Delta}, \qquad g^{(1)}[1] = \frac{-1}{2\Delta}, \tag{3.76}$$
$$g^{(1)}[0] = \frac{3}{2\Delta}, \qquad g^{(1)}[1] = \frac{-2}{\Delta}, \qquad g^{(1)}[2] = \frac{1}{2\Delta}. \tag{3.77}$$

It can be verified that the values in Eq. (3.77) coincide with the ones given by formula (14.28), that is, those derived by the LPA with a rectangular nonsymmetric window, $m = 2$, $h = 2$, and $\Delta = 1$. For the symmetric kernel in Eq. (3.76) we note that it can be derived by using the LPA with $m = 1$ or $m = 2$. The result is identical in both cases because of the symmetry of the kernel support defined by the point selection.

The interpolation technique enables the one-to-one link between the number of points and the order m of the polynomial approximation. In this way we always have the interpolation as well as the differentiation with the minimum number of points.

3.5.1.3 Statistical differentiation

The LPA smoothing and differentiation estimates belong to the class of statistical methods. The order m and the number of points n are independent parameters, and n is selected to be larger than $m + 1$. These estimates are redundant because n is larger than the minimum number of points sufficient for the accurate fitting of the polynomial of degree m. This redundancy is used to smooth errors that can have a place in the data.

The stochastic nature of the observations is essential here. The main objective is to build estimates with the expectation close to the estimated derivative while the variance of the random errors in the estimate is as small as possible.

3.5.2 Multivariate differentiation

3.5.2.1 Gradients

Let us consider some of the classical differentiators used for the 2D case in image processing. The standard differentiator has a square mask of minimum size sufficient for calculation of finite differences or polynomial approximation usually by degree $m = 1$ or $m = 2$.

Because they are small in size, these differentiators perform badly in noisy environments. However, they are very popular because of their simplicity and transparent interpretation. Examples of this type of conventional gradient operator are shown in Table 3.1 ([91], [92]). The boxed elements in this table indicate the location of the estimate reference point.

Table 3.1 Common differentiating operators.

	$\bar{g}^{(r)}$	$\tilde{g}^{(r)}$
Roberts	$\begin{bmatrix} \boxed{0} & 1 \\ -1 & 0 \end{bmatrix}$	$\begin{bmatrix} \boxed{1} & 0 \\ 0 & -1 \end{bmatrix}$
Prewitt	$\begin{bmatrix} -1 & 0 & 1 \\ -1 & \boxed{0} & 1 \\ -1 & 0 & 1 \end{bmatrix}$	$\begin{bmatrix} -1 & -1 & -1 \\ 0 & \boxed{0} & 0 \\ 1 & 1 & 1 \end{bmatrix}$
Sobel	$\begin{bmatrix} -1 & 0 & 1 \\ -2 & \boxed{0} & 2 \\ -1 & 0 & 1 \end{bmatrix}$	$\begin{bmatrix} -1 & -2 & -1 \\ 0 & \boxed{0} & 0 \\ 1 & 2 & 1 \end{bmatrix}$
Isotropic	$\begin{bmatrix} -1 & 0 & 1 \\ -\sqrt{2} & \boxed{0} & \sqrt{2} \\ -1 & 0 & 1 \end{bmatrix}$	$\begin{bmatrix} -1 & -\sqrt{2} & -1 \\ 0 & \boxed{0} & 0 \\ 1 & \sqrt{2} & 1 \end{bmatrix}$

Let us discuss the meaning of these estimates in terms of the differentiating operators introduced in this chapter. Assume that $y(x_1, x_2)$ is defined on the regular grid with $\Delta = 1$. The *Roberts* estimates are calculated as the following finite differences:

$$\hat{y}_{\bar{g}} = y(x_1 + 1, x_2) - y(x_1, x_2 + 1),$$
$$\hat{y}_{\tilde{g}} = y(x_1, x_2) - y(x_1 + 1, x_2 + 1).$$

The subscript of the estimate shows which kernel, $\bar{g}$ or $\tilde{g}$, is used. The *Roberts* estimates are calculated as the diagonal finite differences. The first gives the finite difference for the direction northeast/southwest and the second for the direction northwest/southeast.

The *Prewitt* (*"smoothed" Prewitt*) estimators are defined as the sum of symmetric 1D differences on x_1 and x_2, respectively. In particular,

$$\begin{aligned}\hat{y}_{\bar{g}} = {} & [y(x_1 + 1, x_2 - 1) - y(x_1 - 1, x_2 - 1)] \\ & + [y(x_1 + 1, x_2) - y(x_1 - 1, x_2)] \\ & + [y(x_1 + 1, x_2 + 1) - y(x_1 - 1, x_2 + 1)]. \end{aligned} \tag{3.78}$$

The *Sobel* and *Isotropic* estimates are also defined as sums of the 1D symmetric differences that are different from *Prewitt* estimates only by the weights combining the differences. These weights are equal to 1 for the the *Prewitt* estimates of Eq. (3.78), while in the *Sobel* estimates one of the differences is taken with the weight equal to 2, and in the isotropic estimate this weight is equal to $\sqrt{2}$. We may conclude that these estimates do not go further than the intuitively clear 1D finite differences, which are coordinatewise or diagonalwise. In this respect, the LPA is a much more universal procedure, giving the differentiating kernels of any orders and with nonuniform arbitrary window w.

3.5.2.2 Laplacian operator

By definition, the Laplacian of y is the sum of the second derivatives of y:

$$L\{y\} = y^{(2,0)}(x) + y^{(0,2)}(x). \tag{3.79}$$

In image processing, the Gaussian Laplacian is used for the estimation of Eq. (3.79). The kernel of this operator is defined as the sum of the corresponding second derivatives of the Gaussian density:

$$\begin{aligned} L\left\{\frac{1}{2\pi\sigma^2} e^{-||x||^2/2\sigma^2}\right\} &= (\partial^2_{x_1} + \partial^2_{x_2}) \frac{1}{2\pi\sigma^2} e^{-||x||^2/2\sigma^2} \\ &= \frac{-1}{2\pi\sigma^4}\left(2 - \frac{||x||^2}{\sigma^2}\right) e^{-||x||^2/2\sigma^2}. \end{aligned} \tag{3.80}$$

The LPA allows one to obtain an accurate, discrete version of the Laplacian for any window w. "Accurate" means that it is guaranteed that the derivatives in

Eq. (3.79) are estimated, and these derivatives are accurate for the 2D polynomial $y \in \mathcal{P}^m$, where m is the degree used in the LPA design. The analytical formulas can be also given for these operators for some particular windows w. For example, for the symmetric rectangular window w, using formulas (14.53) and (14.54), we obtain the kernel of the discrete Laplacian as

$$g_h^{(2,0)}[l_1, l_2] + g_h^{(0,2)}[l_1, l_2]$$
$$= \left[\frac{30[3l_1^2 - h_1(h_1+1)]}{h_1(h_1+1)(2h_1+3)(2h_1-1)} + \frac{30[3l_2^2 - h_2(h_2+1)]}{h_2(h_2+1)(2h_2+3)(2h_2-1)} \right] w_{h_1,h_2}[l], \tag{3.81}$$

$$w_{h_1,h_2}[l] = \frac{1}{(2h_1+1)(2h_2+1)}, \qquad |l_1| \le h_1,\ |l_2| \le h_2,\ h_1, h_2 \ge 1.$$

Chapter 4

Integral LPA

Linear continuous filters and integral representations are standard for theory and for many applications. In comparison with discrete filters, continuous ones are simpler in many aspects and have more transparent properties.

An integral version of LPA filters is introduced in two different ways. An integral form of the estimates is derived first directly from the LPA approach. To do this, the sum in the criterion in Eq. (2.24) is replaced by the corresponding integral. Calculations similar to the ones produced in Sec. 2.2 give the continuous estimates and their integral kernel representations.

Another approach starts from the regular-grid discrete filters with the sampling interval Δ approaching zero. This limit gives the LPA kernels and estimates in the integral forms identical to the integral LPA. In this way the integral kernels and the estimates are considered as approximations of the discrete ones. The term "equivalent kernels" is used for the integral kernels obtained in this asymptotic consideration with a reference to the original discrete ones. We start from the 2D signals and then pass to the general d-dimensional case.

4.1 Integral Kernel Estimators

4.1.1 Integral LPA

Assume that a continuous observation model is given as

$$z(x) = y(x) + \varepsilon(x), \qquad x \in X \subset \mathbb{R}^2, \tag{4.1}$$

where $x = (x_1, x_2)$ is a continuous 2D variable belonging to a subspace X and $\varepsilon(x)$ is an additive noise.

4.1.1.1 Criterion

Let x be the "center" of the LPA. Then the estimate for the point $v = (v_1, v_2)$ in the neighborhood of the center x is of the form (2.18):

$$y_h(x, v) = C^T \phi_h(x - v),\ \phi_h(x) = \phi(x/h), \tag{4.2}$$
$$\phi(x) = (\phi_1(x), \phi_2(x), \ldots, \phi_M(x))^T,$$
$$C = (C_1, C_2, \ldots, C_M)^T,$$

where $\phi(x) \in \mathbb{R}^M$ is a vector of linearly independent 2D polynomials as in Eq. (2.20):

$$(-1)^{k_1+k_2} x_1^{k_1} x_2^{k_2} / k_1! k_2! \tag{4.3}$$

of the degrees $k_1 + k_2$ from 0 through m, and $C \in \mathbb{R}^M$ is a vector of parameters. These $\phi_r(x)$, $r = 1, \ldots, M$, form a complete set of polynomials as in Eq. (2.21), that is, for any x and v,

$$\phi(x - v) = A(v)\phi(x), \tag{4.4}$$

where $A(v)$ is an $M \times M$ matrix depending on v only.

For $m = 0$, 1, 2, and 3, the complete sets of 2D linearly independent polynomials are given in Eq. (2.22). The following integral weighted least-square criterion $J_h(x)$ is applied to find the coefficient C in Eq. (4.2):

$$J_h(x) = \int_{\mathbb{R}^2} w_h(x - v)[z(v) - y_h(x, v)]^2 dv, \tag{4.5}$$

where $\int_{\mathbb{R}^2}(\cdot)dv = \int_{-\infty}^{\infty}\int_{-\infty}^{\infty}(\cdot)dv_1 dv_2$ and $w_h(x) = w(x/h)/h^2$ and $h > 0$ is a scale (window size) parameter.

4.1.1.2 Signal estimation

According to the idea of the LPA, the estimate of $y(x)$ at the argument value x is defined as

$$\hat{y}_h(x) = y_h(x, v)|_{v=x} = y_h(x, x),$$

that is, the expansion in Eq. (4.2) is used for $v = x$ only. It gives

$$\hat{y}_h(x) = y_h(x, v)|_{v=x} = C^T \phi(0) \tag{4.6}$$

and, in particular, for the polynomials (2.22),

$$\hat{y}_h(x) = C^T \phi(0) = C_1. \tag{4.7}$$

4.1.1.3 Derivative estimation

The LPA model (Eq. (4.2)) of degree m can be used for the estimation of any derivative of the order $r = (r_1, r_2)$, $r_1 + r_2 \leq m$. The pointwise derivative estimates are of the form

$$\hat{y}_h^{(r)}(x) = \partial_{v_1}^{r_1}\partial_{v_2}^{r_2} y_h(x, v)|_{v=x} = C^T \partial_{v_1}^{r_1}\partial_{v_2}^{r_2}\phi_h(x - v)|_{v=x} = (-1/h)^{r_1+r_2} C^T \phi^{(r)}(0), \tag{4.8}$$

where $\phi^{(r)}(0) = \partial_{x_1}^{r_1}\partial_{x_2}^{r_2}\phi(x)|_{x=0}$. The differentiation in Eq. (4.8) is done with respect to v because $y(x, v)$ is an approximation considered to be a function of v provided that the LPA center x is fixed. For the polynomials in Eq. (2.22), the integral derivative estimates in Eq. (4.8) are exactly as they are in Eq. (2.32).

4.1.1.4 Estimate of C

Minimization of $J_h(x)$ in Eq. (4.5) on C gives a linear normal equation, $\partial_C J_h(x) = 0$, and a solution in the form

$$\begin{aligned} \hat{C}(x, h) &= \Phi_h^{-1} \int_X w_h(x - v)\phi_h(x - v)z(v)dv, \\ \Phi_h &= \int_X w_h(x - v)\phi_h(x - v)\phi_h^T(x - v)dv. \end{aligned} \tag{4.9}$$

It is assumed here that the matrix Φ_h is nonsingular. In general, it depends on x. The formulas (4.9) are an integral version of the solution of Eq. (2.39) obtained for the discrete case.

4.1.1.5 Kernel estimates

Using the estimates $\hat{C}(x, h)$ instead of C, we yield from Eqs. (4.6) and (4.8) that

$$\hat{y}_h(x) = y_h(x, x) = \hat{C}^T(x, h)\phi(0), \tag{4.10}$$

and

$$\hat{y}_h^{(r)}(x) = (-1/h)^{|r|}\hat{C}^T(x, h)\phi^{(r)}(0). \tag{4.11}$$

Inserting $\hat{C}(x, h)$ from Eq. (4.9) into Eqs. (4.10) and (4.11), we find the estimates are outputs of linear filters:

$$\hat{y}_h(x) = \int_X g_h(x, v)z(v)dv, \tag{4.12}$$

$$\hat{y}_h^{(r)}(x) = \int_X g_h^{(r)}(x, v)z(v)dv, \tag{4.13}$$

where

$$g_h(x, v) = w_h(x - v)\phi_h^T(x - v)\Phi_h^{-1}\phi(0), \tag{4.14}$$

$$g_h^{(r)}(x, v) = (-1/h)^{|r|} w_h(x - v)\phi_h^T(x - v)\Phi_h^{-1}\phi^{(r)}(0), \tag{4.15}$$

$$\Phi_h = \int_X w_h(x - v)\phi_h(x - v)\phi_h^T(x - v)dv. \tag{4.16}$$

Formulas (4.12)–(4.13) define the *smoothing and differentiating integral (continuous) filters* of order m. The weights $g_h(x, x_s)$ and $g_h^{(r)}(x, x_s)$ are *smoothing and differentiating kernels (impulse response functions)* of these estimates.

Let integration in Eqs. (4.12–4.13) be produced in infinite limits, $X = \mathbb{R}^2$. Then the matrix Φ_h in Eq. (4.16) becomes invariant and the kernels g_h and $g_h^{(r)}$ depend on the difference of the arguments x and v only. In this case, the estimates in Eqs. (4.12–4.13) can be written as convolutions with *shift-invariant kernels*:

$$\hat{y}_h(x) = \int_{\mathbb{R}^2} g_h(x - v)z(v)dv, \tag{4.17}$$

$$\hat{y}_h^{(r)}(x) = \int_{\mathbb{R}^2} g_h^{(r)}(x - v)z(v)dv, \tag{4.18}$$

where

$$g_h(x) = w_h(x)\phi_h^T(x)\Phi_h^{-1}\phi(0), \tag{4.19}$$

$$g_h^{(r)}(x) = (-1/h)^{|r|} w_h(x)\phi_h^T(x)\Phi_h^{-1}\phi^{(r)}(0), \tag{4.20}$$

$$\Phi_h = \int_{\mathbb{R}^2} w_h(v)\phi_h(v)\phi_h^T(v)dv. \tag{4.21}$$

Using convolution notation, we can rewrite Eqs. (4.17–4.18) in the form

$$\hat{y}_h(x) = (g_h \circledast z)(x), \tag{4.22}$$

$$\hat{y}_h^{(r)}(x) = (g_h^{(r)} \circledast z)(x),$$

where $\circledast$ is used for continuous integral convolution.

The same convolution notation for discrete and continuous cases (Eqs. (3.9) and (4.22), respectively) cannot be confused, because a difference is always clear from the context and the considered problem.

Further transformations show that

$$\begin{aligned}\Phi_h &= \int_{\mathbb{R}^2} w(v/h)\phi(v/h)\phi^T(v/h)d(v/h) \\ &= \int_{\mathbb{R}^2} w(v)\phi(v)\phi^T(v)dv = \Phi,\end{aligned} \tag{4.23}$$

that is, the matrix Φ is constant and independent of h and x.

Substituting Eq. (4.23) into Eqs. (4.19–4.20), we obtain the convolutions in Eq. (4.22) in the following final forms:

$$\begin{aligned}\hat{y}_h(x) &= (g_h \circledast z)(x)\\ &= \frac{1}{h^2}\int_{\mathbb{R}^2} g[(x-v)/h]z(v)dv \qquad (4.24)\\ &= \int_{\mathbb{R}^2} g(v)z(x-hv)dv,\\ \hat{y}_h^{(r)}(x) &= (g_h^{(r)} \circledast z)(x)\\ &= \frac{1}{h^{2+|r|}}\int_{\mathbb{R}^2} g^{(r)}[(x-v)/h]z(v)dv \qquad (4.25)\\ &= \frac{1}{h^{|r|}}\int_{\mathbb{R}^2} g^{(r)}(v)z(x-hv)dv,\end{aligned}$$

where

$$g(x) = w(x)\phi^T(x)\Phi^{-1}\phi(0), \qquad (4.26)$$

$$g^{(r)}(x) = (-1)^{|r|}w(x)\phi^T(x)\Phi^{-1}\phi^{(r)}(0). \qquad (4.27)$$

These formulas are integral counterparts of the discrete estimate formulas (3.1)–(3.2) or (3.10)–(3.11). In comparison with the discrete case, the integral estimates in Eqs. (4.24–4.25) are much simpler because the dependence on the scale h is given analytically.

The scale h in Eqs. (4.24–4.25) can be treated as a rescaling of the kernel (the first expressions in Eqs. (4.24–4.25)) or as a rescaling of the observations (the second expressions in Eqs. (4.24–4.25)). The *scaling* meaning of the parameter h is much more obvious in the integral operator than it is in the case of the discrete ones.

4.1.1.6 Vanishing moments

Consider the *reproducing properties* of the integral estimators with respect to polynomial signals. Let the input y be a polynomial of degree less than or equal to m. It can be proved as it has been in Proposition 1 that the estimates in Eqs. (4.24–4.25) are accurate for this polynomial signal.

For the vector $\phi(x)$ composed from Eq. (4.3), the estimates in Eqs. (4.24–4.25) give accurate results, that is,

$$\int_{\mathbb{R}^2} g(u)\phi(x-hu)du = \phi(x),$$

$$\frac{1}{h^{|r|}}\int_{\mathbb{R}^2} g^{(r)}(u)\phi(x-hu)du = \partial^r\phi(x).$$

These equations are identities that are valid for all x. It follows for $x = 0$ that

$$\int_{\mathbb{R}^2} g(u)\phi(-hu)du = \phi(0), \tag{4.28}$$

$$\frac{1}{h^{|r|}}\int_{\mathbb{R}^2} g^{(r)}(u)\phi(-hu)du = \partial^r\phi(0). \tag{4.29}$$

For the polynomials in Eq. (4.3) one obtains the vanishing moment conditions in a form similar to Eqs. (3.15–3.16):

(a) for the smoothing operator,

$$\int_{\mathbb{R}^2} x^i g(x)dx = \delta_{i,0}, \qquad 0 \le |i| \le m; \tag{4.30}$$

(b) for the differentiating operator,

$$\frac{1}{i!}\int_{\mathbb{R}^2} x^i g^{(r)}(x)dx = (-1)^{|r|}\delta_{i,r}, \qquad 0 \le |i| \le m, \quad 0 \le |r| \le m. \tag{4.31}$$

4.1.2 Multivariate kernels and estimates

Let ϕ used in the LPA in Eq. (4.2) be a vector composed from the polynomials in Eq. (2.59) of powers from 0 through m:

$$(-1)^{|k|}x^k/k!, 0 \le |k| \le m, \qquad x \in \mathbb{R}^d,$$

and let the window $w(x)$ be a function of d variables. Then the formulas (4.12)–(4.16), where $X \in \mathbb{R}^d$, are valid for the general d-dimensional case without any changes.

Consider the shift-invariant estimates. Assume that $X = \mathbb{R}^d$. The integration in Eqs. (4.12–4.16) is produced over $\mathbb{R}^d$ space and we arrive at formulas similar to Eqs. (4.24–4.27) where the dimension d is shown explicitly:

$$\begin{aligned}\hat{y}_h(x) = (g_h \circledast z)(x) &= \frac{1}{h^d}\int_{\mathbb{R}^d} g[(x-v)/h]z(v)dv \\ &= \int_{\mathbb{R}^d} g(v)z(x-hv)dv,\end{aligned} \tag{4.32}$$

$$\begin{aligned}\hat{y}_h^{(r)}(x) = (g_h^{(r)} \circledast z)(x) &= \frac{1}{h^{d+|r|}}\int_{\mathbb{R}^d} g^{(r)}[(x-v)/h]z(v)dv \\ &= \frac{1}{h^{|r|}}\int_{\mathbb{R}^d} g^{(r)}(v)z(x-hv)dv,\end{aligned} \tag{4.33}$$

where

$$g(x) = w(x)\phi^T(x)\Phi^{-1}\phi(0), \tag{4.34}$$

$$g^{(r)}(x) = (-1)^{|r|}w(x)\phi^T(x)\Phi^{-1}\phi^{(r)}(0), \tag{4.35}$$

$$\Phi = \int_{\mathbb{R}^d} w(v)\phi(v)\phi^T(v)dv.$$

The matrix Φ is constant and independent on h and x. These formulas link the kernels g_h, $g_h^{(r)}$ with the kernels $g(x)$, $g^{(r)}(x)$ as follows:

$$g_h(x) = \frac{1}{h^d}g(x/h), \tag{4.36}$$

$$g_h^{(r)}(x) = \frac{1}{h^{d+|r|}}g^{(r)}(x/h). \tag{4.37}$$

The vanishing moment conditions in Eqs. (4.30–4.31) for $x \in \mathbb{R}^d$ are transformed to the obvious forms:

(a) for the smoothing operator,

$$\int_{\mathbb{R}^d} x^i g(x)dx = \delta_{i,0}, \qquad 0 \leq |i| \leq m; \tag{4.38}$$

(b) for the differentiating operator,

$$\frac{1}{i!}\int_{\mathbb{R}^d} x^i g^{(r)}(x)dx = (-1)^{|r|}\delta_{i,r}, \qquad 0 \leq |i| \leq m, \quad 0 \leq |r| \leq m. \tag{4.39}$$

4.1.3 Limit LPA estimates

The above continuous integral estimates can be derived starting from the discrete LPA estimates assuming that the sampling interval Δ is approaching zero.

Proposition 4 *Let the discrete kernel estimates of the signal and the derivatives be defined by Eqs. (3.4–3.8). Then the limit estimates, as $\Delta \to 0$, have the integral forms (4.32)–(4.35).*

The proof is given in Sec. 14.2.

4.1.4 Frequency domain

Classical continuous signal processing algorithms are mostly based on shift-invariant linear operators given by convolutions, e.g., in Eqs. (4.32) and (4.33). The frequency characteristics and the integral Fourier transform are efficient methods for characterizing this type of operator. We start this section with a brief review of the basic concepts of the integral Fourier transform. Details and proofs can be found in many sources (e.g., [18], [173], [229]).

4.1.4.1 Signal estimate

The continuous integral Fourier transform (IFT) is a standard tool for shift-invariant linear systems. Given an absolutely integrable signal $y(x)$, $x \in \mathbb{R}^d$, its IFT is defined by

$$Y(\lambda) = \mathcal{F}\{y(x)\} = \int_{\mathbb{R}^d} y(x) \exp(-j\langle\lambda, x\rangle)dx, \tag{4.40}$$

which is called the Fourier analysis formula, $\langle\lambda, x\rangle = \sum_{l=1}^{d} \lambda_l x_l$.

In contrast to the frequency ω used for the DFT, we exploit λ for the continuous frequency in the IFT. The inverse integral Fourier transform is given by

$$y(x) = \mathcal{F}^{-1}\{Y(\lambda)\} = \frac{1}{(2\pi)^d} \int_{\mathbb{R}^d} Y(\lambda) \exp(j\langle\lambda, x\rangle)d\omega. \tag{4.41}$$

Applying the IFT to the convolution in Eq. (4.32) and assuming that $g(x)$ and $z(x)$ are absolutely integrable, we obtain, after standard manipulations,

$$\begin{aligned}
\hat{Y}_h(\lambda) &= \mathcal{F}\{\hat{y}_h(x)\} \\
&= \int_{\mathbb{R}^d} \left[\int_{\mathbb{R}^d} g(v) z(x - hv)dv\right] \exp(-j\langle\lambda, x\rangle)dx \\
&= \int_{\mathbb{R}^d} g(v) \exp(-j\langle\lambda, hv\rangle)dv \\
&\quad \cdot \int_{\mathbb{R}^d} z(x - hv) \exp(-j\langle\lambda, x - hv\rangle)dx \\
&= \int_{\mathbb{R}^d} g(v) \exp(-j\langle h\lambda, v\rangle)dv \cdot \int_{\mathbb{R}^d} z(x) \exp(-j\langle\lambda, x\rangle)dx \\
&= G(h\lambda)Z(\lambda),
\end{aligned} \tag{4.42}$$

where $\hat{Y}_h(\lambda) = \mathcal{F}\{\hat{y}_h(x)\}$ and $Z(\lambda) = \mathcal{F}\{z(x)\}$ are the IFTs of the corresponding signals, and

$$G_h(\lambda) = G(h\lambda), \qquad G(\lambda) = \mathcal{F}\{g(x)\} \tag{4.43}$$

is the transfer function of the system. In the frequency domain, the parameter h changes only the scale of the frequency λ in the transfer function $G(\lambda)$.

4.1.4.2 Derivative estimate

Applying the IFT to the convolution in Eq. (4.33), we obtain the following, similar to Eqs. (4.42–4.43):

$$\begin{aligned}\hat{Y}_h^{(r)}(\lambda) &= \mathcal{F}\{\hat{y}_h^{(r)}(x)\}\\ &= \frac{1}{h^{|r|}}\int_{\mathbb{R}^d}\left[\int_{\mathbb{R}^d} g^{(r)}(v)z(x-hv)dv\right]\exp(-j\langle\lambda, x\rangle)dx\\ &= \frac{1}{h^{|r|}}\int_{\mathbb{R}^d} g^{(r)}(v)\exp(-j\langle h\lambda, v\rangle)dv\\ &\quad\cdot\int_{\mathbb{R}^d} z(x-hv)\exp(-j\langle\lambda, x-hv\rangle)dx\\ &= \frac{1}{h^{|r|}}G^{(r)}(h\lambda)Z(\lambda),\end{aligned} \tag{4.44}$$

$$G^{(r)}(\lambda) = \mathcal{F}\{g^{(r)}(x)\}. \tag{4.45}$$

Here, $G^{(r)}(\lambda)$ is the transfer function of the differentiating operator. Thus, we have proven that

$$\hat{Y}_h^{(r)}(\lambda) = \frac{1}{h^{|r|}}G^{(r)}(h\lambda)Z(\lambda).$$

4.1.4.3 Vanishing moments

The vanishing moment formulas [Eqs. (4.38–4.39)] in the frequency domain can be presented in a form similar to that given in Eqs. (3.46–3.49) for discrete LPA kernels. We can derive from Eq. (4.40) that

$$G(\lambda)|_0 = \int_{\mathbb{R}^d} g(x)dx,$$

$$\partial^i_{j\lambda}G(\lambda)|_0 = (-1)^{|i|}\int_{\mathbb{R}^d} x^i g(x)dx,$$

where $|_0$ means $\lambda = 0$. This gives the vanishing moment conditions in Eqs. (4.38–4.39) in these equivalent forms:

(a) for the smoothing operator,

$$\begin{aligned}G(\lambda)|_0 &= 1,\\ \partial^i_{j\lambda}G(\lambda)|_0 &= 0, \qquad 0 < |i| \le m;\end{aligned} \tag{4.46}$$

(b) for the differentiating operator,

$$\begin{aligned}G^{(r)}(\lambda)|_0 &= 0,\\ \frac{1}{i!}\partial^i_{j\lambda}G^{(r)}(\lambda)|_0 &= \delta_{i,r}, \qquad 0 \le |i| \le m, \quad 0 < |r| \le m.\end{aligned} \tag{4.47}$$

It follows from Eqs. (4.46–4.47) that the smoothing and differentiating estimators define the *low-pass* and *high-pass* (or *bandpass) filters*, respectively. This is exactly as was discussed in Sec. 3.3.3 with the only difference being that, for the continuous case, the frequency interval is infinite, $\lambda \in \mathbb{R}^d$.

4.2 Analytical Kernels

In this section we use the classical orthonormal polynomials (Legendre, Laguerre, and Hermit) instead of the polynomials in the form of Eq. (4.3). The technique is similar to the one exploited in Sec. 14.1 for the analytical design of the discrete smoothing and differentiating estimators.

4.2.1 1D case

Let the 1D polynomials $\phi(x)$ be orthonormal with the window w. Then the matrix Φ in Eqs. (4.34–4.35) is identical, $\Phi = I$, and the kernel formulas are

$$g(x) = w(x)\phi^T(x)\phi(0), \tag{4.48}$$

and

$$g^{(r)}(x) = (-1)^{|r|} w(x)\phi^T(x)\phi^{(r)}(0). \tag{4.49}$$

The 1D kernels $g(x)$ and $g^{(k)}(x)$ can be derived from the formulas of the orthonormal polynomials. Let us demonstrate how this technique works.

4.2.1.1 *Legendre polynomials I*

Legendre polynomials are orthonormal with the rectangular uniform window. We consider two different cases: symmetric and nonsymmetric windows. The corresponding polynomials are called here the Legendre polynomials I and II, respectively.

The 1D Legendre polynomials I are orthonormal with the symmetric window

$$w_1(x_1) = 1, \qquad |x_1| \le 1. \tag{4.50}$$

These polynomials are

$$\phi_1(x_1) = \sqrt{\tfrac{1}{2}}, \qquad \phi_2(x_1) = \sqrt{\tfrac{3}{2}}x_1, \qquad \phi_3(x_1) = \tfrac{1}{2}\sqrt{\tfrac{5}{2}}(3x_1^2 - 1), \qquad x_1 \in \mathbb{R}. \tag{4.51}$$

The corresponding smoothing kernels of the powers $m = 0, 1, 2$ follow from Eq. (4.48) as

$$g(x_1) = \begin{cases} \tfrac{1}{2} w_1(x_1) & \text{for } m = 0, 1, \\ \tfrac{3}{8}(3 - 5x_1^2) w_1(x_1) & \text{for } m = 2. \end{cases} \tag{4.52}$$

According to Eq. (4.49), the first derivative estimation kernel is

$$g^{(1)}(x_1) = -\tfrac{3}{2}x_1 w_1(x_1), \qquad \text{for } m = 1, 2, \tag{4.53}$$

and the kernel for the second derivative is

$$g^{(2)}(x_1) = \tfrac{15}{4}(3x_1^2 - 1)w_1(x_1), \qquad \text{for } m = 2. \tag{4.54}$$

4.2.1.2 Legendre polynomials II

The Legendre polynomials II are orthonormal with the nonsymmetric rectangular window function

$$w_1(x_1) = 1, \qquad 0 \le x_1 \le 1, \tag{4.55}$$

and have the form

$$\phi_1(x_1) = 1, \qquad \phi_2(x_1) = \sqrt{3}(2x_1 - 1), \qquad \phi_3(x_1) = \tfrac{1}{2}\sqrt{5}[3(2x_1 - 1)^2 - 1]. \tag{4.56}$$

Then, the corresponding smoothing kernels follow from Eq. (4.48) as

$$g(x) = \begin{cases} w_1(x_1) & \text{for } m = 0, \\ (4 - 6x_1)w_1(x_1) & \text{for } m = 1, \\ (9 - 36x_1 + 30x_1^2)w_1(x_1) & \text{for } m = 2. \end{cases} \tag{4.57}$$

The first derivative estimation kernels of different LPA orders follow from Eq. (4.49) and are

$$g^{(1)}(x) = \begin{cases} -6(2x_1 - 1)w_1(x) & \text{for } m = 1, \\ -[6(2x_1 - 1) - 15(3(2x_1 - 1)^2 - 1)]w_1(x_1) & \text{for } m = 2. \end{cases} \tag{4.58}$$

The second derivative estimation kernel follows from Eq. (4.49) as

$$g^{(2)}(x) = 30[3(2x_1 - 1)^2 - 1]w_1(x_1), \qquad \text{for } m = 2. \tag{4.59}$$

4.2.1.3 Laguerre polynomials

Laguerre polynomials, which are orthonormal with the nonsymmetric exponential window function

$$w_1(x) = \exp(-x_1), \qquad 0 \le x_1 < \infty, \tag{4.60}$$

have the form

$$\phi_1(x_1) = 1, \qquad \phi_2(x_1) = 1 - x_1, \qquad \phi_3(x_1) = \tfrac{1}{2}(2 - 4x_1 + x_1^2). \tag{4.61}$$

The corresponding smoothing kernels follow from Eq. (4.48) as

$$g(x_1) = \begin{cases} w_1(x_1) & \text{for } m = 0, \\ (2 - x_1)w_1(x_1) & \text{for } m = 1, \\ \left(3 - 3x_1 + \frac{1}{2}x_1^2\right) w_1(x_1) & \text{for } m = 2. \end{cases} \tag{4.62}$$

The first derivative estimation kernels following from Eq. (4.49) and are

$$g^{(1)}(x_1) = \begin{cases} -(x_1 - 1)w_1(x_1) & \text{for } m = 1, \\ [(1 - x_1)(2 - x_1) + 2 - 4x_1 + x_1^2]w_1(x_1) & \text{for } m = 2. \end{cases} \tag{4.63}$$

The second derivative estimation kernel follows from Eq. (4.49) as

$$g^{(2)}(x_1) = \tfrac{1}{2}(2 - 4x_1 + x_1^2)w_1(x_1), \qquad \text{for } m = 2. \tag{4.64}$$

4.2.1.4 Hermitian polynomials

These polynomials are orthonormal with the symmetric weight

$$w_1(x_1) = \frac{1}{\sqrt{2\pi}} e^{-(x_1^2/2)}, \qquad -\infty < x < \infty, \tag{4.65}$$

and have the form

$$\phi_1(x_1) = 1, \qquad \phi_2(x_1) = x_1, \qquad \phi_3(x_1) = \frac{1}{\sqrt{2}}(x_1^2 - 1). \tag{4.66}$$

The corresponding smoothing kernels of the powers $m = 0, 1, 2$ follow from Eq. (4.48) as

$$g(x_1) = \begin{cases} w_1(x_1) & \text{for } m = 0, 1, \\ \frac{1}{2}(3 - x_1^2)w_1(x_1) & \text{for } m = 2. \end{cases} \tag{4.67}$$

The first derivative estimation kernels follow from Eq. (4.49) and are given by

$$g^{(1)}(x_1) = -x_1 w_1(x_1), \qquad \text{for } m = 1, 2. \tag{4.68}$$

The second derivative estimation kernel follows from Eq. (4.49) and is given by

$$g^{(2)}(x_1) = (x_1^2 - 1)w_1(x_1), \qquad \text{for } m = 2. \tag{4.69}$$

Examples

For $m = 2$, the Legendre polynomials I in Eq. (4.51) give the smoothing in Eq. (4.52) and differentiating kernel in Eq. (4.53) in the form

$$\begin{aligned} g(x_1) &= \tfrac{3}{8}(3 - 5x_1^2)w_1(x_1) \\ g^{(1)}(x_1) &= -\tfrac{3}{2}x_1 w_1(x_1), \end{aligned} \tag{4.70}$$

where $w_1(x_1)$ is defined by Eq. (4.50).

For the same order, $m = 2$, the Hermitian polynomials in Eq. (4.66) give the corresponding kernels in the form

$$\begin{aligned} g(x_1) &= \tfrac{1}{2}(3 - x_1^2)w_1(x_1) \\ g^{(1)}(x_1) &= -x_1 w_1(x_1), \end{aligned} \tag{4.71}$$

where $w(x_1)$ is defined by Eq. (4.65).

Curves of these kernels and their frequency characteristics are shown in Figs. 4.1 and 4.2. The kernels for the Hermitian polynomials are continuous and much smoother than those for the Legendre polynomials. We note that the frequency characteristics of the Hermitian polynomials are smoother and have a narrower low-pass band.

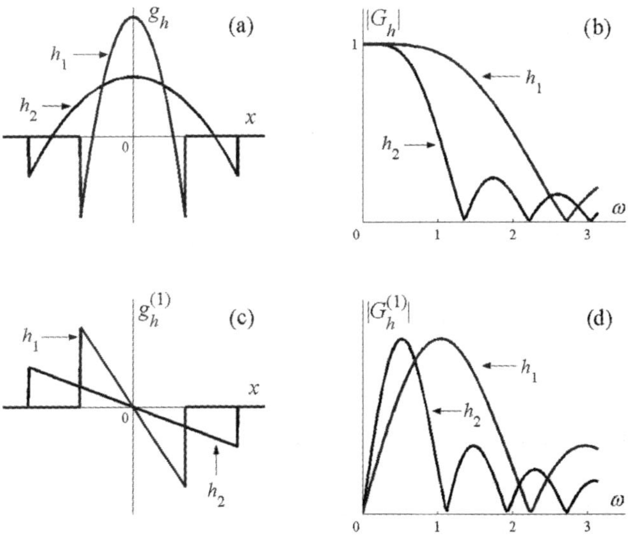

Figure 4.1 Smoothing g_h and differentiating $g_h^{(1)}$ Legendre kernels from Eq. (4.70) and their frequency characteristics shown for the scales $h_1 < h_2$.

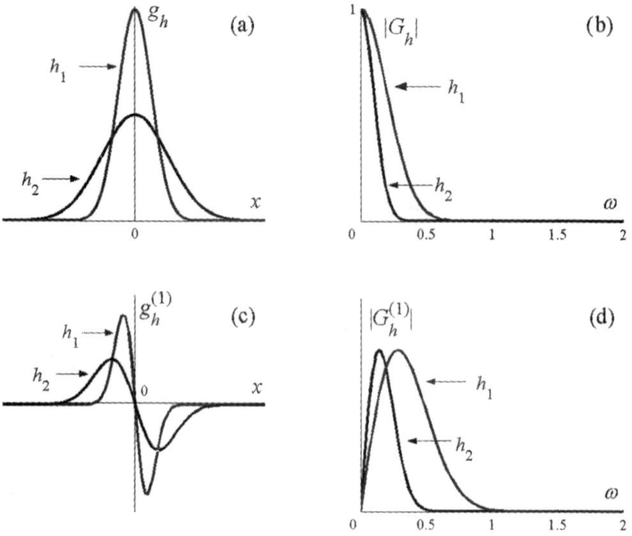

Figure 4.2 Smoothing g_h and differentiating $g_h^{(1)}$ Hermitian kernels of Eq. (4.71) and their frequency characteristics for the scales $h_1 < h_2$.

4.2.2 2D case

Let the 2D polynomials $\phi(x)$ be orthonormal with the window $w(x)$. Then the matrix Φ in Eq. (4.21) is identical and the kernel formulas (4.26) and (4.27) are simplified to

$$g(x) = w(x)\phi^T(x)\phi(0), \tag{4.72}$$

and

$$g^{(r)}(x) = (-1)^{|r|}w(x)\phi(x)\phi^{(r)}(0). \tag{4.73}$$

These formulas look the same as Eqs. (4.48–4.49). However, the difference is that, in the 2D case, $x = (x_1, x_2) \in \mathbb{R}^2$ and the derivative order r is a multi-index $r = (r_1, r_2)$.

Assume that the 2D window $w(x)$ is a product of the 1D windows $w_1(x_1)$ and $w_1(x_2)$:

$$w(x) = w_1(x_1)w_1(x_2). \tag{4.74}$$

Then the 2D polynomials $\phi(x)$ orthonormal with the window $w(x)$ in Eq. (4.74) can be composed from the 1D polynomials orthonormal with the weights $w_1(x_1)$ and $w_1(x_2)$, respectively (see formulas (14.42)). In this case the kernels in Eqs. (4.72) and (4.73) can be derived using these 1D orthonormal polynomials. This is exactly what was done in Sec. 14.1 to obtain the analytical formulas for the kernels of the discrete estimators.

4.2.2.1 Legendre polynomials I

Using the polynomials in Eq. (4.51) which are orthonormal with the weights in Eq. (4.50) and the formulas (4.72), we find

$$g(x) = \begin{cases} w(x), & \text{for } m = 0, 1, \\ \left[\frac{7}{2} - \frac{15}{4}(x_1^2 + x_2^2)\right] w(x), & \text{for } m = 2, \end{cases} \tag{4.75}$$

where

$$w(x) = \tfrac{1}{4} w_1(x_1) w_1(x_2).$$

According to Eq. (4.73), the first derivative kernels are

$$\begin{aligned} g^{(1,0)}(x) &= -3x_1 w(x), && \text{for } m = 1, 2, \\ g^{(0,1)}(x) &= -3x_2 w(x). && \text{for } m = 1, 2. \end{aligned} \tag{4.76}$$

According to Eq. (4.73), the second derivative kernels are

$$\begin{aligned} g^{(1,1)}(x) &= 9x_1 x_2 w(x), && \text{for } m = 2, \\ g^{(2,0)}(x) &= \tfrac{15}{2}(3x_1^2 - 1) w(x), && \text{for } m = 2, \\ g^{(0,2)}(x) &= \tfrac{15}{2}(3x_2^2 - 1) w(x). && \text{for } m = 2. \end{aligned}$$

4.2.2.2 Legendre polynomials II

The 1D Legendre polynomials II are orthonormal with the nonsymmetric rectangular window in Eq. (4.55) and have the form shown in Eq. (4.56).

Let

$$w(x) = w_1(x_1) w_1(x_2).$$

Then, the corresponding smoothing kernels follow from Eq. (4.72) as

$$g(x) = \begin{cases} w(x), & \text{for } m = 0, \\ [7 - 6(x_1 + x_2)] w(x), & \text{for } m = 1, \\ [21 + 30(x_1^2 + x_2^2) - 54(x_1 + x_2) + 36 x_1 x_2] w(x). & \text{for } m = 2. \end{cases}$$

The first derivative kernels follow from Eq. (4.73) and are

$$g^{(1,0)}(x) = \begin{cases} -6(2x_1 - 1) w(x), & \text{for } m = 1, \\ -\left\{12(2x_1 - 1)x_2 - \frac{15}{2}[3(2x_1 - 1)^2 - 1]\right\} w(x), & \text{for } m = 2. \end{cases}$$

$$g^{(0,1)}(x) = \begin{cases} -6(2x_2 - 1) w(x), & \text{for } m = 1, \\ -\left\{12(2x_2 - 1)x_1 - \frac{15}{2}[3(2x_2 - 1)^2 - 1]\right\} w(x). & \text{for } m = 2. \end{cases}$$

The second derivative kernels follow from Eq. (4.73) and are

$$\begin{aligned} g^{(1,1)}(x) &= 36(2x_1-1)(2x_2-1)w(x), && \text{for } m=2,\\ g^{(2,0)}(x) &= 15[3(2x_1-1)^2-1]w(x), && \text{for } m=2,\\ g^{(0,2)}(x) &= 15[3(2x_2-1)^2-1]w(x). && \text{for } m=2. \end{aligned}$$

4.2.2.3 Laguerre polynomials

The 1D Laguerre polynomials, which are orthonormal with the nonsymmetric exponential window Eq. (4.60), have the form shown in Eq. (4.61).

Let

$$w(x) = w_1(x_1)w_1(x_2).$$

The corresponding smoothing kernels follow from Eq. (4.72) as

$$g(x) = \begin{cases} w(x), & \text{for } m=0,\\ [3-(x_1+x_2)]w(x), & \text{for } m=1,\\ \left[6+\frac{1}{2}(x_1^2+x_2^2)-4(x_1+x_2)+x_1x_2\right]w(x). & \text{for } m=2. \end{cases}$$

The first derivative kernels follow from Eq. (4.73) and are

$$g^{(1,0)}(x) = \begin{cases} -(x_1-1)w(x), & \text{for } m=1,\\ [(1-x_1)(3-x_1-x_2)+2-4x_1+x_1^2]w(x), & \text{for } m=2. \end{cases}$$

$$g^{(0,1)}(x) = \begin{cases} -(x_2-1)w(x), & \text{for } m=1,\\ [(1-x_2)(3-x_1-x_2)+2-4x_2+x_2^2]w(x). & \text{for } m=2. \end{cases}$$

The second derivative kernels follow from Eq. (4.73) and are

$$\begin{aligned} g^{(1,1)}(x) &= (1-x_1)(1-x_2)w(x), && \text{for } m=2,\\ g^{(2,0)}(x) &= \tfrac{1}{2}(2-4x_1+x_1^2)w(x), && \text{for } m=2,\\ g^{(0,2)}(x) &= \tfrac{1}{2}(2-4x_2+x_2^2)w(x). && \text{for } m=2. \end{aligned}$$

4.2.2.4 Hermitian polynomials

These 1D polynomials are orthonormal with the symmetric weight in Eq. (4.65) and have the form shown in Eq. (4.66).

The corresponding smoothing kernels of the powers $m = 0, 1, 2$ follow from Eq. (4.72) as

$$g(x) = \begin{cases} w(x), & \text{for } m = 0, 1, \\ \left[2 - \frac{1}{2}(x_1^2 + x_2^2)\right] w(x), & \text{for } m = 2, \end{cases} \tag{4.77}$$

where

$$w(x) = w_1(x_1)w_1(x_2).$$

The first derivative kernels follow from Eq. (4.73) and are

$$\begin{aligned} g^{(1,0)}(x) &= -x_1 w(x), \quad \text{for } m = 1, 2, \\ g^{(0,1)}(x) &= -x_2 w(x). \quad \text{for } m = 1, 2. \end{aligned} \tag{4.78}$$

The second derivative kernels follow from Eq. (4.73) and are

$$\begin{aligned} g^{(1,1)}(x) &= x_1 x_2 w(x), && \text{for } m = 2, \\ g^{(2,0)}(x) &= (x_1^2 - 1)w(x), && \text{for } m = 2, \\ g^{(0,2)}(x) &= (x_2^2 - 1)w(x). && \text{for } m = 2. \end{aligned} \tag{4.79}$$

4.3 Generalized Singular Functions*

There are natural links between the integral LPA kernels and the generalized singular functions. These links and analogies highlight the meanings of the kernel estimates as well as provide constructive methods for the designs of some differentiating kernels.

4.3.1 Gaussian smoothing kernels

Let us start with an example. Consider the integral estimate in Eq. (4.24) as a smoother of the input signal $y(x)$:

$$\hat{y}_h(x) = \int_{\mathbb{R}^d} g(u)y(x - hu)du = \frac{1}{h^d}\int_{\mathbb{R}^d} g[(x - u)/h]y(u)du, \tag{4.80}$$

with the Gaussian kernel

$$g(x) = \frac{1}{(2\pi)^{d/2}} \exp\left(-\tfrac{1}{2}||x||^2\right). \tag{4.81}$$

The following facts can be stated for these $\hat{y}_h(x)$ and $g(x/h)/h^d$ ([62]–[63]):

1. The smoothed $\hat{y}_h(x)$ is continuous for any $h \geq 0$, even for a discontinuous $y(x)$;
2. The smoothed $\hat{y}_h(x)$ is infinitely differentiable regardless of the properties of $y(x)$;
3. If $y(x)$ is continuous at the point x, then
$$\lim_{h \to 0} \hat{y}_h(x) = y(x);$$
4. As $h \to 0$, the kernel $g(x/h)/h^d$ approaches Dirac's delta function,
$$\lim_{h \to 0} g(x/h)/h^d = \delta(x),\ x \in \mathbb{R}^d; \qquad (4.82)$$
5. If $y(x)$, $d = 1$, has a finite jump at the point x, then
$$\lim_{h \to 0} \hat{y}_h(x) = \tfrac{1}{2}[y(x+0) + y(x-0)], \qquad (4.83)$$
where $y(x+0)$ and $y(x-0)$ are values of $y(x)$ on the right- and left-hand sides of the jump point x.

Thus, the estimate $\hat{y}_h(x)$ can be treated as a smoothed $y(x)$ and the properties of this smooth $\hat{y}_h(x)$ can be very different from the properties of the original $y(x)$. At least $\hat{y}_h(x)$ is always continuous, even if the original $y(x)$ is discontinuous. If $y(x)$ is continuous at the point x, the estimate $\hat{y}_h(x) \to y(x)$ as $h \to 0$.

4.3.2 Univariate Dirac delta function

Dirac's delta function $\delta(x)$, $x \in \mathbb{R}$, is an unusual singular function. To include mathematical objects such as the *Dirac delta function* $\delta(x)$ into analyses, the concept of a function should somehow be extended. This is done in terms of the so-called generalized function and distribution theory.

Let us present some brief notes on the relevant mathematical concepts.

The function $\delta(x)$ is called a *generalized function* if, for any infinitely differentiable finite support function $\varphi(x)$,

$$\int_{-\infty}^{\infty} \delta(x)\varphi(x)dx = \varphi(0). \qquad (4.84)$$

The $\varphi(x)$ in Eq. (4.84) is called a *test function*. Thus, by definition, the generalized function is introduced through integration. It is very different from the usual definition of the function, for instance, $y = y(x)$, $x \in X$, $y \in Y$, given as a pointwise mapping of the argument set X into the set Y of function values. The usual function can be given analytically or numerically. However, it is always a type of pointwise mapping.

In engineering disciplines the functional nature of the delta function is often ignored. Instead it is viewed as a special kind of function such that:

$$\delta_{\varepsilon}(x) = 0 \quad \text{if } |x| > \varepsilon, \qquad \int_{-\varepsilon}^{\varepsilon} \delta_{\varepsilon}(x)dx = 1, \tag{4.85}$$

where a small $\varepsilon > 0$ defines a small-sized neighborhood of $x = 0$, where $\delta_{\varepsilon}(x) \to \infty$. This depicts the delta function as an infinitely high pick in a neighborhood of $x = 0$ with the area under the curve equal to 1. For any $\varphi(x)$ continuous at $x = 0$, we obtain

$$\int_{-\infty}^{\infty} \delta_{\varepsilon}(x)\varphi(x)dx = \int_{-\varepsilon}^{\varepsilon} \delta_{\varepsilon}(x)\varphi(x)dx \simeq \varphi(0) \int_{-\varepsilon}^{\varepsilon} \delta_{\varepsilon}(x)dx = \varphi(0).$$

This approximately corresponds to Eq. (4.84). The intuitive definition in Eq. (4.85) works well, but in general it is not sufficient because the properties of the generalized delta function are very different from properties of ordinary functions. As an illustration of this unusualness, note that the generalized function is always differentiable and has derivatives of all orders. Of course, this is not true for ordinary functions.

The *derivative of the Dirac delta function* $\delta^{(r)}(x)$ is defined by the integral

$$\int_{-\infty}^{\infty} \delta^{(r)}(x)\varphi(x)dx = (-1)^r \varphi^{(r)}(0), \tag{4.86}$$

where $\varphi^{(r)}(x) = \partial_x^r \varphi(x)$ is a derivative of the test function $\varphi(x)$.

Formula (4.86) can be verified. It follows from Eq. (4.84) that

$$\int_{-\infty}^{\infty} \delta(x)\varphi^{(r)}(x)dx = \varphi^{(r)}(0). \tag{4.87}$$

By integration by parts we can see that

$$\begin{aligned}\int_{-\infty}^{\infty} \delta^{(1)}(x)\varphi(x)dx &= \delta(x)\varphi(x)|_{-\infty}^{\infty} - \int_{-\infty}^{\infty} \delta(x)\varphi^{(1)}(x)dx \\ &= -\int_{-\infty}^{\infty} \delta(x)\varphi^{(1)}(x)dx = -\varphi^{(1)}(0).\end{aligned}$$

This proves Eq. (4.86) for $r = 1$. Similarly, a recursive integration by parts gives

$$\int_{-\infty}^{\infty} \delta^{(r)}(x)\varphi(x)dx = (-1)^r \int_{-\infty}^{\infty} \delta(x)\varphi^{(r)}(x)dx$$

and proves Eq. (4.86) for $r \geq 0$.

The convolution of $\delta^{(r)}(x)$ and the test function $\varphi(x)$ is

$$\begin{aligned}(\delta^{(r)} \circledast \varphi)(x) &= \int_{-\infty}^{\infty} \delta^{(r)}(x-u)\varphi(u)du \\ &= \int_{-\infty}^{\infty} \delta^{(r)}(u)\varphi(x-u)du.\end{aligned} \tag{4.88}$$

Using Eq. (4.87) we obtain

$$\int_{-\infty}^{\infty} \delta^{(r)}(u)\varphi(x-u)du = (-1)^r \partial_u^r \varphi(x-u)|_{u=0} = \partial_x^r \varphi(x-u)|_{u=0} = \varphi^{(r)}(x), \tag{4.89}$$

which proves that

$$(\delta^{(r)} \circledast \varphi)(x) = \varphi^{(r)}(x). \tag{4.90}$$

This formula shows that the derivative of the delta function can be interpreted as a differentiating kernel.

4.3.3 Multivariate Dirac delta function

The concepts of the Dirac delta function and its derivative are generalized for the multivariate case in the following way. A function $\delta(x)$, $x \in \mathbb{R}^d$, is called a *generalized delta function* if

$$\int_{\mathbb{R}^d} \delta(x)\varphi(x)dx = \varphi(0), \tag{4.91}$$

where $\varphi(x)$, $x \in \mathbb{R}^d$, is a *test function* that is assumed to be infinitely differentiable and has a finite support. The *derivative of the delta function* $\delta^{(r)}(x)$, $x \in \mathbb{R}^d$, is defined as

$$\int_{\mathbb{R}^d} \delta^{(r)}(x)\varphi(x)dx = (-1)^{|r|}\varphi^{(r)}(0), \tag{4.92}$$

where $r = (r_1, \ldots, r_d)$ is a multi-index and $\varphi^{(r)}(x)$ is the rth derivative of the test function $\varphi(x)$.

It can be shown that

$$\int_{\mathbb{R}^d} \delta^{(r)}(x)\varphi(x)dx = (-1)^{|r|} \int_{\mathbb{R}^d} \delta(x)\varphi^{(r)}(x)dx.$$

Using Eq. (4.92) and a transformation similar to Eq. (4.89), we obtain for the convolution

$$(\delta^{(r)} \circledast \varphi)(x) = \int_{\mathbb{R}^d} \delta^{(r)}(u)\varphi(x-u)du = (-1)^{|r|}\partial_u^r \varphi(x-u)|_{u=0} = \varphi^{(r)}(x). \tag{4.93}$$

This last formula proves that the derivative of the delta function can be treated as the differentiating kernel,

$$(\delta^{(r)} \circledast \varphi)(x) = \varphi^{(r)}(x). \tag{4.94}$$

This formula is also of special interest for $r = 0$ because it means that the convolution of the delta function and the test function give the test function

$$(\delta \circledast \varphi)(x) = \varphi(x). \tag{4.95}$$

4.3.4 Dirac's delta sequences

It is conventional to define Dirac's delta function as a limit of a usual function with a parameter. This function with a sequence of parameter values is called a *delta sequence*. Let us use the notation $\delta_h(x)$ for the delta sequence with the parameter $h \to 0$ and define the delta function as the limit of this sequence:

$$\lim_{h\to 0} \delta_h(x) = \delta(x). \tag{4.96}$$

Commonly used equivalent definitions include the following univariate sequences:

$$\begin{aligned} &1)\ \delta_h(x) = \frac{h}{x^2+h^2},\\ &2)\ \delta_h(x) = h \cdot |x|^{h-1},\\ &3)\ \delta_h(x) = \frac{1}{\sqrt{2\pi}h} e^{(-x^2/2h^2)},\\ &4)\ \delta_h(x) = \frac{1}{\pi x} \sin(x/h). \end{aligned} \tag{4.97}$$

These delta sequences are illustrated in Fig. 4.3 for $h_1 < h_2 < h_3$, where it can be seen that the functions $\delta_h(x)$ become narrower and better concentrated at $x = 0$ as $h \to 0$.

Using the delta sequence instead of the delta function in formula (4.95), we have

$$(\delta_h \circledast \varphi)(x) \to_{h\to 0} (\delta \circledast \varphi)(x) = \varphi(x). \tag{4.98}$$

Let us show that the kernel $g(x/h)/h$ of the 1D integral LPA estimate in Eq. (4.32) can be treated as a delta sequence.

Proposition 5 *Assume that the kernel $g(x)$ in Eq. (4.34) satisfies the conditions*

$$\int_{-\infty}^{\infty} g(x)dx = 1 \tag{4.99}$$

and

$$\int_{-\infty}^{\infty} |x| \cdot |g(x)|dx < \infty. \tag{4.100}$$

Then, $g(x/h)/h$ is a delta sequence,

$$\delta_h(x) = g(x/h)/h. \tag{4.101}$$

Proof. Inserting the test function $\varphi(x)$ in Eq. (4.32) we have

$$\hat{\varphi}_h(x) = \frac{1}{h}\int_{-\infty}^{\infty} g[(x-u)/h]\varphi(u)du = \int_{-\infty}^{\infty} g(u)\varphi(x-hu)du.$$

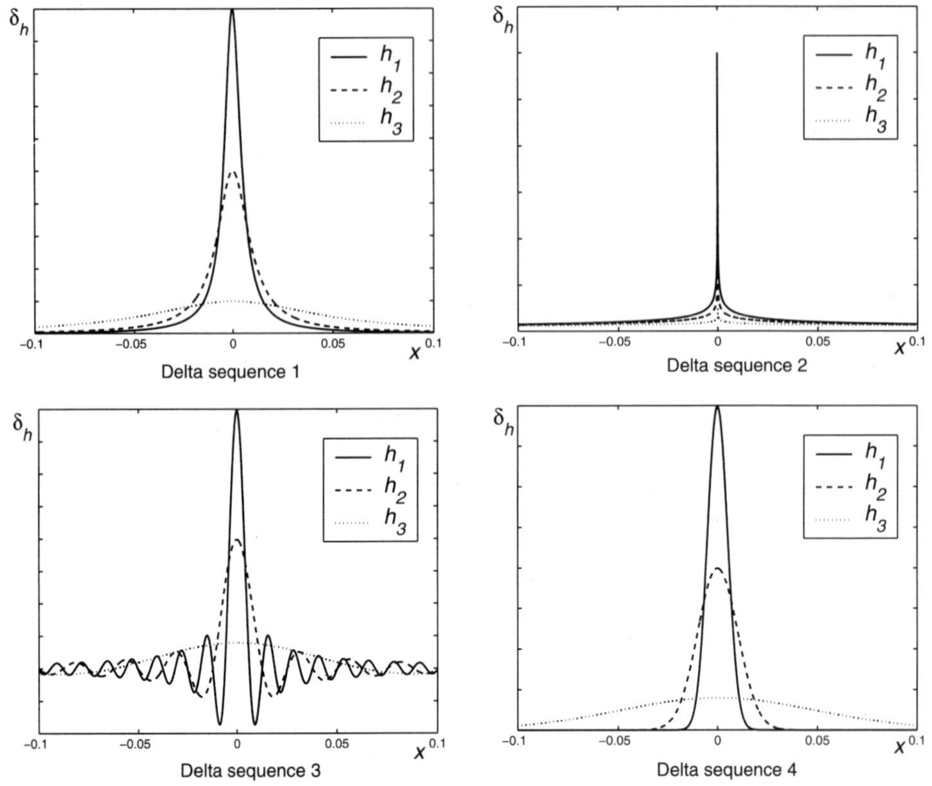

Figure 4.3 Delta sequences given by formulas (4.97).

Considering the difference $\hat{\varphi}_h(x) - \varphi(x)$ and using Eq. (4.99) we have

$$\hat{\varphi}_h(x) - \varphi(x) = \int_{-\infty}^{\infty} g(u)[\varphi(x - hu) - \varphi(x)]du.$$

Because the test function $\varphi(x)$ is differentiable, there exists an L such that $|\varphi(x - hu) - \varphi(x)| < L|u|$. It then follows that

$$\begin{aligned} |\hat{\varphi}_h(x) - \varphi(x)| &\leq \int_{-\infty}^{\infty} |g(u)| \cdot |\varphi(x - hu) - \varphi(x)|du \\ &\leq Lh \int_{-\infty}^{\infty} |g(u)| \cdot |u|du \to 0 \qquad \text{as } h \to 0. \end{aligned}$$

As $h \to 0$ the limit of $\hat{\varphi}_h(x)$ exists and

$$\hat{\varphi}_h(x) = \frac{1}{h} \int_{-\infty}^{\infty} g\left(\frac{x - u}{h}\right) \varphi(u)du \to \varphi(x).$$

The last formula as well as Eq. (4.98) is valid for any test function φ, which means that

$$g(x/h)/h = \delta_h(x), \lim_{h\to 0} g(x/h)/h = \delta(x).$$

This proves Eq. (4.101). □

Note that Proposition 5 is of a general nature because the assumption in Eq. (4.100) is nonrestrictive, that is, it holds for any finite support kernel g.

For any $h > 0$, the scaled kernel $\delta_h(x) = g(x/h)/h$ is a usual function, while the corresponding limit $\delta(x)$ is a singular generalized delta function $\delta(x)$.

The limit of $\hat{y}_h(x)$ in Eq. (4.32) as $h \to 0$ exists under certain assumptions on the input $y(x)$,

$$\hat{y}_0(x) = \lim_{h\to 0} \hat{y}_h(x). \tag{4.102}$$

If $y(x)$ is a continuous function, then $\hat{y}_0(x) = y(x)$. If $y(x)$ is discontinuous at the point x, then the limit $\hat{y}_0(x)$ is not identical to the original $y(x)$. The symmetric $g(x) = g(-x)$ gives this limit as it is in Eq. (4.83) for the Gaussian kernel. However, the symmetry of g is not a compulsory requirement. If g is nonsymmetric and $a_1 = \int_{-\infty}^{0} g(x)dx$, $a_2 = \int_{0}^{\infty} g(x)dx$, then instead of Eq. (4.83) we have for the limit estimate

$$\lim_{h\to 0} \hat{y}_h(x) = a_1 y(x+0) + a_1 y(x-0), a_1 + a_2 = 1,$$

that is, this estimate is a weighted mean of the right- and left-hand side values of y.

A generalization of Proposition 5 for the multivariate case is straightforward.

Proposition 6 *Assume that the kernel $g(x)$, $x \in \mathbb{R}^d$, is normalized,*

$$\int_{\mathbb{R}^d} g(x)dx = 1, \tag{4.103}$$

and

$$\int_{\mathbb{R}^d} ||x|| \cdot |g(x)|dx < \infty. \tag{4.104}$$

Then, $g(x/h)/h^d$ is a delta sequence,

$$\delta_h(x) = g(x/h)/h^d. \tag{4.105}$$

The proof is completely similar to the proof of Proposition 5. Again note that the assumption in Eq. (4.104) is not restrictive. The integral LPA always gives the normalized kernel g and the scaled kernel $g(x/h)/h^d$ defines the delta sequence estimate in Eq. (4.32):

$$\hat{y}_h(x) = \frac{1}{h^d} \int_{\mathbb{R}^d} g[(x-u)/h] y(u)du,$$

where $\delta_h(x) = g(x/h)/h^d$ is a delta sequence kernel with the parameter h.

Further, for the derivative estimate [Eq. (4.33)], we have

$$\hat{y}_h^{(r)}(x) = \frac{1}{h^{d+|r|}} \int_{\mathbb{R}^d} g^{(r)}[(x-u)/h]y(u)du,$$

where the *differentiating delta sequence* is defined as

$$\delta_h^{(r)}(x) = \frac{1}{h^{d+|r|}} g^{(r)}(x/h)$$

and

$$\lim_{h\to 0} \frac{1}{h^{d+|r|}} g^{(r)}(x/h) = \delta^{(r)}(x). \tag{4.106}$$

It can be shown that the above statement is valid for the differentiating delta sequence provided that

$$\int_{\mathbb{R}^d} |u^k| \cdot |g^{(r)}(u)|du < \infty, \; |u^k| = \Pi_{s=1}^d |u_s|^{k_s}$$

where $k_s \geq r_s$ and $\sum_{s=1}^d k_s = \sum_{s=1}^d r_s + 1$. Thus, the kernels designed by the LPA can be interpreted as delta sequences converging to a generalized delta function for the smoothing estimators and to the derivative of the generalized delta function for the differentiating estimators.

To conclude this section we remark that in real function theory and distribution theory, the generalized functions as well as the delta sequences are a special topic. The rigorous mathematical treatment of the field can be seen in the two-volume work by Gel'fand and Shilov [62]–[63] and in the book by Friedlander [59].

4.4 Potential Derivative Estimates

Equation (4.93) says that the derivative of the delta function can be used as a kernel for the derivative estimator. Now we know that the LPA kernels can be treated as delta sequences $g_h(x)$. It follows immediately that the derivative of the delta sequence $g_h(x)$ can be used as a differentiating kernel $g_h^{(r)}(x)$ for the derivative estimate:

$$g_h^{(r)}(x) = \partial_x^r g_h(x). \tag{4.107}$$

Using the formula $g_h(x) = (1/h^d)g(x/h)$ and differentiating this $g_h(x)$, we find that

$$g_h^{(r)}(x) = \partial_x^r g_h(x) = \frac{1}{h^d}\partial_x^r g(x/h) = \frac{1}{h^{d+|r|}}\partial_{x/h}^r g(x/h).$$

Comparing this formula with Eq. (4.37) we obtain the differentiating kernel according to the formula

$$g_h^{(r)}(x) = \frac{1}{h^{d+|r|}} g^{(r)}(x/h), \tag{4.108}$$

$$g^{(r)}(x) = \partial^r g(x), \tag{4.109}$$

where $g(x)$ is defined in Eq. (4.34). Thus, the differentiating kernel $g^{(r)}(x)$ in Eq. (4.108) is calculated as the corresponding derivative of the smoothing kernel $g(x)$. This is a novel definition of the differentiating kernel that is different from the one used in the integral and discrete LPA derivations.

The following expressions have been obtained for the LPA smoothing in Eq. (4.26) and differentiating kernels in Eq. (4.27)

$$g(x) = w(x)\phi^T(x)\Phi^{-1}\phi(0), \tag{4.110}$$

$$g^{(r)}(x) = (-1)^{|r|}w(x)\phi^T(x)\Phi^{-1}\phi^{(r)}(0). \tag{4.111}$$

First, it is obvious that the derivative $\partial^r g(x)$ used in Eq. (4.109), in general, is not equal to $g^{(r)}(x)$ in Eq. (4.111). Thus, in general, formula (4.109) indeed defines a kernel that is different from the one in Eq. (4.111). Second, it is required in the derivation of Eq. (4.111) that the LPA order $m \geq |r|$, but there is no such requirement for the kernel in Eq. (4.109). So, the kernel $g(x)$ can be derived using the LPA of the order $m = 0$ and nevertheless multiple differentiating of this kernel gives the kernel for estimating the corresponding high-order derivatives. Of course, if $g(x)$ is not smooth enough, the derivative $\partial^r g(x)$ will include delta functions. However, the continuous and differentiable $g(x)$ is a good candidate for use in Eq. (4.109) for the kernel design.

The Gaussian kernel

$$g(x) = \frac{1}{(2\pi)^{d/2}} \exp\left(-\frac{1}{2}||x||^2\right) \tag{4.112}$$

is a good example for this sort of procedure. This kernel has the order $m = 1$ and differentiating kernels are obtained as

$$g^{(r)}(x) = \frac{1}{(2\pi)^{d/2}} \partial_x^r \exp\left(-\frac{1}{2}||x||^2\right). \tag{4.113}$$

The differentiation kernels produced according to formula (4.109) are called *potential differentiating kernels* [101]. The "potential" means that for each derivative estimate $\hat{y}_h^{(r)}(x) = (g_h^{(r)} \circledast y)(x)$ there exists a smoothed signal $\hat{y}_h(x) = (g_h \circledast y)(x)$ such that the derivative estimate $\hat{y}_h^{(r)}(x)$ is the accurate derivative of this smoothed signal, i.e.,

$$\hat{y}_h^{(r)}(x) = \partial_x^r \hat{y}_h(x). \tag{4.114}$$

In this way, the smoothed $\hat{y}_h(x)$ serves as a potential function for the estimate $\hat{y}_h^{(r)}(x)$.

The following is important for the potential differentiating estimates. The smoothing estimate in Eq. (4.32) can be given in two different forms:

$$\hat{y}_h(x) = \frac{1}{h^d} \int_{\mathbb{R}^d} g[(x-v)/h]y(v)dv = \int_{\mathbb{R}^d} g(v)y(x-hv)dv.$$

Differentiating both sides of this equation gives

$$\hat{y}_h^{(r)}(x) = \partial_x^r \hat{y}_h(x) = \frac{1}{h^{d+|r|}} \int_{\mathbb{R}^2} g^{(r)}\left(\frac{x-u}{h}\right) y(u)du \tag{4.115}$$

$$= \int_{\mathbb{R}^d} g(v) y^{(r)}(x - hv)dv, \tag{4.116}$$

where $g^{(r)}(x) = \partial^r g(x)$ is the potential differentiating kernel in Eq. (4.109).

Formula (4.116) proves that the potential derivative estimate $\hat{y}_h^{(r)}(x)$ can be treated as an accurate derivative $y^{(r)}(x)$ smoothed by the corresponding smoothing kernel $g(x)$. This fact admits a number of interesting conclusions. To illustrate them, assume that the kernel $g(x), x \in \mathbb{R}^d$, is Gaussian in Eq. (4.112). Then, the properties of the smoothing estimator $\hat{y}_h(x)$ considered in Sec. 4.3.1 can be transferred to the corresponding potential derivative estimator $\hat{y}_h^{(r)}(x)$:

1. The derivative estimate $\hat{y}_h^{(r)}(x)$ is continuous for any $h \geq 0$, even for discontinuous $y^{(r)}(x)$;
2. The derivative estimate $\hat{y}_h^{(r)}(x)$ is infinitely differentiable regardless of the properties of $y^{(r)}(x)$;
3. The derivative estimate $\hat{y}_h^{(r)}(x)$ is a sequence function defining the limit estimate as $\lim_{h \to 0} \hat{y}_h^{(r)}(x)$.

If the derivative $y^{(r)}(x)$ is continuous at the point x, then

$$\lim_{h \to 0} \hat{y}_h^{(r)}(x) = y^{(r)}(x). \tag{4.117}$$

Equation (4.117) is important because it shows explicitly that the differentiation of $\hat{y}_h(x)$ can be used to obtain the derivative estimators. Formula (4.109) gives a simple and constructive way to design the differentiating kernels.

Examples

Gaussian kernel

Let us produce some calculations and derive the differentiating kernels according to formula (4.109) for the Gaussian $g(x)$ in Eq. (4.112) assuming that $d = 2$. It can be seen that

$$\begin{aligned} \partial_{x_1} g(x) &= -x_1 g(x), & \partial_{x_2} g(x) &= -x_2 g(x), \\ \partial_{x_1}\partial_{x_2} g(x) &= x_1 x_2 g(x), & \partial_{x_1}^2 g(x) &= (x_1^2 - 1)g(x), \\ \partial_{x_2}^2 g(x) &= (x_2^2 - 1)g(x). \end{aligned} \tag{4.118}$$

Thus, the differentiating kernels are of the form

$$\begin{aligned} &g^{(1,0)}(x) = -x_1 g(x), && g^{(0,1)}(x) = -x_2 g(x), \\ &g^{(1,1)}(x) = x_1 x_2 g(x), && g^{(2,0)}(x) = (x_1^2 - 1)g(x), \\ &g^{(0,2)}(x) = (x_2^2 - 1)g(x). \end{aligned} \tag{4.119}$$

It is a bit surprising that the equations in (4.119) are identical to the formulas given by the standard LPA procedure in Eqs. (4.78–4.79) provided that the LPA is used with $m = 2$.

Let us make also a few remarks concerning the application of these potential kernels, in particular, for the estimation of the Laplacian in Eq. (3.79):

$$L\{y\} = y^{(2,0)}(x) + y^{(0,2)}(x).$$

Using the kernels in Eq. (4.119), we derive the kernel for the Laplacian estimate in the form of Eq. (3.80) as

$$g_{Lap}(x) = g^{(0,2)}(x) + g^{(2,0)}(x) = (x_1^2 + x_2^2 - 2)g(x). \tag{4.120}$$

The estimate of the Laplacian can be given in a natural convolution form as $\int_{R^2} g_{Lap}(x-u)y(u)$.

The kernels $g^{(0,2)}(x)$ and $g^{(2,0)}(x)$ are potential, and by using these potential differentiating kernels we obtain the corresponding potential derivative estimates. This means that the estimate of the Laplacian is equal to the smoothed accurate Laplacian function:

$$\int_{R^2} g_{Lap}(x-u)y(u)du = \int_{R^2} g(x-u)[y^{(2,0)}(u) + y^{(0,2)}(u)]du, \tag{4.121}$$

where $g(x)$ is defined by Eq. (4.112) with $d = 2$.

The kernel $g_{Lap}(x)$ in Eq. (4.120) and its frequency amplitude characteristic $|G_{Lap}(\lambda)|$ are shown in Fig. 4.4.

With the scale parameter h, the formulas in Eqs. (4.120–4.121) are transformed as follows:

$$\begin{aligned} g_h(x) &= \frac{1}{h^2} g\left(\frac{x}{h}\right), \quad g_{Lap,h}(x) = \frac{1}{h^4} g_{Lap}\left(\frac{x}{h}\right), \\ \int_{R^2} g_{Lap,h}(x-u)y(u)du &= \frac{1}{h^2}\int_{R^2} g_{Lap}(u)y(x-hu)du, \\ \int_{R^2} g_h(x-u)[y^{(2,0)}(u) + y^{(0,2)}(u)]du &= \int_{R^2} g(u)[y^{(2,0)}(x-hu) \\ &\quad + y^{(0,2)}(x-hu)]du. \end{aligned} \tag{4.122}$$

Note that we obtain a factor $1/h^4$ for $g_{Lap,h}(x)$ in Eq. (4.122) because it is a differentiation kernel and the power of h in Eq. (4.108) is defined as $1/h^{d+|r|}$. In the considered case, $d = 2$ and $|r| = 2$.

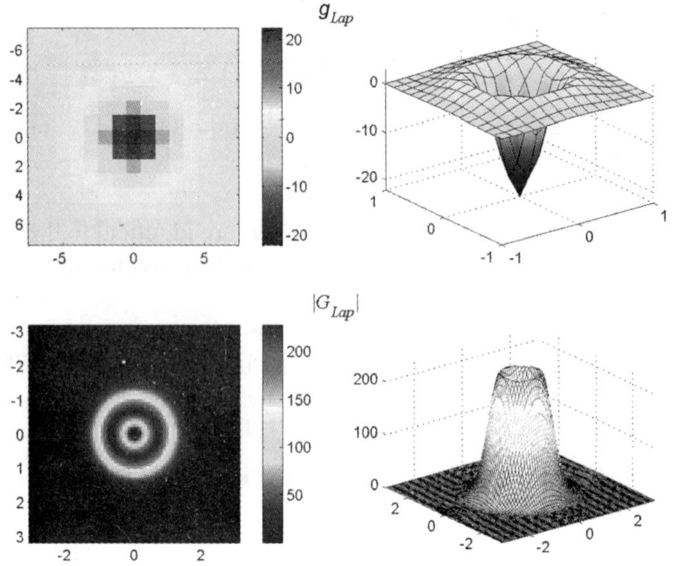

Figure 4.4 The Gaussian Laplacian kernel $g_{Lap}(x)$ in Eq. (4.120) and its frequency amplitude characteristic $|G_{Lap}(\lambda)|$.

Finite-difference differentiation

Consider a smoothed 1D signal

$$\hat{y}_h(x) = \frac{1}{2h}\int_{x-h}^{x+h} y(u)du. \tag{4.123}$$

Differentiate $\hat{y}_h(x)$ on x. According to Eq. (4.115), this gives the potential derivative estimate of y. Simple calculations show that this estimate is the standard symmetric finite difference:

$$\hat{y}_h^{(1)}(x) = \frac{1}{2h}[y(x+h) - y(x-h)]. \tag{4.124}$$

According to the concept of the potential differentiation, this *approximate* derivative estimate of y is an *accurate derivative of the smoothed* $\hat{y}_h(x)$. Taking smaller or larger values of h in the finite-difference estimate in Eq. (4.124), we always obtain the accurate derivative of $\hat{y}_h(x)$ of Eq. (4.123) with the smoothness of this function controlled by h.

Let $y(x)$ demonstrate a sort of irregular behavior with some high-frequency components. Then a larger h in Eq. (4.123) as well as in Eq. (4.124) allows us to filter out these irregular components of y and obtain a derivative corresponding only to the regular, basic part of y.

The curves in Fig. 4.5 illustrate how efficient this large h can be. It is assumed in this simulation that $y(x)$ is composed of a quadratic signal and a high-frequency component. We calculate the finite-difference estimates of the derivatives $\hat{y}_h^{(1)}(x)$ in Eq. (4.124) and the smoothed $\hat{y}_h(x)$ in Eq. (4.123). The smoothed $\hat{y}_h(x)$ are shown

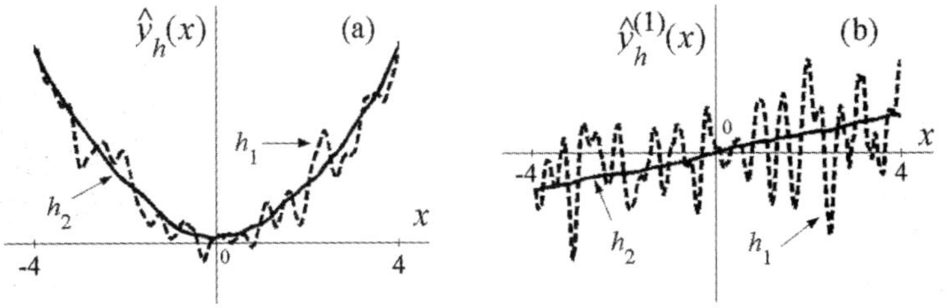

Figure 4.5 The smoothed curves $\hat{y}_h(x)$ and the corresponding finite differencies $\hat{h}_h^{(1)}(x)$ are shown in (a) and (b), respectively, for small $h_1 = 0.01$ and large $h_2 = 1$.

in Fig. 4.5(a) for small $h_1 = 0.01$ and large $h_2 = 1$. The smoothing with large h_2 naturally filters out the high-frequency components of $y(x)$ and reveals a smooth quadratic signal. The derivative estimates calculated according to Eq. (4.124) are shown in Fig. 4.5(b). We observe that with the larger $h_2 = 1$ we set a much smoother (nearly linear) $\hat{y}_{h_2}^{(1)}(x)$ while $\hat{y}_{h_1}^{(1)}(x)$ looks very irregular.

A usual tendency in the finite difference algorithm in Eq. (4.124) is to make h as small as possible with restrictions on the small h defined by the level of the random noise. In the considered example we appeal to a different idea. Using a larger h in the finite difference in Eq. (4.124) allows us to smooth y and obtain a derivative corresponding to a smoothed function in Eq. (4.123). In this way a larger h becomes a useful replacement for small ones, even in situations when there is no random noise.

Chapter 5

Discrete LPA Accuracy

The LPA accuracy is characterized by an error defined as the difference between the true signal value and the estimate:

$$\begin{aligned} e_y(x,h) &= y(x) - \hat{y}_h(x), \\ e_{y^{(r)}}(x,h) &= y^{(r)}(x) - \hat{y}_h^{(r)}(x). \end{aligned} \tag{5.1}$$

Here e_y is the error of estimation of the signal y and $e_{y^{(r)}}$ is the error of estimation of the derivative $y^{(r)}$. For the observation model in Eq. (2.1), these errors are composed of the systematic (bias) and random components corresponding to the deterministic y and the random noise ε, respectively.

The analysis produced in the previous chapters concerns some aspects of the accuracy of the LPA operators. It has been proved that, as the sampling interval approaches zero, the discrete estimators can be replaced by corresponding integral ones. Further, it is shown that the kernels of these integral estimators are delta sequences. It immediately follows that, as $h \to 0$, the approximations of the signal as well as those of the derivatives become accurate. Both of these results jointly yield that there is a convergence of the LPA estimates to the true values of the signal and the derivatives:

$$\hat{y}_h(x) \longrightarrow y(x),\ \hat{y}_h^{(r)}(x) \longrightarrow y^{(r)}(x).$$

However, all of these results are mainly qualitative. In particular, in terms of the delta sequences, this convergence can be claimed only with respect to the test functions. Remember that the test function is continuous smooth with an infinite number of derivatives.

In this chapter we analyze approximation errors of the LPA for more realistic classes of signals with the main goal of evaluating the order of errors as well as the convergence rate of the estimates. Together with the analysis with respect to the random noise, it gives a complete picture of the accuracy of the LPA estimators.

The scale h is a crucial parameter for the accuracy of estimation. When the scale h is relatively small, the LPA gives a good, smooth fit of signals, however, fewer number of observations are used and the estimates are more variable and sensitive with respect to the noise. The best choice of h involves a trade-off between the bias and variance, which depends on the degree of the LPA, a sample period, the noise variance, and the derivatives of y of the orders beyond the degree used in the LPA.

We present an accuracy analysis of the LPA estimates to illustrate these statements. For simplicity we consider only the shift-invariant estimators while the similar results can be obtained for the general case of shift-varying kernel LPA estimates.

We derive the asymptotic formulas for the bias and variance valid for small h and Δ with the main intention of giving the results where a dependence on these parameters is obtained in an analytical form. Further, we use these results to derive the algorithms for data-driven adaptive-scale selection.

5.1 Bias and Variance of Estimates

5.1.1 Main results

5.1.1.1 Estimation errors

The bias of an estimate is the difference between the true signal and the expectation of the estimate:

$$m_{\hat{y}_h}(x,h) = y(x) - E\{\hat{y}_h(x)\},$$
$$m_{\hat{y}_h^{(r)}}(x,h) = y^{(r)}(x) - E\{\hat{y}_h^{(r)}(x)\},$$

where $E\{\cdot\}$ is the mathematical expectation with respect to the random error ε in Eq. (2.1).

The estimates are defined by the convolutions in Eq. (3.9):

$$\begin{aligned} \hat{y}_h(x) &= (g_h \circledast z)(x), \\ \hat{y}_h^{(r)}(x) &= (g_h^{(r)} \circledast z)(x), \end{aligned} \tag{5.2}$$

where the kernels g_h and $g_h^{(r)}$ are of the forms in Eq. (3.5) and Eq. (3.8). Elementary calculations give

$$E\{\hat{y}_h(x)\} = E\{(g_h \circledast z)(x)\} = (g_h \circledast E\{z\})(x) = (g_h \circledast y)(x),$$
$$E\{\hat{y}_h^{(r)}(x)\} = E\{(g_h^{(r)} \circledast z)(x)\} = (g_h^{(r)} \circledast E\{z\})(x) = (g_h^{(r)} \circledast y)(x).$$

Then, the bias of the estimates is

$$\begin{aligned} m_{\hat{y}_h}(x,h) &= y(x) - (g_h \circledast y)(x), \\ m_{\hat{y}_h^{(r)}}(x,h) &= y^{(r)}(x) - (g_h^{(r)} \circledast y)(x), \end{aligned} \tag{5.3}$$

and the random estimation errors are

$$\begin{aligned} e_y^0(x,h) &= -(g_h \circledast \varepsilon)(x), \\ e_{y^{(r)}}^0(x,h) &= -(g_h^{(r)} \circledast \varepsilon)(x). \end{aligned} \tag{5.4}$$

5.1.1.2 Variance

Assuming that the random ε are independent and identically distributed (i.i.d.) with variance σ^2, we obtain for the variance of the estimate $\hat{y}_h(x)$,

$$\begin{aligned}\sigma^2_{\hat{y}_h}(x,h) &= E\{(e^0_y(x,h))^2\} = E\left\{\left(\sum_{l\in\mathbb{Z}^d} g_h(l\Delta)\varepsilon(x-l\Delta)\right)^2\right\}\\ &= E\left\{\sum_{l\in\mathbb{Z}^d} g_h(l\Delta)\varepsilon(x-l\Delta)\cdot\sum_{k\in\mathbb{Z}^d} g_h(k\Delta)\varepsilon(x-k\Delta)\right\}\\ &= \sum_{l\in\mathbb{Z}^d}\sum_{k\in\mathbb{Z}^d} g_h(l\Delta)g_h(k\Delta)\cdot E\{\varepsilon(x-l\Delta)\varepsilon(x-k\Delta)\}\\ &= \sigma^2\sum_{l\in\mathbb{Z}^d}\sum_{k\in\mathbb{Z}^d} g_h(l\Delta)g_h(k\Delta)\cdot\delta_{l,k} = \sigma^2\sum_{k\in\mathbb{Z}^d} g_h^2(k\Delta) = \sigma^2||g_h||^2.\end{aligned} \tag{5.5}$$

A similar formula can be derived for the derivative estimate $\hat{y}_h^{(r)}(x)$, proving that the variance of the smoothing and differentiating estimates can be given as

$$\begin{aligned}\sigma^2_{\hat{y}_h}(x,h) &= \sigma^2||g_h||^2,\\ \sigma^2_{\hat{y}_h^{(r)}}(x,h) &= \sigma^2||g_h^{(r)}||^2,\end{aligned} \tag{5.6}$$

where the Euclidean norm of the kernels is

$$||g_h||^2 = \sum_{k\in\mathbb{Z}^d} g_h^2(\Delta k),\ ||g_h^{(r)}||^2 = \sum_{k\in\mathbb{Z}^d}\left[g_h^{(r)}(\Delta k)\right]^2.$$

In this particular case, the variances $\sigma^2_{\hat{y}_h}(x,h)$ and $\sigma^2_{\hat{y}_h^{(r)}}(x,h)$ do not depend on x. However, for generality, we keep the argument x in the variance notation, because in later chapters we will use estimates with the variance depending on x.

5.1.1.3 Bias assumptions

The bias error in Eq. (5.3) depends on the estimator kernel and the signal y. The properties of y should be specified to evaluate this error. We restrict our analysis to the class of signals in Eq. (2.6) as

$$\mathbb{C}^\alpha = \{y\colon \max_{|r|=\alpha}\left|\partial_x^r y(x)\right| = L_\alpha(x) \le \bar{L}_\alpha, x\in\mathbb{R}^d\}, \tag{5.7}$$

where $L_\alpha(x)$ and $\bar{L}_\alpha$ are finite. This class includes smooth, differentiable signals with bounded derivatives up to the order α. The parameter $L_\alpha(x)$ is the maximum absolute value of the derivatives given as a function of x, while $\bar{L}_\alpha$ is a constant upper-bound uniform on x. We also assume that the derivatives $\partial_x^r y(x)$ in Eq. (5.7) are continuous *Lipshitz functions*,

$$|\partial_x^r y(x) - \partial_{x'}^r y(x')| \leq K_L ||x - x'||^\gamma, 0 < \gamma \leq 1, |r| = \alpha, \tag{5.8}$$

where $K_L > 0$.

Together, assumptions of Eqs. (5.7) and (5.8) define the *Hölder class* of signals with notation $\mathbf{H}^{\alpha,\gamma}$.

We introduce a special parameter

$$M = \min\{m + 1, \alpha\}, \tag{5.9}$$

combining information on the order (degree) m of the LPA and the order α of the derivatives in the class $\mathbb{C}^\alpha$, which gives

$$M = \begin{cases} m + 1, & \text{if } \alpha \geq m + 1, \\ \alpha, & \text{if } \alpha < m + 1. \end{cases} \tag{5.10}$$

The order of the derivatives α in Eq. (5.7) characterizes the smoothness of the signal y. The larger α in Eq. (5.7) means a smoother signal. Thus, if the smoothness of the signal is higher than the order m of the LPA of the estimator, then $M = m + 1$. If the LPA order is higher than the smoothness of the signal, then $M = \alpha$.

5.1.1.4 Accuracy results

Assume that:

(H1) y is deterministic locally smooth belonging to the Hölder class $\mathbf{H}^{M,\gamma}$ of functions.

(H2) The discrete LPA estimates of the signal and the derivatives are defined by Eqs. (3.4–3.8). The order of the LPA is equal to m. The sampling intervals are the same for all variables, $\Delta_i = \Delta, i = 1, \ldots, d$.

Proposition 7 *Let the hypotheses H1–H2 hold, let the noise in the observation model in Eq. (2.1) be white with variance σ^2, and let h, Δ, Δ/h, $\Delta^d/h^{d+2|r|} \to 0$. Then, the accuracy of the multivariate discrete LPA estimates in Eqs. (3.4) and (3.7) is defined for the bias as*

$$m_{\hat{y}_h}(x, h) = (-1)^{M-1} h^M \sum_{|k|=M} y^{(k)}(x) \frac{1}{k!} \int_{\mathbb{R}^d} g(u) u^k du + o(h^M), \tag{5.11}$$

$$m_{\hat{y}_h^{(r)}}(x, h) = (-1)^{M-1} h^{M-|r|} \sum_{|k|=M} y^{(k)}(x) \frac{1}{k!} \int_{\mathbb{R}^d} g^{(r)}(u) u^k du + o(h^{M-|r|}), M > |r|, \tag{5.12}$$

for the variance

$$\frac{1}{\sigma^2} \sigma^2_{\hat{y}_h}(x, h) = \frac{\Delta^d}{h^d} \int_{\mathbb{R}^d} g^2(u) du + o\left(\frac{\Delta^d}{h^d}\right), \tag{5.13}$$

$$\frac{1}{\sigma^2}\sigma^2_{\hat{y}_h^{(r)}}(x,h) = \frac{\Delta^d}{h^{d+2|r|}}\int_{\mathbb{R}^d}\left[g^{(r)}(u)\right]^2 du + o\left(\frac{\Delta^d}{h^{d+2|r|}}\right). \quad (5.14)$$

The kernels g and $g^{(r)}$ in Eqs. (5.11–5.14) are defined in Eqs. (4.34–4.35).

The proof is given in Sec. 14.2.

5.1.2 Discussion

Let us discuss the meaning of the results given in the proposition.

1. *The bias of the signal estimate*
   Formula (5.11) shows that the bias of the signal estimate is proportional to h^M. A smaller h results in a smaller bias error. This result corresponds to the intuitively clear idea that, with a smaller window size, the LPA is able to give a better approximation to a smooth signal with a smaller error.
   A larger α in Eq. (5.7) means that the signal y has a larger number of bounded derivatives, and in this sense the signal is smoother than the signal with a smaller α. If $\alpha \geq m+1$, then, according to Eq. (5.10), $M = m+1$ and the signal estimate error has a bias of the order h^{m+1}. Thus, the estimation bias is defined by the order m of the used LPA kernel. A larger m means a smaller estimation error.
   However, if $\alpha < m+1$, then, according to Eq. (5.10), $M = \alpha$ and the order h^α of the bias error is restricted by the smoothness of the estimated signal, that is, by the order of the bounded derivative of y. This means that the LPA of order larger than α is not able to improve the bias. Even more, usually a larger m gives a larger constant in Eq. (5.11). If, for example, the signal has a bounded first derivative and an unbounded or just as large second derivative, the application of the LPA with $m = 1$ can give a bias that is worse than the one obtained for $m = 0$.
2. *The bias of the derivative estimate*
   Formula (5.12) shows that the bias is proportional to $h^{M-|r|}$. A smaller h results in a smaller bias error. However, the higher-order derivative (larger $|r|$) means that the estimation error becomes larger. It corresponds to the point that the derivative estimation is a more complex problem than the estimation of the signal itself, and a higher-order derivative means a more complex problem with larger bias errors.
   The derivatives of order $|r| \leq m$ can be estimated using an LPA of degree m. Thus the degree of h is always positive in Eq. (5.12). The bias for the derivative estimation is proportional to the derivatives $y^{(k)}(x)$ of y of the same order M as it is for the signal estimation.
   The above comments on the dependence of the bias errors on the order of the LPA and the smoothness of the signal $y(x)$ are valid for the derivative estimation as well.
3. *The variance*
   The ratio h^d/Δ^d is proportional to a number of observations used in the estimate. Thus, according to Eqs. (5.13–5.14), the variance of the estimate

is inversely proportional to the number of observations. This is a standard statistical result for observations corrupted by additive white noise.

For the derivative estimation, the variance is larger than that for the signal estimation because this variance also includes the factor $1/h^{2|r|}$. The estimate of the higher-order derivative has a higher value of the variance.

4. *The signal dimensionality*
The degree of h in the formulas for the bias, Eqs. (5.11–5.12), does not depend on the dimension d of the signal. This means that the order of the bias does not depend on the dimension of the signal. However, the variance in Eqs. (5.13–5.14) depends on d. For small h, a larger d means larger variance values. The integrals in the formulas for the bias and the variance depend on d.
5. *The Lipshitz conditions in Eq. (5.8)*
These are essential to show that the small terms in the bias formulas (5.11)–(5.12) have the orders $o(h^M)$ and $o(h^{M-|r|})$ smaller than the corresponding main terms in these formulas.

5.2 Ideal Scale

5.2.1 Varying scale

5.2.1.1 *Signal estimation*

Let the accuracy of the signal estimation be characterized by the pointwise mean-square risk

$$l_{\hat{y}_h}(x,h) \triangleq E\{(y(x)-\hat{y}_h(x))^2\}.$$

This risk is equal to the sum of the squared bias and the variance of the estimate,

$$\begin{aligned} l_{\hat{y}_h}(x,h) &= (y(x)-E\{\hat{y}_h(x)\})^2 + E\{(\hat{y}_h(x)-E\{\hat{y}_h(x)\})^2\} \\ &= m^2_{\hat{y}_h}(x,h)+\sigma^2_{\hat{y}_h}(x,h). \end{aligned} \tag{5.15}$$

Omitting the higher-order small terms $o(\cdot)$, the following can be derived from Eqs. (5.11) and (5.13):

$$\begin{aligned} |m_{\hat{y}_h}(x,h)| &\le h^M L_M(x) A_g,\ A_g = \sum_{|k|=M} \frac{1}{k!}\left|\int_{\mathbb{R}^d} u^k g(u)du\right|, \\ \sigma^2_{\hat{y}_h}(x,h) &= \sigma^2 \frac{\Delta^d}{h^d} B_g, \quad B_g = \int_{\mathbb{R}^d} g^2(u)du. \end{aligned} \tag{5.16}$$

Denoting the upper-bound of the bias in Eq. (5.16) by

$$|\bar{m}_{\hat{y}_h}(x,h)| = h^M L_M(x) A_g,$$

we obtain for the risk in Eq. (5.15)

$$
\begin{aligned}
&l_{\hat{y}_h}(x,h) \leq \bar{l}_{\hat{y}_h}(x,h),\\
&\bar{l}_{\hat{y}_h}(x,h) = |\bar{m}_{\hat{y}_h}(x,h)|^2 + \sigma^2_{\hat{y}_h}(x,h) = (h^M L_M(x) A_g)^2 + \frac{\Delta^d \sigma^2}{h^d} B_g.
\end{aligned}
\tag{5.17}
$$

The upper-bound risk $\bar{l}_{\hat{y}_h}(x,h)$ is convex on h. Its minimization on h gives

$$
\partial_h \bar{l}_{\hat{y}_h}(x,h) = 0. \tag{5.18}
$$

To proceed, let us derive some auxiliary formulas that will be useful for this sort of minimization.

Auxiliary formulas

Assume that the function to be minimized is of the form

$$
f(h) = Ah^a + Bh^{-b}. \tag{5.19}
$$

Then the equation $\partial f(h)/\partial h = 0$ gives

$$
h^* = \left(\frac{Bb}{Aa}\right)^{1/(a+b)}. \tag{5.20}
$$

For $f(h)$ we can write

$$
f(h) = Bh^{-b}(1 + Ah^{a+b}/B).
$$

Substituting in the last formula h^* from Eq. (5.20) we find for the optimal value of $f(h)$

$$
f^* = f(h^*) = B(h^*)^{-b}(1+\gamma^2), \quad \gamma^2 = \frac{b}{a}, \tag{5.21}
$$

where h^* is the optimal value of h and f^* is the minimum value of the criterion function provided $h = h^*$.

Comparing $\bar{l}_{\hat{y}_h}(x,h)$ in Eq. (5.17) with Eq. (5.19) shows that the parameters of $f(h)$ in Eq. (5.19) for $\bar{l}_{\hat{y}_h}(x,h)$ are

$$
A = \left[L_M(x) A_g\right]^2, \quad B = \Delta^d \sigma^2 B_g, \quad a = 2M, \quad b = d. \tag{5.22}
$$

Then formulas (5.20) and (5.21) give the optimal solution for Eq. (5.18) as

$$
h^*(x) = \left[\Delta^d \frac{\sigma^2 B_g}{A_g^2 L_M^2(x)} \gamma^2\right]^{1/(2M+d)}, \tag{5.23}
$$

and

$$
\bar{l}^*_{\hat{y}_h}(x) = \bar{l}_{\hat{y}_h}(x, h^*(x)) = \sigma^2_{\hat{y}_h}(x, h^*(x))\,(1+\gamma^2), \tag{5.24}
$$

where

$$\gamma = \frac{|\bar{m}_{\hat{y}_h}(x, h^*(x))|}{\sigma_{\hat{y}_h}(x, h^*(x))}. \tag{5.25}$$

Because $\gamma^2 = b/a$ according to Eq. (5.21), and using a and b from Eq. (5.22) we find that

$$\gamma = \sqrt{\frac{d}{2M}} = \sqrt{\frac{d}{2\min(m+1, \alpha)}}. \tag{5.26}$$

The parameter γ is introduced in Eq. (5.25) as a ratio of the upper-bound bias to the standard deviation calculated for the ideal scale $h^*(x)$. However, it follows from Eq. (5.26) that it is a constant, independent on x and h. This conclusion is surprising because the bias and the variance are complex functions of all parameters involved in the calculation of the mean-square risk.

The found optimal value of h is called the "ideal" (or "oracle") because its calculation according to Eq. (5.23) requires an accurate knowledge of the signal derivatives used in $L_M(x)$. Of course, this information is not available in any practical scenario. Formulas (5.23) and (5.24) demonstrate that the ideal scale $h^*(x)$ depends on the Mth-order derivatives of y. The smaller values of these derivatives mean smaller $L_M(x)$ and a larger value of the *ideal* scale $h^*(x)$.

It can be verified that the ideal $h^*(x)$ gives a special balance between the standard deviation and the bias of estimation. This balance can be represented in the form of the following inequalities:

$$|\bar{m}_{\hat{y}_h}(x, h)| \begin{cases} <\gamma \cdot \sigma_{\hat{y}_h}(x, h) & \text{if } h < h^*(x), \\ >\gamma \cdot \sigma_{\hat{y}_h}(x, h) & \text{if } h > h^*(x). \end{cases} \tag{5.27}$$

The ideal scale h^* is defined by Eq. (5.25) as

$$|\bar{m}_{\hat{y}_h}(x, h^*(x))| = \gamma\sigma_{\hat{y}_h}(x, h^*(x)).$$

For the asymptotic defined by Eq. (5.16), Eq. (5.18) has a unique solution.

The optimal balance between the variance and the bias means that the upper-bound of the bias is equal to the standard deviation multiplied by the factor γ. It is seen from Eq. (5.27) that the bias becomes smaller than the ideal, balanced value $|\bar{m}_{\hat{y}_h}(x, h^*(x))|$ if h is smaller than the ideal $h^*(x)$. If h is larger than $h^*(x)$, then the bias error is larger than this balanced $|\bar{m}_{\hat{y}_h}(x, h^*(x))|$.

The equality in Eq. (5.27) is achieved only for

$$h = h^*(x).$$

Thus, there is a one-to-one correspondence between the inequality linking the bias and the standard deviation of the estimate

$$|\bar{m}_{\hat{y}_h}(x, h)| \lesseqgtr \gamma \cdot \sigma_{\hat{y}_h}(x, h),$$

and the inequality

$$h \lesseqgtr h^*(x).$$

These speculations form a base of the ICI rule for the data-driven adaptive-scale selection. In fact, the ICI rule uses the estimate of the signal and the variance of the estimate to test the hypotheses $h \lessgtr h^*(x)$ and to select the adaptive scale close to the ideal $h^*(x)$.

5.2.1.2 Derivative estimation

The analysis of the derivative estimation accuracy is only slightly different from the analysis produced for the signal estimation. It follows from Eqs. (5.12) and (5.14) that

$$\begin{aligned} |m_{\hat{y}_h^{(r)}}(x,h)| &\leq |\bar{m}_{\hat{y}_h^{(r)}}(x,h)| \triangleq h^{M-|r|} L_M(x) A_{g^{(r)}}, \\ \sigma^2_{\hat{y}_h^{(r)}}(x,h) &= \sigma^2 \frac{\Delta^d}{h^{d+2|r|}} B_{g^{(r)}}, \end{aligned} \tag{5.28}$$

where

$$A_{g^{(r)}} = \sum_{|k|=M} \frac{1}{k!} \left| \int_{\mathbb{R}^d} u^k g^{(r)}(u) du \right|, \quad B_{g^{(r)}} = \int_{\mathbb{R}^d} |g^{(r)}(u)|^2 du.$$

The pointwise mean-square risk $l_{\hat{y}_h^{(r)}}(x,h)$ for the derivative estimation can be represented as

$$\begin{aligned} l_{\hat{y}_h^{(r)}}(x,h) &\triangleq E\left[\partial^r y(x) - \hat{y}_h^{(r)}(x)\right]^2 \leq \bar{l}_{\hat{y}_h^{(r)}}(x,h), \\ \bar{l}_{\hat{y}_h^{(r)}}(x,h) &= |\bar{m}_{\hat{y}_h^{(r)}}(x,h)|^2 + \sigma^2_{\hat{y}_h^{(r)}}(x,h) \\ &= \left[h^{M-|r|} L_M(x) A_{g^{(r)}}\right]^2 + \sigma^2 \frac{\Delta^d}{h^{d+2|r|}} B_{g^{(r)}}. \end{aligned} \tag{5.29}$$

Comparing Eqs. (5.29) and (5.19) shows that the parameters of $f(h)$ are

$$A = \left[L_M(x) A_{g^{(r)}}\right]^2, \quad B = \Delta^d \sigma^2 B_e^{(r)}, \quad a = 2(M - |r|), \quad b = d + 2|r|.$$

Using formula (5.20), we obtain for the minimization of $l_{\hat{y}_h^{(r)}}(x,h)$ in Eq. (5.29)

$$h_r^*(x) = \left[\Delta^d \frac{\sigma^2 B_{g^{(r)}}}{A^2_{g^{(r)}} L^2_M(x)} \gamma^2_{|r|}\right]^{1/(2M+d)}, \qquad \gamma^2_{|r|} = \frac{d + 2|r|}{2(M - |r|)}, \tag{5.30}$$

where

$$\gamma_{|r|} = \frac{|\bar{m}_{\hat{y}_h^{(r)}}(x, h_r^*(x))|}{\sigma_{\hat{y}_h^{(r)}}(x, h_r^*(x))}, \tag{5.31}$$

and

$$\bar{l}^*_{\hat{y}_h^{(r)}}(x) = \bar{l}_{\hat{y}_h^{(r)}}(x, h_r^*(x)) = \sigma^2_{\hat{y}_h^{(r)}}(x, h_r^*(x))(1 + \gamma^2_{|r|}). \tag{5.32}$$

The calculations show that

$$\gamma_{|r|} = \sqrt{\frac{d + 2|r|}{2(M - |r|)}} = \sqrt{\frac{d + 2|r|}{2[\min(m+1, \alpha) - |r|]}}. \tag{5.33}$$

The parameter $\gamma_{|r|}$ is introduced as a ratio of the upper-bound bias to the standard deviation of the derivative estimate for the corresponding ideal scale $h_r^*(x)$. It follows from Eq. (5.33) that this parameter is a constant depending only on M and the order $|r|$ of the estimated derivatives.

Formula (5.30) gives the ideal (oracle) scale for the derivative estimation where $L_M(x)$ depends on unknown values of the derivatives of the signal y. The ideal scale depends on x as well as on the order r of the estimated derivative. It can be checked that, similar to Eq. (5.27),

$$|\bar{m}_{\hat{y}_h^{(r)}}(x,h)| \begin{cases} <\gamma_{|r|} \cdot \sigma_{\hat{y}_h^{(r)}}(x,h) & \text{if } h < h_r^*(x), \\ >\gamma_{|r|} \cdot \sigma_{\hat{y}_h^{(r)}}(x,h) & \text{if } h > h_r^*(x). \end{cases} \tag{5.34}$$

The upper-bound of the bias in the derivative estimate is smaller than its standard deviation multiplied by $\gamma_{|r|}$ if h is smaller than the ideal $h_r^*(x)$. If h is larger than $h_r^*(x)$, then the bias error is larger than the standard deviation multiplied by $\gamma_{|r|}$.

According to the definition of $\gamma_{|r|}$ in Eq. (5.31), the equality in Eq. (5.34) is achieved for

$$h = h_r^*(x).$$

As was discussed for signal estimation, there is a one-to-one correspondence between the inequality linking the bias and the standard deviation of the estimation error,

$$|\bar{m}_{\hat{y}_h^{(r)}}(x,h)| \lesseqgtr \gamma_{|r|} \cdot \sigma_{\hat{y}_h^{(r)}}(x,h),$$

and the inequality between the scales

$$h \lesseqgtr h_r^*(x).$$

This fact is used in the ICI rule applied for the data-driven adaptive-scale selection for the derivative estimation.

5.2.1.3 Ideal convergence rate

The formulas derived in the previous section allow us to obtain the *ideal convergence rate*. The "ideal" in this context means that the scale h of the estimates is defined according to formulas (5.23) and (5.30). The calculations required for this derivation are quite routine. In fact, we apply formulas (5.24) and (5.32) for the optimal values of the minimized mean-square error and substitute the corresponding ideal scale $h^*(x)$. The results are formulated as the following proposition.

Proposition 8 *Let the assumptions of Proposition 7 hold, the ideal scales be defined as in Eqs. (5.23) and (5.30), and $\Delta^d \sim 1/n$, where n is a number of observations. Then, in the asymptotic $n \to \infty$, the mean-square errors of the estimates are*

$$\begin{aligned} E\{[y(x) - \hat{y}_h(x)]^2\} &= O(n^{-2M/(2M+d)}), \\ E\{[y^{(r)}(x) - \hat{y}_h^{(r)}(x)]^2\} &= O(n^{-2(M-|r|)/(2M+d)}), \end{aligned} \tag{5.35}$$

Formulas (5.35) show that there is a mean-square convergence of the estimates to the true values of the signal and the derivative as $n \to \infty$. The convergence rate has the form $n^{-\beta}$ with the exponent $\beta = 2M/(2M+d)$ for the estimate of the signal and the exponent $\beta = 2(M - |r|)/(2M+d)$ for the estimate of the derivative of the order $|r|$. Thus, for smoother signals and larger orders of the LPA, we have larger M and a faster convergence with the best possible convergence rate of the order $O(n^{-1})$. The convergence of the derivative estimates is slower for higher-order derivatives with larger values of $|r|$.

5.2.2 Invariant scale

5.2.2.1 *Signal estimation*

Let the accuracy of signal estimation be characterized by the global criterion for $x \in X$, which is defined as the mean over the set X of the pointwise mean-square risk considered above, and let the estimates be used with an invariant scale h. Then, in the notation of Eq. (5.17), we obtain for this global criterion

$$l_{\hat{y}_h}(h) = \frac{1}{\#}\sum_{x \in X} l_{\hat{y}_h}(x,h) \leq \bar{l}_{\hat{y}_h}(h),$$

$$\bar{l}_{\hat{y}_h}(h) = \frac{1}{\#}\sum_{x \in X} \bar{l}_{\hat{y}_h}(x,h),$$

$$\bar{l}_{\hat{y}_h}(x,h) = |\bar{m}_{\hat{y}_h}(x,h)|^2 + \sigma^2_{\hat{y}_h}(x,h),$$

where # is the cardinality of the set X.

Exploiting the formulas for the bias upper-bound $\bar{m}_{\hat{y}_h}(x,h)$ in Eq. (5.17), we can see that

$$\bar{l}_{\hat{y}_h}(h) = \frac{1}{\#}\sum_{x \in X} L_M^2(x) \cdot (h^M A_g)^2 + \frac{\Delta^d \sigma^2}{h^d} B_g. \tag{5.36}$$

Let $\bar{L}^2 = (1/\#)\sum_{x \in X} L_M^2(x)$. Then

$$\bar{l}_{\hat{y}_h}(h) \leq (h^M \bar{L} A_g)^2 + \frac{\Delta^d \sigma^2}{h^d} B_g. \tag{5.37}$$

Minimizing on h the right-hand side upper-bound in Eq. (5.37), we can use formula (5.23) where $L_M(x)$ is replaced by the invariant value $\bar{L}$. This gives the ideal invariant scale in the form

$$h^* = \left(\Delta^d \frac{\sigma^2 B_g}{A_g^2 \bar{L}^2} \gamma^2\right)^{1/(2M+d)}. \tag{5.38}$$

5.2.2.2 Derivative estimation

Similar results are valid for the derivative estimation with the global mean-square risk

$$l_{\hat{y}_h^{(r)}}(h) = \frac{1}{\#}\sum_{x \in X} l_{\hat{y}_h^{(r)}}(x,h) \leq \bar{l}_{\hat{y}_h^{(r)}}(h),$$

$$\bar{l}_{\hat{y}_h^{(r)}}(h) = \frac{1}{\#}\sum_{x \in X} \bar{l}_{\hat{y}_h^{(r)}}(x,h),$$

$$\bar{l}_{\hat{y}_h^{(r)}}(h) \leq (h^{M-|r|}\bar{L}A_{g^{(r)}})^2 + \sigma^2 \frac{\Delta^d}{h^{d+2|r|}} B_{g^{(r)}}.$$

The ideal invariant scale is then defined by formula (5.30), where $L_M(x)$ is replaced by the invariant value $\bar{L}$.

5.2.2.3 Convergence rate

The convergence rate of the estimates with the ideal invariant scales is the same as that for the estimates with the pointwise varying ideal scales. This convergence rate is defined by Proposition 8. The varying scale improves the pointwise accuracy of estimation; however, the convergence rate is identical for the varying and invariant optimal scale selections.

5.2.3 Example

Let us consider the accuracy of the LPA discrete kernel scalar estimate ($d = 1$) using the Gaussian window

$$w(x) = \frac{e^{-x^2/2}}{\sqrt{2\pi}},$$

and an LPA of the degree $m = 0$. The bias defined according to Eq. (5.11) is

$$m_{\hat{y}_h}(x,h) = (-1)^{M-1} h^M \frac{1}{M!} \partial^M y(x) \cdot \int_{-\infty}^{\infty} g(u) u^M du + o(h^M), \tag{5.39}$$

where, according to Eq. (4.67), $g(x) = w(x)$ is the continuous limit kernel obtained from the discrete one as $\Delta, h \to 0$.

Assume that the signal y is smooth enough for $m + 1 < \alpha$. Then

$$M = \min\{m+1, \alpha\} = m + 1.$$

For $m = 0$ we obtain $M = 1$, $\int_{\mathbb{R}} g(u) u du = 0$, and the main term in Eq. (5.39) is equal to zero. What we can see from this formula is

$$m_{\hat{y}_h}(x,h) = o(h).$$

To specify this small term, assume $m = 1$ in Eq. (5.39) to give $M = 2$. Then

$$m_{\hat{y}_h}(x, h) = -h^2 \frac{1}{2} y^{(2)}(x) \cdot \int_{-\infty}^{\infty} g(u)u^2 du + o(h^2). \tag{5.40}$$

Thus we find the first nonzero term in the formula for the bias.

Let us clarify what was done here and why we need to replace $m = 0$ by $m = 1$. It will be discussed in Sec. 14.1 that the symmetric window for even $m = 2s$ and odd $m = 2s + 1$ gives identical LPA kernels. Because the Gaussian window w is symmetric, the kernel g_h (discrete and continuous) obtained by the LPA is the same for $m = 0$ and $m = 1$. In this case, the accuracy formula in Eq. (5.11) gives a nonzero main term only provided that the degree m used in these formulas is the largest m corresponding to the estimation kernel. This means that, for the symmetric window and even m in the accuracy analysis, we always replace m by $m + 1$. In the considered example, we do this by replacing $m = 0$ by $m = 1$. Then the upper-bound of the bias given in the form of Eq. (5.16) is

$$|m_{\hat{y}_h}(x, h)| \leq h^2 L_2(x) A_g,\ A_g = \frac{1}{2} \left| \int_{-\infty}^{\infty} g(u)u^2 du \right|,\ L_2(x) = |y^{(2)}(x)|,$$

and the variance is

$$\sigma^2_{\hat{y}_h}(x, h) = \sigma^2 \frac{\Delta}{h} B_g,\ B_g = \int_{-\infty}^{\infty} g^2(u)du.$$

Formulas (5.23)–(5.26) give a full set of expressions for the ideal scale selection:

$$h^*(x) = \left[\Delta \frac{\sigma^2 B_g}{A_g^2 L_2^2(x)} \gamma^2 \right]^{1/5}, \tag{5.41}$$

and

$$\bar{l}^*_{\hat{y}_h}(x) = \bar{l}_{\hat{y}_h}(x, h^*(x)) = \sigma^2_{\hat{y}_h}(x, h^*(x))(1 + \gamma^2), \tag{5.42}$$

$$\gamma = \frac{|\bar{m}_{\hat{y}_h}(x, h^*(x))|}{\sigma_{\hat{y}_h}(x, h^*(x))}, \quad \gamma = \sqrt{\frac{1}{2 \cdot 2}} = 0.5. \tag{5.43}$$

Elementary calculations show that for the Gaussian $g(x)$

$$A_g = 0.5, \qquad B_g = \frac{1}{\sqrt{4\pi}}.$$

Finally, we find from Eqs. (5.41–5.43) that

$$h^*(x) = \left[\Delta \frac{\sigma^2}{\sqrt{4\pi} |y^{(2)}(x)|^2} \right]^{1/5}, \tag{5.44}$$

$$\sigma^2_{\hat{y}_h}(x, h^*(x)) = \frac{\Delta}{\sqrt{4\pi} h^*(x)} \sigma^2,$$
$$|\bar{m}_{\hat{y}_h}(x, h^*(x))| = 0.5 \cdot \sigma_{\hat{y}_h}(x, h^*(x)). \tag{5.45}$$

A further analysis is numerical. Let us assume that the estimated signal is

$$y(x) = x^2 \cos(3x), \qquad 0 \leq x \leq 5,$$

and

$$\Delta = 0.01, \qquad \sigma = 35.$$

The signal $y(x)$ and the first two derivatives are shown in Fig. 5.1 as curves (a), (b), and (c), respectively. The corresponding ideal varying scale $h^*(x)$ in Eq. (5.44) is given in Fig. 5.2. It is seen that the optimization results in a highly variable behavior of the scale as a function of x. The ideal scale tends to take large values at the points where the second derivative $\partial^2 y(x)$ is equal to zero.

Concerning the meaning of this tendency, note that the points where the second derivative is equal to zero, $y^{(2)}(x) = 0$, are the points of inflection for $y(x)$. In a neighborhood of these points, the signal $y(x)$ is nearly linear. The first-order Gaussian estimator has zero bias error for any linear function. Large ideal scales at these points reflect this property of the estimator.

The squared bias $|\bar{m}_{\hat{y}_h}(x, h^*(x))|^2$ (curve (d)), the variance $\sigma^2_{\hat{y}_h}(x, h^*(x))$ (curve (c)), and the upper-bound of the risk $\bar{l}^*_{\hat{y}_h}(x)$ (curve (b)) are shown in Fig. 5.3. The square bias is equal to one-fourth of the variance ($\gamma^2 = 0.25$) and the risk value is the sum of the squared bias and the variance. We can see very low values of the variance, bias, and risk close to the inflection points of $y(x)$.

For comparison we analyze the accuracy when the scale h is invariant. The corresponding invariant ideal scale is given by Eq. (5.38). In the considered example, similar to Eq. (5.44) we have

$$h^* = \left[\Delta \frac{\sigma^2}{\sqrt{4\pi}\frac{1}{\#}\sum_x L_2^2(x)} \right]^{1/5}. \tag{5.46}$$

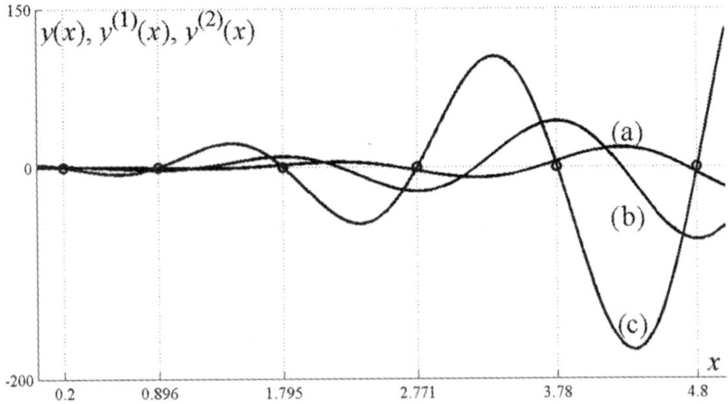

Figure 5.1 The estimated function $y(x)$ and its two derivatives are shown as curves (a), (b), and (c), respectively.

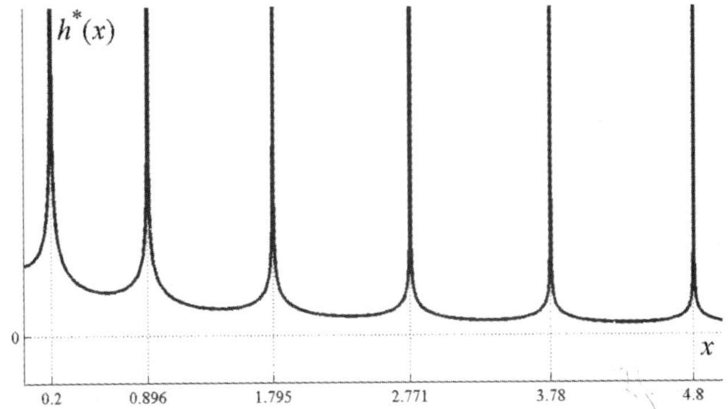

Figure 5.2 The ideal varying scale $h^*(x)$ in Eq. (5.44). Sharp high peaks correspond to inflation points of the estimated $y(x)$.

Let us use this scale value to calculate the upper-bound risk as a function of x. According to Eqs. (5.40) and (5.45), it gives

$$\begin{aligned} \bar{l}_{\hat{y}_h}(x) &= |\bar{m}_{\hat{y}_h}(x, h^*)|^2 + \sigma^2_{\hat{y}_h}(x, h^*), \\ \sigma^2_{\hat{y}_h}(x, h^*) &= \frac{\Delta}{\sqrt{4\pi h^*}}\sigma^2, \\ |\bar{m}_{\hat{y}_h}(x, h^*)|^2 &= (h^*)^4[y^{(2)}(x) \cdot A_g]^2. \end{aligned} \tag{5.47}$$

Curve (a) in Fig. 5.3 shows $\bar{l}_{\hat{y}_h}(x)$ from Eq. (5.47). Naturally, the varying scale gives a uniformly better result with the most striking advantage at the inflation points of the estimated signal, where the risk of the varying-scale estimation is close to zero.

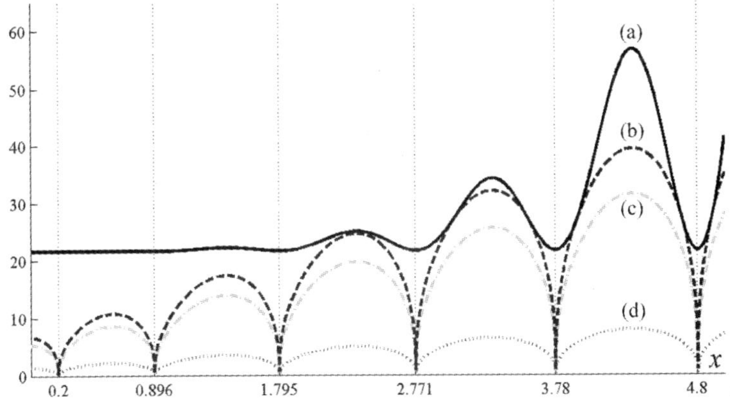

Figure 5.3 The squared bias $|\bar{m}_{\hat{y}_h}(x, h^*(x))|^2$ (curve (d)), the variance $\sigma^2_{\hat{y}_h}(x, h^*(x))$ (curve (c)), and the upper-bound of the risk $\bar{l}^*_{\hat{y}_h}(x)$ (curve (b)) for the ideal varying scale in (5.44). Curve (a) shows the risk $\bar{l}_{\hat{y}_h}(x)$ for the ideal invariant scale h^* in Eq. (5.46).

5.3 Accuracy of the Potential Differentiators*

It can be seen from formulas (4.115)–(4.116) that the integral differentiator with the potential kernel gives the estimate that is the accurate derivative smoothed by the corresponding smoothing kernel.

Let a discrete differentiator be a discrete counterpart (discrete version) of the integral potential differentiator (see Sec. 4.4). We assume that this discrete differentiator is *equivalent* (asymptotically as the sampling interval approaches zero) to the potential differentiating integral estimator. The bias accuracy of this discrete potential differentiator is quite different from the nonpotential ones.

Consider the discrete operators, smoothing (kernel $g_h(\Delta l)$) and differentiation (kernel $g_h^{(r)}(\Delta l)$), and assume that these discrete operators are *equivalent* to the integral ones with the kernels $g(x)$, $g^{(r)}(x)$, respectively, and the kernel $g^{(r)}(x)$ is potential such that $g^{(r)}(x) = \partial^r g(x)$. The accuracy of these discrete estimators is given in the following proposition.

Proposition 9 *Under the conditions of Proposition 7 with* $\alpha \geq m + 1 + |r|$, *the above discrete smoothing and differentiation estimators have the asymptotic bias error defined as*

$$m_{\hat{y}_h}(x, h) = (-1)^m h^{m+1} \sum_{|k|=m+1} y^{(k)}(x) \frac{1}{k!} \int_{\mathbb{R}^d} g(u) u^k du + o(h^{m+1}), \quad (5.48)$$

$$m_{\hat{y}_h^{(r)}}(x, h) = (-1)^m h^{m+1} \sum_{|k|=m+1} y^{(k+r)}(x) \frac{1}{k!} \int_{\mathbb{R}^d} g(u) u^k du + o(h^{m+1}). \quad (5.49)$$

The variance of these estimates is as in Proposition 7.

Proof. Formula (5.48) follows from Eq. (5.11) as soon as $m + 1 \leq \alpha$. Further, according to Eq. (4.116), we have for the bias error of the differentiator

$$m_{\hat{y}_h^{(r)}}(x, h) = y^{(r)}(x) - \hat{y}_h^{(r)}(x) = y^{(r)}(x) - \int_{\mathbb{R}^d} g(u) y^{(r)}(x - hu) du,$$

that is, a differentiation error can be treated as a smoothing error provided that the input signal is the derivative $y^{(r)}(x)$. Then, formula (5.49) follows from Eq. (5.48), where y is replaced by $y^{(r)}$. The bias error of differentiation is defined by the smoothing kernel $g(x)$. For evaluation of this error, formula (5.48) can be used provided that the $y^{(r)}(x)$ derivative belongs to the class of signals $\mathbb{C}^{\alpha}$ with $\alpha \geq m + 1 + |r|$, that is, the derivatives of the orders $m + 1 + |r|$ are bounded. This completes the proof of the proposition. □

The result given by formula (5.49) is nontrivial and surprising because it says that the bias has the same orders on h for the estimates of the signal and the derivatives of any orders r. The only difference between the bias for the signal and the derivative estimation is in the orders of the derivatives of $y(x)$ used in formulas (5.48)–(5.49).

Chapter 6

Adaptive-Scale Selection

Adaptability is a crucial feature of filtering and differentiating methods. It was demonstrated in the previous chapters that the performance of LPA kernel estimates depends strongly on the selection of the scale parameter. In Fig. 2.6 we saw the Nadaraya-Watson estimates for different h. A small h gives the limit estimate $\hat{y}_h(x)$ as a stepwise curve passing exactly through the observations. For large h the same estimator gives a constant value equal to the sample mean of the observations. For intermediate h between these small and large values, we may obtain a large variety of the estimate curves, which are different by their curvature and closeness to the observations.

The idea of the Nadaraya-Watson estimator is natural and fruitful. However, the estimation curve is quite sensitive to h, and the selection of h is a key factor that is able to transform this reasonable idea into an effective working tool. A selection of a proper scale for estimation is a hot topic in both signal processing and nonparametric regression methods. The number of publications in this area is very large and growing quickly.

Two main groups of techniques are exploited in methods based on the nonparametric regression approach. The first one is based on the estimation of the bias and the variance with a scale calculation using the theoretical formulas for the ideal scale (see Chapter 5). These sorts of methods are known as "plug-in" methods.

The second does not deal with the bias and formulas for the ideal scale selection. This is a group of methods based on *quality-of-fit* statistics such as *cross-validation*, *generalized cross-validation*, C_p, *Akaike criteria*, and so on, which are applied for model selection or direct optimization of the estimation accuracy.

Successful implementations of the plug-in approach have been reported by several authors. Overall, these methods give smooth curves with good filtering of random errors. However, the estimate bias depends on unknown high-order derivatives of the signal. As a result, the algorithms are quite complex and have a number of parameters to be tuned, in particular for estimation of these derivatives.

Automatic window size selectors with estimation of the higher-order derivatives of y have been developed and studied in [46]. The estimates at several window sizes are used in [195] to approximate the bias for estimation of the signal and the derivative. Similar ideas have been exploited in adaptive smoothers described in

[198]. Most of publications concerning the quality-of-fit approach are related to a databased global (constant) scale selection (e.g., [77], [87], [102], [199]).

The problem of varying-scale pointwise adaptation has received a powerful impetus in connection with a number of novel ideas developed in mathematical statistics. Various developments of these ideas and various statistical rules for adaptation can be seen in [126], [138], [155], [163], [180], [181], [182], and [208]. These methods are from a class of quality-of-fit statistics applied locally in a pointwise manner.

Methods known under the generic name *Lepski's approach* [135], [136], [137], [138] are of special interest to us because the ICI rule used in this book as a main adaptation tool belongs to this group of methods.

Let us present the idea of *Lepski's approach*. First we introduce a set of scales

$$H = \{h_1 < h_2 < \cdots < h_J\}, \tag{6.1}$$

starting from a small h_1 and increasing to a maximum h_J, and let $\hat{y}_{h_i}$ be the LPA estimate of y calculated for $h_i \in H$.

According to the accuracy analysis produced in Chapter 5, for small h the estimate $\hat{y}_h$ has a small bias and a large standard deviation. The adaptive-scale algorithm compares the estimates of increasing h with the main intention of finding the maximum scale when the estimate's deviation has a nearly ideal balance between the bias and randomness. The meaning of this ideal variance-bias balance is discussed in Chapter 5.

Lepski's approach defines the pointwise adaptive-scale $h^+(x)$ according to the rule

$$h^+(x) = \max\{h_i \colon |\hat{y}_{h_j}(x) - \hat{y}_{h_i}(x)| \le \Gamma(h_j, h_i, x), \quad \forall h_j < h_i, i = 1, \ldots, J\}, \tag{6.2}$$

where $\Gamma(h_j, h_i, x)$ is a given threshold.

The procedure in Eq. (6.2) looks for a largest scale h_i to obtain the maximum smoothing effect for the random errors. However, a large h_i can result in a significant bias error. The estimates $\hat{y}_{h_j}(x)$ with the scale $h_j < h_i$ are compared with the estimate $\hat{y}_{h_i}(x)$ of the scale h_i. If all differences $|\hat{y}_{h_j}(x) - \hat{y}_{h_i}(x)|$, $h_j < h_i$, can be explained by the random errors, the bias is not large and larger h_i can be selected.

The adaptive-scale h^+ is defined as a maximum h_i such that all estimates $\hat{y}_{h_j}(x)$ with $h_j < h_i$ are not different from $\hat{y}_{h_i}(x)$. The threshold $\Gamma(h_j, h_i, x)$ is a key element of the algorithm because it indicates when the difference between the estimates is large or small. The rule in Eq. (6.2) enables a multiple pairwise statistical test on significance of the systematic error in the differences $\hat{y}_{h_j}(x) - \hat{y}_{h_i}(x)$.

Overall, the algorithm searches for the largest local vicinity of the point of estimation x where the LPA assumptions fit well to the data. The estimates are calculated for a few scales and compared. The adaptive scale is defined as the largest one of those for which the estimate does not differ significantly from the estimates corresponding to the smaller scales.

There are a few versions of this concept that are different mainly by estimates of $\hat{y}_h(x)$, sets of scales H, and hypothesis testing rules. The threshold $\Gamma(h_j, h_i, x)$

in the right-hand side of the inequality in Eq. (6.2) is an important parameter of this hypothesis testing.

Theoretical results given in the papers cited above show that the algorithms have exciting asymptotic properties in terms of their adaptivity over wide classes of signals, minimax estimations, and convergence rates. These adaptive algorithms work without special estimation of change points and singularities. The main intention is to estimate the signal with a varying adaptive-scale selection. Near the change points small-scale values are selected automatically, and in this way the accuracy of the local estimation is ensured.

We are mainly concerned with one particular form of this class of methods developed in [66] independently from the mainstream works in this direction. The term "intersection of confidence intervals" (ICI) for this method first appeared in [103].

It is shown in a number of publications that the ICI-based algorithms are practical and efficient for adaptive signal and image processing with applications in median filtering [107], probability density estimation [109], time-frequency analysis [38], [113], beamforming and direction of arrival estimation [112]. A successful application of the ICI rule to image denoising has been reported in [106], [108], and [111]. A number of interesting modifications of the ICI rule for regression problems with Gaussian and non-Gaussian noises are developed in [209].

In this chapter, we give a systematic presentation of the ICI rule, including ideas, algorithms, theoretical performance analysis, and simulation results.

6.1 ICI Rule

Here the ICI rule for the adaptive-scale selection is derived from the accuracy analysis of the LPA estimates.

6.1.1 Foundations

6.1.1.1 Signal estimation

Let $y(x)$ and $\hat{y}_h(x)$, $x \in \mathbb{R}^d$, be the signal and its LPA estimate, respectively, with the estimate defined according to Eq. (2.55) as

$$\hat{y}_h(x) = \sum_s g_h(x, X_s) z_s, \tag{6.3}$$

where $g_h(x, X_s)$ is the kernel given in Eq. (2.56) and the observations z_s are as in Eq. (2.1).

The estimation error e_y in Eq. (5.1) can be represented as a sum of the bias $m_{\hat{y}_h}(x, h)$ and the zero-mean random error $e^0(x, h)$, $e_y(x, h) = m_{\hat{y}_h}(x, h) + e^0(x, h)$, with the natural inequality

$$\left|e_y(x, h)\right| = \left|y(x) - \hat{y}_h(x)\right| \leq |\bar{m}_{\hat{y}_h}(x, h)| + \left|e^0(x, h)\right|, \tag{6.4}$$

where $\bar{m}_{\hat{y}_h}(x,h)$ is a bias upper bound in Eq. (5.17). The random estimation error

$$e_y^0(x,h) = -\sum_s g_h(x, X_s)\varepsilon_s,$$

has zero mean and standard deviation $\sigma_{\hat{y}_h}(x,h)$.

Assuming that the noise ε_s is Gaussian, the random error $e_y^0(x,h)$ is subject to the Gaussian probability density $\mathcal{N}(0, \sigma^2_{\hat{y}_h}(x,h))$. The following inequality holds with probability $p = 1 - \beta$:

$$\left|e_y^0(x,h)\right| \leq \chi_{1-\beta/2} \cdot \sigma_{\hat{y}_h}(x,h), \tag{6.5}$$

where $\chi_{1-\beta/2}$ is the $(1-\beta/2)$th quantile of the standard Gaussian distribution $\mathcal{N}(0,1)$ and

$$\sigma^2_{\hat{y}_h}(x,h) = \sigma^2 \sum_s g_h^2(x, X_s). \tag{6.6}$$

Remember the concept used in the inequality Eq. (6.5). In Fig. 6.1, the curve is the standard Gaussian density $\mathcal{N}(0,1)$. The area of the central shaded space under the curve is equal to p. With probability p, the random $e_y^0(x,h)$ takes values in the interval $[-\chi_{1-\beta/2}, \chi_{1-\beta/2}]$.

In this sort of analysis, p is conventionally large close to 1. The values of the random $e_y^0(x,h)$ belonging to the interval $[-\chi_{1-\beta/2}, \chi_{1-\beta/2}]$ are "usual." The random e_y^0 takes these values often with a large probability p. In contrast, the values of $e_y^0(x,h)$ out of the interval $[-\chi_{1-\beta/2}, \chi_{1-\beta/2}]$ are called "unusual" because they occur rarely with a small probability $1 - p = \beta$. It follows from Eqs. (6.4) and (6.5) that with the same probability p,

$$\left|e_y(x,h)\right| \leq |\bar{m}_{\hat{y}_h}(x,h)| + \chi_{1-\beta/2}\sigma_{\hat{y}_h}(x,h). \tag{6.7}$$

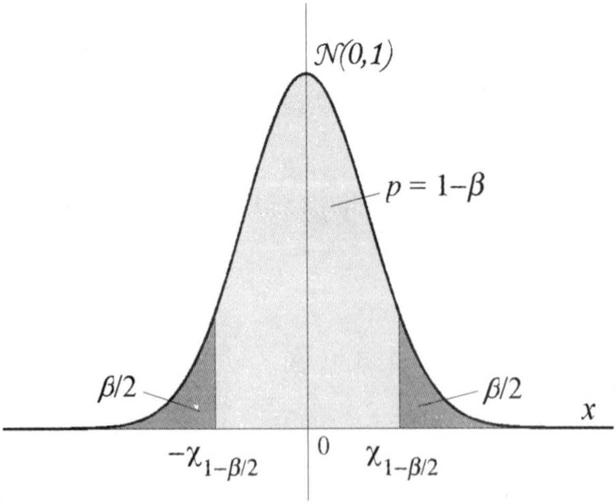

Figure 6.1 $\chi_{1-\beta/2}$ is the $(1-\beta/2)$th quantile of the standard Gaussian distribution $\mathcal{N}(0,1)$.

Let us use the accuracy results derived in Chapter 5 for the shift-invariant kernels. These asymptotic formulas are valid for general shift-variant estimates in Eq. (6.3) with the variance calculated according to Eq. (6.6) and depending on x.

It has been shown that there exists an ideal scale $h^*(x)$ in Eq. (5.23) giving the optimal balance of the bias and the variance. The MSE upper bound in Eq. (5.17) takes its minimum value at $h = h^*(x)$.

It is shown that there exists an optimal balance between the bias and the standard deviation of the random error having a form of the inequality Eq. (5.27):

$$|\bar{m}_{\hat{y}_h}(x,h)| \begin{cases} <\gamma \cdot \sigma_{\hat{y}_h}(x,h) & \text{if } h < h^*(x), \\ >\gamma \cdot \sigma_{\hat{y}_h}(x,h) & \text{if } h > h^*(x), \end{cases} \tag{6.8}$$

where the parameter γ is defined in Eq. (5.26).

Using the first inequality (Eq. (6.8)) in Eq. (6.7), we obtain that for $h \leq h^*(x)$

$$\left|e_y(x,h)\right| \leq \gamma \cdot \sigma_{\hat{y}_h}(x,h) + \chi_{1-\beta/2}\sigma_{\hat{y}_h}(x,h) \leq \Gamma \cdot \sigma_{\hat{y}_h}(x,h), \tag{6.9}$$

$$\Gamma = \gamma + \chi_{1-\beta/2}. \tag{6.10}$$

This inequality is equivalent to

$$|y(x) - \hat{y}_h(x)| \leq \Gamma \cdot \sigma_{\hat{y}_h}(x,h),$$

and can be rewritten in the form

$$\hat{y}_h(x) - \Gamma \cdot \sigma_{\hat{y}_h}(x,h) \leq y(x) \leq \hat{y}_h(x) + \Gamma \cdot \sigma_{\hat{y}_h}(x,h), \quad h \leq h^*(x). \tag{6.11}$$

This determines the confidence interval $Q(h)$ of the estimate $\hat{y}_h(x)$ as

$$Q(h) = [\hat{y}_h(x) - \Gamma \cdot \sigma_{\hat{y}_h}(x,h),\ \hat{y}_h(x) + \Gamma \cdot \sigma_{\hat{y}_h}(x,h)],$$

where Γ in Eq. (6.10) is a parameter. Because inequality Eq. (6.7) is valid with probability p, the same holds for Eq. (6.11) and we can claim that with probability p,

$$y(x) \in Q(h), \tag{6.12}$$

while $h \leq h^*(x)$.

For $h_j \in H$ we define a sequence of confidence intervals $Q_j = Q(h_j)$ as

$$\begin{aligned} Q_j &= [\hat{y}_{h_j}(x) - \Gamma \cdot \sigma_{\hat{y}_h}(x,h_j),\ \hat{y}_{h_j}(x) + \Gamma \cdot \sigma_{\hat{y}_h}(x,h_j)], \\ \Gamma &= \gamma + \chi_{1-\beta/2}, \end{aligned} \tag{6.13}$$

and $y(x) \in Q_j$ for all h_j provided that $h_j \leq h^*(x)$. For this sequence, we may conclude with some probability p that if Eq. (6.9) holds for $h = h_j$, $1 \leq j \leq i$, all of the intervals Q_j, $1 \leq j \leq i$, have a point in common, namely, $y(x)$. This says that the intersection of these confidence intervals is not empty provided that all $h_j \leq h^*(x)$ for $1 \leq j \leq i$. In the opposite case, when the intersection of the confidence intervals is empty, it indicates that $h_j > h^*(x)$. In this way, the intersection of the confidence intervals can be used to verify the inequality

$$h \lessgtr h^*(x). \tag{6.14}$$

The following is the ICI statistic proposed for obtaining the adaptive scales from testing the very existence of the common point of the confidence intervals Q_j [66]:

Consider the intersection of the intervals Q_j, $1 \leq j \leq i$, with increasing i, and let i_+ be the largest of those i for which the intervals Q_j, $1 \leq j \leq i$, have a point in common. This i_+ defines the adaptive scale and the adaptive LPA estimate as follows:

$$\hat{y}^+(x) = \hat{y}_{h^+(x)}(x), \qquad h^+(x) = h_{i_+}. \tag{6.15}$$

6.1.1.2 ICI rule calculations

The following algorithm implements the ICI procedure:

Determine the sequence of the upper and lower bounds of the confidence intervals Q_i as follows:

$$\begin{aligned} Q_i &= [L_i, U_i], \\ U_i &= \hat{y}_{h_i}(x) + \Gamma \cdot \sigma_{\hat{y}_h}(x, h_i), \\ L_i &= \hat{y}_{h_i}(x) - \Gamma \cdot \sigma_{\hat{y}_h}(x, h_i). \end{aligned} \tag{6.16}$$

Let

$$\begin{aligned} \bar{L}_{i+1} &= \max\{\bar{L}_i,\ L_{i+1}\},\ \underline{U}_{i+1} = \min\{\underline{U}_i,\ U_{i+1}\}, \\ i &= 1, 2, \ldots, J-1,\ \bar{L}_1 = L_1,\ \underline{U}_1 = U_1. \end{aligned} \tag{6.17}$$

According to these formulas, $\bar{L}_{i+1}$ and $\underline{U}_{i+1}$ are nondecreasing and nonincreasing sequences, respectively. Find the largest i when

$$\bar{L}_i \leq \underline{U}_i,\ i = 1, 2, \ldots, J, \tag{6.18}$$

is still satisfied. Denote this largest value i_+. This i_+ is the largest of those i for which the confidence intervals Q_i have a point in common, as was discussed above, and the ICI adaptive scale is

$$h^+ = h_{i_+}.$$

The ICI rule is illustrated graphically in Fig. 6.2(a), where the confidence intervals Q_j, $j = 1, 2, 3, \ldots, J$, are shown. The vertical lines with arrows show the successive intersections of the confidence intervals (1, 2) and (1, 2, 3). Assuming that the intersection with the fourth confidence interval (corresponding to $h = h_4$) is empty, we obtain the adaptive scale as $h^+ = h_3$. The algorithm in Eq. (6.18) for the variables L_i, U_i is illustrated in Fig. 6.2(b). These calculations are pointwise repeated for each x.

Let us make some general remarks concerning the ICI rule as a selector of the adaptive scale:

1. The confidence intervals Q_j in Eq. (6.13) are given for the biased estimates. Indeed, we replace the bias in the inequality Eq. (6.7) by the upper bound

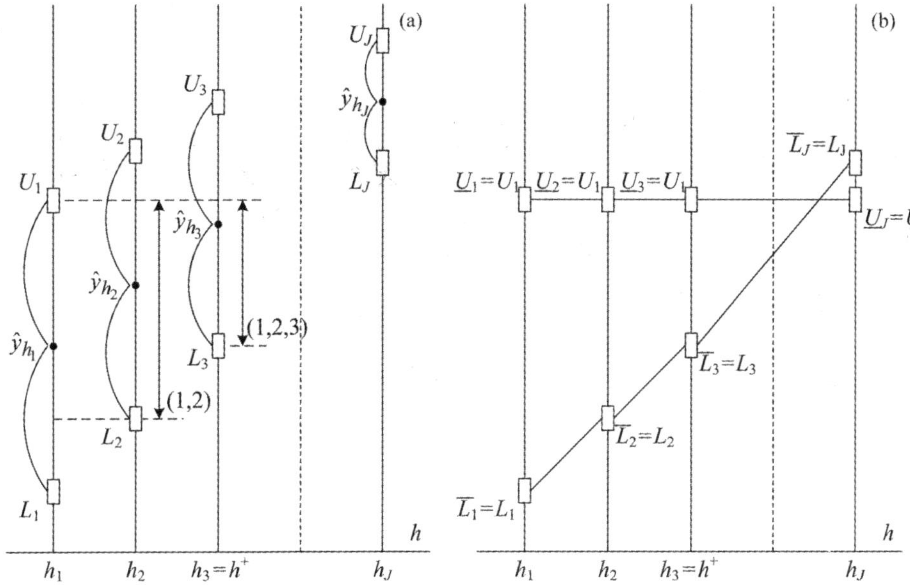

Figure 6.2 The idea (a) and implementation (b) of the ICI rule.

valid for $h \leq h^*$. This bias is replaced by the standard deviation $\sigma_{\hat{y}_h}(x, h_j)$ taken with the factor γ. In this way the confidence intervals in Eq. (6.13) are essentially different from the standard ones defined for unbiased estimates.

2. The intersection of the confidence intervals Q_j depends on the estimates $\hat{y}_{h_j}(x)$ giving a position of the center of the interval and on the width of the interval equal to $2\Gamma\sigma_{\hat{y}_h}(x, h)$. The parameter Γ plays an important role in the ICI algorithm. It is called a threshold parameter of the confidence interval or threshold parameter of the ICI rule.
3. It is assumed in the derivation of the ICI that $h_j \leq h^*$. Then the first inequality in Eq. (6.8) holds, and the intersection of the confidence intervals is not empty and includes the value $y(x)$. The inequality $h_j \leq h^*$ is a sufficient condition for this conclusion.

   However, the inverse statement is not true. The inequality $h_j \leq h^*$ does not follow from the fact that the intersection is not empty. This means that the ICI adaptive-scale h^+ can be smaller or larger than the ideal h^* and the intersection of the confidence intervals may not include the true signal value $y(x)$.
4. It is proved in Lemma 3 (Sec. 14.3.1 of the Appendix) that asymptotically a distance between the adaptive estimate $\hat{y}^+(x)$ and the true signal value $y(x)$ is not larger than 1.5 of the length of the ideal confidence interval $Q(h^*)$, that is, this distance is not larger than $1.5 \times 2\Gamma\sigma_{\hat{y}_h}(x, h^*) = 3\Gamma\sigma_{\hat{y}_h}(x, h^*)$. Then, the accuracy of the ICI adaptive estimate is proportional to the standard deviation $\sigma_{\hat{y}_h}(x, h^*)$ of the ideal scale estimate, that is, the accuracies of the adaptive and ideal scale estimators are close to each other. This guarantees a nearly ideal (asymptotically) performance of the ICI adaptive-scale

estimates. This statement is valid despite the fact that some of the adaptive scales can be larger than h^*.

5. In the ICI algorithm, the estimates of the different scales are compared by their confidence intervals. This is a particular implementation of the general Lepski approach. Indeed, we may conclude that the confidence intervals Q_j and Q_i intersect if and only if

$$|\hat{y}_{h_j}(x) - \hat{y}_{h_j}(x)| \leq \Gamma[\sigma_{\hat{y}_h}(x, h_j) + \sigma_{\hat{y}_h}(x, h_i)]. \tag{6.19}$$

This condition means that the intersection has a place only when a distance between the middle points of the confidence intervals is less than or equal to $\Gamma[\sigma_{\hat{y}_h}(x, h_j) + \sigma_{\hat{y}_h}(x, h_i)]$. In this formalization the ICI rule is identical to Lepski's scheme in Eq. (6.2) with the threshold parameter

$$\Gamma(h_j, h_i, x) = \Gamma[\sigma_{\hat{y}_h}(x, h_j) + \sigma_{\hat{y}_h}(x, h_i)]. \tag{6.20}$$

6.1.1.3 Differentiation

The ICI algorithm is universally valid for estimation of both a signal and any derivative as well as for univariate and multivariate signals. The only difference is in the estimates and formulas that are used for variance. The kernel estimate of the derivative $y^{(r)}(x)$ Eq. (2.57) is of the form

$$\hat{y}_h^{(r)}(x) = \sum_s g_h^{(r)}(x, X_s) z_s, \tag{6.21}$$

where $g_h^{(r)}$ is the kernel in Eq. (2.58) of the estimate.

Similar to Eq. (6.7) we can write for the differentiation error that, with some probability p,

$$\left|e_{y^{(r)}}(x, h)\right| \leq |\bar{m}_{\hat{y}_h^{(r)}}(x, h)| + \left|e^0_{y^{(r)}}(x, h)\right|, \tag{6.22}$$

where the random error satisfies the inequality

$$\left|e^0_{y^{(r)}}(x, h)\right| \leq \chi_{1-\beta/2}\sigma_{\hat{y}_h^{(r)}}(x, h). \tag{6.23}$$

The ideal scale $h_r^*(x)$ in Eq. (5.30) minimizing the least-square error gives the bias-variance balance in Eq. (5.34) as

$$|\bar{m}_{\hat{y}_h^{(r)}}(x, h)| \begin{cases} <\gamma_{|r|} \cdot \sigma_{\hat{y}_h^{(r)}}(x, h) & \text{if } h < h_r^*(x), \\ >\gamma_{|r|} \cdot \sigma_{\hat{y}_h^{(r)}}(x, h) & \text{if } h > h_r^*(x), \end{cases} \tag{6.24}$$

where $\gamma_{|r|}$ is defined in Eq. (5.33) and

$$\sigma^2_{\hat{y}_h^{(r)}}(x, h) = \sigma^2 \sum_s [g_h^{(r)}(x, X_s)]^2.$$

The confidence interval of the estimate in Eq. (6.21) that is similar to Eq. (6.13) is of the form

$$Q_j^{(r)} = [\hat{y}_{h_j}^{(r)}(x) - \Gamma_{|r|} \cdot \sigma_{\hat{y}_h^{(r)}}(x, h_j), \hat{y}_{h_j}^{(r)}(x) + \Gamma_{|r|} \cdot \sigma_{\hat{y}_h^{(r)}}(x, h_j)], \tag{6.25}$$

where

$$\Gamma_{|r|} = \gamma_{|r|} + \chi_{1-\beta/2}. \tag{6.26}$$

Using Eq. (6.25), the ICI algorithm is formulated exactly as it is for signal estimation. This algorithm gives the adaptive varying scales $h_r^+(x)$ that approximate the ideal scales $h_r^*(x)$ in Eq. (5.30) for derivative estimation.

The above remarks given for the adaptive-scale estimates of the signal are valid for the scale-adaptive estimates of the derivatives.

6.1.2 LPA-ICI algorithm

Input variables of the ICI algorithm for signal estimation include:

1. Noise variance σ^2;
2. The estimate $\hat{y}_h(x)$ in Eq. (6.3) given for all $h \in H$ and $x \in X$;
3. The variance $\sigma_{\hat{y}_h}^2(x, h)$ in Eq. (6.6) of the estimates given for all $h \in H$.

6.1.2.1 Noise variance estimation

The following standard methods, which are independent of the ICI, are used for the estimation of σ^2:

- The median of the signal's differences

$$\hat{\sigma} = \{\text{median}_s(|z_s - z_{s-1}| : s = 2, \ldots, n)\}/(\sqrt{2} \cdot 0.6745). \tag{6.27}$$

  The normalized constant $\sqrt{2} \cdot 0.6745$ in Eq. (6.27) makes this estimate unbiased, $E\{\hat{\sigma}\} = \sigma$, for Gaussian $z_s \sim \mathcal{N}(0, \sigma^2)$. The finite differences in Eq. (6.27) are used to eliminate the signal y and have $z_s - z_{s-1} \simeq \varepsilon_s - \varepsilon_{s-1}$. The mean $\sum_{n=2}^{n}(z_s - z_{s-1})^2/(n-1)$ could also be applied as an estimate of σ^2. However, the median has been proven to have better performance because the median is robust with respect to outliers (large rare errors), which, in particular, can appear in the differences $z_s - z_{s-1}$ near change points where discontinuity of the signal has a place.
- Let $\psi_{l,m}(x) = \psi(x/2^l - m)$ be a wavelet of the finest scale l and the wavelet spectrum of the signal z calculated as

$$w_m = \sum_s \psi_{l,m}(x_s) z_s.$$

Then the estimate $\hat{\sigma}$ is calculated as ([150], page 459)

$$\hat{\sigma} = \{\text{median}_m(|w_m| : 0 \leq m < n/2\}/0.6745. \tag{6.28}$$

The estimates in Eqs. (6.27) and (6.28) are different by the linear filter used for suppressing the signal, that is, the finite difference in Eq. (6.27) and the wavelet transform in Eq. (6.28).

6.1.2.2 Calculation of the estimate

If the shift-invariant kernels are applied for estimation, the convolution gives the fast implementation with

$$\hat{y}_h(x) = (g_h \circledast z)(x), h \in H, \tag{6.29}$$

and allows one to get the estimates for all $x \in X$ in parallel.

The shift-variant kernels $g_h(x, x_s)$ can be used, for example, near boundaries of the image/signal, to decrease the boundary effects. There are no fast algorithms for this case, so the general shift-variant estimation formulas in Eq. (6.3) should be used instead.

6.1.2.3 Calculation of estimate variance

While the asymptotic formula (5.13) is useful to clarify the ideas behind the ICI, the accurate formulas for the variance in Eq. (6.6) are recommended for the algorithm. For shift-invariant kernels the variance is independent on x,

$$\sigma^2_{\hat{y}_h}(x, h) = \hat{\sigma}^2||g_h||^2 = \hat{\sigma}^2 \sum_s g_h^2(x_s). \tag{6.30}$$

6.1.2.4 Algorithm organization

A layout of the LPA-ICI adaptive algorithm is shown in Fig. 1.4 of Chapter 1. The algorithm includes the following basic steps:

1. The LPA filter design defining the bank of linear filters $g_h(x), h \in H$;
2. Calculation of the variance of observation noise Eqs. (6.27) or (6.28);
3. Filtering an input signal in Eq. (6.29) with the results given for all $h \in H$, $x \in X$;
4. Calculation of the lower and upper bounds of the confidence intervals in Eq. (6.16);
5. Selection of the ICI adaptive scales $h^+(x)$ and the corresponding adaptive estimates $\hat{y}_{h_+}$ for all $x \in X$.

6.1.2.5 Differentiation

The above algorithm is applicable for derivative estimation if the kernel g_h in formulas (6.29) and (6.30) is replaced by the corresponding differentiating kernel $g_h^{(r)}$.

6.1.3 Complexity

The calculation of the estimate $\hat{y}_h(x)$ for a given h is a linear convolution requiring $N_{conv} \sim n \log n$ operations. If more sophisticated procedures are used for convolution, then $N_{conv} \sim n \log n_{h_J}$, where n_{h_J} is the maximum size of the support of the kernel g_{h_J}.

The ICI algorithm is implemented as a loop on J different scales from the set H. Its complexity is proportional to J. Thus, the overall complexity of the calculations of the ICI adaptive estimates is proportional to $J \cdot N_{conv}$. In this way, the ICI adaptive-scale algorithm belongs to the class of fast algorithms with a number of operations proportional to $n \log n$ or $n \log n_{h_J}$.

6.1.4 Convergence

The asymptotic accuracy analysis of the ICI adaptive algorithms is based on the results and the techniques developed in [66] and [163], where different aspects of the problem are studied, including links with the minimax estimation, lower and upper bounds of the accuracy, as well as analyses for various classes of signals using different local and global risks. We present here results corresponding to the assumptions used in the accuracy analysis of Chapter 5.

It is assumed in this asymptotic that the signal is defined on a regular grid with a small enough sampling interval Δ and with a large number of observations $n \sim 1/\Delta^d$. It is also assumed that the threshold Γ grows as $\log n$. The set of scales H is dense enough starting with a small $h_1 = o(\log n/n)$ and $h_i - h_{i-1} = o(\log n/n)$.

Proposition 10 *Let the assumptions of Proposition 7 hold and the estimates $\hat{y}_h(x)$ and $\hat{y}_h^{(r)}(x)$ be scale-adaptive according to the ICI algorithm. Then, for n large enough and the threshold Γ growing proportionally to $\log n$, the mean-square errors of the estimates are of the following orders*:

$$E\{[y(x) - \hat{y}_h(x)]^2\} = O\left(\left(\frac{\log n}{n}\right)^{\frac{2M}{2M+d}}\right), \tag{6.31}$$

$$E\{[y^{(r)}(x) - \hat{y}_h^{(r)}(x)]^2\} = O\left(\left(\frac{\log n}{n}\right)^{\frac{2(M-|r|)}{2M+d}}\right). \tag{6.32}$$

Let us make some remarks concerning these results:

1. Formulas (6.31)–(6.32) show that there is a mean-square convergence of the adaptive estimates to the true values of the signal and the derivative as $n \to \infty$. The convergence rate is exponential. Comparison with the results given in Proposition 8 shows that the convergence orders obtained for the adaptive scales are different only in the factor $\ln n$ from the convergence orders of the estimates with the ideal scale selection. Naturally, the convergence of the adaptive estimates is slower; that is a price of the adaptivity.
2. It is shown in [135] that if the smoothness of the estimated signal is unknown, the best possible convergence rate differs by this log-factor $\log n$ from the convergence rate of the ideal nonadaptive estimator. For estimating a signal of unknown smoothness, this factor cannot be eliminated, meaning that the mean-square convergence rate given in Eq. (6.31) is the optimal one and cannot be improved.
3. According to Eqs. (5.23) and (5.30), the ideal scales depend on the smoothness of the signal defined by the derivative order α in $M = \min\{m+1, \alpha\}$ and the derivative values at the point x as $L_M(x) = \max_{r_1+\cdots+r_d=M} \left|\partial_{x_1}^{r_1} \cdot \cdots \cdot \partial_{x_d}^{r_d} y(x)\right|$. The adaptive ICI algorithms do not use this information to find the adaptive varying scales, which means that the LPA-ICI algorithms are adaptive with respect to the unknown smoothness of the signal defined by parameter α and with respect to the unknown values of the derivatives $L_M(x)$.

The convergence rates Eqs. (6.31–6.32) for the univariable case $d = 1$ are derived in [66]. For multivariate signals $d > 1$, the convergence rates in Eqs. (6.31–6.32) can be extracted as a special case of more general results given in [163], Theorem 3.3.1.

The proof of Proposition 10 is given in Sec. 14.3 of the Appendix.

6.1.5 Threshold adjustment

6.1.5.1 *Signal estimation*

General recommendations

The threshold parameter Γ in Eqs. (6.13) and (6.16) plays a crucial role in the performance of the algorithm. Simulation shows that too large or too small Γ results in oversmoothing and undersmoothing of the estimates.

Let us start with some figures for Γ following from the theoretical analysis. Assuming $\beta = 0.05$ or 0.01, we have, respectively, $\chi_{1-\beta/2} = 2$ or $\chi_{1-\beta/2} = 3$. It follows from Eqs. (5.26) and (6.10) that

$$\Gamma = \gamma + \chi_{1-\beta/2} = \sqrt{\frac{d}{2\min\{m+1, \alpha\}}} + \chi_{1-\beta/2}. \tag{6.33}$$

For $m + 1 \leq \alpha$ and $d = 2$, this gives

$$\Gamma = \begin{cases} 3.0, & \text{for } \beta = 0.05, \\ 4.0, & \text{for } \beta = 0.01, \quad \text{if } m = 0, \end{cases} \tag{6.34}$$

$$\Gamma = \begin{cases} 2.7, & \text{for } \beta = 0.05, \\ 3.7, & \text{for } \beta = 0.01, \quad \text{if } m = 1. \end{cases}$$

How much can we trust these figures?

A principal ambiguity of the algorithm concerns a value of the confidence level β (probability of first-kind error). This seriously changes the corresponding quantile $\chi_{1-\beta/2}$ in Eq. (6.33). Large values of $\chi_{1-\beta/2}$ (small β) improve the noise resistance of the ICI rule but result in a low sensitivity to changes in y. Small values of $\chi_{1-\beta/2}$ (large β) lead to a large "false alarm" probability β. Further, formula (6.33) is derived for a single-step hypothesis testing, where β is the probability of the first-kind error in this procedure only. Multiple-hypothesis tests used in the ICI require different and more accurate thresholds to control the probability of errors.

Remember also that the formula for γ in Eq. (6.33) is derived for the asymptotic as $\Delta \to 0$. It is assumed in the asymptotic analysis that $y(x)$ is smooth enough in the neighborhood of x, that is, the first- and second-order derivatives exist when the zero-power ($m = 0$) or the first-power ($m = 1$) LPA is applied. In reality, the smoothness of $y(x)$ cannot be guaranteed for every x and the sampling interval Δ cannot be small enough. All these factors influence the probability of errors and, finally, the accuracy of estimation.

It is clear that a proper selection of the threshold Γ is a problem that cannot be resolved in terms of the theoretical analysis only. As a result, it is reasonable to treat the threshold Γ as a design parameter of the ICI algorithm.

Experiments show that simple universal recommendations can be made for Γ that usually enable a stable, good performance of the ICI adaptive algorithms. The following recommendations are a rule of thumb and give smaller values of Γ than the theoretical ones in Eqs. (6.33–6.34):

- The basic interval for Γ is

$$1 \leq \Gamma \leq 2; \tag{6.35}$$

- If the signal y is piecewise smooth with rare and slow variations, larger values of Γ from the interval (6.35) are recommended for better noise reduction;
- If the signal y has multiple change-points and singularities, then smaller values of Γ from the interval (6.35) are recommended for the preservation of small details of the signal.

Data-driven Γ

Another performance improvement can be achieved by adjusting Γ to the particular signal/noise observations or typical scenario. The value of Γ can be optimized by comparing the performance of the algorithm with different Γ. This straightforward idea can be implemented easily for a given test signal y when the evaluation of the performance is trivial. If the signal y is unknown, the performance evaluation is not a simple task, but various statistical procedures can be applied.

Let us show how the cross-validation (CV) statistics (see Sec. 14.4 in the Appendix) can be used for this purpose. Assume that the ICI rule is applied for the adaptive-scale selection and the corresponding pointwise varying scales are found. Let

$$r_+(X_t) = z_t - \hat{y}_{h^+(X_t)}(X_t),$$

be the residuals calculated as differences between observations and adaptive-scale estimates. Then, the CV criteria in Eqs. (14.120) and (14.124) are

$$R_{CV}^2 = \frac{1}{n}\sum_{t=1}^{n} \frac{r_+^2(X_t)}{[1 - g_{h^+(X_t)}(0)]^2}, \tag{6.36}$$

and

$$R_{CV}^2 = \frac{1}{n}\sum_{t=1}^{n} \frac{r_+^2(X_t)}{[1 - g_{h^+(X_t)}(X_t, X_t)]^2}, \tag{6.37}$$

respectively, for the shift-invariant and shift-variant estimates. These R_{CV}^2 are used for a data-adaptive selection of Γ value.

The ICI adaptive estimation is repeated for each $\Gamma \in G$, $G = \{\Gamma_1, \Gamma_2, \ldots, \Gamma_{NG}\}$, and

$$\hat{\Gamma} = \arg\min_{\Gamma \in G} R_{CV}^2, \tag{6.38}$$

gives the CV-adjusted threshold parameter Γ.

The CV in the form of Eqs. (6.37–6.36) presents a reasonable and quite efficient selector for Γ. Our attempts to use other quality-of-fit statistics instead of CV, in particular the C_P [151] and Akaike criteria [3], [87], show no improvement in estimation accuracy.

The CV-adjusted adaptive LPA-ICI estimation consists of the following basic steps:

1. Set $\Gamma = \Gamma_l$, $l = 1, 2, \ldots, N_G$;
2. For $h = h_i$, $i = 1, \ldots, J$, calculate the estimates $\hat{y}_h(X_s)$, the adaptive-scale $h^+(X_s)$, and the ICI adaptive estimate $\hat{y}_{h^+(X_s)}(X_s)$;
3. Repeat Step 2 for all X_s, $s = 1, 2, \ldots, n$, and Γ_l, $l = 1, 2, \ldots, N_G$;
4. Find $\hat{\Gamma}$ from Eq. (6.38) and select estimates $\hat{y}(X_s, h^+(X_s))$ corresponding to this $\hat{\Gamma}$.

6.1.5.2 Derivative estimation

For differentiation, the threshold in the confidence intervals in Eq. (6.25) are defined according to formulas (6.26) and (5.33) as

$$\Gamma_{|r|} = \sqrt{\frac{d + 2|r|}{2(\min\{m + 1, \alpha\} - |r|)}} + \chi_{1-\beta/2}. \tag{6.39}$$

Let $d = 2$ and the estimated derivatives be $r = (1, 0)$ or $r = (0, 1)$, $|r| = 1$, that is, the first derivative on x_1 or x_2 and $|r| \leq m \leq \alpha - 1$. We obtain from Eq. (6.39)

$$\Gamma_{|1|} = \begin{cases} 3.4, & \text{for } \beta = 0.05, \\ 4.4, & \text{for } \beta = 0.01, \quad \text{for } m = 1, \end{cases} \tag{6.40}$$

$$\Gamma_{|1|} = \begin{cases} 3.0, & \text{for } \beta = 0.05, \\ 4.0, & \text{for } \beta = 0.01, \quad \text{for } m = 2. \end{cases}$$

Comparison with the data in Eq. (6.34) for $m = 1$ shows that the derivative estimation requires larger values of the threshold $\Gamma_{|r|}$. However, overall, the situation is similar to the estimation of a signal. A proper selection of the threshold Γ cannot be resolved using theoretical analysis only. As a result, $\Gamma_{|r|}$ is treated as a design parameter of the ICI algorithm.

6.2 Multiple-Window Estimation

Different window functions w_h can be exploited for the LPA filter design. A simple standard idea is to use a window w_h symmetric with respect to the reference point x with a single size (scale) parameter h. A more flexible approach considers varying-size nonsymmetric left/right windows, as illustrated in the univariate example in Sec. 2.2 (Fig. 2.8).

For image processing, quadrant windows shown in Fig. 3.7(b) provide a valuable alternative to the symmetric square window (Fig. 3.7a). Each quadrant window $h \times h$ covers a square part of the corresponding quadrant. The initial point (0, 0) is the center of the LPA estimate for each quadrant. Two two-stage procedures are considered in terms of the *multiple-window estimation*. The first procedure is called a *multiwindow estimate*. First, the adaptive varying sizes for the different windowed estimates are found by the ICI rule independently. In this way we obtain a set of independently calculated adaptive estimates for each x. Second, the obtained estimates are fused (aggregated) into a single, final one. The second procedure is called a *combined-window estimate*. First, the adaptive varying-size windows are found by the ICI rule exactly as is done for the *multiwindow estimate*. Next, we define the *combined-window* and the *combined-window estimate*. The adaptive support of the *combined window* is a union of the supports found for multiple adaptive windowed estimates. The *combined-window estimate* is calculated as the LPA estimates with this adaptive support.

6.2.1 Multiwindow estimate

The adaptive-scale estimates are calculated independently for each window, in particular, for symmetric and quadrant ones and then fused into the final one.

Let $g_{j,h}$ be a kernel of the estimate with the scale h obtained for the jth window. This jth estimate is calculated as

$$\hat{y}_{j,h}(x) = (g_{j,h} \circledast z)(x). \tag{6.41}$$

and has the variance

$$\sigma^2_{\hat{y}_{j,h}} = \sigma^2 \sum_x g^2_{j,h}(x). \tag{6.42}$$

The ICI is used to find the best adaptive-scale h^+ for each j. As a result, we obtain a set of adaptive estimates for each x. These estimates should be fused into a single *final estimate*. This can be done in a number of different ways:

1. *Weighted adaptive mean*
   The weighted mean of these estimates is a reasonable fusing procedure that gives the final estimate in the form

$$\hat{y}(x) = \sum_{j \in I} \lambda_j \hat{y}_{j,h_j^+}(x),$$
$$\lambda_j = \frac{\sigma^{-2}_{\hat{y}_{j,h}}(x, h_j^+)}{\sum_{j \in I} \sigma^{-2}_{\hat{y}_{j,h}}(x, h_j^+)}, \tag{6.43}$$

where $\hat{y}_{j,h_j^+}(x)$ is an estimate with the ICI rule adaptive-scale $h_j^+(x)$, $\sigma_{\hat{y}_{j,h}}(x, h_j^+)$ is the standard deviation of these estimates, and I is a set of the used estimates. In particular, I may consist of the four quadrant windows, $I = \{Q_1, Q_2, Q_3, Q_4\}$, or it may also include the symmetric window Q_0, $I = \{Q_0, Q_1, Q_2, Q_3, Q_4\}$.

The variances $\sigma^2_{\hat{y}_{j,h}}$ in Eq. (6.42) do not depend on x; however, the variances $\sigma^2_{\hat{y}_{j,h}}(x, h_j^+)$ in Eq. (6.43) depend on x because the adaptive scales vary in x. Thus, the weights in Eq. (6.43) are pointwise varying and adaptive.

Assume that the estimates $\hat{y}_{j,h}(x)$ are independent and unbiased with variances $\sigma^2_{\hat{y}_{j,h}}$. Then the fusing algorithm in Eq. (6.43) can be derived as the maximum likelihood estimate provided that the estimates $\hat{y}_{j,h_j^+}(x)$ are Gaussian,

$$\hat{y}_{j,h_j^+}(x) \sim \mathcal{N}(0, \sigma^2_{\hat{y}_{j,h}}(x, h_j^+)).$$

Thus, some optimality properties can be seen behind formula (6.43). This fusing for the multiple ICI estimate originates from [66].

2. *Sample mean*

$$\hat{y}(x) = \frac{1}{\#}\sum_{j\in I}\hat{y}_{j,h_j^+}(x), \tag{6.44}$$

where # is the number of estimates used in the mean.

3. *Median*

$$\hat{y}(x) = median_{j\in I}\,\{\hat{y}_{j,h_j^+}(x)\}. \tag{6.45}$$

4. *Least variance estimate*

$$\begin{aligned} \hat{y}(x) &= \hat{y}_{k,h_k^+}(x), \\ k &= \arg\min_{j\in I}\{\sigma^2_{\hat{y}_{j,h}}(x, h_j^+)\}. \end{aligned} \tag{6.46}$$

The last fusing rule takes as the final estimate the ICI adaptive estimate with the largest window size, that is, the most reliable with the minimum level of random noise.

6.2.2 Combined-window estimate

Let h_j^+ be an adaptive scale found for the jth estimate. Denote the corresponding window function by w_{j,h_j^+} and its support by $supp(w_{j,h_j^+})$. The support of the combined-window estimate is a union of these adaptive supports,

$$U(x) = \cup_{j\in I} supp(w_{j,h_j^+}). \tag{6.47}$$

The LPA technique is used to calculate the corresponding combined-window estimate. The LPA criterion of the form

$$\begin{aligned} J_h(x, C) &= \sum_{X_s\in U(x)} w(x - X_s)[z_s - y(x, X_s)]^2, \\ y(x, v) &= C^T\phi(x - v), \end{aligned} \tag{6.48}$$

is defined on the support $U(x)$ and includes all of the observations X_s belonging to this support.

Minimization of Eq. (6.48) gives the estimate of the vector C,

$$\begin{aligned} \hat{C}(x) &= \Phi^{-1}\sum_{X_s\in U(x)} w(x - X_s)\phi(x - X_s)z_s, \\ \Phi &= \sum_{X_s\in U(x)} w(x - X_s)\phi(x - X_s)\phi^T(x - X_s), \end{aligned} \tag{6.49}$$

with the estimates of the signal and the derivatives defined according to Eq. (2.43) as

$$\hat{y}_h^{(r)}(x) = (-1)^{|r|}\hat{C}^T(x)\phi^{(r)}(0). \tag{6.50}$$

Window functions and LPA orders used to find the adaptive support in Eq. (6.47) and for calculation of the final estimate in Eq. (6.50) can be different. In particular, simple zero-order estimates ($m = 0$) can be used to find the adaptive support $U(x)$ with the final estimate calculated for the higher-order LPA.

In the combined-window estimation, the ICI is exploited as an auxiliary procedure to find a complex-form adaptive neighborhood that is used for the calculation of the final estimate.

6.3 Denoising Experiments

The LPA estimators and the ICI rule for scale selection introduced above define a conceptual framework of adaptive algorithms developed for noise reduction. The major parameters of these estimates are

(a) The threshold Γ;
(b) The power m of the LPA;
(c) The set H defining a grid of scales.

A number of implementations and modifications of these algorithms have been developed. In this section, simulation experiments are presented illustrating the performance of the LPA-ICI algorithms.

6.3.1 Univariate signals

6.3.1.1 *Experiment conditions*

We are given n noisy samples of a univariate signal y:

$$z_s = y_s + \sigma\varepsilon_s, \qquad s = 1, \ldots, n, \tag{6.51}$$

where $y_s = y(X_s)$, $X_s = (s-1)/(n-1)$, $\varepsilon_s \sim \mathcal{N}(0,1)$, $\sigma = 1$, and y is an unknown signal that we would like to recover. The number of observations on the interval $[0, 1]$ is n with a sampling interval $\Delta = 1/(n-1)$. The LPA shift-invariant kernels g_h Eq. (3.4) with $m = 0, 1, 2$ are calculated for the symmetric and nonsymmetric left/right windows.

We use two types of window functions: rectangular and Gaussian (see Table 2.1). The ICI defines the adaptive window sizes for the symmetric and nonsymmetric estimates. The final multiwindow estimate is defined according to the fusing weighted-mean formula (6.43) with the weights given by the variances of the adaptive estimates.

The following set of scales H is used in experiments for the nonsymmetric window function w:

$$H = \{h : h_i = \lfloor 1.45^{i+3} \rfloor, \qquad i = 1, \ldots, 13\}, \tag{6.52}$$

where $\lfloor \cdot \rfloor$ denotes the integer part of the number. The integer parameter $h \in H$ defines the support size of the left/right kernel. The radix 1.45 instead of the standard dyadic 2 allows us to achieve a finer resolution of scale and signal details.

To have more or less equal size supports for the symmetric and nonsymmetric kernels, we use for the symmetric windows the parameter

$$h = \lfloor h/2 \rfloor, \; h \in H. \tag{6.53}$$

The threshold Γ is a design parameter of the ICI rule. We show the results corresponding to the CV-adaptive threshold selected according to Eqs. (6.36–6.38).

As test functions, we exploit the well-known *Blocks*, *Bumps*, *Heavysine*, and *Doppler*, which appear in many publications on signal processing. These signals are used with parameters and in scenarios corresponding to the paper [41], where they have been introduced for experiments with wavelet filters.

The test signals are rescaled to have a fixed SNR, $\sigma_y/\sigma = 7$, where σ_y is the standard deviation of the signal sequence y_s. Thus, the original y is replaced by

$$y \cdot 7/\sigma_y, \sigma_y = \sqrt{\frac{1}{n}\sum_{s=1}^{n}\left(y_s - \frac{1}{n}\sum_{s=1}^{n} y_s\right)^2}. \tag{6.54}$$

The noise standard deviation σ is estimated for the ICI according to Eq. (6.27).

6.3.1.2 Qualitative results

Blocks

Figure 6.3 shows the true and noisy *Blocks* signals and the final multiwindow LPA-ICI estimate. The threshold Γ is CV adaptive and found independently for the symmetric and nonsymmetric estimates. Here we use the LPA estimator of the order $m = 0$ with the rectangular window. The estimate curve demonstrates that the LPA-ICI filter preserves the change-points (jumps) of the signal in an accurate manner with a good smoothing of the piecewise constant parts of the signal.

Figure 6.4 provides details about the algorithm's work. It shows the true signal y and the ICI adaptive scales (window sizes) for the left, right, and symmetric estimates obtained for the noisy signal shown in Fig. 6.3(b). Naturally, the varying adaptive window sizes are noisy. However, overall the ICI gives smaller windows at the discontinuities and larger windows for areas of flat signal behavior.

The different left, right, and symmetric estimators give larger or smaller window sizes for the same jump point depending on which side of the jump is used for estimation. As a result, the adaptive window sizes shown in Fig. 6.4 locate the jumps nearly perfectly and give adaptive varying sizes in a complete agreement with it, as could be done provided that the estimate signal is known in advance.

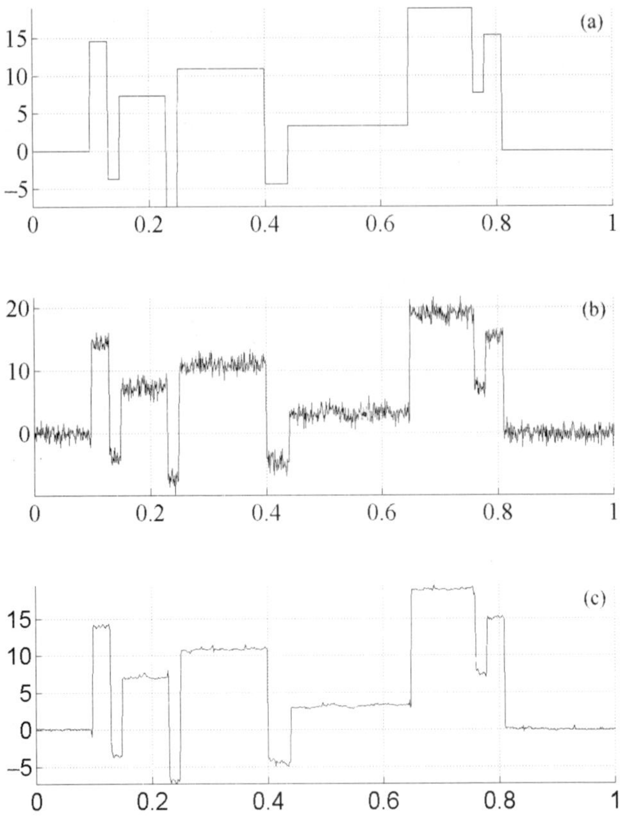

Figure 6.3 *Blocks*: true (a) and noisy (b) signals given by $n = 1024$ samples. (c) The final multiwindow LPA estimate of the order $m = 0$ with the rectangular window w. The criterion value RMSE = 0.2726.

The nonsymmetric window estimates catch the signal jumps from the left and right sides. For each piecewise constant segment of the signal, the left adaptive window size is a growing function of x taking the largest value on the left-hand side of the signal jump point and the smallest one on the right-hand side (Fig. 6.4b). For the right window kernel, the adaptive window sizes demonstrate the opposite behavior (Fig. 6.4c).

The adaptive symmetric window size has a piecewise triangular form. It achieves the largest values at the middle of the segments where the signal is constant. The left and right sides of these triangles are linear decreasing and increasing functions of x having nearly zero values at the signal-jump points (Fig. 6.4d).

Figure 6.5 shows the nearly "ideal" work of the ICI. The adaptive window sizes are shown here with a larger threshold $\Gamma = 2.5$ and for the very fine grid of scales

$$H = \{1, 2, \ldots, 300\}. \tag{6.55}$$

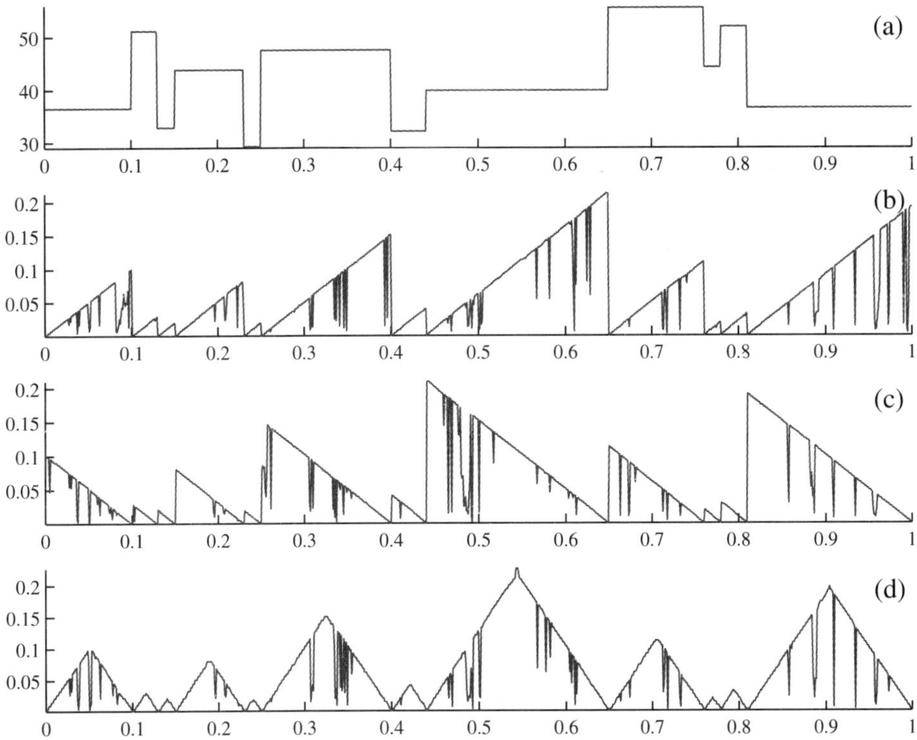

Figure 6.4 *Blocks*: true signal (a), left (b), right (c), and symmetric (d) adaptive window sizes for the noisy signal shown in Fig. 6.3(b).

The larger Γ attenuates the noise effects. The adaptive-scale curves h^+ are piecewise linear with no noise effects, perfectly corresponding to the perception of what the ideal estimation window sizes should look like. However, if we compare the estimates in Fig. 6.3 with the estimates corresponding to the noiseless windows in Fig. 6.5, the comparison is not in favor of the estimates with these noiseless ideal windows. Yes, these latter estimates are smoother, but they are definitely oversmoothed. The edges of the estimate are not as sharp as in Fig. 6.3, and they are slightly shifted.

Comparison based on the numerical criteria also favors the noisier adaptive windows shown in Fig. 6.4.

Larger values of Γ improve the noise reduction properties of the estimator but at the price of detail preservation. The optimal balance on Γ is always at the level when the noise elements are clearly seen in the adaptive window patterns.

We note in Fig. 6.4 that the noise in the adaptive window sizes have specific features. First, the random errors are always directed to the smaller values of the adaptive window sizes. Thus, the ICI with the CV-adaptive Γ overestimates the role of the random components in the observations and gives smaller adaptive

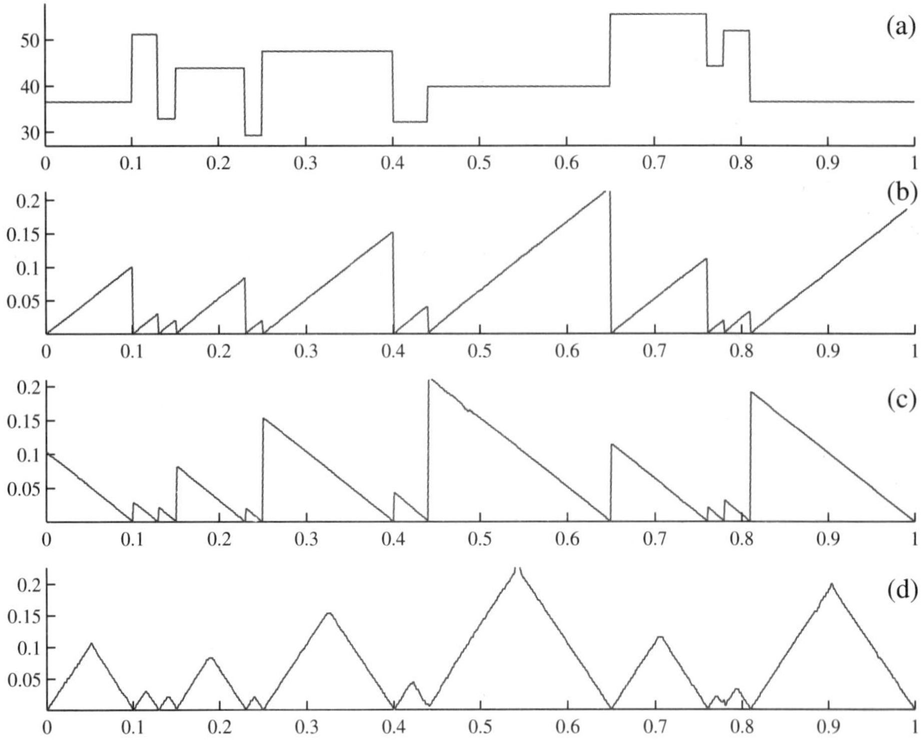

Figure 6.5 *Blocks*: true signal (a), left (b), right (c), and symmetric (d) adaptive window sizes obtained by the ICI with the large threshold Γ = 2.5 and a fine grid of the scales *H* in Eq. (6.55).

window sizes than it should. Second, the noise in the adaptive window sizes has the typical features of random, narrow, but rare spikes when the errors are comparatively large.

The prefiltering of these noisy ICI adaptive $h^+(x)$ is proved to be efficient for improvement of the LPA-ICI adaptive estimates. Our experiments show that the median is a good filter in this case, with the adaptive-scale redefined according to the rule

$$\hat{h}^+(x) = median_{u \in U}\{h^+(x+u)\}. \tag{6.56}$$

When estimating with the prefiltering of the ICI adaptive scales, the sliding median $\hat{h}^+(x)$ in Eq. (6.56) of the values $h^+(x+u)$, $u \in U$, is used instead of $h^+(x)$.

The final estimate in Fig. 6.3 is obtained with this sort of median prefiltering of the ICI-adaptive window sizes. In this estimate, $U = \{-1, 0, 1\}$; thus, the median of three successive values is used as $\hat{h}^+(x)$. The median or weighted median is a relevant tool for the impulse noise appearing as a result of the ICI algorithm.

Bumps

Figure 6.6 shows the true and noisy *Bumps* signals and the final multiwindow LPA-ICI estimate obtained with CV-adaptive thresholds Γ found independently for the symmetric and nonsymmetric windows. Here we use an LPA estimator of order $m = 2$ with Gaussian symmetric and nonsymmetric windows with scales, as in Eqs. (6.52) and (6.53). The final multiwindow estimate demonstrates that the LPA-ICI algorithm is a good filter that preserves the sharp peaks of the signal and removes the noise.

Figure 6.7 provides some details concerning the work that the ICI algorithm does. It shows the true signal y and the ICI adaptive scales for left, right, and symmetric estimates. These varying adaptive scales are noisy. It is seen that overall the ICI rule gives smaller windows at the discontinuities and larger window for areas of flat signal values. Again as it was for *Blocks*, the different (left, right, symmetric)

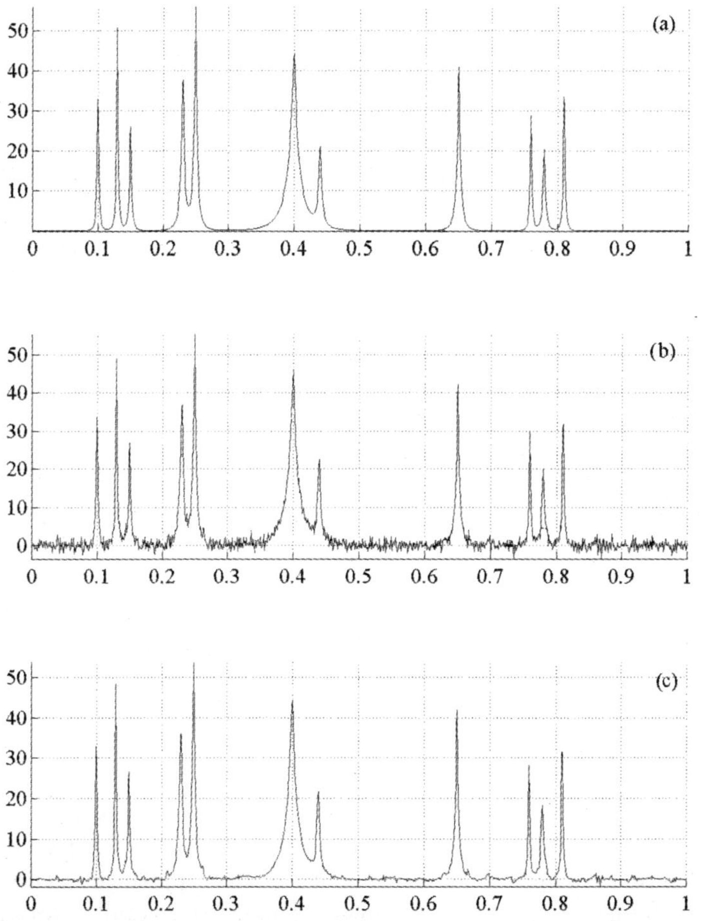

Figure 6.6 *Bumps*: true (a) and noisy (b) signals given by n = 1024 samples. Curve (c): The final multiwindow LPA estimates with m = 2 and the Gaussian window, RMSE = 0.4890.

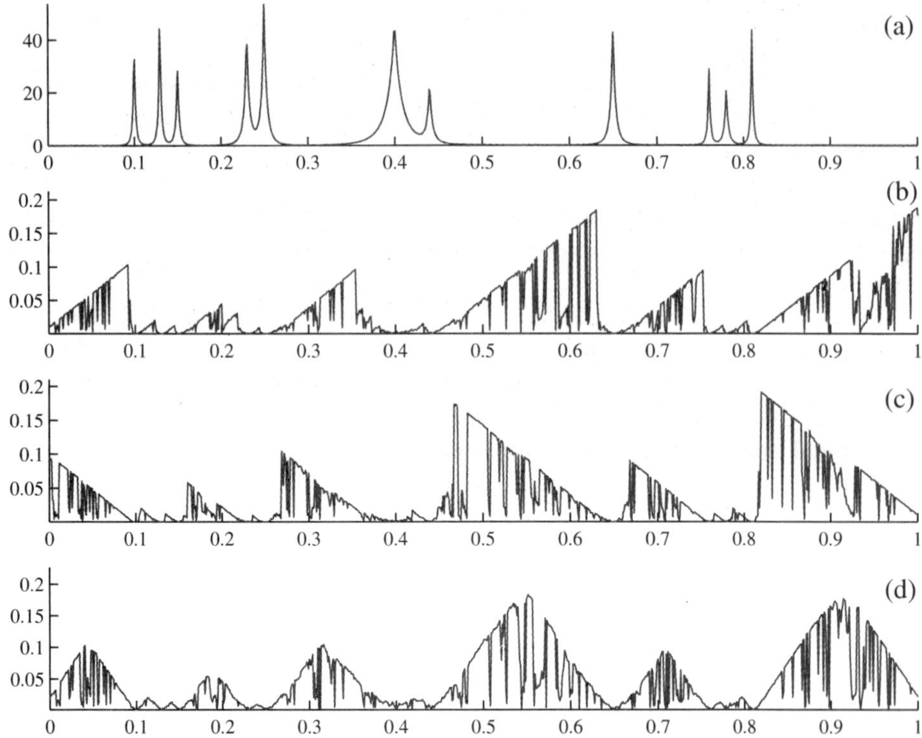

Figure 6.7 *Bumps*: true signal (a), left (b), right (c), and symmetric (d) adaptive window sizes obtained by the ICI with the CV-adaptive thresholds and the scales given in Eq. (6.52).

estimators give larger or smaller window sizes for the same change-point, depending on which side of the change-point neighborhood is used for estimation. The adaptive window sizes shown in Fig. 6.7 locate the change-points nearly perfectly and give reasonable adaptive varying sizes.

Figure 6.8 shows the adaptive window sizes when the ICI is used with a larger value of $\Gamma = 2.5$ and a finer grid of scales Eq. (6.55). It is seen that the noise in the adaptive window sizes is automatically filtered out. However, it does not mean that the quality of filtering is improved. As is discussed for *Blocks*, the estimate becomes oversmoothed with worse values of accuracy criteria.

Heavisine and Doppler

Figures 6.9 and 6.10 show the true and noisy *Heavisine* and *Doppler* signals, and the final multiwindow LPA-ICI estimates obtained with CV-adaptive thresholds Γ. The LPA estimators are of the order $m = 2$ with Gaussian symmetric and nonsymmetric windows. The final multiwindow estimates demonstrate good performance of the adaptive filters tracking the varying curvatures of these signals.

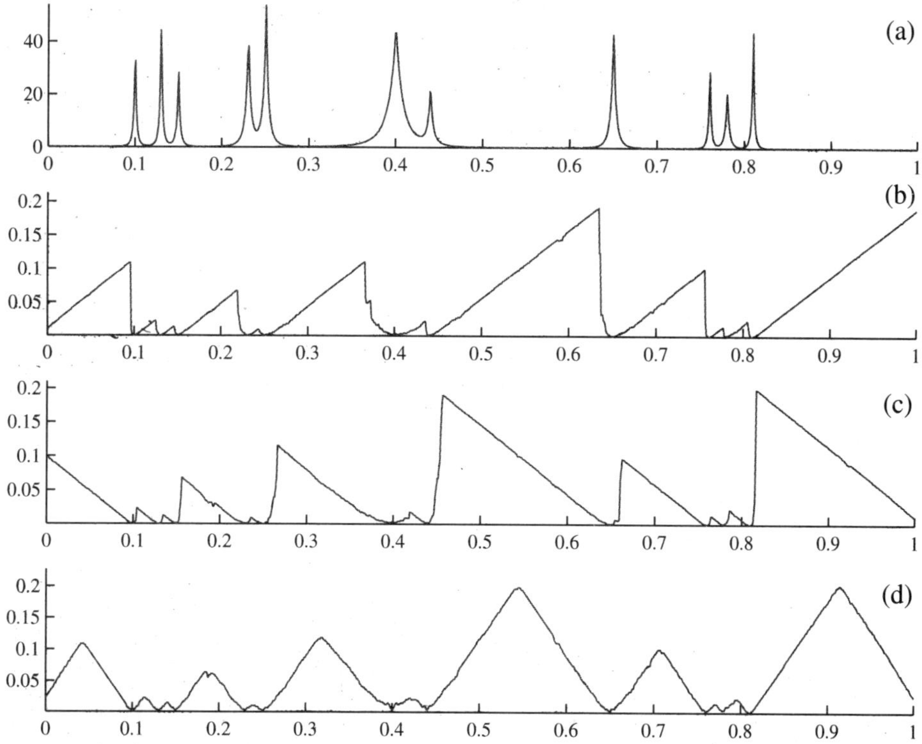

Figure 6.8 *Bumps*: true signal (a), left (b), right (c), and symmetric (d) adaptive window sizes obtained by the ICI with the large threshold $\Gamma = 2.5$ and a fine grid of the scales (Eq. (6.55)).

Figure 6.11 demonstrates the work of the CV-adaptive threshold selection for *Heavisine*. The dashed lines correspond to the CV curves and the solid ones to the MSE as functions of Γ. The main idea is to use a minimization of the CV criteria to find Γ optimal for the MSE. Comparing the curves in Fig. 6.11, we may conclude that the minima of the CV curves give a good indication of the optimal Γ values.

We remark that CV works well for a large sample size n, but this cannot be guaranteed for smaller n and all kinds of signals.

6.3.1.3 Quantitative results

In this section the accuracy analysis of the adaptive estimators is produced. The comparison is done versus the wavelet filters applied to the considered test signals. For comparison, we use the results of the accuracy study presented in [41] for a family of adaptive wavelet filters (*Haar*, *DB4*, *Coif3*, *Symm8*, *WvJS*, *LPJS*, *ThrC3*, *ThrS8*, *VisS8*).

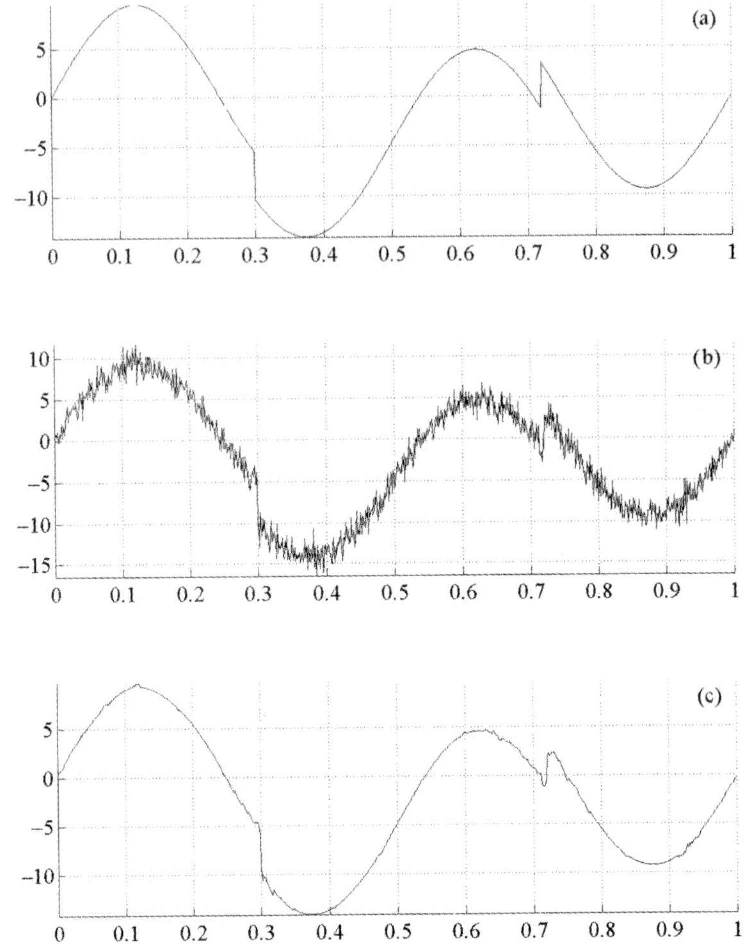

Figure 6.9 *Heavisine*: true (a) and noisy (b) signals given by $n = 1024$ samples. Curve (c): The final multiwindow LPA estimators with $m = 2$ and a Gaussian window, RMSE = 0.1768.

These results are completed by *translation-invariant wavelets* (TI) using *Haar* wavelets for *Blocks* and *Symmlet 8* for other signals with the universal threshold calculated as $\sigma\sqrt{2\log n}$. Monte-Carlo statistical modeling of the LPA-ICI estimates is done with $L = 100$ independent simulation runs for signal lengths $n = 128, 256, 512, 1024$ and with $L = 50$ for $n = 4096$. As the accuracy criterion we use the root-mean-square error (RMSE):

$$\mathrm{RMSE} = \sqrt{\frac{1}{Ln}\sum_{j=1}^{L}\sum_{i=1}^{n}[y(X_i) - \widehat{y}^{[j]}(X_i)]^2}, \tag{6.57}$$

where $[j]$ means the jth simulation run.

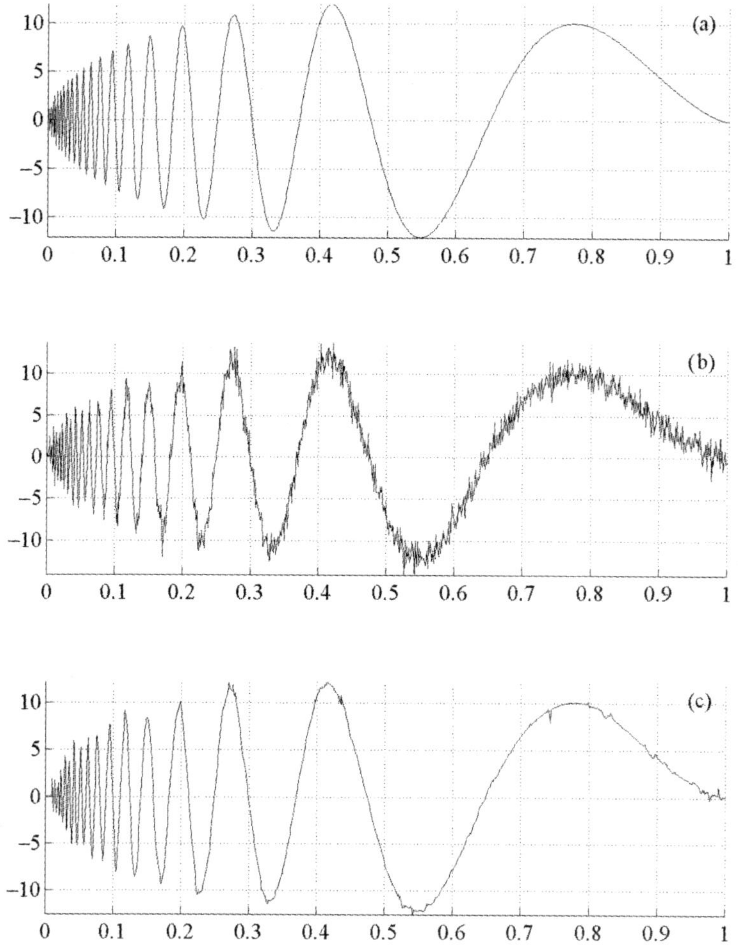

Figure 6.10 *Doppler*: true (a) and noisy (b) signals given by n = 1024 samples. (c) The final multiwindow LPA estimators with m = 2 and Gaussian window, RMSE = 0.3868.

The results are presented in Table 6.1. The columns marked by *I* and *II* correspond to the LPA-ICI algorithms with and without median prefiltering of the adaptive scales, respectively.

The median filter in Eq. (6.56) of the length $N_U = 3$ is used for *Blocks* for all n. For other signals, the length N_U of this filter is different for different n:

1. $N_U = 3$ for $n = 128, 256$;
2. $N_U = 5$ for $n = 512$;
3. $N_U = 7$ for $n = 1024$, 2048 and 4096.

The results are given for LPAs of orders $m = 0, 1, 2$. The threshold Γ is CV-adaptive and takes different values in independent simulation runs. The adaptive

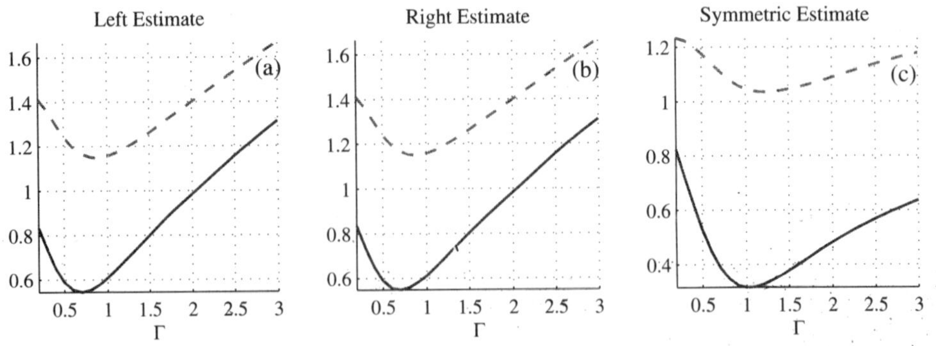

Figure 6.11 The CV criterion (dashed curve) for a *Heavisine* signal, $\Gamma \in \{0.2i : i = 1, \ldots, 15\}$. The CV criterion is calculated according to Eq. (6.36). The solid curve is for the MSE. The minimums of the CV and MSE curves are achieved at nearly equal values of Γ. These optimal Γ are different for symmetric and nonsymmetric windows.

Γ values are selected according to Eqs. (6.36–6.38) using the set of values $G = \{0.2i,\ i = 1, 2, \ldots, 15\}$.

The last column (Wavelets) in Table 6.1 presents results achieved by the wavelets. There are two subcolumns in this wavelet column. The first subcolumn, marked by [41], shows the lower and upper boundary values of the RMSE achieved by the algorithms studied in [41]. The second subcolumn shows the RMSE obtained by the TI wavelets.

Let us comment further on Table 6.1. Comparing the estimates of the orders $m = 0, 1, 2$, we conclude that the zero order $m = 0$ is the best one for the piecewise constant *Blocks*. For the smoother signals, *Bumps*, *Heavysine*, and *Doppler*, better results are obtained by the higher-order approximations.

Comparing column I versus column II, we note that the median prefiltering mainly improves the results. However, this improvement is not large in terms of the average RMSE, and sometimes the prefiltering even makes the criterion values worse. Visually, the prefiltering results in smoother estimates.

The LPA-ICI estimates are more accurate versus the wavelet filters from [41], at least for $n > 128$, which means that the RMSE for the LPA-ICI estimates takes smaller values than the lower bounds of the RMSE corresponding to the wavelet algorithms considered in [41].

A numerical comparison versus the *TI* wavelets is not so definite. For *Blocks*, the zero-order LPA-ICI and *TI* wavelets give very close RMSE values. We may say that the algorithms are equivalent for this piecewise constant signal.

For *Heavisine*, the LPA-ICI of the order $m = 2$ with the prefiltering gives figures that are uniformly better than those obtained by the *TI* wavelet algorithm. For *Bumps*, some of the LPA-ICI estimates give results that are better or slightly worse than those obtained by the *TI* wavelets. *Doppler* is the most difficult signal for the LPA-ICI mainly because of its high-frequency start, where the LPA does not work. This explains the higher values of the RMSE for the LPA-ICI algorithm.

Table 6.1 RMSE for LPA-ICI technique with CV-adaptive threshold Γ.

LPA-ICI	$m = 0$		$m = 1$		$m = 2$		Wavelets	
n	I	II	I	II	I	II	[41]	TI
Blocks								
128	.64	1.15	.78	.84	.97	1.0	.81-1.18	**.53**
256	.49	.61	.63	.60	.76	.88	.68-1.20	**.41**
512	.32	.39	.44	.44	.60	.75	.59-1.12	**.30**
1024	.27	.25	.42	.43	.47	.49	.47-1.03	**.23**
2048	.30	**.16**	.23	.25	.38	.38	.41-.85	.17
4096	.18	.15	.21	.18	.24	.23	.29-.72	**.14**
Bumps								
128	2.25	3.49	2.23	3.56	2.19	3.22	.93-1.35	**.87**
256	1.23	**.70**	1.37	1.83	1.43	1.68	.85-1.41	.79
512	.64	**.61**	.72	.80	.83	1.08	.73-1.38	.62
1024	.56	.55	.54	.54	.51	.53	.63-1.16	**.50**
2048	.48	.44	.45	.43	.40	.39	.51-.96	**.38**
4096	.40	.35	.35	.32	.32	**.30**	.41-.77	**.30**
Heavisine								
128	.72	.60	.65	.58	.57	**.52**	.61-.77	.58
256	.59	.51	.52	.45	.44	**.40**	.49-.62	.45
512	.48	.39	.41	.35	.35	**.31**	.40-.55	.34
1024	.38	.32	.31	.27	.24	**.22**	.32-.46	.26
2048	.30	.26	.24	.21	.18	**.16**	.25-.35	.20
4096	.24	.21	.19	.17	.14	**.13**	.20-.28	.15
Doppler								
128	.86	.82	1.39	1.82	1.80	2.19	.80-.93	**.65**
256	.79	.71	.89	.94	.89	1.14	.73-.98	**.54**
512	.68	.57	.63	.54	.57	.53	.58-.89	**.42**
1024	.59	.48	.52	.44	.44	.39	.49-.72	**.32**
2048	.48	.40	.42	.36	.35	.31	.35-.61	**.23**
4096	.38	.32	.33	.28	.27	.24	.28-.50	**.16**

Overall, the adaptive LPA-ICI demonstrates a good filtering performance visually and numerically.

The CV adaptation for Γ works well with the *Heavisine* signal. Figure 6.11 shows a clear similarity between the curves for the RMSE and CV. It is not always as good for other signals, however. In particular, for small n, the CV curves can be multimode and quite different from the RMSE curve. This results in wrong values selected for the threshold Γ.

In Table 6.2 we show the results obtained from a best possible selection of threshold values. We assume that the true signal is known and Γ is found by direct minimization of the corresponding RMSE. Thus, in Table 6.2 we can see the ideal oracle Γ figures.

Table 6.2 with the oracle Γ highlights interesting aspects of the considered adaptive estimation. First, the oracle threshold improves the accuracy. The comparison versus the wavelets is mainly in favor of the LPA-ICI (excluding *Doppler*).

Table 6.2 RMSE for LPA-ICI technique with oracle-adaptive threshold Γ.

LPA-ICI	Γ	$m=0$		$m=1$		$m=2$		Wavelets	
n	*var*	*I*	*II*	*I*	*II*	*I*	*II*	[41]	*TI*
Blocks									
128	.7–1.3	**.40**	1.0	.60	.64	.79	.80	.81–1.18	.53
256	.8–1.6	**.34**	.51	.47	.48	.66	.67	.68–1.20	.41
512	1.1–1.5	**.25**	.30	.35	.37	.52	.52	.59–1.12	.30
1024	1.3–1.9	.20	**.17**	.27	.26	.38	.38	.47–1.03	.23
2048	1.4–1.7	.18	**.14**	.26	.23	.29	.27	.41–.85	.17
4096	1.5–2.0	.18	.15	.18	.17	.24	.22	.29–.72	**.14**
Bumps									
128	.4–.6	.78	**.77**	.80	.80	.81	.81	.93–1.35	.87
256	.4–.8	**.71**	**.71**	**.71**	.72	**.71**	.73	.85–1.41	.79
512	.5–.9	.63	**.61**	.62	.63	.60	.61	.73–1.38	.62
1024	.6–1.3	.54	.55	.51	.54	**.50**	**.50**	.63–1.16	**.50**
2048	.8–1.5	.48	.44	.41	.43	.41	.40	.51–.96	**.38**
4096	1.1–1.7	.37	.34	.32	.31	.32	**.30**	.41–.77	**.30**
Heavisine									
128	.6–1.2	.72	.60	.63	.58	.55	**.51**	.61–.77	.58
256	.9–1.4	.60	.52	.51	.45	.42	**.39**	.49–.62	.45
512	1–1.7	.49	.40	.40	.35	.33	**.30**	.40–.55	.34
1024	1.2–1.8	.39	.33	.30	.27	.23	**.21**	.32–.46	.26
2048	1.4–2.0	.31	.26	.23	.20	.18	**.15**	.25–.35	.20
4096	1.4–2.2	.24	.21	.17	.15	.13	**.12**	.20–.28	.15
Doppler									
128	.2–.4	.86	.81	.80	.77	.75	.76	.80–.93	**.65**
256	.4–.7	.79	.70	.71	.66	.65	.63	.73–.98	**.54**
512	.6–1.0	.69	.57	.62	.53	.55	.50	.58–.89	**.42**
1024	.8–1.2	.60	.49	.51	.44	.44	.39	.49–.72	**.32**
2048	.9–15	.50	.41	.41	.35	.34	.30	.35–.61	**.23**
4096	1.0–1.8	.39	.33	.32	.28	.26	.23	.28–.50	**.16**

Comparing the results for *Heavisine* we see that the oracle figures in Table 6.2 are very close to the corresponding data in Table 6.1. This confirms the nearly perfect performance of the CV threshold selection for this signal.

For all signals, the ideal oracle results with large $n = 2048, 4096$ are very close to the CV-adaptive ones. Thus, for a larger number of samples, the CV works very well for all signals.

The column Γ in Table 6.2 shows the ranges of the ideal Γ obtained by different Monte-Carlo simulation runs. We can see that these ideal Γ grow with n.

Adjusting the ICI for data of different lengths, we can use the model $\Gamma(n) = \alpha_0 + \alpha_1(\ln n)^{M/(2M+d)}$, $d = 1$, corresponding to formula (14.97) obtained for the convergence rate analysis in Proposition 10. Finding α_0 and α_1 from simulation data, we obtain the threshold as a function of n. The results with the ideal threshold illustrate the potential of the considered adaptive estimates as well as demonstrate the ideal values of Γ.

6.3.2 Binary imaging

6.3.2.1 *Experiment conditions*

Suppose we are given n samples $\{z_s, X_s\}$ of a bivariate 2D signal y defined on a regular square grid with a sample interval $\Delta = 1$. The samples are corrupted by additive noise according to Eq. (6.51).

A signal y is a binary image intensity taking values 0 or 1. As a test signal we use a *Cheese* image (128×128) shown in Fig. 6.12.

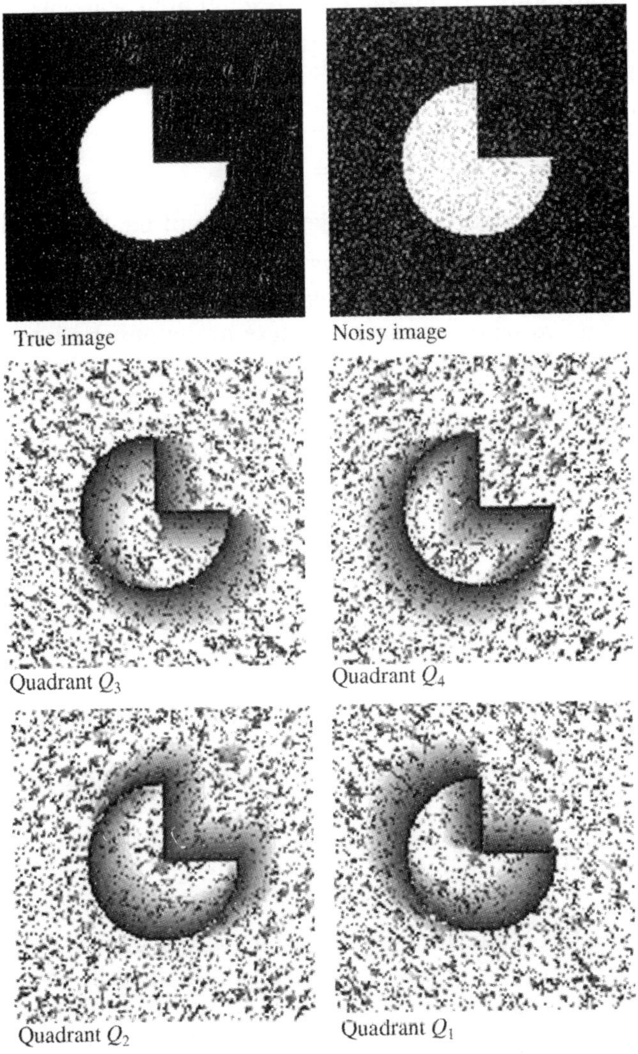

Figure 6.12 *Cheese*: true and noise images; LPA-ICI adaptive window sizes without prefiltering.

The nonsymmetric quadrant LPA filters of order $m = 0$ are used as the estimators. These four quadrants—Q_1, Q_2, Q_3, Q_4—are shown in Fig. 3.7 and give the windows of the LPA as

$$w(x) = \begin{cases} 1, & 0 \le x_1 \le 1, \quad -1 \le x_2 \le 0, \quad \text{quadrant } Q_1, \\ 1, & -1 \le x_1 \le 0, \quad -1 \le x_2 \le 0, \quad \text{quadrant } Q_2, \\ 1, & -1 \le x_1 \le 0, \quad 0 \le x_2 \le 1, \quad \text{quadrant } Q_3, \\ 1, & 0 \le x_1 \le 1, \quad 0 \le x_2 \le 1, \quad \text{quadrant } Q_4. \end{cases} \tag{6.58}$$

The estimators are sample means of observations in these quadrants. The scale h defines the sizes of these quadrant windows with the number of observations in the quadrant equal to h^2. To obtain a fine resolution in scale, we take a set H of the form

$$H = \{1, 2, 3, 4, 5, 7, 9, 11, 13, 15, 17, 19\}. \tag{6.59}$$

The ICI threshold has a fixed value

$$\Gamma = 1.4. \tag{6.60}$$

The ICI gives the adaptive varying scales for each of the four quadrant estimates $h_j^+(x)$ with the adaptive estimates $\hat{y}_{j,h_j^+(x)}(x)$, $j = 1, 2, 3, 4$. Here and in what follows, j refers to the number of the quadrant Q_j.

The ICI adaptive scales are noisy because of the observation noise. We filter these $h_j^+(x)$ by median filters:

$$\hat{h}_j^+(x) = median_{u \in U}\{h_j^+(x+u)\}, \tag{6.61}$$

where U is the square 3×3 window. As a result, the quadrant estimates with the adaptive-scale prefiltering are defined as $\hat{y}_{j,\hat{h}_j^+(x)}(x)$, with the variances of these estimates calculated as

$$\sigma_j^2(x) = \frac{\sigma^2}{[\hat{h}_j^+(x)]^2}, \qquad j = 1, 2, 3, 4, \tag{6.62}$$

where $[h_j^+(x)]^2$ is a number of observations in the quadrant j for the estimate at pixel x.

The final fused multiwindow estimate is defined according to Eqs. (6.43) and (6.62) as

$$\hat{y}(x) = \sum_{j \in I} \lambda_j \hat{y}_{j,\hat{h}_j^+(x)}(x), \ \lambda_j = \frac{[\hat{h}_j^+(x)]^2}{\sum_{j=1}^4 [\hat{h}_j^+(x)]^2}. \tag{6.63}$$

To evaluate the effect of the median prefiltering, we calculate the estimate with and without prefiltering from Eq. (6.61). Without prefiltering, the final fused estimate instead of Eq. (6.63) has the form

$$\hat{y}(x) = \sum_{j \in I} \lambda_j \hat{y}_{j,h_j^+(x)}(x), \qquad \lambda_j = \frac{[h_j^+(x)]^2}{\sum_{j=1}^4 [h_j^+(x)]^2}. \tag{6.64}$$

As criteria we use RMSE:

$$\text{RMSE} = \sqrt{\frac{1}{n}\sum_{s}[y(X_s) - \hat{y}(X_s)]^2}, \tag{6.65}$$

and the improvement in the signal-to-noise ratio (ISNR) is calculated in decibels as:

$$\text{ISNR} = 20\log_{10}(\hat{\sigma}/\text{RMSE}), \tag{6.66}$$

$$\hat{\sigma}^2 = \frac{1}{n}\sum_{s=1}^{n}[z_s - y(X_s)]^2.$$

6.3.2.2 Results

Figure 6.12 shows the noisy image with $\sigma = 0.1$. Black-and-white points on the white-and-black backgrounds in the noisy image show the noise-affected observations.

The other four images in Fig. 6.12 show the four adaptive varying window sizes for the quadrant window estimates. Small and large window sizes are shown by black and white, respectively. The ICI rule mainly selects large adaptive window sizes for flat areas where the true image has constant (0 or 1) values. Smaller windows are obtained in a neighborhood of edges where the image intensity is discontinuous and changes its values from zero to one or from one to zero.

Actually, the ICI measures the distance between the estimate point x to the closest change-point. This measurement is directional because the ICI compares the estimates of the different increasing h. For the first quadrant Q_1 this means that the growing size quadrants are directed (grows) to the southeast and produce a measurement of the distance in this direction. When x is far from the change-point, the adaptive h^+ is large. When x is closer to the change-point, the adaptive h^+ becomes smaller and a gray color appears in the scale imaging. The closer x is to the change-point, the smaller h^+ is and the darker is the color of h^+ in the adaptive-scale imaging. As a result, the h^+ imaging for the quadrant Q_1 has shadows from the northwest side of the image edges. A similar shadowing of the edges can be seen for all quadrant-adaptive scales. Naturally, different quadrants produce shadowing of the edges from different directions.

The quadrants in Fig. 6.12 are arranged according to their location around the estimation point x, as was shown in Fig. 3.7. The different quadrant estimators give larger or smaller values of the window size for the same estimation point x, depending on which side of the edge neighborhood is used for estimation. As a result, the adaptive window sizes shown in Fig. 6.12 delineate the image intensity quite accurately and from different directions.

The noise effects are clearly seen in these adaptive h^+ as spread-out black, isolated points. These are small erroneous window sizes given by the ICI on the base of the noisy data. As was discussed for the univariate case, the ICI produces this sort of isolated random error, decreasing the adaptive-scale values with respect to their ideal values. The median filtering is a good tool for this sort of random error.

Figure 6.13 demonstrates the results. We show the quadrant estimates $\hat{y}_{j,\hat{h}_j^+(x)}(x)$ obtained with the prefiltering of the adaptive window sizes and the final fused multiwindow estimates. These fused estimates are shown in two versions. The first one is obtained according to Eq. (6.63) from the shown quadrant estimates and corresponds to fusing the prefiltered adaptive window estimates. The second one is obtained assuming that the adaptive window sizes are used for quadrant

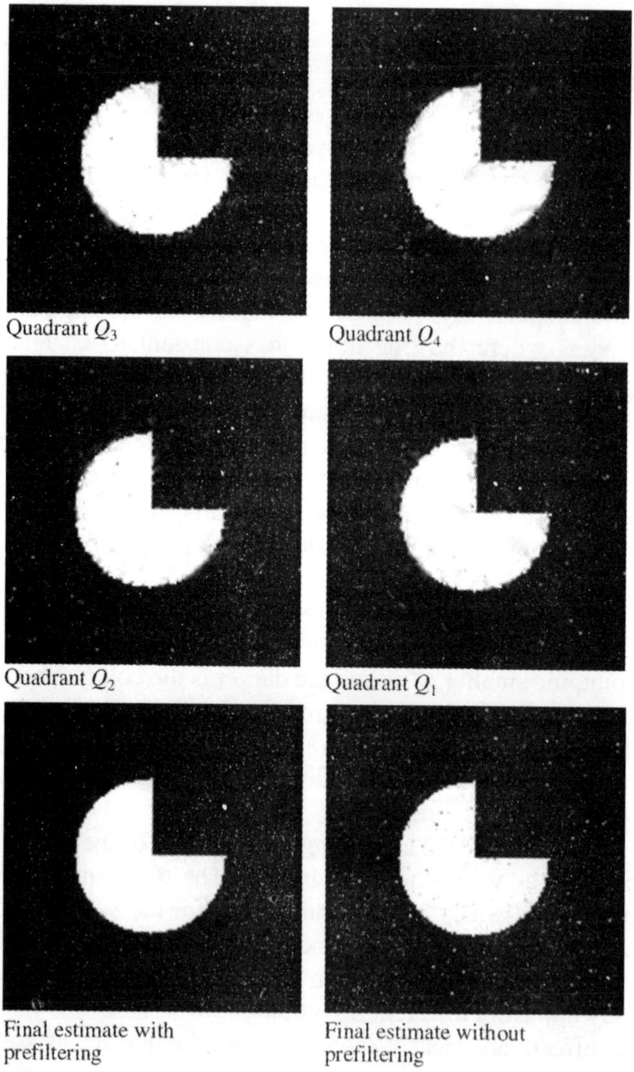

Figure 6.13 *Cheese*: LPA-ICI estimates for quadrants and final multiple-window estimates with and without prefiltering.

Table 6.3 RMSE and ISNR for LPA-ICI estimates of *Cheese* image.

	RMSE prefiltering	RMSE no prefiltering	ISNR prefiltering	ISNR no prefiltering
Q_1	0.0609	0.0856	10.42	7.47
Q_2	0.0598	0.0864	10.59	7.38
Q_3	0.0624	0.0856	10.20	7.47
Q_4	0.0607	0.0866	10.44	7.36
Final	**0.0169**	0.0559	**21.53**	11.17

estimation without median prefiltering and the final estimate is calculated according to Eq. (6.64).

The final estimate with the prefiltered adaptive window sizes is nearly perfect. The edges are well preserved and the noise is suppressed completely.

Some noise effects are clearly seen in the final image obtained without prefiltering. Comparing the final fused estimates with the quadrant ones, we can also see that the fusing produced in the multiwindow estimates really improves the estimation.

The numerical results shown in Table 6.3 demonstrate a number of interesting effects. The performance of the four quadrant estimates for each scenario is more or less equivalent. In all cases, the multiwindow fusing gives results that are dramatically better for the final estimate compared to the quadrant estimates. It gives 0.0169 for RMSE with prefiltering (versus about 0.06 for quadrants) and 0.0559 without prefiltering (versus about 0.086 for quadrants). For ISNR it gives 21.53 dB with prefiltering (versus about 10.4 dB for quadrants) and 11.17 dB without prefiltering (versus about 7.4 dB for quadrants). The source of this improvement is obvious if we recall that the quadrant estimates actually work differently depending on which side of the edge the estimate quadrant is located.

Another important point is the prefiltering. We note that prefiltering improves the criteria values in an essential manner. For the final estimates one gets RMSE = 0.0169 with prefiltering versus RMSE = 0.0559 without prefiltering. For the final estimates, we have ISNR = 21.53 dB with prefiltering versus ISNR = 11.17 dB without prefiltering. Thus, prefiltering improves the final estimate by about 10 dB. Note that the improvement as a result of prefiltering for the quadrant estimates is only about 3 dB.

Chapter 7

Anisotropic LPA

7.1 Directional Signal Processing

7.1.1 Recent developments

Intuitively, an isotropy means that a signal $y(x)$ has more or less the same behavior in all directions in a neighborhood of x. A constant $y(x) = const$ is an ideal isotropic signal with no variation in the signal at all. A more complex linear model $y(x) = \sum_{j=1}^{d} a_j x_j$ might have different derivatives with respect to the directions/variables x_j. However, this is again a trivial case because this signal is invariant in the space of the parameters a_j.

In a nonparametric context, anisotropy mainly denotes unpredictable nonparametric changes in the behavior of a signal. Examples are a univariate y with jumps (*Blocks* in Fig. 6.3) or the piecewise constant/linear two-variable signals shown in Figs. 2.10 and 2.11. These signals have different values for x on different sides of the jump points in *Blocks* or change-lines in Figs. 2.10 and 2.11.

Points, lines, edges, and textures are present in all images. They are locally defined by position, orientation, size, or scale. Often being of small size, these specific features encode a great proportion of the information contained in images. An image intensity y is a typical example of an anisotropic nonparametric signal. In what follows, we mainly consider anisotropic methods in the context of 2D imaging.

The idea that biological visual systems analyze images by orientation and scale dates back to the 1960s, when two points of view became quite popular. The first was that visual systems compute directional derivatives along space and time, and the second was that biological visual systems implement the concept of "matched filtering." In the computational vision and image-recognition literature, corresponding approaches of analyzing images along multiple orientations appeared at the beginning of the 1970s with works in early vision systems. Respectively, methods have been proposed for differentiation and matching typical visual elements in images.

7.1.1.1 Directional differentiation

Oriented differentiating filters are used in many vision and image processing tasks, such as edge detection, texture and motion analysis. Recent developments have shown the impressive success of methods for directional image/multivariate signal processing.

A kernel estimate can be directionally sensitive only when the corresponding kernel is nonsymmetric. Let us consider an integral smoothing estimate with a Gaussian kernel Eq. (4.81),

$$g(x) = \frac{1}{2\pi} \exp\left(-\frac{1}{2}||x||^2\right). \tag{7.1}$$

This kernel is circularly symmetric and nondirectional because it has a constant value for any x_1 and x_2 on a circle $||x||^2 = const.$

The differentiating kernels for the derivatives $\partial_{x_1}^{r_1}$ and $\partial_{x_2}^{r_2}$ on x_1 and x_2 can be obtained from Eq. (7.1) using the potential differentiating kernels,

$$g^{(r)}(x) = \frac{1}{2\pi} \partial_{x_1}^{r_1} \partial_{x_2}^{r_2} \exp\left(-\frac{1}{2}||x||^2\right).$$

For the first- and second-order derivatives this gives

$$\begin{aligned} &g^{(1,0)}(x) = -x_1 g(x), && g^{(0,1)}(x) = -x_2 g(x), \\ &g^{(1,1)}(x) = x_1 x_2 g(x), && g^{(2,0)}(x) = (x_1^2 - 1)g(x), \\ &g^{(0,2)}(x) = (x_2^2 - 1)g(x), && \end{aligned} \tag{7.2}$$

where $g(x)$ is defined by Eq. (7.1). These differentiating kernels are definitely nonsymmetric and directional, as are the corresponding derivatives. For example, the kernel $g^{(1,0)}(x)$ is symmetric (even) on x_2 and nonsymmetric (odd) on x_1. This makes a difference between the arguments x_1 and x_2 as well as between the positive and negative directions along x_1. The convolution of this kernel with a signal gives the corresponding derivative estimate.

A differentiation of kernels is exploited in directional signal processing to obtain nonsymmetric directional kernels. However, not all differentiating operators have nonsymmetric kernels. For instance, the Laplacian operator in Eq. (4.120) has a symmetrical kernel and cannot be used as a directional estimator.

How do we design a kernel for a desired direction? The concept of the *directional derivative* gives a good initial idea. By definition, the directional derivative for the direction defined by the angle θ is

$$\partial_\theta y(x) = \lim_{\rho \to 0} \frac{1}{\rho}[y(x_1 + \rho\cos\theta, x_2 + \rho\sin\theta) - y(x_1, x_2)]. \tag{7.3}$$

First, a deviation of the signal in the direction θ is calculated as a difference and then divided by the distance $\rho \geq 0$ between the arguments defining the initial and final points of this difference. The derivative is a limit of this ratio as $\rho \to 0$.

If y is continuous and differentiable, then elementary calculations give a well-known result,

$$\partial_\theta y(x) = \cos\theta \cdot \partial_{x_1} y(x) + \sin\theta \cdot \partial_{x_2} y(x). \tag{7.4}$$

To find the derivative for any given θ, it suffices to estimate the derivatives on x_1 and x_2 and calculate the directional derivative according to Eq. (7.4). This statement can be quite misleading in image processing, for instance, for the edge detection problem. Indeed, it looks like the derivatives $\partial_{x_1} y(x)$ and $\partial_{x_2} y(x)$ carry all information concerning the behavior of $y(x)$ in any direction in a neighborhood of a given x. However, the image intensity can be nearly constant, say in the directions of the axes x_1 and x_2, and have a sharp ridge along the diagonal between these axes. Then, of course, formula (7.4) gives a wrong result, saying that the derivative in the diagonal direction is also nearly zero.

What is actually wrong in this speculation? Without a doubt, formula (7.4) is accurate. However, it holds only for a continuous x and provided that y is a continuous and differentiable signal.

In imaging, the intensity $y(x)$ is a discrete signal defined on a discrete grid of x values. For this discrete signal the derivatives are replaced by the finite differences

$$\partial_{x_1} y(x) \simeq \frac{[y(x_1+\rho, x_2) - y(x_1, x_2)]}{\rho}, \quad \partial_{x_2} y(x) \simeq \frac{[y(x_1, x_2+\rho) - y(x_1, x_2)]}{\rho}.$$

Inserting these finite differences into Eq. (7.4), we have

$$\partial_\theta y(x) \simeq \frac{\{\cos\theta \cdot [y(x_1+\rho, x_2) - y(x_1, x_2)] + \sin\theta \cdot [y(x_1, x_2+\rho) - y(x_1, x_2)]\}}{\rho}. \tag{7.5}$$

According to this formula, the derivative is calculated on the basis of three values of the signal: $y(x_1, x_2)$, $y(x_1+\rho, x_2)$, and $y(x_1, x_2+\rho)$ for any θ, including $\theta = \pi/4$, when

$$\partial_{\pi/4} y(x) \simeq \frac{1}{\sqrt{2}\rho}\big\{[y(x_1+\rho, x_2) - y(x_1, x_2)] + [y(x_1, x_2+\rho) - y(x_1, x_2)]\big\}. \tag{7.6}$$

However, for this diagonal direction the corresponding finite difference can be defined in a different and more natural way as

$$\partial_{\pi/4} y(x) \simeq \frac{1}{\sqrt{2}\rho}[y(x_1+\rho, x_2+\rho) - y(x_1, x_2)]. \tag{7.7}$$

For a discrete signal, the value $y(x_1+\rho, x_2+\rho)$ in Eq. (7.7) can be quite different from $y(x_1+\rho, x_2)$ and $y(x_1, x_2+\rho)$ used in Eq. (7.6). Thus, the discrete estimates in Eqs. (7.6) and (7.7) can be very different.

Continuous models of the discrete image intensity are commonly used for image processing. However, the estimate of the directional sensitivity in the simple form of Eq. (7.4) is too rough to be useful for those applications in imaging where the sharpness and details are of top priority.

The only reliable way to obtain true directional information for discrete image intensity is to calculate variations of $y(x)$ in the desirable direction θ and estimate the directional derivative by finite differences similar to Eq. (7.7). In more general terms, this means that the direction-sensitive image analysis should be based on narrow support kernels well concentrated in desirable directions.

7.1.1.2 Steerable filters

One of the popular approaches to directional filtering that originated in [58] is based on extracting orientation information using the rotation of directional differentiating kernels. It is proposed to synthesize an arbitrarily oriented differentiating filter as a linear combination of a fixed set of differentiating filters for some basic directions.

Let $g_{\theta_i}(x)$ be a basis kernel for differentiation in the direction θ_i. Define the kernel $g_\theta(x)$ of the derivative in the desirable direction θ as a linear combination of $g_{\theta_i}(x)$ with the coefficients $k_i(\theta)$:

$$g_\theta(x) = \sum_{i=1}^{K} k_i(\theta) g_{\theta_i}(x). \tag{7.8}$$

The estimate of the directional derivative is calculated as a convolution,

$$\hat{y}_\theta(x) = (g_\theta \circledast y)(x). \tag{7.9}$$

Formula (7.4) illustrates the idea behind the steering approach because in this case we have in terms of Eq. (7.8): $K = 2, \theta_1 = 0$, and $\theta_2 = \pi/2$, $g_{\theta_1}(x) = g^{(1,0)}(x)$, $g_{\theta_2}(x) = g^{(0,1)}(x)$, $k_1(\theta) = \cos\theta$, $k_2(\theta) = \sin\theta$.

The representation in Eq. (7.8) is a design formula that defines the kernel for the direction θ. It is based on a fixed number K of directional kernels and allows one to calculate the derivatives for any direction.

The estimates can be calculated as

$$\begin{aligned} \hat{y}_\theta(x) &= \sum_{i=1}^{K} k_i(\theta) \hat{y}_{\theta_i}(x), \\ \hat{y}_{\theta_i}(x) &= (g_{\theta_i} \circledast y)(x). \end{aligned} \tag{7.10}$$

The kernel $g_\theta(x)$ is called *steerable* if there are weights $k_i(\theta)$ in Eq. (7.8) such that the differentiating kernel for any direction θ can be represented as a linear combination of its values for $\theta = \theta_i$, that is,

$$g_{\theta_i}(x) = g_\theta(x)|_{\theta=\theta_i}. \tag{7.11}$$

This is a linear interpolation formula defining the kernel function for any θ through its values at $\theta = \theta_i$, $i = 1, \ldots, K$.

Equation (7.11) is called a steering condition [58]. It claims that the derivative for an arbitrary direction $\hat{y}_\theta(x)$ can be calculated as a linear combination of the derivative for the basic directions $\hat{y}_{\theta_i}(x)$:

$$\hat{y}_\theta(x) = \sum_{i=1}^{K} k_i(\theta)\hat{y}_{\theta_i}(x).$$

The kernel steerability in Eq. (7.11) has a place provided that the kernel $g_\theta(x)$ is sufficiently smooth in the space domain x.

The Gaussian differentiating kernels in Eq. (7.2) are commonly used for calculating derivatives. One of the steerability properties presented in [58] is summarized as follows.

Let y be given in polar coordinates as a Fourier series:

$$y(r, \phi) = \sum_{n=-N}^{N} a_n(r)e^{in\phi}, \tag{7.12}$$

where $r = \sqrt{x_1^2 + x_2^2}$, $\phi = \arg(x, y)$, $a_n(r)$ are real functions.

If the weights $k_i(\theta)$ in Eq. (7.11) are solutions of the equation

$$\begin{pmatrix} 1 \\ \exp(i\theta) \\ \cdots \\ \exp(i\theta N) \end{pmatrix} = \begin{bmatrix} 1 & 1 & \cdots & 1 \\ \exp(i\theta_1) & \exp(i\theta_2) & \cdots & \exp(i\theta_K) \\ \cdots & \cdots & \cdots & \cdots \\ \exp(i\theta_1 N) & \exp(i\theta_2 N) & \cdots & \exp(i\theta_K N) \end{bmatrix} \begin{pmatrix} k_1(\theta) \\ k_2(\theta) \\ \cdots \\ k_K(\theta) \end{pmatrix}, \tag{7.13}$$

then the steering condition in Eq. (7.10) holds.

One is interested in a minimum number of summands in Eq. (7.10). It is shown that this minimum K is defined by a number of nonzero $a_n(r)$ in Eq. (7.12) [58]. The steering property has a place only for an angularly sufficiently smooth y allowing a small number of terms in the Fourier series of Eq. (7.12).

If the assumption in Eq. (7.12) is not fulfilled, Eq. (7.10) can be treated only as an approximation. For a nonsmooth y, the N in Eq. (7.12) and K in Eq. (7.10) become large for a good accuracy approximation. In this case, the usefulness of the representation in Eq. (7.11) is quite questionable because the basic idea is to exploit derivative estimates for a few directions to restore the derivatives for all directions.

The steerability is not restricted to the first derivative only. It can be extended to derivatives of arbitrary orders. There are a number of papers reporting successful applications of these interpolation techniques mainly for edge detection problems (e.g., [178], [201], [202]).

We wish to emphasize that there is a principal difference between the steerability of the kernel and the steerability of the estimate. These are two very different concepts. The first problem is related to the directional kernel design and requires a smoothness for the kernel only, regardless of the properties of y. The second problem is mainly restricted by the properties of the signal, because it assumes that the true directional derivative y_θ for any θ can be represented as a linear combination of the derivatives y_{θ_i} for the fixed directions θ_i.

Thus, the steerable filter approach requires not only smoothness of the kernel but definite smoothness of the analyzed signal. The latter requirement is definitely not fulfilled at discontinuous points and change-points of images.

7.1.1.3 Directional wavelets

It is recognized that the usual wavelets exhibit an inherent limit performance in image denoising. We quote from [210]: "This limit arises from the well-known and frequently depicted fact that the 2D wavelet transform of images exhibits large wavelet coefficients even at fine scales, all along the important edges of the images, so that in a map of the large coefficients one sees the edges of images repeated at scale after scale. With so many coefficients to estimate, denoising faces certain difficulties ... which lead to relatively high mean-square error."

The continuous *ridgelet transform* was introduced as a multidirectional alternative to the usual wavelets. In two dimensions it allows the representation of an arbitrary signal $y(x_1, x_2)$ by superpositions of elements (*ridgelets*) of the form

$$\psi_{h,\theta,b}(x) = h^{-1/2}\psi[(x_1 \cos\theta + x_2 \sin\theta - b)/h]. \tag{7.14}$$

Here ψ is a univariate wavelet function, $h > 0$ is a scale parameter, θ is an orientation angle, and b is a scalar location parameter.

The ridgelet is constant along the ridge line $x_1 \cos\theta + x_2 \sin\theta - b = 0$, and it is a wavelet along the orthogonal direction. For small h, the ridgelet is quite sharp and localized along the ridge line. For larger h, the ridgelet can be a smooth function with smooth waves. Thus, different scale ridgelets can be used for directional approximations in smooth and straight-line edge areas. As an example we show in Fig. 7.1 a ridgelet based on a Daubechies wavelet ψ of length 10.

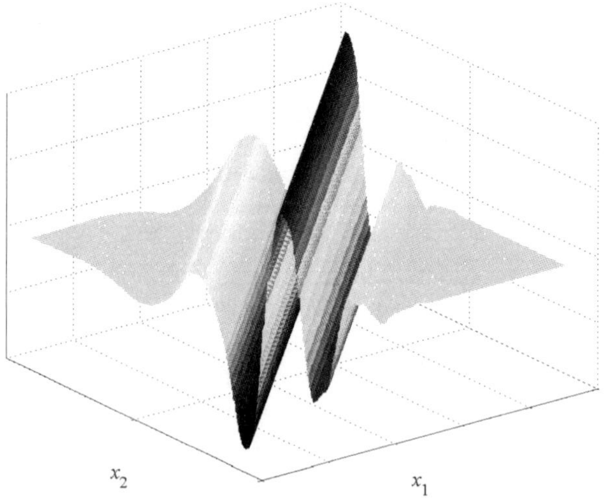

Figure 7.1 The ridgelet in Eq. (7.14) is depicted as a function of x. For smaller h, the waves of the ridgelet become sharper, and for larger h they become smoother.

The wavelet ψ used in the ridgelet in Eq. (7.14) fulfills the condition

$$\int_{\mathbb{R}^2} \psi(x)dx = 0.$$

According to the first condition in Eq. (4.31), this means that it is a differentiating 2D kernel. Varying the parameters θ, b, and h, we obtain ridgelets in all directions, all locations, and all scales.

The ridgelets in the definition in Eq. (7.14) are nonlocal because they are infinitely elongated along the direction θ. In image processing, the length of the ridgelet is fixed. Note that the scale h works as a width parameter of the ridgelet. The sharp linewise ridgelets are not steerable functions, because they cannot be presented in the form of Eq. (7.11) with a small K.

The *curvelet transform* was proposed as a further development of the idea of the directional approximation. Based on the representation in Eq. (7.14), it uses ridgelets of varying lengths and widths. It is assumed that the length is dyadic and the length and width are linked as $width \sim length^2$. If $width = h$, this means that

$$length = \sqrt{h}. \tag{7.15}$$

For small h, the width-length anisotropy increases and the curvelets form a set of needle-shaped elements.

Conceptually, ridgelets and curvelets employ a 1D analysis function (wavelet ψ) to generate a 2D elementary function. These transforms have been proposed in the continuous domain and discretized for sampled data. They require a rotation operation, making the construction simple in the continuous domain but causing problems in the discrete domain because it assumes some interpolation to the regular image pixel grid and this interpolation is not trivial.

7.1.1.4 Anisotropic kernel estimates

In the framework of denoising anisotropic signals, a novel nonparametric regression procedure was recently constructed with a local adaptive selection of the multi-dimensional scale parameter [118, 119]. The method is a generalization of Lepski's approach for multiscale multivariate estimates and roughly consists of choosing the "coarsest" scales such that the estimated bias is negligible. The continuous integral estimates are introduced and studied for smoothing in d-dimensional space, $x \in \mathbb{R}^d$, with the vector scale parameter $h \in \mathbb{R}^d$.

The proposed d-dimensional kernel $g_h(x)$ is a product of the univariate kernels:

$$g_h(x) = \prod_{i=1}^{d} k(x_i/h_i)/h_i, \tag{7.16}$$

$$\int_{\mathbb{R}^1} k(x_1)dx_1 = 1, \qquad \int_{\mathbb{R}^1} k(x_1)x_1^j dx_1 = 0, \qquad j = 1, \ldots, m, \tag{7.17}$$

where the scale parameters h_i are exploited for independent scaling of the arguments x_i. As is clear from Eq. (7.17), this univariate kernel $k(x_1)$ has the order m. Symmetric finite support $[-1, 1]$ is assumed for this kernel,

$$k(x_1) = 0 \qquad \text{if } x_1 \notin [-1, 1]. \tag{7.18}$$

The scales h_i are discretized into dyadic ones, $h_i = 2^{-j_i}$, and then used in tests: biasedness versus the standard deviation of the estimates. A multidimensional version of the adaptive-scale selection algorithm in Eq. (6.2) is proposed.

The works [118] and [119] concentrate on the theoretical aspects of the problem and show, in particular, that the adaptive method attains exactly the minimax convergence rate and adapts to an unknown degree of anisotropy for a large class of anisotropic Besov spaces.

7.1.1.5 *Adaptive weight smoothing*

A new development in spatially adaptive nonparametric filtering based on the *structural adaptation* and the *adaptive weights smoothing* (AWS) is proposed in [180] and [182]. The idea is to define the estimates in the form of a weighted mean,

$$\hat{y}(x) = \frac{1}{\sum_s w_h(x, X_s)} \sum_s w_h(x, X_s) z_s, \tag{7.19}$$

where the size and form of a neighborhood of x is defined by the adaptive weights $w_h(x, X_s)$.

These weights are observation-dependent and calculated according to a recursive procedure where the previous estimates of y are used for weight updating and refining the size and shape of the pointwise neighborhood of x. The estimator has the form of the Nadaraya-Watson algorithm with the weight w depending on the estimated signal.

In [181] and [183], the AWS algorithm is extended to a broad class of nonparametric local likelihood models including Poisson and a general class of exponential models.

These new algorithms are based on the local likelihood for estimation and the Kullbac-Leibler statistics for measuring a distance between the local models. Reported numerical experiments and applications demonstrate a good performance of the algorithms.

7.1.2 Directional LPA-ICI approach

7.1.2.1 *Idea*

Consider a ball (disk) of radius h defining a spherical neighborhood of x,

$$\mathcal{B}_h = \{u : ||x - u|| \leq h\}.$$

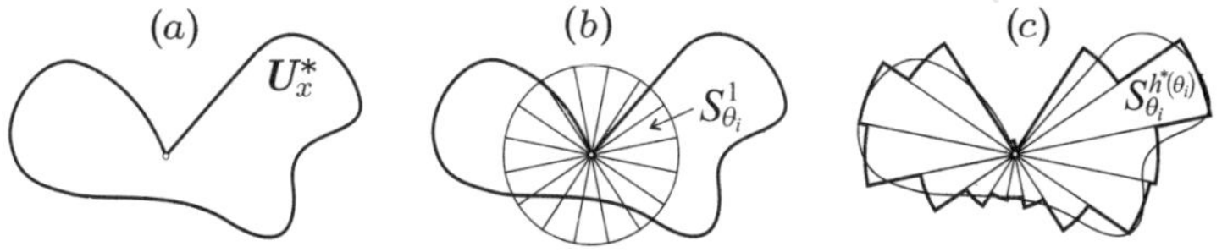

Figure 7.2 A neighborhood of the estimation point x: (a) the best estimation set U_x^*, (b) the unit ball segmentation, and (c) sectorial approximation of U_x^*.

Introduce a sectorial partition of this ball with K nonoverlapping sectors having x as a common vertex. Exploit the LPA for the design of the estimate kernels using these sectors as the supports and the vertex x as the LPA center. In this way, we obtain K estimators with K nonoverlapping supports covering the ball $\mathcal{B}_h$.

Use the ICI to find the adaptive estimation scale for each sector independently. Thus, we arrive at K adaptive-scale estimators that have different adaptive scales/sizes. The union of supports of these sectors is no longer a ball $\mathcal{B}_h$ but a star-shaped set that can have quite a sophisticated configuration (see Fig. 7.2(c)).

Because it is found from the ICI rule, this star-shaped set is a pointwise adaptive neighborhood of x used for estimation. The final estimate can be produced in one of the ways discussed in Sec. 6.2, as a multiwindow or a combined-window estimate. In this way we obtain a highly anisotropic adaptive-scale estimator with the angular resolution depending on a number of sectors K. In what follows we explain this idea in detail and link the directional LPA with an estimation support optimization and an approximation of this optimal solution by a star-shaped set.

7.1.2.2 Support optimization

It is convenient to start from a *continuous observation* model

$$z(x) = y(x) + \sigma\varepsilon(x), \tag{7.20}$$

where $x \in \mathbb{R}^2$ is continuous and $\varepsilon(x)$ is a standard Gaussian white noise, $\varepsilon \sim \mathcal{N}(0, 1)$.

Consider a shift-invariant kernel estimator

$$\hat{y}(x) = \int_U g(u)z(x-u)du,$$

where the kernel g is given and has a finite support $U \subset \mathbb{R}^2$. The bias and variance of this estimate are of the forms, respectively,

$$m_{\hat{y}_h}(x) = y(x) - \int_U g(u)y(x-u)du,$$

$$\sigma^2_{\hat{y}_h} = \sigma^2 \int_U g^2(u)du,$$

and the pointwise mean-square risk is

$$l_{\hat{y}}(x) = E\{(y(x) - \hat{y}(x))^2\} = m_{\hat{y}_h}^2(x) + \sigma_{\hat{y}_h}^2.$$

Let there be an optimal support $U^*(x)$ minimizing this risk,

$$U^*(x) = \arg\min_U l_{\hat{y}}(x). \tag{7.21}$$

This optimal $U^*(x)$ gives both the size and shape of the neighborhood of x, enabling the best possible mean-square error of estimation (see Fig. 7.2(a)).

7.1.2.3 Scale optimization

Optimization as in Eq. (7.21) can be quite a problem because in the LPA the kernel g itself depends on U. To make the idea more practical, a specification of the support is required. Let us use the kernels equipped with the usual scale parameter h:

$$\begin{aligned}\hat{y}(x) &= \int_{\mathbb{R}^2} g(u)z(x - hu)du = \frac{1}{h^2}\int_{\mathbb{R}^2} g(u/h)z(x - u)du \\ &= \int_{\mathbb{R}^2} g_h(u)z(x - u)du,\end{aligned}$$

where $g_h(x) = g(x/h)/h^2$.

The support of this estimate is equal to the support of the kernel $g(x/h)$:

$$U_h = supp[g(x/h)].$$

It is assumed that the support of g is finite and the scaling with h means a smaller support U_h for a smaller h.

It is natural to make the assumption that the support sets U_h are nested,

$$U_{h_1} \subset U_{h_2} \text{ for } h_2 > h_1. \tag{7.22}$$

Then, for the sequence of $h \in H$ in Eq. (6.1) we have

$$U_{h_1} \subset U_{h_2} \subset \cdots \subset U_{h_J}. \tag{7.23}$$

We restrict the class of kernels and say that the support of g is star-shaped. Then all supports in the sequence of Eq. (7.23) are star-shaped.

How general is this assumption? By definition, the set is *star-shaped* if a segment linking the origin with any point of the set belongs to this set.

The disk and square are star-shaped supports, as is the tricky set shown in Fig. 7.3(a). The set shown in Fig. 7.3(b) is not star-shaped. In the image given in Fig. 7.3(a), the segments $A0$, $B0$, and $C0$, as well as all others linking the origin with any point of the set, belong to the set. This set is star-shaped. In the image given in Fig. 7.3(b), not all points of the segments $C0$ belong to the set. This set is not star-shaped.

If the support of the window w is star-shaped, the LPA kernel g_h is also star-shaped for all h. The supports of these kernels are nested according to Eq. (7.23). If

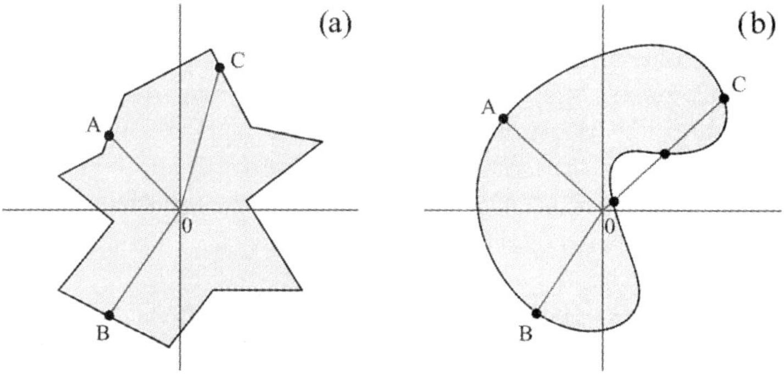

Figure 7.3 (a) Star-shaped and (b) non-star-shaped sets.

the kernel g is fixed, then the scale parameter h is used to control the size of the star-shaped neighborhood of the estimation point x. In this way we arrive at the kernel estimates studied in Chapters 2 and 3 with the ideal scales (supports) found as

$$h^*(x) = \arg\min_h l_{\hat{y}_h}(x),$$

where the risk $l_{\hat{y}_h}(x)$ is considered as a function of the scale h. For every fixed star-shaped kernel g, the support's scale $h^*(x)$ defines the size of the star-shaped neighborhood of x as optimal for estimation.

7.1.2.4 Star-shaped approximation of support

Let us look at what types of optimal solutions can appear in the context of general support optimization in Eq. (7.21). Consider a binary image y defined as equal to 1 inside the unit disk and equal to 0 outside of this disk, as shown in Fig. 7.4 (top row, left). It is obvious that the best possible estimate is the sample mean of the observations belonging to the disk for $\|x\| \leq 1$ (the white area in this image) and the

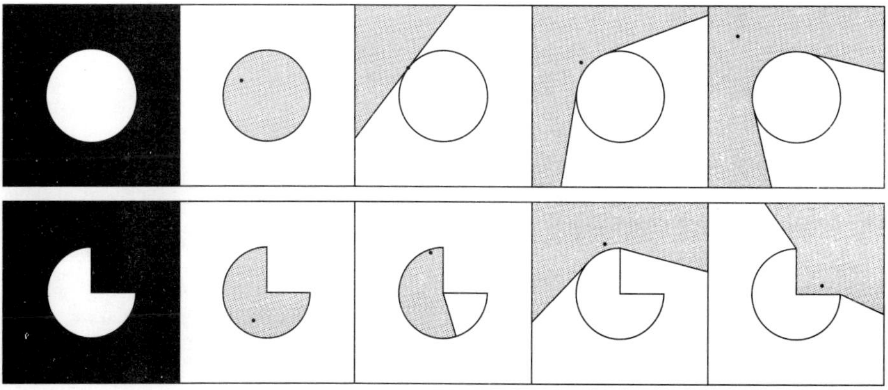

Figure 7.4 Examples of the ideal star-shaped supports of the sample mean estimates for binary images.

sample mean of the observations outside of the disc for $\|x\| > 1$ (the black area in this image). This solution gives the optimal support $U^*(x)$ provided that the problem in Eq. (7.21) is unconstrained, that is, we are looking for an arbitrary-shaped $U^*(x)$.

Now, consider the constrained optimization. Assume that the solution to Eq. (7.21) we seek is a star-shaped set. These optimal sets $U^*(x)$ for different x are shown by the gray color in the other images in the top row of Fig. 7.4, where the estimation point x is denoted by a black point. For all x inside the unit disc, the optimal sets are the same and are equal to this unit disc.

However, when x is outside of the unit disc, the solution is totally different. For $\|x\| = 1$, the optimal star-shaped set is a half-plane support tangent to the circle at x. For $\|x\| > 1$, the optimal star-shaped set is the union of all half-plane supports tangent to the circle $||x|| = 1$ and containing x.

For the binary *Cheese* image (bottom row, left in Fig. 7.4) the optimal unconstrained sets $U^*(x)$ are shown by black and white in the first image, respectively, for x belonging to *Cheese* and outside of *Cheese*. The optimal estimate is the sample mean calculated over the observations in the corresponding areas. These optimal unconstrained U^* are similar to what we have for the disc image.

Some constrained optimal star-shaped sets for the *Cheese* image are shown in other images of Fig. 7.4 (bottom row). Because the support of *Cheese* is a nonconvex set, the corresponding optimal sets U^* can have an exotic shape. Even when x is inside of *Cheese*, the optimal U^* is only a part of the *Cheese* support.

Thus, regardless of the simplicity of these examples, each point x has its own different optimal star-shaped neighborhood $U^*(x)$. These neighborhoods fit perfectly to the image edges and typically are nonconvex. Roughly speaking, the class of star-shaped sets implies a *line-of-sight* model in which the estimation point "cannot see" beyond obstacles. Therefore $U^*(x)$ never contains points that are not "directly visible" from x. This leads to an important aspect that we exploit in what follows.

Let $U^*(x)$ be an optimal neighborhood of x as shown in Fig. 7.2(a). Introduce a sectorial (angular) partition of the unit disc centered at the estimation point x (Fig. 7.2(b)), where the angle θ_i names the corresponding sector, and use the lengths of these sectors to approximate $U^*(x)$ (Fig. 7.2(c)). If the set $U^*(x)$ is a star-shaped body as in Fig. 7.2(a), then, provided one has a sufficient number of angular sectors, a good approximation of $U^*(x)$ can be achieved.

Figure 7.2 illustrates the basic idea of the directional scale optimization approach introduced in this section.

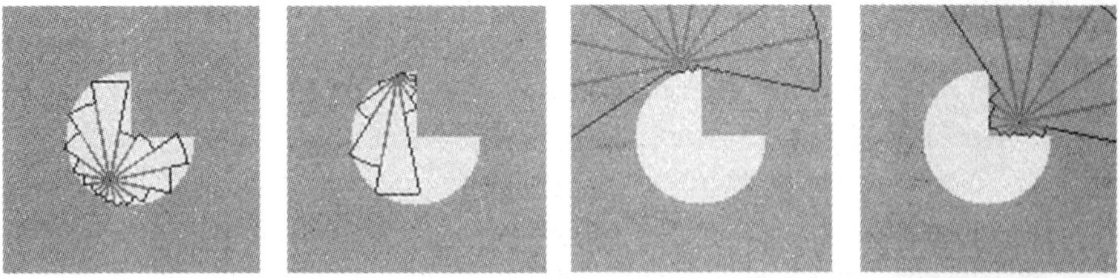

Figure 7.5 *Cheese*: adaptive sectorial neighborhoods obtained by the LPA-ICI algorithm.

Figure 7.6 *Cameraman*: Adaptive sectorial neighborhoods obtained by LPA-ICI algorithm.

Figures 7.5 and 7.6 show estimation neighborhoods obtained using the LPA-ICI algorithm for adaptive varying-scale selections for the sectors. The images are noisy and an LPA estimator of the power $m = 0$ is used for the sectors.

For the binary *Cheese* image, a similarity between these sectorial adaptive scales and the optimal scales shown in Fig. 7.4 is evident. Thus, the adaptive-scale selection gives a good approximation of this star-shaped optimal support.

The adaptive sectorial scales shown in Fig. 7.6 for a textured *Cameraman* image appear to be reasonable approximations of the star-shaped optimal neighborhoods that could be used for estimation.

In all cases the adaptive shape neighborhoods are sensitive to image edges and have shapes/sizes giving the pointwise segmentation of the image into maximum-sized subareas of flat behavior of the corresponding intensities.

7.1.2.5 Directional scale optimization

It is assumed in what follows that the optimal estimation neighborhoods are star-shaped or well approximated by star-shaped sets. This assumption satisfies two important ideas. First, for many practical problems, the use of line-of-sight modeling to support the LPA corresponds not only to common sense but prior information about the problem at hand. Second, this assumption fits well to the LPA technique modified to the sectorial supports.

Let $g_{h,\theta_i}(x)$ be a kernel with a support belonging to the sector θ_i for all h. Thus, $g_{h,\theta_i}(x) = 0$ for x out of the sector θ_i. The sectorial estimate is defined as

$$\hat{y}_{h,\theta_i}(x) = \int_{\mathbb{R}^2} g_{h,\theta_i}(u) z(x-u) du.$$

Thus, the estimate $\hat{y}_{h,\theta_i}(x)$ uses the observations from the sector θ_i only with the scale h responsible for the size of the support in this sector.

The kernel $g_{h,\theta_i}(u)$ satisfies some vanishing moment conditions, for example,

$$\int_{\mathbb{R}^2} g_{h,\theta_i}(u)dt = 1, \qquad \int_{\mathbb{R}^2} g_{h,\theta_i}(u) u_1^{t_1} u_2^{t_2} du = 0, \qquad 0 < t_1 + t_2 \leq m,$$

where m denotes the order of the kernel.

According to the idea of the multiwindow estimation, the final estimate is defined as the linear adaptive combination in Eq. (6.43) of the corresponding partial directional estimates:

$$\hat{y}(x) = \sum_{i=1}^{K} \lambda(\theta_i) \int_{\mathbb{R}^2} g_{h,\theta_i}(u) z(x-u) du \tag{7.24}$$

with the weights $\lambda(\theta_i)$ such that

$$\lambda(\theta_i) \geq 0, \sum_{i=1}^{K} \lambda(\theta_i) = 1.$$

The selection of these weights is a special topic. However, the weights defined as inverse variances of the directional estimates have proved to be efficient. The properties of the directional estimates for the general multivariate case are a subject for further analysis.

Formula (7.24) makes clear the basic idea of directional algorithms.

We introduce the directional estimates $\hat{y}_{h,\theta_i}(x)$; optimize the scale parameter for each direction; and fuse these directional estimates into the final one $\hat{y}(x)$ using the weights $\lambda(\theta)$.

Two key points are important here. First, we are able to find good approximations of the optimal unknown estimation supports that can be of a complex form. Second, for each direction, we are looking for a scalar parameter h, and, overall, the complexity of the problem is linear with respect to the number K of the used sectors. The importance of these properties is clear if we look at the approach developed in [118] and [119] as one possible alternative.

Figure 7.7(a) shows the support of the kernel in Eqs. (7.16–7.18) for $d = 2$ with different scales h_1 and h_2 for the variables. If $h_1 = h_2$, this support is square. For $h_1 \gg h_2$ or $h_1 \ll h_2$, we have supports narrowing along axes x_1 and x_2, respectively.

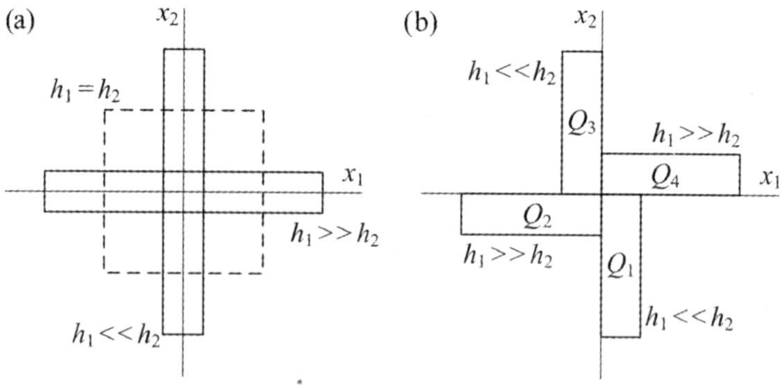

Figure 7.7 Examples of the directional kernel supports: (a) The supports of the separable kernel in Eqs. (7.16–7.18) for $d = 2$ with different scales h_1 and h_2. If $h_1 = h_2$, this support is square. For $h_1 \gg h_2$ or $h_1 \ll h_2$, we have supports narrowing along axes x_1 and x_2, respectively. (b) Quadrant supports transformed in the narrowed ones directed along the main axes by the corresponding selection of the scales h_1 and h_2.

There are no such h_1 and h_2 that are able to give a support narrowed and oriented in the diagonal direction. This means that the multivariate scaling used in [118] and [119] is restricted to dealing with the anisotropy of $y(x)$ along axes x_1 and x_2 only. The kernels in Eq. (7.24) can be steered to any θ and in this way are able to deal with anisotropic variations of $y(x)$ in any direction.

Another point concerns a multidimensional version of the adaptive-scale selection in general. First, the kernel estimates considered as functions of the vector scale parameter h as it is in Eqs. (7.16–7.18) cannot be uniquely ordered according to the values of their variances, which are basically exploited in the ICI algorithm. This happens mainly because the variance is defined by the number of points in the support, and this number is proportional to the product $h_1 h_2$. We have the same value for this product and the same variance of the estimate for multiple combinations of h_1 and h_2.

Further, the multidimensional scale selection requires a multiple comparison of the estimates for different values of the elements of the vector scale parameter h. It assumes a multidimensional optimization on $h \in \mathbb{R}^d$.

In the sectorial approach, this multidimensional search is replaced by 1D ones for each of the directions.

What follows in this chapter mainly concerns some pragmatic aspects of the developed approach and includes:

- Design of discrete directional kernels $g_{h,\theta}$;
- Application of the ICI rule for the adaptive varying-scale selection for each direction;
- Fusing the directional estimates in the final one using the data-driven weights $\lambda(\theta)$;
- Accuracy and convergence analyses highlighting the theoretical background of the directional algorithms.

7.2 Directional LPA

The standard LPA technique can be modified to meet specific requirements of directional estimation. The modifications mainly concern the window function w, which becomes nonsymmetric directional, and the orders of the LPA, which are independently specified for each of the variables. We name this modified method the *directional* LPA.

If the LPA window w is circularly symmetric, then the kernels g_h derived according to Eq. (2.56) are nondirectional. Note that the LPA-differentiating kernels are mainly nonsymmetric, because they calculate the derivatives on x_1 and x_2, and in this way they make a difference between these axes as well as between the positive and negative directions along these axes.

We wish to obtain both nonsymmetric well-oriented smoothing and differentiating kernels. The conventional approaches to directional filtering are mainly concentrated on differentiating estimates.

The first step to the directional estimation was been done in Chapter 2, where we introduced nonsymmetric left/right windows. From the very beginning of this book we demonstrated that on many occasions the nonsymmetry of the estimator is a key tool for the efficient estimation of signals with discontinuities and singularities.

For bivariate imaging we replace the symmetric window Q_0 by the square quadrant windows $Q_i, i = 1, 2, 3, 4$ (see Fig. 3.7) actually directed to the southeast, southwest, northwest, and northeast. However, these kernels are too wide to be well oriented in 2D space, and the next idea is to exploit narrowed oriented windows. These types of kernels demonstrate a remarkable improvement in the denoising and deblurring of images as compared with the estimates based on symmetric and quadrant windows [108, 111]. In this chapter, the directional kernel design is presented in a refined and modified form.

The technique provides a number of valuable benefits:

- Unlike many other methods, which start from the continuous domain and then discretize the continuous kernels, this technique works directly in the discrete space domain.
- The designed kernels are multivariate, nonseparable, and anisotropic with arbitrary orientations, widths, lengths, and orders.
- The required smoothness of the kernels along and across the main directions is enabled by the corresponding vanishing moment conditions.
- The kernel support can be flexibly shaped to any desirable geometry to capture geometrical structure and pictorial information. A special design can actually be done for special-form objects.
- The smoothing and differentiating directional kernels can be designed in terms of this universal approach.

By default, the standard LPA considered in Chapter 2 assumes that the orthogonal horizontal and vertical axes x_1 and x_2 are treated as main directions with the best approximation of the signal along these main axes. In addition, the support of the designed estimators is assumed to be square symmetric Q_0 or a nonsymmetric square quadrant.

We modify this technique in a number of ways. First, we use different-order approximations for x_1 and x_2. Second, we exploit different scales h_1 and h_2 and, respectively, different sizes of the support on the variables x_1 and x_2. Third, we rotate the kernel support to direct it to the desired direction θ. In what follows we restrict our consideration to the 2D case where the directionality can be presented in a very transparent way. A generalization to any dimension is straightforward.

7.2.1 Polynomials

Let us introduce a set of 2D polynomials in a form similar to the one in Eq. (2.45)

$$(-1)^{k_1+k_2} x_1^{k_1} x_2^{k_2} / k_1! k_2!, \qquad k_1 = 0, \ldots, m_1, \qquad k_2 = 0, \ldots, m_2, \qquad (7.25$$

where the powers m_1 and m_2 can be selected independently.

These polynomials are linearly independent with a total number of polynomials

$$M = (m_1 + 1)(m_2 + 1),$$

and the maximum power of the polynomials equal to $m_1 + m_2$.

Note that this system is very different from the one used in Chapter 2, where the scalar m defines the maximum power of the 2D polynomials as $0 \le k_1 + k_2 \le m$.

In Eq. (7.25) we use m as a multi-index

$$m = (m_1, m_2),$$

with elements showing the powers on the corresponding variables. The LPA estimates are defined according to the model in Eq. (2.18), where the vector $\phi(x)$ is composed of M polynomials, as in Eq. (7.25). We also use a multi-index parameter $h = (h_1, h_2)$ to define independent scales on the variables x_1 and x_2.

In this multi-index notation, formula (2.18) is applicable and defines the LPA model in the form

$$\begin{aligned} y_h(x, v) &= C^T \phi_h(x - v),\ \phi_h(x) = \phi(x/h), \\ \phi(x) &= [\phi_1(x), \phi_2(x), \ldots, \phi_M(x)]^T, \\ C \in \mathbb{R}^M,&\quad x, v \in \mathbb{R}^2,\quad x/h = (x_1/h_1, x_2/h_2). \end{aligned} \tag{7.26}$$

For example, for $m = (2, 1)$, $M = (m_1 + 1)(m_2 + 1) = 6$ and the set of polynomials $\phi_k(x)$, $k = 1, \ldots, 6$, can be given in the form

$$\phi_1 = 1,\ \phi_2 = -x_1,\ \phi_3 = -x_2,\ \phi_4 = x_1^2/2,\ \phi_5 = -x_1 x_2,\ \phi_6 = -x_1^2 x_2/2. \tag{7.27}$$

7.2.2 Directional windowing

Setting different scales h_1, h_2 for the arguments x_1 and x_2 is a first step to directional estimation because it makes it possible to obtain the kernels elongated along one of the axes x_1 or x_2. Figure 7.7(a) shows the windows $w_h(x)$ obtained from the uniform symmetric square $w(x) = 1, |x_1| \le 1$ and $|x_2| \le 1$, with $h_1 \gg h_2$ and $h_1 \ll h_2$. For $h_1 \gg h_2$ the window $w_h(x)$ is a narrow rectangle elongated along the axis x_1, and for $h_1 \ll h_2$ the window $w_h(x)$ is a rectangle elongated along the axis x_2. Different scales transform a window symmetric for $h_1 = h_2$ into nonsymmetric ones.

In a similar way, the use of $h_1 \gg h_2$ or $h_1 \ll h_2$ for the quadrant windows Q_i, $i = 1, 2, 3, 4$, produces the narrowed rectangular windows shown in Fig. 7.7(b). We treat the narrowness and orientation of the support of w along some direction as the key elements of the directional LPA. The weights of the windows and the polynomials used in the LPA provide a further specification of the kernel, while the orientation and directionality are mainly defined by the geometry of the window support and by the scale parameters.

We call the window *directional* if it is narrowed and oriented along one of the radial directions. The windows in Fig. 7.7 are directional because they point to the positive or negative directions of the axes x_1 and x_2.

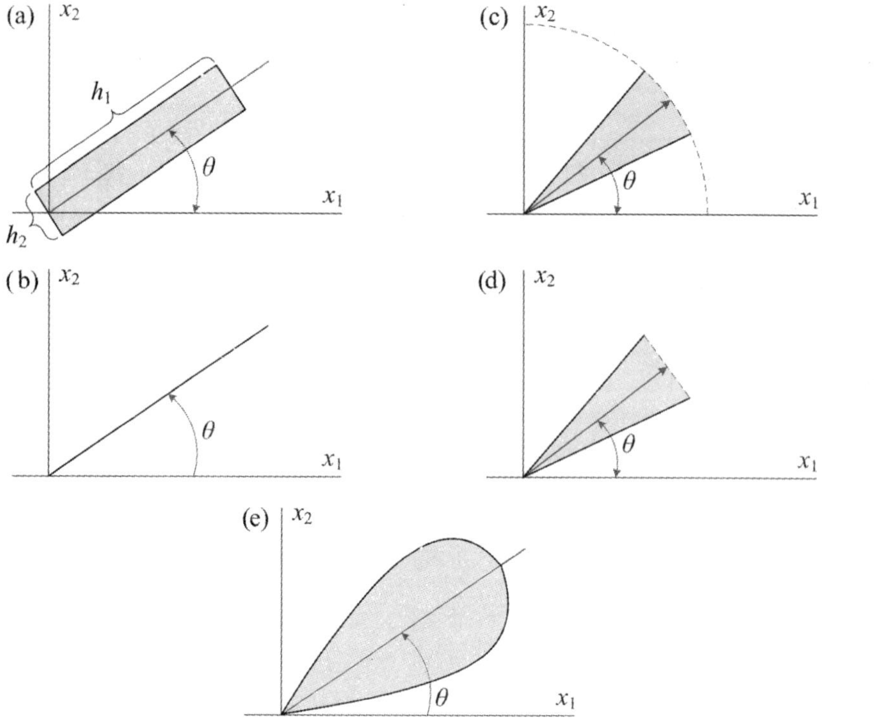

Figure 7.8 Directional supports: (a) rectangular, (b) sectorial, (c) line-wise, (d) triangular, and (e) beam.

The window w has to satisfy the standard LPA assumptions in Eq. (2.34). The reference point always corresponds to the maximum value of w. The directionality is guaranteed by the shape of the window support. It can be defined geometrically by using a finite support w narrowed in a given direction. Figure 7.8 illustrates the possible directional supports of w, which vary from rectangular (a) and straight-line (b) to sectors (c, d) and beam supports (e).

The only requirement restricting the LPA procedure and the window selection is that the matrix Φ_h in Eq. (2.44) should be nonsingular. The windows corresponding to this condition are called *admissible*. We also may talk about admissible scales as too-small scales can result in a singular matrix Φ_h. When we use the zero-order model $m = (0, 0)$, every nonempty window w_h is admissible.

In our design we start from a *basic directional kernel* directed along the axis x_1. The parameters h_1 and h_2 define, respectively, the length and the width of this kernel.

The kernel directed to the direction θ is obtained by a special rotation of the basic kernel.

7.2.3 Rotation

The design of the directional kernel consists of two successive steps. The first one is a rotation of the support of the basic window w_h. The finite discrete grid support is found for the new direction as a result of this step. The second step is the LPA design of the kernel for this new support. We call the kernels obtained in this way "*directional LPA kernels.*"

Let us introduce the "*basic*" and "*rotated*" coordinate systems. The rotation of the original (x_1, x_2) coordinates on the angle θ is defined by the geometrical transform

$$\begin{aligned} u_1 &= x_1 \cos\theta + x_2 \sin\theta, \\ u_2 &= -x_1 \sin\theta + x_2 \cos\theta, \end{aligned} \tag{7.28}$$

where (u_1, u_2) are new rotated coordinates depending on x_1, x_2 and a fixed angle θ. The inverse of Eq. (7.28) is

$$\begin{aligned} x_1 &= u_1 \cos\theta - u_2 \sin\theta, \\ x_2 &= u_1 \sin\theta + u_2 \cos\theta. \end{aligned} \tag{7.29}$$

In matrix notation we have for Eqs. (7.28–7.29)

$$\begin{gathered} u = U(\theta)x,\; x = U^T(\theta)u, \\ U(\theta) = \begin{pmatrix} \cos\theta & \sin\theta \\ -\sin\theta & \cos\theta \end{pmatrix}. \end{gathered} \tag{7.30}$$

When a dependence of u on θ is emphasized we show θ as one of the arguments of u, that is,

$$u(x, \theta) = U(\theta)x. \tag{7.31}$$

The rounding-off effects of rotation on the discrete grid are illustrated in Fig. 7.9. The nodes of the window after rotation are found as an intersection of the basic regular rectangular grid and the rotated support. In Fig. 7.9 the grid of the initial basic support is a straight line of six nodes. After rotation, the new support consists of eight nodes that form an irregular grid in the rotated coordinates u. Depending

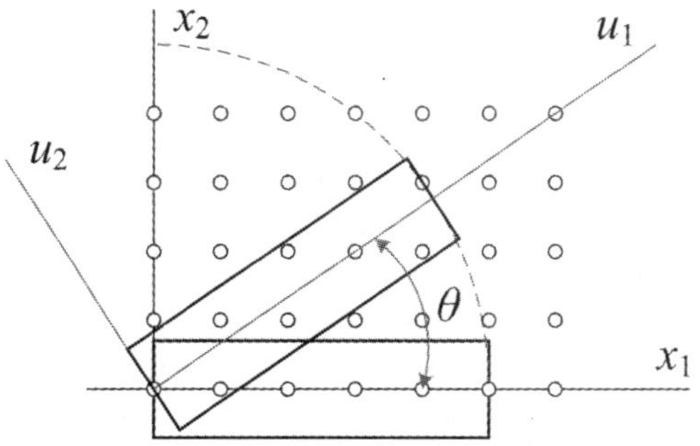

Figure 7.9 A rotation of the rectangular support shows that the number of pixels as well as their location inside the support depends on the rotation angle θ. For the LPA design of the kernels, the new variables u_1 and u_2 are used as arguments for the calculation of the window function and the polynomials.

on the angle θ and the scale h, the rotated support can be very different in its configuration as well as in the number of nodes.

It can be seen in Fig. 7.9 that there are three nodes just outside of the rotated rectangular support that are near its boundary. These nodes can be included in the support. With these three extra nodes included, the discrete rotated support would have a more complete form closer to the full rectangular one. This sort of morphological repairing of the shape of the rotated support is quite useful.

Consider the new variable u as an argument of the LPA with $u = 0$ as a reference point of the approximation. Replacing $x = (x_1, x_2)$ by $u = (u_1, u_2)$ in the formulas for the LPA, we may apply all formulas of Chapters 2 and 3 for the LPA kernel design in the new variables. This straightforward technique defines the directional LPA design.

The directional LPA kernels obtained in this way are *accurate polynomials* for any discrete support obtained after rotation. This means that the properties of the rotated kernels in the new variable u are identical to the properties of the usual basic kernels in the original variable x. The rotated kernels are reproducing with respect to the polynomial signals in the variable u and satisfy the corresponding vanishing moment conditions also in the variables u. This allows us to preserve the desired smoothness of the discrete rotated kernel (orders of the polynomial approximation and vanishing moment conditions) for any θ because the LPA is applicable for any particular set of support nodes included in the rotated support.

To emphasize the special features of this procedure we now discuss briefly the rotation based on interpolation techniques as one of the possible alternative techniques for the directional kernel design. Let a basic directional kernel, say $g_{h,\theta_0}(x)$, be LPA-designed for the *main direction* along axis x_1 with $\theta_0 = 0$. This *basic kernel* is rotated geometrically to the desired direction θ. Both kernels, the basic and the rotated, $g_{h,\theta_0}(x)$ and $g_{h,\theta}(x)$, are discretely defined on the regular x grid. In geometric rotation the nodes of the basic kernel do not coincide with the nodes of the regular grid. Various interpolation methods (linear, bilinear, nearest neighborhood, etc.) can be applied to calculate the values of the rotated kernel for the new grid argument values. Let us call the kernel obtained in this way a *directional interpolation LPA kernel*.

For high-accuracy image reconstruction we need the estimators with good sensitivity with respect to singularities. The window/kernel should have narrow and sharp supports, which enable good angular resolution of the estimator. Usually, the windows/kernels are discontinuous at the edges of the support. The ICI compares the estimates of the different support sizes. The discontinuity of the windows at the narrow, sharp apex makes the estimates very sensitive to any variations in the signal when the kernel with increasing h crosses image change-points.

However, the directional interpolation can be efficient for smooth kernels only. For nonsmooth, sharp-support LPA kernels, the interpolation results in strong blurring effects that smooth the rotated kernel. The interpolation-based rotation results in a severe degradation of this sort of kernel. Note also that the interpolation does not preserve the order of the polynomial approximation and vanishing moment conditions. Thus, in this rotation we are not able to control the polynomial properties of the kernel.

In the above-described design of *directional LPA kernels*, the kernels are specially calculated (constructed) for each desirable direction θ_i and for each $h \in H$. In this way we obtain a set of kernels $g_{h,\theta_i}(x)$ that can be treated as a filter bank of the directional LPA. This directional LPA design procedure is applicable for differentiating kernels and gives a filter bank of the directional differentiating kernels $g^{(r)}_{h,\theta_i}(x)$.

7.2.4 Calculations

Let the observations be given on the grid $\{X_s\}$ and the estimates be needed for a desired x. The LPA model in Eq. (7.26) is used for the variables u,

$$y_h(u, v_s) = C^T \phi_h(u - v_s), \tag{7.32}$$

where

$$u = u(x, \theta), \qquad v_s = u(X_s, \theta) \tag{7.33}$$

are calculated using Eq. (7.31).

The weighted least-square criterion J_h in Eq. (2.24) is

$$J_h(u, C) = \sum_s w_h(u - v_s)[z_s - y_h(u, v_s)]^2, \tag{7.34}$$

where u is the center (reference point) of the LPA in the rotated coordinates and

$$w_h(u) = \frac{1}{h_1 h_2} w(u/h).$$

We produce a fit of the observations $z_s = z(X_s)$ by the polynomials defined in the rotated variables.

Let $\hat{y}_{h,\theta}(x)$ denote the LPA estimate of $y(x)$, where the index θ shows the dependence of the estimate on the direction. This estimate is defined according to the basic idea of the local approximation as

$$\hat{y}_{h,\theta}(x) = y_h(u, v_s)|_{u=v_S} = y_h(u(x, \theta), u(X_s, \theta))|_{x=X_s} = \hat{C}^T \phi(0).$$

For the polynomials in Eq. (7.27), this yields

$$\hat{y}_{h,\theta}(x) = \hat{C}_1. \tag{7.35}$$

Following Eq. (2.30) we obtain for the derivative estimates

$$\hat{y}^{(r)}_{h,\theta}(x) = \partial^r_v y_h(u, v)|_{u=v} = \hat{C}^T \partial^r_v \phi_h(u - v)|_{u=v} = \frac{(-1)^{|r|}}{h_1^{r_1} h_2^{r_2}} \hat{C}^T \phi^{(r)}(0). \tag{7.36}$$

It is emphasized that $\hat{y}^{(r)}_{h,\theta}(x)$ is the estimate of the *directional derivative*. This directional derivative of order $r = (r_1, r_2)$ is calculated on the variables u_1 and u_2, $\partial^r_u y(u) = \partial^{r_1}_{u_1} \partial^{r_2}_{u_2} y(u)$. Minimizing $J_h(u, C)$ on C and substituting it in formulas

(7.35) and (7.36), we obtain the formulas for the kernel representation of the LPA estimates analogous to Eqs. (2.55–2.58):

$$\hat{y}_{h,\theta}(x) = \sum_s g_h(u, v_s) z_s, \tag{7.37}$$

$$\begin{aligned} g_h(u, v_s) &= w_h(u - v_s)\phi_h^T(u - v_s)\Phi_h^{-1}\phi(0), \\ \Phi_h &= \sum_s w_h(u - v_s)\phi_h(u - v_s)\phi_h^T(u - v_s), \end{aligned} \tag{7.38}$$

$$\hat{y}_{h,\theta}^{(r)}(x) = \sum_s g_h^{(r)}(u, v_s) z_s, \tag{7.39}$$

$$g_h^{(r)}(u, v_s) = \frac{(-1)^{|r|}}{h_1^{r_1} h_2^{r_2}} w_h(u - v_s)\phi_h^T(u - v_s)\Phi_h^{-1}\phi^{(r)}(0). \tag{7.40}$$

The dependence of the kernels on x and θ is imbedded in the variables u and v_s. Substituting $u = u(x, \theta)$ and $v_s = u(X_s, \theta)$, we can represent these estimates in the forms

$$\begin{aligned} \hat{y}_{h,\theta}(x) &= \sum_s g_{h,\theta}(x, X_s) z_s, \\ g_{h,\theta}(x, X_s) &= g_h(u(x, \theta), u(X_s, \theta)), \end{aligned} \tag{7.41}$$

$$\begin{aligned} \hat{y}_{h,\theta}^{(r)}(x) &= \sum_s g_{h,\theta}^{(r)}(x, X_s) z_s, \\ g_{h,\theta}^{(r)}(x, X_s) &= g_h^{(r)}(u(x, \theta), u(X_s, \theta)), \end{aligned} \tag{7.42}$$

where we use the summation over the original grid $\{X_s\}$ of the observations.

The directional LPA design gives the kernels defined on the basic data grid to calculate the estimates for any desirable θ and x.

It is emphasized once more that the number of nodes u_s used in the calculations of the kernels depends on θ and h, and in this way the kernels can be quite different for different directions and different scales. The orders of this estimate $m = (m_1, m_2)$ and the derivatives that can be estimated in this way are defined by the polynomials used in ϕ. In particular, for the example in Eq. (7.27), these derivatives are $\partial_{u_1}, \partial_{u_2}, \partial_{u_1}^2, \partial_{u_1}\partial_{u_2}, \partial_{u_1}^2\partial_{u_2}$.

7.2.5 Shift-invariant kernels

The formulas for the directional LPA kernels considered above are quite general. They are applicable for regular and irregular grids and for finite and infinite numbers of observations. Let the LPA estimator be applied to an image defined on the rectangular infinite regular grid:

$$X = \{\Delta_1 k_1, \Delta_2 k_2 : k_1, k_2 \in \mathbb{Z}\}. \tag{7.43}$$

The rotation formulas (7.29) can be represented as

$$\begin{aligned} u_1(\Delta k_1, \Delta k_2, \theta) &= \Delta_1 k_1 \cos\theta + \Delta_2 k_2 \sin\theta, \\ u_2(\Delta k_1, \Delta k_2, \theta) &= -\Delta_1 k_1 \sin\theta + \Delta_2 k_2 \cos\theta, \end{aligned} \tag{7.44}$$

or in the compact form as

$$u = u(\Delta k, \theta) \quad \text{or} \quad u = U(\theta)\Delta k. \tag{7.45}$$

For a fixed θ,

$$u - v_s = u(\Delta k, \theta) - u(\Delta s, \theta) = U(\theta)(\Delta k - \Delta s).$$

Then, the matrix Φ_h in Eqs. (7.38) and (7.40) is shift-invariant:

$$\begin{aligned} \Phi_h &= \sum_{s\in\mathbb{Z}^2} w_h(U(\theta)(\Delta k - \Delta s))\phi_h(U(\theta)(\Delta k - \Delta s))\phi_h^T(U(\theta)(\Delta k - \Delta s)) \\ &= \sum_{s\in\mathbb{Z}^2} w_h(U(\theta)\Delta s)\phi_h(U(\theta)\Delta s)\phi_h^T(U(\theta)\Delta s), \end{aligned}$$

and the estimates in Eqs. (7.41) and (7.42) also become shift-invariant.

In other words, we arrive at *shift-invariant directional* LPA estimators similar to those considered in Sec. 3.1. In this case, the kernels $g_h(u, v_s)$, $g_h^{(r)}(u, v_s)$ become shift-invariant with respect to the shift on the grid in Eq. (7.43) and can be essentially simplified because they depend on the difference between the arguments Δk and Δs only:

$$\begin{aligned} g_h(u, v_s) &= g_h(U(\theta)(\Delta k - \Delta s)), \\ g_h^{(r)}(u, v_s) &= g_h^{(r)}(U(\theta)(\Delta k - \Delta s)). \end{aligned} \tag{7.46}$$

The shift-invariant estimates have the convolution form

$$\hat{y}_{h,\theta}(\Delta k) = \sum_{s\in\mathbb{Z}^2} g_{h,\theta}[k-s]z_s = \sum_{s\in\mathbb{Z}^2} g_{h,\theta}[s]z_{k-s}, \tag{7.47}$$

where

$$\begin{aligned} g_{h,\theta}[s] &= w_h(U(\theta)\Delta s)\phi_h^T(U(\theta)\Delta s)\Phi_h^{-1}\phi(0), \\ \Phi_h &= \sum_{s\in\mathbb{Z}^2} w_h(U(\theta)\Delta s)\phi_h(U(\theta)\Delta s)\phi_h^T(U(\theta)\Delta s). \end{aligned} \tag{7.48}$$

In a similar way, the derivative estimate in Eq. (7.39) can be written in the form

$$\hat{y}_{h,\theta}^{(r)}(\Delta k) = \sum_{s\in\mathbb{Z}^2} g_{h,\theta}^{(r)}[k-s]z_s = \sum_{s\in\mathbb{Z}^2} g_{h,\theta}^{(r)}[s]z_{k-s}, \tag{7.49}$$

where

$$g_{h,\theta}^{(r)}[s] = \frac{(-1)^{|r|}}{h_1^{r_1} h_2^{r_2}} w_h(U(\theta)\Delta s)\phi_h^T(U(\theta)\Delta s)\Phi_h^{-1}\phi^{(r)}(0). \tag{7.50}$$

Finally, we can represent these estimates in the form of convolutions similar to the standard ones for the discrete sequences in Eqs. (3.10–3.11). In this way, we have

$$\hat{y}_{h,\theta}[k] = \sum_{s\in\mathbb{Z}^2} g_{h,\theta}[s] z_{k-s} \tag{7.51}$$

and

$$\hat{y}_{h,\theta}^{(r)}[k] = \sum_{s\in\mathbb{Z}^2} g_{h,\theta}^{(r)}[s] z_{k-s}. \tag{7.52}$$

Further, the general discrete variable convolution notation is applicable:

$$\begin{aligned} \hat{y}_{h,\theta}[k] &= (g_{h,\theta} \circledast z)[k], \\ \hat{y}_{h,\theta}^{(r)}[k] &= (g_{h,\theta}^{(r)} \circledast z)[k], \; k \in \mathbb{Z}^2, \end{aligned} \tag{7.53}$$

where $\circledast$ denotes discrete convolution. The estimates in Eqs. (3.12) and (7.53) differ only by the kernels used.

Let us comment on the obtained directional kernels and estimates.

1. The directional kernels $g_{h,\theta}$ and $g_{h,\theta}^{(r)}$ are given by formulas (7.48) and (7.50), which are valid for any direction θ. The support of these kernels is defined by the grid points where $w_h(U(\theta)\Delta s) > 0$. This corresponds to the accurate geometric rotation of the kernel support, for instance, shown in Fig. 7.9. However, this support can be repaired (corrected) as discussed above. Then, formulas (7.48) and (7.50) should be used with the correction nodes included in these formulas.
2. The polynomials in the rotated variables mean that the polynomial approximations of the orders m_1 and m_2 are applied along the rotated axes u_1 and u_2. They define the directional smoothness of this LPA. The directionality of the approximation is accurately preserved for any direction θ and does not depend on variations of the node number in the support caused by the rotation. Of course, this is valid provided that the total number of nodes is sufficient for the polynomial fit.

7.2.6 Frequency domain

In the frequency domain (see Sec. 3.3) the convolution in Eq. (7.51) is of the form

$$Y_{h,\theta}(e^{-j\bar{\omega}}) = G_{h,\theta}(e^{-j\bar{\omega}}) Z(e^{-j\bar{\omega}}), \tag{7.54}$$

where

$$Y_{h,\theta}(e^{-j\bar{\omega}}) = \sum_k \hat{y}_{h,\theta}[k] \exp(-j \langle k, \bar{\omega} \rangle),$$

$$G_{h,\theta}(e^{-j\bar{\omega}}) = \sum_k g_{h,\theta}[k] \exp(-j \langle k, \bar{\omega} \rangle), \tag{7.55}$$

$$Z(e^{-j\bar{\omega}}) = \sum_k z[k] \exp(-j \langle k, \bar{\omega} \rangle),$$

and $\bar{\omega} = \Delta\omega$ is the normalized frequency in Eq. (3.27). Formula (7.55) defines the transfer function $G_{h,\theta}(e^{-j\bar{\omega}})$ of the directional kernel.

According to Eq. (7.48),

$$g_{h,\theta}[k] = g_{h,\theta}(\Delta k) = g_h(U(\theta)\Delta k).$$

Inserting it into Eq. (7.55), we have

$$G_{h,\theta}(e^{-j\bar{\omega}}) = \sum_k g_h(U(\theta)\Delta k) \exp(-j \langle \Delta k, \omega \rangle)$$

and changing the variables in the summation as

$$\Delta l = U(\theta)\Delta k,$$

where to be accurate l is not always integer, we can write

$$G_{h,\theta}(e^{-j\bar{\omega}}) = \sum_l g_h(\Delta l) \exp(-j \langle U^T(\theta)\Delta l, \omega \rangle)$$
$$= \sum_l g_h(\Delta l) \exp(-j \langle l, U(\theta)\bar{\omega} \rangle). \tag{7.56}$$

The frequency characteristic of the basic ($\theta = 0$) kernel in Eq. (3.25) is

$$G_{h,0}(e^{-j\bar{\omega}}) = \sum_k g_h(\Delta k) \exp(-j \langle k, \bar{\omega} \rangle), \tag{7.57}$$

that is,

$$G_h(e^{-j\bar{\omega}}) = G_{h,0}(e^{-j\bar{\omega}}).$$

Compare Eq. (7.56) with Eq. (7.57). First, the frequency $\bar{\omega}$ in Eq. (7.57) is replaced in Eq. (7.56) by its rotated counterpart $U(\theta)\bar{\omega}$. Second, the grid Δk in Eq. (7.57) is replaced in Eq. (7.56) by the grid Δl.

If the sampling interval Δ is small, then the discrete operator g_h in Eqs. (7.56) and (7.57) as well as the sums defining the discrete Fourier transforms can be replaced by the corresponding integrals. In this limit passage to the integrals, the difference between the grids Δk and Δl disappears and we may conclude that there is the following approximate link between the frequency characteristics of the basic and rotated kernels:

$$G_{h,\theta}(e^{-j\bar{\omega}}) \simeq G_{h,0}(e^{-jU(\theta)\bar{\omega}}). \tag{7.58}$$

The frequency characteristic of the rotated kernel $G_{h,\theta}(e^{-j\bar{\omega}})$ is approximately equal to the rotated frequency characteristic of the basic $G_{h,0}(e^{-j\bar{\omega}})$.

7.3 Numerical Directional Kernels

Here we show some examples of numerical shift-invariant discrete kernels for smoothing and differentiation with the main intention of demonstrating the effects of directionality and rotation. The directional basic kernel is always elongated along the x_1 axis. We distinguish between the following three types of these kernels [108]:

- Kernels with support even with respect to both variables x_1 and x_2, and $w(x_1, x_2) = w(-x_1, x_2)$, $w(x_1, x_2) = w(x_1, -x_2)$; this is a kernel of type 00.
- Kernels with support even with respect to variable x_2, and $w(x_1, x_2) = w(x_1, -x_2)$, and nonnegative $x_1 \geq 0$; this is a kernel of type 10,
- A kernel with support with nonnegative x_1, $x_2 \geq 0$; this is a kernel of type 11.

7.3.1 Sectorial kernels

7.3.1.1 Sectorial kernels of type 10

A uniform binary window function w for a basic sectorial kernel of the type 10 is defined as follows

$$w = 1, \quad \text{if} \left| \frac{x_1}{\sqrt{x_1^2 + x_2^2}} \right| \leq \sin\alpha, \qquad ||x|| \leq 1, x_1 \geq 0. \tag{7.59}$$

The support of this window and of the corresponding kernels is a sector of the unit disc symmetric with respect to the x_1 axis. The angular width of the sector is equal to 2α. This window defines a kernel of type 10, as $w = 0$ for $x_1 < 0$.

Figure 7.10 shows the basic kernel $g_{h,\theta}(\theta = 0)$ of order $m = (0, 0)$ with the window in Eq. (7.59), where $\alpha = \pi/16$. The figures in the first row present g_h as a function of two variables, x_1 and x_2. The first and second in the row images show g_h as a flat 2D binary image and as a 3D surface, respectively. The second row presents the amplitude frequency characteristic of the kernel $|G_{h,\theta}|$ as a function of two normalized frequencies, $\bar{\omega}_1$ and $\bar{\omega}_2$. Similar to those the first row, these frequency characteristics are also given by 2D imaging and by a 3D surface. The vertical bars between 2D and 3D images give the scales for the 2D imaging.

The kernel scale parameters $h_1 = h_2 = 21$ define the size of the symmetric square mask where the kernel g_h is defined. The stepwise shape of g_h, clearly seen in the first image in the first row, shows that the kernel is discretely defined on the regular discrete grid. The directionality of the kernel is obvious in the images

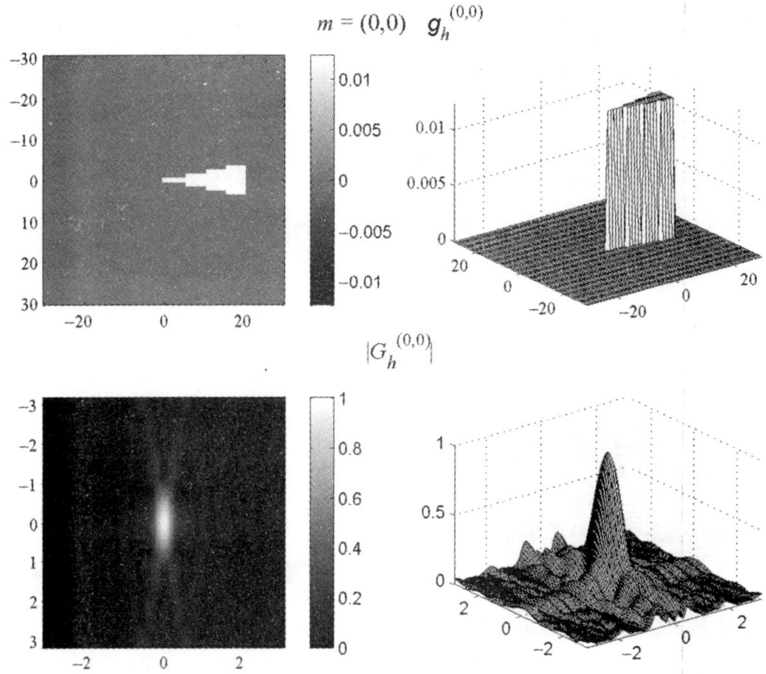

Figure 7.10 The basic kernel ($\theta = 0$) of order $m = (0, 0)$, the type 10 with the sectorial window support in Eq. (7.59). The kernel g_h is shown as a 2D image and a 3D surface. The amplitude frequency response of the kernel is also given in these two forms. The scale parameters $h_1 = h_2 = 21$ define the size of the kernel mask.

of the kernel as well as in its frequency characteristic. The frequency characteristic $|G_h|$ is elongated in the direction orthogonal to the symmetry axis x_1 of the kernel g_h.

Figure 7.11 demonstrates the rotation effects for the sectorial kernel in Eq. (7.59) with $\theta = \pi/4$. This rotation can be seen in both the kernel itself as well as in its frequency characteristic. The shape of g_h is slightly blurred compared with the sharp binary shape of this kernel in Fig. 7.10. This blur is a result of the rotation of the kernel support from $\theta = 0$ to $\theta = \pi/4$.

Let us use some higher-order polynomial models. Consider the kernel with the same sectorial support in Eq. (7.59) and of the order $m = (2, 0)$. The second order essentially changes a character of the kernel g_h in the direction of the x_1 axis. The second-order polynomial approximation allows us to not only estimate the signal but also to differentiate up to the order 2. This differentiation is directional and directed along the x_1 axis. Figures 7.12–7.14 illustrate the basic directional smoothing $g_h^{(0,0)}$ and differentiating kernels $g_h^{(1,0)}$, $g_h^{(2,0)}$ corresponding to $\theta = 0$. The rotation effects for these higher-order kernels are very similar to those shown in Fig. 7.11. Note that the frequency characteristics have zero values at the zero frequency for the differentiating kernels and are equal to one for the smoothing kernel. These polynomial properties of the LPA kernels are accurately preserved for directional versions of the kernels for all θ.

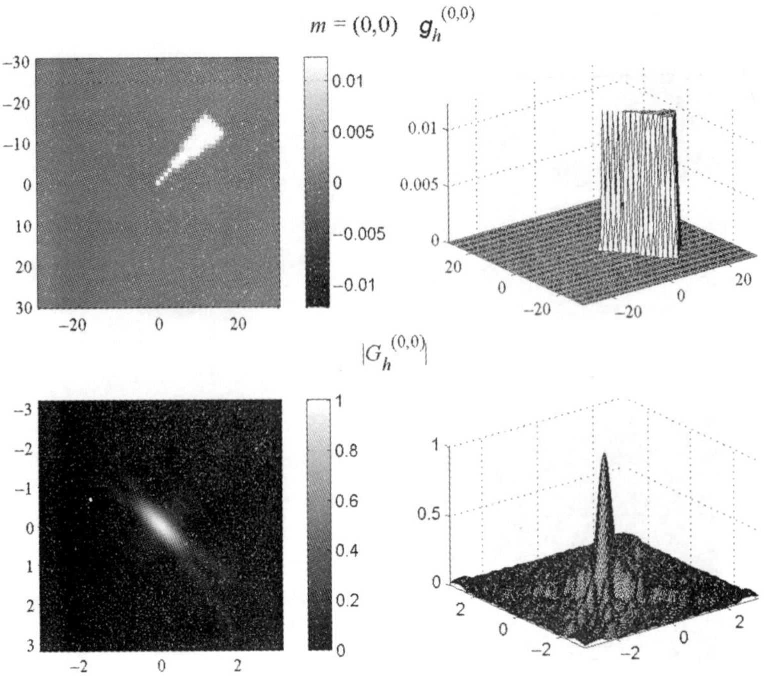

Figure 7.11 A rotation of the basic kernel $g_h^{(0,0)}$ (see Fig. 7.10) is shown, $\theta = 45^0$. The support of the kernel is slightly blurred compared with the sharp support in Fig. 7.10.

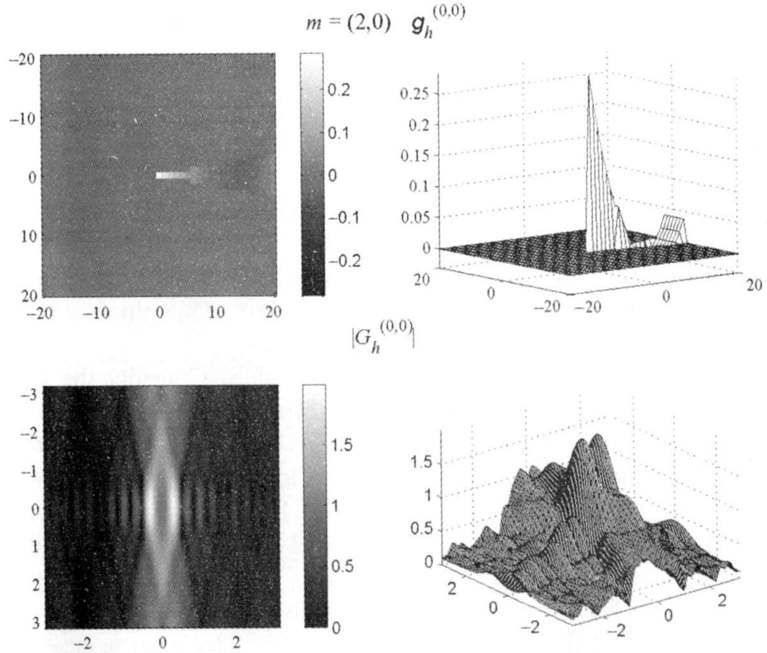

Figure 7.12 The basic smoothing kernel $g_h^{(0,0)}$, $\theta = 0$, type 10, sectorial support. The order $m = (2, 0)$ results in a quadratic shape of $g_h^{(0,0)}$ clearly seen in the 3D curve.

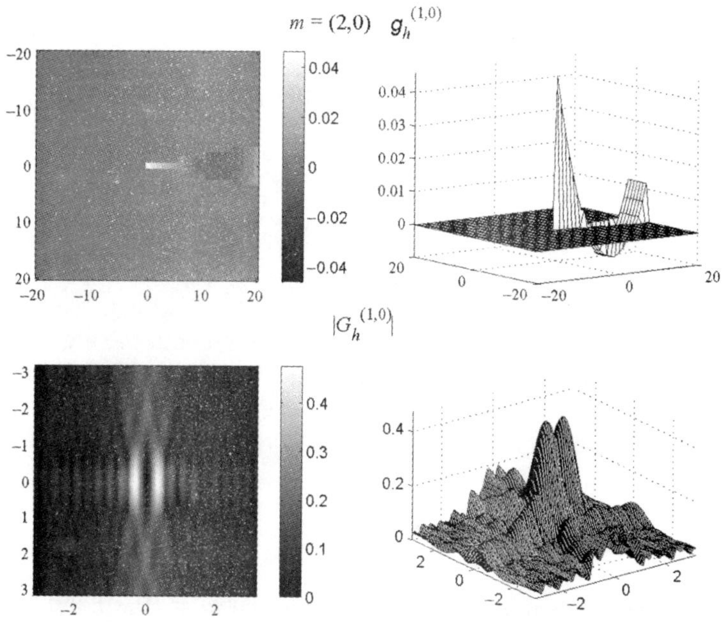

Figure 7.13 The differentiating basic kernel $g_h^{(1,0)}$, θ = 0, type 10, sectorial support. The first-order differentiated on x_1, $\partial^{(1,0)} = \partial_{x_1}$. It is seen in the frequency characteristic that $|G_h^{(1,0)}(0)| = 0$.

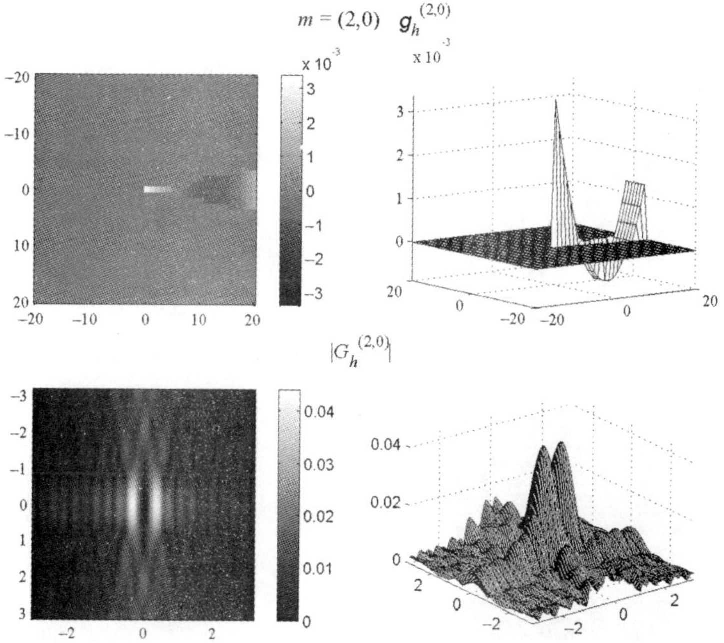

Figure 7.14 The differentiating basic kernel $g_h^{(2,0)}$, θ = 0, type 10, the sectorial support. The second-order differentiating on x_1, $\partial^{(2,0)} = \partial_{x_1}^2$. It is seen in the frequency characteristic that $|G_h^{(2,0)}(0)| = 0$.

7.3.1.2 Sectorial kernels of type 00

Figure 7.15 corresponds to the basic kernel of order $m = (0, 0)$ and type 00 with the sectorial windows

$$w = 1, \quad \text{if} \left| \frac{x_1}{\sqrt{x_1^2 + x_2^2}} \right| \leq \sin\alpha, \qquad \alpha = \pi/16, ||x|| \leq 1. \tag{7.60}$$

This support is different from the one in Eq. (7.59) by assumptions for x_1. In Eq. (7.59), $x_1 \geq 0$ and in Eq. (7.60), x_1 takes positive and negative values. As a result, the support of Eq. (7.60) is an even function on both the arguments x_1 and x_2. The kernel is shown for $\theta = 0$. The directionality is clearly seen in the figures given for the kernel as well as for the frequency characteristic. The frequency characteristic $|G_h|$ is elongated in the direction orthogonal to the x_1 axis.

7.3.1.3 Sectorial kernels of type 11

Figure 7.16 corresponds to the basic kernel of order $m = (0, 0)$ and type 11 with the sectorial window support

$$w = 1, \quad \text{if} \left| \frac{x_1}{\sqrt{x_1^2 + x_2^2}} \right| \leq \sin\alpha, ||x|| \leq 1, \qquad x_1 \geq 0, \quad x_2 \geq 0, \quad \alpha = \pi/16. \tag{7.61}$$

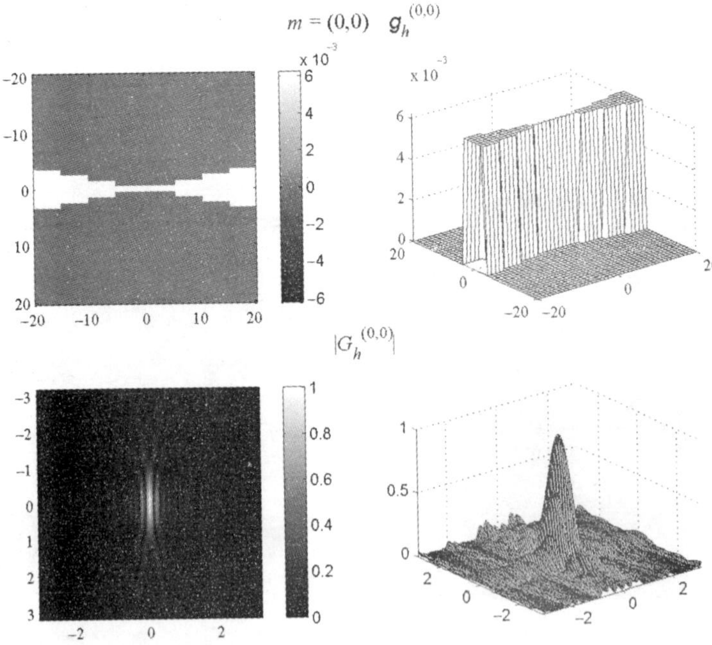

Figure 7.15 The basic kernel ($\theta = 0$) of order $m = (0, 0)$. A type of 00 kernel with the sector window as in Eq. (7.60). The only difference between this and Fig. 7.10 is that this kernel symmetric with respect to the vertical axis.

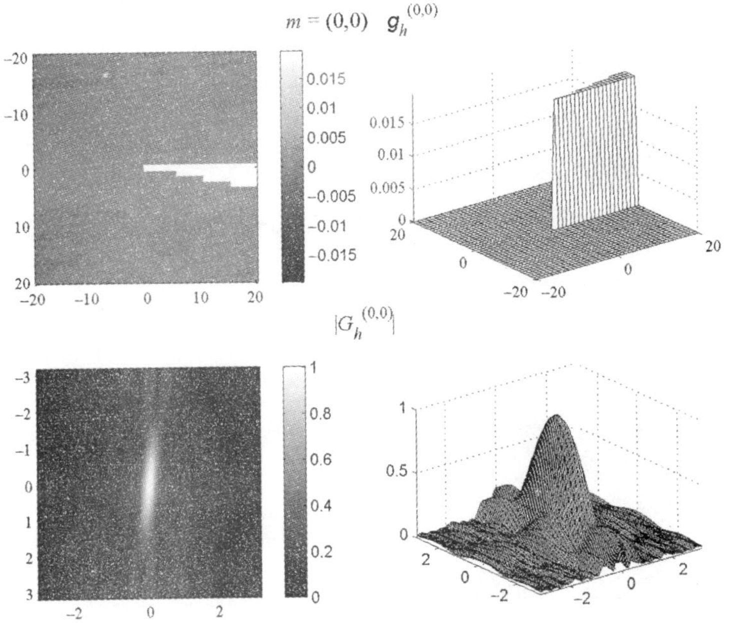

Figure 7.16 The basic kernel with sectorial support as in Eq. (7.61), order $m = (0, 0)$, and type 11.

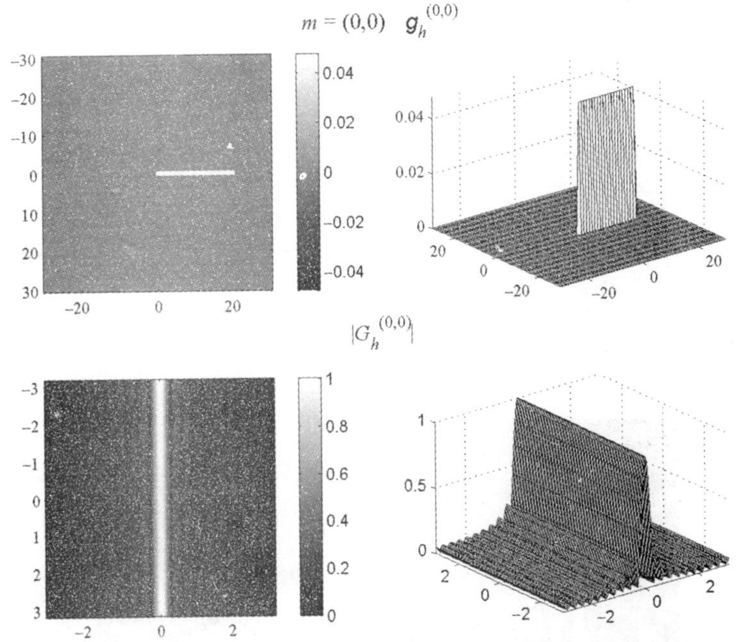

Figure 7.17 The basic kernel ($\theta = 0$) of order $m = (0, 0)$. A type of 10 kernel with the linewise window as in Eq. (7.62). The kernel g_h is shown as the image and 3D surface. The amplitude frequency response of the kernel is given in these two forms. The scale parameters $h_1 = h_2 = 21$ define the size of the kernel mask.

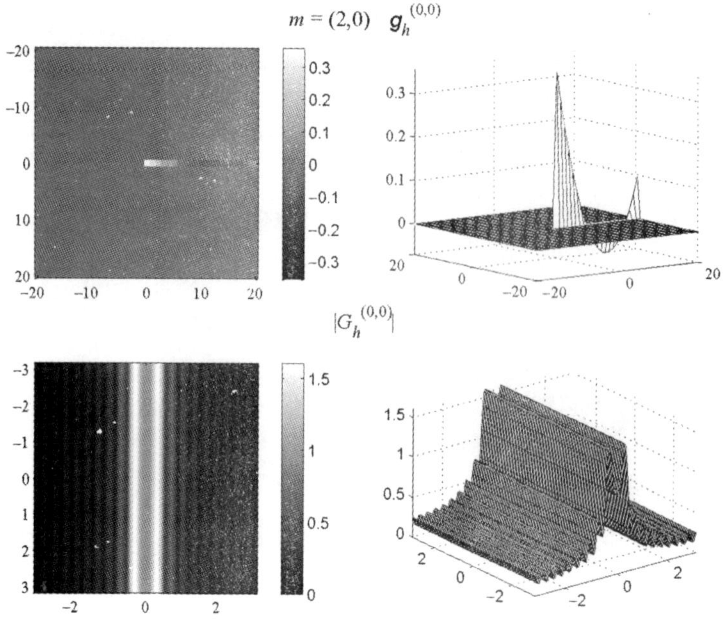

Figure 7.18 The basic smoothing kernel $g_h^{(0,0)}$, θ = 0, linewise support. The order m = (2, 0) results in the quadratic shape of $g_h^{(0,0)}$ clearly seen in 3D curve.

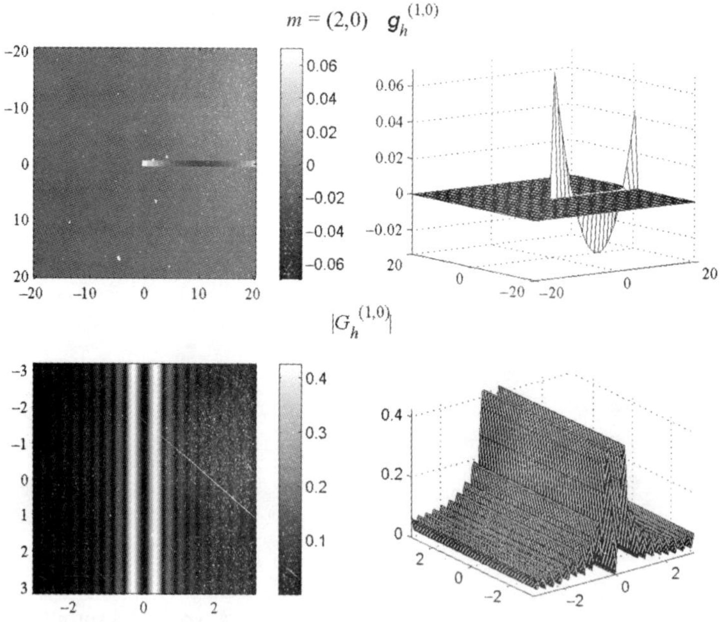

Figure 7.19 The differentiating basic kernel $g_h^{(1,0)}$, θ = 0, linewise support. The first-order differentiated on x_1, $\partial^{(1,0)} = \partial_{x_1}$. It is seen in the frequency characteristic that $|G_h^{(1,0)}(0)| = 0$.

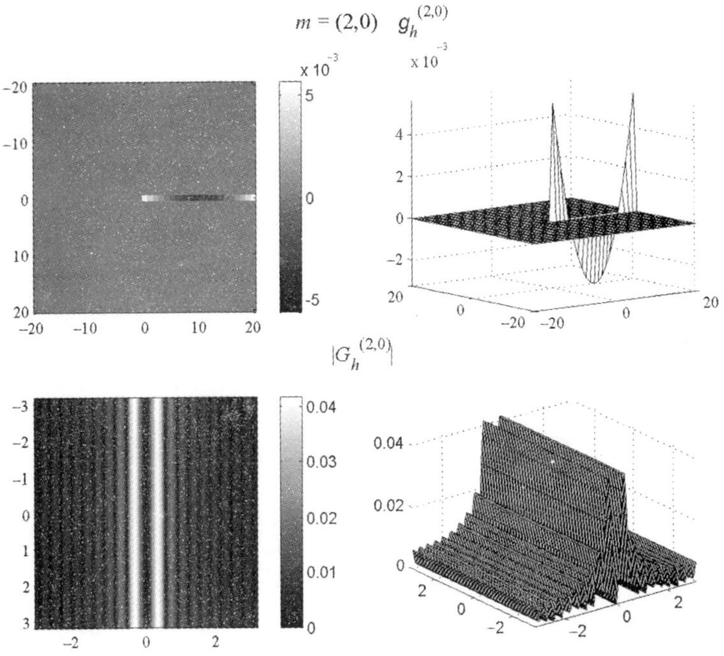

Figure 7.20 The differentiating basic kernel $g_h^{(2,0)}$, $\theta = 0$, linewise support. The second-order differentiated on x_1, $\partial^{(2,0)} = \partial_{x_1}^2$. It is seen in the frequency characteristic that $|G_h^{(2,0)}(0)| = 0$.

This support is different from the one in Eq. (7.59) by having different assumptions for x_1 and x_2. In Eq. (7.59), $x_1 \geq 0$, $x_2 \lesseqgtr 0$ while in Eq. (7.61), $x_1 \geq 0$, $x_2 \geq 0$. As a result, for type 11 the support is one-half of the support in Eq. (7.59) and one-quarter of the support in Eq. (7.60). The kernel is shown for $\theta = 0$.

7.3.2 Line-wise kernels

Figure 7.17 shows the basic kernel of the order $m = (0, 0)$, type 10, with the linewise window

$$w = 1, \qquad \text{if } 0 \leq x_1 \leq 1, \quad x_2 = 0. \tag{7.62}$$

We can see the imaging of this kernel and its 3D surface. The amplitude frequency response of the kernel is given in the two similar forms. The directionality of the kernel is obvious. The scale parameters $h_1 = h_2 = 21$.

Let us exploit the higher-order polynomial approximation on x_1 along the line support, that is, we use the LPA with $m = (2, 0)$. The second-order polynomial approximation allows us to produce differentiation up to the order 2. This differentiation is directional and directed along the x_1 axis. We show in Figs. 7.18–7.20 the smoothing $g_h^{(0,0)}$ and differentiating kernels $g_h^{(1,0)}$, $g_h^{(2,0)}$ for $\theta = 0$ only.

The frequency characteristics have zero values at zero frequency for the differentiating kernels and are equal to one for the smoothing kernel. These properties of the LPA kernels are automatically preserved in the directional kernels for all θ.

Chapter 8

Anisotropic LPA-ICI Algorithms

In this chapter, we start with accuracy analyses with the main emphasis on the specific features of the directional estimates: different LPA orders m and different scales h of the variables. The directionality of the supports and the estimates is specified as corresponding to the anisotropy properties of the object signal y.

The estimation accuracy is characterized by errors:

$$\begin{aligned} e_{y,\theta}(x,h) &= y(x) - \hat{y}_{h,\theta}(x), \\ e_{y^{(r)},\theta}(x,h) &= y^{(r)}(x) - \hat{y}^{(r)}_{h,\theta}(x), \end{aligned} \tag{8.1}$$

where θ indicates the directionality of the estimators.

As usual, these errors are composed of systematic (bias) and random components corresponding to the deterministic y and the random noise ε, respectively.

We derive the asymptotic formulas for the bias and variance valid for small h and Δ. The obtained ideal scales allow us to determine a ratio of scales (width-to-length ratio) that minimizes the quadratic risk. These results are used to justify the ICI algorithms for anisotropic data-driven adaptive-scale selection. In this chapter we present and use the results mainly for 2D image processing.

8.1 Accuracy Analysis*

8.1.1 Polynomial smoothness

The directional LPA produces kernels that give accurate reconstruction and estimates for polynomial signals. It follows from the design procedure, which is the moving least-square method, and the polynomials as models of the signal imbedded in the LPA. However, the directional kernels are different from those introduced in Chapter 2 in two important respects. First, the orders m_1 and m_2 of the LPA polynomials are specified for variables x_1 and x_2 independently. Second, these variables enter into the kernels with different scales h_1 and h_2.

We wish to reformulate for the directional estimates the statements given in Chapter 5 for the nondirectional estimates. Assume that the angle θ is fixed and consider the estimate kernels in the rotated coordinate system. The LPA method uses the polynomials in the rotated variables $u = U(\theta)x$, $x = \Delta k$, where k are integers and Δ is a sampling period. Let us consider these kernels as functions of the variables u and note that the grid of these u depends on θ and, in general, is irregular.

The shift-invariant kernel estimate in Eq. (7.51) can be written in the rotated variables as

$$\begin{aligned} \hat{y}_{h,\theta}[k] &= \hat{y}_h(u) = \sum_v g_h(v) z(u - v), \\ g_h(v) &= w_h(v)\phi^T(0)\Phi_h^{-1}\phi_h(v), \\ \Phi_h &= \sum_v w_h(v)\phi_h(v)\phi_h^T(v), \end{aligned} \tag{8.2}$$

where

$$u = U(\theta)\Delta k, \qquad v = U(\theta)\Delta s.$$

Let us prove that the directional LPA filters give a perfect reconstruction of the signal and a perfect derivative estimation for input signals of the appropriate degrees.

It can be verified that, for the vector ϕ_h composed of the polynomials in Eq. (7.25), $(-1)^{k_1+k_2}(u_1^{k_1}u_2^{k_2}/k_1!k_2!)$, we have

$$\phi_h(u - v) = A_h(u)\phi_h(v), \tag{8.3}$$

where the matrix $A_h(u)$ is square and does not depend on v, and

$$\phi_h(u) = A_h(u)\phi(0). \tag{8.4}$$

Consider a response of the filter in Eq. (8.2) with respect to the vector input signal $z(u) = \phi_h(u)$. Transformations using Eqs. (8.3) and (8.4) show that

$$\begin{aligned} \sum_v g_h(v)\phi_h(u - v) &= \sum_v w_h(v)\phi_h(u - v)\phi^T(0)\Phi_h^{-1}\phi_h(v) \\ &= \sum_v w_h(v)A_h(u)\phi_h(v)\phi_h^T(v)\Phi_h^{-1}\phi(0) \\ &= A_h(u)\left[\sum_v w_h(v)\phi_h(v)\phi_h^T(v)\right]\Phi_h^{-1}\phi(0) \\ &= A_h(u)\Phi_h\Phi_h^{-1}\phi(0) \\ &= A_h(u)\phi(0) = \phi_h(u). \end{aligned}$$

This proves the perfect reconstruction for the polynomial vector ϕ_h,

$$\sum_v g_h(v)\phi_h(u - v) = \phi_h(u). \tag{8.5}$$

In a similar way, we have for the differentiating kernels Eqs. (7.49–7.50)

$$\hat{y}^{(r)}_{h\theta}[k] = \hat{y}^{(r)}_{h}(u) = \sum_{v} g^{(r)}_{h}(v)y(u-v),$$
$$g^{(r)}_{h}(v) = \frac{(-1)^{|r|}}{h_1^{r_1}h_2^{r_2}} w_h(v)\phi^T_h(v)\Phi^{-1}_h\phi^{(r)}(0). \tag{8.6}$$

Differentiation of $\phi_h(u-v)$ gives

$$\partial^r_v\phi_h(u-v) = \partial^r_v\phi\left(\frac{u-v}{h}\right) = \frac{(-1)^{|r|}}{h_1^{r_1}h_2^{r_2}}\phi^{(r)}\left(\frac{u-v}{h}\right). \tag{8.7}$$

where

$$\phi^{(r)}(u) = \partial^r_u\phi(u),$$

and differentiation of Eq. (8.3) on v gives

$$\partial^r_v\phi_h(u-v) = A_h(u)\partial^r_v\phi_h(v) = \frac{1}{h_1^{r_1}h_2^{r_2}}A_h(u)\phi^{(r)}_h(v).$$

Substituting Eq. (8.7) in the left-hand side of this equation, we have

$$\frac{(-1)^{|r|}}{h_1^{r_1}h_2^{r_2}}\phi^{(r)}\left(\frac{u-v}{h}\right) = \frac{1}{h_1^{r_1}h_2^{r_2}}A_h(u)\phi^{(r)}(v/h), \tag{8.8}$$

and for $v = 0$ it gives

$$(-1)^{|r|}\phi^{(r)}(u/h) = A_h(u)\phi^{(r)}(0). \tag{8.9}$$

Let the input signal of the differentiating filter in Eq. (8.6) be the vector $z(u) = \phi_h(u)$. We obtain for the response of this differentiator

$$\begin{aligned}\sum_{v} g^{(r)}_{h}(v)\phi_h(u-v) &= \frac{(-1)^{|r|}}{h_1^{r_1}h_2^{r_2}}\sum_{v} w_h(v)\phi_h(u-v)\phi^T_h(v)\Phi^{-1}_h\phi^{(r)}(0)\\ &= \frac{(-1)^{|r|}}{h_1^{r_1}h_2^{r_2}}\sum_{v} w_h(v)A_h(u)\phi_h(v)\phi^T_h(v)\Phi^{-1}_h\phi^{(r)}(0)\\ &= \frac{(-1)^{|r|}}{h_1^{r_1}h_2^{r_2}}A_h(u)\phi^{(r)}(0) = \frac{1}{h_1^{r_1}h_2^{r_2}}\phi^{(r)}(u/h) = \partial^r_u\phi_h(u).\end{aligned} \tag{8.10}$$

In this derivation we use Eq. (8.3) to replace $\phi_h(u-v)$ by $A_h(u)\phi_h(v)$ and Eq. (8.9) in the last line of Eq. (8.10). This proves that

$$\sum_{v} g^{(r)}_{h}(v)\phi_h(u-v) = \partial^r_u\phi_h(u). \tag{8.11}$$

Comparing Eqs. (8.5) and (8.11) versus Eqs. (3.17) and (3.18), we see the similarity of these formulas, and as a result the identity of the reproducing properties of the directional and nondirectional estimators.

Note that in the rotated variables, the order of the polynomials depends on θ. For example, let

$$y(x) = \varphi^T(x)p, \varphi(x) = (1, x_1)^T, \qquad p = (p_1, p_2)^T, \qquad p \in \mathbb{R}^2,$$

that is, for $x \in \mathbb{R}^2$, the polynomial $y(x)$ has the powers $m_1 = 1$ and $m_2 = 0$. After the rotation,

$$\begin{aligned} y_\theta(u) &= y(U^T(\theta)u) = \varphi(U^T(\theta)u)p \\ &= (1, u_1 \cos\theta - u_2 \sin\theta)p = p_1 + (u_1 \cos\theta - u_2 \sin\theta)p_2, \end{aligned}$$

and $y_\theta(u)$ has the powers $m_1 = 1$ and $m_2 = 1$ for all θ such that $\cos\theta \neq 0$ and $\sin\theta \neq 0$.

Let $\mathcal{P}^m$, $m = (m_1, m_2)$, be a class of 2D polynomials, where m is a multi-index showing the maximal powers on the polynomials on the corresponding variables x_1 and x_2. Assume further that a direction θ is fixed and $u = U(\theta)x$ are variables in the rotated coordinates. Then y as a function of u is

$$y_\theta(u) = y(U^T(\theta)u). \tag{8.12}$$

In order to specify the class of directional polynomials for a particular direction θ, we use the notation $\mathcal{P}^m_\theta$. Then $y(x) \in \mathcal{P}^m_\theta$ means that, in the rotated coordinates, as a function of u, the polynomial $y(U^T(\theta)u) \in \mathcal{P}^m$.

Let $y(x)$ be polynomial belonging to the class $\mathcal{P}^m_\theta$ or, equivalently, $y_\theta(u) \in \mathcal{P}^m$. The polynomials in the vector $\phi_h(u)$ belong to the class $\mathcal{P}^m$ and form a complete set of the polynomials of power m. The completeness means that any polynomial $y(u) \in \mathcal{P}^m$ can be written in the form

$$y(u) = p^T\phi(u) \in \mathcal{P}^m, \tag{8.13}$$

where $p \in \mathbb{R}^M$ is a constant vector.

Proposition 11 *Let the estimate $\hat{y}_{h,\theta}[k]$ in Eq. (7.51) of the signal $y(x)$ and the estimate $\hat{y}^{(r)}_{h,\theta}[k]$ in Eq. (7.52) of the derivative $y^{(r)}(x)$ be designed for the direction θ. Assume that $y(x) \in \mathcal{P}^m_\theta$. Then, for any $h > 0$ such that* $\det \Phi_h \neq 0$,

$$\hat{y}_{h,\theta}[k] = y[k], \tag{8.14}$$

$$\hat{y}^{(r)}_{h,\theta}[k] = y^{(r)}[k]. \tag{8.15}$$

The proof follows from Eqs. (8.5, 8.11), and (8.13).

8.1.2 Vanishing moments

The conditions in Eqs. (8.5) and (8.11) for $u = 0$ can be written as

$$\sum_v g_h(v)\phi_h(-v) = \phi_h(0),$$
$$\sum_v g_h^{(r)}(v)\phi_h(-v) = \partial_u^r \phi_h(u)|_{u=0}. \tag{8.16}$$

The *vanishing moment conditions* similar to Eqs. (3.15) and (3.16) follow from Eq. (8.16):

(a) For the smoothing,

$$\sum_v g_h(v) = 1,$$
$$\sum_v g_h(v)v_1^{l_1} v_2^{l_2} = 0, \tag{8.17}$$
$$0 \le l_1 \le m_1, \qquad 0 \le l_2 \le m_2, \qquad l_1 + l_2 \ne 0;$$

(b) For the differentiating,

$$\frac{1}{r!}\sum_v g_h^{(r)}(v)v_1^{l_1} v_2^{l_2} = \begin{cases} (-1)^{|r|}, & \text{if } l_1 = r_1, \quad l_2 = r_2, \\ 0, & \text{otherwise,} \end{cases} \tag{8.18}$$
$$0 \le l_1 \le m_1, \qquad 0 \le l_2 \le m_2.$$

Remember that the variables v in Eqs. (8.16–8.18) are defined as $v = U(\theta)\Delta s$. With the notation $v_1(\Delta s)$ and $v_2(\Delta s)$ for these variables, in the basic coordinate system the conditions in Eqs. (8.17–8.18) have the following forms:

(a) For the smoothing,

$$\sum_{s\in\mathbb{Z}^2} g_{h,\theta}[s] = 1,$$
$$\sum_{s\in\mathbb{Z}^2} g_{h,\theta}[s]v_1^{l_1}(\Delta s)v_2^{l_2}(\Delta s) = 0, \tag{8.19}$$
$$0 \le l_1 \le m_1, \qquad 0 \le l_2 \le m_2, \qquad \text{or } l_1 + l_2 \ne 0,$$

where the kernel $g_{h,\theta}$ is defined in Eq. (7.48);

(b) For the differentiating,

$$\frac{1}{r!}\sum_{s\in\mathbb{Z}^2} g_{h,\theta}^{(r)}[s]v_1^{l_1}(\Delta s)v_2^{l_2}(\Delta s) = \begin{cases} (-1)^{|r|}, & \text{if } l_1 = r_1, \quad l_2 = r_2, \\ 0, & \text{otherwise,} \end{cases} \tag{8.20}$$

where $0 \le l_1 \le m_1$, $0 \le l_2 \le m_2$, and the kernels $g_{h,\theta}^{(r)}$ are defined in Eq. (7.50).

8.1.3 Taylor series

The directional kernels may use different-order approximations on the variables and different scales on these variables. This gives more flexibility in kernel design while the accuracy analysis and the scale selection are a bit different from those given in Chapter 5.

Let us start from some notes on the Taylor series used in this analysis. In the multi-index notation for $x \in \mathbb{R}^d$, the standard Taylor series exploited in the previous chapters has the form (see the Appendix, formulas (14.64)):

$$\begin{aligned} f(x-u) &= \sum_{t=0}^{|Q|}(-1)^t \sum_{|k|=t} \frac{1}{k!} u^k \partial^k f(x) + R_{|Q|}, \\ R_{|Q|} &= (-1)^{|Q|+1} \sum_{|k|=|Q|+1} \frac{1}{k!} u^k \partial^k f(x-\lambda u), \qquad 0 \le \lambda \le 1, \end{aligned} \tag{8.21}$$

where the power of the polynomial $u^k = u_1^{k_1} \cdot \ldots \cdot u_d^{k_d}$ is calculated as the sum $|k| = \sum_{j=1}^{d} k_j$, and $|Q|$ is a maximal power in the main term of the series. The parameter λ in Eq. (8.21) can take different values in different items of the Lagrange residual term $R_{|Q|}$. This form of the Taylor series does not fit to the modeling of the directional kernels. To be relevant to these models, the main term of the Taylor series should be given in a form where the powers are different for variables.

We start from the series in Eq. (8.21) to introduce this *separable* (*anisotropic*) form of the Taylor series. Let the multi-index $Q = (Q_1, \ldots, Q_d)$ denote the maximal powers of the Taylor series for the corresponding variables, and the main term of the series with different powers on the variables have the form

$$\sum_{t=0}^{Q}(-1)^{|t|}\frac{u^t}{t!}\partial_x^t f(x) \triangleq \sum_{t_1=0}^{Q_1} \cdots \sum_{t_d=0}^{Q_d}(-1)^{|t|}\frac{u_1^{t_1} \cdot \ldots \cdot u_d^{t_d}}{t_1! \cdot \ldots \cdot t_d!}\partial_{x_1,\ldots,x_d}^{(t_1,\ldots,t_d)} f(x_1, \ldots, x_d).$$

Reordering the items of the series Eq. (8.21), we represent it in the form

$$f(x-u) = \sum_{t=0}^{Q}(-1)^{|t|}\frac{u^t}{t!}\partial_x^t f(x) + R_Q. \tag{8.22}$$

Assume in Eq. (8.21) that $|Q| = \sum_{i=1}^{d} Q_i$. Then the residual R_Q in Eq. (8.22) is

$$R_Q = \sum_{t=0}^{|Q|}(-1)^{|t|}\frac{u^t}{t!}\partial_x^t f(x) - \sum_{t=0}^{Q}(-1)^{|t|}\frac{u^t}{t!}\partial_x^t f(x) + R_{|Q|}, \tag{8.23}$$

which says that the representation in Eq. (8.22) uses the standard Taylor series in Eq. (8.21) of the power $|Q|$, where all items excluding the main term

$\sum_{t=0}^{Q}(-1)^{|t|}\frac{u^t}{t!}\partial_x^t f(x)$ are collected in the residual R_Q. Let us call the representation in Eq. (8.22) the *anisotropic Taylor series*, where the different powers are used for the variables in the main term.

Consider a simple example illustrating the difference between the standard and anisotropic Taylor series. Assume that $d = 2$ and the powers are $Q_1 = 2$, $Q_2 = 1$.

In the main part of the standard Taylor series in Eq. (8.21) of the power $|Q| = Q_1 + Q_2 = 3$, we have the polynomials $u_1^{t_1}u_2^{t_2}$ of the powers $|t_1 + t_2| \leq 3$, that is,

$$1,\ u_1,\ u_1^2,\ u_1^3,\ u_2,\ u_2^2,\ u_2^3,\ u_1u_2,\ u_1^2u_2,\ u_1u_2^2. \tag{8.24}$$

The residual term $R_{|Q|}$ includes polynomials of the power $|Q| + 1 = 4$. In the main part of the anisotropic series in Eq. (8.22), we have the polynomials $u_1^{t_1}u_2^{t_2}$, of the powers $0 \leq t_1 \leq 2$, $0 \leq t_2 \leq 1$, that is,

$$1,\ u_1,\ u_1^2,\ u_2,\ u_1u_2,\ u_1^2u_2. \tag{8.25}$$

Comparing the sets of polynomials in Eqs. (8.24) and (8.25), we may realize that the residual R_Q in Eq. (8.23) of the anisotropic Taylor series, besides the terms of $R_{|Q|}$, includes the polynomials

$$u_1^3, u_2^2, u_2^3, u_1u_2^2. \tag{8.26}$$

For small u_1 and u_2, the order of the residual in Eq. (8.23) of the anisotropic series is defined by the polynomials in Eq. (8.26) as the terms of $R_{|Q|}$ that are smaller. This means that

$$R_Q = O(u_1^3) + O(u_2^2).$$

For the directional estimation, we are interested in the main terms of the Taylor series separately for each of the variables. In general, for the arbitrary orders Q_1 and Q_2, we have the orders of these main terms of the residual as

$$R_Q = O(u_1^{Q_1+1}) + O(u_2^{Q_2+1}). \tag{8.27}$$

The anisotropic Taylor series has different maximum powers with respect to the variables, and the residual term is of the corresponding different orders for these variables. Contrary to it, the standard Taylor series has the same maximum powers and the residual composed from the items having the same order with respect to all variables.

8.1.4 Basic estimates

In this section we consider the accuracy of the directional basic estimates for the anisotropic class of signals defined as relevant to the models used in the directional LPA. The basic estimate assumes that $\theta = 0$ and the directionality means that the different scales and LPA orders are used for the variables in the kernel design. We will use the notation x for the estimation point Δk.

For the general multivariate case $x \in \mathbb{R}^d$, the regular grid estimates in Eqs. (7.47–7.48) and Eqs. (7.49–7.50) can be rewritten as the following convolutions:

For the signal estimate,

$$\begin{aligned} \hat{y}_h(x) &= (g_h \circledast z)(x), \\ g_h(x) &= w_h(x)\phi_h^T(x)\Phi_h^{-1}\phi(0), \\ \Phi_h &= \sum_{l\in\mathbb{Z}^d} w_h(\Delta l)\phi_h(\Delta l)\phi_h^T(\Delta l), \end{aligned} \tag{8.28}$$

and for the derivative estimate,

$$\begin{aligned} \hat{y}_h^{(r)}(x) &= (g_h^{(r)} \circledast z)(x), \\ g_h^{(r)}(x) &= \frac{(-1)^{|r|}}{\Pi_{i=1}^d h_i^{r_i}} w_h(x)\phi_h^T(x)\Phi_h^{-1}\phi^{(r)}(0), \\ w_h(x) &= \frac{1}{\Pi_{i=1}^d h_i} w(x/h). \end{aligned} \tag{8.29}$$

The estimation errors are then given as

$$\begin{aligned} e_y(x,h) &= y(x) - \hat{y}_h(x), \\ e_{y^{(r)}}(x,h) &= y^{(r)}(x) - \hat{y}_h^{(r)}(x). \end{aligned} \tag{8.30}$$

Because the new kernels have different orders with respect to the different argument variables, we specify the corresponding class of anisotropic signals as follows:

$$\mathbb{C}^{(\alpha)} = \{y : |\partial^\alpha y(x)| = L_\alpha(x) \le \bar{L}_\alpha, x \in \mathbb{R}^d\}, \tag{8.31}$$

where α is the multi-index showing that the orders of the derivatives on the arguments x; $\bar{L}_\alpha$ is finite.

The class $\mathbb{C}^{(\alpha)}$ is obviously different from the one in Eq. (2.6), where α is a scalar denoting the total order of the derivatives. This class also includes the polynomial functions. If $\bar{L}_{m+1} = 0$, then all derivatives $\partial^{m+1}y(x) = 0$, $m+1 = (m_1+1, \ldots, m_d+1)$, which means that $y(x)$ is a polynomial of degree m, that is, $y(x) \in \mathcal{P}^m$, where m is the multi-index.

Besides, we assume that the derivatives $\partial^\alpha y(x)$ in Eq. (8.31) are continuous *Lipshitz functions*,

$$|\partial^\alpha y(x) - \partial^\alpha y(x')| \le K_L||x - x'||^\gamma, \qquad 0 < \gamma \le 1, \tag{8.32}$$

where $K_L > 0$ and α is a multi-index.

Together, the assumptions in Eqs. (8.31) and (8.32) define the *Hölder class* of signals with notation $\mathbf{H}^{\alpha,\gamma}$. Assume that:

(D1) $y(x)$ is deterministic locally smooth and belongs to the Hölder class $\mathbf{H}^{M,\gamma}$ of M differentiable functions in Eq. (8.31), where $M = (M_1, \ldots, M_d)$ and

$$M_i = \min\{m_i + 1, \alpha_i\};$$

(D2) The LPA estimates of the signal and the derivatives are defined by Eqs. (8.28) and (8.29).

The following proposition gives the accuracy of these basic estimates.

Proposition 12 *Let the hypotheses D1 and D2 hold; the noise in the observation model in Eq. (2.1) be white with the variance σ^2; and h, Δ, $\Delta/h \to 0$. Then, the accuracy of the multivariate discrete regular grid estimates in Eqs. (8.28–8.29) is defined by the following formulas for the bias:*

$$
\begin{aligned}
m_{\hat{y}_h}(x,h) &= \sum_{i=1}^{d}(-1)^{M_i-1}h_i^{M_i}y^{(M_i)}(x)A_{g,i} + o\left(\sum_{i=1}^{d}h_i^{M_i}\right),\\
A_{g,i} &= \frac{1}{M_i!}\int_{\mathbb{R}^d} g(u)u_i^{M_i}du,
\end{aligned}
\tag{8.33}
$$

$$
\begin{aligned}
m_{\hat{y}_h^{(r)}}(x,h) &= \sum_{i=1}^{d}(-1)^{M_i-1}h^{M_i-r_i}y^{(M_i)}(x)A_{g^{(r)},i} + o\left(\sum_{i=1}^{d}h_i^{M_i-r_i}\right),\\
A_{g^{(r)},i} &= \frac{1}{M_i!}\int_{\mathbb{R}^d} g^{(r)}(u)u_i^{M_i}du
\end{aligned}
\tag{8.34}
$$

and the following formulas for the variances:

$$
\begin{aligned}
\frac{1}{\sigma^2}\sigma^2_{\hat{y}_h}(x,h) &= \Pi_{i=1}^{d}\left(\frac{\Delta_i}{h_i}\right)B_g + o\left(\Pi_{i=1}^{d}\frac{\Delta_i}{h_i}\right),\\
B_g &= \int_{\mathbb{R}^d}[g(u)]^2du,
\end{aligned}
\tag{8.35}
$$

$$
\begin{aligned}
\frac{1}{\sigma^2}\sigma^2_{\hat{y}_h^{(r)}}(x,h) &= \Pi_{i=1}^{d}\left(\frac{\Delta_i}{h_i^{1+2r_i}}\right)B_{g^{(r)}} + o\left(\Pi_{i=1}^{d}\frac{\Delta_i}{h_i^{1+2r_i}}\right),\\
B_{g^{(r)}} &= \int_{\mathbb{R}^d}[g^{(r)}(u)]^2du,
\end{aligned}
\tag{8.36}
$$

The kernels g and $g^{(r)}$ are defined in Eqs. (4.34–4.35), where the polynomials of the corresponding multi-index power m are assumed in the vector functions ϕ.

The proof of the proposition is given in Appendix 14.5. This proposition is analogue to Proposition 7 in Chapter 5.

Omitting the small terms $o(\cdot)$, the following upper bounds for the bias can be derived from Eqs. (8.33) and (8.35):

$$
\begin{aligned}
l_{\hat{y}_h}(x,h) &\triangleq E(y(x)-\hat{y}_h(x))^2 \leq \bar{l}_{\hat{y}_h}(x,h),\\
\bar{l}_{\hat{y}_h}(x,h) &= \left(\sum_{i=1}^{d}h_i^{M_i}L_{M_i}(x)A_{g,i}\right)^2 + \sigma^2\Pi_{i=1}^{d}\frac{\Delta_i}{h_i}B_g.
\end{aligned}
\tag{8.37}
$$

In a similar way, for the derivative estimation we have from Eqs. (8.34) and (8.36)

$$l_{\hat{y}_h^{(r)}}(x,h) \triangleq E[y^{(r)}(x) - \hat{y}_h^{(r)}(x)]^2 \leq \bar{l}_{\hat{y}_h^{(r)}}(x,h),$$

$$\bar{l}_{\hat{y}_h^{(r)}}(x,h) = \left(\sum_{i=1}^{d} h^{M_i - r_i} L_{M_i}(x) A_{g^{(r)},i}\right)^2 + \sigma^2 \Pi_{i=1}^{d}\left(\frac{\Delta_i}{h_i^{1+2r_1}}\right) B_{g^{(r)}}. \tag{8.38}$$

Minimizing $\bar{l}_{\hat{y}_h}$ and $\bar{l}_{\hat{y}_h^{(r)}}$ on the vector h, the ideal scales can be found. These ideal scales h^* for the signal and derivative estimations satisfy the equations

$$\partial_h \bar{l}_{\hat{y}_h}(x,h) = 0, \tag{8.39}$$

$$\partial_h \bar{l}_{\hat{y}_h^{(r)}}(x,h) = 0. \tag{8.40}$$

The corresponding optimal solutions are studied in Propositions 19 and 20 of Sec. 14.5 of the Appendix. Besides the optimal scales and convergence rates of the estimates, these propositions show that there are special links between the ideal directional scales.

Let us discuss the corresponding corollaries from Propositions 19 and 20.

- Assume that all derivatives of y used in the propositions are bounded and of more or less the same magnitude, and let h_k^* be the optimal scale for the variable x_k for the signal estimation. Then the following link of the optimal scales is valid (see Eq. (14.148)):

$$(h_k^*)^{m_k+1} \sim (h_l^*)^{m_l+1}, l, k = 1, \ldots, d. \tag{8.41}$$

For $d = 2$, this can be interpreted in the following way. Let the kernel be elongated along x_1. Then h_1 and h_2 define the length and width of the kernel, respectively. It is natural to assume that we can use a larger-order approximation for the length than for the width, that is, $m_1 > m_2$.

Formula (8.41) installs the following correspondence between the length and the width:

$$(h_2^*)^{m_2+1} \sim (h_1^*)^{m_1+1}, \tag{8.42}$$

where m_1 and m_2 are the orders of the polynomials in the LPA on x_1 and x_2, respectively. In particular, for $m_1 = 1$ and $m_2 = 0$, we have

$$h_2^* \sim (h_1^*)^2. \tag{8.43}$$

This is *the scaling law width* $\sim$ *length*^{2} in Eq. (7.15) exploited in the curvelet transform [210]. This relation is derived from a general accuracy analysis. It shows that this law is a type of optimal relation between the scales in the directional LPA provided the degrees $m_1 = 1$ and $m_2 = 0$ are used for the approximation along and across the directional window.

- The convergence rate for the signal estimation according to Eqs. (14.150–14.151) has the form

$$\bar{l}_{\hat{y}_h}(x, h^*) = O\left(\left(\frac{\sigma^2}{n}\right)^{\beta_1}\right), \quad \beta_1 = \frac{1}{1 + (1/2)\sum_{k=1}^{d}[1/(m_k + 1)]}. \tag{8.44}$$

  If we assume that all of these orders are equal to each other, that is,

$$m_k = m_0, \alpha_k > m_0, \qquad k = 1, \ldots, d,$$

  then

$$\beta_1 = \frac{2(m_0 + 1)}{2(m_0 + 1) + d},$$

  and the convergence rate is identical to that observed in Eq. (5.35) provided that $M = m_0 + 1$ in these formulas.
- Assume again that all derivatives of y used in the propositions are bounded and of more or less the same magnitude, and let h_k^* be the optimal scale of the variable x_k for the derivative estimation. We obtain from Eq. (14.165) a link between the optimal scales that is different from Eq. (8.41). This link is of the form

$$(h_k^*)^{m_k+1-r_k} \sim (h_l^*)^{m_l+1-r_l}, l, k = 1, \ldots, d. \tag{8.45}$$

  We may conclude that the links between the length and width of the kernels depend on the used orders of the LPA and on the orders of the estimated derivatives.
- The convergence rate for the derivative estimation according to Eqs. (14.168–14.169) has the form

$$\bar{l}_{\hat{y}_h^{(r)}}(x, h^*) = O\left(\left(\frac{\sigma^2}{n}\right)^{\beta_2}\right),$$
$$\beta_2 = \frac{1}{1 + (1/2)\sum_{k=1}^{d}[(1 + 2r_k)/(m_k + 1 - r_k)]}. \tag{8.46}$$

For $r_k = 0$, formula (8.46) naturally coincides with Eq. (8.44).

8.1.5 Rotated estimates

It is natural to analyze the directional estimates in the rotated coordinates using the signal model as well as the estimates given in these coordinates. The assumptions in Eqs. (8.31) and (8.32) concerning the anisotropic properties of the object signal are replaced by the corresponding directional ones:

$$\mathbb{C}_\theta^{(\alpha)} = \{y_\theta : |\partial_u^\alpha y_\theta(u)| = L_{\alpha,\theta}(u) \le \bar{L}_{\alpha,\theta}, u \in \mathbb{R}^2\}, \tag{8.47}$$

$$|\partial^\alpha y_\theta(u) - \partial^\alpha y_\theta(v)| \le K_{L_\theta}||u - v||^{\gamma_\theta}, 0 < \gamma_\theta < 1. \tag{8.48}$$

Here, $y_\theta(u)$ is a signal given in the rotated coordinates in Eq. (8.12) as

$$y_\theta(u) = y(U^T(\theta)u).$$

A class of directionally anisotropic signals is defined by Eqs. (8.47–8.48) as the derivative values as well as the smoothness α of y and can be direction-dependent.

Represent the directional estimates in Eqs. (7.47–7.50) also in the rotated variables as Eqs. (7.37–7.40) using the shift-invariant kernels Eq. (7.46):

$$\hat{y}_{h,\theta}(u) = \sum_s g_h(u - v_s)z_s, \tag{8.49}$$

$$\begin{aligned} g_h(u) &= w_h(u)\phi_h^T(u)\Phi_h^{-1}\phi(0), \\ \Phi_h &= \sum_s w_h(v_s)\phi_h(v_s)\phi_h^T(v_s), \end{aligned} \tag{8.50}$$

$$\begin{aligned} \hat{y}_{h,\theta}^{(r)}(u) &= \sum_s g_h^{(r)}(x - v_s)z_s, \\ g_h^{(r)}(u) &= \frac{(-1)^{|r|}}{h_1^{r_1}h_2^{r_2}} w_h(u)\phi_h^T(u)\Phi_h^{-1}\phi^{(r)}(0). \end{aligned} \tag{8.51}$$

In these formulas, the kernels $g_h(u)$, $g_h^{(r)}(u)$ are defined only by the given nodes v_s and the argument u. Note that the kernels g_h and $g_h^{(r)}$ are shift-invariant on the regular grid in Eq. (7.43) and in general are shift-varying in variables u. In general, the grid of v_s is irregular (see Fig. 7.9).

For the accuracy analysis we need to represent the observations z_s in the rotated coordinates. Using the notation in Eq. (8.12), we obtain for the signal and observations

$$y_s = z(U^T v_s) \triangleq y_\theta(v_s), \qquad z_s = z(U^T v_s) \triangleq z_\theta(v_s). \tag{8.52}$$

The estimation errors in Eq. (8.1) in the rotated variables can be given as

$$\begin{aligned} e_{y,\theta}(u,h) &= y_\theta(u) - \hat{y}_{h,\theta}(u), \\ e_{y^{(r)},\theta}(u,h) &= y_\theta^{(r)}(u) - \hat{y}_{h,\theta}^{(r)}(u). \end{aligned} \tag{8.53}$$

Inserting Eqs. (8.52) in (8.49) and (8.51), we may note that, provided fixed nodes v_s, the kernels of the estimates do not depend on θ and the directionality of the estimate is ensured by the rotation of the observations z_s.

Comparing Eqs. (8.49–8.53) with Eqs. (8.28–8.30), we may see that these formulas are identical provided that x, X_s, z, y are replaced by u, v_s, z_θ, y_θ, respectively.

The bias error is equal to

$$m_{\hat{y}_{h,\theta}^{(r)}}(u,h) = \partial_u^r y_\theta(u) - E\{\hat{y}_{h,\theta}^{(r)}(u)\}, \tag{8.54}$$

$$E\{\hat{y}_{h,\theta}^{(r)}(u)\} = (g_h^{(r)} \circledast y_\theta)(u). \tag{8.55}$$

The properties of the directional y_θ used for the bias evaluation are defined by the assumptions Eqs. (8.47–8.48). The random errors are

$$e^0_{y^{(r)},\theta}(u,h) = -(g_h^{(r)} \circledast \varepsilon)(u).$$

Assuming that the random ε are independent and identically distributed with the variance σ^2, the variance of the estimates is given in the form

$$\sigma^2_{\hat{y}^{(r)}_{h,\theta}}(u,h) = \sigma^2 ||g_h^{(r)}||^2$$

and does not depend on u.

It is quite obvious that Propositions 12 and 20 (Sec. 14.5 of the Appendix), which give the accuracy results for the basic anisotropic estimates, are valid for the directional estimate in the form of Eqs. (8.49–8.51) with a fixed arbitrary θ provided that the following replacements are produced in the formulas of the propositions:

$$x \Longrightarrow u,\ y \Longrightarrow y_\theta,\ y^{(M_i)}(x) \Longrightarrow \partial_u^{M_i} y_\theta(u), \quad L_\alpha(\cdot) \Longrightarrow L_{\alpha,\theta}(\cdot), \quad \bar{L}_\alpha \Longrightarrow \bar{L}_{\alpha,\theta}$$

and

$$m_{\hat{y}_h^{(r)}}(x,h) \Longrightarrow m_{\hat{y}^{(r)}_{h,\theta}}(u,h), \qquad \sigma^2_{\hat{y}_h^{(r)}}(x,h) \Longrightarrow \sigma^2_{\hat{y}^{(r)}_{h,\theta}}(u,h).$$

In particular, this means that the optimal scaling law in Eq. (8.41) holds for any θ.

8.2 Adaptive-Scale Algorithms

8.2.1 Calculations

8.2.1.1 *Scale ratio and ICI*

The scaling multi-index h controls the size as well as the shape of the directional kernel. The link between h_1 and h_2 in Eq. (8.42), found from the accuracy optimization, can be used to define the width as a function of the length and to simplify the scale selection by reducing it to a selection of the length only.

In what follows, we always consider the multi-index h as an ordered sequence of the pairs "*length-width*":

$$H = \{h^{\langle 1\rangle}, h^{\langle 2\rangle}, \ldots, h^{\langle J\rangle}\}, \tag{8.56}$$

where $h^{\langle k\rangle} = (h_1^{\langle k\rangle}, h_2^{\langle k\rangle})$ is the kth pair of length $h_1^{\langle k\rangle}$ and width $h_2^{\langle k\rangle}$.

In the set H, the width $h_2^{\langle k\rangle}$ can be considered as a function of the length $h_1^{\langle k\rangle}$ or just as a function of the number $^{\langle k\rangle}$. The ICI is applied for the selection of the best pair from H. The pairs are ordered and tested by the ICI according to their number $^{\langle k\rangle}$. It is a principal point that the set H has a unique parameter $^{\langle k\rangle}$ defining the scales, and in this way the scale selection continues to be a univariate problem. The unique parameter $^{\langle k\rangle}$ ordering the scales makes it possible to apply the standard ICI technique for a pointwise data-driven selection of $h^+(x,\theta)$ for each of the directional estimates $\hat{y}_{h,\theta}(x)$ and for each direction θ.

The width $h_2^{\langle k\rangle}$ in H can be selected according to the rule in Eq. (8.42) or from different speculations because, practically, the set H is one of the design parameters of the algorithm. One of the principal restrictions on the selection and ordering of the elements in H is that the variance of the estimates calculated for $h^{\langle k\rangle}$ should be a decreasing function of the number $\langle k\rangle$.

8.2.1.2 Estimates and estimate fusion

A sectorial scale optimization results in multiple adaptive estimates for each x. Using K sectors we obtain K independently derived estimates. This raises the problem of fusing or aggregating these directional scale-adaptive estimates.

Recall the original idea of the direction estimation discussed in Sec. 7.1.2. The scales $h^{+}(x,\theta_j)$ shape the adaptive estimation neighborhood as a union of the supports for the directions θ_j, that is, $\bigcup_j \operatorname{supp} g_{h+(x,\theta_j),\theta_j}$. In this way we arrive at the spatially adaptive varying-scale approximation of the best neighborhood U^* as it is given in Fig. 7.2.

The multiwindow and combined window estimates (see Sec. 6.2) are exploited to obtain the unique final estimate $\hat{y}(x)$ from the partial directional ones. In particular, the multiwindow estimate with the inverse variances as the weights for the linear fusing gives

$$\begin{aligned} \hat{y}(x) &= \sum\nolimits_{j=1}^{K} \lambda_j \hat{y}_{\theta_j}(x), \quad \hat{y}_{\theta_j}(x) = \hat{y}_{h,\theta_j}(x)|_{h=h_+(x,\theta_j)}, \\ \lambda_j &= \frac{\sigma_j^{-2}(x)}{\sum_{i=1}^{K} \sigma_i^{-2}(x)}, \end{aligned} \tag{8.57}$$

where

$$\hat{y}_{h,\theta_j}(x) = (g_{h,\theta_j} \circledast z)(x), \quad \sigma_j^{-2}(x) = \sigma^2_{\hat{y},\theta_j}(x,h)|_{h=h_+(x,\theta_j)},$$

and the variance of the adaptive-scale estimate is calculated as

$$\sigma^2_{\hat{y},\theta_j}(x,h) = \sigma^2 \sum_x g^2_{h,\theta_j}(x). \tag{8.58}$$

Note that the weights λ_j in Eq. (8.57) are data-driven adaptive as $\sigma_j^{-2}(x)$ depend on the adaptive pointwise $h^{+}(x,\theta_j)$. Assuming that the supports of the kernels $g_{\theta_j,h}$ are not overlapping, and neglecting the fact that the kernels have the point x as a common one, we obtain for the variance of the fused estimate in Eq. (8.57)

$$\sigma^2_{\hat{y}}(x) = \sum\nolimits_{j=1}^{K} \lambda_j^2 \sigma_j^2(x) = \frac{1}{\sum_{j=1}^{K} \sigma_j^{-2}(x)}. \tag{8.59}$$

8.2.2 Recursive algorithms

The above calculations can be exploited in standard and recursive modes. The flowchart of these algorithms is presented in Fig. 8.1. In the *standard mode* (Fig. 8.1(a)), the LPA, ICI, and fusing blocks work successively and only once:

- *The LPA estimation block.* The inputs are the observations z and the observation variance σ_z. The outputs are the estimates $\hat{y}_{h,\theta_j}$ calculated for all scales $h \in H$ and for all directions θ_j in the multiple-window version of the algorithm. In addition, the variances $\sigma^2_{\hat{y}_{h,\theta_j}}$ of these estimates are also calculated. The block includes the LPA filter design corresponding to the scales h and to the given directional windows.
- *The ICI block.* The inputs are the estimates $\hat{y}_{h,\theta_j}$ and the variances $\sigma^2_{\hat{y}_{h,\theta_j}}$. The outputs are the adaptive-scale estimates $\hat{y}_{h^+,\theta_j}$ and the variances $\sigma^2_{\hat{y}_{h^+,\theta_j}}$ of these estimates.
- *The fusing block.* The inputs are the adaptive-scale estimates $\hat{y}_{h^+,\theta_j}$ and the variances $\sigma^2_{\hat{y}_{h^+,\theta_j}}$ of these estimates. The outputs are the final estimates $\hat{y}(x)$ and the variances $\sigma^2_{\hat{y}}(x)$ of these estimates.

It is convenient to treat this complex LPA-ICI multidirectional algorithm as an adaptive filter with two inputs z and σ_z and two outputs $\hat{y}$ and $\sigma_{\hat{y}}$ (Fig. 8.1(b)).

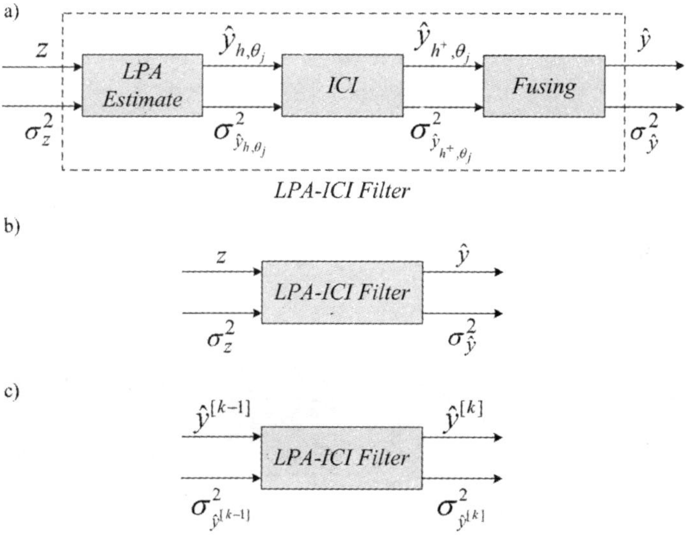

Figure 8.1 Flowcharts of the LPA-ICI-adaptive-scale algorithm: (a) standard algorithm, (b) input-output model of the standard LPA-ICI algorithm, and (c) the LPA-ICI as an element of the recursive algorithm.

Denoting the calculations imbedded in this algorithm as an $\mathcal{LI}$ operator, the input-output equation can be written as

$$(\hat{y}, \sigma_{\hat{y}}) = \mathcal{LI}\{z, \sigma_z\}. \tag{8.60}$$

This input-output link is defined by formulas (8.57)–(8.59).

In the *recursive mode*, the LPA-ICI multidirectional algorithm is applied repeatedly to the data obtained as outputs of the previous filtering (Fig. 8.1c). In this way we arrive at the *recursive directional LPA-ICI algorithm.*

The equations for the recursive algorithm can be written in the form

$$(\hat{y}^{[k]}, \sigma_{\hat{y}}^{[k]}) = \mathcal{LI}\{z^{[k-1]}, \sigma_z^{[k-1]}\}, \qquad k = 1, 2, \ldots, \tag{8.61}$$

where the index $^{[k]}$ numbers the iterations.

Two iterations of this recursive LPA-ICI filter are visualized in Fig. 8.2.2. At the natural initialization $\hat{y}^{[0]} = z$, $\sigma_{\hat{y}[0]} = \sigma_z$, we have the noisy image and the constant standard deviation of the additive input noise. After the first iteration, the image is filtered. The standard deviation becomes smaller and depends on x with a larger level of the noise near the edges of the image. After the second iteration, the standard deviation of the noise becomes significantly smaller overall and we obtain a good filtered image.

The estimation variance in Eq. (8.59) now depends on x. This means that in the recursive algorithm, starting from the second iteration, we cannot use the variance formula (8.58), assuming that the variance of the input signal is the same for all x. Also, starting from the second iteration, the errors in the estimate $\hat{y}$ can be correlated because the resulting estimation neighborhoods may overlap each other.

For each x and each iteration, we know the adaptive-scale size used in the estimate in Eq. (8.57). This makes it possible to reconstruct all kernels linking the estimate for a given x with the initial observations z, where the errors are uncorrelated. Using these kernels, we can accurately calculate the variances of the estimates. However, these calculations are complex and too demanding for computer memory to be considered as practical ones. We implement simplified variance calculations for the recursive algorithm that mimic the ones used in the standard scheme.

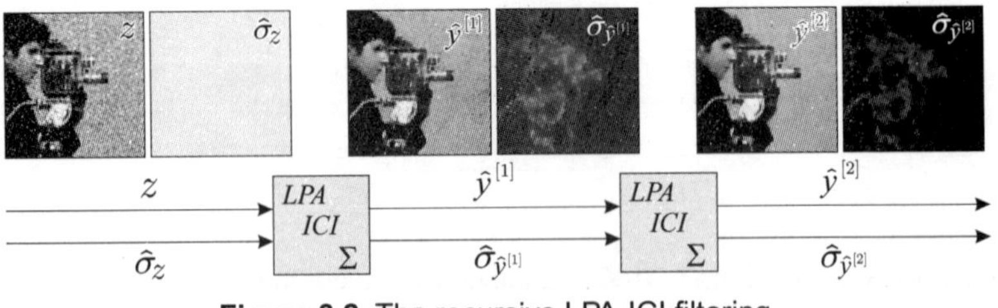

Figure 8.2 The recursive LPA-ICI filtering.

The estimate calculation and fusing are defined according to Eq. (8.57) as follows [54]:

$$\begin{aligned} \hat{y}^{[k]}(x) &= \sum\nolimits_{i=1}^{K} \lambda_i \hat{y}_{\theta_i}^{[k]}(x), \\ \hat{y}_{\theta_i}^{[k]}(x) &= (g_{h,\theta_i} \circledast \hat{y}^{[k-1]})(x)|_{h=h^+}, \qquad k = 1, \ldots, \end{aligned} \tag{8.62}$$

where

$$\lambda_i = \frac{\sigma_i^{-2}(x)}{\sum_{j=1}^{K} \sigma_j^{-2}(x)}, \sigma_i^2(x) = \sigma^2_{\hat{y}_{h,\theta_i}^{[k]}}(x, h^+). \tag{8.63}$$

These formulas are completely similar to Eq. (8.57) provided that the variances $\sigma_i^2(x)$ are known. The only difference is that the input $\hat{y}^{[k-1]}$ of the directional estimate $\hat{y}_{\theta_i}^{[k]}(x)$ is a result of the previous iteration.

The most serious changes concern the calculations of variance of the estimate $\hat{y}_{\theta_i}^{[k]}(x)$. We assume for these calculations that the errors of the estimate $\hat{y}^{[k-1]}(x)$ are uncorrelated for different x. Then the variance of the estimate $\hat{y}_{h,\theta_i}^{[k]}(x) = (g_{\theta_i,h} \circledast \hat{y}^{[k-1]})(x)$ is calculated as the convolution

$$\sigma^2_{\hat{y}_{\theta_i,h}^{[k]}}(x, h) = (g^2_{h,\theta_i} \circledast \sigma^2_{\hat{y}^{[k-1]}})(x), \tag{8.64}$$

where $\sigma^2_{\hat{y}^{[k-1]}}$ is the variance of the fused estimate of the previous iteration.

To calculate the variance of the fused estimate $\sigma^2_{\hat{y}^{[k-1]}}$, we again assume that the directional estimates have uncorrelated errors for different x and use formula Eq. (8.59):

$$\sigma^2_{\hat{y}(k-1]}(x) = \frac{1}{\sum_{j=1}^{K} \sigma_j^{-2}(x)}, \qquad \sigma_i^2(x) = \sigma^2_{\hat{y}_{h,\theta_i}^{[k-1]}}(x, h^+). \tag{8.65}$$

However, because the assumption that the random errors are uncorrelated for different x does not hold, the quality of the estimate deteriorates and typically results in oversmoothing details in the image.

It turns out that, for zero- and first-order LPA kernels, a simple compensating factor for the variance effectively reduces this deterioration. We use this variance compensation in the simple form of a constant factor in formula Eq. (8.64), replacing it by

$$\sigma^2_{\hat{y}_{h,\theta_i}^{[k]}}(x, h) = \gamma^2 \left(g^2_{h,\theta_i} \circledast \sigma^2_{\hat{y}_{h,\theta_i}^{[k-1]}} \right)(x), \tag{8.66}$$

where $0 < \gamma < 1$ is a compensation factor. In our simulation experiments, we use $\gamma = 2/3$.

Thus, the operator $\mathcal{LI}$ in the recursive algorithm in Eq. (8.61) is defined by formulas (8.62)–(8.63) and Eqs. (8.65–8.66). This recursive algorithm has been used for detail-preserving image denoising, segmentation, and edge detection applications.

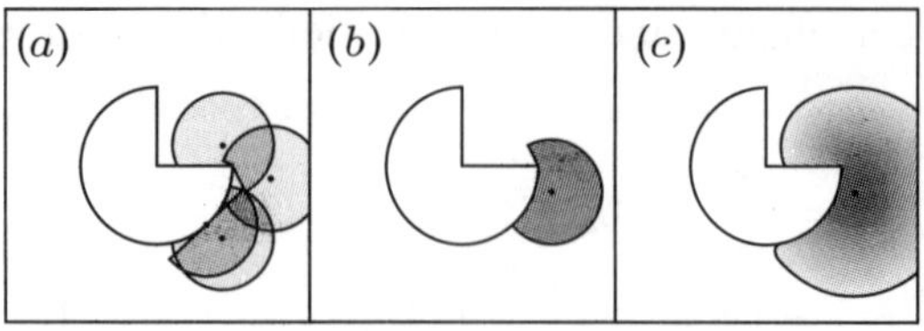

Figure 8.3 The estimation neighborhood enlargement: (b) ICI adaptive neighborhood of the point *x*, (a) ICI adaptive neighborhoods of the points included in the neighborhood of *x*, and (c) an effective neighborhood for the point *x* in the next iteration of estimation.

The images in Fig. 8.3 give a general idea why the recursive algorithm is more efficient than the standard ones. Let the adaptive estimation neighborhood for the point x obtained during the first iteration be as shown in Fig. 8.3(b). Note that each point from this neighborhood has its own ICI adaptive neighborhood, shown in Fig. 8.3(a). These neighborhoods are explicitly exploited in the first iterative estimate.

At the second iteration, the estimate at x, using the first iteration estimates as the input of the LPA-ICI filter, in fact, has an enlarged support combining the supports of all of the first iteration estimates shown in Fig. 8.3(a). This enlarged support, illustrated in Fig. 8.3(c), is not the compulsory star shape.

The efficiency of the recursive LPA-ICI algorithm will discussed later, in the simulation results.

8.2.3 Algorithm complexity

The complexity of the algorithm is proportional to the number of directions (sectors) used. According to Sec. 6.1.3, we note that the algorithm is fast based on the fast convolution operations. The calculation of the estimate $\hat{y}_{h,\theta_i}(x)$ for a given scale h is a linear convolution requiring $N_{conv} \sim n \log n$ operations, where n is the size of the signal. Because the sectorial kernels usually have small supports, the complexity can be reduced to $N_{conv} \sim n \log n_{h_J}$, where n_{h_J} is the maximum size of the mask of the kernels $g_{h_j,\theta}$. This procedure is repeated $J \cdot K$ times, where K is the number of sectors in the estimator and J is the number of pair-scales in H in Eq. (8.56). Overall, the algorithm complexity is proportional to $J \cdot K \cdot N_{conv}$. For the recursive algorithm, this complexity is multiplied by the number of iterations.

8.3 Directional Image Denoising

8.3.1 Criteria and algorithms

In this section we present simulation results demonstrating the performance of the developed directional algorithms in image denoising. We use the following standard criteria:

(1) Root-mean-square error (RMSE):

$$\mathrm{RMSE} = \sqrt{\frac{1}{n}\sum_x (y(x) - \hat{y}(x))^2}; \tag{8.67}$$

(2) Signal-to-noise ratio (SNR) in decibels (dB):

$$\mathrm{SNR} = 10\log_{10}\frac{\sum_x |y(x)|^2}{\sum_x |y(x) - \hat{y}(x)|^2}; \tag{8.68}$$

(3) Improvement in SNR (ISNR) in decibels (dB):

$$\mathrm{ISNR} = 20\log_{10}\left(\frac{\hat{\sigma}_z}{RMSE}\right), \tag{8.69}$$

where $\hat{\sigma}_z$ is the estimate of the observation standard deviation;

(4) Peak signal-to-noise ratio (PSNR) in decibels (dB):

$$\mathrm{PSNR} = 20log_{10}\left(\frac{max_x|y(x)|}{RMSE}\right); \tag{8.70}$$

(5) Mean absolute error (MAE)

$$\mathrm{MAE} = \frac{1}{n}\sum_x \left|y(x) - \hat{y}(x)\right|; \tag{8.71}$$

(6) Maximum absolute error (MAXDIF):

$$\mathrm{MAXDIF} = \max_x |y(x) - \hat{y}(x)|. \tag{8.72}$$

his set of criteria was completed by the universal image quality index (UIQI) roposed recently in [231]. It provides a criterion normalized to the interval $[-1, 1]$ ith larger values implying better image quality.

There is no one-to-one link between the visual image quality and the above riteria. It has been noticed that different criteria show different optimal values for e design parameters. Thus, a visual inspection, which of course is quite subjective, ontinues to be the most important final performance criterion.

A test signal is the 256×256 *Cameraman* image (8-bit grayscale) corrupted y an additive zero mean Gaussian noise. The image intensity takes integer values $\leq y \leq 255$ with the noise standard deviation $\sigma = 25.5$ (SNR = 14.39 dB). The oisy image can be seen in Fig. 8.4.

To emphasize the directional character of the estimates, we use narrow nonoverpping sectorial kernels of type 10 (Sec. 7.3) of order $m = 0$ with a uniform indow function w. The kernel supports are defined by the 2D scale $h = (h_1, h_2)$ ith $h \in H$,

$$H = \left\{\begin{pmatrix}1\\1\end{pmatrix}, \begin{pmatrix}2\\1\end{pmatrix}, \begin{pmatrix}3\\1\end{pmatrix}, \begin{pmatrix}5\\1\end{pmatrix}, \begin{pmatrix}7\\2\end{pmatrix}, \begin{pmatrix}11\\3\end{pmatrix}\right\}. \tag{8.73}$$

Figure 8.4 Noisy *Cameraman* image, SNR = 14.39 dB, PSNR = 19.97 dB, RMSE = 25.57 MAE = 20.38, MAXDIF = 106.73.

The scales h_1 and h_2 form pairs and are used for estimation in these pairs. The supports of the kernels are line-wise of the width and equal $h_2 = 1$ for $h_1 = 1, 2, 3, 5$ and the width is increased to 2 and 3 pixels for $h_1 = 7$ and $h_1 = 11$, respectively. This set of basic kernel supports is depicted in Fig. 8.5 for $\theta = 0$.

The estimates are calculated as the sample means of observations included in the kernel supports. The ICI rule is used with the threshold $\Gamma = 1$. The estimates and the adaptive scales $h_j^+(x)$ are found for eight directions:

$$\theta_j = (j - 1)\pi/4, \qquad j = 1, \ldots, 8. \tag{8.74}$$

These ICI-adaptive directional estimates are fused in the final one using the multiwindow method in Eq. (8.57).

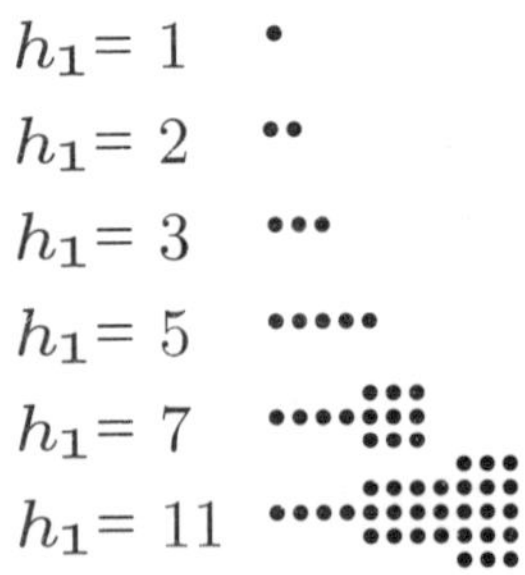

Figure 8.5 Discrete sectorial kernel supports for $\theta = 0$, where h_1 and h_2 are defined according to Eq. (8.73).

8.3.2 Experiments

8.3.2.1 *Results for* Cameraman

The central panel of Fig. 8.6 shows the true image. Eight surrounding panels show the ICI adaptive scales $h_j^+(x)$ for the corresponding eight directions θ_j. Thus, we can see the adaptive scales for directional estimates looking at the horizontal and vertical directions, i.e., to east, north, west, and south, as well as to four diagonal directions: northeast, northwest, southwest, and southeast.

White and black correspond to large and small scale values, respectively. The adaptive scales delineate the image intensity very well, as is the case provided that the intensity function is known in advance. This delineation is obviously directional because the contours of the image are shadowed from the corresponding directions.

Recall that, in Sec. 6.3.2, we use four quadrant estimates for imaging. The eight narrowed kernels exploited here allow one to make the estimates highly sensitive to image details and essentially improve the quality of denoising.

Figure 8.6 ICI-adaptive directional scales $\hat{h}_j^+(x)$, $\theta_j = (j-1)\pi/4$, $j = 1, \ldots, 8$, for the *Cameraman* test image. The true image is shown in the center panel.

Figure 8.7 demonstrates the estimates obtained. The center panel shows the final fused estimate calculated from the sectorial ones according to the multiwindow estimate formula, Eq. (8.57). The surrounding panels show the sectorial directional adaptive-scale estimates $\hat{y}_{\hat{h}_j^+(x),\theta_j}(x)$, $j = 1, \ldots, 8$, corresponding to the adaptive scales given in Fig. 8.6 for the relevant directions.

The noise effects are clearly seen in the adaptive scales h_j^+ as spread-out black isolated points. The black means that the ICI erroneously takes smaller values of the scale.

The directional nature of the adaptive estimates $\hat{y}_{\hat{h}_j^+(x),\theta_j}(x)$ is obvious since the corresponding directions are seen as a line-wise background of this imaging. The multiwindow fusing allows us to delete and smooth these directional line effects and obtain a good-quality final estimate. Overall, the multiwindow estimation allows us to reveal and preserve a tiny detail of the image and at the same time efficiently suppress the noise.

Figure 8.7 LPA-ICI adaptive-scale directional estimates $\hat{y}_{j,\hat{h}_j^+(x)}(x)$, $\theta_j = (j-1)\pi/4$, $j = 1, \ldots, 8$. The center panel shows the the multiwindow fused estimate with ISNR = 8.07 dB and RMSE = 10.10.

Table 8.1 *Cameraman* image: criteria values for the eight directional and final multiwindow estimates.

	θ_1	θ_2	θ_3	θ_4	θ_5	θ_6	θ_7	θ_8	Final Est.
ISNR, dB	4.13	3.57	4.08	3.55	4.11	3.44	4.07	3.55	**8.07**
SNR, dB	18.52	17.96	18.46	17.94	18.50	17.83	18.45	17.94	**22.46**
PSNR, dB	24.10	23.54	24.05	23.53	24.08	23.41	24.04	23.52	**28.05**
RMSE	15.89	16.95	15.99	16.98	15.93	17.20	16.01	16.98	**10.10**
MAE	10.67	1.54	10.79	11.59	10.68	11.70	10.81	11.57	**6.66**
MAXDIF	133.0	114.72	124.22	117.0	112.62	142.45	114.42	125.88	**85.26**
UIQI	0.34	0.29	0.30	0.29	0.33	0.29	0.30	0.29	**0.41**

The numerical results in Table 8.1 are given for the eight directional estimates. They show the criteria values for the all eight directional and final estimates. The figures were obtained by Monte-Carlo simulation with averages over 100 runs.

The criterion values for the final estimate compared with the directional sectorial ones show a strong improvement in the fused estimate. In particular, we have for ISNR values of about 4 dB for the sectorial estimates while for the fused estimate ISNR $\simeq$ 8.07 dB. The adaptive weight fusing is an efficient tool for improving accuracy. Visually, the improvement effects are clear from a comparison of the directional and final estimates in Fig. 8.7. The fusing works very well for all criteria.

Table 8.2 shows similar results obtained using the four quadrant estimates as they were presented in Sec. 6.3: $h_1 \in H_1$, $h_1 = h_2$, $\Gamma = 1.2$. These four quadrant estimates compared with the eight sectorial estimates from Table 8.1 show that each of the four quadrant estimates mainly performs better than each of the eight sectorial estimates. In particular, we can see the values of ISNR $\simeq 4.55, 4.58, 4.55, 4.46$ for the quadrant estimates and ISNR $\simeq$ $4.13, 3.57, 4.08, 3.55, 4.11, 3.44, 4.07, 3.55$ for the sectorial estimates. However, for the final estimate we obtain ISNR $=$ 8.07 dB for the eight sectorial estimates

Table 8.2 *Cameraman* image: criteria values for the four quadrant and final multi-window estimates.

	Q_1	Q_2	Q_3	Q_4	Final Est.
ISNR, dB	4.55	4.58	4.55	4.46	**7.25**
SNR, dB	18.94	18.97	118.94	18.85	**21.64**
PSNR, dB	24.52	24.55	24.52	24.44	**27.22**
RMSE	15.14	15.09	15.14	15.29	**11.09**
MAE	9.48	9.42	9.44	9.51	**6.65**
MAXDIF	132.99	120.59	119.71	112.25	**98.93**
UIQI	0.33	0.32	0.33	0.32	**0.38**

versus ISNR = 7.25 dB for the four quadrant estimates. This confirms a significantly better performance of the multiwindow directional narrow sector estimates versus the four quadrant estimates using the wide quadrant kernels.

Of course, this sort of conclusion is not absolute and depends on the character of the image. In particular, it is valid for texture images with multiple varying directional edges typical for the *Cameraman* test image. The quadrant estimates give better results for internal points of images with large flat areas while the narrowed multidirectional estimates perform better near edges and change-points.

8.3.2.2 Comparison against more conventional algorithms

We compare the performance of the directional estimates versus the more conventional estimates demonstrating a good performance for image denoising. As counterparts we take the translation-invariant (TI) Haar and Daubechies wavelets (length 4, two vanishing moments, three decomposition levels). Further, we use for comparison the discrete cosine transform (DCT) with a sliding 8×8 window. This gives a number of estimates for each x. These estimates are accumulated in a buffer and averaged in order to produce a final estimate for every pixel. The thresholds in wavelets and DCT filtering are equal to 2.7σ. The results are given in Fig. 8.8, where the ISNR ratio is shown as a function of the SNR of the observations. The ISNR is calculated by Monte-Carlo simulation as an average over 100 runs.

In order to make the comparison more transparent, we show in Fig. 8.9 the ISNR achieved by the considered algorithms minus the ISNR for the directional

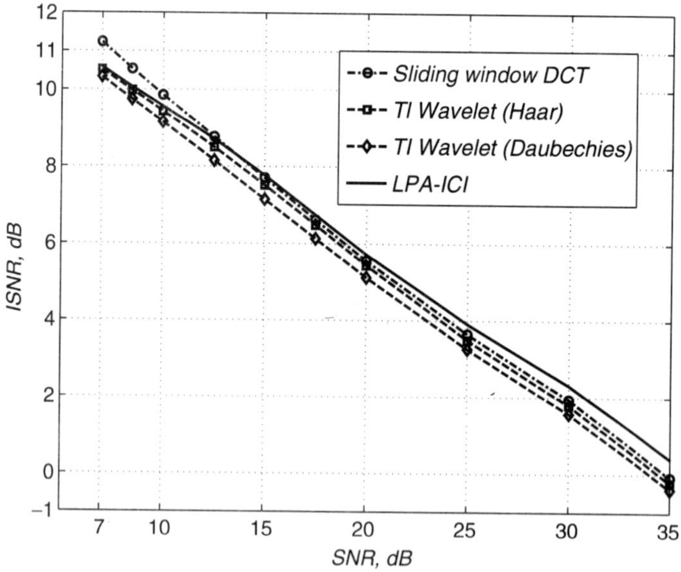

Figure 8.8 *Cameraman* image: ISNR versus SNR of the observations. The ISNR is calculated by Monte-Carlo simulation as an average over 100 runs.

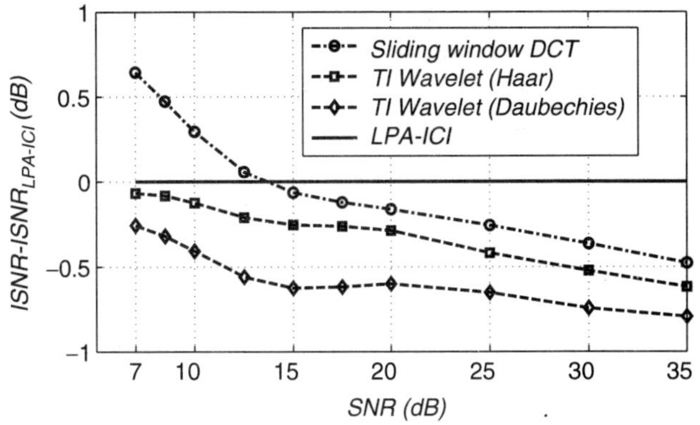

Figure 8.9 *Cameraman* image: differences of ISNR versus SNR of the observations. The differences are calculated as ISNR for the compared algorithm minus the ISNR for the LPA-ICI algorithm.

LPA-ICI algorithm. Thus, the horizontal nil line corresponds to the LPA-ICI algorithm. Negative values of these differences are in favor of the directional LPA-ICI algorithm while the positive differences are in favor of the counterparts.

The directional LPA-ICI algorithm is always better versus both the TI wavelets. This advantage increases for larger SNR. The DCT algorithm is always better than the wavelet algorithms and also outperforms the directional LPA-ICI algorithm for large levels of noise with SNR < 12.5 dB.

Here we mention once more that comparisons on the basis of ISNR as well as on the basis of all other conventional numerical criteria on many occasions do not correspond to visual inspection. In particular, while the DCT shows better values of ISNR for small SNR, the denoised images suffer from multiple artifacts typical for the estimates based on the orthogonal series with thresholding. The LPA-ICI algorithm shows worse ISNR for SNR < 12.5 dB; however, visually it demonstrates better imaging.

8.3.2.3 ICI pilot DCT algorithm

Here we introduce an unusual use of the ICI rule. It serves as an auxiliary *pilot estimator* defining the adaptive varying window sizes for each x. The DCT is exploited for filtering in these adaptive windows. The final estimate is a pointwise sample mean of the estimates obtained by these varying sized DCT filters. The algorithm is composed of the following steps:

1. The square symmetric uniform window w and the zero-order LPA define the estimator as a sample mean of the observations in this window.
2. The ICI is applied with this simple estimator to find the corresponding adaptive pointwise window sizes $h^+(x)$. These $h^+(x)$ give a pointwise segmentation defining the adaptive varying sized subsets for each x.

3. The discrete cosine transform (DCT) is applied to filter observations in these adaptive varying-sized subsets. The filtering is produced by the standard hard-thresholding procedure (Sec. 13.3).
4. The DCT filter gives estimates for all points of the adaptive-sized subsets. Because these subsets have different sizes and overlapping, we obtain different numbers of estimates for each x.
5. The estimates that are obtained are accumulated in a buffer. The final estimate for a given x is calculated as a mean of the DCT-filtered observations available for each x.

In this strategy the ICI is exploited as an auxiliary *pilot estimator* of the window size only, and the DCT filter that is different from the sample mean used in the ICI gives the final estimate. In this manner, as a pilot tool the ICI can be exploited with any orthogonal series-type filter.

We apply this algorithm for the 256×256 *Cameraman* image in the scenario described above. The following square windows are used for the DCT: 4×4, 8×8, 16×16, and 32×32. The window-size-dependent threshold t_N has the following values: $t_4 = 1.8$, $t_8 = 2.2$, $t_{16} = 2.5$, and $t_{32} = 2.7$, found from optimization of the procedure. The results, collected in Table 8.3, are shown for four types of algorithms: ICI pilot DCT, the sliding window DCT with a fixed window size of 8×8, LPA-ICI recursive (three iterations) (Sec. 8.2.2), and standard LPA-ICI with the eight directional multiwindow estimations described above. The ICI pilot DCT demonstrates very good numbers and the best for nearly all criteria. However, visual comparison does not favor this algorithm. Ringing effects around the edges and other artifacts are clearly seen in the estimates.

Figure 8.10 shows a fragment (75×75) of the *Cameraman* image for the algorithms compared. The artifacts mentioned previously are clearly seen in the images obtained by the ICI pilot DCT and the fixed-window size DCT. Visually, the LPA-ICI recursive algorithm demonstrates the best performance.

Comparing the recursive LPA-ICI versus the standard LPA-ICI algorithm with eight directional multiwindow estimates, we can see that the figures for the former algorithm are better. For the ISNR, we have the value 8.47 dB for the recursive

Table 8.3 *Cameraman* image: criterion values for adaptive ICI pilot DCT, fixed-size DCT, LPA-ICI recursive, and LPA-ICI eight directional multiwindow estimates.

	ICI pilot DCT	DCT (8×8)	LPA-ICI (Recursive)	LPA-ICI (8 sectors)
ISNR, dB	**8.60**	8.03	8.47	8.07
SNR, dB	**23.0**	22.42	22.86	22.46
PSNR, dB	**28.58**	28.01	28.44	28.05
RMSE	**9.49**	10.14	9.65	10.09
MAE	9.49	6.55	**6.12**	6.44
MAXDIF	87.00	85.55	111.10	**85.26**
UIQI	**0.43**	0.42	**0.43**	0.41

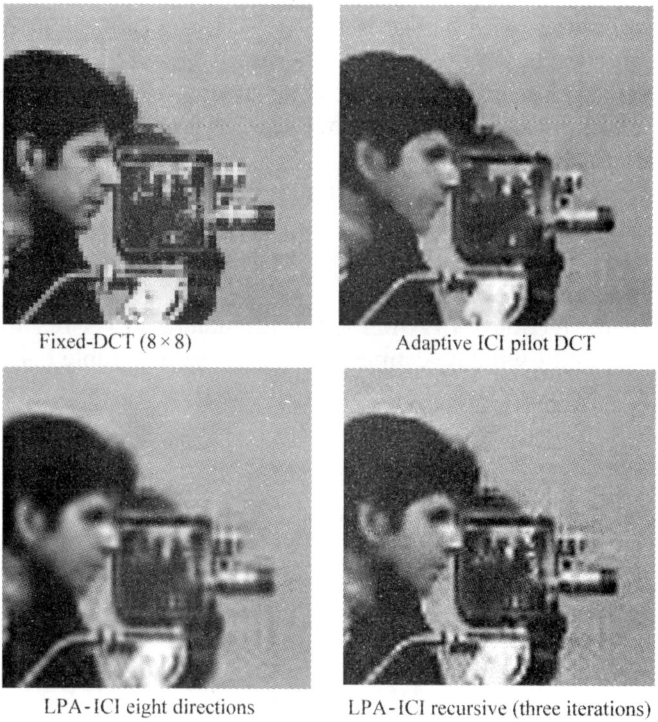

Figure 8.10 *Cameraman* image for four algorithms.

algorithm versus 8.07 dB for the standard nonrecursive one. Visually, the comparison is also in favor of the recursive LPA-ICI algorithm. The recursive algorithm demonstrates a fast convergence. Usually, two or three iterations are enough to achieve stable values of the criteria.

Application of the above four algorithms to the 256 × 256 test-image *Lena* is illustrated in Table 8.4. The observations are corrupted by an additive zero-mean

Table 8.4 *Lena* image: criteria values for adaptive ICI pilot DCT, fixed-size DCT, LPA-ICI recursive, and LPA-ICI eight directional multiwindow estimates.

	DCT (ICI pilot)	DCT (8 × 8)	LPA-ICI (Recursive)	LPA-ICI (8 sectors)
ISNR, dB	**11.08**	11.00	10.77	9.41
SNR, dB	**25.41**	25.32	25.09	23.73
PSNR, dB	**31.09**	31.00	30.77	29.41
RMSE	**7.11**	7.18	7.37	8.62
MAE	**5.05**	5.11	5.22	5.99
MAXDIF	67.00	**58.87**	63.64	88.45
UIQI	**0.55**	**0.55**	**0.55**	0.47

Gaussian noise with $\sigma = 25.5$. The ICI pilot DCT algorithm demonstrates good performance with the best values for all criteria but MAXDIF.

Ringing effects and artifacts again can be seen in the image restoration using the adaptive ICI pilot DCT algorithm, however, they are not so strong and obvious as for the *Cameraman* image.

Overall we may conclude that the ICI pilot DCT algorithm demonstrates a very good performance and can be treated as a highly competitive algorithm. Note that the algorithm is very fast since it uses only a set of the square windows and estimates calculations concerning the fast DCT.

Comparing the recursive LPA-ICI versus the standard LPA-ICI algorithm with eight directional multi-window estimates, we can see that Table 8.4 demonstrates the advantage of the recursive algorithm in all criterion values.

8.3.2.4 Results for a piecewise smooth signal

An interesting example of a piecewise smooth signal that is difficult for denoising is proposed in [190]. This signal, shown in Fig. 8.11(a), is given by

$$y(x_1, x_2) = \begin{cases} -4[(x_1 - 0.5)^2 + (x_2 - 0.5)^2], & 0 \leq x_1 \leq 0.5, \\ & 0 \leq x_2 \leq 0.75, \\ -4[(x_1 - 0.5)^2 + (x_2 - 0.5)^2] + 1, & \text{otherwise}, \end{cases}$$

The noisy surface is shown in Fig. 8.11(b) with the additive noise $\sigma = 0.25$. For this noisy image, SNR $= 7.4$. The LPA-ICI is used with the following parameters: $m = (0{,}0)$, $\Gamma = 0.95$,

$$H = \left\{ \begin{pmatrix} 1 \\ 1 \end{pmatrix}, \begin{pmatrix} 2 \\ 1 \end{pmatrix}, \begin{pmatrix} 3 \\ 1 \end{pmatrix}, \begin{pmatrix} 5 \\ 1 \end{pmatrix} \right\},$$

and the eight directions defined by Eq. (8.74).

The filtered surfaces in Figs. 8.11(c) and (d) are different by the fusing used for the directional estimates. The fusing in Fig. 8.11(c) is done by exploiting the *multiwindow* estimation with inverse variances as the weights of the estimates according to formula Eq. (8.57).

The estimate in Fig. 8.11(d) is a *combined window* estimate, as in Eqs. (6.48–6.50), where the eight directional varying-sized windows are used only to obtain the adaptive neighborhood of x. The estimate is calculated as the sample mean of the observations in this neighborhood.

Comparison of the surfaces in Figs. 8.11(c) and (d) definitely favors the combined window estimate, where the surface $\hat{y}(x_1, x_2)$ is smoother and there are no multiple erroneous peaks clearly seen in the multiwindow estimate.

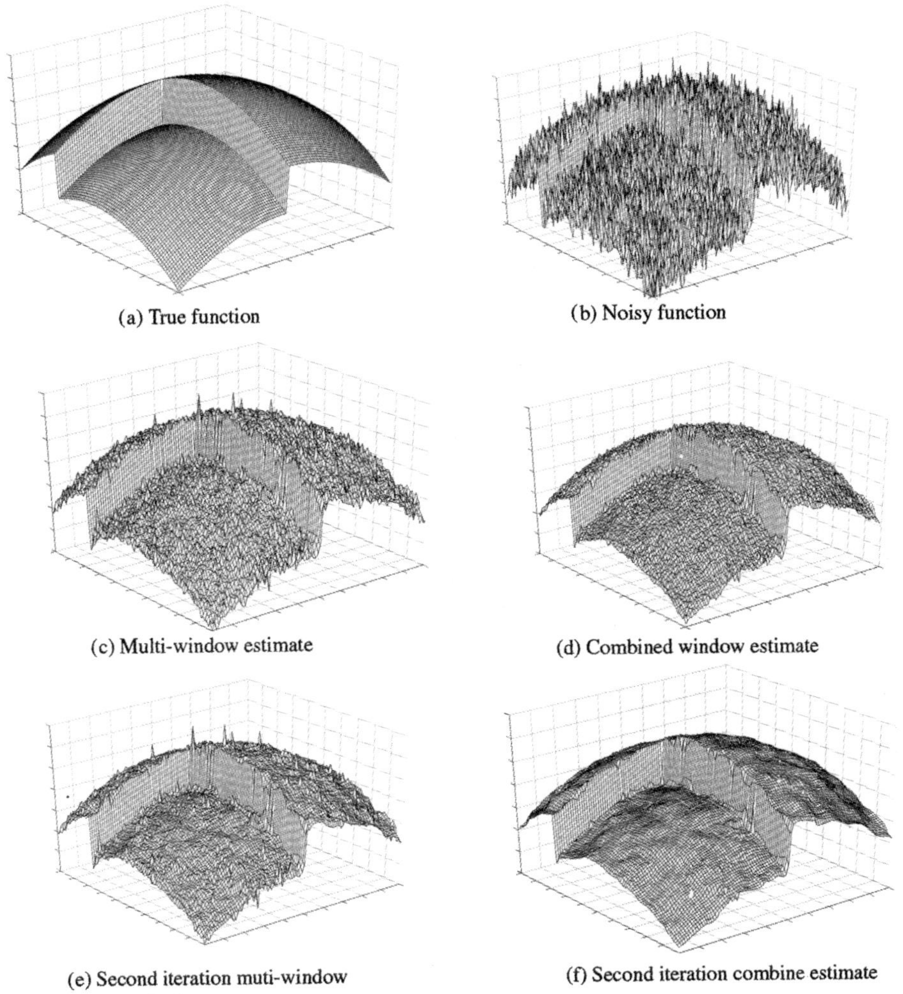

(a) True function
(b) Noisy function
(c) Multi-window estimate
(d) Combined window estimate
(e) Second iteration muti-window estimate
(f) Second iteration combine estimate

Figure 8.11 Piecewise quadratic function: data and estimates.

Further, the curves in Figs. 8.11(e) and (f) show what happens if the LPA-ICI recursive filtering is used. This recursive filtering demonstrates an essential filtering improvement in the first few steps. We may evaluate this improvement visually in Figs. 8.11(e) and (f).

The ISNR sizes given in Table 8.5 demonstrate two things. First, they confirm the advantage of the combined window estimate versus the multiwindow estimate. The ISNR of the former is about 3 dB higher than the latter. Second, the second iteration of the recurrent filtering improves the ISNR by about 2 dB for both methods. If we go up to ten iterations, the further slight improvement amounts to 0.1 dB for the multiwindow estimate and there is no improvement for the combined estimate.

Table 8.5 ISNR for the recursive application of the multiwindow and combined window filters.

Filter/Iterations	1	2	10
Multiwindow	10.3	12.7	12.8
Combined	13.5	15.2	15.2

8.4 Directional Differentiation

In a number of applications one is interested not in the signal itself but rather in the signal's derivative. The LPA provides flexible tools for designing the linear differentiating kernels. In particular, for the shift-invariant kernel, the differentiating estimate and the differentiating kernel can be given in the form of Eq. (3.7). These estimates, when generalized to directional differentiation in the convolution form of Eqs. (8.49–8.51), are defined by formulas (7.49)–(7.50) as

$$\hat{y}_{h,\theta}^{(r)}(\Delta k) = \sum_l g_{h,\theta}^{(r)}[k-l]z(\Delta l) = \sum_l g_{h,\theta}^{(r)}[l]z(\Delta(k-l)), \tag{8.75}$$

$$\begin{aligned} g_{h,\theta}^{(r)}[k] &= \frac{(-1)^{|r|}}{h_1^{r_1}h_2^{r_2}} w_h(u(\Delta k,\theta))\phi_h^T(u(\Delta k,\theta))\Phi_h^{-1}\phi^{(r)}(0), \\ \Phi_h &= \sum_k w_h(u(\Delta k,\theta))\phi_h(u(\Delta k,\theta))\phi_h^T(u(\Delta k,\theta)), \\ u(\Delta k,\theta) &= U(\theta)\Delta k, \end{aligned} \tag{8.76}$$

where θ indicates the direction and the scale parameters h_1 and h_2 define the "length" (along the direction θ) and the "width" of the kernel.

What sort of derivatives do we calculate using Eqs. (8.75–8.76)? Assume $r = (r_1, r_2)$. Then $\hat{y}_{h,\theta}^{(r)}$ is the estimate of the mixed derivative of order r_1 along the u_1 axis and of order r_2 along the u_2 axis, where u_1 and u_2 are the axes rotated by the angle θ with respect to basic Cartesian axes x_1, x_2 (Fig. 7.9). These estimates have the scale parameter h, which defines the support of the estimate and the weights of the observations used for estimation.

The ICI rule is applicable for the adaptive varying-scale selection for all estimates of this type where there is a standard bias-variance trade-off. We wish to demonstrate the performance of the adaptive-scale directional derivative estimates as well as discuss some specific problems arising from the signal anisotropy—in particular, the concept of an anisotropic differentiation. This sort of derivative is a generalization of the directional finite differences. The generalization concerns possible use of higher-order polynomials on both rotated axes u_1 and u_2 and an arbitrary shape of the window support w. For clarity, we restrict our consideration to the first-order derivative with $r = (1, 0)$ in Eqs. (8.75–8.76). It is calculated along the positive semiaxis u_1 in the direction θ.

8.4.1 Anisotropic gradient

Here we introduce the concept of a *directional anisotropic gradient* based on directional nonsymmetric LPA differentiation and the ICI adaptive-scale selection.

8.4.1.1 Directional estimate of gradient

Consider a spherical neighborhood (unit ball) of the point x:

$$\mathcal{B}_h = \{v : ||x - v|| \leq h\}, \tag{8.77}$$

where h is the radius of this neighborhood. As was done in Sec. 7.1.2 for signal estimation, we introduce a sectorial partition of this ball with K nonoverlapping sectors having x as the common vertex. Let $g_{h,\theta}^{(1,0)}$ be differentiating kernels defined on these sectors with the corresponding estimates $\hat{y}_{h,\theta}^{(1,0)}$. In this way, we obtain K estimators with K nonoverlapping supports covering the ball $\mathcal{B}_h$. When we estimate the signal, all sectorials $\hat{y}_{h,\theta_i}(x)$ estimate the same variable $y(x)$. In this way, all of these estimates have the same meaning and can be naturally fused in a single estimate, say as a weighted mean.

The situation is different for differentiation, because for each sector θ we estimate the directional derivative $\hat{y}_{h,\theta_i}^{(1,0)}(x)$ and these derivatives are different for each direction. Thus, using a fusing similar to the one used for signal estimation makes no sense for the derivatives. Therefore, something common should be found in the directional derivatives in order to use them jointly.

Assume for the moment that the signal y is differentiable at the point x with a vector gradient

$$\nabla y = (\partial_{x_1} y, \partial_{x_2} y)^T.$$

According to Eq. (7.29), the directional derivative (on the definition $\partial_\theta y \triangleq \partial_{u_1} y$) can be written as

$$\partial_\theta y = \partial_{x_1} y \cdot \partial_{u_1} x_1 + \partial_{x_2} y \cdot \partial_{u_1} x_2 = \partial_{x_1} y \cdot \cos\theta + \partial_{x_2} y \cdot \sin\theta \tag{8.78}$$

and can be treated as an indirect observation of the gradient components $\partial_{x_1} y$ and $\partial_{x_2} y$.

Provided that a number of the directional derivative estimates $\hat{y}_{h,\theta_i}^{(1,0)}$ is given, the gradient ∇y can be found by minimizing the weighted mean-square criterion:

$$J = \sum_{i=1}^{K} \frac{1}{\sigma_{\theta_i}^2} [\hat{y}_{h,\theta_i}^{(1,0)} - (\partial_{x_1} y \cdot \cos\theta_i + \partial_{x_2} y \cdot \sin\theta_i)]^2,$$

where the weights $\sigma^2_{\theta_i}$ are the variances of the estimates $\hat{y}^{(1,0)}_{h,\theta_i}$. In the vector-matrix notation, J can be rewritten as

$$J = (\hat{\mathbf{y}}^{(1,0)}_h - B\nabla y)^T \Lambda (\hat{\mathbf{y}}^{(1,0)}_h - B\nabla y), \tag{8.79}$$

where $\hat{\mathbf{y}}^{(1,0)}_h = (\hat{y}^{(1,0)}_{h,\theta_1}, \ldots, \hat{y}^{(1,0)}_{h,\theta_K})^T$ is a vector of the estimates, $\Lambda = diag\{1/\sigma^2_{\theta_1}, \ldots, 1/\sigma^2_{\theta_K}\}$ is a diagonal matrix, and $B = (B_i)_{K\times 2}$, $B_i = (\cos\theta_i, \sin\theta_i)$.

Minimization of Eq. (8.79) gives the estimate of the gradient as

$$\hat{\nabla} y = (B^T \Lambda B)^{-1} B^T \Lambda \hat{\mathbf{y}}^{(1,0)}_h. \tag{8.80}$$

Thus, if the gradient of y exists, it can be estimated using a set of directional derivatives $\hat{y}^{(1,0)}_{h,\theta_i}$. Formula (8.80) gives one of the possible ways to jointly use this directional information.

Further, assume that the ICI rule is used in each directional estimate $\hat{y}^{(1,0)}_{h,\theta_k}$ for the adaptive window size selection. The union of the support of these adaptive estimates is a star-shaped set that allows the following interesting interpretation.

This set is the largest star-shaped neighborhood of x where the signal y is differentiable. In more precise terms, we can say that this star-shaped set is the largest star-shaped area where the estimate of the vector gradient $\hat{\nabla} y$ fits signal observations provided that y is differentiable at the point x.

Thus, the ICI adaptive estimate of the directional derivatives used in Eq. (8.80) gives two complementary important results. The first is the estimate of the gradient, and the second is the largest star-shaped neighborhood of x, where y is differentiable.

8.4.1.2 Anisotropic gradient

A signal y is differentiable at x if there exists a linear functional $\mathcal{L}_x$ such that

$$y(x+v) - y(x) - \mathcal{L}_x(v) = o(|v|), \; v \in \mathcal{B}_h, \tag{8.81}$$

where $\mathcal{B}_h$ is a ball neighborhood as in Eq. (8.77). If such a functional $\mathcal{L}_x$ exists, it is linear on v and allows the following expression:

$$\mathcal{L}_x(v) = v^T \nabla y(x), \tag{8.82}$$

where ∇y is the gradient vector of the signal y and v is a vector deviation from the point x. When the signal is differentiable, essential information about the local behavior of y in $\mathcal{B}_h$ is expressed by this approximating linear functional $\mathcal{L}_x$. Most notably, the direction of steepest ascent for y at x is equal to the direction of the gradient vector $\nabla y(x)$, and the rate of ascent is given by the modulus $|\nabla y(x)|$. If the graph of the signal y is thought of as a surface in the 3D space, then $\mathbf{n}_y(x) = \left(-\partial_{x_1} y, -\partial_{x_2} y, 1\right)^T$ is the normal vector to this surface, that is, the normal vector to the tangent plane. Such information is often of critical importance in many theoretical problems (optimization theory, algorithm convergence, ordinary and partial differential equations) and practical applications (geophysics, imaging, computer graphics).

The aforementioned properties of the gradient do not hold, in general, when the signal is not differentiable and no functional $\mathcal{L}_x$ satisfying Eq. (8.81) exists. Checking the differentiability hypothesis can be an extremely difficult—if not impossible—task, especially when the signal y is known only partially (by sampling) or with limited precision (quantization, noise). Nevertheless, in practical applications, the differentiability hypothesis is often given implicitly, and formula (8.82) is used to derive a functional that may not satisfy Eq. (8.81). The most striking example is probably the edge detection problem: The gradient is used to detect edges that are clearly points of non-differentiability.

A traditional approach to estimation of the normal $\mathbf{n}_y(x)$ is looking for surface-fitting observations. After the surface is found, the normal vector $\mathbf{n}_y(x)$ at x is taken from the analytic approximation of the surface. Although this technique produces very good results when the underlying data are smooth, particular care must be taken in the vicinity of edges, where surface fitting may not be suitable.

More sophisticated approaches (e.g., in [239]) first segment the image into regions of smoothness and then evaluate the normals within these regions using surface-fitting or kernel estimates. Other authors (e.g., [26], [203]) estimate the gradient by using both forward and backward differences. In the spirit of edge detection, edges of the regions are found by exploiting thresholding schemes for the magnitude of the gradient and/or on the discrepancy between forward and backward derivative estimates. Such schemes are very complicated and their robustness to noise is questionable. Instead, we propose a different approach based on the notion of *an anisotropic gradient.*

The standard definitions of the differentiability and the derivative mean that

$$\partial_{x_1} y(x_1, x_2) = \lim_{v_1, v_2 \to 0} \{[y(x_1 + v_1, x_2 + v_2) - y(x_1, x_2 + v_2)]/v_1\}, \tag{8.83}$$

and

$$\partial_{x_2} y(x_1, x_2) = \lim_{v_1, v_2 \to 0} \{[y(x_1 + v_1, x_2 + v_2) - y(x_1 + v_1, x_2)]/v_2\},$$

where the limits do not depend on how small v_1, v_2 get as they approach zero. These small parameters can take positive or negative values and have different orders of smallness; in any case, the result is the same. Defaults of these limits assume that the signal y is defined in any small ball neighborhood of Eq. (8.77) of the point x, and it is a continuous function in this neighborhood.

Now let us introduce the idea of the anisotropic gradient [54]. Denote a sector of the ball as in Eq. (8.77) as $V_\theta(x)$, where θ indicates the directionality of the sector. Assume that the limits in Eq. (8.83) exist and do not depend on how the sequence $v_1, v_2 \to 0$ provided that v_1, v_2 belong to the sector $V_\theta(x)$, that is,

$$\begin{aligned} \partial_{x_1} y(x_1, x_2) &= \lim_{v_1, v_2 \to 0, v_1, v_2 \in V_\theta(x)} \{[y(x_1 + v_1, x_2 + v_2) - y(x_1, x_2 + v_2)]/v_1\}, \\ \partial_{x_2} y(x_1, x_2) &= \lim_{v_1, v_2 \to 0, v_1, v_2 \in V_\theta(x)} \{[y(x_1 + v_1, x_2 + v_2) - y(x_1 + v_1, x_2)]/v_2\}. \end{aligned} \tag{8.84}$$

It may happen that the same limit in Eq. (8.84) is achieved for some adjacent sectors $V_\theta(x)$. Let these adjacent sectors be joined in a single one. Further assume that $V_\theta(x)$ is the largest sector where Eq. (8.84) holds with the same limits. The linear approximation Eq. (8.81) is valid in this sector in terms of

$$y\left(x+v\right)-y\left(x\right)-v^{T}\nabla y\left(x\right)=o\left(|v|\right),\, v\in V_{\theta}(x), \tag{8.85}$$

where the gradient's $\nabla y(x)$ components are defined by Eq. (8.84).

It is obvious that if the gradient in Eq. (8.84) is the same for all $V_\theta(x) \subset \mathcal{B}_h$, the signal y is differentiable. Let $V_\theta(x)$ be a sector of the ball in Eq. (8.77) and let $\nabla y(x)$ be the gradient of $y(x+v)$, $v \in V_\theta(x)$, with components defined according to Eqs. (8.84–8.85).

The gradient is called *anisotropic* if at least two sectors exist such that the corresponding gradients are different. We call the gradient *isotropic* if the limit in Eq. (8.84) does not depend on the selection of the sector.

Conceptually, the anisotropic gradient is linked to the isotropic neighborhood of x. If the signal is not differentiable, it is possible that there are a few anisotropic gradients and a few corresponding neighborhoods at the point x.

Figure 8.12 illustrates the concept of the anisotropic gradient for two cases—one where the signal y is discontinuous, and one where the signal is continuous. Figure 8.12(a) shows a piecewise linear discontinuous y composed of two linear fragments. It is assumed that the point x^0 stays at the discontinuity point. There are left and right neighborhoods of x^0 belonging to the corresponding argument areas $V(x^0_-)$ and $V(x^0_+)$, where the signal is regular with a unique derivative and a unique normal vector $\mathbf{n}_y$. Thus, at the point x^0 we have two neighborhoods, $V(x^0_-)$ and $V(x^0_+)$, giving two different derivatives $\partial_x y$ and two different normal vectors.

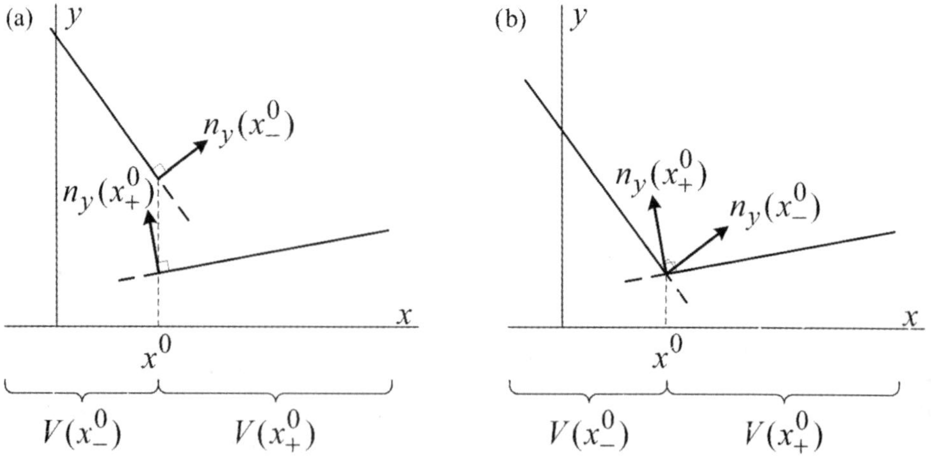

Figure 8.12 The concept of the anisotropic gradient. A signal $y(x)$ is piecewise linear composed of two linear fragments. There are two gradients and two normal vectors for the left- and right-hand side neighborhoods of the point x. (a) continuous signal, (b) discontinuous signal.

Compare this situation with another one depicted in Fig. 8.12(b), where y is again composed of two linear functions and the point x^0 belongs to intersection of these two lines. In this case, the function is continuous at the point x^0 and there are two different normal vectors $\mathbf{n}_y\left(x^0\right)$ and two different left and right derivatives of y.

8.4.1.3 Ideal neighborhood and adaptive estimation

Consider a sectorial partition of the ball in Eq. (8.77) as

$$V = \{V_\theta(x) : \cup_\theta V_\theta(x) = \mathcal{B}_h\}. \tag{8.86}$$

Assume that the support of the differentiating kernel $g_{h,\theta}^{(1,0)}$ in Eqs. (8.79–8.80) belongs to the sector $V_\theta(x)$ in Eq. (8.85), and assume that for each direction θ the directional estimate is optimized by the selection of the ideal scale $h^*(\theta)$ minimizing the corresponding least-square error.

The union of these optimally-sized supports defines the ideal neighborhood $U_g^*(x)$ for estimation of the gradient $\nabla y(x)$. This is the largest neighborhood of x that can be used for gradient estimation. It is obvious that this differentiation ideal neighborhood is similar to the ideal neighborhood for the signal estimation in Eq. (7.21).

Using the ICI for the adaptive-scale derivative estimation allows us to get some approximation of this largest ideal differentiation neighborhood. Let the LPA directional derivative estimates be calculated and the ICI be exploited to find the adaptive scale (sector size) for each θ_i. Then, the ICI adaptive gradient estimate is calculated according to Eq. (8.80), where the estimate vector is replaced by the vector of the estimates with the adaptive scales $\hat{\mathbf{y}}_{h^+}^{(1,0)} = \left(\hat{y}_{h^+(\theta_1),\theta_1}^{(1,0)}, \ldots, \hat{y}_{h^+(\theta_K),\theta_K}^{(1,0)}\right)^T$ and $\sigma_{\theta_k}^2$ in Λ are variances of these estimates.

The derivative estimates $\hat{y}_{h^+(\theta_i),\theta_i}^{(1,0)}$ in Eq. (8.80) are calculated for all directions. For those θ_i directed to a change-point, the ICI gives a minimum value to the corresponding adaptive size $h^+(\theta_i)$. This means a larger estimate variance $\sigma_{\theta_i}^2$ and a smaller weight of the corresponding residual in the criterion in Eq. (8.79) and in the estimate in Eq. (8.80). The estimates $\hat{y}_{h^+(\theta_i),\theta_i}^{(1,0)}$ for x close to the change-points and with θ_i directed to a change-point are attenuated or completely eliminated from the gradient estimate. In other words, the adaptive gradient estimate in Eq. (8.80) uses the derivative estimates $\hat{y}_{h^+(\theta_i),\theta_i}^{(1,0)}$ only belonging to the isotropic neighborhood of the point x. The union of the ICI adaptive supports $V_{h^+(\theta_i),\theta_i}(x)$ gives a star-shaped approximation of the ideal set $U_g^*(x)$ defining the isotropic neighborhood of the point x. The vector $\hat{\nabla} y$ defined in Eq. (8.80) is the ICI adaptive estimate of the anisotropic gradient.

The following examples provide further illustrations of the anisotropic gradient concept as well as show how the ICI works when solving two different problems: segmentation of the neighborhood of x with identification of the anisotropy neighborhoods, and estimation of the corresponding gradients.

8.4.2 Examples

8.4.2.1 Example 1: Horizontal derivative for pyramid

Figures 8.13–8.15 show estimates of the horizontal derivative ∂_{x_1} for a signal y corresponding to the pyramid shown in the first row of Fig. 8.13 by the true and noisy surfaces. Because the derivative estimates are sensitive to the noise, we assume that the level of the noise is small to have a clear visual demonstration. The line-wise directional 1D differentiating kernels are used in the estimate. These kernels $g_h^{(1)}$ can be given in the analytical form as in (14.26), with $h = 3$ for the short and $h = 11$ for the longer kernel.

The second row of images in Fig. 8.13 shows the estimates obtained with the fixed-size short and longer kernels. The short kernel gives very noisy estimates, while the longer one attenuates the noise as well as oversmooths the estimate. For longer kernels we obtained noiseless but blurred imaging of the derivative.

Figure 8.14 shows the work of the ICI rule for the derivative estimation. We use the kernel in Eq. (14.26) for estimation of the left and right derivatives, that is, directional derivatives with $\theta = 0$ and $\theta = \pi$. The window sizes are defined by the set $H = \{3, 5, 9, 11, 15, 21\}$. The ICI rule threshold parameter is $\Gamma = 1$.

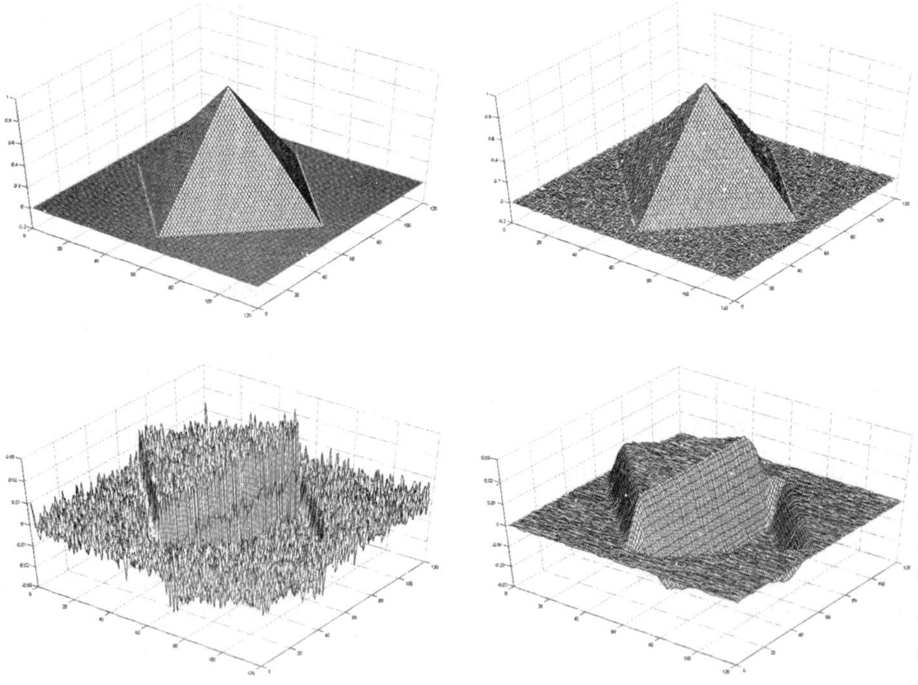

Figure 8.13 Nonadaptive derivative estimation. The original pyramidal signal (top left) is contaminated by additive noise (top right). A short differentiating kernel of length 3 gives an extremely noisy result (bottom left). A larger kernel of length 11 reduces the noise but blurs the sharp transitions in the derivative (bottom right).

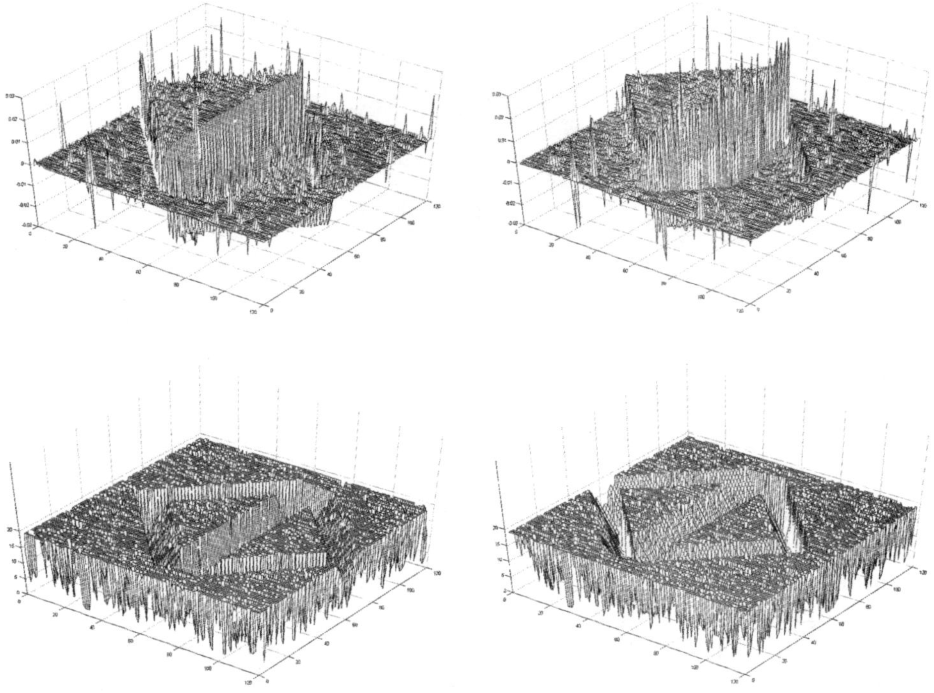

Figure 8.14 Adaptive nonsymmetric LPA derivative estimation: left and right derivatives (first row); the corresponding ICI adaptive scales h^{+} are shown in the second row.

The first row of the images in Fig. 8.14 shows the left ($\theta = \pi$) and right ($\theta = 0$) adaptive estimates. The advantage of these estimates is quite clear if we compare them with the estimates shown in Fig. 8.13. The noise is mainly suppressed and sharp edges of the derivatives are well preserved. However, the derivative at the change-points corresponding to the edges of the pyramid is quite noisy, because the ICI uses small window sizes for these areas.

The second row of images in Fig. 8.14 shows the used ICI adaptive window sizes h^{+}. Far from the pyramid edges the window sizes take the maximum value $h = 21$. However, near the edges the adaptive window sizes become small.

The pyramid is a typical example of a nondifferentiable signal having different left- and right-hand side derivatives at the edges.

Let us apply the above-described anisotropic procedure for the derivative calculation using the left- and right-hand side adaptive directional derivative estimates. In this case, the general formula Eq. (8.80) is reduced to the weighted mean,

$$\hat{y}^{(1,0)}(x) = \frac{1/[\sigma^2_{\hat{y}_h^{(1,0)}}(x, h^{+}(0))]\hat{y}^{(1,0)}_{h^{+}(0),0} - 1/[\sigma^2_{\hat{y}_h^{(1,0)}}(x, h^{+}(\pi))]\hat{y}^{(1,0)}_{h^{+}(\pi),\pi}}{1/[\sigma^2_{\hat{y}_h^{(1,0)}}(x, h^{+}(0))] + 1/[\sigma^2_{\hat{y}_h^{(1,0)}}(x, h^{+}(\pi))]}, \quad (8.87)$$

where the weights are given by the derivative estimate variance, $\sigma^2_{\hat{y}_h^{(1)}}(x, h) = \sigma^2 ||g_h^{(1)}||^2$.

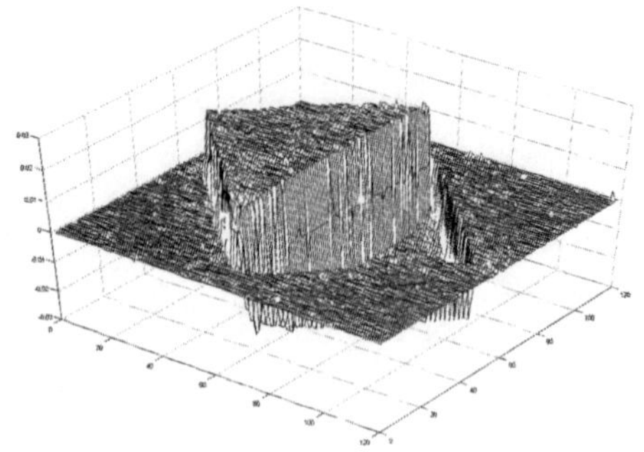

Figure 8.15 The estimate of the derivative ∂_{x_1} obtained using Eq. (8.87), the ICI-adaptive nonsymmetric left- and right-differentiating kernels, $\theta_1 = 0$, $\theta_2 = \pi$.

At the point of differentiability, the left and right derivatives are the same and the average in Eq. (8.87) improves the estimation accuracy of this well-defined derivative. The minus sign in Eq. (8.87) allows us to agree different signs of these left and right directional derivatives.

Near the edge points, the estimate in Eq. (8.87) works as the anisotropic gradient (derivative) estimate in Eq. (8.80) with automatic selection of the largest (left or right) neighborhood. Formula (8.87) is applicable as a generalized estimator for all x, regardless of the differentiability of the signal.

The obtained anisotropic derivative estimate is shown in Fig. 8.15. We see a significant improvement in the quality of differentiation when we compare this result to the ones shown in Fig. 8.13 and 8.14.

8.4.2.2 *Example 2: Discontinuous surface [53]*

Consider the following 2D piecewise smooth function given by

$$y\left(x_1, x_2\right) = \begin{cases} 0, & \text{if } r \leq \frac{1}{10}, \\ \varphi(x_1, x_2), & \text{if } r > \frac{1}{10}, \end{cases} \tag{8.88}$$

where r is a module and φ is an argument of the complex-valued function $x_1 + ix_2$, that is,

$$r = \sqrt{x_1^2 + x_2^2} \quad \text{and} \quad \varphi(x_1, x_2) = \arg(x_1 + ix_2), \quad i = \sqrt{-1}, \quad x_1, x_2 \in \mathbb{R}.$$

The 3D imaging of this function is shown in Fig. 8.16.

This surface can be thought of as being obtained from a horizontal plane having zero value for the square $|x_1| \leq 1/2$ and $|x_2| \leq 1/2$. Cut a circle through this plane of radius $r = 1/10$ with its center at the origin $x_1 = x_2 = 0$. Next, cut a

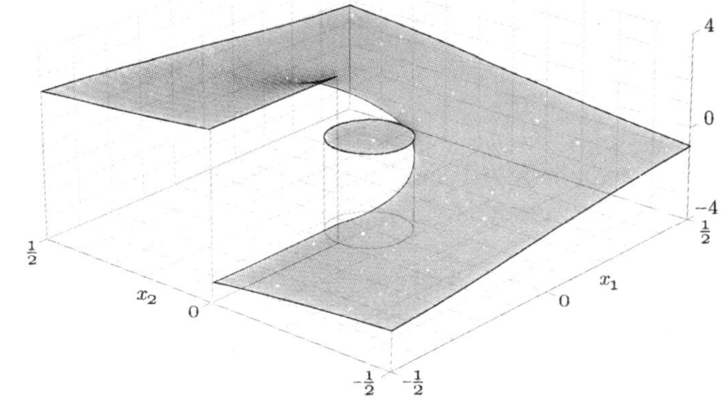

Figure 8.16 Three-dimensional imaging of the function in Eq. (8.88).

straight line parallel to the x_1 axis starting from the circle (point with the coordinates $x_1 = r, x_2 = 0$) and go to the end of the plane with the coordinates $x_1 = -1/2$, $x_2 = 0$.

We then take two neighboring points corresponding to the end of this slit straight line (with the coordinates $x_1 = -1/2$, $x_2 = 0$) and move one of them up and the other one down. Then the slit plane is separated along the cut line and we can recognize the surface shown in Fig. 8.16. The slit disc of radius r (with the value $y = 0$) is hanging in the air and touching another part of the surface at a single point $x_1 = 1/10$, $x_2 = 0$.

The function y is smooth and differentiable everywhere for $|x_1| \leq 1/2$ and $|x_2| \leq 1/2$, excluding the points where the cuts have been produced. Thus, the function is discontinuous on the set

$$E = \{x_1, x_2 : r = 1/10\} \cup \{x_2 = 0, -1/2 \leq x_1 < -1/10\}.$$

We take this function for numerical experiments as a difficult object for differentiation and some specific features of the anisotropic gradient can be demonstrated.

It can be verified by differentiation of Eq. (8.88) that for $x \notin E$ the gradient $\nabla y = \left(\partial_{x_1} y, \partial_{x_2} y\right)^T$ can be calculated analytically:

$$\nabla y = \begin{cases} (0, 0)^T, & \text{for } r < \frac{1}{10}, \\ \left(-x_2/x_1^2 + x_2^2, x_1/x_1^2 + x_2^2\right)^T, & \text{for } x \notin E. \end{cases} \tag{8.89}$$

It is not difficult to realize from Fig. 8.16 that for every discontinuity point $x \in E$ there exists a neighborhood $V_\theta(x)$ such that $y(x)$ is differentiable for $x \in V_\theta(x)$. This means that, while the function y is discontinuous at $x \subset E$ and, of course, nondifferentiable in the classical meaning of this concept, the anisotropic gradient can be calculated for each such x.

By restricting to a particular neighborhood of the point x, we can differentiate $y(x)$ for any $|x_1| \leq 1/2$ and $|x_2| \leq 1/2$. It follows that the gradient generalized as the anisotropic one can be defined by formula (8.89) for any x as

$$\nabla y = \begin{cases} (0, 0)^T, & \text{for } r \leq \frac{1}{10}, \\ \left(-x_2/x_1^2 + x_2^2, x_1/x_1^2 + x_2^2\right)^T, & \text{for } r > \frac{1}{10}. \end{cases} \quad (8.90)$$

This anisotropic gradient ∇y is standard for the points where $y(x)$ is continuous and the anisotropic gradient for the points $x \in E$ where the function $y(x)$ is discontinuous.

The vector gradient in Eq. (8.89) of the function in Eq. (8.88) is considered a classical example of a curious discontinuous function in some mathematical analysis textbooks. The anisotropy concept that is used here is different from the traditional analysis of this function. At the very least, it allows us to define a unique gradient calculated according to Eq. (8.90) for each x whether the function is continuous or discontinuous. The only point where the anisotropic gradient is not univalent is $x_1 = 1/10, x_2 = 0$.

The main objective of this example is to demonstrate that the adaptive LPA-ICI differentiation allows us to numerically obtain the gradient values corresponding to the analytical formula (8.90) for each x but the point $x_1 = 1/10, x_2 = 0$, where there exist two different anisotropic gradients.

In terms of this procedure, the algorithm automatically finds the isotropic sets $V_\theta(x)$ where the function is regular. These results demonstrate the unique abilities of the developed differentiation procedure to obtain the correct results in extremely complex situations. For the critical point $x_1 = 1/10, x_2 = 0$, the LPA-ICI differentiation gives the weighted mean of the two anisotropic gradients.

Numerical experiments

Let the function y be sampled on a regular grid of size 200×200 and with a Gaussian additive noise with a small variance $\sigma = 0.01$ added to the observations. The noisy data are imaged in Fig. 8.17 (right). The set E, where the function is discontinuous and nondifferentiable according to the classical concepts, is shown in Fig. 8.17 (left).

Figure 8.18 shows the true anisotropic gradient ∇y calculated using the analytical formulas (8.90).

The two images in the first row show the derivatives $\partial_{x_1} y$ and $\partial_{x_2} y$, respectively. The two images in the second row show the length of the vector-gradient $||\nabla y|| = \sqrt{(\partial_{x_1} y)^2 + (\partial_{x_2} y)^2}$ and the direction (angle φ) of the vector gradient calculated from

$$\cos\varphi = \frac{\partial_{x_1} y}{||\nabla y||}, \quad \sin\varphi = \frac{\partial_{x_2} y}{||\nabla y||}.$$

The discontinuity lines shown in Fig. 8.17 (left) have no direct influence on the true gradient images. The central circle is seen in Fig. 8.18 only because along this circle the gradient takes different values inside and outside of the disc.

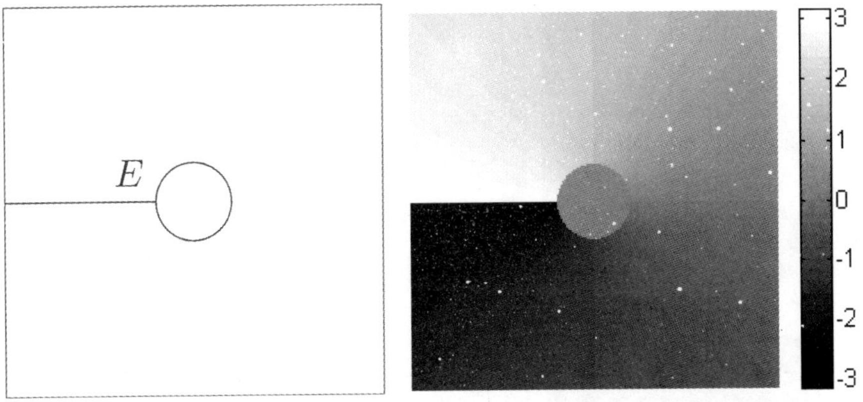

Figure 8.17 The set of nondifferentiability E (left) and the function y sampled on a discrete 200×200 grid (right).

We do not see the horizontal discontinuous line, $-1/2 \leq x_1 \leq -1/10$, $x_2 = 0$ in Fig. 8.18, because the gradient has equal values on both sides of this line.

In Figs. 8.19 and 8.20, the gradient estimates are obtained by standard differentiating filters of a fixed nonadaptive size. We use here 1D symmetric differentiation filters of length 3 (small scale) and length 7 (large scale). The kernels of the filters are defined by Eq. (14.14) with $h = 1$ for length 3 and $h = 3$ for the length 7.

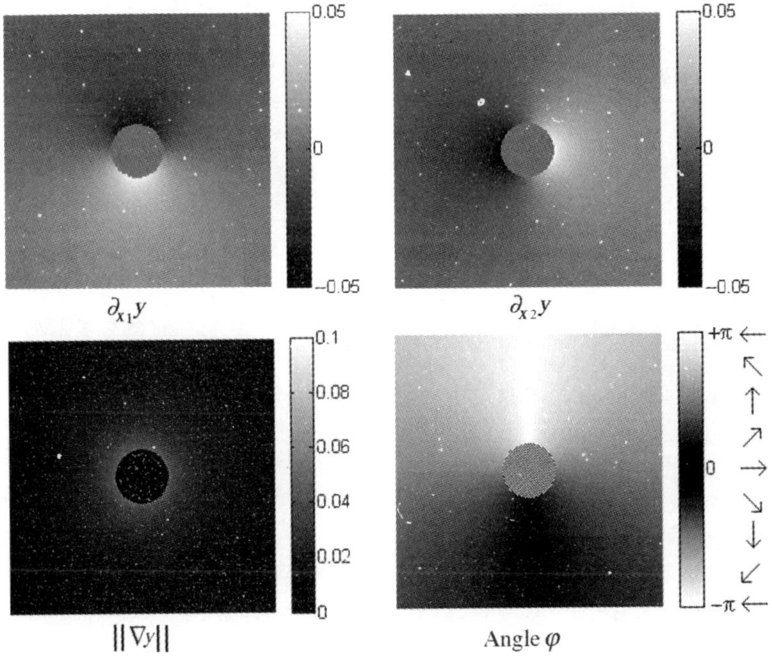

Figure 8.18 True gradient of the function in Eq. (8.88).

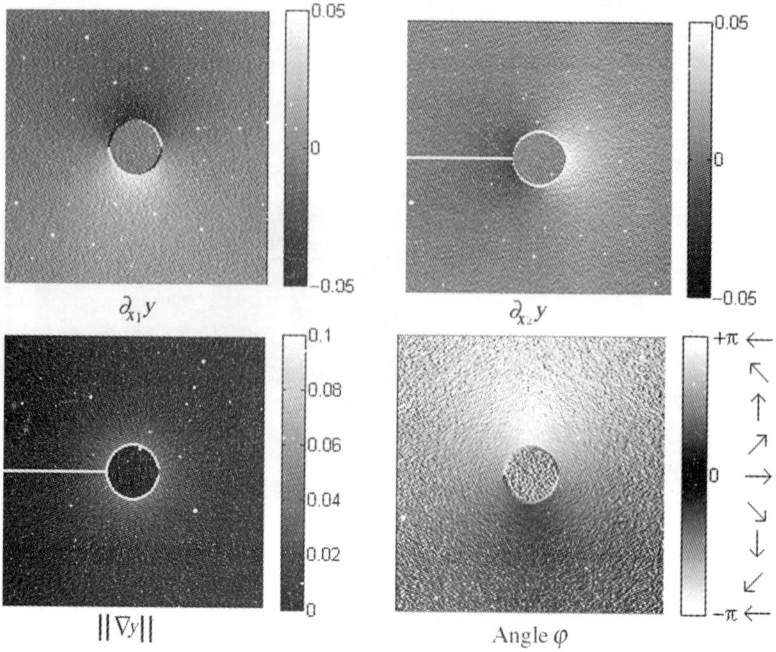

Figure 8.19 Fixed small size kernel estimate of the gradient of the function in Eq. (8.88).

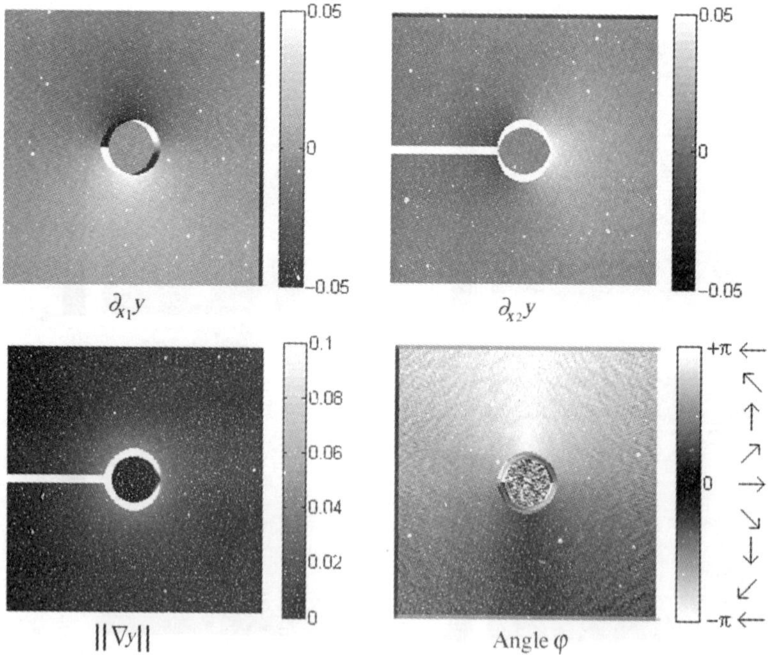

Figure 8.20 Fixed large size kernel estimate of the gradient of the function in Eq. (8.88).

This straightforward approach gives results where the discontinuity circle and straight line are seen clearly. These discontinuities affect sensibly the estimated gradient, in particular, at large $h = 3$.

The developed adaptive anisotropic gradient estimate is shown in Fig. 8.21. In these estimates we use the asymmetric kernels in Eq. (14.26) with an adaptive varying scale $h \in H, H = \{2 : 6\}$. It is evident that the adaptive estimate in Fig. 8.21 gives a much better fit to the accurate gradient images in Fig. 8.18. At least for this adaptive-scale estimate the discontinuity lines are invisible exactly as it is for the true gradient.

Comments

Those from the signal processing community could find this example quite unusual and perhaps even puzzling. In fact, here we present a technique for gradient/derivative estimation where the edges in the image become invisible and the estimates are not sensible to the edges.

This contradicts the common practice of using derivatives to identify edges. Two facts should be clarified here. First, the very notion of the derivative across the edge where the derivative does not exist is contradictory. This notion is rejected in the concept of an anisotropic gradient, because the edge clearly introduces a discontinuity that destroys the differentiability of the signal. We choose instead nonsymmetric directional derivatives intrinsically compatible with edges.

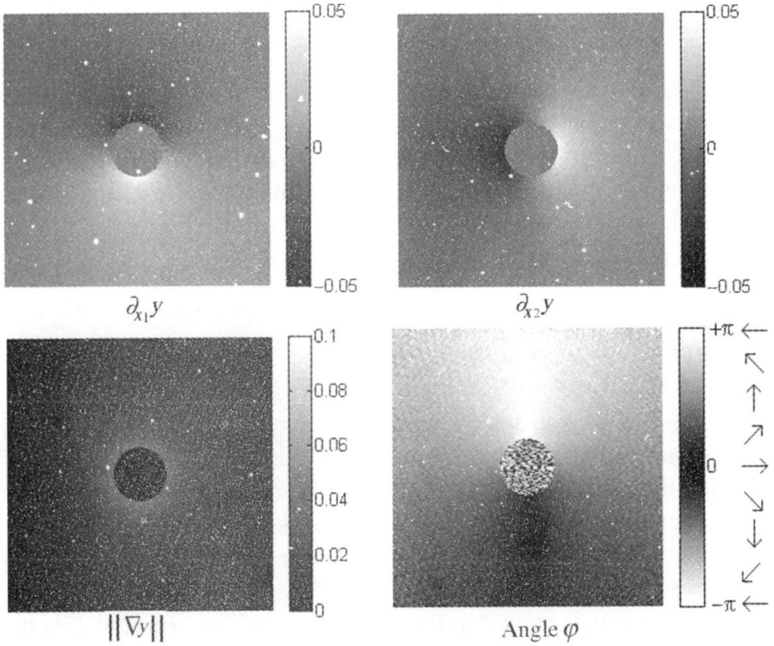

Figure 8.21 The anisotropic gradient estimate of the function in Eq. (8.88).

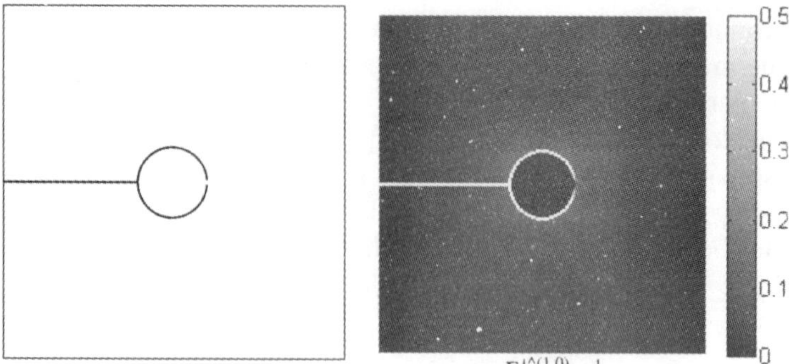

Discontinuity lines $\quad \sum_i \left|\hat{y}^{(1,0)}_{h^+(\theta_i),\,\theta_i}\right|$

Figure 8.22 Discontinuity lines of the function in Eq. (8.88) detected by the sum of the absolute values of the derivatives.

Second, within our approach the edges are implicitly detected by the LPA-ICI algorithm. Indeed, in the ICI algorithm the partial derivatives are calculated for all scales (window sizes) as well as for the scales using the signal values across the discontinuity borders.

The ICI algorithm identifies the derivatives across the border as having too large a bias. In this way the derivatives across the border lines are rejected and the derivatives are calculated only over the areas where the signal is continuous and differentiable.

We stress that this method does not use a thresholding of the derivative magnitude. Instead, it is based on the notion of the coherence of derivative estimation with varying scales, which is closely connected with the analytical continuous-domain differentiability.

Figure 8.22 (right) shows an imaging of the sum of the absolute values of the derivatives calculated with adaptive scales for all (horizontal and vertical) directions. Discontinuity lines are clearly seen and easily identified. These types of functionals of estimates can be used in the edge-detection procedure. We use the multiple-scale estimates, which make them different from the traditional edge-detection procedures where the scale is usually fixed.

More examples

Application of the anisotropic gradient technique to a texture image called *Peppers* is illustrated in Figs. 8.23 and 8.24. Figure 8.23 shows the ICI adaptive scales for the horizontal left and horizontal right derivatives (middle row of images), respectively, in the left and right images. The bottom line of the images is given for the vertical derivative. This shows the up and down vertical (left) and right derivatives (right), respectively. The mean value of the adaptive scales computed for all directions is shown in the right image of the first row. It is interesting to note that, as for signal estimation, the adaptive scales for derivative estimation

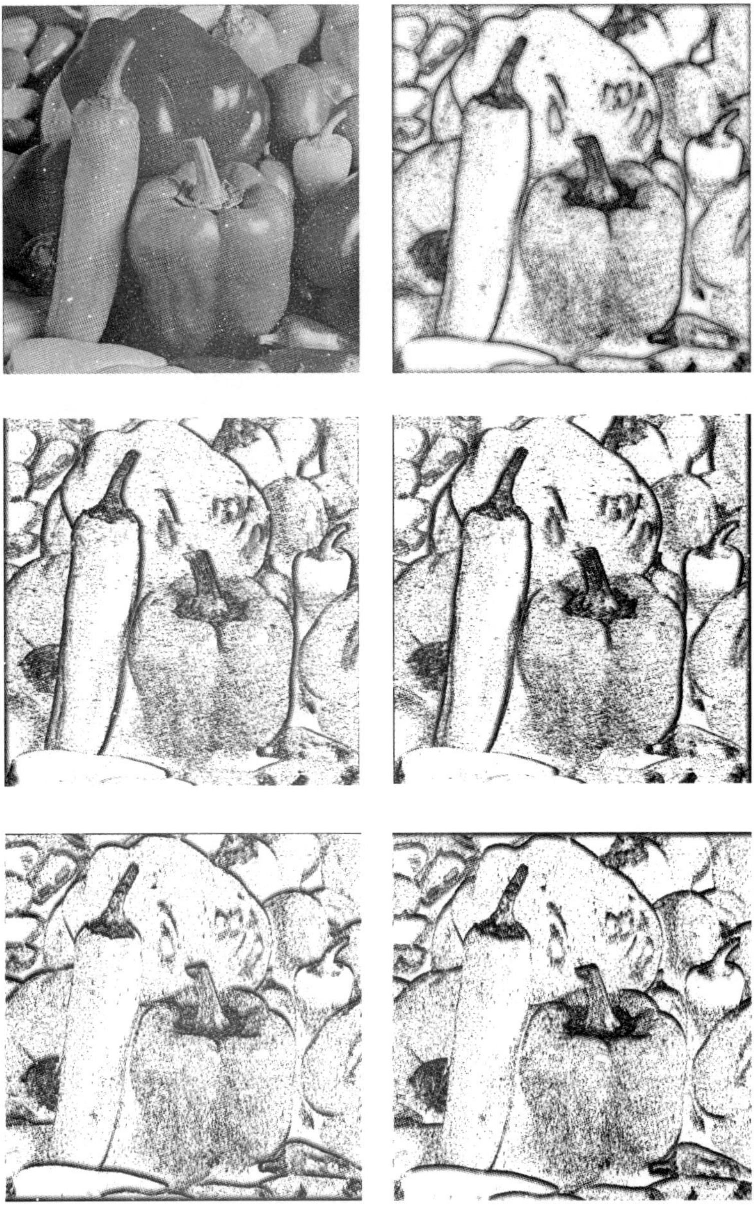

Figure 8.23 *Peppers* image (top left): adaptive varying scales for differentiation. In the middle row, the adaptive scales for the horizontal left and horizontal right derivatives are shown in the left and right images, respectively. In the bottom row, the adaptive scales for the vertical up and vertical down derivatives are shown in the left and right images, respectively. The mean value of the adaptive scales computed among all directions is shown in the top right image. Observe how the adaptive scales reveal the edges and contours of the image.

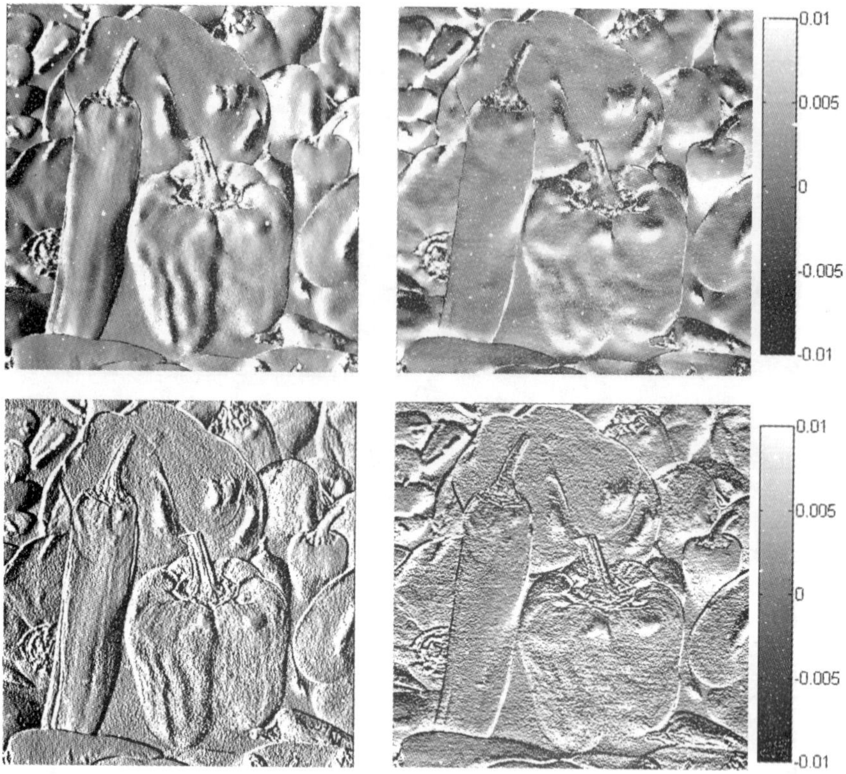

Figure 8.24 Anisotropic gradient: $\partial_{x_1}y$ (top left) and $\partial_{x_2}y$ (top right). The same derivatives estimated by the Sobel derivative filters are shown in the bottom row.

accurately indicate the edges and contours of the image details. Thus the adaptive varying scales can be exploited for edge detection.

Figure 8.24 shows the partial derivatives calculated using the ICI adaptive varying scales and the anisotropic gradient technique for fusing these estimates.

The directional derivatives are calculated for sixteen directions as

$$\theta_i = i \cdot \pi/8, \qquad i = 0, 1, \ldots, 15, \tag{8.91}$$

and the estimation of the gradient is produced according to formulas (8.79)–(8.80).

The adaptive-scale anisotropic derivatives are shown in the first line of images in Fig. 8.24. When estimating we do not go through the image edges and use only the areas where the intensity signal is regular. In this way the anisotropic differentiation assumes a pointwise segmentation of the image defining a regular neighborhood for each pixel x.

We can see some edges in the first line of images of Fig. 8.24 only because the derivative values are different for the different sides of these edges.

The nature of the edges in this anisotropic gradient imaging is quite different from those that we obtain in the standard edge detection methods, where the derivatives are calculated across the edges and the edges identified by large values of these derivatives. The amplitudes of the anisotropic derivatives shown in Fig. 8.24 are very small because they are calculated over the areas where the intensity signal is continuous. For comparison, Sobel's differentiation is shown in the second line of images in Fig. 8.24.

8.5 Shading from Depth

Volumetric objects can be generated from measured images, simulated data, or by discretizing analytic functions defining objects or surfaces. Volumetric data visualization appears in computer graphics and some medical and technical applications. When rendering these objects, shading provides visual cues about object shape and texture and can have a strong effect on the visual realism of a rendered scene.

There are a number of image-space and object-space shading methods that have been used for shading volumetric objects. Most of them require an estimate of the object surface normal for shading calculations. Errors in contact positions and surface normals create artifacts and blurring effects in renderings and inconsistencies in physical modeling [115], [116].

In what follows we use one of these techniques where the gradient of the depth map is exploited. It is shown that the concept of the anisotropic gradient naturally fits to specific requirements of this problem, and that the LPA-ICI estimates of the gradient allow us to get good results.

Let $y(x)$ be the volumetric data for some object obtained in the following measuring procedure. Rays emanate from a viewpoint, pass through an image plane, and then intersect the object surface (Fig. 8.25). The distance $y(x)$ to the surface is recorded. The recorded data $y(x)$ form a depth map of the object.

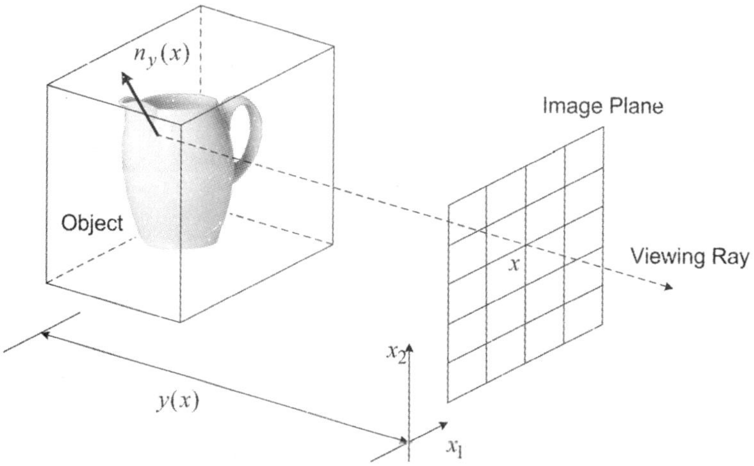

Figure 8.25 Rays emanate from a viewpoint, pass through an image plane, and then intersect the object surface. The distance $y(x)$ to the surface is recorded. The recorded data form a depth map of the object.

Let $\mathbf{n}_y(x) = \left(-\partial_{x_1} y, -\partial_{x_2} y, 1\right)^T$ be a normal vector to a surface $y(x)$. We assume that a reflected luminous intensity l_y in any direction from an element of the surface y is proportional to the cosine of the angle between the vector $\mathbf{n}_y(x)$ and the direction of the illumination $\boldsymbol{v} = (v_1, v_2, v_z)^T$,

$$l_y \sim \frac{\boldsymbol{v}^T \mathbf{n}_y(x)}{|\boldsymbol{v}| \left|\mathbf{n}_y(x)\right|}. \tag{8.92}$$

This is Lambert's law for diffusion reflection advanced by Lambert over 200 years ago. It is used extensively in geometrical optics and modern computational vision systems [159].

As we discussed previously, the estimation of the correct gradient (and then the correct normal vector) can be compromised by the failure of the differentiability assumption as well as by noise. When the normal vector is used for the realistic visualization (rendering) of a 3D surface, even weak noise can produce a dramatic loss of quality. Indeed, noise can be dealt with by means of larger size derivative estimation kernels. However, these larger kernels are unable to fit discontinuities in the underlying signal.

Figure 8.26 illustrates the differentiability problem at hand. The estimator of the normal vector uses the volumetric data from the set $V(x_-^0)$ for the argument value $x < x^0$ and from the set $V(x_+^0)$ for the argument value $x > x^0$. The depth map

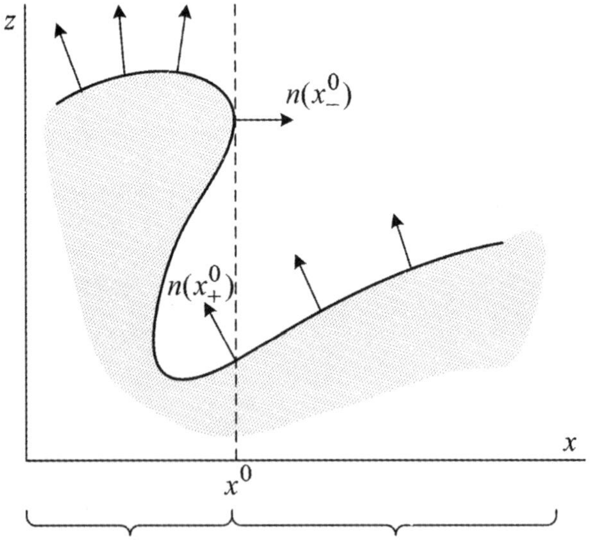

Figure 8.26 The estimator of the normal vector uses the volumetric data for the argument value $x < x^0$ from the vicinity $V(x_-^0)$ and from the vicinity $V(x_+^0)$ if $x > x^0$. The depth map exhibits a clear discontinuity at the point x^0 for both the intensity and the normal vector. An error in estimation of this normal vector can result in erroneous shading in a neighborhood of the point x^0.

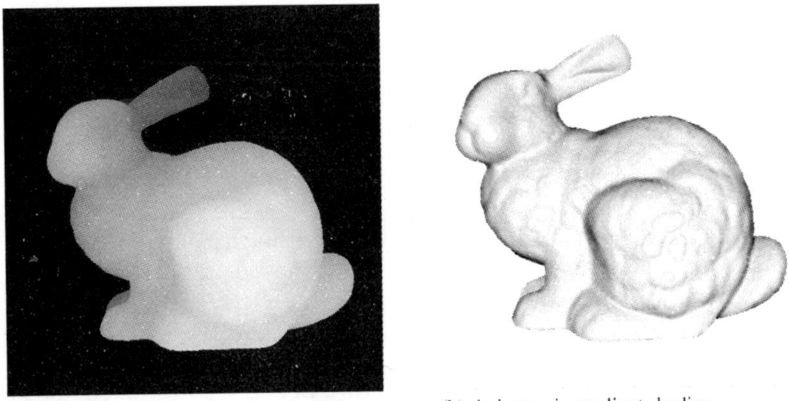

(a) rabbit: noisy depth map (z-buffer) (b) Anisotropic gradient shading

Figure 8.27 Shading obtained from the anisotropic gradient: (a) input data-noisy depth map (z-buffer), (b) shading obtained using the anisotropic gradient concept for estimation of the normal vactor.

exhibits a clear discontinuity at the point x^0 for both the intensity and the normal vector. An error in estimation of this normal vector can result in erroneous shading in a neighborhood of the point x.

The concept of an anisotropic gradient based on the relevant adaptive segmentation of the estimation neighborhood accurately corresponds to the considered shading from depth problem.

Figure 8.27(a) presents the depth map, also called a *z-buffer*. Components of the anisotropic gradient are shown in Fig. 8.28. Here we use the line-wise nonsymmetric differentiating filters in Eq. (14.26) for the scales $H = \{2, 3, 7\}$ for 32 sectorial directions $\theta_i = i\pi/16$, $i = 0, 1, \ldots, 31$.

The shading of the depth map with the illumination direction perpendicular to the image plane is calculated according to formula (8.92) and shown in Fig. 8.27(b).

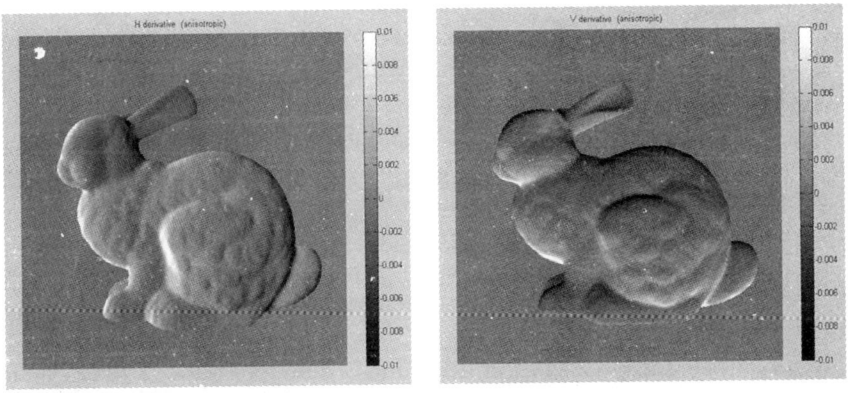

Figure 8.28 Anisotropic gradient: $\partial_{x_1} y$ (left) and $\partial_{x_2} y$ (right).

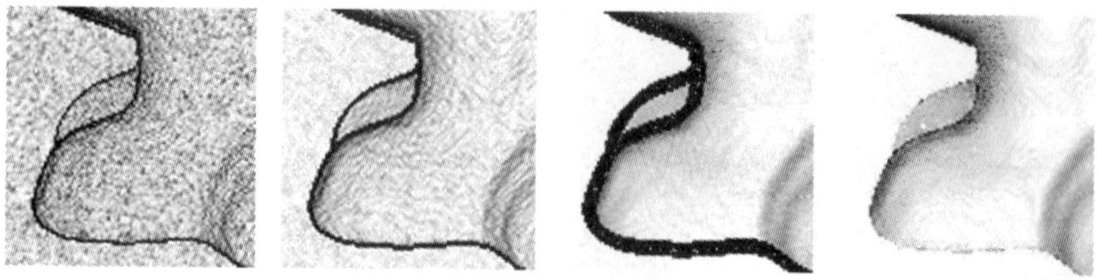

Figure 8.29 Detail of the rabbit's paws: from left to right, shading from gradient, kernel lengths 2, 3, 7, and anisotropic gradient.

Details of this shading (rabbit's paws) are shown in Fig. 8.29 for fixed nonadaptive window sizes and for the scale-adaptive anisotropic differentiator.

The nonadaptive estimator is not able to differentiate between the two paws of the rabbit, which are in different planes, one behind the other. It differentiates the depth map as a continuous 2D signal. As a result, in the corresponding images the rabbit's paws have black contours that become wider for larger h. A more intelligent anisotropic adaptive estimate of the gradient differentiates between the visible and obscured parts of the object, and for the rabbit's paws these black contours are practically eliminated.

8.6 Optical Flow Estimation

The relative motion of objects and camera is an obvious source of temporal variations in recorded image sequences. Parameters of 3D motion (rotation, translation) play an important role in image motion modeling. If these parameters are known precisely, the object motion can be accurately predicted and used in such problems as image compression, motion blur removing, and high-resolution imaging from an image sequence. The 3D relative motion of objects and camera induces 2D motion on the image plane via a suitable projection system.

The corresponding 2D motion projections, called an *apparent motion* or *optical flow*, form observation data used for sequence image processing. Naturally, these data are degraded by noise and disturbances of different natures. A basic problem in signal processing is to reconstruct 2D movements in the image plane from intensity and color variations in the sequence of the recorded 2D image plane projections.

The problem of estimating the object motion has received much attention because of its many different applications. Tasks such as scene interpretation and segmentation, high-resolution imaging from a sequence of images, target tracking and recognition, and so on, all use the optical flow as input information.

A variety of techniques dealing with original space-time data or going to transform domains has been developed as general theoretical methodology as well as for applications.

The image intensity can be considered as a function of the spatial x and time t variables with notation $y(x, t)$. For a sequence of time instants $\{t_j\}$ we have a sequence of intensities $\{y(x, t_j)\}$, where for each pixel x the signal $y(x, t_j)$ gives a variation of the intensity in time. Thus, the time-varying sequence of the images is described by

$$\{y(x, t_j), x \in X \quad \text{and} \quad j = 1, \ldots, K\},$$

where K is the size of the sequence and X is a grid of the image pixels.

Every motion estimation technique starts from assumptions about the relationship between the motion of objects and the image intensity. One of the simplest and practically efficient hypotheses is that the optical flow (or image intensity y) is invariant along the motion trajectory [83], [213]. This implies, in particular, that the scene illumination is invariant during motion and the object surface is opaque. This optical flow hypothesis approximately describes the dominant properties of natural images and usually works well, but is not satisfied exactly.

8.6.1 Optical flow equation

Let s be a parameter along a motion trajectory. Then the invariance of y along the trajectory means that

$$\partial_s y = 0.$$

Using the chain rule for differentiation, we obtain

$$\partial_s y = (\partial_x y)^T \cdot \partial_s x + \partial_t y \cdot \partial_s t = 0.$$

Dividing both parts of the last equation by $\partial_s t$, we find $(\partial_x y)^T \cdot \partial_s x / \partial_s t + \partial_t y = 0$, which gives

$$(\partial_x y)^T \cdot \partial_t x + \partial_t y = 0.$$

Let us use the notation $\mathbf{v} = \partial_t x$ for the vector velocity and ∇y for the spatial vector gradient. Then the *optical flow equation* can be given in the form

$$\mathbf{v}^T \nabla y + \partial_t y = 0. \tag{8.93}$$

If the gradient ∇y and time derivative $\partial_t y$ are given, then the only unknown in the flow equation is the component of the vector velocity $\mathbf{v}$.

Equation (8.93) holds for each x, but two components of $\mathbf{v}$ cannot be found from a single equation. Because of this, extra constraints on the estimated movement should be used to find a unique solution and to resolve this ambiguity [213].

8.6.1.1 Global movement

Let the vector $\mathbf{v}$ be the same for all $x \in X$. Then it can be found by minimizing the quadratic residual criterion that

$$J = \sum_{x \in X} [\mathbf{v}^T \nabla y(x) \cdot \mathbf{v} + \partial_t y(x)]^2.$$

The minimum conditions give

$$\sum_{x \in X} \nabla y(x)[\nabla y(x)]^{T \cdot \hat{\mathbf{v}}} + \sum_{x \in X} \nabla y(x) \partial_t y(x) = 0 \tag{8.94}$$

with the solution in the form

$$\hat{\mathbf{v}} = -\left\{ \sum_{x \in X} \nabla y(x)[\nabla y(x)]^T \right\}^{-1} \sum_{x \in X} \nabla y(x) \partial_t y(x). \tag{8.95}$$

8.6.1.2 *Local movement*

Assume that the differences between velocities are negligible for neighboring pixels. Then the global estimate can be replaced by the local one minimizing the quadratic residual criterion calculated for a neighborhood of x^0:

$$J(x) = \sum_{x \in U(x^0)} (\mathbf{v}^T \nabla y + \partial_t y)^2$$

with the solution in the form

$$\mathbf{v}(x^0) = -\left\{ \sum_{x \in U(x^0)} \nabla y(x)[\nabla y(x)]^T \right\}^{-1} \sum_{x \in U(x^0)} \nabla y(x) \partial_t y(x). \tag{8.96}$$

The solution $\mathbf{v}(x^0)$ is unique if the matrix $\sum_{x \in U(x^0)} \nabla y(x)[\nabla y(x)]^T$ is nonsingular, that is, there exists at least one pair of vector gradients linearly independent in the neighborhood $U(x^0)$. This restriction means that the image must be varying in the neighborhood $U(x^0)$ and have details about the movement can be seen in a sequence of images.

8.6.1.3 *Pixel-wise color image movement*

Image diversity can be improved by using color imaging. In this way we obtain at least three different image intensities y_i, where $i = 1, 2, 3$, for red, green, and blue. Assuming that the velocity is the same for all colors, we arrive at the following pixel-wise criterion:

$$J(x) = \sum_i [\mathbf{v}^T \nabla y_i(x) + \partial_t y_i(x)]^2,$$

which gives the least-square estimate in the form

$$\hat{\mathbf{v}}(x) = -\left\{ \sum_i \nabla y_i(x)[\nabla y_i(x)]^T \right\}^{-1} \sum_i \nabla y_i(x) \partial_t y_i(x). \tag{8.97}$$

8.6.1.4 Directional differentiation

Using multidirectional differentiation of y instead of the standard components of the gradient with the derivatives on x_1 and x_2 may be one of the more interesting alternatives. Applied to an image intensity defined on a discrete grid, this allows one to find y diversities that are invisible by the standard differentiation on x_1 and x_2. In the following, we consider the multidirectional approaches to estimation of the pixel-wise movements in image sequences.

8.6.2 Velocity estimation

Let u_1 and u_2 be variables defined in rotated coordinates as shown in Fig. 7.9. The link between the unmoving and rotated variables is defined by formula (7.30) as

$$u = U(\theta)x, \qquad x = U^T(\theta)u,$$
$$U(\theta) = \begin{pmatrix} \cos\theta & \sin\theta \\ -\sin\theta & \cos\theta \end{pmatrix}. \tag{8.98}$$

Consider the image intensity in the original and rotated variables as $y(x)$ and $y_\theta(u) = y(U^T(\theta)u)$. Using the chain rule for vector differentiation, we obtain for the gradient of $y_\theta(u)$ calculated on u

$$\partial_u y_\theta(u) = U(\theta)\partial_x y(x) \tag{8.99}$$

and

$$\partial_x y(x) = U^T(\theta)\partial_u y_\theta(u). \tag{8.100}$$

The derivatives in the different directions are linked through the rotation matrix $U(\theta)$.

Inserting the gradient $\nabla y(x) \triangleq \partial_x y(x)$ into Eq. (8.93), we obtain the optical flow equation in the directional form

$$[U(\theta)\mathbf{v}]^T\partial_u y_\theta(u) + \partial_t y = 0, \tag{8.101}$$

where $U(\theta)\mathbf{v}$ is a projection of the vector velocity $\mathbf{v}$ on the axes u_1, u_2. The directional derivative estimates $\hat{y}^{(r)}_{h,\theta}(x)$ for $r = (1, 0)$ and $r = (0, 1)$ give estimates of the components of the vector $\partial_u y_\theta$, that is,

$$\partial_{u_1} y_\theta \simeq \hat{y}^{(1,0)}_{h,\theta}(x),\ \partial_{u_2} y_\theta \simeq \hat{y}^{(0,1)}_{h,\theta}(x).$$

Denote this vector estimate as

$$\partial_u \hat{y}_{h,\theta}(x) = [\hat{y}^{(1,0)}_{h,\theta}(x),\ \hat{y}^{(0,1)}_{h,\theta}(x)]^T. \tag{8.102}$$

Using this vector in Eq. (8.101) for multiple directions of θ_i, we can find an estimate of the vector velocity by minimizing the quadratic criterion

$$J = \sum_{\theta_i}\{[\partial_u \hat{y}_{h,\theta_i}(x)]^T U(\theta_i)\mathbf{v} + \partial_t y\}^2.$$

The minimum conditions $\partial_{\mathbf{v}} J = 0$ give the equation for $\mathbf{v}$

$$\sum_{\theta_i} U^T(\theta_i)\partial_u \hat{y}_{h,\theta_i}(x)\{[\partial_u \hat{y}_{h,\theta_i}(x)]^T U(\theta_i)\hat{\mathbf{v}} + \partial_t y\}$$
$$= \left\{\sum_{\theta_i} U^T(\theta_i)\partial_u \hat{y}_{h,\theta_i}(x)[\partial_u \hat{y}_{h,\theta_i}(x)]^T U(\theta_i)\right\}\hat{\mathbf{v}} + \sum_{\theta_i} U^T(\theta_i)\partial_u \hat{y}_{h,\theta_i}(x)\partial_t y = 0. \tag{8.103}$$

Then the estimate of $\mathbf{v}$ is of the form

$$\hat{\mathbf{v}}(x) = -\Xi^{-1}(x)b(x)\cdot\partial_t y,\ b(x) = \sum_{\theta_i} U^T(\theta_i)\partial_u \hat{y}_{h,\theta_i}(x),$$
$$\Xi(x) = \sum_{\theta_i} U^T(\theta_i)\partial_u \hat{y}_{h,\theta_i}(x)[\partial_u \hat{y}_{h,\theta_i}(x)]^T U(\theta_i).$$

These equations look very similar to those in Eqs. (8.94–8.96) with the only difference being that we start from the directional derivatives and project them onto the x_1 and x_2 axes.

Remember that, according to Eq. (8.100), the product $U^T(\theta_i)\partial_u \hat{y}_{h,\theta_i}(x)$ is a projection of the directional vector $\partial_u \hat{y}_{h,\theta_i}(x)$ on the original axes x_1, x_2.

For continuous variables and differentiable functions, all these directional estimates carry no extra information about the intensity y. However, for discrete variables and anisotropic functions, the situation is different because derivatives calculated for various directions reveal different features of the function.

These directional derivative estimates can be exploited in three different modes: with a *single invariant scale* as in Eq. (8.103); with *multiple invariant scales* in the form

$$\left\{\sum_h \sum_{\theta_i} U^T(\theta_i)\partial_u \hat{y}_{h,\theta_i}(x)[\partial_u \hat{y}_{h,\theta_i}(x)]^T U(\theta_i)\right\}\hat{\mathbf{v}}(x)$$
$$+ \sum_h \sum_{\theta_i} U^T(\theta_i)\partial_u \hat{y}_{h,\theta_i}(x)\partial_t y = 0, \tag{8.104}$$

where the summation on the used scales is included; and with the ICI adaptive scales,

$$\left\{\sum_{\theta_i} U^T(\theta_i)\partial_u \hat{y}_{h^+(x),\theta_i}(x)[\partial_u \hat{y}_{h^+(x),\theta_i}(x)]^T U(\theta_i)\right\}\hat{\mathbf{v}}(x)$$
$$+ \sum_{\theta_i} U^T(\theta_i)\partial_u \hat{y}_{h^+(x),\theta_i}(x)\partial_t y = 0, \tag{8.105}$$

where the ICI adaptive varying scales $h^+(x)$ are used for the estimates. The following numerical experiments illustrate the performance of the pixel-wise LPA-ICI adaptive estimates in Eq. (8.105).

8.6.3 Experiments

We assume that the set of data consists of two successive images for each experiment and the time velocity $\partial_t y$ is calculated as the finite difference of these two images.

Let us start from a simple geometrical figure called *Square*. It is a binary image with image intensity equal to zero and one inside and outside of the figure, respectively. The optical flows obtained are shown in Fig. 8.30.

The three figures show the optical flows for the horizontal, vertical, and diagonal shifts of *Square*. The shift sizes are equal to two pixels, respectively, in the right and up directions for the horizontal and vertical movements. For the diagonal movement, the object is moved by two pixels in the right and up directions.

The image is binary. It then follows that it is not possible to reveal a movement of the areas where the image intensity is constant, that is, for the internal pixels of the square as well as along the edges directed in the direction of the movement. For the horizontal movement we can see only the movements of the vertical left and right edges of the square (left image in Fig. 8.30). For the middle pixels of these edges we can accurately observe horizontal arrows showing the correct direction of the movement. Near the vertex of the square the picture is different, because the arrows are not horizontal and show some vertical components of the velocity. It can be realized that this effect accurately corresponds to changes of the image intensity as a result of the horizontal shift of the binary object image.

For the vertical movement we can observe only the movements of the horizontal top and bottom horizontal edges of the square (middle image in Fig. 8.30). For the middle pixels of these edges we can accurately observe vertical arrows showing the correct movement of the square. Near the corners of the square the picture is disturbed because the arrows objectively seem to have some horizontal components of the velocity.

For the diagonal movement (right image in Fig. 8.30) the middle pixels of the horizontal edges show vertical movement and the middle pixels of the vertical edges

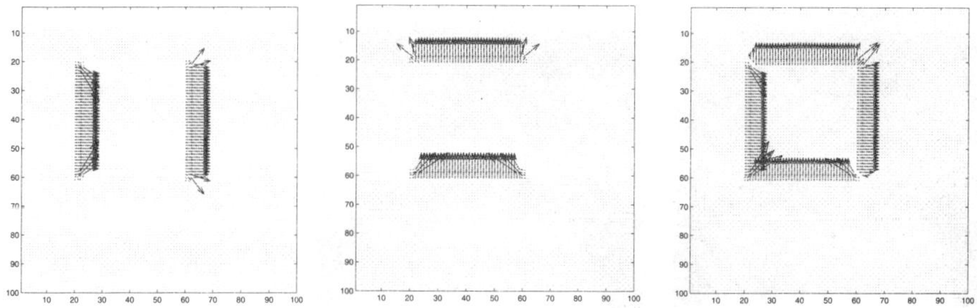

Figure 8.30 *Square*: optical flow for horizontal, vertical, and diagonal movements.

Figure 8.31 *Akyoi*: optical flow for eyes and mouth movement.

show the horizontal movement. Only near the vertexes do some of the pixels show accurate diagonal movement.

Figure 8.31 shows two images of *Akyoi*. Akyoi's eyes open in the first image and close in the second. Some mouth movement also differs in the images. The right image in Fig. 8.31 demonstrates the obtained optical flow. The arrows in this image represent the velocity vectors $\boldsymbol{v} = (\partial_t x_1, \partial_t x_2)^T$ of the optical flow, that is, they show the size and the direction of the velocity as they are defined by the horizontal $\partial_t x_1$ and vertical $\partial_t x_2$ components of the velocity.

Figure 8.32 shows two images of *Akyoi* that are different in the more complex movements of the head and body. The left image in Fig. 8.32 demonstrates the estimate of the corresponding optical flow.

Both sets of images in *Square* and *Akyoi* show that the proposed algorithm is very sensitive with respect to even tiny movements and allows one to correctly detect movements and estimate the map of the corresponding velocities.

The details of the algorithm used in these experiments are as follows. The differentiating kernels have the order $m = (1, 1)$ with the uniform narrowed supports of type 10 (Fig. 7.9) defined by the scales

$$H = \left\{ \begin{pmatrix} 2 \\ 1 \end{pmatrix}, \begin{pmatrix} 4 \\ 2 \end{pmatrix}, \begin{pmatrix} 8 \\ 2 \end{pmatrix}, \begin{pmatrix} 16 \\ 4 \end{pmatrix}, \begin{pmatrix} 32 \\ 4 \end{pmatrix} \right\}.$$

The directional resolution is equal to 16 in Eq. (8.91), and the threshold parameter $\Gamma = 1$.

Figure 8.32 *Akyoi*: optical flow for head and body movement.

Chapter 9

Image Reconstruction

One fundamental branch in image processing concerns image reconstruction. A digital image or a sequence of digital images can be corrupted optically and electronically. These image data can be results of indirect observations (experiments) where linear or nonlinear transformations of an image of interest are possible. The problem is to reconstruct a clear and sharp high-resolution image of the original image-object relevant to the observed phenomenon and correctly understood by humans.

The reconstruction of quantities that we want is called an *inverse problem*. In an inverse problem we measure an effect and want to determine the cause.

Modeling is a key starting element of any reconstruction method because it allows one to present and formalize information about signals and systems defining a signal distortion. Linear models are the most popular, and they define a link between the input signal of interest y and the measurement z in integral form as an integral convolution,

$$z(x) = \int_{\mathbb{R}^2} v(x-u)y(u)du + \varepsilon(x) = (v \circledast y)(x) + \varepsilon(x). \tag{9.1}$$

The kernel $v(x)$ in Eq. (9.1) is called a *point spread function* (*PSF*) of the system. It can be *shift-invariant (homogeneous)*, depending only on a single argument as in Eq. (9.1), or *shift-variant (heterogeneous)*, as in the model

$$z(x) = \int_{\mathbb{R}^2} v(x,u)y(u)du + \varepsilon(x), \tag{9.2}$$

where v depends on two arguments x and u.

The blurring phenomenon, modeled by the kernel v, is very evident in many image processing applications. The essence of the deblurring is to invert the input-output link in Eqs. (9.1–9.2) and find y from z. Here are two examples of some common continuous variable optical PSFs [129]:

- Atmospheric turbulence blur (Gaussian PSF):

  $$v(x) = \mathcal{N}(0, \sigma^2),$$

  where the standard deviation parameter σ controls the level of the blur.

- Uniform out-of-focus blur, used in geometric optics, in the form

$$v(x_1, x_2) = \begin{cases} \dfrac{1}{\pi R^2}, & \text{if } \sqrt{x_1^2 + x_2^2} \le R, \\ 0, & \text{otherwise,} \end{cases}$$

where the radius R depends on the parameters of the lens and the distance of the lens from the object to the image sensor.

The term "inverse problem" seems to have slightly different meanings in different communities. It is strongly connected to the operation of convolution/deconvolution (blurring/deblurring) in digital signal processing, while in general it denotes a much broader class of problems. Mathematically, it is a solution of the integral Eqs. (9.1–9.2).

Many approaches use discretization of the integral equations as a first step. In this way we arrive at a set of linear algebraic equations similar to Eqs. (9.1–9.2), where integration is replaced by summation. Sensitive aspects of correct sampling should be addressed at this step so as to not lose important information in the signal y.

Discretization means that a function of continuous argument $y(x)$ is replaced by a sequence $y(\Delta k)$ and the inverse problem is reduced to the estimation of this sequence. Alternative approaches, e.g., Galerkin's methods, focus on the reconstruction of y as a function of the continuous argument x replacing y in Eqs. (9.1–9.2) by some series.

The inverse problem formalization is typical for optical systems, satellite imaging, radiometry, and ultrasonic and magnetic resonance imaging. Details, further references, and applications can be seen in the following books and reviews: [10], [24], [71], [93], [129], [211].

In this chapter we are concerned with discrete inverse problems as they appear in digital imaging. Algorithmic solutions are focused on the LPA-ICI adaptive-scale methods. It is assumed that the PSF is shift-invariant and known.

9.1 Image Deblurring

9.1.1 Blur modeling and deblurring

Suppose that we wish to reconstruct a 2D image intensity function y, but we are able to observe only the convolution $(y \circledast v)(x)$, where v is a given PSF. It is assumed that the convolution is discrete and noisy, so we observe z given as

$$z(x) = (y \circledast v)(x) + \varepsilon(x), \tag{9.3}$$

where x is defined on the regular $n_1 \times n_2$ lattice $X = \{k_1, k_2 : k_1 = 1, 2, \ldots, n_1, k_2 = 1, 2, \ldots, n_2\}$, and ε is a zero-mean white Gaussian noise with variance σ^2.

In the following we will use DFT techniques (see Sec. 3.3.4). Applying the DFT to both parts of Eq. (9.3), we obtain the equation in the frequency domain,

$$Z(f) = Y(f)V(f) + \varepsilon(f), \tag{9.4}$$

where $Z(f)$, $Y(f)$, $V(f)$, and $\varepsilon(f)$ are the DFTs of the corresponding signals, $f = (f_1, f_2)$, $f \in F$, and

$$F = \{f_1,\ f_2 : f_i = k_i,\ k_i = 0, 1, \ldots, n_i - 1, i = 1, 2\}$$

is the 2D discrete frequency.

Recall that Eq. (9.4) by default assumes the periodicity of $y(x)$, $v(x)$, and $z(x)$ with the periods n_1 and n_2 on the variables x_1 and x_2, respectively. This means that, strictly speaking, the equation in the spatial domain in Eq. (9.3) is not equivalent to the equation in the frequency domain in Eq. (9.4). The accurate equivalent to the frequency domain Eq. (9.4) can be written using formula (3.54), which is different from Eq. (9.3). However, these differences mainly concern the areas close to the image boundaries and may be neglected in many cases.

In principle, an unbiased estimate of $Y(f)$ can be obtained as a straightforward "naive inverse" solution of Eq. (9.4),

$$\hat{Y}(f) = \frac{Z(f)}{V(f)}. \tag{9.5}$$

However, in the cases of interest, the convolution operator v is not invertible, in the sense that the inverse of v does not exist as a bounded linear operator. Such inverse problems are called *ill-posed* or *ill-conditioned* [220].

In the frequency domain, being ill-posed means that $V(f)$ may take zero or close to zero values. If the system is ill-conditioned, the random components in the observations are enormously amplified in the inversion in Eq. (9.5) to yield an extremely noisy and useless estimate.

It is now standard to approach the inverse problem by the method of regularization, in which one applies, rather than the inversion, a regularized inverse operator [220]. In the frequency domain this is defined by

$$\hat{Y}(f) = \frac{Z(f)V^*(f)}{|V(f)|^2 + \alpha^2}, \tag{9.6}$$

where $\alpha^2 > 0$ is a regularization parameter.

In addition, there are plenty of different ideas and methods exploited for image reconstruction. Here we briefly review some of the recent tendencies. One of the typical approaches to the image reconstruction problem is based on *projection methods*, which assume that the observed image can be decomposed as a sum of basis functions:

- A natural way of projecting for ill-posed problems is associated with singular value decomposition (SVD). Denote by v^* the adjoint of the PSF v and assume that v^*v is an operator with positive eigenvalues and some eigenfunctions. These eigenfunctions are used as a basis in SVD methods (see, e.g.,

[22], [23], and the references therein). Despite the great popularity of schemes based explicitly or implicitly on SVD, this method suffers from natural performance limitations that are rooted in the fact that the basis eigenfunctions are derived from the convolution operator v and not from the object y. Thus, efficient representation of the object y by the eigenfunctions of v^*v cannot be guaranteed for all applications [22].

- There are important alternative strategies based on developing basis functions for the object y rather than for the convolution operator v. Wavelets provide an effective solution to the problem. Many real-world images and signals have economical representations in which few large wavelet coefficients capture most of the signal energy. Signal and image reconstructions using wavelet shrinkage are spatially adaptive processes well-suited to signals with edges and other singularities [2], [42], [97], [150].

  The wavelet-vaguelette decomposition has been proposed in [43] for the deconvolution problem. In this approach, wavelet expansion is applied to the signal obtained as an output of the pure deconvolution. A scale-dependent shrinkage is employed to estimate the signal wavelet coefficients.

  Further development of the wavelet approach is proposed in [22], with a new decomposition that is much better adapted to the type of edge-dominated object while providing an almost diagonal representation of the Radon convolution operator. This is done by introducing new basis functions, the so-called curvelets: smooth, anisotropic directional elements that are well-suited for detecting and synthesizing curved edges.

  A new, efficient hybrid wavelet-based deconvolution algorithm was developed in [161]. This algorithm comprises Fourier-domain regularized inversion followed by wavelet-domain noise suppression. It is shown that the developed algorithm outperforms the conventional Wiener filter as well as some other wavelet-based image restoration algorithms.

- The expectation-maximization (EM) algorithm for image deblurring, based on a penalized maximum likelihood setting, is proposed in [50]. The algorithm uses the EM technique in the wavelet domain. It is shown that the algorithm is very efficient and outperforms many of the best methods.

A common point of all of the above methods is that the basis functions are applied for parametric approximations of the object function y in the form of a series with coefficients (parameters) defined from the observations. These functions may be Fourier harmonics (as in the frequency domain Eq. (9.4)), eigenfunctions of the convolution operator in SVD methods, or wavelets in wavelet-based decompositions. There exist many deconvolution techniques based on this type of approach.

Fundamentally different ideas and methods arise from the nonparametric regression approach studied in this book. The main idea is to build a pointwise approximation of y in the spatial domain using observations from local neighborhoods.

The estimation is produced in a sliding window, meaning that every pixel of the object is treated in a special way to achieve the best possible performance. The best size of this neighborhood (scale) is selected for each pixel, and this single-scale estimate is used in the estimation. This serves as the base of the LPA-ICI adaptive approach.

The first application of this kind of adaptive nonparametric regression for blurred 1D observations was reported in [67], where the ICI window size adaptive kernel estimate was applied to a signal obtained by the pure inverse. The continuous observation model was used for the algorithm presentation as well as for the analysis.

The idea for this sort of a nonparametric deconvolution technique originated in probability density estimation. If a desired random signal is observed with an additive random noise, the observation density is a convolution of the noise and the desired signal densities. If the noise density is known, the estimation of the desired signal density is the deconvolution operation with a known PSF (e.g., [45]).

In this chapter we show how the adaptive nonparametric approach can be extended to discrete image deblurring. New adaptive algorithms are developed in forms of the regularized inverse and more efficient Wiener inverse LPA-ICI spatially adaptive filters. Experiments show that the scale-adaptive Wiener inverse allows a further significant improvement in the accuracy of image deblurring.

9.1.2 LPA deblurring

To derive the solutions for the regularized inverse and Wiener filtering, we need some technical results concerning the spectrum transformation of random processes and minimization of the quadratic lost function in the frequency domain. For completeness, we present these techniques in Sec. 14.6 of the Appendix.

9.1.2.1 Naive inverse

The naive inverse Eq. (9.5) gives a solution that is unbiased for the observation model Eq. (9.4),

$$E\{\hat{Y}(f)\} = E\{Z(f)\}/V(f) = E\{V(f)Y(f) + \varepsilon(f)\}/V(f) = Y(f), \quad (9.7)$$

because $E\{\varepsilon(f)\} = 0$. The spectrum of the white noise $\varepsilon(x)$ is invariant and equal to the variance σ^2. Then, the variance of this naive estimate $\hat{y}$ in the frequency domain is calculated according to Eq. (14.189) as

$$\sigma_{\hat{y}}^2 = var\{\hat{y}(x)\} = \frac{\sigma^2}{n_1 n_2}||1/V(f)||^2, \quad (9.8)$$

where the Euclidean norm means $||1/V(f)||^2 = \sum_{f \in F} 1/|V(f)|^2$.

For the ill-conditioned convolution the operator $|V(f)|$ takes values close to zero, which gives a large value of the noise variance Eq. (9.8). This is a well-known effect of the noise amplification that is typical for the naive deconvolution. Usually, this results in the high-level noise completely destroying the image and making this solution unacceptable.

Filtering of the naive solution is one way to manage the problem. The regularization and Wiener filtering are two of the more classical approaches. We exploit both of these ideas and complete them by LPA-ICI adaptive smoothing.

9.1.2.2 LPA smoothing

The basic idea of the proposed deblurring algorithm is to estimate the smoothed image intensity y_h instead of the original one y with the link between them defined by the smoothing kernel g_h:

$$y_h(x) = (g_h \circledast y)(x). \tag{9.9}$$

This smoothing kernel is designed according to the LPA procedure and can be defined according to the general formulas (3.5) or (7.47) if the directional version of the smoothing is used.

The scale parameter h in g_h controls the level of the smoothing and can be used to suppress the noise as much as possible while preserving the details of the original object signal $y(x)$. Applying the kernel operator g_h to both sides of Eq. (9.3) yields the following equivalent representations:

$$z_h(x) = [g_h \circledast (y \circledast v)](x) + \varepsilon_h(x) \tag{9.10}$$

$$= (y_h \circledast v)(x) + \varepsilon_h(x) = (y \circledast v_h)(x) + \varepsilon_h(x), \tag{9.11}$$

where

$$v_h(x) = (g_h \circledast v)(x)$$

is the smoothed PSF and

$$\varepsilon_h(x) = (g_h \circledast \varepsilon)(x)$$

is the smoothed observation noise.

In the frequency domain, Eqs. (9.10–9.11) can be presented in two equivalent forms:

$$Z_h(f) = Y_h(f)V(f) + \varepsilon_h(f), \; Z_h(f) = Y(f)V_h(f) + \varepsilon_h(f), \tag{9.12}$$

where $Z_h(f)$, $Y_h(f)$, $V_h(f)$, and $\varepsilon_h(f)$ stand for the DFTs of the corresponding smoothed signals. It is clear that $Y_h(f) = G_h(f)Y(f)$ as well as $V_h(f) = G_h(f)V(f)$ and $Z_h(f) = G_h(f)Z(f)$. The operator g_h smooths the observations, giving $Z_h(f)$ instead of $Z(f)$.

The naive deconvolution for this smoothed signal can be found from the equation

$$Z_h(f) = \hat{Y}_h(f)V(f), \tag{9.13}$$

with the smoothed naive inverse

$$\widehat{Y}_h(f) = \frac{Z_h(f)}{V(f)} = G_h(f)\frac{Z(f)}{V(f)} = G_h(f)\widehat{Y}(f), \tag{9.14}$$

where $\widehat{Y}(f)$ is given in Eq. (9.5).

Using Eq. (14.186) we can see that the variance of the estimate in Eq. (9.14) is

$$\sigma^2_{\hat{y}_h} = \frac{\sigma^2}{n_1 n_2}\left\| \frac{G_h(f)}{V(f)} \right\|^2.$$

With the appropriate choice of G_h, the variance $\sigma^2_{\hat{y}_h}$ is smaller than the one given by the pure inverse Eq. (9.8). However, if v is ill-conditioned and $|V(f)|$ has small values close to zero, this sort of filtering alone cannot be sufficient. We must combine it with regularization and Wiener filtering.

9.1.2.3 Regularization

Let us look for a solution $\hat{y}_h$ for y_h from the observations in Eq. (9.10) based on the criterion

$$\begin{aligned} &R^2_\psi = R^2 + \alpha^2\psi, \qquad \alpha > 0, \\ &R^2 = \sum_x r^2(x), \qquad r(x) = z_h(x) - \hat{z}_h(x), \qquad \hat{z}_h(x) = (y_h \circledast v)(x), \\ &\psi = \sum_x r_d^2(x), \qquad r_d(x) = (y_h \circledast d)(x). \end{aligned} \tag{9.15}$$

The combined criterion R^2_ψ consists of two parts. The first one, R^2, is a residual functional computed as a mean-square difference between the smoothed observation $z_h(x)$ and the prediction $\hat{z}_h(x)$. The second part, ψ, is a *penalty* characterizing the "total energy" of the convolution of y_h with a regularization operator d. The parameter α defines the weight of this penalty in the criterion.

It is convenient to rewrite R^2_ψ in the frequency domain. Using the Parseval formula (3.57) we have

$$R^2_\psi = \frac{1}{n_1 n_2}||Z_h(f) - V(f)Y_h(f)||^2 + \frac{\alpha^2}{n_1 n_2}||Y_h(f)D(f)||^2, \tag{9.16}$$

where $D(f)$ is the DFT for $d(x)$.

The Euclidian norm used in Eq. (9.16) means that

$$\begin{aligned} R^2_\psi = \frac{1}{n_1 n_2}\sum_{f\in F}\{&[Z_h(f) - V(f)Y_h(f)][Z^*_h(f) - V^*(f)Y^*_h(f)]\} \\ &+ \alpha^2 Y_h(f)|D(f)|^2 Y^*_h(f)], \end{aligned} \tag{9.17}$$

where the star (*) stands for complex conjugate variables. Let us also rewrite R_{ψ}^2 as

$$R_{\psi}^2 = \frac{1}{n_1 n_2} \sum_{f \in F} r_{\psi}^2(f), \tag{9.18}$$

$$r_{\psi}^2(f) = [Z_h(f) - V(f)Y_h(f)][Z_h^*(f) - V^*(f)Y_h^*(f)] + \alpha^2 Y_h(f)|D(f)|^2 Y_h^*(f). \tag{9.19}$$

Formula (14.197) gives the minimum conditions for R_{ψ}^2 in the form

$$\partial_{Y_h^*(f)} R_{\psi}^2 = 0, \qquad f \in F. \tag{9.20}$$

Thus, the derivation of the equation for the optimal solution is reduced to the calculation of the derivative in Eq. (9.20). It results in the linear equation for $\widehat{Y}_h(f)$:

$$-[Z_h(f) - V(f)\widehat{Y}_h(f)]V^*(f) + \alpha^2 \widehat{Y}_h(f)|D(f)|^2 = 0$$

with the solution in the frequency domain as

$$\widehat{Y}_h(f) = \frac{Z_h(f)V^*(f)}{|V(f)|^2 + \alpha^2|D(f)|^2} = \frac{G_h(f)V^*(f)}{|V(f)|^2 + \alpha^2|D(f)|^2} Z(f). \tag{9.21}$$

In the spatial domain,

$$\hat{y}_h(x) = \mathcal{F}^{-1}[\widehat{Y}_h(f)].$$

It is a typical regularized inverse solution with a regularization parameter α^2 and a regularization operator given in the frequency domain by $D(f)$.

If the penalty ψ is defined as the energy of $\hat{y}_h$, then $d(x)$ is the identity operator, $D(f) = 1$, and

$$\psi = \sum_x \hat{y}_h^2(x). \tag{9.22}$$

Differentiating operators is usually recommended for $d(x)$, enabling the smoothness of the solution. The gradient and Laplacian are typical examples of differentiating operators. The differentiating kernels $g_h^{(r)}$ can be used for such differentiating operators $d(x)$.

Large values of α^2 in Eq. (9.15) mean that the main emphasis in this criteria is carried out on the minimization of the penalty term. The solution then becomes smoother. Smaller α^2 means that the smoothness requirements are sacrificed. These are standard speculations for the regularization techniques overall.

In the adaptive-scale approach, we use the penalty term in Eq. (9.22) with a small regularization parameter to avoid any possible oversmoothing. The main filtering is done by the LPA-ICI adaptive-scale algorithms.

In our development we use the scale h as a basic tool to control the smoothness of the solution, and the regularization is exploited only to improve the conditioning of the system. For this goal it is enough to use $D(f) = 1$ giving the regularized inversion in Eq. (9.21) in the form

$$\widehat{Y}_h^{RI}(f) = \frac{Z_h(f)V^*(f)}{|V(f)|^2 + \alpha^2} = \frac{G_h(f)V^*(f)}{|V(f)|^2 + \alpha^2} Z(f). \tag{9.23}$$

Denoting the transfer function of this regularized estimator as

$$Q_{RI}(f, \alpha) = \frac{G_h(f)V^*(f)}{|V(f)|^2 + \alpha^2}, \tag{9.24}$$

we have

$$\widehat{Y}_h^{RI}(f) = Q_{RI}(f, \alpha)Z(f). \tag{9.25}$$

This regularization guarantees that

$$\min_{f \in F}\{|V(f)|^2 + \alpha^2\} \geq \alpha^2$$

and improves the conditioning of the system.

9.1.2.4 Wiener filtering

In Wiener filtering we are looking for an optimal linear estimate of the smoothed signal $y_h(x)$ in the form of the linear shift-invariant operator

$$\hat{y}_h(x) = (z \circledast q_{WI})(x), \tag{9.26}$$

with the kernel q_{WI} to be found. The index WI means "Wiener inverse" because the considered Wiener filtering includes the inverse operation.

The signal z in Eq. (9.26) is defined by the observation model in Eq. (9.10) and the quadratic risk is

$$R^2 = E\left\{\sum_x [y_h(x) - \hat{y}_h(x)]^2\right\}. \tag{9.27}$$

It is assumed that the signal y is deterministic while the noise ξ is random. Thus, the expectation in R^2 is calculated with respect to the random noise.

According to the Parseval theorem in Eq. (3.57), the sum in R^2 is equal to

$$\sum_x [y_h(x) - \hat{y}_h(x)]^2 = \frac{1}{n_1 n_2} \sum_{f \in F} |Y_h(f) - \widehat{Y}_h(f)|^2, \tag{9.28}$$

where

$$\widehat{Y}_h(f) = Q_{WI}(f)Z(f),$$
$$Z(f) = V(f)Y(f) + \varepsilon(f),$$
$$Q_{WI}(f) = \mathcal{F}\{q_{WI}(x)\}.$$

Substituting Eq. (9.28) in Eq. (9.27), we have

$$R^2 = E\left\{\frac{1}{n_1 n_2}\sum_{f\in F}|Y_h(f) - Q_{WI}(f)Z(f)|^2\right\}.$$

The necessary minimum conditions for R^2 in a form similar to Eq. (9.20) give

$$\partial_{Q^*_{WI}(f)} R^2 = 0, \qquad f \in F.$$

Computing this derivative we arrive at

$$E\{[Y_h(f) - Q_{WI}(f)Z(f)] \cdot Z^*(f)\} = 0, \qquad f \in F.$$

This gives a transfer function of the optimal Wiener filter in the form

$$Q_{WI}(f) = \frac{E\{Y_h(f)Z^*(f)\}}{E\{|Z(f)|^2\}}, \qquad f \in F. \tag{9.29}$$

Assume a more general model for the random noise than the one used in Eq. (9.3). Let ε be a zero-mean stationary random process with the spectrum $S_\varepsilon(f)$, that is, $E\{\varepsilon(f)\} = E\{\varepsilon^*(f)\} = 0$ and $E\{|\varepsilon(f)|^2\} = n_1 n_2 S_\varepsilon(f)$ (see formula (14.187)). For the white noise in Eq. (9.3), $S_\varepsilon(f) = \sigma^2$.

The expectation over this random ε gives for the numerator in Eq. (9.29)

$$\begin{aligned} E\{Y_h(f)Z^*(f)\} &= E\{Y_h(f)[V^*(f)Y^*(f) + \varepsilon^*(f)]\} \\ &= Y_h(f)V^*(f)Y^*(f) = |Y(f)|^2 V^*(f)G_h(f), \end{aligned}$$

and for the denominator in Eq. (9.29)

$$\begin{aligned} E\{|Z(f)|^2\} &= E\{[V(f)Y(f) + \varepsilon(f)][V^*(f)Y^*(f) + \varepsilon^*(f)]\} \\ &= |V(f)|^2|Y(f)|^2 + n_1 n_2 S_\varepsilon(f). \end{aligned}$$

Then the transfer function of the Wiener filter is

$$Q_{WI}(f) = \frac{|Y(f)|^2 V^*(f)G_h(f)}{|V(f)|^2|Y(f)|^2 + n_1 n_2 S_\varepsilon(f)}, \qquad f \in F.$$

which can also be written as

$$Q_{WI}(f) = \frac{V^*(f)G_h(f)}{|V(f)|^2 + n_1 n_2 S_\varepsilon(f)/|Y(f)|^2}, \qquad f \in F. \tag{9.30}$$

The optimal estimate of y_h is

$$\hat{Y}_h^{WI}(f) = Q_{WI}(f)Z(f) = \frac{V^*(f)G_h(f)}{|V(f)|^2 + n_1 n_2 S_\varepsilon(f)/|Y(f)|^2} Z(f). \qquad (9.31)$$

Comparing the transfer functions of the regularized inverse estimator in Eq. (9.24) and the Wiener estimator in Eq. (9.30), we see that they are different by the second summand in the denominator. We have invariant α^2 for the regularized inverse estimator and a varying frequency $n_1 n_2 S_\varepsilon(f)/|Y(f)|^2$ for the Wiener inverse. This last item can be treated as an optimal "regularizator" because the Wiener filter is an optimal estimator. This optimal regularizator is a ratio of the noise and signal spectrums. The last point is a drawback of this optimal solution because it requires knowledge of the unknown estimated signal y.

Both the regularized inverse and the Wiener inverse operators simultaneously solve two tasks. First, they invert the convolution operator, and, second, they filter the result of this inversion. If we assume $\alpha = 0$ in Eq. (9.24) and $S_\varepsilon = 0$ in Eq. (9.30), both of the transfer functions have an identical form:

$$Q_{WI}(f) = Q_{RI}(f, 0) = \frac{G_h(f)}{V(f)}$$

with the output signal

$$\hat{Y}_h(f) = \frac{1}{V(f)} G_h(f)Z(f).$$

In this case we obtain the naive inverse of the convolution and filtering of this result by the smoothing filter $G_h(f)$.

9.2 LPA-ICI Deblurring Algorithms

9.2.1 ICI adaptive scale

Consider the scale selection problem for the regularized inverse and Wiener inverse estimators. Represent the estimates in Eq. (9.23) and Eq. (9.31) in the forms

$$\hat{Y}_h^{RI}(f) = G_h(f)Z_R(f), \; Z_R(f) = \frac{V^*(f)}{|V(f)|^2 + \alpha^2|D(f)|^2} Z(f),$$
$$\hat{Y}_h^{WI}(f) = G_h(f)Z_W(f), \; Z_W(f) = \frac{V^*(f)}{|V(f)|^2 + n_1 n_2 S_\varepsilon(f)/|Y(f)|^2} Z(f). \qquad (9.32)$$

Both estimators have the smoothing transfer function $G_h(f)$ as a factor of their total transfer functions. This means that application of $G_h(f)$ can be treated as a filtering of the signals $Z_R(f)$ and $Z_W(f)$.

The signals $Z_R(f)$ and $Z_W(f)$ are the regularized inverse and the Wiener inverse for estimation of the original signal y (or $Y(f)$) from the observation z. The inversion of $V(f)$ and $G_h(f)$ filtering can be treated as two independent successive operations.

The analysis of the bias and the variance of these estimates shows that the scale parameter h in $G_h(f)$ continues to play a role typical for the LPA estimates considered in the previous chapters. The most important point is that there exists an optimal balance between the variance and the bias of these estimates that is controlled by the scale parameter h. The usual arguments leading to the ICI rule for the adaptive-scale selection continue to be valid for these estimates.

The variances of the estimates in Eq. (9.32) exploited in the ICI are calculated according to the following formulas: for the regularized inverse,

$$\sigma^2_{\hat{y}_h^{RI}} = \frac{\sigma^2}{n_1 n_2} \left\| \frac{V^*(f)G_h(f)}{|V(f)|^2 + \alpha^2 |D(f)|^2} \right\|^2, \tag{9.33}$$

and for the Wiener inverse,

$$\sigma^2_{\hat{y}_h^{WI}} = \frac{\sigma^2}{n_1 n_2} \left\| \frac{V^*(f)G_h(f)}{|V(f)|^2 + n_1 n_2 \sigma^2 / |Y(f)|^2} \right\|^2. \tag{9.34}$$

In these formulas we assume that the noise is white with variance σ^2.

The properties of the filter kernel g_h should agree with the PSF convolution kernel v so that the variances in Eqs. (9.33–9.34) are bounded and not too large. Roughly speaking, the ratio $G_h(f)/V(f)$ and the corresponding variances should decay as h grows. This assumes that, to suppress the high-frequency components of the noise, the low-pass filter g_h should be stronger than the inverse of the operator v.

9.2.2 Algorithms

In the forthcoming simulation experiments we use the regularized inverse and regularized Wiener inverse estimates, specified as follows:

(A) Regularized inverse (RI)

$$\widehat{Y}_h^{RI}(f) = \frac{V^*(f)G_h(f)}{|V(f)|^2 + \alpha_1^2} Z(f); \tag{9.35}$$

(B) Regularized Wiener inverse (RWI),

$$\widehat{Y}_h^{RWI}(f) = \frac{V^*(f)G_h(f)}{|V(f)|^2 + \alpha_2^2 n_1 n_2 \sigma^2 / |Y(f)|^2} Z(f). \tag{9.36}$$

The standard Wiener inverse corresponds to $\alpha_2^2 = 1$ in Eq. (9.36). The parameter α_2 in the denominator of Eq. (9.36) is proved to be efficient for improving the filtering. Because $n_1 n_2 \sigma^2 / |Y(f)|^2$ plays the role of the optimal regularizator, we call α_2 the regularization parameter of the Wiener filter. Respectively, for the estimate and the transfer functions of the Wiener filter with the parameter α_2, we use the index RWI (regularized Wiener inverse) instead of the index *WI* that is used for the Wiener filter.

The estimates in Eqs. (9.35) and (9.36) are biased with respect to $y_h(x)$ if α_1^2, $\alpha_2^2 > 0$. The variance of these estimates is defined by

$$\sigma^2_{\hat{y}_h^{RI}}(h) = \frac{\sigma^2}{n_1 n_2}\left\| \frac{V(f)G_h(f)}{|V(f)|^2 + \alpha_1^2} \right\|^2, \tag{9.37}$$

$$\sigma^2_{\hat{y}_h^{RWI}}(h) = \frac{\sigma^2}{n_1 n_2}\left\| \frac{V(f)G_h(f)}{|V(f)|^2 + \alpha_2^2 n_1 n_2 \sigma^2/|Y(f)|^2} \right\|^2. \tag{9.38}$$

In our algorithm, we use the RI estimates in Eq. (9.35) with α_1 as small as possible, leaving the main filtering to be done by the LPA with the spatially adaptive smoothing parameter h.

The output of the RI is used in the RWI as a reference signal to estimate the power $|Y(f)|^2$ of the true image. Thus, the RI filter serves as an auxiliary to the RWI estimator. The ICI rule is exploited in the algorithm for the adaptive-scale selection for both the RI and the RWI estimates.

9.2.2.1 Basic steps of the algorithm

The main steps of the RI estimator can be summarized as follows (see Fig. 9.1):

1. Compute the transfer function for the RI in Eq. (9.35) as

$$Q_{RI}(f, \alpha_1) = \frac{V^*(f)G_h(f)}{|V(f)|^2 + \alpha_1^2}$$

   and the variances in Eq. (9.37) as

$$\sigma^2_{\hat{y}_h^{RI}}(h) = \frac{\sigma^2}{n_1 n_2}||Q_{RI}(f, \alpha_1)||^2.$$

2. Compute the estimates $\hat{y}_h^{RI}(x)$ for all $h \in H$ as the inverse DFT of Eq. (9.35).

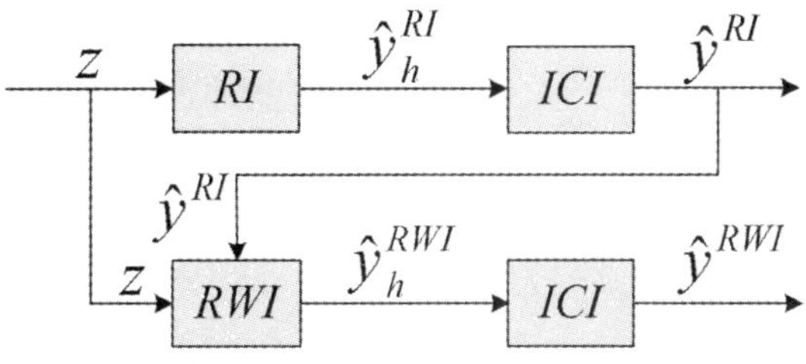

Figure 9.1 The LPA-ICI regularized Wiener inverse algorithm. The first line of the flowchart shows that the RI estimates $\hat{y}_h^{RI}(x)$ are calculated for a set of scales $h \in H$ and the ICI algorithm is used for varying the adaptive-scale selection, giving the RI image estimate $\hat{y}^{RI}(x)$. In the second line of the flowchart, the RWI estimates $\hat{y}_h^{RWI}(x)$, $h \in H$, are calculated using $\hat{y}^{RI}(x)$ as a reference signal in the Wiener filtering. After that, the ICI algorithm gives the final adaptive-scale RWI estimate $\hat{y}^{RWI}(x)$.

3. Apply the ICI algorithm to find the adaptive scales $h^{+}(x)$.
4. Fix the adaptive RI estimate according to the rule $\hat{y}^{RI}(x) = \hat{y}^{RI}_{h^{+}(x)}(x)$.

The RWI consists of the same steps as the RI algorithm with the corresponding repiacements of the transfer function and the formula for the estimation variance. These are the main steps of the RWI algorithm:

1. Compute the transfer function for the RWI in Eq. (9.36) as

$$Q_{RWI}(f, \alpha_2) = \frac{V^{*}(f)G_h(f)}{|V(f)|^2 + \alpha_2^2 n_1 n_2 \sigma^2/|\widehat{Y}_{RI}(f)|^2}$$

and the variance according to Eq. (9.38) as

$$\sigma^2_{\hat{y}_h^{RWI}}(h) = \frac{\sigma^2}{n_1 n_2}||Q_{RWI}(f, \alpha_2)||^2.$$

2. Compute the estimates $\hat{y}_h^{RWI}(x)$ as the inverse DFT of Eq. (9.36) for all $h \in H$.
3. Apply the ICI algorithm to find the adaptive window sizes $h^{+}(x)$.
4. Fix the adaptive RWI estimates according to $\hat{y}^{RWI}(x) = \hat{y}^{RWI}_{h^{+}(x)}(x)$.

9.2.2.2 *Implementation: directional estimation*

A symmetric window w in the LPA kernels g_h can be a good choice if y is isotropic in a neighborhood of the estimation point. However, as was discussed in Chapter 7, if y is anisotropic, a nonsymmetric approximation of y becomes much more reasonable.

To deal with the anisotropy of y, multiple nonsymmetric directional-window LPA estimates are exploited. It is assumed that the neighborhood of the pixel x is separated in a number of nonoverlapping sectorial subareas. The LPA-ICI adaptive estimates are calculated for each of these subareas and fused together to yield the final estimate. Experiments prove that the narrowed sectorial estimates demonstrate a much more efficient performance than can be achieved using the symmetric and even four-quadrant estimates.

The technique that we develop here is based on the following directional version of the above algorithm:

$$\widehat{Y}^{RI}_{h,\theta}(f) = \frac{V^{*}(f)G_{h,\theta}(f)}{|V(f)|^2 + \alpha_1^2}Z(f), \tag{9.39}$$

$$\widehat{Y}^{RWI}_{h,\theta}(f) = \frac{V^{*}(f)G_{h,\theta}(f)}{|V(f)|^2 + \alpha_2^2 n_1 n_2 \sigma^2/|\widehat{Y}^{RI}(f)|^2}Z(f). \tag{9.40}$$

The estimate of y given by the RWI deconvolution scheme uses the ICI-based RI estimate $\hat{y}^{RI}(x)$ as the reference signal. Thus, we have the two-step multidirectional procedure shown in Fig. 9.2.

The adaptive procedure assumes that the estimates $\{\hat{y}^{RI}_{h,\theta_k}\}_{h \in H}$ are calculated according to Eq. (9.39) for a set of scales H, and the ICI rule selects the best scales

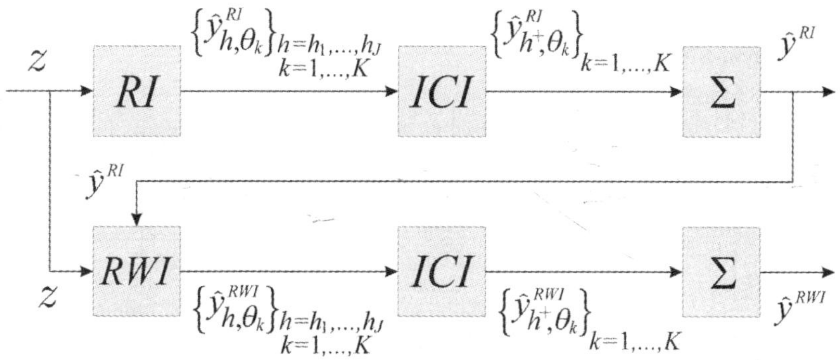

Figure 9.2 The directional LPA-ICI regularized Wiener inverse algorithm. In the first line of the flowchart, the RI estimates are calculated for a set of scales h and directions θ, and the ICI is used to obtain the pointwise adaptive-scale directional estimates that are then fused into the $\hat{y}^{RI}$ estimate. In the second line the RWI estimates are calculated using $\hat{y}^{RI}$ as a reference signal in Wiener filtering, and again ICI and fusing are performed to obtain the final adaptive $\hat{y}^{RWI}$ estimate.

for each direction and for each pixel. In this way we obtain the directional varying-scale adaptive estimates $\hat{y}^{RI}_{h^+_{\theta_k}(x),\theta_k}(x)$, $k = 1, \ldots, K$, which are fused in the final $\hat{y}^{RI}(x)$ according to Eq. (8.57).

This $\hat{y}^{RI}(x)$ serves as the reference signal in the RWI procedure in Eq. (9.40). The adaptive RWI algorithm is similar and gives the ICI adaptive varying-scale estimates $\hat{y}^{RWI}_{h^+_{\theta_k}(x),\theta_k}(x)$ for each direction and each x. Then, the final estimate $\hat{y}^{RWI}(x)$ is obtained by again fusing these directional ones according to Eq. (8.57).

The ICI adaptive varying scales $h^+_{\theta_k}$ represent the distribution of image features across the directions θ_k. As was discussed in Sec. 6.3, the randomness in the ICI adaptive scales has the special feature of the impulse noise. The prefiltering of these noisy ICI adaptive $h^+_{\theta_k}$ is useful for improving the adaptive estimates. We use weighted medians for prefiltering the adaptive scales because they are defined by Eq. (6.56). These specially designed median filters effectively remove noise from $h^+_{\theta_k}$ and preserve the accurate edge adaptation.

9.2.3 Performance study

In this section we present simulation results demonstrating the performance of the developed algorithm in comparison with some recent advanced wavelet-based methods [50], [161]. The term "state-of-the art" has been used in [50] for these methods, which demonstrate high values of imaging performance criteria.

The estimation kernels g_h have been designed using the narrowed conically supported uniform windows w. A scale set H for the length h_1 and the width h_2 are given in Eq. (8.73). The LPA orders are $m = (1, 0)$ and $m = (0, 0)$ for the RI and RWI estimators, respectively.

The level of the noise in the observations is characterized in decibels by the BSNR (blurred SNR) as

$$BSNR = 10\log_{10}\left\{\frac{1}{\sigma^2 n_1 n_2}\sum_x\left[(v \circledast y)(x) - \frac{1}{n_1 n_2}\sum_x (v \circledast y)(x)\right]^2\right\}$$

$$\simeq 10\log_{10}\frac{\sum_x\left[(v \circledast y)(x) - \frac{1}{n_1 n_2}\sum_x (v \circledast y)(x)\right]^2}{\sum_x [z(x) - (v \circledast y)(x)]^2}. \qquad (9.41)$$

Thus, the ratio of the standard deviation of the blurred original image to the standard deviation of the noise additive to the blurred image is used in the BSNR.

The design parameters of the LPA-ICI deblurring are the ICI thresholds Γ_1, Γ_2 and the regularization parameters α_1, α_2, respectively, for the RI in Eq. (9.39) and the RWI in Eq. (9.40) algorithms. In multiple experiments with different images we found the following values of these parameters, which overall enable a good performance:

$$\Gamma_1 = 1.6, \qquad \Gamma_2 = 1.3, \qquad \alpha_1^2 = 1.44 \cdot 10^{-4}, \qquad \alpha_2^2 = 0.13. \qquad (9.42)$$

Table 9.1 presents the results for the following four experiments:

1. *Cameraman* image, 9×9 uniform (box-car) PSF, BSNR $= 40$ dB (Exp. 1);
2. *Cameraman* image, $v(x_1, x_2) = (1 + x_1^2 + x_2^2)^{-1}$, $x_1, x_2 = -7, \ldots, 7$, $\sigma^2 = 2$ (Exp. 2);
3. *Cameraman* image, $v(x_1, x_2) = (1 + x_1^2 + x_2^2)^{-1}$, $x_1, x_2 = -7, \ldots, 7$, $\sigma^2 = 8$ (Exp. 3);
4. *Lena* image, v is a 5×5 separable filter with weights $[1, 4, 6, 4, 1]/16$ in the horizontal and vertical directions, BSNR $= 15.93$ dB (Exp. 4).

For these experiments, a set of eight directions as in Eq. (8.74) is used.

Overall, the values of ISNR in Table 9.1 show that the RWI algorithm demonstrates a good performance and outperforms the state-of-the-art techniques developed in [37], [50], and [161]. In Table 9.1, for the first three experiments we also show the figures for the RI algorithm. A comparison of these ISNR values with corresponding values for the RWI algorithm gives a clear demonstration of

Table 9.1 ISNR (dB) of the LPA-ICI algorithm for four experiments considered in [37], [50], and [161].

Method | Experiment	1	2	3	4
RWI multidirectional	**8.23**	**7.78**	**6.04**	**3.76**
RI multidirectional	7.70	7.18	5.30	—
GEM [37]	8.10	7.47	5.17	—
EM [50]	7.59	6.93	4.88	2.94
ForWaRD [161]	7.30	6.75	5.07	2.98

Figure 9.3 Original *Cameraman* image (left) and noisy blurred observations (Exp. 1) (right).

the advantage of the RWI algorithm. Thus, the RI estimates working as the pilot ones in the RWI algorithm allow essential improvement of the results.

Visual inspection also favors the RWI multidirectional algorithm. Figure 9.3 shows a fragment of the original and noisy blurred *Cameraman* image for experiment 1. This image restoration as well as the adaptive varying scales for $\theta = \pi/4$ are shown in Fig. 9.4.

The conical narrow kernels are an important element of this good performance. For example, in the same algorithm, the four-quadrant estimate gives ISNR = 7.52 dB for experiment 1 versus ISNR = 8.23 dB in Table 9.1.

As a further illustration of the flexibility of the used technique, we present two examples of differentiation of y using the noisy blurred observations. First, we replace the smoothing kernels $g_{h,\theta_k}^{(0,0)}$ by the discrete differentiation kernels $g_{h,\theta_k}^{(1,0)}$

Figure 9.4 LPA-ICI algorithm performance: restored image SNR = 8.23 dB (left) and adaptive scales $h^{+}_{\pi/4}$ (right).

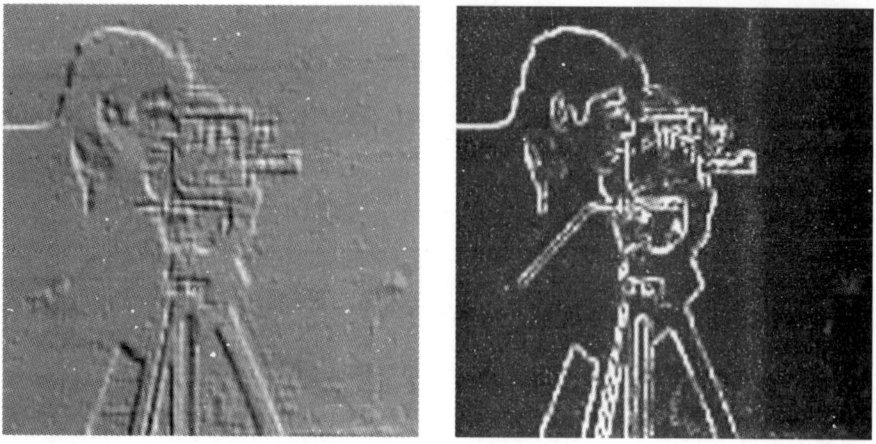

Figure 9.5 Directional derivative (left) and edge detection (right).

of orders $m = (1, 0)$ in the RWI stage of the algorithm. Then the output $\hat{y}^{RWI}_{h^+,\theta_k}$ of this two-stage algorithm gives the estimate of the directional derivative $\partial_{\theta_k} y$. Figure 9.5 (left) shows the diagonal derivative estimate for $\theta = \pi/4$ as the mean of the two one-sided directional derivatives with $\theta = \pi/4$ and $\theta = -\pi/4$. These derivative estimates are fused as $(\hat{y}^{RWI}_{h^+,\pi/4} - \hat{y}^{RWI}_{h^+,-\pi/4})/2$.

Further, for the edge detection we calculate the sum of the absolute values of these adaptive-scale derivatives $\sum_{k=1}^{4} |\hat{y}^{RWI}_{h^+,\theta_k} - \hat{y}^{RWI}_{h^+,-\theta_k}|/2$. The image of this sum is given in Fig. 9.5 (right). It shows a very accurate recovery of the image's edges from the blurred noisy image data.

9.3 Motion Deblurring

A motion blur due to camera or object motion during the exposure time can significantly degrade the quality of the image. The simplest model of this blur can be given in the form of a discrete convolution with the line-wise PSF as

$$v(x_1, x_2) = \begin{cases} 1/L, & \text{if } x_1 = x_2 \tan\beta, \quad \sqrt{x_1^2 + x_2^2} \leq L/2, \\ 0, & \text{otherwise}, \end{cases} \tag{9.43}$$

where the length L of the kernel depends on the movement velocity, and the slope $\tan\beta$ is defined by the movement direction β. This model assumes the same movement for all pixels of the image.

The LPA-ICI deblurring algorithm is sensitive to edges, change-points and other details and demonstrates a very good performance in deconvolving the motion blur PSF in Eq. (9.43), even when the velocity of the movement and the noise make the blurred image completely unrecognizable.

Figure 9.6 shows text images degraded by the kernel in Eq. (9.43) with $L = 41$ pixels and a vertical direction of movement. The images are given for different values of the BSNR. The reconstructed images are shown in Fig. 9.7. Comparing

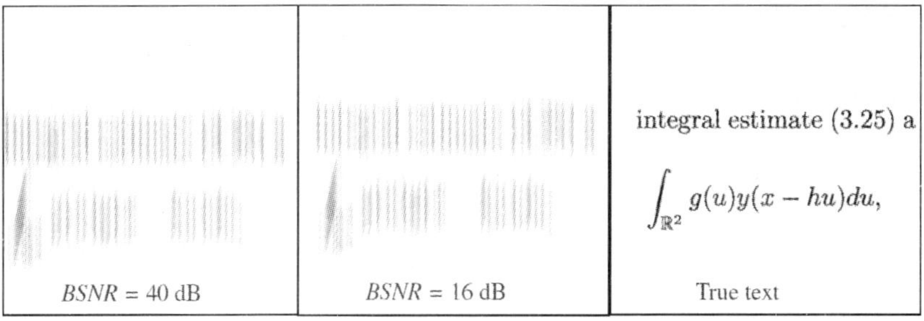

Figure 9.6 True text and noisy images degraded by vertical motion blur.

these reconstructions with the original image in Fig. 9.6, we may conclude that, for the low level of noise (BSNR = 40 dB), the reconstruction is nearly perfect. Even for the larger level of noise (BSNR = 16 dB), the reconstruction is quite clear and readable. For these experiments we used the scale-adaptive RI and RWI filters with the parameters given in the previous section.

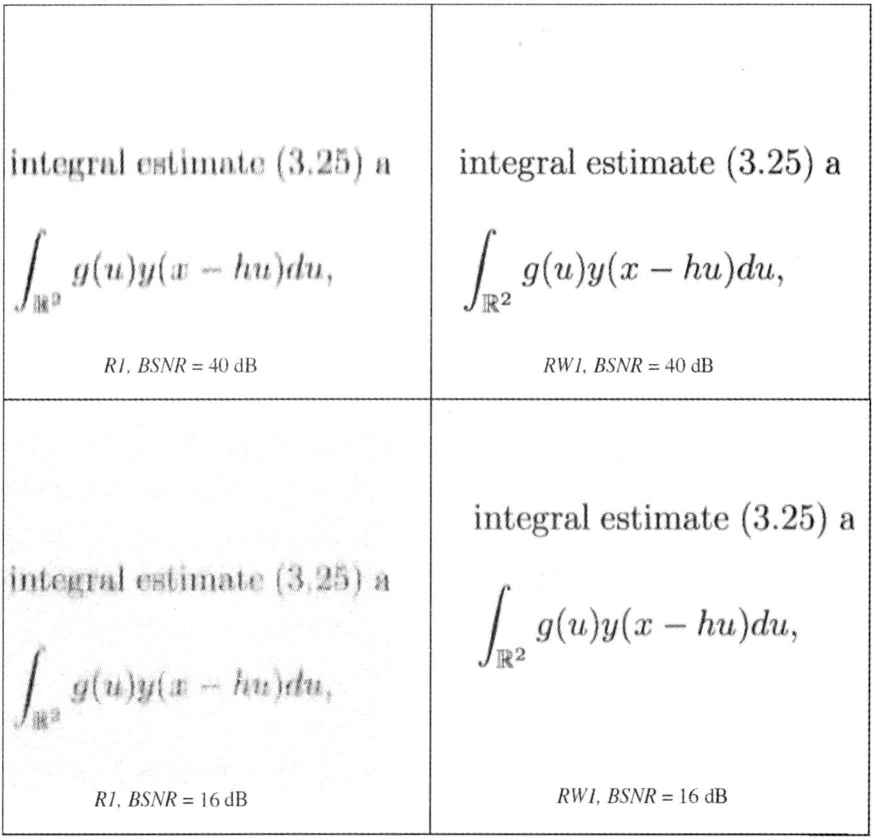

Figure 9.7 Reconstruction of the motion-blurred images using the RI and RWI adaptive-scale algorithms.

9.4 Super-Resolution Imaging

The image spatial resolution (the number of pixels per unit area of the image) is a principal factor in determining the quality of imaging. The current strategy of obtaining high-resolution images depends on sensor technology that tries to reduce the size of the sensor's pixels. However, this has a principal limitation because the number of photons registered by the sensor pixel becomes too small, resulting in an unacceptable level of random noisy components in the signal. The cost of the system with the increased sensor density and required high-precision optics also becomes inappropriate for general-purpose commercial applications.

Therefore, resolution enhancement based on signal processing becomes of great concern. One idea that could prove efficient is a method of obtaining high-resolution imaging from a set of low-resolution images. This technique is called *super-resolution*, where the term "super" refers to the intention to extract subpixel information from multiple images and overcome the natural limitations of low-resolution images.

The low-resolution images are considered to be samples of the high-resolution image of interest. Naturally, subpixel information can be available only if frames of low-resolution images are shifted with respect to each other and aliased. Then each image cannot be obtained from the others (e.g., [177]).

The idea of shifted subpixel sampling is illustrated in Fig. 9.8. A few low-resolution frames together form the high-resolution frame. The simplest scenario occurs when sampled data sets (low-resolution frames) have the same period and are shifted by the sampling period with respect to each other.

In general, low-resolution data sets can be of different sampling periods and have different shifts. Samples of the shifted frames fill the space between the samples of one of the low-resolution frames, which is treated as the reference frame.

When low-resolution frames are shifted with respect to the pixels of the high-resolution image, some interpolation is used to obtain a blurred version of the high-resolution image from the available low-resolution ones. If the shifts are known, the

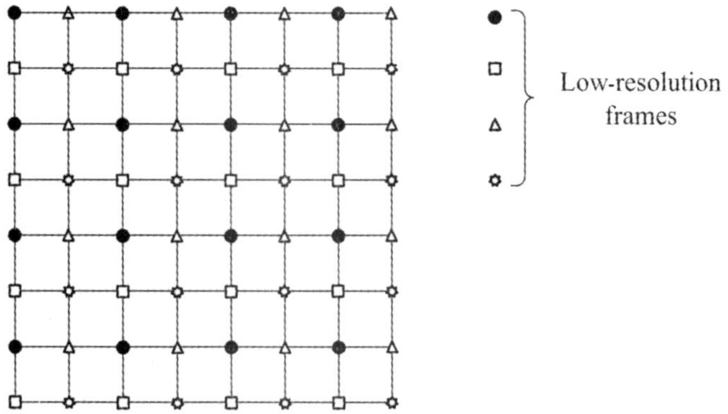

Figure 9.8 The idea of shifted subpixel sampling: four low-resolution frames together form the high-resolution data grid.

observation model of the low-resolution images follows from the high-resolution model in Eq. (9.3) as its sampled version:

$$z(x) = (y \circledast v)(x) + \varepsilon(x), \qquad (9.44)$$
$$x \in X_l, \qquad l = 1, \dots, L, \qquad L \geq 2.$$

where X_l is a grid of lth low-resolution images and L is the number of low-resolution images.

For the integer high-resolution grid $X = \{k_1, k_2 : k_1 = 1, 2, \dots, n_1,\ k_2 = 1, 2, \dots, n_2\}$, the grid X_l of the low-resolution subsampled image can be defined as

$$X_l = \{k_1 L + l,\ k_1 L + l : 0 \leq k_1 \leq (n_1 - l)/L, 0 \leq k_2 \leq (n_2 - l)/L\}, \qquad (9.45)$$
$$l = 1, 2, \dots, L.$$

The sets X_l do not have common elements and give an accurate decomposition of the set X,

$$X_l \cap X_k = \varnothing, \qquad k \neq l, \qquad (9.46)$$
$$\cup_{l=1}^{L} X_l = X.$$

According to Eq. (9.46), the blurred version of the high-resolution image $y(x)$, $x \in X$, can be obtained from the low-resolution observations in Eq. (9.44) using the coordinates in Eq. (9.45) of the lower resolution pixels. Thus, if the shifts are given and the subsamples form the regular grids corresponding to the high-resolution image, super-resolution imaging is reduced to the deblurring problem with the PSF of the low-resolution images.

The deconvolution algorithms of Sec. 9.2 are applicable to this $z(x)$, $x \in X$. The frequency domain technique exploited in these algorithms has a strong advantage over the spatial domain recursive algorithms that are commonly used for this sort of problem (e.g., [165]).

Let us illustrate how this super-resolution imaging technique works. Low-resolution maps of Hervanta (a suburb of Tampere, Finland) are freely available from the Internet. A data acquisition system allows one to obtain these low-resolution maps with an arbitrary shift of the frames. The size of the shift is multiple to quite a small fixed value. We assume that this small fixed-size shift is equal to the pixel size of a underlying high-resolution map.

The acquisition of the low-resolution maps can be treated as a subsampling of this high-resolution map produced with some shifts of the sampling grids. The procedure corresponds to the subsampling model in Eq. (9.45) with the integer shifts between the frames of the low-resolution images equal to the sampling period (pixel size) of the high-resolution image.

A study of the obtained shifted images shows that the low-resolution images are blurred and the corresponding PSF is well approximated by the uniform box-car of the size 5×5.

Besides, the intensities of these low-resolution images are available with a small amplitude resolution. The number of quantization levels of the image intensity is equal to 43. Thus, the data are available with high quantization errors.

It was found also from the analysis of the low-resolution images that they are periodical with respect to the shifts, with the periods equal to 5, and these periods are the same for both the horizontal and the vertical shifts. This shows that, in the periodical sampling model in Eq. (9.45), the parameter $L = 5$.

For the high-resolution image reconstruction we use an $L \times L = 25$ low-resolution image shifted according to Eq. (9.45). The high-resolution image has a number of pixels equal to the size of the low-resolution image multiplied by 25. This says that for each pixel of the reference low-resolution image we obtain 24 subpixels for a resolution enhancement.

Figure 9.9 illustrates the work of the ICI adaptive deconvolution technique for high-resolution imaging. The top image shows one of the 25 low-resolution images. It is given of the size of the reconstructed high-resolution image to demonstrate clearly the coarseness of these low-resolution data.

The middle image shows the blurred image built from the original 25 low-resolution ones. It is interpreted as a blurred high-resolution image. For deblurring of

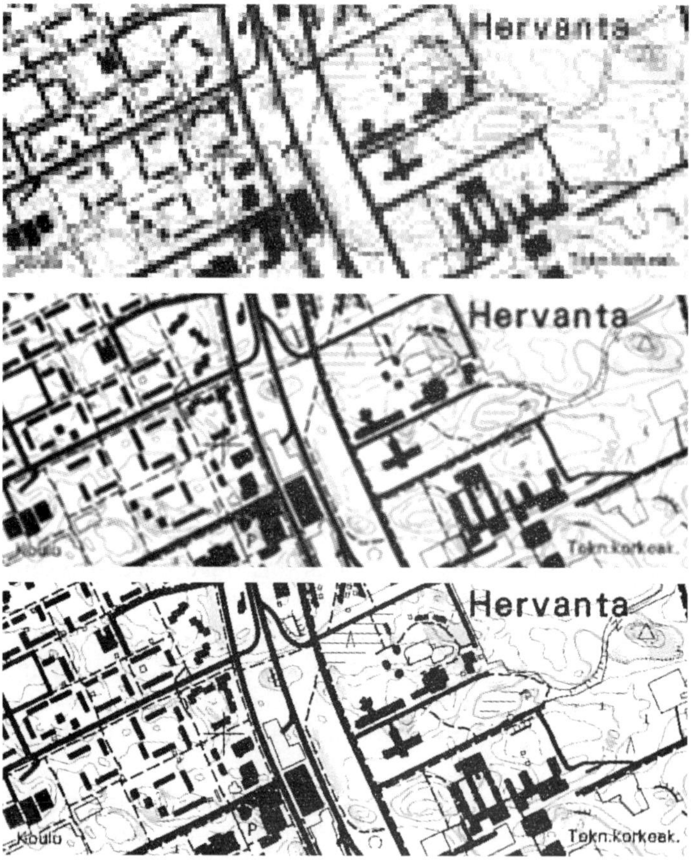

Figure 9.9 Super-resolution imaging. From the top to the bottom: low-resolution original data (one of 25 images), accurately shifted and combined lower resolution images, high-resolution image obtained by deblurring of the combined images.

the last image we use the adaptive-scale RI and RWI algorithms with the parameters given in Sec. 9.2.3. Finally, the third image in Fig. 9.9 shows the reconstruction. It is obvious that it is indeed a high-resolution map with easily seen details that are quite smoothed in the low-resolution maps.

9.5 Inverse Halftoning

In the last two decades the color depth of digital images, graphic cards, computer displays, and digital cameras has steadily increased. The current standard for consumer devices is 8 or more bits for each color channel. In particular, for grayscale images this is equivalent to 256 or more different intensity values. Since the human eye is usually not able to distinguish between such close shades of gray, such grayscale images are often called *continuous-tone* images. Coarser palettes nowadays are considered only for lossy image/video compression applications.

Despite this progress, many output and rendition devices are still unable to reproduce these continuous-tone shades and can provide only binary (black-and-white) output. Typical examples of such devices are office and industrial printers as well as low-cost displays for mobile devices.

Digital halftoning is the rendition process of a continuous tone into a binary image. Although the naive approach where shades lighter or darker than a 50% gray level are thresholded, respectively, to white or black, is the simplest to implement, it is almost never used because of its poor visual results. Taking into account the characteristics of the human visual system, which acts as a low-pass filter, halftones are generated in such a way that the difference between the halftoned binary image and the original grayscale image is compacted into the high-frequency part of a Fourier spectrum.

Halftoning techniques include ordered dithering or screening (dispersed-dot and clustered-dot), error diffusion, blue-noise dithering [224], and direct binary search [5]. The latter is known to provide the highest quality halftones. However, the most widely used methods, because of their computational efficiency, are order dithering and error diffusion. Figure 9.10 illustrates the halftoning process.

Halftoned images may look good printed on paper; however, due to their high-frequency characteristics they cannot be used in many situations. For example,

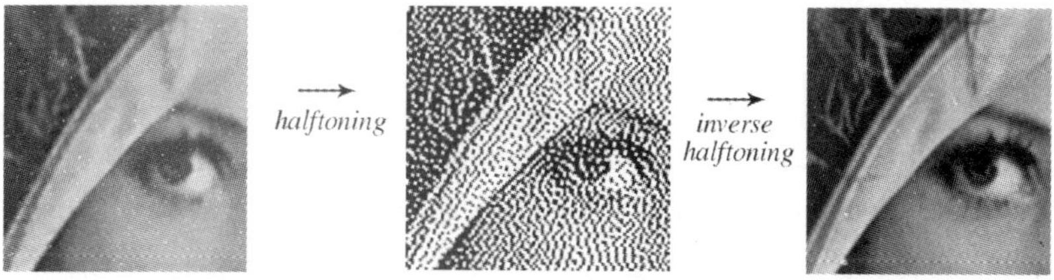

Figure 9.10 An illustration of halftoning and inverse halftoning. Detail of the *Lena* image: original (left), Jarvis error diffusion halftone (center), and inverse halftoning LPA-ICI estimate (PSNR = 33.0 dB) (right).

scanning and reprinting a high-resolution halftone would result in a poor quality output (see, e.g., the left column of Fig. 9.13). When displayed on a computer screen, halftones present evident aliasing artifacts. Further processing, such as resizing or contrast enhancement, can severely degrade the image quality. Moreover, standard compression techniques do not process halftones efficiently. The development of applications such as high-quality digital archives of old newspapers or scientific journals can thus still be considered a challenging task. In all of these cases it would be desirable to process, whenever available, the original grayscale image rather than the black-and-white halftones.

Inverse halftoning is the reconstruction process of a continuous-tone image from its binary halftone, as illustrated in the last image in Fig. 9.10. It is clear from the above discussion that inverse halftoning should mimic the human visual system. Thus, all inverse halftoning techniques perform some sort of low-pass filtering.

A fixed-kernel low-pass filtering is simple to implement, nevertheless it very seldom yields satisfactory results. In recent years, inverse halftoning has gained renewed interest and several new adaptive methods have been proposed [236]. They include thresholding in the transform domain [160, 162], projections onto convex sets (POCS) [79, 225], MAP projection [235], and anisotropic diffusion [124] and look-up tables (LUT) based on learning/training [120], [156], [157].

In this section we present a novel inverse halftoning technique [55]. It combines a linear model for error diffusion [124] and anisotropic scale-adaptive RI and RWI filtering. We assume that the error diffusion kernel is known. We show simulation results obtained for the Floyd-Steinberg [52] and Jarvis *et al.* [94] error diffusion kernels.

Just as for the traditional image deblurring problem, the anisotropic LPA-ICI-based deconvolution yields state-of-the-art performance through a two-stage, noniterative, filtering procedure where blur and noise are simultaneously removed. The anisotropy of the proposed estimator allows one to accurately restore edges and details, producing a result that is quite faithful to the original.

9.5.1 Error diffusion

Roughly speaking, error diffusion halftoning works by raster-scanning the continuous-tone image and recursively distributing, or "diffusing," the quantization errors due to binarization on the neighboring pixels.

Let y be the original continuous-tone image, $X_s = (x_{1,s}, x_{2,s})$ be the integer pixel coordinate Eq. (2.1), $s = 1, \ldots, n$, and z be the halftoned image (to be generated). With the initial conditions [52]

$$y'(X_1) = y(X_1),$$

$$X_1 = (1, 1) \qquad \text{(start from the top-left pixel)},$$

the error diffusion is defined by the following iterative procedure:

$$\begin{aligned} z(X_s) &= q[y'(X_s)], \\ e(X_s) &= z(X_s) - y'(X_s), \\ y'(X_s + x) &= y(X_s + x) - e(X_s) \cdot w^{ed}(x), \ \forall x, \\ X_{s+1} &= successor\ (X_s), \end{aligned} \tag{9.47}$$

where w^{ed} is the weight kernel; $q[\tilde{y}]$ is the binarization (or rounding) operation (i.e., $q[\tilde{y}] = 1$ iff $\tilde{y} \geq 1/2$, otherwise $q[\tilde{y}] = 0$); and *successor* denotes the next pixel to be processed in the raster-scanning. In other words, at every step, the error diffusion algorithm

- binarizes the current s pixel (i.e., rounding it to 0 or 1);
- computes the quantization error $e(X_s)$;
- diffuses the error $e(X_s)$ on neighboring pixels with the weights w^{ed};
- moves to the next $s + 1$ pixel.

The variable $y'(X_s + x)$ in Eq. (9.47) differs from $y(X_s + x)$ by the errors $e(X_s)$ spread with the weights $w^{ed}(x)$ in a neighborhood of X_s. The kernel w^{ed} is called the *error filter*. Examples of the error filter kernels w^{ed} are shown in Fig. 9.11. Observe that the weights are nonzero only for those pixels that have not already been scanned. The diffusion never goes backward with respect to the scanning direction, and after a pixel has been binarized its value is not modified by future iterations.

The algorithm ends when the bottom-right pixel has been processed. The diffusion of the quantization error guarantees that the local means of the halftoned z are close to the corresponding local averages of the continuous tone y. Several modifications to the above procedure (such as different pixel-scan ordering or threshold modulations) are possible [224].

9.5.2 Linear model of error diffusion

We may approximate the nonlinear quantization operation by a linearized stochastic model [123],

$$q[y] \simeq K \cdot y + \varepsilon, \tag{9.48}$$

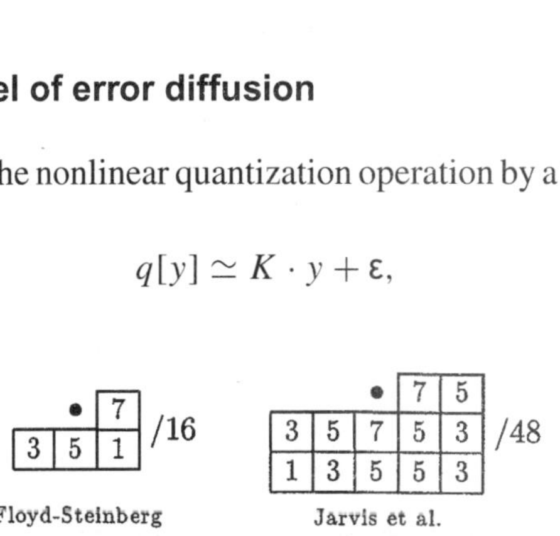

Figure 9.11 Error filters w^{ed} for the Floyd-Steinberg [52] (left) and Jarvis [94] error diffusions. The black dot indicates the center of the kernel.

where K is a constant and ε is random white noise. Inserting Eq. (9.48) in Eq. (9.47) and using the convolution for the error spread operation, we can represent Eq. (9.47) in the nonrecursive form equivalent to Eq. (9.47) as

$$\begin{aligned} z(X_s) &= Ky'(X_s) + \varepsilon(X_s), \\ e(X_s) &= z(X_s) - y'(X_s), \\ y'(X_s) &= y(X_s) - (e \circledast w^{ed})(X_s). \end{aligned} \tag{9.49}$$

Let Y, Y', Z, and W^{ed} be the DFTs of y, y', z, and w^{ed}, respectively. Then

$$\begin{aligned} Z(f) &= KY'(f) + \varepsilon(f), \\ e(f) &= Z(f) - Y'(f), \\ Y'(f) &= Y(f) - e(f)W^{ed}(f). \end{aligned} \tag{9.50}$$

Solving these equations, we arrive at

$$Z(f) = P(f)Y(f) + Q(f)\varepsilon(f), \tag{9.51}$$

where the transfer functions P and Q are

$$P(f) = \frac{K(f)}{1 + [K(f) - 1]\, W^{ed}(f)}, \qquad Q(f) = \frac{1 - W^{ed}(f)}{1 + [K(f) - 1]\, W^{ed}(f)}, \tag{9.52}$$

and ε is white Gaussian noise (with standard deviation σ).

Here W^{ed} is the frequency response of the error filter w^{ed} and K is a gain constant from Eq. (9.48). K is found to be essentially independent of y and depends instead only on the used error filter; for example, $K = 2.0$ and $K = 4.5$ for the Floyd-Steinberg and Jarvis error filters, respectively [122], [123]. Since typically W^{ed} is a low-pass filter, Q is a high-pass filter. This is consistent with the fact that error diffusion halftoned images differ from the continuous-tone original mostly in the high-frequency components of the spectrum.

The model in Eqs. (9.51–9.52) has been proved to be quite accurate [123], and has been exploited in a number of algorithms (e.g., [125], [160], [162]).

In the spatial domain, multiplications in Eq. (9.51) are replaced by the convolutions and Eq. (9.51) becomes

$$z(x) = (p \circledast y)(x) + (q \circledast \varepsilon)(x), \tag{9.53}$$

where p and q are the impulse responses of P and Q, respectively. According to this model, the inverse halftoning process, estimation of y from z, can be formulated as a deconvolution problem where p is the given point-spread function and the observations z are contaminated by colored noise $(q \circledast \varepsilon)$.

9.5.3 Anisotropic deconvolution

A unbiased solution of the deconvolution problem in Eq. (9.51) can be obtained in a straightforward manner by first inverting the convolution operator P and then removing the noise $P^{-1}Q\eta$. However, this naive inverse procedure gives a very noisy solution. The approach with regularization of the inverse and anisotropic ICI adaptive smoothing kernels is very reasonable here.

A collection of compactly supported directional LPA kernels $\{g_{h,\theta_k}\}_{h\in H,k=1,\ldots,K}$ is designed specifically for this problem. Each kernel is characterized by a direction θ_k and a scale parameter h. The notation $\hat{y}_{h,\theta_k}(x)$ is used for the estimate obtained by filtering with the kernel g_{h,θ_k}.

The directionality of the kernels as detailed in Chapter 7 arises from two independent facts:

- The supports (i.e., the windowing function w) of the kernels g_{h,θ_k} are asymmetric and oriented along the directions θ_k;
- The vanishing moment conditions and the accurate polynomial reproducing properties are satisfied for each kernel g_{h,θ_k} and for each corresponding direction θ_k.

In contrast to sectorial segmentation, the designed kernels have rectangular supports as in Fig. 7.8(a). We use kernels of type 10 (see Sec. 7.3) with the Gaussian window w and the scales $h = (h_1, h_2)$. Type 10 means here that the window w_h is a segment of the Gaussian probability density function defined as

$$\begin{aligned} w_h(x) &= f(x_1/h_1\gamma_1)\cdot f(x_2/h_2\gamma_2), \\ x_1 &= 0, 1, \ldots, h_1 - 1, \\ x_2 &= -h_2 + 1, \ldots, -1, 0, 1, \ldots, h_2 - 1, \end{aligned} \tag{9.54}$$

where $f(x_1) = \exp(-x_1^2/2)$ and $\gamma_1, \gamma_2 > 0$.

The parameters h_1 and h_2 are integers, $h_1 \geq h_2$, defining simultaneously the scale and the size of the finite support. These windows are essentially overlapping at the origin $x = 0$.

We use the RI and RWI filters in directional estimates. The corresponding inverse regularized estimates, similar to Eqs. (9.39–9.40), are defined by

$$\widehat{Y}^{RI}_{h,\theta}(f) = \frac{P^*(f)G_{h,\theta}(f)}{|P(f)|^2 + \alpha_1^2\,|Q(f)|^2}Z(f), \tag{9.55}$$

$$\widehat{Y}^{RWI}_{h,\theta}(f) = \frac{P^*(f)G_{h,\theta}(f)}{|P(f)|^2 + \alpha_2^2 n_1 n_2 \sigma^2\,|Q(f)|^2/|\widehat{Y}^{RI}(f)|^2}Z(f). \tag{9.56}$$

These estimates are different from Eqs. (9.39) and (9.40) by the regularization terms, including the factor $|Q(f)|^2$ defining the spectrum of the color noise in the model in Eq. (9.51). The RWI in the form of Eq. (9.56) follows from Eq. (9.31) as the spectrum of the noise $S_\varepsilon(f) = \sigma^2\,|Q(f)|^2$. The RI in the form of Eq. (9.55) is derived from Eq. (9.21) provided $D = Q$. For the setting considered in Eq. (9.16),

$$R_{\psi}^2 = \frac{1}{n_1 n_2}||Z_{h,\theta}(f) - P(f)Y_{h,\theta}(f)||^2 + \frac{\alpha_1^2}{n_1 n_2}||Y_{h,\theta}(f)Q(f)||^2.$$

This means that we minimize residuals of the RI estimator provided that the penalization term $||Y_{h,\theta}(f)Q(f)||^2$ is bounded.

The adaptive procedure assumes that the estimates $\{\hat{y}_{h,\theta}^{RI}\}_{h\in H}$ are calculated for a set of scales and $\theta = \theta_k$, and the ICI rule selects the best scale for each direction and for each pixel. In this way we obtain the directional varying-scale adaptive estimates $\hat{y}_{h^+,\theta_k}^{RI}$, $k = 1, \ldots, K$, which are fused in the final one $\hat{y}^{RI}$ according to the rule

$$\hat{y}^{RI}(x) = \sum_{k=1}^{K} \lambda_k^{RI} \hat{y}_{h^+,\theta_k}^{RI}(x), \tag{9.57}$$
$$\lambda_k^{RI} = [\sigma_k^{RI}(x)]^{-2} / \sum_{i=1}^{K} [\sigma_i^{RI}(x)]^{-2},$$

where σ_k^{RI} is the standard deviation of the adaptive-scale estimate $\hat{y}_{h^+,\theta_k}^{RI}$.

The fused RI estimate $\hat{y}^{RI}$ serves as the reference signal in the RWI procedure in Eq. (9.56) (see Fig. 9.2). The adaptive RWI algorithm is similar and gives the ICI adaptive varying-scale estimates $\hat{y}_{h^+,\theta_k}^{RWI}$ for each direction and each x. Then, the final estimate $\hat{y}^{RWI}$ is obtained by fusing these directional ones again as in Eq. (9.57):

$$\hat{y}^{RWI}(x) = \sum_{k=1}^{K} \lambda_k^{RWI} \cdot \hat{y}_{h^+,\theta_k}^{RWI}(x), \tag{9.58}$$
$$\lambda_k^{RWI} = [\sigma_k^{RWI}(x)]^{-2} / \sum_{i=1}^{K} [\sigma_i^{RWI}(x)]^{-2},$$

where σ_k^{RWI} is the standard deviation of the adaptive-scale estimate $\hat{y}_{h^+,\theta_k}^{RWI}$.

The final estimate of y of the proposed inverse halftoning algorithm is the output $\hat{y}^{RWI}$ given by the RWI deconvolution scheme that uses the ICI-based RI estimate as a reference signal Y. Thus, we implement the two-step procedure shown in Fig. 9.2.

The weights λ^{RI} and λ^{RWI} in Eqs. (9.57–9.58) are data-driven adaptive because the variances depend on the adaptive-scale $h_{\theta_k}^{+}(x)$. The variances of the estimates $\hat{y}_{h,\theta_k}^{RI}$ and $\hat{y}_{h,\theta_k}^{RWI}$ are obtained, respectively, as

$$\sigma_{\hat{y}_h^{RI},\theta_k}^2 = \frac{\sigma^2}{n_1 n_2}\left\| \frac{P^*(f)G_{h,\theta_k}(f)Q(f)}{|P(f)|^2 + \alpha_1^2\,|Q(f)|^2} \right\|^2, \tag{9.59}$$
$$\sigma_{\hat{y}_h^{RWI},\theta_k}^2 = \frac{\sigma^2}{n_1 n_2}\left\| \frac{P^*(f)G_{h,\theta_k}(f)Q(f)}{|P(f)|^2 + n_1 n_2 \alpha_2^2 \sigma^2\,|Q(f)|^2/|\widehat{Y}^{RI}(f)|^2} \right\|^2.$$

σ^2 in the above formulas is the variance of the noise ε in formula (9.51).

9.5.4 Simulation results

Directional LPA kernels have been designed for asymmetric windows oriented along the eight directions in Eq. (8.74). The LPA orders are $m = (1, 0)$ and $m = (0, 0)$ for the RI and RWI filters, respectively. The parameters γ_1 and γ_2 in the Gaussian window functions in Eq. (9.54) are equal to 0.6.

The LPA kernels g_{h,θ_k}, the scales h_1 and h_2, and the ICI thresholds Γ_1 and Γ_2 are considered as the design parameters of the proposed inverse halftoning algorithm. We optimize and fix the values of these parameters. For all results and figures presented in this sections, these fixed values of the design parameters are exploited.

The sets of 6 and 9 scales are used for the RI and RWI filters, respectively:

(a) RI estimator:

$$H = \left\{ \begin{pmatrix} 2 \\ 1 \end{pmatrix}, \begin{pmatrix} 3 \\ 2 \end{pmatrix}, \begin{pmatrix} 4 \\ 2 \end{pmatrix}, \begin{pmatrix} 5 \\ 2 \end{pmatrix}, \begin{pmatrix} 6 \\ 3 \end{pmatrix}, \begin{pmatrix} 8 \\ 3 \end{pmatrix} \right\}; \tag{9.60}$$

(b) RWI estimator:

$$H = \left\{ \begin{pmatrix} 1 \\ 1 \end{pmatrix}, \begin{pmatrix} 2 \\ 1 \end{pmatrix}, \begin{pmatrix} 3 \\ 1 \end{pmatrix}, \begin{pmatrix} 5 \\ 1 \end{pmatrix}, \begin{pmatrix} 4 \\ 2 \end{pmatrix}, \begin{pmatrix} 4 \\ 3 \end{pmatrix}, \begin{pmatrix} 6 \\ 2 \end{pmatrix}, \begin{pmatrix} 5 \\ 3 \end{pmatrix}, \begin{pmatrix} 8 \\ 4 \end{pmatrix} \right\}. \tag{9.61}$$

The first smallest scale kernel for the RWI estimator is the discrete Dirac delta function.

Small scales are mainly used to preserve small details in images. It is surprising to note that for RWI estimators h_1 and h_2 are not monotonically growing sequences in Eq. (9.61). The point here is that the size and shape of the kernels are defined by both parameters h_1 and h_2 jointly.

One of the basic characteristics of the kernel that defines its filtering ability is the quadratic norm giving the variances in Eq. (9.59) for the output signal. This norm, which is calculated for the scales ordered in the set H, should be a decreasing sequence. This is one of the basic requirements for the ICI adaptive-scale selection. This property holds for the sequences in Eqs. (9.60) and (9.61). The regularization parameters are $\alpha_1^2 = 3.24$ and $\alpha_2^2 = 0.002$, and the thresholds are $\Gamma_1 = 0.45$ and $\Gamma_2 = 0.4$ for the RI and RWI algorithms, respectively.

Figure 9.12 shows how the ICI rule works. The directional adaptive scales delineate and reveal all small details of the image. Although a little investigation has been done in the optimization of the algorithm's design parameters, the anisotropic LPA-ICI inverse halftoning already yields more than satisfactory results (see Fig. 9.13). Overall, the PSNR values in Table 9.2 show that the developed spatially adaptive algorithm demonstrates good performance and outperforms some of the state-of-the-art techniques.

Visual inspection also favors the new algorithm. Figure 9.14 shows a fragment of the restored *Lena* image. When compared to the similar wavelet-based inverse halftoning method [162], the proposed LPA-ICI procedure shows its superiority by restoring finer details without introducing any visible artifacts.

Recently it has been claimed in [120] that a decision tree learning LUT algorithm can yield a PSNR of 34.75 dB for the *Lena* image, sensibly outperforming all previous records of other authors, in particular that of the other LUT-based algorithm [157]. We do not include this result in Table 9.2 because it is achieved for a unusually sized 1050×1050 image. The results for this table, as well as all of the figures in this section, correspond to standard 512×512 images. We also tested our algorithm for this "oversized *Lena*," obtaining PSNR $= 37.75$ dB versus PSNR $= 34.75$ dB given in [120].

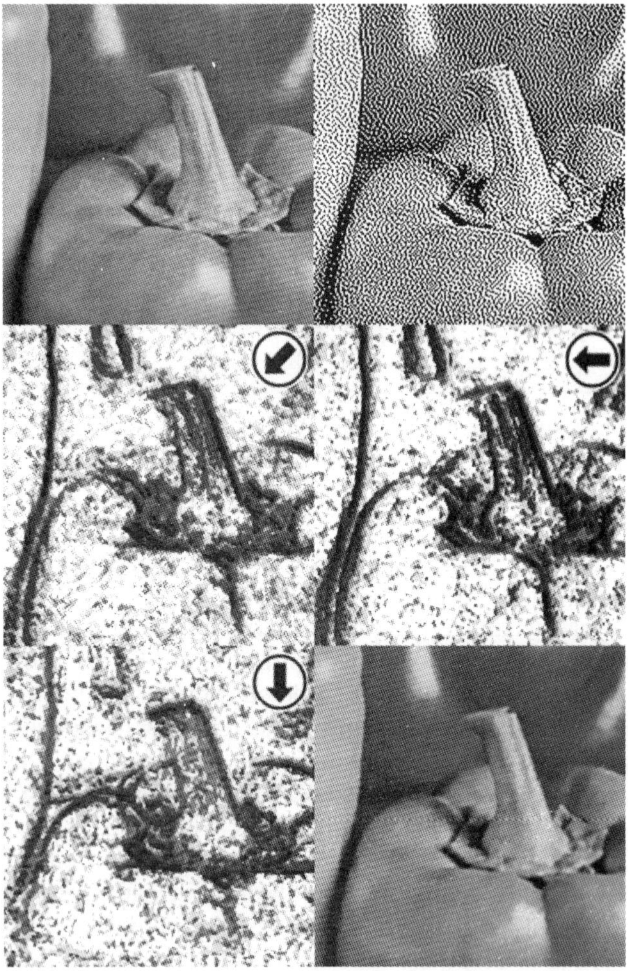

Figure 9.12 *Peppers* (detail): original image (top row, left), Jarvis error diffusion halftone (top row, right), adaptive scales $h^+_{\pi/4}$ (middle row, left), $h^+_0(x)$ (middle row, right), $h^+_{\pi/2}(x)$ (bottom row, left), and LPA-ICI estimate (PSNR = 31.6 dB) (bottom row, right). The arrows indicate the orientation of the kernels.

Figure 9.13 Examples of Floyd-Steinberg error diffusion LPA-ICI inverse halftoning: *Peppers* (PSNR = 31.6 dB), *Lena* (PSNR = 32.4 dB), and *Boats* (PSNR = 29.5 dB). The pictures in the right column are the estimates obtained by the proposed procedure from the binary halftone images shown in the left column.

Table 9.2 PSNR (dB) performance of the proposed algorithm and of other methods for restoration from Floyd-Steinberg error diffusion.

Inverse halftoning technique	*Lena*	*Peppers*
Anisotropic LPA-ICI	**32.4**	**31.6**
WinHD [162]	32.1	31.2
Wavelet-Vaguelette [160]	31.9	31.0
Wavelet [238]	31.7	30.7
Gradient [124]	31.3	31.4
Kernel [235]	32.0	30.3
LUT [157]	31.0	—
LMS–MMSE [27]	31.4	31.2
Gaussian (POCS) [79]	28.6	27.6
POCS [225]	32.2	30.9

9.6 3D Inverse

9.6.1 Sectioning microscopy

When we look through a microscope, there is one plane that appears in focus. However, specimens are not flat but are 3D structures. Therefore, some portions of the specimen are out of focus. Nevertheless, these out-of-focus structures are in the field of view and thus obscure the in-focus plane.

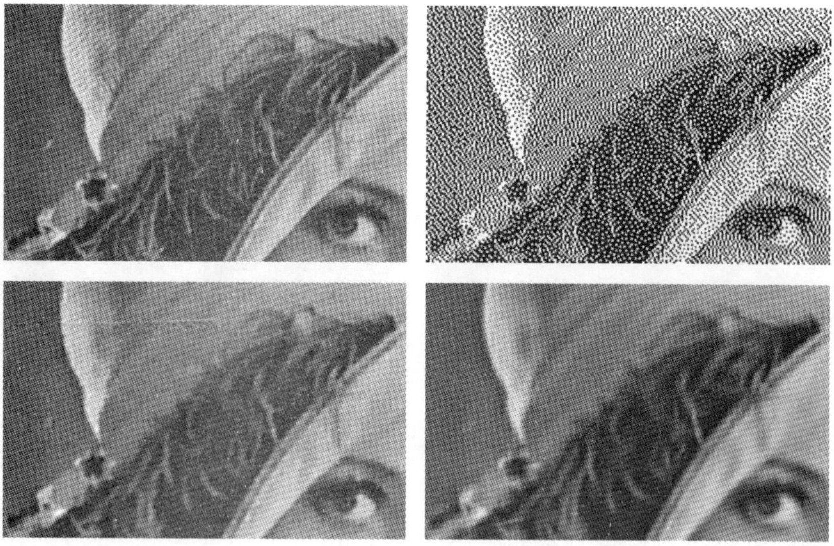

Figure 9.14 Visual comparison for a detail of the *Lena* image: from left to right, original and Floyd-Steinberg halftone (top), the proposed anisotropic LPA-ICI estimate (PSNR = 32.4 dB), and the "*WInHD*" estimate [162] (PSNR = 32.1 dB) based on deconvolution and filtering in the complex wavelet domain [121].

To obtain a 3D image from a specimen, it is common to use a method called optical sectioning. The microscope is focused at a given plane in the specimen and the image is recorded. This image is an optical slice. Then, the microscope is refocused to another plane and another image is recorded. This process is repeated until the whole specimen is covered.

There are two approaches in fluorescence microscopy to reducing out-of-focus contributions, optical and computational. In the optical approach, a confocal microscope is used to reduce the contributions from out-of-focus fluorescence.

The decision to use confocal or widefield microscopy depends on the application. The former produces high-quality images because only the light from the region near the in-focus plane is detected. However, the cost of such equipment is high. Widefield microscopes are cheaper than confocal ones. They are particularly valuable for work with living specimens, because they avoid specimen damage due to the laser light used in confocal microscopy, and also because they allow imaging with wavelengths not provided by standard lasers.

In the computational approach, image processing is applied to process the 2D optical slices to reduce the out-of-focus contributions. These methods require information about the process of image formation. The most severe degradation is often caused by diffraction at the objective and condenser lenses. This degradation is characterized by the PSF of the microscope optical system. The computational approach is used for both widefield and confocal microscopy to enhance the image quality.

3D sectioning microscopy equipped with digital deblurring algorithms is a powerful modern tool for visualizing specimens in biology, medicine, mineralogy, and so on.

There exist many publications concerning computational 3D microscopy. Computational restoration methods applied to slice images are quite an efficient prospective tool. Reviews of the field, applications, and further references can be found in [12], [16], [93], [154], and [188].

In this section we generalize the developed spatially adaptive 2D deblurring algorithm to 3D imaging.

9.6.2 Observation model

Suppose that we wish to reconstruct a 3D image intensity function $y(x)$, $x \in \mathbb{R}^3$. In the argument $x = (x_1, x_2, x_3)$, the first two variables x_1 and x_2 define the pixel's coordinates in the 2D image obtained from $y(x)$ with a fixed-depth coordinate x_3. The x_3 axis is parallel to the optical axis of the microscope optical system and perpendicular to the 2D image plane. For image formation we use the following general 3D integral model [188]:

$$z(\tilde{x}, x_3) = \int_{\mathbb{R}^3} v(\tilde{x} - \tilde{u},\ x_3, u_3) y(\tilde{u}, u_3) d\tilde{u} du_3, \tag{9.62}$$

where v is the PSF of the optical system and y is a 3D image object function. To make a difference between the depth x_3 and the 2D image coordinates (x_1, x_2), we use in Eq. (9.62) 2D variables,

$$\tilde{x} = (x_1, x_2), \qquad \tilde{u} = (u_1, u_2).$$

In optical sectioning microscopy, the image records are obtained slice-by-slice for a set of different focal planes or different values of x_3 in Eq. (9.62). We number the slices of the observations in the form

$$z_i(\tilde{x}) = \int_{\mathbb{R}^3} v(\tilde{x} - \tilde{u},\ x_3(i),\ u_3) y(\tilde{u}, u_3) d\tilde{u} du_3, \qquad i = 1, \ldots, n_3, \tag{9.63}$$

where $\tilde{x}$ and $\tilde{u}$ are 2D coordinates in the plane slices, and n_3 is the number of observed optical slices with depth coordinates $x_3(i)$, $i = 1, \ldots, n_3$.

It is assumed in the model in Eqs. (9.62–9.63) that the PSF is shift-invariant (homogeneous) in $\tilde{x}$ and can be shift-varying (nonhomogeneous) in the depth x_3. It is a much more realistic assumption in comparison with the convolution 3D model when the PSF is assumed to be shift-invariant over all three variables [188].

Assuming that $v(u)$ does not change rapidly in depth, the integral in Eq. (9.63) is replaced by an approximate sum with respect to the integration variable u_3,

$$z_i(\tilde{x}) = \sum_{j=1}^{l} \int_{\mathbb{R}^2} \beta(u_3(j)) v(\tilde{x} - \tilde{u},\ x_3(i), u_3(j)) y(\tilde{u}, u_3(j)) d\tilde{u},\ i = 1, \ldots, n_3. \tag{9.64}$$

Usually, the grid for $u_3(j)$ is uniform,

$$\{u_3 : u_3(j) = u_{3,0} + j\Delta_3,\ j = 0, 1, \ldots, l - 1\},$$

with the initial value $u_{3,0}$, the sampling period Δ_3, and the number of slices l covering the support of the PSF with respect to the argument u_3. The weights $\beta(u_3(j))$ appear from the used approximation for the integral. In the simplest case, $\beta(u_3(j)) = \Delta_3$.

In Eq. (9.64), the object function y is partitioned into l nonoverlapping strata:

$$y_j(\tilde{u}) = y(\tilde{u}, u_3(j)).$$

The 3D image model in Eq. (9.64) can be represented as the sum of l 2D convolutions,

$$z_i(\tilde{x}) \simeq \sum_{j=1}^{l} \int_{\mathbb{R}^2} v_{i,j}(\tilde{x} - \tilde{u}) y_j(\tilde{u}) d\tilde{u}, \qquad i = 1, \ldots, n_3, \tag{9.65}$$

where the convolutions are calculated over the strata and the PSF, $v_{i,j}$, is defined according to

$$v_{i,j}(\tilde{x}) = v(\tilde{x},\ x_3(i), u_3(j)) \beta(u_3(j)). \tag{9.66}$$

Assume that the observations $z_i(\tilde{x})$ are given on the integer regular $n_1 \times n_2$ lattice, $\tilde{x} \in X = \{k_1, k_2 : k_1 = 1, 2, \ldots, n_1,\ k_2 = 1, 2, \ldots, n_2\}$, and replace the 2D integrals in Eq. (9.65) by the approximate discrete convolutions similar to those used in Eq. (9.3):

$$z_i(\tilde{x}) \simeq \sum_{j=1}^{l} (v_{i,j} \circledast y_j)(\tilde{x}), \qquad i = 1, \ldots, n_3.$$

In this way we arrive at the discrete 3D observation model in the form

$$z_i(\tilde{x}) = \sum_{j=1}^{l} (v_{i,j} \circledast y_j)(\tilde{x}) + \varepsilon_i(\tilde{x}), \qquad i = 1, \ldots, n_3, \tag{9.67}$$

where the residuals $\varepsilon_i(\tilde{x})$ include observation errors as well as errors of discretization of the original integral model in Eq. (9.62). It is assumed for simplicity that it is zero-mean white Gaussian noise with variance σ_i^2. All weights appear from replacement of the integration by summations are included in the PSF $v_{i,j}$.

The 3D Gaussian model for $v_{i,j}$ used in our simulation experiments has the form

$$\begin{aligned} v_{i,j}(\tilde{x}) &= v_{i,j}^{(1)}(\tilde{x}) v_{i,j}^{(2)}, \\ v_{i,j}^{(1)}(\tilde{x}) &= \frac{1}{2\pi\sigma_x^2(x_3(i) - x_3(j))} \exp\left\{ -\frac{\tilde{x}_1^2 + \tilde{x}_2^2}{2\sigma_x^2(x_3(i) - x_3(j))} \right\}, \\ v_{i,j}^{(2)} &= \frac{1}{\sqrt{2\pi}\sigma_z} \exp\left\{ -\frac{[x_3(i) - x_3(j)]^2}{2\sigma_z^2} \right\}, \qquad \tilde{x} \in X. \end{aligned} \tag{9.68}$$

This model is presented as a product of two functions. The first one, $v_{i,j}^{(1)}(\tilde{x})$, models the shift-invariant Gaussian blur in the image (slice) plane with variance $\sigma_x^2(x_3(i) - x_3(j))$ depending on the distance $x_3(i) - x_3(j)$ between the focal plane i and the stratum j. This variance function $\sigma_x^2(x_3(i) - x_3(j)) \to 0$ for $i = j$ because the image of the accurately focused stratum has no blur. It is also assumed that $\sigma_x^2(x_3(i) - x_3(j))$ is a growing function of $|x_3(i) - x_3(j)|$; that is, a larger distance between the stratum and the focal plane means an increasing blur for the out-of-focus stratum.

The second function, $v_{i,j}^{(2)}(\tilde{x})$, models a contribution (magnitude) of the out-of-focus strata to the focal plane image. If $\sigma_z^2 \to 0$, the only stratum corresponding to $i = j$ is imaged in the focal plane. If $\sigma_z^2 \to \infty$, all blurred and sharp strata are imaged in the focal plane with equal weights.

The vector-matrix notation for Eq. (9.67) gives a compact form of this model as

$$\mathbf{z}(\tilde{x}) = (\mathbf{v} \circledast \mathbf{y})(\tilde{x}) + \boldsymbol{\varepsilon}(\tilde{x}), \tag{9.69}$$

where $\mathbf{z}$ is an n_3-vector function of 2D observations, $\mathbf{z}(\tilde{x}) = [z_1(\tilde{x}), z_2(\tilde{x}), \ldots, z_{n_3}(\tilde{x})]^T$, $\mathbf{v}(\tilde{x}) = [v_{i,j}(\tilde{x})]_{n_3 \times l}$ is an $n_3 \times l$ matrix of PSFs, $\mathbf{y}(\tilde{x}) = [y_1(\tilde{x}), y_2(\tilde{x}), \ldots, y_l(\tilde{x})]^T$ is an object function. The element $v_{i,j}(\tilde{x})$ of the matrix $\mathbf{v}(\tilde{x})$

is the PSF linking the jth slice of the object function $\mathbf{y}$ with the measurements produced in the ith optical focus plane.

The discrete observation model in Eq. (9.69) is used for the reconstruction of $\mathbf{y}$ from the observation $\mathbf{z}$. Note that in this discrete model the object function is defined by its l strata $y_j(\tilde{x})$, $j = 1, \ldots, l$. We use this approximate discrete model to estimate the unknowns y_j from the observations given by the focus planes in z.

The set of equations in Eq. (9.69) has $n_1 n_2 l$ unknowns $y_j(k_1, k_2)$ with the total number of equations (observations) equal to $n_1 n_2 n_3$. The natural assumption for estimation is that $n_3 \geq l$, that is, the number of observed focus planes is equal to or larger than the number of strata in question. These strata are used as discrete approximations of the continuous object 3D function $y(x)$. We cannot find more strata than the number of observed optical slices. If one wishes to have a finer in-depth approximation of y, then the parameter l should be large as well as the number of optical slices $n_3 \geq l$.

In the frequency domain, the equation for Eq. (9.69) is of the form

$$\mathbf{Z}(f) = \mathbf{V}(f)\mathbf{Y}(f) + \boldsymbol{\varepsilon}(f), \tag{9.70}$$

with $\mathbf{Z}(f) = \mathcal{F}\{\mathbf{z}(\tilde{x})\}$, $\mathbf{Y}(f) = \mathcal{F}\{\mathbf{y}(\tilde{x})\}$, and $\boldsymbol{\varepsilon}(f) = \mathcal{F}\{\boldsymbol{\varepsilon}(\tilde{x})\}$ being the vectors of the respective sizes. Here $f \in F$, $F = \{(f_1, f_2), f_i = k_i, k_i = 0, 1, \ldots, n_i - 1, i = 1, 2\}$, is the 2D normalized frequency. The PSF transfer matrix $\mathbf{V}(f) = \mathcal{F}\{\mathbf{v}(\tilde{x})\}$ has size $n_3 \times l$.

Equation 9.70 gives the observation model in the frequency domain. It is a vector analog of Eq. (9.4).

Figures 9.15 and 9.16 illustrate the setting of the problem. A phantom object is modeled by five balls. Each of these balls has a fluorescence intensity distributed uniformly in the volume of the ball. Figure 9.15 shows the distribution of the balls in the space and the cross-section slices (strata) of the object function. Each stratum includes the balls appearing only in the corresponding cross-section. The sharp images of the balls in the stratum correspond to these cross-sections.

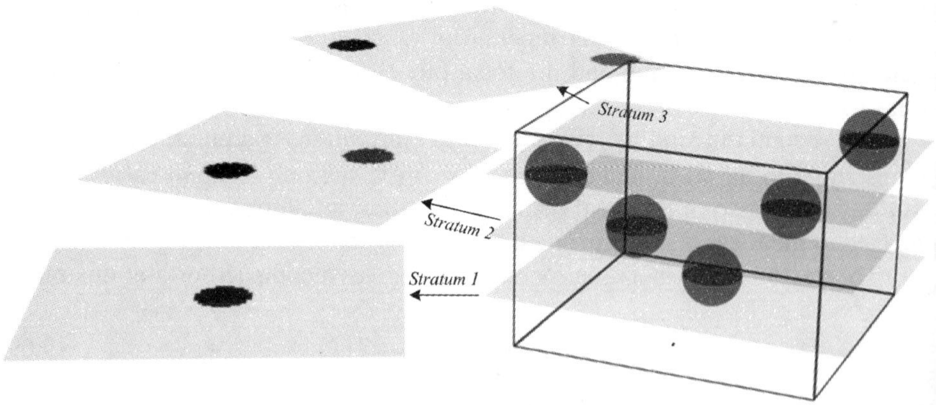

Figure 9.15 A phantom object of five balls. Three cross-sections (strata) of the object are shown.

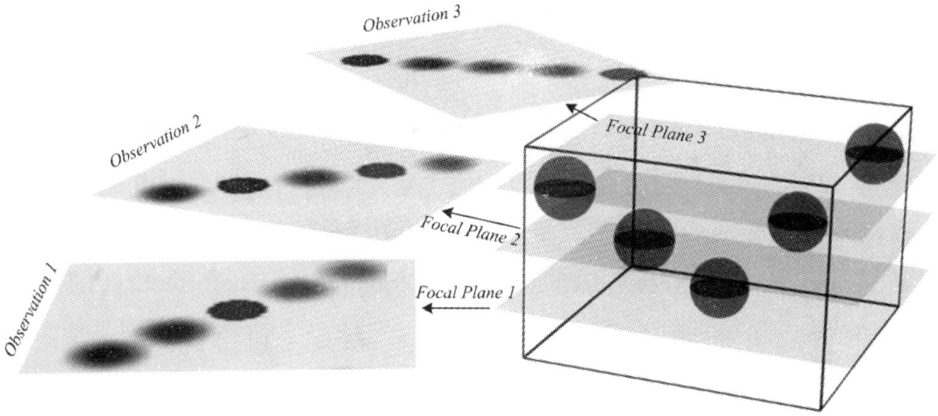

Figure 9.16 A phantom object of five balls: optical sectioning. Three focal planes (observation slices) correspond to the cross sections shown in Fig. 9.15.

Figure 9.16 illustrates the observation model. In each focal plane we observe projections of all five balls. However, only the cross-sections appearing in this particular focal plane give sharp, clear images. Other balls and parts of the balls are out-of-focus and give blurred images.

These focal plane images correspond to the observations $\mathbf{z}(\tilde{x})$ in Eq. (9.69). The problem is to restore the sharp stratum images shown in Fig. 9.15 from the observed focal plane images shown in Fig. 9.16. Thus, it is required to obtain a clear picture for the stratum of interest and eliminate the interference from all other strata in the scene.

9.6.3 3D spatially adaptive inverse

9.6.3.1 Naive inverse

The least-square method can be used to solve the matrix Eq. (9.70). This solution is defined as a minimizer for the quadratic loss function calculated in the frequency domain as the sum of all the residuals in Eq. (9.70):

$$R^2 = \sum_{f \in F} \sum_{i=1}^{n_3} |E_i(f)|^2 = \sum_{f \in F} \mathbf{E}^H(f)\mathbf{E}(f), \mathbf{E}(f) = \mathbf{Z}(f) - \mathbf{V}(f)\mathbf{Y}(f). \quad (9.71)$$

Using a vector version of Eq. (9.20), we obtain that the minimum of R^2 is achieved when

$$\partial_{\mathbf{Y}^H(f)} R^2 = 0. \quad (9.72)$$

Represent R^2 in the form

$$R^2 = \sum_{f \in F} R_f^2,$$

where

$$R_f^2 = [\mathbf{Z}^H(f) - \mathbf{Y}^H(f)\mathbf{V}^H(f)][\mathbf{Z}(f) - \mathbf{V}(f)\mathbf{Y}(f)].$$

Then the minimum conditions in Eq. (9.72) are

$$\partial_{\mathbf{Y}^H(f)} R_f^2 = 0 \Longrightarrow \partial_{\mathbf{Y}^H(f)}[\mathbf{Z}^H(f) - \mathbf{Y}^H(f)\mathbf{V}^H(f)][\mathbf{Z}(f) - \mathbf{V}(f)\mathbf{Y}(f)] = 0. \tag{9.73}$$

For calculation of the vector derivative in this equation we use formula (14.198):

$$\begin{aligned}\partial_{\mathbf{Y}^H(f)} R_f^2 &= -\partial_{\mathbf{Y}^H(f)}\{\mathbf{Y}^H(f)\mathbf{V}^H(f)\cdot[\mathbf{Z}(f) - \mathbf{V}(f)\mathbf{Y}(f)]\}\\ &= -\mathbf{V}^H(f)\cdot[\mathbf{Z}(f) - \mathbf{V}(f)\mathbf{Y}(f)].\end{aligned}$$

This gives the equation for the estimate

$$\mathbf{V}^H(f)\mathbf{V}(f)\hat{\mathbf{Y}}(f) = \mathbf{V}^H(f)\mathbf{Z}(f). \tag{9.74}$$

If $n_3 \geq l$ and the $l \times l$ matrix $\mathbf{V}^H(f)\mathbf{V}(f)$ is nonsingular, the solution of Eq. (9.74) has the form

$$\hat{\mathbf{Y}}(f) = [\mathbf{V}^H(f)\mathbf{V}(f)]^{-1}\mathbf{V}^H(f)\mathbf{Z}(f). \tag{9.75}$$

The estimate $\hat{\mathbf{Y}}(f)$ is random because the observation $\mathbf{Z}(f)$ is random. Inserting Eq. (9.70) in Eq. (9.75), we find for the expectation

$$\begin{aligned}E\{\hat{\mathbf{Y}}(f)\} &= [\mathbf{V}^H(f)\mathbf{V}(f)]^{-1}\mathbf{V}^H(f)E\{\mathbf{Z}(f)\}\\ &= [\mathbf{V}^H(f)\mathbf{V}(f)]^{-1}\mathbf{V}^H(f)\mathbf{V}(f)\mathbf{Y}(f) = \mathbf{Y}(f).\end{aligned}$$

Thus, the estimate $\hat{\mathbf{Y}}(f)$ is unbiased.

The spectrum of the random components of the vector $\hat{\mathbf{Y}}(f)$ is defined as

$$\mathbf{S}_{\hat{\mathbf{Y}}}(f) = E\{(\hat{\mathbf{Y}}(f) - E\{\hat{\mathbf{Y}}(f)\})(\hat{\mathbf{Y}}(f) - E\{\hat{\mathbf{Y}}(f)\})^H\}. \tag{9.76}$$

We use it to calculate the variance of the estimate on the base of formula (14.214):

$$\sigma_{\hat{y}_k}^2 = var\{\hat{y}_k(x)\} = \frac{1}{n_1 n_2}\sum_{f\in F}[\mathbf{S}_{\hat{\mathbf{Y}}}(f)]_{k,k}, \tag{9.77}$$

where $[\mathbf{S}_{\hat{\mathbf{Y}}}(f)]_{k,k}$ is the kth diagonal item of the matrix $\mathbf{S}_{\hat{\mathbf{Y}}}(f)$.

Let us derive the formula for the spectrum matrix $\mathbf{S}_{\hat{\mathbf{Y}}}(f)$. Inserting Eq. (9.75) and Eq. (9.70) in Eq. (9.76), we obtain

$$\begin{aligned}\mathbf{S}_{\hat{\mathbf{Y}}}(f) &= [\mathbf{V}^H(f)\mathbf{V}(f)]^{-1}\mathbf{V}^H(f)E\{\varepsilon(f)\varepsilon^H(f)\}\mathbf{V}(f)[\mathbf{V}^H(f)\mathbf{V}(f)]^{-1}\\ &= [\mathbf{V}^H(f)\mathbf{V}(f)]^{-1}\mathbf{V}^H(f)\sigma^2\mathbf{V}(f)[\mathbf{V}^H(f)\mathbf{V}(f)]^{-1},\end{aligned}$$

where

$$\sigma^2 = diag\{\sigma_1^2, \ldots, \sigma_l^2\}. \tag{9.78}$$

Then the variance $\sigma^2_{\hat{y}_k}$ is

$$\sigma^2_{\hat{y}_k} = \frac{1}{n_1 n_2} \sum_{f \in F} \{[\mathbf{V}^H(f)\mathbf{V}(f)]^{-1}\mathbf{V}^H(f)\sigma^2\mathbf{V}(f)[\mathbf{V}^H(f)\mathbf{V}(f)]^{-1}\}_{k,k}, \tag{9.79}$$

where $(\cdot)_{k,k}$ means the (k,k)th element oi the corresponding matrix. If $\sigma_1^2 = \sigma_2^2 = \cdots = \sigma_l^2$, we arrive at the simpler formula

$$\sigma^2_{\hat{y}_k} = \frac{\sigma_1^2}{n_1 n_2} \sum_{f \in F} \{[\mathbf{V}^H(f)\mathbf{V}(f)]^{-1}\}_{k,k}. \tag{9.80}$$

Comparing Eqs. (9.75) and (9.80) with Eqs. (9.7) and (9.8), we realize that the solution in Eq. (9.75) is the naive inverse generalized for the multivariate case.

9.6.3.2 Conditioning

Ill conditioning is a principal difficulty of all kinds of inverse problems. It results in noise amplification as well as in a high sensitivity of the solution to small variations in the parameters of the system. In the 2D inverse problems considered in Sec. 9.1.2, we use the regularized inverse, the regularized Wiener inverse, as well as the LPA-ICI spatially adaptive filters to suppress these undesirable effects.

In the 3D case we encounter the ill conditioning in a much more complex form. The basic step of the considered 3D inverse is a solution of Eq. (9.74). It is a set of l equations with l unknowns $\widehat{Y}_j(f)$, $j = 1, \ldots, l$, to be found. These equations should be solved for all $f \in F$.

The matrix $\mathbf{V}^H(f)\mathbf{V}(f)$ of the set in Eq. (9.74) is Hermitian, that is, $\mathbf{V}^H(f)\mathbf{V}(f) = [\mathbf{V}^H(f)\mathbf{V}(f)]^H$. It follows that the eigenvalues of this matrix are real and nonnegative. Let $\lambda_j(f)$, $j = 1, \ldots, l$, denote these eigenvalues. By definition, the conditioning number of the Hermitian matrix $\mathbf{V}^H(f)\mathbf{V}(f)$ is a ratio of the maximum eigenvalue to the minimum one,

$$\gamma(f) = \frac{\max_j \lambda_j(f)}{\min_j \lambda_j(f)}. \tag{9.81}$$

For the linear operator given by the matrix $\mathbf{V}^H(f)\mathbf{V}(f)$, the conditioning number includes the minimization and maximization of these eigenvalues also over the frequency f with the conditioning number calculated as

$$\gamma = \frac{\max_{f \in F} \max_j \lambda_j(f)}{\min_{f \in F} \min_j \lambda_j(f)}. \tag{9.82}$$

A large γ means that the system is ill-conditioned.

For the 2D problem from Sec. 9.1.2, $V^H(f)V(f) = |V(f)|^2$ is a scalar function of f. Then $\lambda(f) = |V(f)|^2$ and the conditioning number in Eq. (9.82) is calculated as

$$\gamma = \frac{\max_{f \in F} |V(f)|^2}{\min_{f \in F} |V(f)|^2}. \tag{9.83}$$

If for some of the frequencies f the function $|V(f)|^2$ takes values close to zero, the conditioning number γ becomes large and says that the system is ill-conditioned.

In the general 3D case, the elements of the matrix $\mathbf{V}^H(f)\mathbf{V}(f)$ are calculated as

$$[\mathbf{V}^H(f)\mathbf{V}(f)]_{k,l} = \sum_{j=1}^{n_3} V_{k,j}^H(f)V_{j,l}(f) = \sum_{j=1}^{n_3} V_{j,k}^*(f)V_{j,l}(f).$$

Thus, $\mathbf{V}^H(f)\mathbf{V}(f)$ is a combination of all elements of the matrix PSF and the conditioning number, in particular, depends on the interference (overlapping) of $V_{j,l}(f)$ elements of this matrix PSF.

The regularization and Wiener filterings are two of the classical approaches to dealing with ill conditioning. For the 3D inverse problem, we exploit matrix versions of these methods joined with the LPA-ICI adaptive filtering.

The mathematical work for the derivation of the corresponding filters is done in Sec. 14.7 of the Appendix. Here we present and discuss the final results only.

9.6.3.3 LPA regularized inverse

The regularized inverse, including the LPA filtering, is defined by (see Eqs. (14.227–14.229) in Sec. 14.7 of the Appendix):

$$\hat{\mathbf{Y}}_h^{RI}(f) = \mathbf{Q}_{RI}(f, \alpha)\mathbf{Z}(f), \tag{9.84}$$

where $\hat{\mathbf{Y}}_h^{RI}(f)$ is a vector of the regularized inverse estimates of $\mathbf{Y}(f)$, and the transfer matrix of the regularized inverse filter is

$$\mathbf{Q}_{RI}(f, \alpha_1) = G_h(f)[\mathbf{V}^H(f)\mathbf{V}(f) + \alpha_1^2 I_{l\times l}]^{-1}\mathbf{V}^H(f). \tag{9.85}$$

Here α_1^2 is a regularization parameter and $I_{l\times l}$ is the $l \times l$ identity matrix. The scalar LPA filter $G_h(f)$ in Eq. (9.85) with the scale parameter h can be used independently for each component of the vector estimate $\hat{\mathbf{Y}}_h^{RI}(f)$. It is convenient to calculate this estimate in two successive steps:

$$\tilde{\mathbf{Y}}^{RI}(f) = [\mathbf{V}^H(f)\mathbf{V}(f) + \alpha_1^2 I_{l\times l}]^{-1}\mathbf{V}^H(f)\mathbf{Z}(f), \tag{9.86}$$

$$\hat{\mathbf{Y}}_h^{RI}(f) = G_h(f)\tilde{\mathbf{Y}}^{RI}(f). \tag{9.87}$$

The regularized inverse estimate $\tilde{\mathbf{Y}}^{RI}(f)$ does not depend on h and is calculated only once. The LPA filtering in Eq. (9.87) has the vector $\tilde{\mathbf{Y}}^{RI}(f)$ as a given input and the same filter $G_h(f)$ for each component of this vector. This two-step procedure essentially simplifies the calculations because the complex matrix manipulations in Eq. (9.86) are produced only once and are used for all scales $h \in H$. The variance of the regularized inverse estimate $\hat{y}_{h,k}^{RI}(x) = \mathcal{F}^{-1}\{\widehat{Y}_{h,k}^{RI}(f)\}$ is calculated as

$$\sigma_{\hat{y}_{h,k}^{RI}}^2 = var\{\hat{y}_{h,k}^{RI}(x)\} = \frac{1}{n_1 n_2}\sum_{f\in F}[\mathbf{Q}_{RI}(f, \alpha_1)\sigma^2\mathbf{Q}_{RI}^H(f,\alpha_1)]_{k,k}. \tag{9.88}$$

9.6.3.4 LPA-regularized Wiener inverse

The regularized Wiener inverse filter including the LPA filtering, is defined by (see Eqs. (14.237–14.240) in Sec. 14.7 of the Appendix):

$$\hat{\mathbf{Y}}_h^{RWI}(f) = \mathbf{Q}_{RWI}(f, \alpha_2)\mathbf{Z}(f), \tag{9.89}$$

with the transfer matrix

$$\mathbf{Q}_{RWI}(f, \alpha_2) = G_h(f)\mathbf{Y}(f)\mathbf{Y}^H(f)\mathbf{V}^H(f)[\mathbf{V}(f)\mathbf{Y}(f)\mathbf{Y}^H(f)\mathbf{V}^H(f)+n_1n_2\alpha_2^2\boldsymbol{\sigma}^2]^{-1},$$
$$f \in F, \tag{9.90}$$

where α_2^2 is a regularization parameter and $\boldsymbol{\sigma}^2$ is a diagonal matrix of the noise variances. The variances of this estimate are calculated as

$$\sigma^2_{\hat{y}^{RWI}_{h,k}}(x) = var\{\hat{y}_{h,k}^{RWI}(x)\} = \frac{1}{n_1n_2}\sum_{f\in F}[\mathbf{Q}_{RWI}(f, \alpha_2)\boldsymbol{\sigma}^2\mathbf{Q}^H_{RWI}(f, \alpha_2)]_{k,k}. \tag{9.91}$$

It is also convenient to calculate this estimate in two successive steps, similar to Eqs. (9.86–9.87):

$$\tilde{\mathbf{Y}}^{RWI}(f) = \mathbf{Y}(f)\mathbf{Y}^H(f)\mathbf{V}^H(f)[\mathbf{V}(f)\mathbf{Y}(f)\mathbf{Y}^H(f)\mathbf{V}^H(f) + n_1n_2\alpha_2^2\boldsymbol{\sigma}^2]^{-1}\mathbf{Z}(f),$$
$$f \in F, \tag{9.92}$$

$$\hat{\mathbf{Y}}_h^{RWI}(f) = G_h(f)\tilde{\mathbf{Y}}^{RWI}(f). \tag{9.93}$$

Here the regularized Wiener inverse $\tilde{\mathbf{Y}}^{RWI}(f)$ is independent of h and used for calculation of $\hat{\mathbf{Y}}_h^{RWI}(f)$ for all $h \in H$. Thus, the complex matrix calculations in Eq. (9.92) are produced only once for all $h \in H$.

9.6.3.5 ICI adaptive scale

Formulas (9.84) and (9.89) define the varying-scale RI and RWI estimates of the vector function $\mathbf{y}(x)$. The variances of these estimates are given, respectively, by formulas (9.88) and (9.91). We can calculate the estimates with the different scales h and use the ICI for the pixel-wise adaptive-scale selection.

For the deblurring we use the algorithm developed in Sec. 9.2. This algorithm is based on the regularized inverse and regularized Wiener inverse estimates. Exploiting the directional estimators, we improve the adaptive-scale selection and quality of filtering overall. The structure of the directional deblurring filter is as shown in Fig. 9.2.

Let $\hat{\mathbf{Y}}^{RI}(f)$ be the final RI adaptive-scale estimate obtained after the fusing of the multidirectional estimates (see Fig. 9.2). Then the RWI estimates

Eqs. (9.92–9.93) are defined as

$$\tilde{\mathbf{Y}}^{RWI}(f) = \hat{\mathbf{Y}}^{RI}(f)[\hat{\mathbf{Y}}^{RI}(f)]^H\mathbf{V}^H(f)[\mathbf{V}(f)\hat{\mathbf{Y}}^{RI}(f)(\hat{\mathbf{Y}}^{RI}(f))^H\mathbf{V}^H(f) + n_1n_2\alpha_2^2\sigma^2]^{-1}\mathbf{Z}(f), \quad f \in F, \tag{9.94}$$

$$\hat{\mathbf{Y}}_h^{RWI}(f) = G_h(f)\tilde{\mathbf{Y}}^{RWI}(f). \tag{9.95}$$

The ICI adaptive-scale selection is used independently for each component of the vector $\hat{\mathbf{Y}}_h^{RI}(f)$ in the RI algorithm and for the vector $\hat{\mathbf{Y}}_h^{RWI}(f)$ in the RWI algorithm. Thus, roughly speaking, the number of calculations is proportional to the length l of the vectors $\hat{\mathbf{Y}}_h^{RI}(f)$ and $\hat{\mathbf{Y}}_h^{RWI}(f)$.

The complexity of the matrix operations heavily depends on the depth dimension of the problem, that is, on the parameters n_3 and l. Remember that we are dealing with vectors and matrices of the following sizes: $\mathbf{Y}(f)$ has length l, $\mathbf{V}(f)$ is an $n_3 \times l$ matrix, and $\mathbf{S}_\varepsilon(f)$ is an $n_3 \times n_3$ matrix. However, all of these matrix operations are produced in the frequency domain only once while the multiscale filtering for each component of the vectors $\tilde{\mathbf{Y}}^{RI}(f)$ and $\tilde{\mathbf{Y}}^{RWI}(f)$ is identical to the one used in Sec. 9.2.

The matrix nature of the 3D inversion is very demanding for the memory of the computer and requires a manipulation with large data arrays. Recursive frequency domain algorithms can be developed to simplify this computing.

9.6.4 Simulation

9.6.4.1 Perfect reconstruction

The first row of images in Fig. 9.17 shows the observations obtained by focusing in planes given in Fig. 9.16. We use these observations to reconstruct the phantom density in stratum 1 (Fig. 9.15).

The second row of images in Fig. 9.17 shows the reconstructions. In the left image we use the data from focal plane 1 only. Naturally, this reconstruction is identical to the used observations, that is, to the observation image of this plane. Thus, the left images in the first and second rows in Fig. 9.16 are identical.

In the middle image of the second row we show the reconstruction obtained using the observations from focal planes 1 and 2 (left and middle images from the first row). We see that the level of interference from other strata becomes lower.

The last right image in the second row corresponds to the reconstruction using all three focal plane data. It gives a perfect reconstruction of the object of interest, that is, the first stratum image in Fig. 9.16. The interference of the images from other strata is completely eliminated.

Thus, for this noiseless experiment we show how much the number of focal plane observations improves the reconstruction and eliminates the interference of the images from the parts of the phantom that are not in the stratum of interest. The algorithm gives a perfect reconstruction when a sufficient number of focal planes is available.

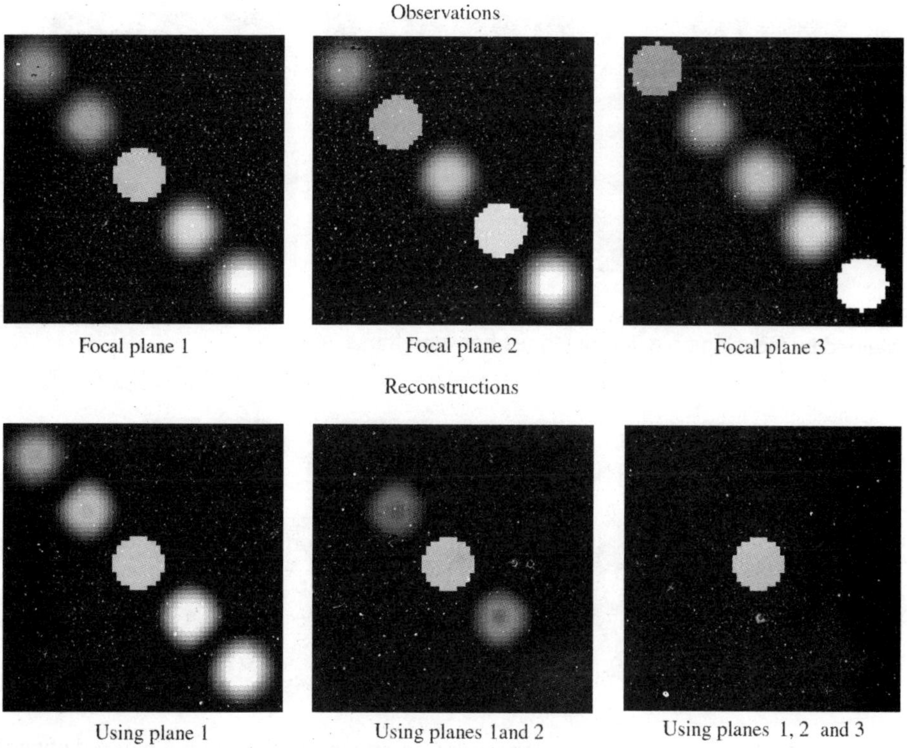

Figure 9.17 The observations obtained in focal planes 1, 2, and 3 are shown in the first row of images. The reconstruction is produced for stratum 1. Results are given in the second row of images: In the left image we use focal plane 1 only; in the middle image we use focal planes 1 and 2; in right image we use all three focal planes and achieve a perfect reconstruction of the stratum 1 image.

For simplicity in this simulation we assume that the phantom consists of only three 2D cross-section (stratum) planes. Then, the demonstrated perfect reconstruction is an expected result.

9.6.4.2 *Noisy data*

When the data are noisy the deconvolution becomes a more difficult task because of the ill-conditioned nature of the inversion. A redundant number of observations ($n_3 > l$) and proper filtering become essential issues.

For a demonstration of the ability of the proposed technique to decrease the interference and to separate the images from different strata, we consider the observations shown in Fig. 9.18.

It is assumed that there are two 2D plane images (*Cameraman* and *Lena*) that are transparent and form two strata objects similar to those presented in Fig. 9.15. The observations in Fig. 9.18 were obtained by focusing on different planes out

Figure 9.18 The mixed noisy observations of transparent *Cameraman* and *Lena* images obtained by focusing on different planes between these images.

Figure 9.19 The reconstruction of *Lena* and *Cameraman* images from the mixed observations given in Fig. 9.18.

of strata. The first image in Fig. 9.18 corresponds to the focusing plane close to *Cameraman*, the last one is close to *Lena*, and the middle image was obtained by focusing between the *Cameraman* and *Lena* strata.

These observation images are given with the additive Gaussian noise with $\sigma = 0.01$. It corresponds more or less to the level of noise in the deblurring experiments discussed in Sec. 9.2.3. For denoising we use the ICI adaptive RI and RWI algorithms presented above. The filter G_h has a conical narrowed support with the scale parameters as in Sec. 9.2.3.

The reconstruction results are shown in Fig. 9.19. The images mixed in the observations are nearly perfectly separated and well denoised. These results were obtained for reconstruction of two images from three image observations. The reconstructed images are noisier and not separated very well when only two of the three image observations are used.

Chapter 10

Nonlinear Methods

10.1 Why Nonlinear Methods?

10.1.1 Basic hypotheses

The LPA estimates studied in the previous chapters give universal and flexible tools for signals given with noise. However, there are a number of good reasons to go beyond these linear estimators to nonlinear ones.

If we look at the estimation problem in general, we see that the selection of an algorithm depends mainly on three hypotheses:

1. Signal model;
2. Observation model;
3. Criterion.

For example, the LPA estimators assume the following:

1) A signal model is linear with respect to some parameters, specifically, the signal y can be approximated by a truncated Taylor series with unknown coefficients.
2) Observations are linear with respect to the signal, and additive noise has invariant characteristics independent of the signal. For direct observations with Gaussian noise, this model has the form

$$z_s = y(X_s) + \varepsilon_s, \qquad s = 1, \ldots, n, \tag{10.1}$$

where the ε_s are zero-mean Gaussian with variance σ^2.
3) The estimation criterion is quadratic.

As a result of these assumptions, the LPA estimates are linear with respect to the observations z_s. This linearity is an important moment defining the transparency of all elements of signal processing starting from the algorithm derivation and ending with its implementation and adjustment. If we are looking for optimal or nearly optimal procedures, strictly speaking, a violation of any of the above hypotheses will result in it being necessary to use nonlinear algorithms.

The role of linearity is obvious. If the signal y is nonlinear with respect to unknown parameters or the observation z is nonlinear with respect to y, then the observation z is a nonlinear function of the signal parameters, and for estimation we need to solve nonlinear equations. This automatically means that the estimation procedure becomes nonlinear.

The LPA criterion in Eq. (2.24) is a sum of the weighted squared residuals e_s

$$\sum_s w_h(x - X_s)e_s^2, \; e_s = z_s - \hat{y}_h(x, X_s), \; \hat{y}_h(x, X_s) = \phi^T(x - X_s)C.$$

This criterion is quadratic on C, and the estimates of C as well as the signal estimates are linear on observations z_s.

Let us replace the squared residuals by a nonquadratic convex function of the residuals, say ψ, with the criterion

$$J_{\psi,h} = \sum_s w_h(x - X_s)\psi(e_s). \qquad (10.2)$$

The criterion becomes nonquadratic on C, and the estimate of C and the signal are nonlinear on the observations z_s.

Why do we need to use something other than our comfortable quadratic residuals?

The criterion in Eq. (10.2) is a measure of the quality of fit of the observations. It is used as a design tool to obtain a signal estimate that approximates and predicts data or, in other words, it explains or properly interprets the data.

A discrepancy between the data and the model can be measured in different ways, depending on the problem at hand. When we need to avoid large errors we use the maximum error criterion, $\max_s |e_s|$. Then the estimation is a minimax procedure with estimates selected by minimizing the maximum errors and ignoring all errors smaller than the maximal ones.

Criteria using mean values take into consideration all errors weighted according to their magnitude. For instance, consider the exponential loss function $\psi = |e_s|^p$, with a parameter $p > 0$ and the l_p norm of the residuals defined as

$$l_p = \left(\sum_{s=1}^{n} |e_s|^p\right)^{1/p}, \qquad p > 0. \qquad (10.3)$$

What is a role of the parameter p in the criterion in Eq. (10.3)? Figure 10.1 shows the sets $\{e : l_p \leq 1\}$, called l_p-balls, for p equal to and smaller and larger than 2. This figure illustrates the 2D case $n = 2$ while the conclusions can be induced for general multivariate data $n > 2$.

The standard quadratic criterion corresponds to $p = 2$. Then, the l_2-ball is a disc with errors $\sum_{s=1}^{n} |e_s|^2 \leq 1$. The l_p-balls become progressively smaller and clustered around the coordinate axis as p decreases. For a larger p, as $p \to \infty$ the ball takes a maximal size and becomes the unit square for the limit maximum error criterion

$$l_\infty = \lim_{p \to \infty} l_p = \max_s\{|e_s|\}.$$

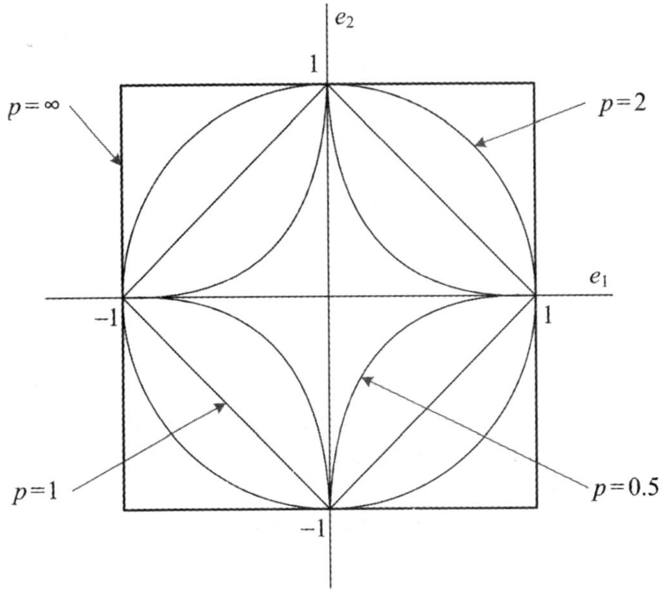

Figure 10.1 l_p-balls, the sets $\{e : l_p \leq 1\}$, illustrating the distribution of elements of vectors e for different powers p.

Compare the areas of the l_p-balls in Fig. 10.1. It is obvious that these areas become smaller for smaller p. The ball with the smaller p is imbedded in the balls with the larger p.

For $p < 1$, the area of the ball is mainly defined by small errors close to the origin, and the contribution of the parts of the ball with large errors clustered around the axis becomes negligible as p takes smaller values.

This allows one to conclude that the l_p-ball (or l_p-norm) is dominated by small errors (the majority of the errors are small) for $p < 1$. Minimization of the l_p-ball with $p < 1$ mainly means the minimization of majority of the small errors, while a few large errors are not very essential for the area of the ball and for the criterion value. This criterion is insensitive to a small number of larger errors.

For $p > 1$, the l_p-ball (or l_p-norm) is dominated by larger errors. Minimization of this l_p-ball means the minimization of larger errors because this criterion is very sensitive to large errors and small errors are neglected.

Depending on p, the residuals in the exponential criterion are treated (weighted) in different ways according to their magnitude, and the usual quadratic criterion with $p = 2$ is just one of the possible choices for the parameter p. There are no reasons to believe that $p = 2$ is a universally good selection.

10.1.2 *M*-estimation

The term *M-estimate* (*M-estimation*) is used for estimates obtained by minimizing an empirical mean of a nonquadratic residual loss function.

The M-estimation nonlinear version of the linear LPA estimate (Sec. 1.3.2) is introduced in the form

$$\hat{C}(x,h) = \arg\min_{C \in \mathbb{R}^M} J_{\psi,h}(x,C), \tag{10.4}$$

$$J_{\psi,h}(x,C) = \sum_s w_h(x - X_s)\psi(e_s), \tag{10.5}$$

$$e_s = z_s - y_h(x,X_s),\ y_h(x,X_s) = C^T\phi(x - X_s),$$

$$\hat{y}_h(x) = \hat{C}^T(x,h)\phi(0). \tag{10.6}$$

Here $y_h(x,X_s) = C^T\phi(x - X_s)$ is the LPA of the signal y as it was exploited in Eq. (2.18). The vector C is an unknown parameter of this model found by the minimization of $J_{\psi,h}(x,C)$. The estimate of the signal for the point x is defined in the standard form in Eq. (10.6). The only difference between this estimate and the linear LPA is the nonquadratic loss function ψ in Eq. (10.5). All other elements of this estimate are completely identical to the ones considered and interpreted in the previous chapters.

Formulas (10.4)–(10.6) define an LPA nonparametric *M-estimate* of the regression $y(x)$.

There exist different ideas concerning the selection of a loss function ψ. One of them is to obtain an estimate that is insensitive (robust) with respect to variations in noise distribution. The minimax *Huber's* estimation theory discussed in Sec. 10.2.4 implements this idea.

In this introductory section we concern ourselves only with a family of the exponential loss functions:

$$\psi(x) = |x|^p, \qquad p > 0. \tag{10.7}$$

Using $\psi(x)$ in Eq. (10.3), we see that $l_p = \left[\sum_{s=1}^{n}\psi(e_s)\right]^{1/p}$. The criterion in Eq. (10.5) is a weighted l_p-norm of the residuals.

The optimal $\hat{C}(x,h)$ in Eq. (10.4) satisfies

$$\partial_C J_h(x,C) = \sum_s w_h(x - X_s)\psi^{(1)}(e_s)\phi_h(x - X_s) = 0, \tag{10.8}$$

where $\psi^{(1)}(x) = \partial_x\psi(x)$. For $x \neq 0$ we have $\psi^{(1)}(x) = p|x|^{p-1}sign(x)$, and $\psi(x)$ is nondifferentiable at $x = 0$ if $p \leq 1$.

In Fig. 10.2 we show $\psi(x) = |x|^p$, and $\psi^{(1)}(x) = p|x|^{p-1}sign(x)$, of different powers p. The quadratic loss function, $p = 2$, gives the linear $\psi^{(1)}(x) = x$, which shows that in Eq. (10.8) a larger residual gives a larger term with $|\psi^{(1)}(e_s)| = |e_s|$.

For $p = 1$, the derivative $|\psi^{(1)}(e_s)| = 1$, and then all residuals, no matter what their absolute values, have the same contribution in Eq. (10.8).

It is obvious that, for $p > 1$, we have the loss function where the role of the larger residuals is stronger than in the quadratic case with $p = 2$ as $|\psi^{(1)}(e_s)|$ grows faster than $|e_s|$. For $p < 1$, the situation is inverse because $|\psi^{(1)}(e_s)|$ is a monotone decreasing function of $|e_s|$. This means that the larger residuals have a smaller

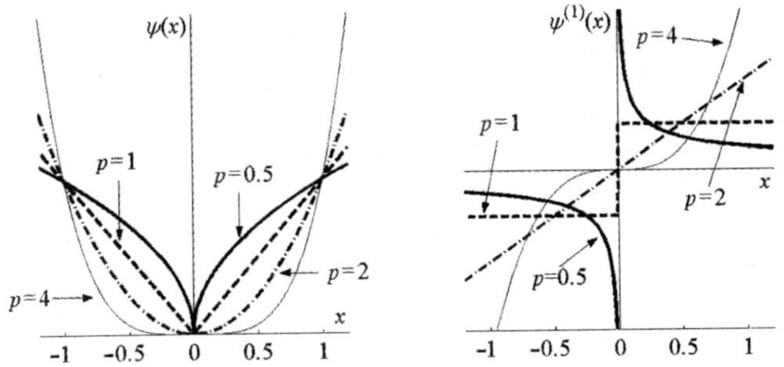

Figure 10.2 Loss functions $\psi(x) = |x|^p$, and derivatives $\psi^{(1)}(x) = p|x|^{p-1} sign(x)$ of different powers p.

influence on the estimate than the smaller ones. This fact is used in robust estimations when we wish to build estimates that are insensitive to rare but large errors in observations.

10.1.3 Statistical approach

Let the observation arguments X_s be given and the corresponding observations z_s be random. In statistics, the observation model usually means a joint probability distribution of the observation pairs $\{z_s, X_s, s = 1, \ldots, n\}$ that is given within some unknown parameters. The main problem is the estimation of these parameters.

A specification of the distribution requires a complete description of the deterministic and random components of the signal as well as of the observation model. In this way the first two hypotheses mentioned above (signal and observation models) are explicitly imbedded in the distribution function.

The maximum likelihood (ML) is one of the statistical techniques universally applicable for many distributions and enables good optimal properties of estimates. The ML is a criterion (third of three hypotheses discussed above) defining an estimation method. The distribution and the ML together completely specify an estimate.

The optimal statistical properties of the ML estimate mean that this method can extract all information about the estimated unknowns available in given observations. The ML assumes that the distribution function of the observations is given within some unknown parameters. In many circumstances the accurate distribution is unknown and cannot be given. Then, the standard ML is not applicable.

It is known that some of the ML estimates (in particular, the popular Gaussian least-square method) are very sensitive with respect to even small variations of distributions. These ML estimates, which are optimal for nominal distributions, can be practically useless when a small variation of this distribution significantly degrades the estimate. A large variety of techniques has been developed in statistics to deal with these drawbacks of the standard ML.

We consider two efficient alternative approaches: the quasi-likelihood (QL) method and the minimax robust M-estimation. The QL is a semiheuristic method developed for problems where the observation variance is a function of the signal. The QL enables good properties of the estimates close to the optimal properties of the ML.

The QL is a distribution-free method that does not require accurate knowledge of the observation distribution. It makes a serious difference versus the ML.

The minimax M-estimation is an approach that deals with uncertainty in the observation model. It starts with the idea that the distribution is unknown but belongs to a class of distributions. The worst distribution from this class is selected and the estimate is the ML estimate for this worst distribution. This leads to algorithms that are insensitive (*robust*) to variations in the distributions.

This chapter is organized as follows. We begin with the robust M-estimation and demonstrate its efficiency for median filters.

Further, the elements of the minimax estimation theory are given for the general type of estimate to present the basic ideas and links with the ML. The emphasis, as everywhere in the book, is on the local nonparametric and adaptive-scale estimation. We show that with some modifications the ICI can be exploited for these nonlinear estimators.

The local ML and QL are presented in the next chapter as general concepts for the adaptive-scale nonparametric regression estimate. The asymptotic accuracy results are used for the design of varying-scale adaptive LPA estimators. However, first we need the statistical distributions presented in the next section.

10.1.4 Some distributions

The observation model can always be given in the additive form

$$z_s = y(X_s) + \varepsilon_s, \qquad i = 1, \ldots, n, \tag{10.9}$$

where the random ε is zero-mean and $E\{z(x)\} = y(x)$.

By definition, the regression $y(x)$ is a conditional mean of z provided that is a fixed x. In the model in Eq. (10.9), the expectation of z considered as a function of x is a *regression* of the random z on x. We consider the following distributions of random observations leading to nonlinear regression estimates.

10.1.4.1 Gaussian distribution with varying variance

The random ε in Eq. (10.9) is zero-mean independent Gaussian with the variance depending on the signal y, $\varepsilon \sim \mathcal{N}(0, \sigma^2(y))$. The observations z are also Gaussian with mean $y = E\{z\}$ and variance $\sigma^2(y)$,

$$z \sim \mathcal{N}(y, \sigma^2(y)).$$

The probability density of z with a parameter y is

$$f(z, y) = \frac{1}{\sqrt{2\pi}\sigma(y)} \exp\left[-\frac{(z-y)^2}{2\sigma^2(y)}\right]. \tag{10.10}$$

The standard Gaussian model is a special case of Eq. (10.10) assuming that the observations are independent identically distributed (i.i.d.), that is, the variance σ^2 is the same for all $y(x)$.

10.1.4.2 Laplacian distribution

The Laplace probability density function (p.d.f.) for z in Eq. (10.9) has the form

$$f(z, y) = \frac{1}{2a} \exp\left(-\frac{|z - y|}{a}\right), \qquad a > 0, \tag{10.11}$$

where a is a parameter of the Laplace distribution. The parameter a defines the width (sharpness) of the Laplace probability density. A smaller a results in a sharper probability density. The mean and the variance for the Laplacian random variable are

$$E\{z\} = y, \qquad var\{z\} = 2a^2. \tag{10.12}$$

Figure 10.3 shows the p.d.f. of the Gaussian and Laplace distributions for the same variances. The Laplace density is sharper than the Gaussian and has a heavier tail. This means that the Laplace p.d.f. approaches zero more slowly than the Gaussian p.d.f. as $|z| \to \infty$. The expectation y as a function of x is called a *Laplacian regression.*

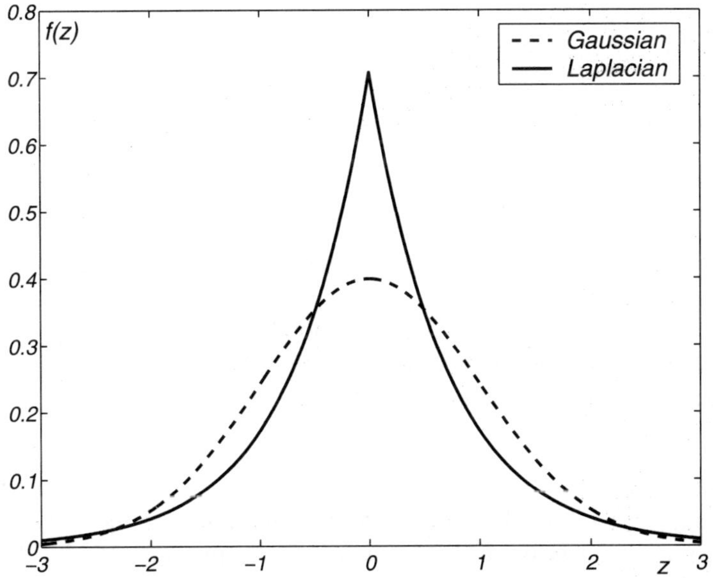

Figure 10.3 Gaussian ($\sigma = 1$) and Laplacian ($a = 1/\sqrt{2}$) probability densities.

10.1.4.3 Binary (Bernoulli) distribution

Assume that z is binary random taking the values 1 and 0 with probabilities p and $1-p$:

$$P\{z=k\}=\begin{cases}p, & k=1,\\ 1-p, & k=0.\end{cases} \tag{10.13}$$

Then

$$E\{z\}=p, \qquad var\{z\}=p(1-p). \tag{10.14}$$

Assume also that the distribution depends on the argument x and this dependence is given in the form $p=p(x)$. The function $p(x)$ is called the *Bernoulli regression*. Note that it is a regression because

$$p(x)=E\{z(x)\},$$

that is, $p(x)$ is the conditional expectation of random z provided that x is fixed.

Such models arise in many economics applications such as reliability and risk research; they are also widely used in classification problems and digital imaging.

10.1.4.4 Poisson distribution

The Poisson distribution is a distribution of discrete-integer random variables while the Gaussian and Laplacian are distributions of continuous random variables.

A random variable z has a Poisson distribution with the parameter $\lambda>0$ if z takes integer values $0, 1, \ldots,$ with probability

$$P\{z=k\}=\exp(-\lambda)\frac{\lambda^k}{k!}, \qquad k=0,1,\ldots. \tag{10.15}$$

The moments of the random Poisson z are

$$E\{z\}=\lambda, \qquad var\{z\}=\lambda. \tag{10.16}$$

This distribution is used to model on count data. In particular, this model is commonly exploited in queueing theory, digital photography, and positron emission tomography.

The parameter λ is the intensity of randomly counted events. This intensity is counted in time and can be varying in space x. We assume that $\lambda=\lambda(x)$. This dependence is called a *Poisson regression*.

We will use the notation

$$z\sim\mathcal{P}(\lambda) \tag{10.17}$$

for the random z having a Poisson distribution with the parameter λ.

For large λ, the Poisson distribution is well approximated by the Gaussian one [152]:

$$(z-\lambda)/\lambda^{1/2}\sim\mathcal{N}(0,1)+O(\lambda^{-1/2}). \tag{10.18}$$

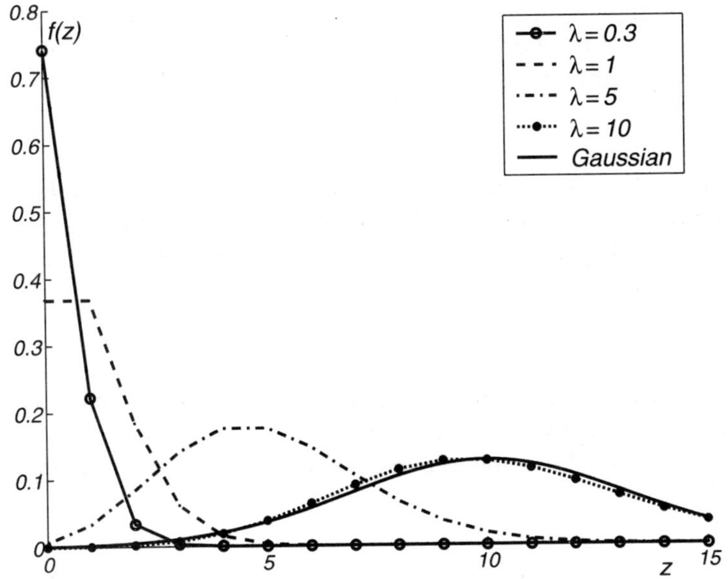

Figure 10.4 The Poisson p.d.f. with parameters λ = 0.3, 1, 5, 10. The last curves nearly coincide with the Gaussian *N*(10, 10).

The Poisson p.d.f. is shown in Fig. 10.4 for a few values of λ. For $\lambda = 10$, it is nearly identical to the Gaussian probability density $\mathcal{N}(m, \sigma^2)$ with $m = 10$ and $\sigma = \sqrt{10}$.

10.1.4.5 Exponential distribution*

The above distributions can be considered as special cases of a general exponential family of distributions [46], [145]. A one-parameter exponential distribution has the following canonical form:

$$f_\theta(z) = \exp\{[\theta z - b(\theta)]/a(\varphi) + c(z, \varphi)\}, \tag{10.19}$$

where $a(\cdot)$, $b(\cdot)$, $c(\cdot, \cdot)$ are some known functions and θ is a *canonical parameter*. The variable φ is called dispersion.

The following holds for the exponential distribution

$$\begin{aligned} m_z &= E\{z\} = \partial_\theta b(\theta) \triangleq b^{(1)}(\theta), \\ var\{z\} &= a(\varphi)\partial_\theta^2 b(\theta) \triangleq a(\varphi) b^{(2)}(\theta). \end{aligned} \tag{10.20}$$

If the parameter θ depends on x, then $E\{z\}$ also depends on x and $m_z(x)$ is a regression of the observations z on x. The function $\theta(x)$ is called a *canonical regression*. Formulas (10.20) show that the expectation and the variance of z are expressed through θ.

In linear regression analysis, the expectation m_z is assumed to be dependent on the argument x and is modeled as a function (regression) of x linear with respect to an unknown vector parameter C, in the form

$$m_z(x) = C^T \phi(x). \tag{10.21}$$

A class of the so-called generalized linear models introduced for the exponential distribution is essentially different. The regression $m_z(x)$ in Eq. (10.20) is built not in the linear form in Eq. (10.21) but as a nonlinear function of such a linear form.

The expectation $m_z(x)$ is defined by the equation

$$g(m_z(x)) = C^T \phi(x), \tag{10.22}$$

where g is called a *link function.* A selection of this sort of link function is not unique and is dictated by the nature of the considered problem.

For an exponential distribution, the *link function* g is called *canonical* if

$$g(m_z) = \theta$$

and

$$g(\cdot) = (b^{(1)})^{-1}(\cdot), \tag{10.23}$$

where $(\cdot)^{-1}$ stands for the inverse function. The linear model $C^T \phi(x)$ in Eq. (10.22) is used for the canonical regression $\theta(x)$ in Eq. (10.22).

For the canonical link function,

$$m_z = g^{-1}(\theta) = [(b^{(1)})^{-1}(\theta)]^{-1} = b^{(1)}(\theta), \tag{10.24}$$

which corresponds to Eq. (10.20). It gives the nonlinear model for m_z in the form

$$m_z(x) = b^{(1)}(C^T \phi(x)). \tag{10.25}$$

If the canonical regression $\theta(x)$ is linear on x, in general, the regression $m_z(x)$ is nonlinear on x. Thus, in general, the generalized linear models are nonlinear on the parameter C.

The generalized linear models (theory and methods) are thoroughly studied in statistical literature and are an important class of nonlinear estimators relevant to many applications. More information on generalized linear models and the exponential distribution and further references can be found in [78], [152].

The canonical parameter θ in the class in Eq. (10.19) is a convenient object for estimation for a number of reasons:

- The exponential p.d.f. and the corresponding ML are concave on θ. Then, the ML gives a unique estimate for θ and the estimate has good accuracy and convergence properties.
- There are no restrictions on the canonical parameter θ even if such restrictions exist for m_z. Thus the estimation can always be given as an unconstrained maximization procedure.
- If some p.d.f.s belong to the exponential family, they can be transformed to the canonical form and the canonical parameter θ can be specified. Then,

estimation of the regression $m_z(x)$ can always be started from the estimation of the canonical regression $\theta(x)$ with the following recalculation of this estimate into the estimate of $m_z(x)$ using the link in Eq. (10.24).

The meaning of the functions $a(\cdot)$, $b(\cdot)$, $c(\cdot,\cdot)$ $g(\cdot)$ in the canonical exponential p.d.f. in Eq. (10.19) for the some distributions are illustrated as follows.

1. The *Gaussian density*

$$f(z) = \frac{1}{\sqrt{2\pi}\sigma} \exp\left[-\frac{(z-m)^2}{2\sigma^2}\right]$$

can be represented in the form

$$f(z) = \exp\left[\frac{mz - m^2/2}{\sigma^2} - \frac{z^2}{2\sigma^2} - \ln(\sqrt{2\pi}\sigma)\right]. \qquad (10.26)$$

Comparing this with Eq. (10.19), we see that, in terms of the exponential distribution,

$$\theta = m, \quad a(\varphi) = \varphi, \quad \varphi = \sigma^2, \quad b(m) = m^2/2,$$
$$c(z,\varphi) = -\frac{z^2}{2\varphi} - \ln(\sqrt{2\pi\varphi}).$$

The equality $\theta = m$ shows that the expectation m is the canonical parameter θ of the Gaussian p.d.f. The estimation of this mean does not require any transformation. It can be done directly from observations of z.

Let us check the identities in Eq. (10.20). Calculation of the derivatives $b^{(1)}(m) = m$ and $b^{(2)}(m) = 1$ gives

$$b^{(1)}(m) = m, \quad a(\varphi)b^{(2)}(m) = \sigma^2.$$

This proves Eq. (10.20) for the Gaussian distribution. Because $b^{(1)}(m) = m$, the canonical link is

$$g(x) = (b^{(1)}(x))^{-1} = x.$$

2. The *Poisson distribution* in Eq. (10.15) can be transformed to the exponential form as

$$f(z) = \exp[z \ln\lambda - \lambda - \ln(z!)]. \qquad (10.27)$$

In the notation of Eq. (10.19),

$$\theta = \ln\lambda, \quad a(\varphi) = 1, \quad b(\theta) = \lambda, \quad c(z,\varphi) = -\ln(z!). \qquad (10.28)$$

The logarithm transform equation

$$\theta = \ln\lambda \qquad (10.29)$$

defines the canonical parameter θ as a logarithm of λ. Thus, the canonical link function g is a logarithmic function. It shows that the estimation of λ

can be started from the estimation of θ. If $\hat{\theta}$ is the estimate of θ, then the estimate of λ is the inverse of the canonical link function with the estimate of λ in the form

$$\hat{\lambda} = \exp \hat{\theta}. \tag{10.30}$$

For the Poisson regression estimation this means that

$$\hat{\lambda}(x) = \exp \hat{\theta}(x). \tag{10.31}$$

The parameter λ of the Poisson p.d.f. is nonnegative, $\lambda \geq 0$. A direct estimation of λ or the regression $\lambda(x)$ would require the constrained estimation because the corresponding estimates should be nonnegative. The nonlinear transform in Eq. (10.30) automatically gives nonnegative values for $\hat{\lambda}$ for any $\hat{\theta}$. The unconstrained estimate of the canonical parameter θ enables nonnegativity for the estimate of the parameter λ.

Calculate the derivative $\partial_\theta b(\theta)$ to find the canonical transform from Eq. (10.23):

$$\partial_\theta b(\theta) = \partial_\lambda b(\theta) \cdot \partial_\theta \lambda = \partial_\lambda \lambda \cdot \partial_\theta \exp \theta = \partial_\theta \exp \theta = \exp \theta.$$

Then the canonical link function in Eq. (10.23) is

$$g(x) = (b^{(1)}(x))^{-1} = (\exp(x))^{-1} = \ln x,$$

and according to Eq. (10.24),

$$\hat{m}_z = g^{-1}(\hat{\theta}) = \exp \hat{\theta}.$$

This estimate for the expectation m_z naturally coincides with Eq. (10.31) as $\lambda(x) = m_z(x)$.

3. The exponential form of the *binary Bernoulli distribution* in Eq. (10.13) is

$$f(z) = \exp\left[z \ln \frac{p}{1-p} + \ln(1-p)\right], \qquad z = 0, 1.$$

For the parameters of the canonical form in Eq. (10.19) this means that

$$\theta = \ln \frac{p}{1-p}, \quad a(\varphi) = 1, \quad b(\theta) = -\ln(1-p) = \ln[1 + \exp(\theta)],$$
$$c(z, \varphi) = 0. \tag{10.32}$$

The first equation in these formulas defines the canonical link function as

$$\theta = \ln \frac{p}{1-p}. \tag{10.33}$$

This function is called the *logit function* (*logit transform*). With the canonical parameter θ instead of p, the distribution $f(z)$ has the form

$$f(z) = \exp[z\theta - \ln(1 + \exp \theta)]. \tag{10.34}$$

Estimation of the parameter p could be started from the estimation of θ. If $\hat{\theta}$ is found, then the estimate of p is the inverse logit transform of $\hat{\theta}$,

$$\hat{p} = \frac{1}{1 + \exp(-\hat{\theta})}. \tag{10.35}$$

For the regression estimation, this means that

$$\hat{p}(x) = \frac{1}{1 + \exp[-\hat{\theta}(x)]}. \tag{10.36}$$

The probability parameter p of the Bernoulli distribution is constrained by the natural conditions

$$0 \leq p \leq 1. \tag{10.37}$$

A direct estimation of p or the regression $p(x)$ from the observation z would require the constrained estimation. The nonlinear logit transform reduces the problem to an unconstrained estimation of the canonical parameter θ. The inverse logit transforms in Eqs. (10.35–10.36) automatically give the estimates satisfying Eq. (10.37).

For the Bernoulli distribution we can verify from Eq. (10.32) that

$$b^{(1)}(\theta) = \frac{1}{1 + \exp(-\theta)}$$

and the canonical link function Eq. (10.23) is

$$g(x) = (b^{(1)}(x))^{-1} = \ln \frac{x}{1-x}.$$

This coincides with Eq. (10.33).

10.2 Robust *M*-Estimation

A general class of nonparametric LPA robust M-estimates is defined by formulas (10.4)–(10.6). The *median* is an estimate from this class corresponding to $\psi(x) = |x|$ and the zero-order LPA in Eqs. (10.4–10.6). It is a very popular and well-studied estimator.

We consider the *median* and its generalization, the *weighted median,* and we do so in the general context of the robust M-estimation. It is well known that the *sliding median* estimate with respect to additive Gaussian noise is nearly as good as the sliding sample mean, while the median estimate demonstrates a good resistance to random non-Gaussian noise.

Another important feature of the median estimates is their ability to preserve edges and discontinuities in curves, which are usually smoothed by linear filtering. These properties, as well as some advantages of implementation, explain the great interest in median filters, especially those used in image, speech, and communication signal processing [9].

First, we demonstrate that the ICI rule can be used with median estimators. In this way the median becomes a varying-window-size adaptive estimator. Second, a general class of nonlinear l_1-norm estimators can be interpreted as a generalization of this adaptive median.

10.2.1 Median estimates

Suppose we are given noisy observations of a univariate signal y with a unit sampling period $\Delta = 1$:

$$z(X_s) = y(X_s) + \varepsilon_s, \qquad X_s \in \mathbb{Z}, \tag{10.38}$$

where ε_s are i.i.d. random errors with $E(\varepsilon_s) = 0,\ E(\varepsilon_s^2) = \sigma^2$. The noise in Eq. (10.38) is non-Gaussian.

The underlying signal y belongs to a nonparametric class of piecewise continuous differentiable functions. The goal is to reconstruct y from the observations in Eq. (10.38) in such a way that certain desirable features such as jumps or instantaneous slope changes are preserved.

We consider median-based filters unusual in two important aspects [107]:

- The signal estimate is determined as a linear combination of the medians calculated in symmetric and nonsymmetric left/right windows;
- The window sizes of the medians are pointwise varying and adaptive. The ICI rule is used for a data-driven window size selection.

The absolute error (modulus) loss function can be applied to derive the *weighted median* filter as a solution of the optimization problem.

This optimization can be given as

$$\begin{aligned} \hat{y}_h(x) &= \arg\min_C \ J_h(x, C), \\ J_h(x, C) &= \sum_s w_h(X_s)|z(x - X_s) - C|, \qquad x \in \mathbb{Z}, \end{aligned} \tag{10.39}$$

where optimization on a real C gives the weighted sliding median $\hat{y}_h$ as an estimate of $y(x)$. For integer x, $X_s \in \mathbb{Z}$, this criterion can be rewritten in the equivalent form

$$J_h(x, C) = \sum_s w_h(x - X_s)|z_s - C|. \tag{10.40}$$

Actually, Eqs. (10.39–10.40) provide an accurate definition of the weighted median concept. Recently this concept has been generalized to weights w_h that are not necessarily positive. It is shown that estimates derived in this way demonstrate advanced performance [8], [205].

The scale parameter h determines the estimate window size. For symmetric and nonsymmetric left/right rectangular windows (invariant weights of the residuals in Eqs. (10.39–10.40), $w(x) = 1$ for $|x| \leq 1/2$, $0 \leq x \leq 1$, and $-1 \leq x \leq 0$, respectively. Then, the symmetric and nonsymmetric left/right weighted medians become the corresponding usual median estimates having the forms

$$\hat{y}_h(x) = \begin{cases} median[z(x - \lfloor h/2 \rfloor), \ldots, z(x), & \\ \quad \ldots, z(x + \lfloor h/2 \rfloor)], & \text{symmetric median } (S), \\ median[z(x - \lfloor h \rfloor), \ldots, z(x)], & \text{nonsymmetric left median } (L), \\ median[z(x), \ldots, z(x + \lfloor h \rfloor)], & \text{nonsymmetric right median } (R), \end{cases} \quad (10.41)$$

where $\lfloor \cdot \rfloor$ stands for the integer part of number.

Comparing Eq. (10.39) with Eq. (10.5), we note that the median can be treated as a particular case of the general M-estimate provided that the loss function is

$$\psi(x) = |x| \quad (10.42)$$

and the zero-power polynomial is used as for $y_h(x, X_s) = C$.

Let $z_{(r)}$ be ordered observations $z(x - X_s)$ from Eq. (10.39):

$$z_{(1)} \leq z_{(2)} \leq z_{(3)} \leq \cdots. \quad (10.43)$$

Then the following computation of the weighted median is efficient for any weights $w_h > 0$:

1. Introduce the normalized weights

$$v_s = \frac{w_h(X_s)}{\sum_k w_h(X_k)} \quad (10.44)$$

and order these weights according to the ordered sequence in Eq. (10.43) as

$$v_{(1)} \leq v_{(2)} \leq v_{(3)} \cdots. \quad (10.45)$$

This ordering means that if $z(x - X_s)$ from Eq. (10.39) takes the place r' in Eq. (10.43), the corresponding normalized weight v_s takes the place r' in Eq. (10.45).

2. Consider the sum $S_q = \sum_{r=1}^{q} v_{(r)}$ and find the minimal q from the condition $S_q \geq 1/2$, that is,

$$\hat{q} = \min\{q : S_q \geq 1/2\}.$$

The weighted median is given as

$$\hat{y}_h = z_{(\hat{q})}. \quad (10.46)$$

The weighted median algorithm in Eqs. (10.44–10.46) is valid for any integer and noninteger weights w_h. It is different from the standard calculations having for Eq. (10.39) the form (e.g., [9])

$$\hat{q} = median\{\ldots, w_h(-1)\Diamond z(x+1), w_h(0)\Diamond z(x), w(1)\Diamond z(x-1), \ldots\}, \quad (10.47)$$

where the weights $w_h(s) = w_h(X_s)$ are integer positive and $\diamondsuit$ is a replication operator, for example,

$$w_h(0) \diamondsuit z(x) = \overbrace{z(x), \ldots, z(x)}^{w_h(0)\text{ times}}.$$

Any rational weights w_h in Eq. (10.39) can be equivalently replaced by integers and we can always use Eq. (10.47), but these calculations are different from the compact form in Eqs. (10.44–10.46).

The window size h is crucial in the efficiency of the median estimators for noisy observations. When the window size is relatively small, the estimator $\hat{y}_h$ gives a good approximation of a smooth y and the estimation error has a small bias, but then fewer data are used and $\hat{y}_h$ is more variable and sensitive to the noise. The best choice of h involves a trade-off between the bias and the variance exactly as it is for the sample mean or general linear estimates. This trade-off depends on the sampling period, the standard deviation of the noise, and the smoothness (derivatives) of the signal y.

Figure 10.5 illustrates the smoothing effects of the median filter for a stepwise signal and a symmetric median. The median estimate exactly delineates the signal without noise ($\sigma = 0$). The smoothing effects appear when noise is added to the signal. The curves in Fig. 10.5(a) show the expectation of the median estimate to demonstrate its bias. The noise is zero-mean Gaussian $N(0, \sigma^2)$. The smoothing effect increases as the standard deviation of the noise becomes larger. The larger h (larger number of samples in the median) corresponds to a stronger smoothing, as is illustrated in Fig. 10.5(b).

There exist a number of proposals for median filters with adaptive window lengths (e.g., [90], [143], [218]) based on changes in distance between the mean and the median estimates, edge detection, local entropy, detection of a nonstationary behavior of the signal analyzing its temporal (spatial) correlation, and so on. The adaptive medians using the ICI rule originated in [107]. These nonlinear adaptive estimates, which are similar to their linear adaptive counterparts, are based on the paradigm of direct adaptive accuracy optimization.

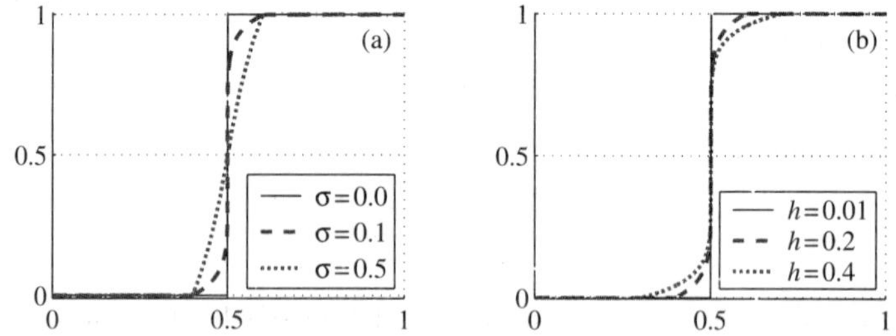

Figure 10.5 Smoothing effects of the median filter for noisy signals. A stepwise signal with additive Gaussian noise is estimated by the symmetric median. The curves show the expectation of the median estimate calculated for: (a) $h = 0.2$ and $\sigma = 0, 0.1, 0.5$; and (b) $\sigma = 0.1$ and $h = 0.01, 0.2, 0.4$. Larger values of σ or h result in stronger smoothing.

10.2.2 Performance of M-estimates

It is convenient to present the accuracy analysis in a form that is more general than is necessary for the median estimation. This gives the results for a general class of M-estimates.

Let the estimate be defined as a solution of the following optimization problem:

$$\hat{y}_h(x) = \arg\min_C J_h(x, C), \qquad J_h(x, C) = \sum_s w_h(x - X_s)\psi(z_s - C). \quad (10.48)$$

The necessary minimum condition for Eq. (10.48) has the form

$$\sum_s w_h(x - X_s)\psi^{(1)}(z_s - C) = 0. \quad (10.49)$$

The weighted median in Eq. (10.39) is a special case of this class of estimates with $\psi(x) = |x|$.

The following proposition can be extracted from the M-estimate accuracy analysis given in ([223], Theorem V, and [46], Sec. 5.5.1). Note that the results given in [223] and [46] are obtained provided that the argument X_s in observations $z_s = z(X_s)$ is random with a given probability density. We adapt these results for a periodic sampling with the sampling interval Δ.

We make the following assumptions:

($N1$) The loss function ψ is convex and symmetric,

$$\psi(x) = \psi(-x), \quad (10.50)$$

and almost everywhere there exist two derivatives $\psi^{(1)}(x) = \partial_x\psi(x)$ and $\psi^{(2)}(x) = \partial_x^2\psi(x)$.

($N2$) F and $f = \partial_x F$ are the symmetric distribution and probability density of the noise ε,

$$f(x) = f(-x), \qquad F(x) = 1 - F(-x). \quad (10.51)$$

($N3$) The following moments of ψ are finite and positive:

$$\begin{aligned} 0 < E\{(\psi^{(1)})^2\} &= \int_{\mathbb{R}} [\psi^{(1)}(x)]^2 dF(x) < \infty, \\ 0 < E\{\psi^{(2)}\} &= \int_{\mathbb{R}} \psi^{(2)}(x) dF(x) < \infty. \end{aligned} \quad (10.52)$$

Note that it follows from the symmetry of the loss function and the noisy distribution that

$$E\{\psi\} = \int_{\mathbb{R}} \psi^{(1)}(x) dF(x) = 0. \quad (10.53)$$

Denote

$$V(\psi, F) = \frac{E\{(\psi^{(1)})^2\}}{(E\{\psi^{(2)}\})^2}, \quad (10.54)$$

where the arguments ψ and F of V emphasize that V depends on the loss function ψ and the noise distribution F.

($N4$) The signal y is continuous and has the first two derivatives bounded at the argument value x.

Proposition 13 1. *Let the sampling interval Δ and the scale parameter h be small such that Δ, $h \to 0$ and $h/\Delta \to \infty$. Then the error of the M-estimate in Eq. (10.48) can be represented in the form*

$$e_y(x,h) = y(x) - \hat{y}_h(x) = m_{\hat{y}_h}(x,h) + e_y^0(x,h), \tag{10.55}$$

where $e_y^0(x,h)$ is random with

$$\begin{aligned} &E\{e_y^0(x,h)\} = 0, \qquad \sigma^2_{\hat{y}_h}(x,h) = V\frac{\Delta}{h}B + o\left(\frac{\Delta}{h}\right), \\ &B = \int_{\mathbb{R}} w^2(u)du, \end{aligned} \tag{10.56}$$

and $m_{\hat{y}_h}(x,h)$ is a systematic error of estimation,

$$m_{\hat{y}_h}(x,h) = \begin{cases} -\frac{1}{2}h^2 y^{(2)}(x)\int_{\mathbb{R}} w(u)u^2du + o(h^2), & \text{if } \int_{\mathbb{R}} w(u)udu = 0, \\ hy^{(1)}(x)\int_{\mathbb{R}} w(u)udu + o(h), & \text{if } \int_{\mathbb{R}} w(u)udu \neq 0. \end{cases} \tag{10.57}$$

2. *Provided some unrestrictive additional assumptions, the estimation error in Eq. (10.55) is asymptotically Gaussian with mean $m_{\hat{y}_h}(x,h)$ and the variance $\sigma^2_{\hat{y}_h}(x,h)$:*

$$e_y(x,h) \sim \mathcal{N}\left(m_{\hat{y}_h}(x,h), V\frac{\Delta}{h}B\right). \tag{10.58}$$

The proof of the proposition is outlined in Sec. 14.8.1.

Comments on Proposition 13.

1. The assumptions of the proposition mean that the window size h and the sampling period Δ should be small, while the number of observations $\lfloor h/\Delta \rfloor$ in the window is large. These assumptions, as is clear from Eqs. (10.56) and (10.57), enable the estimation errors to be small.

   The estimation variance $Vb\Delta/h$ is small when the number of observations proportional to h/Δ is large. The estimation bias is small when $|h^2y^{(2)}(x)|$ or $|hy^{(1)}(x)|$ is small, that is, when the window size h is small.

   Two formulas for the bias are used in Eq. (10.57). If $w(u) = w(-u)$, then $\int w(u)udu = 0$ and the bias is defined by the first line formula in Eq. (10.57), while if the $w(u)$ is nonsymmetric, then the bias is defined by the second formula in Eq. (10.57). It is assumed in the proposition formulas that

$$\int_{\mathbb{R}} w(u)du = 1.$$

2. For the median we have

$$\psi(x) = |x|, \qquad \psi^{(1)}(x) = sign(x), \qquad \psi^{(2)}(x) = 2\delta(x).$$

It follows then that

$$E\{(\psi^{(1)})^2\} = \int_{\mathbb{R}} [\psi^{(1)}(x)]^2 dF(x) = \int_{\mathbb{R}} dF(x) = 1,$$

$$E\{\psi^{(2)}\} = \int_{\mathbb{R}} 2\delta(x) dF(x) = 2f(0),$$

and we have for $V(\psi, F)$ in Eq. (10.54) that

$$V(\psi, F) = \frac{1}{4f^2(0)}. \tag{10.59}$$

For the Gaussian noise $f(0) = 1/(\sigma\sqrt{2\pi})$, and then we obtain from Eq. (10.59) that

$$V(\psi, F) = \frac{\pi}{2}\sigma^2. \tag{10.60}$$

3. The proposition is valid for any symmetric residual function ψ and general assumptions on the probability density of the noise, which can be different from the Gaussian. Note that the loss function ψ and the probability distribution F (density f) influence the value of the parameter V only.
4. For the linear estimates, $\psi = x^2/2$ and $V = \sigma^2$. Then we arrive at the known results for the linear sliding mean estimate with $g(x) = w(x)/\int_{\mathbb{R}} w(x)dx = w(x)$ as $\int_{\mathbb{R}} w(x)dx = 1$.

   If the window is symmetric, then this $g(x)$ corresponds to an LPA of order $m = 1$. If the window is nonsymmetric, then $g(x)$ corresponds to the order $m = 0$.

   It is also assumed in Proposition 13 that the signal is twice differentiable. This means that $M = m + 1$ in Proposition 7.

   Compare the accuracy formulas (10.56)–(10.57) with the corresponding results given in Proposition 7 for the linear LPA estimate.

   If $\int w(u)udu = 0$, then the first formula in Eq. (10.57) coincides with Eq. (5.11) for $M = 2$. If $\int w(u)udu \neq 0$, then the second formula in Eq. (10.57) coincides with Eq. (5.11) for $M = 1$.

   Because $V = \sigma^2$ and $w(x) = g(x)$, the formulas for the variances in Eqs. (10.56) and (5.13) for $d = 1$ coincide. Thus, the results for the linear and nonlinear M-estimates are identical within the scalar factor V, which is equal to σ^2 for the linear case and can take different values for the nonlinear estimates.
5. The accuracy results in Eqs. (10.56–10.57) can be exploited for the scale parameter optimization as has been done for the linear estimates. We can use the corresponding formulas from Sec. 5.2 for the general nonlinear case provided that the noise variance is replaced by V.

   The reader will want to remember these results because we are going to use them for the adaptive-scale selection in median (nonlinear) estimation.

Consider the pointwise MSE and the ideal window size selection. It follows from Proposition 13 that the MSE depends on the location x of the estimator and has the form:

$$E[e_y(x,h)]^2 \simeq [a_q h^q y^{(q)}(x)]^2 + VB\Delta/h, \tag{10.61}$$
$$B = \int_{\mathbb{R}} w^2(u)du, \; a_q = \frac{1}{q!}\int_{\mathbb{R}} w(u)u^q du,$$

where $q = 2$ for the symmetric window w and $q = 1$ otherwise. Minimizing the right-hand side of Eq. (10.61) on h and using formulas Eqs. (5.19–5.21), we get the ideal window size $h^*(x)$:

$$h^*(x) = \left\{\Delta \frac{VB}{a_q^2[y^{(q)}(x)]^2} \frac{1}{2q}\right\}^{1/(2q+1)}. \tag{10.62}$$

The ideal window size $h^*(x)$ depends on the smoothness of the signal as it is defined by the derivative $y^{(q)}(x)$. This window size should be varying if the derivative of the signal is varying.

For $h = h^*(x)$, the ratio of the asymptotic bias to the asymptotic standard deviation $\sigma_{\hat{y}_h}(x,h)$ is a constant that does not depend on the derivative $y^{(q)}(x)$ or on x, and it is given by

$$|m_{\hat{y}_h}(x, h^*(x))|/\sigma_{y_h}(x, h^*(x)) = \gamma, \qquad \gamma = \sqrt{\frac{1}{2q}}, \tag{10.63}$$

$$\sigma_{\hat{y}_h}(x,h) = \sqrt{V\frac{\Delta}{h}\int_{\mathbb{R}} w^2(u)du}.$$

It follows from the formulas for the bias and the standard deviation that

$$|m_{\hat{y}_h}(x,h)| \leq \gamma \cdot \sigma_{\hat{y}_h}(x,h), \qquad \text{for } h \leq h^*(x). \tag{10.64}$$

In general, the equality is achieved in Eq. (10.64) for $h = h^*(x)$ only.

This inequality is used in the ICI rule derivation to test hypotheses $h \lesseqgtr h^*(x)$ exactly as was done for the linear estimates.

10.2.3 Median in Gaussian noise (simulation)

All considered univariate signals are of the length $n = 1024$ and given on the segment $[0, 1]$, $\Delta = 1/(n-1)$. We use the ICI rule for the adaptive varying window size selection as it was presented in Sec. 6.1.2 with the estimate variance $\sigma^2_{\hat{y}_h}(x,h)$ calculated according to formula (10.60). The noise is Gaussian and its standard deviation σ is estimated by Eq. (6.27).

The ICI rule threshold parameter Γ plays an important role in the performance of the algorithm. Too large or too small Γ results in oversmoothing or undersmoothing estimates, respectively. It is treated as a design parameter with a fixed value $\Gamma = 2.5$ in these experiments.

The window sizes are given through the number of observations in the window. A set of the number of observations is defined as

$$N = \{N_j = \lfloor 1.45^j \rfloor, j = 1, \ldots, 17\} \tag{10.65}$$

with $h = N\Delta$.

The asymptotic results of Proposition 13 are valid for a large number of observations in the window. However, the simulation shows that better performance is achieved when small window sizes are included in N. In particular, we use $N_1 = 1$.

The ICI is applied to find the adaptive varying window sizes for the left (N_L^+), right (N_R^+), and symmetric (N_S^+) medians in Eq. (10.41). The estimates of the signal are finalized in three forms of multiple-window estimates (Sec. 6.2):

1. Multiple-window left-right (LR) estimates,
2. Multiple-window left-right-symmetric (LRS) estimates,
3. Combined-window left-right weighted median (W) estimate.

The multiple-window LR and LRS estimates are obtained as linear combinations of the prime L, R, and S adaptive estimates with weights inversely proportional to the variances of the estimates as in Eq. (6.43).

The combined-window left-right weighted median (W-median) is defined in a different way. The adaptive left N_L^+ and right N_R^+ window sizes are found using the left and right median estimators.

The W-median is calculated according to the general formula (10.39), where the following triangle window is used:

$$w_h(X_s) = 1 - \frac{|s|}{1 + \max(N_L^+, N_R^+)}, \quad -N_R^+ \le s \le N_L^+. \tag{10.66}$$

This is a discrete and nonsymmetric version of a more usual triangular window $w(x) = 1 - |x|$, $|x| < 1$. The adaptive varying window Eq. (10.66) is nonsymmetric as, in general, $N_R^+ \neq N_L^+$.

10.2.3.1 Stepwise signal

We consider a stepwise signal: $y(x) = 0, 0 \le x \le 1/2$, and $y(x) = 1, 1/2 < x \le 1$, with the noise standard deviation $\sigma = 0.1$ (Fig. 10.6a). Figs. 10.6(b) and (c) show the adaptive varying window sizes $h_L^+ = \Delta N_L^+, h_R^+ = \Delta N_R^+$ obtained by the ICI rule for the left and right medians. Note that these curves are in complete agreement with what could be selected provided that the signal as well as the jump location are known in advance.

In Fig. 10.6(b), the first part of this curve, $0 \le x \le 1/2$, for the left median is a linear growing function of x. It happens because the left median uses for

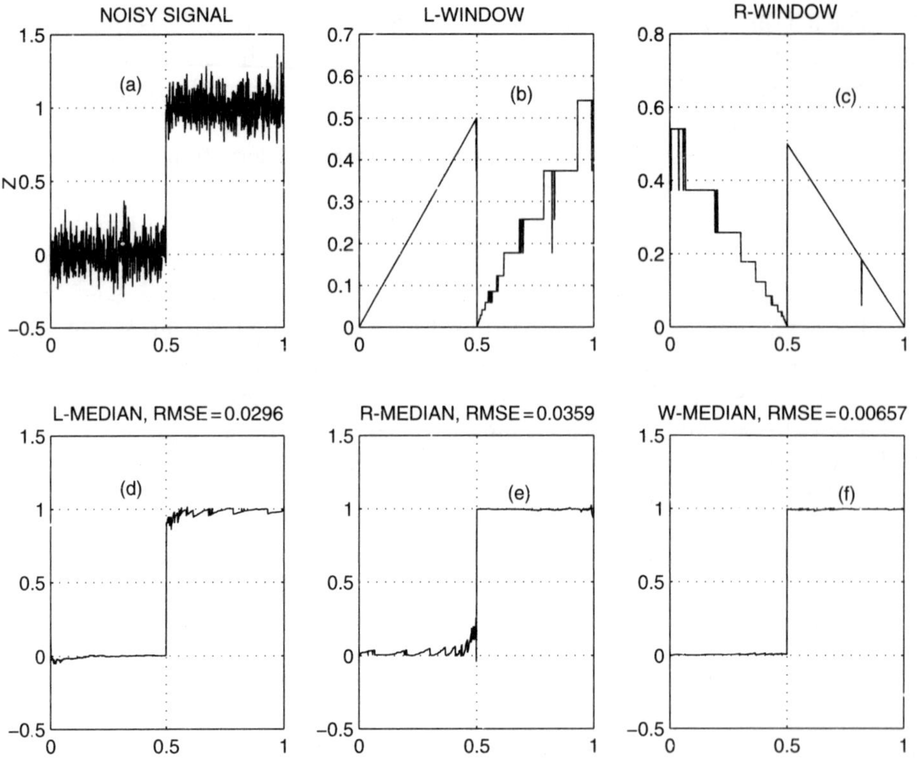

Figure 10.6 (a) Noisy stepwise signal; (b) and (c) left and right ICI adaptive windows; (d)–(f) left, right, and combined-window *W*-median adaptive estimates, respectively.

estimation only the observations that are on the left-hand side of the estimation point x. Immediately after $x = 1/2$, the window size drops to the minimum value as the signal changes from $y = 0$ to $y = 1$ and only a small number of observations carry useful information and can be used for the left-side estimates near $x = 1/2$. While x is increasing toward the value 1, the window sizes are also increasing proportionally since a larger number of observations can be used for estimation. A stepwise form of this adaptive window size is a round-off effect because H consists of the finite set of window sizes.

The left median estimate in Fig. 10.6(d) accurately delineates the jump in the signal from its left-hand side while suppressing the noise perfectly. The performance of this estimate is not as good on the right-hand side of the jump because only a small number of observations can be used by the left median for estimation for values of x that are close to $x = 1/2$. Similar interpretation can be given for the curves in Figs. 10.6(c) and (e), which show the results obtained by using the right-hand side median estimator.

Figure 10.6(f) demonstrates an advantage and a nearly perfect performance of the combined-window W-median estimate. What we see here is very similar to the behavior of the LPA-ICI adaptive estimates considered in Sec. 6.3.

10.2.3.2 Blocks

A performance of the algorithms for more complex piecewise constant *Blocks* is demonstrated in Fig. 10.7. A noisy signal is shown in Fig. 10.7(a). The ICI adaptive left and right median windows shown in Figs. 10.7(b) and (c) demonstrate the reasonably good performance of the ICI.

In Figs. 10.7(d), (e), and (f) the adaptive estimates are shown. The comparison favors the combined-window W-median, which outperforms the multiwindow counterparts in both the lower values of the RMSE as well as visually. It gives a smoother estimate with better noise suppression.

RMSE values for the signal *Blocks* are given in Table 10.1. These RMSE values are calculated as the sample mean of 50 Monte-Carlo simulation runs. The multiple-window estimates LR and LRS essentially show better accuracy than the adaptive prime estimates L, R, and S. The best accuracy is achieved by the combined-window W-median.

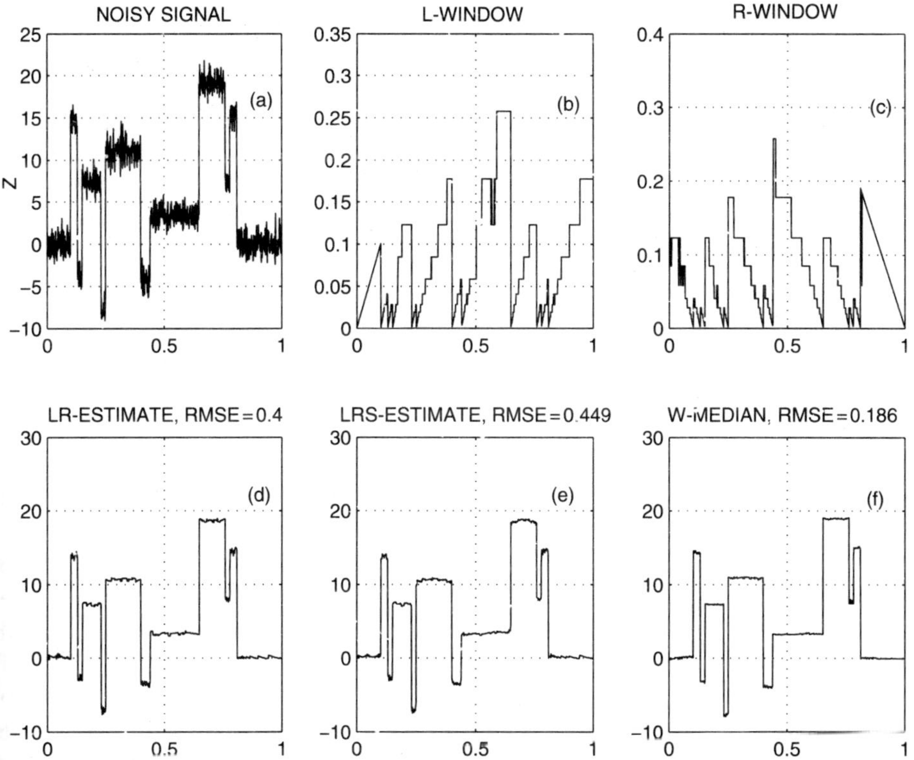

Figure 10.7 (a) Noisy *Blocks* signal; (b) and (c) left and right ICI adaptive windows; (d) and (e) multiwindow left-right and left-right-symmetric adaptive window size estimates; (f) combined-window weighted median with adaptive left and right window sizes.

Table 10.1 RMSE for the ICI adaptive median algorithms and *Blocks* signal (SNR = 7).

Algorithm	L	R	S	LR	LRS	W
RMSE	0.89	0.88	0.55	0.35	0.39	0.20

10.2.4 Theory of minimax estimation*

10.2.4.1 Sensitivity of ML

For many typical applications, observation noise can be modeled by the Gaussian distribution. The corresponding optimal filters are known to be linear and can be found by using the universal ML approach. However, what is also well known is that if the noise becomes non-Gaussian, these linear estimates degrade quickly, and the nonlinear estimation is a valuable alternative.

Median filters comprise a class of simple tools that can be efficient on many occasions. However, if we wish to have optimal estimates, the ML again can be used and gives nonlinear estimates that are optimal for a given noise distribution.

When dealing with optimization, we can always ask how sensitive the optimization result will be with respect to the exploited hypothesis. A sensitivity analysis is important because optimization models are approximations of reality and small or large variations are unavoidable. If the optimal solution is very sensitive to hypothesis variations, then these types of results will have little practical use.

Let us say that in the observations in Eq. (10.1) the noise ε mainly is Gaussian. However, from time to time non-Gaussian ε also can happen. Such observations, which do not satisfy some basic noise distribution, are called *outliers* in statistics. It is assumed that outliers are rare, and the following mix distribution model is traditional:

$$f(x) = (1-\alpha) f_0(x) + \alpha f_1(x), \qquad 0 < \alpha < 1. \tag{10.67}$$

Here f_0 and f_1 are the probability densities of the basic and the outlier. We model the noise distribution as an additive mix of two distributions with the parameter α which is the probability that the outlier occurs. This parameter α defines the level (proportion) of contamination of the basic random errors by outliers.

The sensitivity of the estimates is analyzed assuming that the contamination parameter α is small. One says that the estimate is sensitive or insensitive to the disturbance distribution f_1 if the estimate degradation caused by f_1 is large or small.

Actually, Eq. (10.67) is a standard model for analyzing the sensitivity of estimates. A special class of contamination distributions f_1 is given by the so-called heavy-tailed distribution. A distribution is a heavy-tailed one if the ratio $|f_1(x)/f_0(x)| \gg 1$ as $|x| \to \infty$. In other words, the probability for a random variable to take a large value is much higher for a heavy-tailed distribution $f_1(x)$ than it is for a basic $f_0(x)$.

Heavy-tailed distributions are typical models of impulsive noises or outliers, that is, random errors that occur rarely but with a large magnitude. Linear estimates are known to be quite sensitive with respect to the heavy-tailed distribution contamination. Even a small number of outliers can destroy a linear estimate completely.

Impulsive noise requires sophisticated techniques that can handle additional information about the noise distribution. However, use of the standard ML is not desirable because it can result in estimates that can again be sensitive with respect to the noise distribution.

Many books and papers have been published in this area concerning the practical and theoretical aspects of the problem as well as supplying conceptually different models and approaches (e.g., [6], [64], [86], [100], [114], [166], [194], [204]).

The minimax Huber's estimation theory [86] appears to be a fruitful and clear alternative to the standard ML, where the noise distribution is replaced by the concept of a class of distributions. It was shown that using a class instead of a specific noise distribution allows one to design estimates that are characterized by decreased sensitivity with respect to unknown variations of the distribution functions. The term "robust" estimation, which is actually quite overused in science, in the considered approach, has a definite meaning for estimates that insensitive with respect to variations in noise distribution.

10.2.4.2 Minimax estimation

Let us discuss an optimal selection of the loss function ψ in the M-estimation criterion that highlights links of the proposed methods with the minimax robust estimation theory [86]. It is shown in Proposition 13 that the M-estimate of y is asymptotically unbiased, the bias approaches zero as $h \to 0$, and the estimate variance has $V(\psi, F)$ in Eq. (10.54) as a factor. Note that this factor is the only term of the variance that depends on the loss function ψ and the noise distribution F. The bias does not depend on ψ and F. Thus, the estimation accuracy depends on the noise distribution F and the loss function ψ only through $V(\psi, F)$. The analysis and optimization of V can be used to agree on a selection of the loss function ψ with the noise distribution F.

Known noise distribution

If the noise distribution F is known and differentiable, then the minimum of

$$V(\psi, F) = \frac{E\{(\psi^{(1)})^2\}}{(E\{\psi^{(2)}\})^2} \tag{10.68}$$

s achieved for $\psi(x) = \psi_0(x)$ [86] by

$$\psi_0(x) = -\ln f(x). \tag{10.69}$$

The minimum value of $V(\psi, F)$ is calculated as

$$V(-\ln(f(x)), F) = 1/I(f), \tag{10.70}$$
$$I(f) = \int_{\mathbb{R}} [f^{(1)}(x)]^2/f(x)dx,$$

where $I(f)$ is *Fisher's information.*

Unknown noise distribution

Let the accurate distribution F be unknown and belong to a certain class $\mathcal{A}$. The worst density $f \in \mathcal{A}$ is naturally the one for which $1/I(f)$ in Eq. (10.70) is maximal or $I(f)$ is minimal. It can be defined as a solution of the problem $\min_{f\in\mathcal{A}} I(f)$:

$$f_0(x) = \arg\min_{f\in\mathcal{A}} I(f). \tag{10.71}$$

The robust Huber's estimator minimax over the density class $\mathcal{A}$ according to Eq. (10.69) is

$$\psi_0(x) = -\log f_0(x). \tag{10.72}$$

Thus, to obtain the minimax robust estimator, first a class $\mathcal{A}$ should be specified, and after that the loss function ψ can be given by Eqs. (10.71–10.72).

Let the pairs (ψ_0, f), (ψ_0, f_0), and (ψ, f_0) of the loss function ψ and densities f be given. Then the following is true for $V(\psi, f)$:

$$V(\psi_0, f) \leq V(\psi_0, f_0) \leq V(\psi, f_0),$$
$$V(\psi_0, f_0) = 1/I(f_0). \tag{10.73}$$

These inequalities hold provided we have some general assumptions about (ψ, f) ([86]; see also Theorem 4 in [185]). Formula (10.73) presents a basic idea and the mathematical results of the minimax approach: The best residual function ψ_0 taken according to Eqs. (10.71–10.72) minimizes the variance of the estimates with respect to ψ, while the worst density f_0 maximizes the variance, which has the corresponding $V(\psi, f)$ value as a factor.

Finally, the minimax optimization over the class $\mathcal{A}$ is formulated as

$$(\psi_0, F_0) = \arg\left(\min_{\psi}\max_{F\in\mathcal{A}} V(\psi, F)\right). \tag{10.74}$$

Here minimization on ψ says that the loss function $\psi_0(x)$ is selected as the best, giving the minimal variance of $\max_{F\in\mathcal{A}} V(\psi, F)$, while the maximization says that the noise distribution F is selected as the worst from class $\mathcal{A}$. The minimax approach selects the best loss function $\psi_0(x)$ for the worst noise distribution $F_0(x)$ from the class $\mathcal{A}$.

It is shown that the order of maximization and minimization in Eq. (10.74) can be changed provided we have some unrestrictive assumptions:

$$(\psi_0, F_0) = \arg\left(\max_{F\in\mathcal{A}}\min_{\psi} V(\psi, F)\right). \tag{10.75}$$

The robustness property of the minimax loss functions $\psi_0(x)$ can be achieved for some special distribution classes $\mathcal{A}$.

More about these classes of robust distributions, optimal loss functions ψ_0, techniques used for the solution of the minimax problem in Eq. (10.74), as well as more details about the minimax robustness concept can be found in [64], [86], [114], [185], and [204].

10.2.4.3 Some minimax solutions

Let us present some classes $\mathcal{A}$ that enable the estimate to be robust and the corresponding minimax loss functions ψ_0:

1. Class of approximate Gaussian distributions [86]:

$$\mathcal{A}_1 = \{f(x) : f(x) = (1-\alpha)f_0(x) + \alpha f_1(x),\ 0 < \alpha < 1\}, \qquad (10.76)$$

where $f_0(x) \in \mathcal{N}(0, \sigma^2)$ and $f_1(x)$ is an arbitrary density with $\int_{\mathbb{R}} x f_1(x)dx = 0$. The approximate Gaussian means that for small α the distribution f is nearly Gaussian while the disturbance f_1 is arbitrary. The minimax ψ_0 is as follows [86]:

$$\psi_0(x) = \begin{cases} \frac{1}{2}|x|^2, & |x| \le \delta, \\ \delta|x| - \delta^2/2, & |x| \ge \delta, \end{cases} \qquad (10.77)$$

where $\delta > 0$ is a parameter that can be found from

$$\frac{1}{1-\alpha} = \int_{-\delta}^{\delta} f_0(x)dx + 2f_0\left(\frac{\delta}{\sigma}\right)\frac{\sigma}{\delta}. \qquad (10.78)$$

2. Class of nonsingular distributions [185]:

$$\mathcal{A}_2 = \{f(x) : f(0) \ge \frac{1}{2a} > 0, a > 0\}. \qquad (10.79)$$

The minimax ψ_0 is found as

$$\psi_0(x) = |x|. \qquad (10.80)$$

3. Class of approximate exponential distributions [185]:

$$\mathcal{A}_3 = \{f(x) = (1-\alpha)f_0(x) + \alpha f_1(x),\ 0 < \alpha < 1\}, \qquad (10.81)$$

where

$$f_0(x) = \frac{1}{2a}\exp\left(-\frac{|x|}{a}\right) \qquad (10.82)$$

is the Laplace distribution and $f_1(x)$ is an arbitrary distribution $\int_{\mathbb{R}} x f_1(x)dx = 0$. Then the minimax ψ_0 is defined by Eq. (10.80).

The first example in Eq. (10.76) with $\psi_0(x)$ of the form in Eq. (10.77) shows that even a small disturbance of the Gaussian distribution by an arbitrary f_1 results in a serious modification of the quadratic $\psi(x) = x^2$, which is the best loss function for the Gaussian f_0.

The linear-quadratic loss function in Eq. (10.77) is quadratic for small x and linear for x having absolute values larger than δ. It decreases the influence of large errors in estimation and in this way makes the estimates robust with respect to impulse noise. Note that δ is an important parameter of this loss function that has to be adjusted. Equation (10.78) gives the optimal solution for δ.

A similar disturbance of the Laplace distribution by an arbitrary distribution f_1 in Eq. (10.81) gives as a result the module $\psi_0(x) = |x|$, which is the best ML loss function for the Laplacian distribution f_0. The Laplacian distribution is robust with respect to a small disturbance because both the basic Laplacian distribution and its contaminated version in Eq. (10.81) give the same optimal loss function as in Eq. (10.80). A small contamination of the Laplacian distribution does not change the form of the corresponding optimal loss function.

The distribution in Eq. (10.79) shows that the module function in Eq. (10.80) is also optimal provided that nothing is known about the distribution except that $f(0) > 0$.

The module loss function in Eq. (10.80) results in median estimates (see Sec. 10.2.1). Thus, the median gives estimates that are robust in the minimax sense for the Laplacian distribution as well as for any nonsingular distribution.

In conclusion, we emphasize the following important fact about robust estimations on the whole: Nonlinear robust estimates outperform the conventional linear ones in heavy-tailed error distribution environment in a very impressive way. At the same time, for the Gaussian distribution, the advantage of the quadratic loss function, which is optimal in this case, is not strong.

Example

Let us compare the least-squared (LS) estimates versus the mean absolute error (MAE) estimates, that is, $\psi = x^2$ versus $\psi = |x|$, provided that errors have the following three types of distribution: Gaussian, Laplace, and Cauchy. The p.d.f. for the last distribution is

$$f(x) = \frac{a}{\pi[1 + (ax)^2]}, \qquad a > 0. \tag{10.83}$$

The columns in Table 10.2 are given respectively for the ML estimates (with loss function (10.69)), for the MAE estimates (with $\psi = |x|$), and for the mean-square error (MSE) estimates (with $\psi = x^2/2$). The relative accuracy of the MAE and MSE with respect to the ML are determined by the ratios $\sqrt{\text{MAE}/\text{ML}}$ and $\sqrt{\text{MSE}/\text{ML}}$ and given in the corresponding columns of the table. The least $V(\psi, F)$ in every row are given naturally by the ML.

For Cauchy noise, the relative loss of the efficiency of the MAE versus the ML is equal to $\sqrt{\pi^2/2} \simeq 2.22$, while the relative loss of the efficiency of the MSE

Table 10.2 $V(\psi, F)$ values and comparative efficiencies of the ML, MAE, and MSE estimates.

Distribution\Estimation	ML	MAE	$\sqrt{\frac{MAE}{ML}}$	MSE	$\sqrt{\frac{MSE}{ML}}$
Cauchy	$\frac{1}{2a^2}$	$\frac{\pi^2}{4a^2}$	$\sqrt{\frac{\pi^2}{2}}$	∞	∞
Laplace	a^2	a^2	1	$2a^2$	$\sqrt{2}$
Gaussian	σ^2	$\sigma^2\frac{\pi}{2}$	$\sqrt{\frac{\pi}{2}}$	σ^2	1

versus the ML is infinite. Remember that the Cauchy distribution has very heavy tails and even an expectation does not exist for this distribution.

The formulas given in the row for the Gaussian noise show that the efficiency loss when we use the MAE instead of the MSE is equal to $\sqrt{\pi/2} \simeq 1.25$. Thus the MAE works in an effective way for the difficult Cauchy distribution and the efficiency loss is not large, even when we use the MAE instead of the MSE for the Gaussian noise.

Table 10.2 illustrates the following important facts about minimax robust estimation: The robust estimates outperform the conventional (MSE) ones obtained from the quadratic loss function in an environment of heavy-tailed error distributions in a very impressive way. At the same time, the advantage of the optimal ML estimates is not strong versus the MAE estimates [64], [185].

The curves in Fig. 10.8 demonstrate examples of the robust loss functions ψ and their first derivatives $\psi^{(1)}$. Some are derived as solutions of the minimax problem and thus are optimal for some classes of the distribution, while others are proposed from heuristic speculations. A major property of this type of robust loss function is decreasing values of the large residuals.

In the quadratic loss function $\psi(x) = x^2$, larger residuals dominate in the estimate because $|x_1| > |x_2|$ means that $\psi(x_1) > \psi(x_2)$ and this domination is quadratic, $|x_1|^2 > |x_2|^2$. For the absolute error loss function $\psi(x) = |x|$, this domination is only linear because we have for this loss function $|x_1| > |x_2|$.

Assuming that very large residuals are caused by outliers, we can reduce the weight of these large residuals in estimates by selecting as $\psi(x)$ the loss function that decreases starting from some values of x.

In Fig. 10.8(I) we can see the absolute error loss function. The stepwise constant derivative $\psi^{(1)}(x) = sign(x)$ shows that all nonzero residuals are treated as having equal influence on the estimate ($|sign(x)| = 1, x \neq 0$). Figure 10.8(II) shows the loss function nulling the residuals with absolute values larger than δ. For this loss function $|\psi^{(1)}(x)| = 0$ for $|x| > \delta$. Figure 10.8(III) shows the truncated absolute value ψ when the residuals larger than δ are equal. In this way the observations with these large residuals are dropped from the estimate and their influence is eliminated.

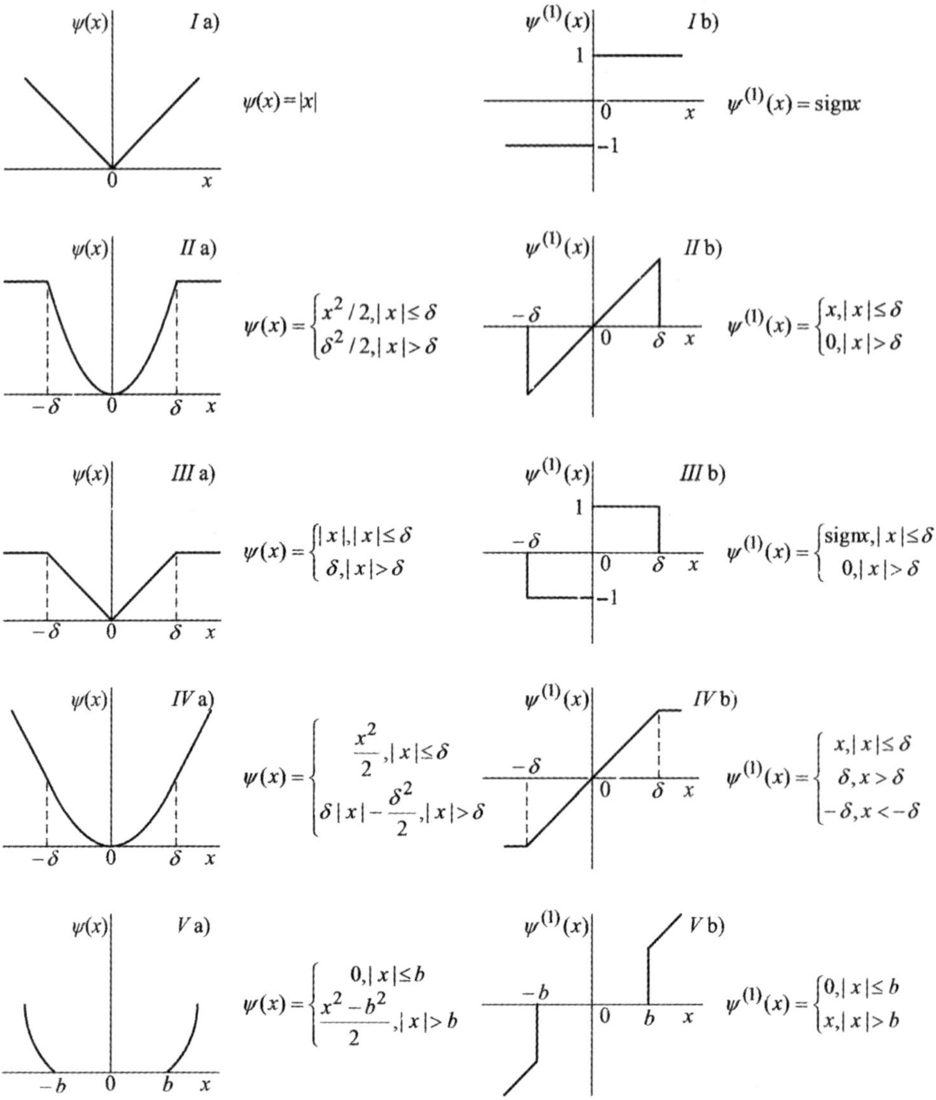

Figure 10.8 Examples of robust loss functions ψ and their first derivatives $\psi^{(1)}$.

Figure 10.8(IV) shows Huber's optimal loss function in Eq. (10.77), which is quadratic for $|x| < \delta$ and linear for $|x| \geq \delta$. The quadratic residuals for small residuals enable good estimation accuracy for observations with small errors while observations with large errors have decreased influence on the estimate because $|\psi^{(1)}|$ is invariant for all $|x| \geq \delta$.

The loss function with a zero zone around $x = 0$ [Fig. 10.8(V)] is used to stop an estimation procedure when absolute values of the residuals are smaller than b. This event is treated as a sign that the desired estimation accuracy has been achieved.

10.2.5 Median in non-Gaussian noise (simulation)

The estimated signals and algorithms are the same as in Sec. 10.2.3 except that the noise is non-Gaussian. It is assumed that the noise distribution is a mix of two Gaussian distributions that differ by their variances:

$$f(x) = (1 - \alpha) f_0(x) + \alpha f_1(x), \tag{10.84}$$

where $f_0(x) = \mathcal{N}(0, \sigma_0^2)$, $f_1(x) = \mathcal{N}(0, \sigma_1^2)$, $\sigma_1 \gg \sigma_0$, and $0 \leq \alpha \leq 1$.

In the modeling according to Eq. (10.84), noise values are generated with distributions $\mathcal{N}(0, \sigma_0^2)$ or $\mathcal{N}(0, \sigma_1^2)$. These distributions are selected randomly with probability α for $\mathcal{N}(0, \sigma_1^2)$ and probability $(1 - \alpha)$ for $\mathcal{N}(0, \sigma_0^2)$. The parameter α defines the proportion of high-variance random impulses/outliers in the sequence of these experiments.

The stepwise and *Blocks* noisy signals are shown in Fig. 10.9. The impulse-style components of the noise, which are essentially different from the background Gaussian noise, are clearly identified by their high narrowed peaks.

We use the following parameters of noise in this simulation: For the stepwise signal $\sigma_0 = 0.02$ and $\sigma_1 = 0.2$, and for *Blocks* $\sigma_0 = 1$ and $\sigma_1 = 5$. For both signals $\alpha = 0.05$.

The left and right medians as well as the combined-window W-median are applied for filtering. The result for the stepwise signal can be viewed in the first row of Fig. 10.10. The curves in panels (b) and (c) show the ICI adaptive window sizes for the left (h_L^+) and right (h_R^+) median estimates.

We observe that, overall, these adaptive window sizes are small near and large far from the discontinuous change-points of the signal. This is quite similar to the behavior of the adaptive window sizes shown in Fig. 10.6 for Gaussian noise. However, this correct picture is disturbed severely by multiple erroneous small window sizes.

The source of these errors is obvious. The ICI, being sensitive with respect to sharp variations in the signal, interprets noise peaks as signal change-points and gives corresponding small values for the adaptive window sizes. In this way, the ICI isolates the impulse-noise peaks in the adaptive window sizes h^+. Using these noisy

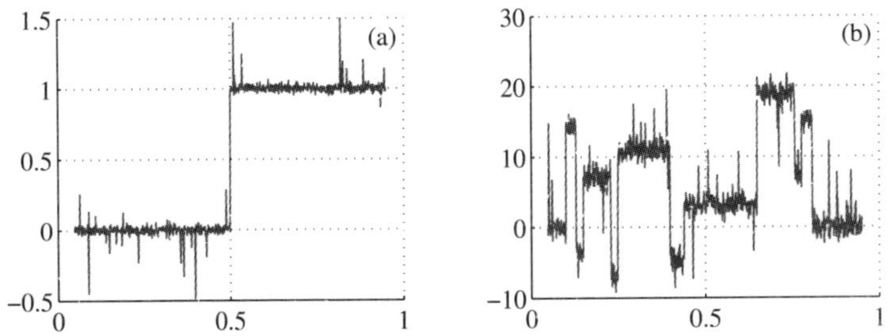

Figure 10.9 Stepwise and *Blocks* signals with impulsive noise.

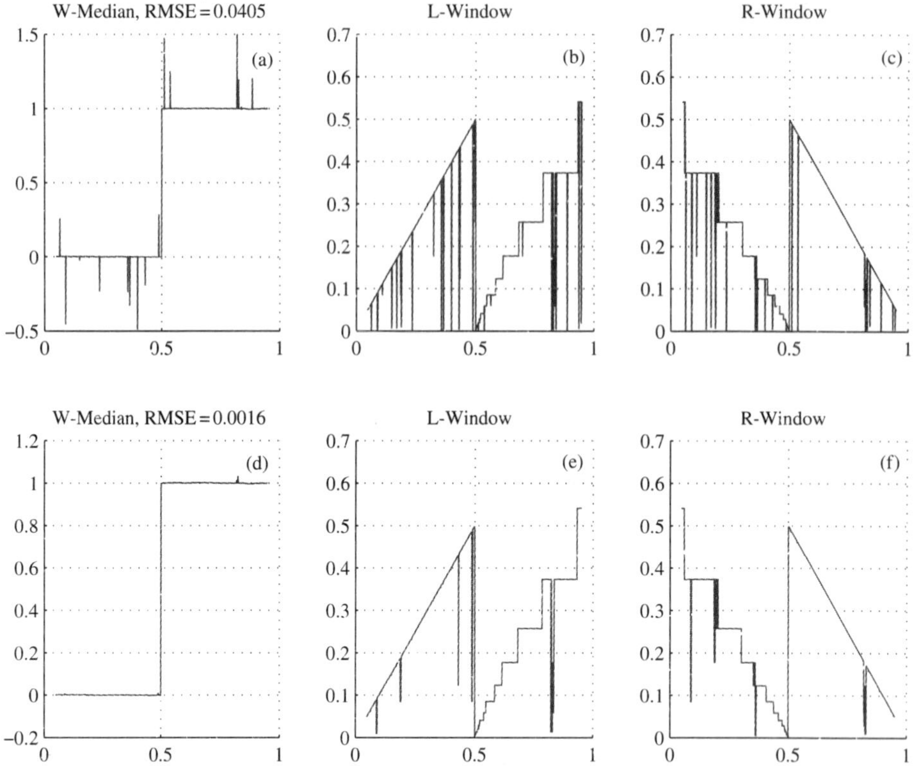

Figure 10.10 The ICI adaptive median estimate for a stepwise signal and impulsive noise observations. First row: (a) combined-window *W*-median estimate without prefiltering; (b) and (c) adaptive window sizes for left and right median estimates. Second row: (d) the combined-window *W*-median estimate with prefiltering of adaptive windows; (e) and (f) obtained by the median prefiltering the ICI adaptive windows (b) and (c).

adaptive window sizes gives the estimate shown in Fig. 10.10(a). Clear erroneous peaks in the estimate correspond to impulsive noise in the observations.

The isolated character of these peaks allows a number of ways to correct them. One is by prefiltering the ICI adaptive window sizes before they are used for estimation. The results of this are shown in the second row of curves in Fig. 10.10. The symmetric median filter of size 5 is used for filtering the adaptive windows h_L^+ and h_R^+. The filtered windows and the corresponding combined-window W-median estimate are shown in Figs. 10.10(e), (f), and (d), respectively. The improvement of the estimates is obvious.

Similar results for *Blocks* can be viewed in Fig. 10.11. In the first row of this figure, the curves are given for the ICI adaptive left/right window sizes in panels (b) and (c), respectively, and for the W-median estimate in panel (a). In general, the behavior of the adaptive window sizes agrees with the results shown for Gaussian noise in Fig. 10.7. However, the correct picture is disturbed by impulsive errors clearly seen in the adaptive window sizes as well as in the estimate of the signal.

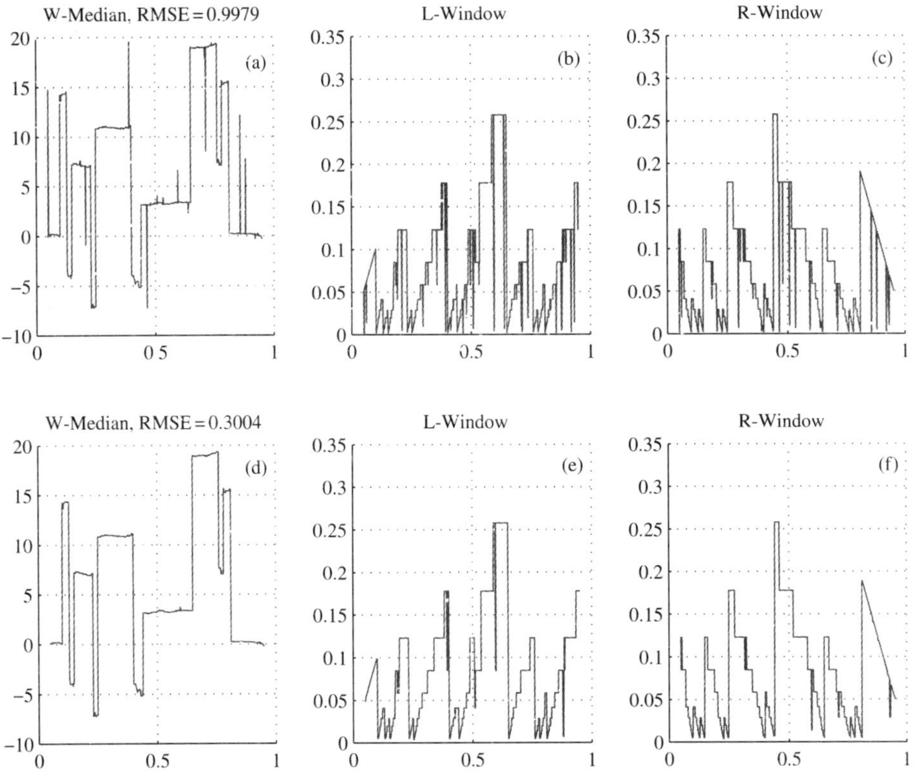

Figure 10.11 The ICI adaptive median estimate for the *Blocks* and impulsive noise observations. First row: (a) combined-window *W*-median estimate without prefiltering; (b) and (c) adaptive window sizes for left and right median estimates. Second row: (d) the combined-window *W*-median estimate with prefiltering of adaptive windows (e) and (f) obtained by median prefiltering of the ICI adaptive windows (b) and (c).

The second row of panels in Fig. 10.11 shows the adaptive window sizes filtered using the symmetric median filter of size 5 and the corresponding combined-window W-median estimate. The improvement of these estimates is clearly seen.

Finally, note that the linear LPA-ICI adaptive estimates are not applicable for impulsive noise. The linear estimates plus ICI do not isolate the effects of impulsive noise and show a complete degradation of the estimates.

10.3 LPA-ICI Robust *M*-Estimates

Exploiting the LPA of the order m for $y_h(x, X_s)$ in the criterion in Eq. (10.5), we arrive at the LPA M-estimates. Here we can use the LPA of any order m as well as apply any window w, in particular, the directional windows considered in Chapter 7. In this way we obtain directional robust M-estimates.

When a vector C is found from Eq. (10.4), the local estimates have the conventional form

$$\hat{y}_h(x) = \hat{C}^T \phi(0), \qquad \hat{y}_h^{(r)}(x) = \left(-\frac{1}{h}\right)^r \hat{C}^T \phi^{(r)}(0). \tag{10.85}$$

The module loss function $\psi(x) = |x|$ corresponds to the considered median estimates if $m = 0$ and defines a more general class of estimates if $m > 0$. These are called the *local polynomial l_1-norm* estimates, where l_1-norm says that the absolute error loss function is the one used in Eq. (10.5). These estimates are a higher-order generalization of the weighted median. We can use in Eq. (10.5) any of the previously discussed loss functions ψ to obtain robust estimates. The robust LPA nonparametric regression estimates originated in papers [29] and [100].

10.3.1 Performance of LPA *M*-estimates

The asymptotic accuracy analysis given in Proposition 13 is extended to the general M-estimates based on the LPA of order m. This extension concerns two points. First, we consider LPA estimates of order $m > 0$, and, second, the estimates are defined for the general multivariate case of dimension d.

The asymptotic formulas for the bias and the variance can be given in forms that are completely similar to those derived for the linear LPA estimates in Chapter 5 (Proposition 7). The only difference between these formulas is that σ^2 in the formulas for the linear estimates is replaced by $V(\psi, F)$ in Eq. (10.54) in the formulas for nonlinear M-estimates. These results are summarized in the following proposition.

Proposition 14 *Let the hypotheses H1–H2 of Proposition 7 hold, $m + 1 \leq \alpha$, and the random errors in the observation model be i.i.d. with probability density f. Assume also that the loss function ψ and the density f satisfy conditions Eqs. (10.50–10.54), and h, Δ, Δ/h, $\Delta^d/h^{d+2|r|} \to 0$. Then,*

1) *The accuracy of the multivariate* M*-estimate in Eqs. (10.4) and (10.85), using the LPA of order m, is defined by the following expressions:*

for the estimation bias,

$$m_{\hat{y}_h^{(r)}}(x, h) = (-1)^m h^{m+1-|r|} \sum_{|k|=m+1} y^{(k)}(x) \frac{1}{k!} \int_{\mathbb{R}^d} g^{(r)}(u) u^k du + o(h^{m+1-|r|}),$$
$$m + 1 > |r|, \tag{10.86}$$

and for the estimation variance,

$$\sigma^2_{\hat{y}_h^{(r)}}(x, h) = V(\psi, F) \frac{\Delta^d}{h^{d+2|r|}} B_{g^{(r)}} + o\left(\frac{\Delta^d}{h^{d+2|r|}}\right),$$
$$B_{g^{(r)}} = \int_{\mathbb{R}^d} \left[g^{(r)}(u)\right]^2 du. \tag{10.87}$$

where the kernels are defined as in Eq. (4.35):

$$g^{(r)}(x) = (-1)^{|r|} w(x)\phi^T(x)\Phi^{-1}\phi^{(r)}(0),$$
$$\Phi = \int_{\mathbb{R}^d} w(v)\phi(v)\phi^T(v)dv. \tag{10.88}$$

2) *Provided some unrestrictive additional assumptions, the estimation error is asymptotically Gaussian with mean value $m_{y^{(r)}}(x, h)$ and variance $\sigma^2_{\hat{y}_h^{(r)}}(x, h)$:*

$$e_{y^{(r)}}(x, h) \sim \mathcal{N}\left(m_{\hat{y}_h^{(r)}}(x, h),\ V(\psi, F)\frac{\Delta^d}{h^{d+2|r|}} B_{g^{(r)}}\right). \tag{10.89}$$

The proof of the proposition can be seen in Sec. 14.8.2 of the Appendix.

Comments on the proposition

1. The similarity with the results given for the linear LPA estimates using the quadratic loss function ψ is surprising, in particular, when we see in formulas (10.86) for the bias and Eq. (10.87) for the variance the kernel function $g_h^{(r)}(x)$ that is natural for linear estimates and quite unnatural for nonlinear ones. Note that the results for the M-estimates in Proposition 14 are given for sufficiently smooth signals y, $\alpha \geq m + 1$, and they can be compared with the results for the linear LPA estimates only under this assumption.
2. The results of the proposition are formulated in a general form that is applicable for estimates of the signal ($r = 0$) as well as for differentiation. The dependence of the variance and the bias on h is the same as for the linear estimates. We have here the bias-variance trade-off, which says that the optimal scale h can be selected to minimize the pointwise MSE. This scale selection is an important tool for the accuracy optimization of M-estimates.
3. The bias error as defined in Eq. (10.86) does not depend on the loss function ψ. It is important to note that, for polynomial signals of power less than or equal to m, the M-estimates are accurate for the noiseless case and asymptotically unbiased for the noisy data. The polynomial reproducing property discussed for the linear estimates has a place for general nonlinear M-estimates.
4. The asymptotic gaussivity of the errors in Eq. (10.89) is important for adaptive estimation. It is exploited in the ICI rule because the standard Gaussian distribution can be used to calculate the quantiles in Eq. (6.5) for any noise distribution in the observations.
5. For the "directional version" of M-estimates using the directional kernels and different order LPA for different directions, the accuracy results can be extracted from Proposition 12 where σ^2 is replaced by V and M_i is replaced by $m_i + 1$.

10.3.2 ICI algorithm for *M*-estimates

The similarity of the accuracy results for linear and nonlinear M-estimates means that all speculations concerning the selection of the ideal and adaptive-scale parameters h from the bias-variance trade-off are completely valid for the nonlinear M-estimates exactly as thay are for the linear ones.

The usual ICI rule can be used for the adaptive-scale selection for nonlinear M-estimates. This adaptive-scale algorithm has the standard structure shown in Fig. 1.4.

However, the calculation of the nonlinear estimates is quite different. They cannot be obtained using the fast convolution algorithms that give estimates for all x simultaneously. Instead, in general, recursive procedures are used for the calculation of $\hat{C}(x, h)$ in Eq. (10.4). This assumes that the calculation is performed in a pointwise manner for each x.

The parameter V in Eq. (10.54) is used in the ICI instead of σ^2 to calculate the corresponding confidence intervals. It can be estimated using the residuals of the estimates of y as follows:

$$\hat{V} = \hat{V}_1/\hat{V}_2^2, \tag{10.90}$$

$$\hat{V}_1 = \frac{1}{\#}\sum_s \{\psi^{(1)}(\hat{e}(X_s))\}^2,$$

$$\hat{V}_2 = \frac{1}{\#}\sum_s \psi^{(2)}(\hat{e}(X_s)), \qquad \hat{e}_s = z_s - \hat{y}_h(X_s),$$

where $\hat{V}_1$ and $\hat{V}_2$ are sample mean estimates of the corresponding expectations $E\{(\psi^{(1)})^2\}$ and $E\{(\psi^{(2)})\}$. The estimation procedure can be thought of as a recursive one. At the first step, the variance of the signal is estimated from the observations, say, using the algorithm in Eq. (6.27). The second step starts with the estimation of the parameter V according to formulas (10.90), where $\hat{y}_h(X_s)$ are the estimates obtained in the first step. The updating of the value of V can be repeated up to the convergence of the estimate.

Good asymptotic properties of the estimates as well as the robustness with respect to the impulsive noise are guaranteed for the M-estimates only for large h, that is, for a large number of observations taken into consideration in the estimate. For small h, the nonlinear M-estimates plus the nonlinear ICI rule can demonstrate a complex behavior.

The simulation results given above for impulsive noise are an example of a simple median filtering of adaptive window sizes doing a good job of improving the results. The LPA M-estimate is a strong tool that is universally applicable in many circumstances. However, it requires special care for small h, which we have to use to preserve the fine details in reconstructed signals.

10.4 Nonlinear Transform Methods

Nonlinear transformations such as logarithms or square roots are often used in statistics to simplify the structure of observations so that they follow a convenient statistical model. These transformations have one or more objectives, including:

a) a simpler input-output relationship;
b) stabilizing a variance, that is, inducing a constant variance in transformed observations provided that the variance of the initial one is varying;
c) inducing a particular type of distribution (usually Gaussian) of random errors in transformed observations.

Transformations can be applied to the dependent y (or z) and/or independent x variables to convert a nonlinear model to a linear regression one. For example, a log transformation of a dependent variable would change an exponential growth model into a linear regression, thereby simplifying its estimation procedure. Because transformations have multiple effects, their use requires care. A good introduction to transformations, including the Box-Cox transformation, can be found in [164].

Data transformations were developed at a time when far fewer tools were available in statistics and signal processing than there are now. Many problems that in the past would have been tackled by data transformation with a linear model are now solved using more sophisticated techniques, for example, generalized linear models, nonlinear models, smoothing, and variance function models. Nonetheless, model simplification is still important and data transformation is still an effective and widely used tool.

10.4.1 Model simplification

10.4.1.1 Nonlinear observations

Suppose that a nonlinear observation model has the form

$$z_s = \chi(y(X_s)) + \varepsilon_s, \tag{10.91}$$

where $\chi(\cdot)$ and y are known and unknown nonlinear functions, respectively. The nonlinearity of the first function χ models the nonlinear indirect observation effects reflecting some physical phenomenon, sensor properties, and so on. We can exploit the basic ideas of the LPA approach to derive a nonparametric estimate in the following straightforward manner.

First, introduce the LPA criterion as weighted residuals between the observations z_s and their nonlinear predictions defined as $\chi(y_h(x, X_s))$:

$$J_h(x, C) = \sum_s w_h(x - X_s)[z_s - \chi(y_h(x, X_s))]^2, \tag{10.92}$$

$$y_h(x, v) = C^T \phi_h(x - v).$$

Second, estimate y naturally in the form

$$\hat{C} = \arg\min_{C} J_h(x, C), \quad (10.93)$$
$$\hat{y}_h(x) = \hat{C}^T \phi(0).$$

We minimize the criterion $J_h(x, C)$ on C to find the pointwise estimate of the parameters C and then use them for the estimation of y. While the formulas for $\hat{y}_h$ look exactly like those for the standard LPA, the estimate of C defined by Eq. (10.93) cannot be obtained in the analytical form because the criterion $J_h(x, C)$ can be nonquadratic with respect to C.

The idea of the estimates in Eqs. (10.92–10.93) is accurately and completely relevant to the observation model in Eq. (10.91). The technical drawback of these estimates is their nonlinearity with respect to C and, as a result, the corresponding computing difficulties. The LPA of the nonlinear transform $\chi(y)$ instead of the direct LPA of the signal y can be exploited to simplify the calculations.

Assume that χ is a one-to-one function, that is, the inverse of this function exists. The LPA can be applied to $\chi(y)$ considered as a function of x in the form

$$\chi_h(x, v) = C^T \phi_h(x - v)$$

with the parameters C found from the quadratic criterion

$$J_h(x, C) = \sum_s w_h(x - X_s)[z_s - \chi_h(x, X_s)]^2$$

and the estimates $\hat{\chi}_h(x) = \hat{C}^T \phi(0)$ for $\chi(x)$.

The estimate of the signal in question y is defined as the inverse of the estimate of χ:

$$\hat{y}_h(x) = \chi^{-1}(\hat{\chi}_h(x)), \quad (10.94)$$

where χ^{-1} is the inverse of χ. Computationally, the estimate in Eq. (10.94) is much simpler than the one in Eq. (10.93). However, at least two points define a weakness of this approach.

First, y and χ as functions of x can be significantly different and, as a result, the ability of the LPA to approximate these functions also can be different. Second, the signal independent noise additive in the initial model in Eq. (10.91) after the inverse becomes nonadditive with the variance depending on the signal y. Nevertheless, in applications this accuracy degradation can be treated as a secondary effect when compared with the advantage of computational simplicity.

10.4.1.2 Transformation of both sides of the model

If the signal y is parametric, then a nonlinear transformation can be used to simplify the model. For example, assume that

$$y(x) = \alpha x_1^{\gamma_1} \cdots x_d^{\gamma_d}. \quad (10.95)$$

The logarithm of y makes this model linear with respect to the parameters α, γ_k, $k = 1, \ldots, d$,

$$\ln y(x) = \gamma_0 + \gamma_1 \ln x_1 + \cdots + \gamma_d \ln x_d,\ \gamma_0 = \ln \alpha. \tag{10.96}$$

Using $\ln x_k$ as new arguments, we have the logarithm of the observations $\ln y(x)$ as a linear function of the new arguments. In this way, the complex nonlinear function in Eq. (10.95) is transformed to a simpler one in Eq. (10.96) that is linear with respect to the unknown coefficients γ_k.

This transform is applied to both sides of the observation Eq. (10.95). For noisy data

$$z(x) = y(x) + \varepsilon(x), \tag{10.97}$$

which means that

$$\ln z(x) = \ln(y(x) + \varepsilon(x)). \tag{10.98}$$

The right-hand side of this equation can be simplified as

$$\ln z(x) = \ln y(x) + \varepsilon_1(x) = \gamma_0\alpha + \gamma_1 \ln x_1 + \cdots + \gamma_d \ln x_d + \varepsilon_1(x), \tag{10.99}$$

where $\varepsilon_1(x)$ are residuals of this new linear model,

$$\varepsilon_1(x) = \ln z(x) - \left(\gamma_0 + \sum_{k=1}^{d} \gamma_k \ln x_k\right). \tag{10.100}$$

Formally, we replace the originally highly nonlinear system in Eq. (10.95) by the linear one in Eq. (10.99). However, to be more accurate, we note that the new residuals $\varepsilon_1(x)$ depend on the signal y. This means that the model in Eq. (10.99) is not exactly equivalent to our standard model with a constant noise variance.

For small $\varepsilon(x)$ we can write

$$\varepsilon_1(x) \simeq \partial_\varepsilon \ln[y(x) + \varepsilon(x)]|_{\varepsilon=0} \times \varepsilon(x) = \varepsilon(x)/y(x).$$

Thus, if $E\{\varepsilon\} = 0$, $var\{\varepsilon\} = \sigma^2$, we have for the new residuals $E\{\varepsilon_1\} = 0$ and $var\{\varepsilon_1\} = \sigma^2/y^2(x)$.

The main objective of the logarithm transformation in Eq. (10.99) is a simpler linear model for a systematic part of the observations. However, it also increases the complexity of the random error ε_1. This error depends on the unknown parameters and becomes *heteroscedastic*, that is, varying with x. The variance of this random error ε_1 is dependent on the signal y.

10.4.2 Variance stabilization

It has been noted that some nonlinear transformations result in data with approximately homoscedastic (invariant) variance while the initial data are essentially heteroscedastic with a varying variance. Assume that in the additive model $z(x) = y(x) + \varepsilon(x)$ the variance $\sigma_\varepsilon^2(x)$ of the noise depends on x and this variance is small. Consider the nonlinear transform of both sides of this equation.

The Taylor linearization of the transformation gives

$$\chi(z) \simeq \chi(y) + \chi^{(1)}(y)\varepsilon$$

and the variance of $\chi(z)$ is

$$\sigma_\chi^2(x) \simeq |\chi^{(1)}(y(x))|^2 \sigma_\varepsilon^2(x) \tag{10.101}$$

The transformation χ stabilizes the variance of $\chi(z)$ if $\sigma_\chi^2(x)$ is approximately constant.

For many important cases the noise variance $\sigma_\varepsilon^2(x)$ depends on $y(x)$. Simple conclusions concerning the recommended nonlinear transformations can be made. For example, if $\sigma_\varepsilon^2(x) \sim y^\lambda$, then it can be seen from Eq. (10.101) that $\chi(y) \sim y^{1-\lambda/2}$ is variance-stabilizing for $\lambda \neq 2$. A number of classical probability distributions satisfy these assumptions.

Let z in $z(x) = y(x) + \varepsilon(x)$ be random and have a Poisson distribution. Then $y = E\{z\}$, $E\{\varepsilon\} = 0$, and $var\{z\} = var\{\varepsilon\} = y$. In this case, $\sigma_\varepsilon^2(x) \sim y^\lambda$ with $\lambda = 1$ and thus the nonlinear square root transform

$$\chi(y) = \sqrt{y} \tag{10.102}$$

is variance-stabilizing.

For $\lambda = 2$, $\sigma_\varepsilon^2(x) \sim y^2$, it follows from Eq. (10.101) that the logarithm is the stabilizing transform

$$\chi(y) = \ln y. \tag{10.103}$$

10.4.2.1 Box-Cox transform

There are several universal transformations that are in common use [17], [164]. The most important is the power *Box-Cox transform*. It is defined for positive z as

$$\chi(z,\lambda) = \begin{cases} (z^\lambda - 1)/\lambda, & \text{if } \lambda \neq 0, \\ \ln z, & \text{if } \lambda = 0, \end{cases} \tag{10.104}$$

with λ as a parameter of this transform. It is obvious that the transforms Eqs. (10.102–10.103) are special cases of Eq. (10.104).

The polynomial $(z^\lambda - 1)/\lambda$ in Eq. (10.104) gives $\ln z$ as the limit for $\lambda \to 0$. In this way the logarithm transform is embedded into the power transformation $(z^\lambda - 1)/\lambda$. Figure 10.12 shows $\chi(z,\lambda)$ for $\lambda = -2, 0, 0.5, 1, 3$. Note that the curves $\chi(z,\lambda)$ are convex (downward bending) for $\lambda > 1$ and concave (upward bending) for $\lambda < 1$. The Box-Cox transform can serve as a variance-stabilizer when the variance of the signal is a power of the mean of z.

The shifted power transform is $\chi(z + a, \lambda)$, $a > 0$, where $\chi(z,\lambda)$ is defined in Eq. (10.104) and a is used in order to enable the positive sign for the argument in $\chi(z,\lambda)$. Sometimes parameters λ and a can be selected based on prior information as in Eqs. (10.102–10.103). In other situations, they can be treated as parameters of the observation model and as subjects of estimation from data. The ML, in particular, can be used for these purposes.

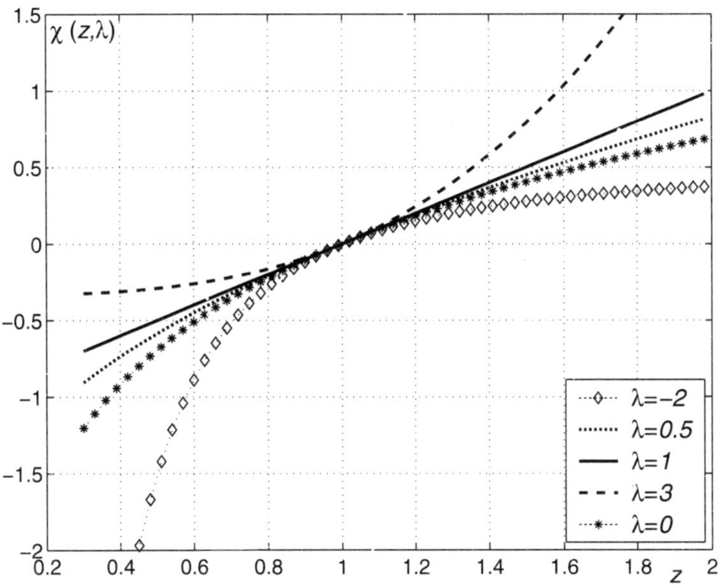

Figure 10.12 Curves of the Box-Cox transform (Eq. (10.104)).

10.4.2.2 Anscombe transform

Here we present some results concerning a nonlinear transform of Poisson random variables that is very useful for applications. This transform is derived in [7] and known as the *Anscombe* transform.

Let z be a Poisson random variable with the parameter λ as in Eq. (10.15). The *Anscombe* transform is defined as

$$\chi(z) = 2\sqrt{z + 3/8}. \tag{10.105}$$

It is shown in [7] that, for large λ,

$$var\{\chi(z)\} \simeq 1 + \frac{1}{16\lambda^2}, \tag{10.106}$$

$$E\{\chi(z)\} \simeq 2\sqrt{\lambda + 3/8} - \frac{1}{4\lambda^{1/2}}. \tag{10.107}$$

Thus, for large λ, the variance of $\chi(z)$ is approximately equal to one and the expectation is approximately equal to the Anscombe transform of the expectation of z, that is, $E\{\chi(z)\} \simeq 2\sqrt{\lambda + 3/8}$. It is important to note that the variance of the transformed variable is more or less independent of λ. For the regression estimation when $\lambda = \lambda(x)$, the variance of the transformed random variables does not depend on x.

10.4.2.3 LPA-ICI estimates

The heteroscedastic variance significantly diminishes the performance of LPA estimates. The variance stabilizing transforms can be very useful for this. The algorithm includes the following three steps:

1. Transform data;
2. Filter the transformed data;
3. Inverse transform the filtered data.

Assume that the observations z have a Poisson distribution:

$$z(x) = \lambda(x) + \varepsilon(x), \qquad (10.108)$$
$$E\{z(x)\} = \lambda(x), \qquad var\{z(x)\} = \lambda(x).$$

One is required to restore $\lambda(x)$ from the observations $z(x)$. As the first step of the procedure, the Anscombe transform in Eq. (10.105) replaces the original data z by

$$\xi(x) = 2\sqrt{z(x) + 3/8}. \qquad (10.109)$$

In the second step, the LPA (with the ICI adaptive-scale selection) is used to filter these new data and the estimation of $E\{\xi(x)\}$. We denote the corresponding estimate by $\hat{\xi}(x)$. The final step is to obtain the inverse of the Anscombe transform with the estimate of the target signal $\lambda(x)$ as

$$\hat{\lambda}(x) = \hat{\xi}^2(x)/4 - 3/8. \qquad (10.110)$$

We note that these forward and inverse nonlinear transforms are not as harmless as one would assume. The biasedness of the final estimate $\hat{\lambda}(x)$ is the usual price. Formula (10.107) shows the term defining this biasedness in the inverse transform. It is essential only for small $\lambda(x)$.

In most applications, the biasedness of the Anscombe transform estimates is ignored, but it can cause some degradation in signal processing. This biasedness can be corrected in part by defining the inverse estimate $\hat{\lambda}(x)$ as a solution of the nonlinear equation Eq. (10.107) as

$$\hat{\xi}(x) = 2\sqrt{\hat{\lambda}(x) + 3/8} - \frac{1}{4\sqrt{\hat{\lambda}(x)}}. \qquad (10.111)$$

Example: 1D filtering

Consider a random signal z having a Poisson distribution with the varying parameter $\lambda(x)$. Let this parameter be the stepwise function shown in Fig. 10.13(a). Then according to Eq. (10.108), the expectation and variance of this random z are varying on x:

$$E\{z\} = \lambda(x), \qquad var\{z\} = \lambda(x).$$

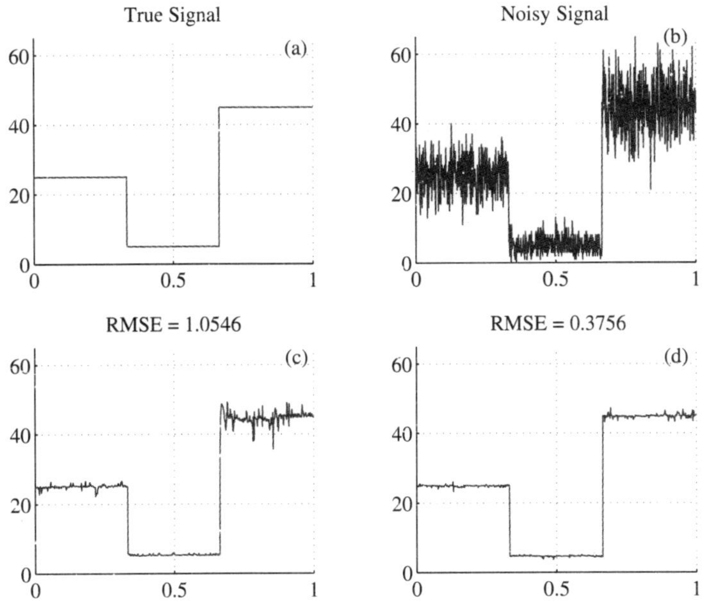

Figure 10.13 Filtering of Poisson signal: (a) a varying intensity of the Poisson signal; (b) Poisson observations; (c) direct LPA-ICI filtering of the Poisson observations; (d) LPA-ICI filtering using the Anscombe transform.

The independent samples of z are generated for $x \in [0, 1]$ with the sampling interval $\Delta = 1/(n-1)$. In total, $n = 1024$ samples of z are shown in Fig. 10.13(b).

The observations $z(x)$ are random, and for each x the variance and the expectation are varying in x according to $\lambda(x)$. It is clearly seen in Fig. 10.13(b) that the variance of $z(x)$ is smaller and larger for x where $\lambda(x)$ takes smaller and larger values, respectively.

The estimation problem at hand is to estimate $\lambda(x)$ from the varying variance data shown in Fig. 10.13(b). The curves in Figs. 10.13(c) and (d) show the scale-adaptive LPA-ICI estimates obtained in two different ways.

First, we apply the LPA-ICI algorithm directly to the observed Poisson data $z(x)$ ignoring the fact that the variance is varying. The filtering results are shown in Fig. 10.13(c).

Second, we start from the nonlinear Anscombe transform Eq. (10.109) to stabilize the variance of the data. Then we use the LPA-ICI filtering and finally apply the inverse Anscombe transform. The results are shown in Fig. 10.13(d).

The estimator is the symmetric sample mean in both cases. The ICI rule is exploited with the parameter $\Gamma = 1.6$, and the window size set H is defined as in Eq. (6.52).

The visual and numerical comparisons are definitely in favor of the algorithm using the stabilizing Anscombe transform. The RMSE value is nearly threefold lower for this algorithm. Application of the Anscombe transform to the imaging of Poisson data will be discussed later, in Sec. 12.1.4.

Chapter 11

Likelihood and Quasi-Likelihood

11.1 Local Maximum Likelihood

In this section the local maximum likelihood approach is introduced as a generalization of the linear LPA. It provides universal tools for designing methods and algorithms for a variety of stochastic models that are different from the standard Gaussian one. This approach combines two different ideas: generalized linear modeling and localization of parametric fitting.

In the generalized linear model the parameters of distributions are assumed to be varying and the linear regression is used to fit them. The ML approach is exploited to make the estimation of these parameters relevant to the stochastic properties of the observations [78]. Localization of the ML allows one to relax restrictions imposed by the standard parametric ML. The theory and applications of the local maximum likelihood can be seen in [46], [47], [145], and [219].

11.1.1 Parametric ML

The ML gives a general constructive idea of how to design optimal nonlinear estimators corresponding to a given distribution. Let us recall briefly the basic concepts of this powerful statistical approach [164], [174].

Assume that the signal y depends on the argument x, and, provided a fixed $x = X_s$, an observed z_s is random with probability density f. This stochastic model can be written in the general form

$$z_s \sim f(z, y_s), \tag{11.1}$$

where $f(z, y)$ is a conditional p.d.f. provided that x is fixed and the signal y is a parameter of this p.d.f. with $y_s = y(X_s)$. Recall that in Eq. (11.1), z_s is random and z is a deterministic argument of the p.d.f.

Let the observations be given by the pairs $\{z_s, X_s\}$, $s = 1, \ldots, n$. The dependence on X_s for z_s goes through $y_s = y(X_s)$. The estimation problem that we are interested in is a reconstruction of the signal $y(x)$ from the observations $\{z_s, X_s\}$, $s = 1, \ldots, n$.

Let us start from the standard parametric ML. The signal y is assumed to have a parametric representation within a vector coefficient C. Denote this parametric form of y by $y_C(x)$:

$$y_C(x) = C^T \phi(x), \tag{11.2}$$

where $\phi \in \mathbb{R}^M$ is a vector of basis functions.

The p.d.f. of the random z_s is $f(z, y_C(X_s))$. Note again that z in $f(z, y_C(X_s))$ is a possible value of the random z_s. This difference between the observed random z_s and the possible value of this random variable is essential in the concept of the likelihood.

Assume that the observations $\{z_s, X_s\}$ are independent for different s. The joint p.d.f. of the observation pairs is a product of the densities for each pair. Replace the possible values of z in this joint p.d.f. by the observed z_s.

The value obtained in this way is called a *likelihood.* Considered as a function of the parameter C, the likelihood is represented in the form

$$l(C) = \Pi_s f(z_s, y_C(X_s)). \tag{11.3}$$

Thus, the likelihood is the value of the joint p.d.f. where possible values of random variables are replaced by observed ones. Loosely speaking, the likelihood is the probability that the random variables take the observed values.

The value of C that maximizes the likelihood is known as the *maximum likelihood estimate*:

$$\widehat{C} = \arg\max_C l(C). \tag{11.4}$$

The *log-likelihood* is defined as the logarithm of the likelihood:

$$L(C) = \sum_s \ln f(z_s, y_C(X_s)). \tag{11.5}$$

This is more convenient in calculations as well as in analyses. Because the logarithm is a one-to-one monotonic function, the ML estimate in Eq. (11.4) can be defined as

$$\widehat{C} = \arg\max_C L(C). \tag{11.6}$$

Provided there are some relevant regularity conditions on f, the ML estimates have the following desirable optimality properties:

1. These estimates are asymptotically unbiased and efficient. The efficiency means that the estimator has the smallest variance, and thus the narrowest confidence interval, of all estimators;
2. They have approximate normal distributions and approximate sample variances that can be used to generate confidence bounds and hypothesis tests for the parameters;

3. More specifically,

$$\widehat{C} \sim \mathcal{N}\left(C, \frac{1}{n}I^{-1}\right), \qquad I = (I_{kl})_{M\times M},$$
$$I_{kl} = E\{\partial_{C_k} \ln f(z_s, y_C(X_s)) \cdot \partial_{C_l} \ln f(z_s, y_C(X_s))\}$$
$$= -E\{\partial_{C_k}\partial_{C_l} \ln f(z_s, y_C(X_s))\}, \tag{11.7}$$

where n is the sample size and I is the Fisher information matrix.

The expectations in these formulas are calculated over the random (z_s, X_s), and it is assumed that these observation pairs are independent for different s. The approximate Fisher information matrix (approximate variances for $\widehat{C}$) can be calculated using the empirical mean estimates of I_{kl}:

$$\hat{I}_{kl} = \frac{1}{n}\sum_s \partial_{C_k} \ln f(z_s, y_C(X_s)) \cdot \partial_{C_l} \ln f(z_s, y_C(X_s)). \tag{11.8}$$

11.1.2 Nonparametric ML

11.1.2.1 General approach

The local ML is a nonparametric counterpart of the parametric ML technique. It extends the scope of the parametric ML to a broader class of functions.

The local likelihood and local log-likelihood use the window function w for localization of estimation in a neighborhood of the point of interest (center) x and replace the global parametric model $y_C(x)$ by a local one.

The linear models in Eq. (11.2) used in $l(C)$ and $L(C)$, which are known in statistics as *generalized linear models*, become *local (nonparametric) generalized linear models*.

In this context, "*generalized*" refers to at least two different things. First, it is not compulsory to use the conditional mean of the observation z as in the standard definition of the regression in Eq. (2.69). It can be a nonlinear function of this conditional mean of z. Second, it is exploited for general types of distributions. In general, the generalized linear models result in estimates that are nonlinear with respect to the observations.

For the polynomial linear model $y_h(x, v)$ in Eq. (2.23) introduced in Chapter 2, this local log-likelihood criterion can be presented in the form

$$L_h(x, C) = \sum_s \ln f(z_s, y_h(x, X_s))w_h(x - X_s),$$
$$y_h(x, X_s) = C^T\phi_h(x - X_s), \qquad \phi_h(x) = \phi(x/h), \tag{11.9}$$
$$w_h(x) = w(x/h)/h^d, \qquad C \in \mathbb{R}^M,$$

where h stands for the scale parameter used both in the window function w and for the scaling of the polynomials ϕ.

What is the difference between a *local* and *non-local likelihood*? The latter uses a global parametric model that is valid for all x. The local likelihood relaxes this assumption and uses a linear model that is valid only locally in some neighborhood of the estimation point x. In this way, the likelihood becomes a nonparametric technique.

Following the idea of the local nonparametric estimation, we use the criterion $L_h(x, C)$ for computing C and the model $y_h(x, v)$ is exploited in a pointwise manner for $x = v$ only as it is in Eq. (2.43). In this way we arrive at the following local ML version of the LPA:

$$\widehat{C} = \arg\max_{C \in \mathbb{R}^M} L_h(x, C),$$
$$\hat{y}_h(x) = \widehat{C}^T \phi(0), \; \hat{y}_h^{(r)}(x) = \left(-\frac{1}{h}\right)^{|r|} \widehat{C}^T \phi^{(r)}(0). \tag{11.10}$$

All ideas of the standard LPAs concerning the window w (shape, anisotropy, directionality, etc.), the scaling h (scalar, multivariate), and the estimation of the signal and derivatives are naturally valid in this nonparametric pointwise estimation.

The estimate $\widehat{C}$ in Eq. (11.10) is a solution of the equation $\partial_C L_h(x, C) = 0$, which has the form

$$\sum_s \partial_y \ln f(z_s, y_h(x, X_s)) w_h(x - X_s) \phi_h(x - X_s) = 0. \tag{11.11}$$

obtained by differentiating $L_h(x, C)$ on C.

Consider how the local ML works for some particular cases.

11.1.2.2 Additive i.i.d. noise

First, the observation model is changed from the global $y_C(x)$ to a local one:

$$z_s = y_h(x, X_s) + \varepsilon_s, \qquad s = 1, \ldots, n,$$
$$y_h(x, X_s) = C^T \phi_h(x - X_s), \tag{11.12}$$

where the random ε_s are i.i.d. with probability density f. Here the deterministic part $y_h(x, X_s)$ of the model is local depending on x. Thus, it is a model for a single fixed x only. Remember that in the LPA the argument x is treated as a fixed parameter of the model. Further, the local observation model depends on the scale parameter h.

The probability density of z_s is $f(z_s - y_h(x, X_s))$, and for the set of observations we have the local log-likelihood in Eq. (11.9) in the form:

$$L_h(x, C) = \sum_s w_h(x - X_s) \ln f(e_s), \, e_s = z_s - y_h(x, X_s). \tag{11.13}$$

Denote

$$\psi(v) = -\ln f(v). \tag{11.14}$$

Then $\sum_s \ln f(e_s) = -\sum_s \psi(e_s)$ and the local log-likelihood estimates are found from the problem in Eq. (11.10) as

$$\widehat{C} = \arg\min_{C \in \mathbb{R}^M} \sum_s \psi(e_s) w_h(x - X_s), \qquad e_s = z_s - y_h(x, X_s), \tag{11.15}$$

where ψ is a loss function of the residuals.

Because of the minus sign in Eq. (11.14), the maximization in the ML is replaced by minimization in Eq. (11.15). In this way we derive the local nonparametric M-estimates in Eq. (10.4) with the loss function ψ defined through the probability density f.

11.1.2.3 Gaussian noise

Let the observation noise be Gaussian with the invariant variance, $\varepsilon \sim \mathcal{N}(0, \sigma^2)$. Omitting the invariant part $\ln(\sqrt{2\pi}\sigma)$ in $\ln f = -\ln(\sqrt{2\pi}\sigma) - v^2/(2\sigma^2)$, we can define the loss function in Eq. (11.14) as $\psi(v) = v^2$ with the weighted mean-square residual criterion

$$\sum_s w_h(x - X_s) e_s^2, \qquad e_s = z_s - y_h(x, X_s). \tag{11.16}$$

Then, the local ML estimate in Eq. (11.15) in the form

$$\widehat{C} = \arg\min_{C \in \mathbb{R}^M} \sum_s w_h(x - X_s) e_s^2 \tag{11.17}$$

gives the standard linear LPA.

11.1.2.4 Laplacian noise

Let the observation noise in the additive noise model be independent Laplacian, that is, $f(v) = (1/2a)\exp(-|v|/a)$. Then $\ln f(v) = \ln[1/(2a)] - |v|/a$ and the loss function in Eq. (11.14) is defined as $\psi(v) = |v|$ with the modulus residual criterion

$$\sum_s |e_s| w_h(x - X_s), \qquad e_s = z_s - y_h(x, X_s). \tag{11.18}$$

In this way we derive the M-estimates in Eq. (10.4) with the loss function $\psi(x) = |x|$ as the local ML estimate.

11.1.3 Theory for local ML

In this section we present theoretical results concerning the accuracy of the local ML estimates for general likelihood functions. These results evaluate the bias and variance of the estimates and claim asymptotic convergence conditions. The variance is of a special practical concern because it is used in the ICI algorithm for the varying-scale selection for general nonlinear ML estimates.

We begin with some preliminary remarks.

- For the observation model in Eq. (11.1), $\int_{\mathbb{R}} f(z,y)dz = 1$ and as a result

$$\partial_y \int_{\mathbb{R}} f(z,y)dz = \int_{\mathbb{R}} \partial_y f(z,y)dz = 0, \tag{11.19}$$

provided that the order of integration and differentiation can be changed.
- The Fisher information value for $f(z,y)$ with scalar random z and scalar given y is defined as

$$I_y = E\{[\partial_y \ln f(z,y)]^2\} \tag{11.20}$$

or, equivalently, as

$$I_y = -E\{\partial_y^2 \ln f(z,y)\}. \tag{11.21}$$

Surprisingly, the main results considered in Chapter 5 for linear estimates are also valid for the general local nonlinear ML estimates provided that in the formulas for the variance σ^2 is replaced by $1/I_y$.

Proposition 15 *Let the p.d.f. $f(z,y)$ be twice differentiable on y with the finite Fisher information I_y. Under the assumptions of Proposition 7, where $M = m+1$, the accuracy of the multivariate discrete local ML estimates in Eqs. (11.9–11.10) are defined by the following expressions:*

for the estimation bias,

$$m_{\hat{y}_h^{(r)}}(x,h) = (-1)^m h^{m+1-|r|} \sum_{|k|=m+1} y^{(k)}(x)\frac{1}{k!} \times \int_{\mathbb{R}^d} g^{(r)}(u)u^k du + o(h^{m+1-|r|}), \qquad m \geq |r|, \tag{11.22}$$

and for the estimation variance,

$$\sigma^2_{\hat{y}_h^{(r)}}(x,h) = \frac{1}{I_y(x)}\frac{\Delta^d}{h^{d+2|r|}} \int_{\mathbb{R}^d} [g^{(r)}(u)]^2 du + o\left(\frac{\Delta^d}{h^{d+2|r|}}\right),$$
$$I_y(x) = E\{[\partial_y \ln f(z,y(x))]^2\} = -E\{\partial_y^2 \ln f(z,y(x))\}. \tag{11.23}$$

The proof of the proposition is given in Sec. 14.8.3 of the Appendix.

This result is quite general. It covers, for example, the linear LPA estimates as well as the robust M-estimates considered in Chapter 10.

The LPA of degree m is assumed for $y_h(x, X_s)$ in Eq. (11.9). For simplicity we also assume here that the signal y is continuous and differentiable up to the derivatives of order $m+1$. Then, the bias of the estimate is defined by these derivatives.

In the local likelihood in Eq. (11.9) we can use the directional window w and obtain the anisotropic sectorial ML estimates, which generalize the linear directional estimates (Chapter 7) to non-Gaussian distributions.

Different orders of the polynomials and different scales can be used in these anisotropic nonlinear estimates for different elements of $x \in \mathbb{R}^d$. The bias and variance results given in Proposition 12 for the linear directional estimates can be translated to the local directional ML estimates.

The estimation bias is identical for the linear and nonlinear ML estimates, while the variance σ^2 in the formulas of Proposition 12 is replaced by $1/I_y$ for the nonlinear ML estimates.

The Fisher information I_y, which is an important parameter in statistics, is calculated for all basic distributions:

1. *For the Gaussian distribution* we have:

$$f(z, y) = f(z - y) = \frac{1}{\sqrt{2\pi}\sigma} \exp[-(z - y)^2/(2\sigma^2)],$$

$$\ln f(z - y) = -\ln(\sqrt{2\pi}\sigma) - (z - y)^2/(2\sigma^2),$$

$$\partial_y \ln f(z - y) = (z - y)/\sigma^2, \qquad \partial_y^2 \ln f(z - y) = -1/\sigma^2.$$

Then

$$I_y(x) = E\{[\partial_y \ln f(z, y(x))]^2\} = E\{(z - y)^2/\sigma^4\} = 1/\sigma^2$$

or

$$I_y(x) = -E\{\partial_{yy}^2 \ln f(z - y)\} = E\{1/\sigma^2\} = 1/\sigma^2. \tag{11.24}$$

Thus,

$$I_y(x) = 1/\sigma^2. \tag{11.25}$$

With this $1/I_y(x) = \sigma^2$, Proposition 15 is identical to Proposition 7.

2. *For the Laplacian distribution* in Eq. (10.11) we have

$$I_y(x) = 1/a^2. \tag{11.26}$$

3. *For the Bernoulli distribution* in Eq. (10.13), using the result from [174], page 312, we obtain

$$I_y(x) = \frac{1}{y(x)[1 - y(x)]}. \tag{11.27}$$

4. *For the Poisson distribution* in Eq. (10.15), using the result from [174], page 312, we have

$$I_y(x) = \frac{1}{y(x)}. \tag{11.28}$$

Formulas (11.23) give the asymptotic accuracy when the sampling interval and the scale are small.

11.1.4 LPA-ICI algorithms

It follows from Proposition 15 that the asymptotic properties of the nonlinear LPA-ML estimates are similar to the properties of the corresponding linear LPA estimates. The formula for the bias in Eq. (11.22) demonstrates that the bias is small for small h and large for large h. The asymptotic formula for the variance in Eq. (11.23) shows that the variance is large for small h and small for a large one. This means that all ideas concerning the selection of the adaptive-scale parameter h from the bias-variance trade-off are completely valid for the general nonlinear ML estimates exactly as they are for the linear LPA estimates.

In principle, the ICI rule is applicable in this nonlinear estimation scenario for the adaptive-scale selection. However, the ML estimates are nonlinear and, in general, require recursive calculations. The variance depends on the estimated regression y and, in this way, also can be calculated using updated estimates of y.

The main difference between the nonlinear ML and linear LPA estimates is in the complexity of the required calculations only.

11.1.4.1 Estimate calculation

The maximization of $L_h(x, C)$ nonquadratic with respect to C cannot be expressed in an explicit analytical form and requires numerical recursive calculations. In this section we recall some basic methods applicable for the considered nonlinear estimates (e.g., [11], [51], [184]).

Mainly, this sort of maximization is based on the vector gradient and the matrix of second derivatives of L_h on $C \in \mathbb{R}^M$. The following notations are usual for the vector gradient:

$$\partial_C L_h(x, C) = [\partial_{C_i} L_h(x, C)]_{M\times 1},$$

and the matrix of the second derivatives (*Hessian* matrix),

$$\partial_C \partial_{C^T} L_h(x, C) = [\partial_{C_i} \partial_{C_j} L_h(x, C)]_{M\times M}.$$

The derivatives for $L_h(x, C)$ in Eq. (11.9) can be given in the form

$$\partial_C L_h(x, C) = \sum_s \partial_{y_h} \ln f(z_s, y_h(x, X_s)) w_h(x - X_s) \phi_h(x - X_s), \tag{11.29}$$

$$\partial_C \partial_{C^T} L_h(x, C) = \sum_s \partial^2_{y_h} \ln f(z_s, y_h(x, X_s)) w_h(x - X_s) \phi_h(x - X_s) \phi_h^T(x - X_s), \tag{11.30}$$

where

$$y_h(x, X_s) = C^T \phi_h(x - X_s). \tag{11.31}$$

Maximization of the criterion $L_h(x, C)$, concave on $C \in \mathbb{R}^M$, can be produced using a gradient-type recursive procedure:

$$C^{(k+1)} = C^{(k)} + \gamma_k A^{(k)} \partial_C L_h(x, C^{(k)}), \qquad k = 0, 1, \ldots, \tag{11.32}$$

where $C^{(k+1)}$ are successive iterations of C, and the gradient $\partial_C L_h$ is calculated for $C = C^{(k)}$.

The methods are different by an $M \times M$ weight matrix $A^{(k)}$ and a step size parameter $0 < \gamma_k \leq 1$:

1. *Simple gradient ascent.* The identity matrix is used for $A^{(k)}$, $A^{(k)} = I_{M \times M}$. The convergence rate is linear,

$$||C^* - C^{(k+1)}|| \leq q_k ||C^* - C^{(k)}||, \tag{11.33}$$

   and characterized by the parameter q_k, $0 \leq q_k < 1$.

   Here C^* is a vector of the optimal values of C maximizing $L_h(x, C)$. An invariant step size parameter $\gamma_k = \gamma$ is selected to enable the convergence of the iterations, $q_k < 1$. A main drawback of the algorithm is a low convergence rate because q_k is close to 1 if the Hessian matrices $\partial_C \partial_{C^T} L_h(x, C^{(k)})$ are ill-conditioned.
2. *Steepest gradient ascent.* This is defined by Eq. (11.32), where γ_k is selected by the maximization of the criterion function

$$\gamma_k = \arg\max_{\gamma} L_h(x, C^{(k)} + \gamma A^{(k)} \partial_C L_h(x, C^{(k)})). \tag{11.34}$$

   In this method the gradient defines the direction of the ascent and Eq. (11.34) gives the best value of the step size in this direction. The convergence is linear. For a complex criterion function $L_h(x, C)$ (in particular, ill-conditioned Hessian matrices) the convergence can be slow.
3. *Newton method.* The inverse Hessian matrix is used for $A^{(k)}$,

$$A^{(k)} = [\partial_C \partial_{C^T} L_h(x, C^{(k)})]^{-1}. \tag{11.35}$$

   Here we assume that $A^{(k)}$ is inverse or pseudo-inverse of the $M \times M$ matrix of $\nabla_C^2 L_h$. The convergence is quadratic,

$$||C^* - C^{(k+1)}|| \leq q_k ||C^* - C^{(k)}||^2, \; q_k > 0. \tag{11.36}$$

   This convergence rate is very good, but the algorithm is sensitive with respect to initialization. For the quadratic convergence, a good initial guess $C^{(0)}$ is required.
4. *Gauss-Newton method.* The $A^{(k)}$ in Eq. (11.32) is an approximation of the inverse Hessian matrix that is computed using the first derivatives of y only. The convergence rate is linear in Eq. (11.33) but with small q_k. The convergence rate is insensitive with respect to the initialization.

There are many modifications of the above algorithms that are different by calculations of the gradient and the matrix A.

11.1.4.2 Variance calculation

Formula (11.23) gives the variance of the local ML estimates. The Fisher information in these formulas depends on unknown values of the regression $y(x)$. Thus, the variance calculation requires some preliminary estimates of the regression.

11.1.4.3 Algorithm organization

Open loop and feedback algorithms are used for nonlinear estimate calculations. *An open loop algorithm* assumes that the estimates $\widehat{C}$ of the vectors C are calculated using some recursive algorithms, e.g., Eq. (11.32) for $h \in H$. The estimates of the regression are

$$\hat{y}_h(x, X_s) = \widehat{C}^T \phi_h(x - X_s), \qquad \hat{y}_h(x) = \widehat{C}^T \phi(0). \tag{11.37}$$

The estimate variances are computed according to Eq. (11.23), where the Fisher information $I_y(x)$ depends on estimated $y(x)$. The corresponding y_s are defined according to Eq. (11.37) as $y_s = \hat{y}_h(x, X_s)$. The estimates of the regressions and their variances are exploited for the ICI adaptive-scale selection. If the multi-directional estimation is used, the adaptive directional estimates are fused in the final one, $\hat{y}(x)$. This is a straightforward implementation of the standard LPA-ICI technique.

A *feedback algorithm* differs from the open loop algorithm in its use of recursive updating of the regression estimates $\hat{y}(x)$. These estimates can be used in the formulas for the variances as well as for the estimate calculation. The gradient in Eq. (11.29) and the Hessian in Eq. (11.30) depend on $\hat{y}_h(x, X_s)$. The final signal estimate

$$\hat{y}(X_s) = \hat{y}(x)|_{x=X_s},$$

obtained after fusing can be exploited in these formulas as well as in the formulas for the variance instead of $\hat{y}_h(x, X_s)$ as more accurate estimates of the regression.

Figure 11.1 shows flowcharts of these *open loop* and *feedback* algorithms for multidirectional estimates. The open loop algorithm shown in Fig. 11.1(a) includes the following blocks:

- ML estimation. The inputs are the observations z and the initial estimate of the LPA parameter $\widehat{C}$. The outputs are the directional estimates $\hat{y}_{h,\theta}$, $h \in H$, and the corresponding variances $\sigma^2_{\hat{y}_{h,\theta}}$.
- ICI rule. The inputs are the estimates $\hat{y}_{h,\theta}$ and their variances $\sigma^2_{\hat{y}_{h,\theta}}$. The outputs are the adaptive-scale estimates $\hat{y}_{h^+,\theta}$ and their variances $\sigma^2_{\hat{y}_{h^+,\theta}}$.
- Fusing. The inputs are the adaptive-scale estimates $\hat{y}_{h^+,\theta}(x)$ and their variances $\sigma^2_{\hat{y}_{h^+,\theta}}(x, h)$. The outputs are the fused estimates $\hat{y}(x)$ and their variances $\sigma^2_{\hat{y}}(x)$.

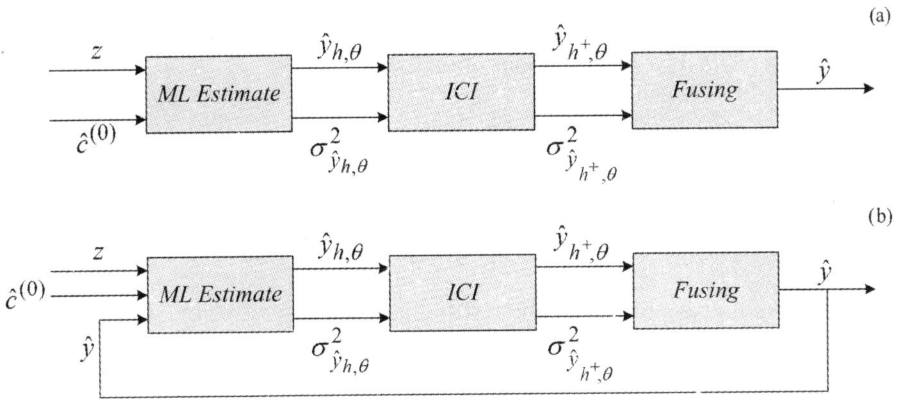

Figure 11.1 Flowcharts of the ML algorithm: (a) open loop algorithm; (b) feedback algorithm.

The feedback algorithm is shown in Fig. 11.1(b). The fused estimates $\hat{y}(x)$ are used in ML estimation blocks for calculation of the gradient in Eq. (11.29) and the Hessian in Eq. (11.30) instead of $y_h(x, X_s)$ as more accurate estimates of $y_s = y(X_s)$.

In a similar way, these final estimates of $\hat{y}(x)$ are used in $I_y(x)$ for calculation of the estimate variance according to formula (11.23). The feedback algorithm should be initiated by some estimates of $y(x)$. The open loop algorithm can be used for this initialization.

11.2 Binary and Counting Observations

Consider the specifications of the general ML framework to important discrete Bernoulli and Poisson distributions.

11.2.1 Bernoulli regression

Assume that z is a binary random variable subject to the Bernoulli distribution in Eq. (10.13). It takes values 1 and 0 with probabilities p and $1 - p$, respectively. Assume also that the performance depends on the argument x and this dependence, $p = p(x)$, is the subject of estimation.

To be in line with the notation exploited in the previous chapters, here and in what follows we use $y(x)$ for the regression regardless of the meaning of this regression function. In particular, for the Bernoulli distribution we replace $p(x)$ by $y(x)$. Thus, $y(x)$ is the probability that a random z takes a value 1, and this probability depends on x.

The function $y(x)$ is a *Bernoulli regression*. It is a real regression because $y(x)$ is the conditional expectation of random z provided that x is fixed,

$$E\{z(x)\} = 1 \cdot y(x) + 0 \cdot [1 - y(x)] = y(x). \tag{11.38}$$

The experiments are produced for different x, and we are interested in revealing the function $y(x)$ from independent observations:

$$z_s = y(X_s) + \varepsilon(X_s), \qquad s = 1, \ldots, n, \tag{11.39}$$

where $\varepsilon(x) = z(x) - y(x)$ is a zero-mean random error. The moments of this random z in Eq. (10.14) are

$$E\{z\} = y, \qquad var\{z\} = E\{z^2\} - E^2\{z\} = y(x)[1 - y(x)]. \tag{11.40}$$

The random variable z and Bernoulli distribution are discrete. In the likelihood instead of p.d.f. we use the probability

$$f(z, y) = y^z(1 - y)^{1-z}, \qquad z = 0, 1. \tag{11.41}$$

Assume for the moment that the regression y has the parametric form $y_C = C^T\phi(x)$. The likelihood for the Bernoulli discrete random variable similar to Eq. (11.3) is defined as

$$l(C) = \Pi_s y_C^{z_s}(X_s)[1 - y_C(X_s)]^{1-z_s},$$

with the log-likelihood

$$L(C) = \ln l(C) = \sum_s \{z_s \ln y_C(X_s) + (1 - z_s)\ln[1 - y_C(X_s)]\}. \tag{11.42}$$

Formally, the estimate of C can be obtained by a straightforward maximizing of $L(C)$ on C. However, the regression y has a meaning of probability, and it is constrained by the natural conditions

$$0 \leq y(x) \leq 1. \tag{11.43}$$

These conditions cannot be guaranteed for the regression approximation $y_C(x)$ with C found from the standard unconstrained ML. In general, the Bernoulli regression cannot be built starting from the model in Eq. (11.42).

The alternative standard idea in statistics is to change the variable y for the canonical parameter θ using the *logit transform* in Eq. (10.33), as

$$\theta(x) = \ln \frac{y(x)}{1 - y(x)}, \tag{11.44}$$

and fit the regression for this new variable as

$$\theta_C(x) = C^T\phi(x). \tag{11.45}$$

The regression $\theta(x)$ defined by Eq. (11.44) is called a *logistic regression.*

The estimate of y as the inverse logit transform in Eq. (10.35) gives the regression for y in the form

$$\hat{y}(x) = \frac{1}{1 + \exp(-\theta_C(x))}.$$

It is obvious that the conditions in Eq. (11.43) hold for this $\hat{y}(x)$ with any $\theta_C(x)$. In this way, the Bernoulli regression can be obtained by unconstrained maximization of the likelihood.

This is an example of the generalized linear regression technique where the linear regression is fit for an intermediate variable θ and then recalculated to the variable of the interest y.

For one important particular case this nonlinear logit transformation can be avoided. Let $y_C(X_s) = C$ be a constant in Eq. (11.42). Then

$$\partial_C L(C) = \sum_s [z_s/C - (1 - z_s)/(1 - C)] = 0,$$

gives the sample mean as an estimate of C,

$$\widehat{C} = \frac{1}{n} \sum_{s=1}^{n} z_s,$$

where n is the number of observations. This solution always satisfies the condition in Eq. (11.43) that

$$0 \leq \widehat{C} \leq 1.$$

In what follows we consider a localized nonparametric version of the Bernoulli regression and we do it in a different manner for the zero ($m = 0$) and higher-order ($m > 0$) LPA.

11.2.1.1 Zero-order nonparametric regression

The localized version of the log-likelihood is obtained by replacing L in Eq. (11.42) by

$$L_h(x, C) = \sum_s [z_s \ln C + (1 - z_s) \ln(1 - C)] w_h(x - X_s), \tag{11.46}$$

where C is a constant and the window w_h is nonnegative. The local ML regression estimate is defined as

$$\hat{y}_h(x) = \widehat{C}, \quad \widehat{C} = \arg\max_C L_h(x, C). \tag{11.47}$$

The maximum of $L_h(x, C)$ on C gives the regression estimate in the form

$$\hat{y}_h(x) = \frac{\sum_s z_s w_h(x - X_s)}{\sum_s w_h(x - X_s)}. \tag{11.48}$$

Thus, the zero-order nonparametric Bernoulli regression has the form of the linear Nadaraya-Watson estimate in Eq. (2.70). For the binary observations $z_s = 0, 1$ and $w_h > 0$, the conditions in Eq. (11.43) hold:

$$0 \leq \hat{y}_h(x) \leq 1. \tag{11.49}$$

Note that

$$\hat{y}_h(x) - E\{\hat{y}_h(x)\} = \frac{\sum_s (z_s - y_s) w_h(x - X_s)}{\sum_s w_h(x - X_s)}.$$

Using that $[z_s - y(X_s)]$ and $[z_{s'} - y(X_{s'})]$ are independent for $s \neq s'$ and $E\{(z-y)^2\} = y(1-y)$ in Eq. (11.40), we find for the variance of the estimate in Eq. (11.48) that

$$\sigma^2_{\hat{y}_h} = \frac{1}{\left[\sum_s w_h(x - X_s)\right]^2} \sum_s (1 - y_s) y_s w_h^2(x - X_s). \tag{11.50}$$

11.2.1.2 High-order nonparametric regression

The local version of the likelihood in Eq. (11.42) can be rewritten as

$$L_h(x, C) = \sum_s \left\{ z_s \ln \frac{y(X_s)}{1 - y(X_s)} + \ln[1 - y(X_s)] \right\} w_h(x - X_s). \tag{11.51}$$

The logit transform in Eq. (11.44) for y gives this likelihood as

$$L_h(x, C, \theta) = \sum_s \{z_s \theta(X_s) - \ln[1 + \theta(X_s)]\} w_h(x - X_s). \tag{11.52}$$

Using the LPA fit for the canonical variable $\theta(x)$ in Eq. (11.52), we obtain this likelihood in the following form:

$$L_h(x, C, \theta) = \sum_s \{z_s \theta_h(x, X_s) - \ln[1 + \theta_h(x, X_s)]\} w_h(x - X_s),$$
$$\theta_h(x, v) = C^T \phi_h(x - X_s), \qquad \phi_h(x) = \phi(x/h). \tag{11.53}$$

The parameter C is found by maximizing $L_h(x, C, \theta)$,

$$\widehat{C} = \arg\max_C L_h(x, C, \theta), \tag{11.54}$$

with the estimate of the logistic regression

$$\hat{\theta}_h(x) = C^T \phi(0). \tag{11.55}$$

The estimate of the Bernoulli regression $y(x)$ requires the inverse logit transform with the Bernoulli regression estimate

$$\hat{y}(x) = \frac{1}{1 + \exp[-\hat{\theta}_h(x)]}. \tag{11.56}$$

It is obvious that $\hat{y}(x)$ satisfies the conditions in Eq. (11.49).

11.2.2 Poisson regression

Let z be a random variable subject to the Poisson distribution in Eq. (10.15). Assume that the Poisson intensity λ depends on the argument x and this dependence, $\lambda = y(x)$, is the subject of estimation. The function $y(x)$ is called a *Poisson regression*

For a fixed x, the moments of the Poisson z are

$$E\{z\} = y(x), \qquad var\{z\} = y(x). \tag{11.57}$$

The regression $y(x)$ has a sense of the intensity of the count flow and should be nonnegative:

$$y(x) \geq 0. \tag{11.58}$$

We use the notation in Eq. (10.17),

$$z \sim \mathcal{P}[y(x)],$$

for the random z having the Poisson distribution with the parameter depending on x, where, according to Eq. (10.15),

$$\mathcal{P}(y) = \exp(-y)\frac{y^z}{z!}.$$

The random variable z is nonnegative discrete, $z_s \in \mathbb{Z}_+$. The corresponding likelihood is defined as

$$l = \Pi_s \exp[-y(X_s)]\frac{y^{z_s}(X_s)}{z_s!}$$

with the log-likelihood

$$L = \ln(l) = \sum_s \{-y(X_s) + z_s \ln[y(X_s)]\}, \tag{11.59}$$

where the additive term $-\ln(z_s!)$, independent on y, is omitted.

We consider the localized nonparametric estimate of the Poisson regression and do it differently for the zero ($m = 0$) and higher ($m > 0$) orders of the LPA.

11.2.2.1 Zero-order nonparametric regression

The zero-order nonparametric regression $\hat{y}_h(x)$ assumes that $y(x)$ is approximated locally by a constant and the windowed likelihood is used. For Eq. (11.59) it has the form

$$L_h(x, C) = \sum_s (-C + z_s \ln C) w_h(x - X_s),$$
$$\hat{y}_h(x) = \widehat{C}, \qquad \widehat{C} = \arg\max_C L_h(x, C). \tag{11.60}$$

The equation $\partial_C L_h(x, C) = 0$ gives as a solution

$$\hat{y}_h(x) = \frac{\sum_s z_s w_h(x - X_s)}{\sum_s w_h(x - X_s)}. \tag{11.61}$$

Thus, again as it was for the Bernoulli regression in Eq. (11.48), we arrive at the Nadaraya-Watson linear nonparametric estimate. It is clear that, for nonnegative w_h and z_s, the requirement in Eq. (11.58) always holds.

11.2.2.2 High-order nonparametric regression

Use the logarithmic transform of y as the canonical parameter θ in Eq. (10.29) of the likelihood Eq. (11.59):

$$\theta = \ln y. \tag{11.62}$$

This gives the local log-likelihood with respect to θ in the form

$$\begin{aligned} L_h(x, C, \theta) &= -\sum_s \{-\exp[\theta_h(x, X_s)] + z_s\theta_h(x, X_s)\}w_h(x - X_s), \\ \theta_h(x, X_s) &= C^T\phi_h(x - X_s). \end{aligned} \tag{11.63}$$

The parameter C is found by maximizing Eq. (11.63) with the canonical regression estimate,

$$\widehat{C} = \arg\max_C L_h(x, C, \theta), \qquad \hat{\theta}_h(x) = \hat{C}^T\phi(0). \tag{11.64}$$

Finally, the estimate of the Poisson regression is obtained by the inverse of Eq. (11.62):

$$\hat{y}_h(x) = \exp[\hat{\theta}_h(x)]. \tag{11.65}$$

The nonnegativity condition in Eq. (11.58) always holds for the estimate in Eq. (11.65).

11.3 Local Quasi-Likelihood

There are many practical circumstances in which even though the full likelihood is unknown, one can specify the relationship between the mean and the variance. In a situation such as this, estimation of the mean (regression, local regression) can be achieved by replacing the log-likelihood, $\ln f(z, y)$, by a quasi-likelihood function $Q(z, y)$. The quasi-likelihood has been proposed as a distribution-free method (note that, contrary to it, the ML is a distribution-based method).

In many applications, the noise that corrupts the signal is signal-dependent, with the most widely encountered types being multiplicative, Poisson, film-grain, and speckle noise. Their common feature is that the power of the noise is related to the magnitude of the signal.

There are many adaptive-neighborhood approaches to filtering signals and images corrupted by signal-dependent noise. Instead of using fixed-size, fixed-shape neighborhoods, statistics of the noise and the signal are computed within variable-size, variable-shape neighborhoods that are selected for every point of the estimation. In particular, the adaptive filters developed by Lee [133], [134] and Kuan [128] are well-known in the field.

In the quasi-likelihood approach, we are concerned with the same class of problems and approaches but based on a different technique. The quasi-likelihood is used to derive the estimate relevant to the observation model, and the ICI enables the neighborhood adaptivity close to the optimal one.

11.3.1 Parametric quasi-likelihood

One assumes that we have independent observations z with expectations $y = E\{z\}$ and the variance of these observations is modeled as

$$\sigma^2(y) = \sigma_0^2 \rho(y), \tag{11.66}$$

where σ^2 is unknown and ρ is a given positive function of y called a *variance function.* Thus, the variance is considered to be product of an unknown invariant scalar σ_0^2 and a given function $\rho(y)$ of unknown y. For example, we have for the Bernoulli in Eq. (11.40) and the Poisson in Eq. (11.57) models: $\sigma_0 = 1$, $\rho(y) = (1 - y)y$ and $\sigma_0 = 1$, $\rho(y) = y$, respectively. In theory, the σ_0 is accurately specified for these models. However, in practice, data never have the accurate Poisson distribution. At best, it can be an approximate Poisson distribution.

Statisticians differentiate between *overdispersed* and *underdispersed Poisson* data if the σ_0 in Eq. (11.66) are, respectively, larger or smaller than 1. For this sort of situation, σ_0^2 in Eq. (11.66) is considered to be a useful parameter.

The estimation problem is to reconstruct $y(x)$ from the observation pairs $\{z_s, X_s\}$, $s = 1, \ldots, n$. The varying variance in Eq. (11.66) gives the idea that the residuals in estimation should be weighted by these variances. The standard LPA technique in Chapter 2, based on the observations with invariant variances, can be quite misleading in this case.

The quasi-likelihood function $Q(z, y)$ is defined by the relation [232]:

$$\partial_y Q(z, y) = (z - u)/\sigma^2(y). \tag{11.67}$$

Naturally,

$$Q(z, y) = \int_{-\infty}^{y} \frac{z - u}{\sigma^2(u)} du + \text{function of } z, \tag{11.68}$$

but practically, the score function $\partial_y Q(z, y)$ is used in estimation and we do not need to know the quasi-likelihood $Q(z, y)$.

There are quasi-likelihoods $Q(z, y)$ in Eq. (11.68) with particular variance functions $\sigma^2(y)$ coinciding with the logarithm of some distributions,

$$\ln f(z, y) = Q(z, y). \tag{11.69}$$

Examples of such quasi-likelihoods and their associated distributions are shown in Table 11.1.

The second column of Table 11.1 shows the quasi-likelihoods obtained according to Eq. (11.68) and corresponding to the variance function from the first column. The third column gives the canonical parameter θ as a function of y. The corresponding distributions are indicated in the forth column.

Indeed, a comparison of the second row of Table 11.1 with the log-likelihoods for the Gaussian, Poisson, and Bernoulli distributions shows that they are identical.

If y allows the parametric representation

$$y_C(x) = C^T \phi(x), \tag{11.70}$$

Table 11.1 Quasi-likelihoods, associated variance functions, and distributions.

Variance function $\sigma(y)$	Quasi-likelihood $Q(z, y)$	Canonical parameter θ	Distribution
1	$-(z-y)^2/2$	y	Gaussian
$y, y > 0$	$z \ln y - y, z \geq 0$	$\ln y$	Poisson
$y(1-y), 0 < y < 1$	$z \ln \frac{y}{1-y} + \ln(1-y), 0 \leq z \leq 1$	$\ln \frac{y}{1-y}$	Bernoulli
$y^2, y > 0$	$-z/y - \ln y, z > 0$	$-1/y$	Gamma
$y^3, y > 0$	$-z/(2y^2) + 1/y, z > 0$	$-1/(2y^2)$	Inverse Gaussian
$y^2(1-y)^2, 0 < y < 1$	$(2z-1)\ln \frac{y}{1-y} - \frac{z}{y} - \frac{1-z}{1-y}$, $0 < z < 1$	–	–

the coefficient C is found by maximization of the quasi-likelihood

$$\widehat{C} = \max_{C} \sum_{s} Q(z_s, y_C(X_s)). \tag{11.71}$$

Under general assumptions, the estimates of the quasi-likelihood behave like the log-likelihood estimates even if there is no density $f(z, y)$ satisfying Eq. (11.69).

With the factor σ_0^2 in Eq. (11.66), the quasi-likelihood becomes different from the log-likelihood. The conventional estimate of σ_0^2 is a second moment based on the residuals, namely,

$$\hat{\sigma}_0^2 = \frac{1}{n} \sum_{s} \frac{[z_s - \hat{y}_h(X_s)]^2}{\rho(\hat{y}_h(X_s))}. \tag{11.72}$$

Because the maximum conditions for Eq. (11.71) are

$$\sum_{s} \partial_C Q(z_s, y_C(X_s)) = 0,$$

we do not need to know the quasi-likelihood function Q since $\partial_C Q$ is defined in Eq. (11.67). The calculations give

$$\begin{aligned}\sum_{s} \partial_C Q(z_s, y_C(X_s)) &= \sum_{s} \partial_y Q(z_s, y_C(X_s)) \partial_C y(X_s) \\ &= -\sum_{s} \frac{z_s - y_C(X_s)}{\sigma^2(y_C(X_s))} \phi(X_s) = 0.\end{aligned} \tag{11.73}$$

Inserting Eq. (11.70) into Eq. (11.73), we arrive at an equation quasi-linear on C:

$$\sum_{s} \frac{1}{\sigma^2(y_C(X_s))} \phi(X_s)\phi^T(X_s)\hat{C} = \sum_{s} \frac{z_s}{\sigma^2(y_C(X_s))} \phi(X_s). \tag{11.74}$$

We use the term quasi-linear to emphasize the fact that Eq. (11.74) is nonlinear on C because the weights $1/\sigma^2(y_C(X_s))$ depend on the unknown C.

The solution for C can be obtained according to the following recursive procedure. Provided fixed weights $1/\sigma^2(y_C(X_s))$, the vector $\widehat{C}$ is found from the linear Eq. (11.74) and these estimates are used to update the weights. In this form, it is a well-known iteratively reweighted least-squares [197].

11.3.2 Nonparametric quasi-likelihood

A local version of the quasi-likelihood model can be defined using the window function w in Eq. (11.73) as weights of the residuals and the LPA model $y_h(x, X_s)$ for the estimated $y(x)$. This gives the local quasi-likelihood equations in the form

$$\begin{aligned} &\sum_s \frac{z_s - y_h(x, X_s)}{\sigma^2(y_h(x, X_s))} w_h(x - X_s)\phi_h(x - X_s) = 0, \\ &y_h(x, X_s) = C^T \phi_h(x - X_s). \end{aligned} \tag{11.75}$$

or in the equivalent form similar to Eq. (11.74) as

$$\sum_s \frac{w_h(x - X_s)}{\sigma^2(y_h(x, X_s))}\phi_h(x - X_s)\phi_h^T(x - X_s)\widehat{C} = \sum_s z_s \frac{w_h(x - X_s)}{\sigma^2(y_h(x, X_s))}\phi_h(x - X_s). \tag{11.76}$$

The calculations are completely as in the parametric case. The estimates of C are found from Eq. (11.76) and the variances $\sigma^2(y(X_s))$ in the weights are updated as

$$\sigma^2(y_h(x, X_s)) = \sigma^2(\widehat{C}^T \phi_h(x, X_s)).$$

Thus, we arrive at estimates that are nonlinear with respect to the observations. However, each iteration requires that only the linear equations in Eq. (11.76) be solved.

When the estimates of C are found, they are used for the estimation of y and the derivatives in the standard LPA style as

$$\begin{aligned} &\hat{y}_h(x) = \widehat{C}^T \phi(0), \\ &\hat{y}_h^{(r)}(x) = \left(\frac{-1}{h}\right)^{|r|} \widehat{C}^T \phi^{(r)}(0). \end{aligned} \tag{11.77}$$

The local quasi-likelihood estimates in Eqs. (11.76–11.77) can be used for any orders of the LPA and for any windows w directional or nondirectional. This is a general construction that is applicable for observations with varying variances.

The estimates in Eq. (11.76) also can be treated like those obtained by the minimization of the quadratic loss function with fixed weights:

$$\widehat{C} = \arg\min_{C} Q_h(C), \tag{11.78}$$

$$Q_h(C) = \sum_s \frac{e_s^2}{\sigma_s^2} w_h(x - X_s),$$
$$e_s = z_s - y_h(x, X_s), \qquad y_h(x, X_s) = C^T \phi_h(x - X_s), \tag{11.79}$$

where the variances σ_s^2 are fixed in minimization on C and updated in terms of the previously discussed recursive procedure with

$$\sigma_s^2 = \sigma_0^2 \rho(y_h(x, X_s)). \tag{11.80}$$

The nonparametric quasi-likelihood estimates are calculated by the following recursive procedure:

1. Initialization: Use equal weights $\sigma_s^2 = 1$ in Eq. (11.79) and find the initial estimate $\hat{y}_h(x, X_s)$from Eq. (11.76);
2. Update the variances $\sigma_s^2 = \rho(y_h(x, X_s))$ according to Eq. (11.80) with $\sigma_0 = 1$;
3. Find estimates of C from Eq. (11.76) and the estimates $y_h(x, X_s)$, and update the weights $1/\sigma_s^2$ using Eq. (11.80);
4. Repeat steps (2) and (3) until convergence of the estimates of C is obtained;
5. Use these estimates of C for calculations of the estimates $\hat{y}_h(x)$ and $\hat{y}_h^{(r)}(x)$ according to Eq. (11.77);
6. Repeat these pointwise calculations for all x;
7. Calculate $\hat{\sigma}_0^2$ by Eq. (11.72) and the estimates of the noise variances σ_s^2 using Eq. (11.80) for $x = X_s$, that is, according to the formula

$$\sigma_s^2 = \hat{\sigma}_0^2 \rho(\hat{y}_h(X_s)). \tag{11.81}$$

We use the term "variance-adaptive" for this type of procedure since the recursive estimation of the unknown σ_s^2 and the weights $1/\sigma_s^2$ is a key point here. In a similar manner, the corresponding estimates in Eq. (11.77) are called *variance-adaptive quasi-likelihood estimates.*

Comment 1

In general, the quasi-likelihood is developed in connection with a class of exponential distributions and allows one to estimate y when the variance of observations is a given nonlinear function of y. The quasi-likelihood and the ML are identical for a one-parameter exponential class of distributions provided there is a corresponding definition of the variance function $\sigma^2(y)$ in Eq. (11.67). However, in general, the quasi-likelihood and maximum likelihood are based on different statistical models and techniques. More details and further references on the quasi-likelihood can be found in [46], [78], [80], [145], [152], and [232].

Comment 2

This comment concerns the natural restrictions in Eq. (11.66),

$$\rho(y_h(x, X_s)) > 0. \tag{11.82}$$

For the Poisson regression, $\rho(y) = y > 0$ means that the estimate $y_h(x, X_s)$ should satisfy the condition $y_h(x, X_s) > 0$. For the Bernoulli regression with $\rho(y) = y(1 - y)$, it means that $0 < y_h(x, X_s) < 1$.

11.4 Quasi-Likelihood LPA-ICI Algorithms

We develop local adaptive window size estimates with the neighborhood size given by the ICI rule. The ICI adaptive-scale estimates are considered separately for the zero and higher-order LPAs.

11.4.1 Zero-order regression

Let us use the quasi-likelihood Eq. (11.76) for $m = 0$. Since the estimate $\hat{y}_h(x, X_s) = \widehat{C}$ is constant, the variance $\sigma^2(y_h(x, X_s))$ is also constant and the solution of Eq. (11.75) can be given in an explicit form as the weighted mean

$$\widehat{C} = \sum_s z_s g_h(x - X_s), \; g_h(x - X_s) = \frac{w_h(x - X_s)}{\sum_s w_h(x - X_s)}, \tag{11.83}$$

with the regression estimate

$$\hat{y}_h(x) = \widehat{C}. \tag{11.84}$$

The observation variance is calculated as $\sigma^2 \rho(\hat{y}_h(x))$ with the variance for the estimate in Eq. (11.83) as

$$\sigma^2_{\hat{y}_h}(x) = \sigma_0^2 \sum_s \rho(\hat{y}_h(X_s)) g_h^2(x - X_s) = \sigma_0^2 \cdot (\rho(\hat{y}_h) \circledast g_h^2)(x). \tag{11.85}$$

Formulas (11.83) define the conventional linear Nadaraya-Watson estimate. The fast convolution algorithms can be applied for the calculation of the estimate in Eq. (11.84) and the variance in Eq. (11.85). This regression estimate does not depend on the variance and does not require any variance adaptation. The unknown σ_0 in Eq. (11.85) is estimated according to Eq. (11.72).

11.4.1.1 LPA-ICI(AV) algorithm

In this algorithm we use the recursive update of the variance exploited in the ICI. The algorithm is called the LPA with the adaptive variance ICI or, in short, the LPA-ICI(AV), and consists of the following steps:

1. The estimate and variance calculation: The estimates $\hat{y}_h(x)$ in Eq. (11.84) and the variances $\sigma^2_{\hat{y}_h}(x)$ in Eq. (11.85) are calculated for $h \in H$, where H is a set of the scales.
2. Adaptive-scale selection: The ICI rule is used for the adaptive-scale selection with the output results $\hat{y}_{h^+}(x)$ and $\sigma^2_{\hat{y}_{h^+}}(x)$.
3. ICI with the variance update: The estimates $\hat{y}_{h^+}(x)$ are used for the calculation of $\rho(\hat{y}_{h^+}(x))$ in Eq. (11.85) and for the updating of the variances in Eq. (11.85) as

$$\sigma^2_{\hat{y}_h}(x) = \hat{\sigma}_0^2 \cdot (\rho(\hat{y}_{h^+}) \circledast g_h^2)(x). \tag{11.86}$$

Calculate $\hat{\sigma}_0^2$ as

$$\hat{\sigma}_0^2 = \frac{1}{n}\sum_s \frac{[z_s - \hat{y}_{h^+}(X_s)]^2}{\rho(\hat{y}_{h^+}(X_s))}. \tag{11.87}$$

Apply the ICI scale selection with these adjusted values of the variance. Repeat steps (2) and (3) up to convergence of the estimates.

The variances in Eq. (11.85) and Eq. (11.86) depend on x, which makes an essential difference versus the linear estimation with a shift-invariant kernel where the variance $\sigma^2_{\hat{y}_{h^+}}(x)$ is invariant with respect to x.

Note that, in this algorithm, for all iterations we use the same set of estimates obtained in step 1.

If the multi-window (Sec. 6.2) or directional (Sec. 8.2) estimation is exploited, then the final estimate $\hat{y}(x)$ is obtained by fusing the estimates for different windows and different directions. The layout of the directional version of the algorithm is shown in Fig. 11.2(a). In this case, the variance update formula (11.86) is replaced with

$$\sigma^2_{\hat{y}_h}(x) = \hat{\sigma}_0^2(\rho(\hat{y}) \circledast g_h^2)(x), \tag{11.88}$$

where the fused estimate $\hat{y}$ of the signal is used instead of the scale-adaptive $\hat{y}_{h^+}$ in Eq. (11.86) and $\hat{\sigma}_0^2$ is calculated as

$$\hat{\sigma}_0^2 = \frac{1}{n}\sum_s \frac{[z_s - \hat{y}(X_s)]^2}{\rho(\hat{y}(X_s))}. \tag{11.89}$$

11.4.2 High-order regression

11.4.2.1 LPA(AV)-ICI(AV) algorithm

There is a principal difference between the zero-order, $m = 0$, and high-order ($m > 0$) LPA regressions, as the latter has an estimate depending on the observation variance. The signal estimate is nonlinear because the weights in the estimate Eq. (11.76) are signal-dependent.

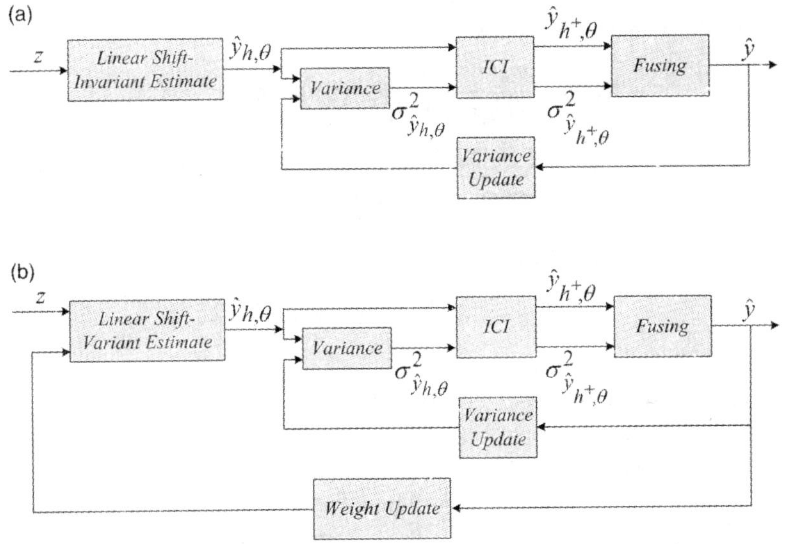

Figure 11.2 The variance-adaptive LPA-ICI algorithms: (a) LPA-ICI(AV) algorithm with the adaptive variance used in the ICI rule; (b) LPA(AV)-ICI(AV) algorithm with adaptive variance used in both the LPA by updating the weights of the estimator and in the ICI rule.

We use the recursive update of the variance for both the weights in the estimate and for the variance in the ICI. The algorithm is called the LPA(AV)-ICI(AV) algorithm, where "AV" emphasizes that the LPA procedure as well as the ICI rule include recursive calculations based on the variance update.The algorithm consists of the following steps:

1. Initialization: the estimate and variance calculation. The parameter C is found by the recursive reweighted least-square procedure for Eq. (11.76):

$$\sum_s \frac{w_h(x - X_s)}{\sigma^2(\hat{y}_h^{[l-1]}(x, X_s))} \phi_h(x - X_s)\phi_h^T(x - X_s)\widehat{C}^{[l]}$$

$$= \sum_s z_s \frac{w_h(x - X_s)}{\sigma^2(\hat{y}_h^{[l-1]}(x, X_s))} \phi_h(x - X_s), \tag{11.90}$$

$$\hat{y}_h^{[l-1]}(x, X_s) = \phi_h^T(x - X_s)\widehat{C}^{[l-1]}, \qquad l = 1, \ldots, \quad \sigma^2(\hat{y}_h^{[0]}(x, X_s)) = const,$$

with the estimate of the regression as

$$\hat{y}_h^{[l]}(x) = \phi^T(0)\widehat{C}^{[l]}. \tag{11.91}$$

This estimate can be represented in kernel form as

$$\hat{y}_h(x) = \sum_s z_s g_{h,s}(x - X_s), \tag{11.92}$$

with the shift-variant kernels

$$g_{h,s}(x - X_s) = \frac{w_h(x - X_s)}{\sigma^2(\hat{y}_h(x, X_s))}\phi^T(0)\Phi^{-1}\phi_h(x - X_s),$$
$$\Phi = \sum_s \frac{w_h(x - X_s)}{\sigma^2(\hat{y}_h(x, X_s))}\phi_h(x - X_s)\phi_h^T(x - X_s). \tag{11.93}$$

Then the variance of the estimates $\hat{y}_h(x)$ in Eq. (11.92) are defined as

$$\sigma^2_{\hat{y}_h}(x) = \hat{\sigma}_0^2 \sum_s \rho(\hat{y}_h(X_s))g^2_{h,s}(x - X_s). \tag{11.94}$$

This estimation is repeated for all x and $\hat{\sigma}_0^2$ is calculated according to Eq. (11.72).

2. ICI scale selection. The estimates $\hat{y}_h(x)$ in Eq. (11.91) and the variances $\hat{\sigma}^2_{\hat{y}_h}(x)$ in Eq. (11.94) are calculated for $h \in H$, where H is a set of the scales. The ICI rule is used for the adaptive-scale selection with the output results $\hat{y}_{h^+}(x)$ and $\hat{\sigma}^2_{\hat{y}_{h^+}}(x)$.
3. LPA(AV) estimation. Replace $\hat{y}_h(x, X_s)$ in $\sigma^2(\hat{y}_h(x, X_s))$ by the scale-adaptive estimate $\hat{y}_{h^+}(x)$in formulas (11.90):

$$\sum_s \frac{w_h(x - X_s)}{\sigma^2(\hat{y}_{h^+}(x))}\phi_h(x - X_s)\phi_h^T(x - X_s)\widehat{C}$$
$$= \sum_s z_s \frac{w_h(x - X_s)}{\sigma^2(\hat{y}_{h^+}(x))}\phi_h(x - X_s). \tag{11.95}$$

Then solve these linear equations and calculate the estimates $\hat{y}_h(x, X_s) = \phi_h^T(x - X_s)\widehat{C}$.

4. ICI(AV) scale selection. Update the variances of the observations and calculate the estimate variances as

$$\sigma_h^2(x) = \hat{\sigma}_0^2 \sum_s \rho(\hat{y}_{h^+}(X_s))g^2_{h,s}(x - X_s), \tag{11.96}$$

where

$$g_{h,s}(x - X_s) = \frac{w_h(x - X_s)}{\sigma^2(\hat{y}_{h^+}(X_s))}\phi^T(0)\Phi^{-1}\phi_h(x - X_s), \tag{11.97}$$
$$\Phi = \sum_s \frac{w_h(x - X_s)}{\sigma^2(\hat{y}_{h^+}(X_s))}\phi_h(x - X_s)\phi_h^T(x - X_s).$$

Calculate the estimate of $\hat{\sigma}_0^2$ as in Eq. (11.87). Use these variances in the ICI for the adaptive-scale selection.

5. Repeat steps (3) and (4) up to convergence of the estimates.

If the multiwindow (Sec. 6.2) or directional (Sec. 8.2) estimation is exploited, then the final estimate $\hat{y}(x)$ is obtained by fusing the estimates for different windows and different directions. This final estimate $\hat{y}(x)$ is used instead of the scale-adaptive $\hat{y}_{h^+}$ in the formulas for the variance update in the estimation in Eq. (11.95) and in the ICI in Eq. (11.96).

The flowchart of the directional version of the high-order algorithm is shown in Fig. 11.2(b). It differs from the flowchart of the zero-order algorithm by two feedbacks showing the use of the fused signal estimate for the update of the weight in the estimation algorithm in Eq. (11.95) and in the estimate of the variance calculation in Eq. (11.96) for the ICI adaptive-scale selection. The zero-order algorithm in Fig. 11.2(a) has a single feedback for the updated observation variance used in the ICI algorithm.

The shift-variant kernel in the estimation procedure means that the estimates are calculated in a pointwise manner. Assuming that the variance in the estimate Eq. (11.90) is constant, we may significantly simplify the calculations because the estimation kernel $g_{h,s}(x - X_s)$ becomes shift-invariant and the convolution can be used for fast calculations. In this case the feedback for the LPA(AV) is not used and the LPA(AV)-ICI(AV) algorithm is different from the zero-order LPA(AV)-ICI(AV) only by the high-order estimation kernel used instead of the zero-order Nadaraya-Watson estimate in Eqs. (11.83–11.84).

11.4.2.2 Variance function calculation

It is assumed by default in the quasi-likelihood approach that $\sigma^2(\cdot) > 0$. Therefore, the denominators of the estimation equations are different from zero. However, say for the Poisson regression, the variance function in Eq. (11.66), $\rho(y) = y$, and $\sigma^2(y) = \sigma_0^2 \rho(y) = \sigma_0^2 y$. When the estimate $\hat{y}$ is used instead of y, it is a natural requirement for the estimate to be positive, $\hat{y} > 0$, or at least nonnegative. As was discussed for ML estimates, this requirement is automatically fulfilled for zero-order Nadaray-Watson estimates and cannot be fulfilled for high-order estimates. Thus, when using the high-order LPA, special care should be taken when calculating the variance function ρ.

One straightforward idea is to project the estimate $\hat{y}$ on a domain of nonnegative values:

$$[\hat{y}]_+ = \begin{cases} \hat{y}, & \text{if } \hat{y} \geq 0, \\ 0, & \text{if } \hat{y} < 0. \end{cases} \tag{11.98}$$

This gives nonnegative estimates for the signal; however, zero values of the variance $\sigma^2(\hat{y}) = \sigma_0^2[\hat{y}]_+$ if $\hat{y} < 0$ are not acceptable in quasi-likelihood equations. We suggest replacing $\hat{y}$ by the posteriori Bayes mean, which is always nonnegative, and inserting it into the variance function ρ.

The following mathematics can be used here. Represent an estimate of the signal y in the form $\hat{y} = y + e_y^0$, assuming that this estimate is unbiased with Gaussian zero-mean error $e_{\hat{y}}^0 \sim \mathcal{N}(0, \sigma^2)$. Let us use the notation $f_{e_y^0}$ for this p.d.f. Then the conditional p.d.f. of the observation $\hat{y}$ (provided we have a fixed y) is

$$f_{\hat{y}}(\hat{y}|y) = f_{e_y^0}(\hat{y} - y),$$

and the posteriori p.d.f. for y provided we have a fixed estimate $\hat{y}$ is given by Bayes' formula

$$f(y|\hat{y}) = \frac{f_{\hat{y}}(\hat{y}|y)f_0(y)}{\int f_{\hat{y}}(\hat{y}|y)f_0(y)dy}, \tag{11.99}$$

where $f_0(y)$ stands for the prior p.d.f. of y.

Assuming the uniform prior distribution for the area $y \geq 0$, that is $f_0(y) = const$, if $y \geq 0$ and $f_0(y) = 0$ otherwise, we derive from Eq. (11.99)

$$f(y|\hat{y}) = \begin{cases} f_{\hat{y}}(\hat{y}|y)/\int_0^\infty f_{\hat{y}}(\hat{y}|y)dy, & y \geq 0, \\ 0, & y < 0. \end{cases}$$

Then the conditional expectation of y provided a given $\hat{y}$ is calculated as

$$E\{y|\hat{y}\} = \int_0^\infty yf(y|\hat{y})dy. \tag{11.100}$$

For a Gaussian $f_{e_y^0}$ we obtain the conditional p.d.f. as

$$f(y|\hat{y}) = \frac{\exp[-(y-\hat{y})^2/(2\sigma^2)]}{\int_0^\infty \exp[-(v-\hat{y})^2/(2\sigma^2)]dv}, \qquad y \geq 0. \tag{11.101}$$

The expectation in Eq. (11.100) is in the form

$$E\{y|\hat{y}\} = \int_0^\infty y \frac{\exp[-(y-\hat{y})^2/(2\sigma^2)]}{\int_0^\infty \exp[-(v-\hat{y})^2/(2\sigma^2)]dv} dy. \tag{11.102}$$

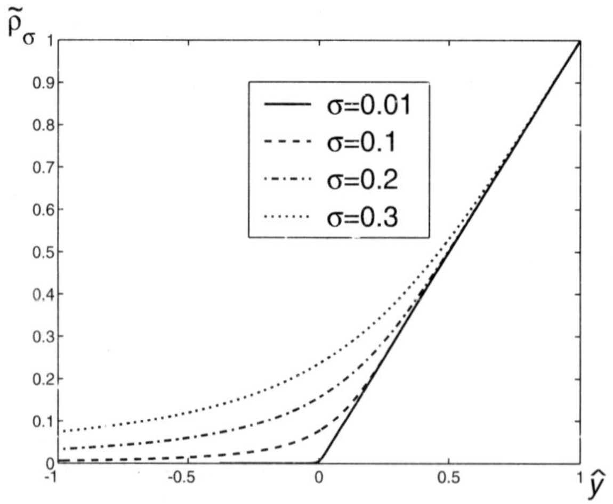

Figure 11.3 The variance function $\tilde{\rho}_\sigma$. For small σ it is close to the piecewise linear function in Eq. (11.98). For larger σ, the increasing smoothing effects are obvious.

Inserting the expectation $E\{y|\hat{y}\}$ into the formula for the variance function $\rho(y) = y$, $y \geq 0$, we obtain

$$\tilde{\rho}_\sigma(\hat{y}) \triangleq \rho(E\{y|\hat{y}\}) = E\{y|\hat{y}\}, \tag{11.103}$$

as $E\{y|\hat{y}\} > 0$. Note that the expectation $E\{y|\hat{y}\}$ as well as the variance function $\tilde{\rho}_\sigma$ depend on the estimate $\hat{y}$. The subscript σ in Eq. (11.103) indicates the dependence of the variance function on the standard deviation of this estimate.

Figure 11.3 illustrates the curves $\tilde{\rho}_\sigma(\hat{y}) = E\{y|\hat{y}\}$ for different σ. These curves are smoothed versions of the projection in Eq. (11.98) with the level of smoothing depending on the parameter σ, and $\tilde{\rho}_\sigma(\hat{y}) \rightarrow [\hat{y}]_+$ as $\sigma \rightarrow 0$.

The function $\tilde{\rho}_\sigma(\hat{y})$ is always positive for $\sigma > 0$. It is a good replacement for the projection variance function ρ in Eq. (11.98).

A similar problem with the sign of the variance function ρ occurs in Bernoulli regressions with $\rho(y) = y(1 - y)$. It follows that the estimate $\hat{y}$ should belong to the interval [0, 1].

Replace the estimate $\hat{y}$ by the posteriori Bayes mean assuming that the prior p.d.f. is uniform on the interval [0, 1]. Then we obtain the posteriori mean that always belongs to the interval [0, 1]. Calculations similar to those above give the following formula for this posteriori Bayes' mean:

$$E\{y|\hat{y}\} = \int_0^1 y \frac{\exp[-(y - \hat{y})^2/(2\sigma^2)]}{\int_0^1 \exp[-(v - \hat{y})^2/(2\sigma^2)]dv} dy. \tag{11.104}$$

Chapter 12

Photon Imaging

In many imaging systems, recorded observations have the physical meaning of a quantity of detected photons. The photons are counted at different spatial locations and in this way form an image of the object. This sort of scenario is typical for imaging in digital photography and for many problems in medicine, including positron emission tomography (PET) and single-photon emission tomography (SPET) [131], [170]; in gamma astronomy; and in microscopy [95], [207].

For instance, confocal microscopy is used to obtain volume images of small fluorescent objects with high-spatial resolutions [176]. The microscope performs a 3D scan of the object, and at each point of the scan a photon multiplier tube measures the emission of fluorescence from the object, essentially acting as a photon counter. Due to the geometry of these microscopes, a "blurring" is introduced into the measurement process. The Poisson distribution is a conventional probabilistic model used for a random number of photons appearing during an exposure time.

In this chapter we concentrate on Poissonian models and simulation experiments demonstrating the performance of the developed algorithms. In Sec. 12.1 we consider the direct observations where the filtering of Poissonian data is the main goal. The remaining sections deal with the complex problems associated with indirect observations when the filtering is complicated by the required deconvolution.

Let us specify the considered problem once more. For direct photon observations, when blurring effects are negligible, observed data can be presented in standard additive form as

$$z(x) = y(x) + \varepsilon(x), \tag{12.1}$$

where with a fixed x the observed random z has the Poisson distribution $z \sim \mathcal{P}(y)$ in Eq. (10.15). This means that

$$\begin{aligned} P\{z = k\} &= \exp(-y)\frac{y^k}{k!}, \qquad k = 0, 1, \ldots, \\ E\{z\} &= y, \qquad var\{z\} = y, \end{aligned} \tag{12.2}$$

where y is the intensity or the mean value of the Poisson distribution.

In the model in Eq. (12.1) we consider the Poisson parameter y as a regression that depends on the argument x. The Poisson observations can be quite noisy. The mean-to-standard-deviation ratio is expressed as

$$\frac{E\{z\}}{\sqrt{var\{z\}}} = \frac{y(x)}{\sqrt{y(x)}} = \sqrt{y(x)}. \tag{12.3}$$

This says that, for a small y, this ratio is small. Thus, for a small y, the signal is noisy and can be very noisy if $y \to 0$.

For a fixed x, estimation of the parameter y of the distribution is not a problem because a sample mean gives the best estimate of y as

$$\hat{y}(x) = \frac{1}{n} \sum_{i=1}^{n} z_i(x), \tag{12.4}$$

where n is the number of independent observations $z_i(x)$ obtained during nonoverlapping time intervals of the unit length. For large n, the distribution of the estimate $\hat{y}(x)$ is well approximated by the Gaussian as $\mathcal{N}(\hat{y}(x), \hat{y}(x)/n)$. Obviously, the accuracy can be improved for longer observation times and for larger n in Eq. (12.4).

The problem occurs when we wish to reconstruct an image or a surface $y(x)$ using the discrete observations given for the grid of X_s. The straightforward idea of estimating $y(X_s)$ using Eq. (12.4) and then interpolating these estimates for other x usually does not work because it requires too many observations and too much time to achieve an acceptable accuracy.

The only workable alternative is to use some model of the signal of interest $y(x)$ and to estimate the signal using this model. The LPA is a nonparametric technique developed for this sort of problem. Overall, the situation is standard for most problems concerning signal reconstruction from noisy data. The specific point to be made here is that in photon imaging we often have to count observations characterized by a high level of randomness and by the essential variability of the noise variance.

Further, we consider an indirect observation model in a form similar to Eq. (9.3) and assume that a measurement at position x is obscured by effects from nearby points (tissues, bones, objects, etc.). Sometimes these effects are modeled by the convolution kernel v, which can be treated as a PSF of the system.

Let v be shift-invariant and the observation mean (not the observations themselves) presented in the convolution form $(y \circledast v)(x)$. We arrive at the Poisson indirect observation model

$$z \sim \mathcal{P}((y \circledast v)(x)), \tag{12.5}$$

or

$$z(x) = (y \circledast v)(x) + \varepsilon(x), \tag{12.6}$$

where $x \in X$ is defined on a regular $n_1 \times n_2$ lattice $X = \{k_1, k_2 : k_1 = 1, 2, \ldots, n_1$ $k_2 = 1, 2, \ldots, n_2\}$, and ε is zero-mean with variance $var\{\varepsilon(x)\} = (y \circledast v)(x)$.

We have for the mean of z

$$E\{z(x)\} = (y \circledast v)(x), \tag{12.7}$$

and for the variance of z

$$\sigma_z^2(x) = var\{z(x)\} = (y \circledast v)(x).$$

Note that z in Eq. (12.5) is a random counting variable that takes nonnegative integer values.

The convolution model allows an interesting and useful probabilistic interpretation. To make it clearer, rewrite the model in Eq. (12.6) in an explicit form as

$$z(x) = \sum_k y(x-k)v(k) + \varepsilon(x).$$

Let the PSF be nonnegative and normalized,

$$v(k) \geq 0 \qquad \text{and} \qquad \sum_k v(k) = 1.$$

Then $v(k)$ can be considered as the probability that a photon originated from the point $x - k$ is counted at the coordinate x.

In nuclear medical applications, the values of $v(k)$ depend on various factors, including the geometry of the detection system, the activity of the isotopes, the exposure time, and the extent of attenuation and scattering between the source and the detector [69]. The PSF works as a general modeling concept for various complex systems.

12.1 Direct Poisson Observations

Let us start by assuming that the sensor is ideal. Then, $v(x) = \delta_{x,0}$ and the direct observation model in Eq. (12.1) is valid,

$$z(x) \sim \mathcal{P}(y(x)). \tag{12.8}$$

The following two methods considered in Chapters 10 and 11 can be exploited as the basis for nonparametric estimations of y from Eq. (12.8):

- Anscombe nonlinear transform (Sec. 10.4);
- Local quasi-likelihood (Sec. 11.4).

12.1.1 Anscombe transform

Using Eq. (10.109) replaces the observation z in Eq. (12.8) by

$$\xi = 2\sqrt{z + 3/8}. \tag{12.9}$$

The Anscombe transform enables the variance stabilizing effect and the new observations $\xi(x)$ have an approximately invariant variance for all x. The adaptive varying-scale LPA-ICI filtering techniques developed in Chapters 6–8 are applicable to the transformed observations in Eq. (12.9).

If the LPA-ICI estimate $\hat{\xi}$ of ξ is found, the final step of the algorithm is to apply the inverse Anscombe transform with the estimate for y as

$$\hat{y} = \hat{\xi}^2/4 - 3/8. \tag{12.10}$$

This method is as complex as the LPA-ICI algorithm. However, it does not take into consideration that the counting nature of the observations and the variance stabilization is approximate, as is clear from Eqs. (10.106–10.107). This means that the local likelihood or quasi-likelihood-based techniques are promising for better accuracy results.

12.1.2 Local quasi-likelihood

The quasi-likelihood algorithms from Sec. 11.4 with $\rho(y) = y$ are applicable for direct Poisson observation problems. The estimation algorithm has the standard form of the reweighted LSM. The flowchart of the algorithms is shown in Fig. 11.2.

To use fast convolution computing, we assume that the variances $\sigma^2(\hat{y}_h(x, X_s))$ are invariant in the estimates. This assumption is supported by the ICI algorithm, which automatically restricts the size of the estimation neighborhood and also restricts the range of the observation variance exploited in the estimates.

In this way we drop the variance adaptation from the estimate calculations. However, we use this variance adaptation in the ICI rule for the threshold variance adaptations. This means that the LPA-ICI(AV) algorithm with the variance-adaptive ICI rule (Sec. 11.4) is exploited for filtering Poissonian signals.

12.1.3 Recursive LPA-ICI(AV) algorithm

We also propose a recursive version of the LPA-ICI(AV) algorithm that is similar to the recursive LPA-ICI algorithm described in Sec. 8.2 but modified for Poissonian observations and the variance update in the ICI rule. The algorithm is based on the LPA-ICI(AV) procedures presented in Sec. 11.4 (Fig. 11.2a). The adaptive-scale LPA-ICI(AV) filtering is applied a few times using as inputs the results of the previous filterings. The parameters of the algorithm can be different for each step of this blockwise recursive procedure.

12.1.3.1 Recursive LPA-ICI(AV) filter

For each step, the LPA-ICI filter is multidirectional with narrow conical supports of the estimation kernels. The linear LPA estimates of zero-order $m = 0$ are calculated for all scales H and for all K directions:

$$\hat{y}_{h,\theta_i}(x) = (g_{h,\theta_i} \circledast z)(x), \qquad i = 1, \ldots, K, \quad h \in H, \tag{12.11}$$

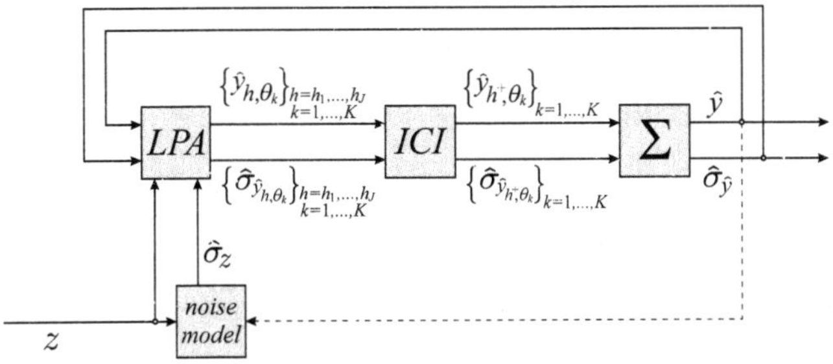

Figure 12.1 Recursive LPA-ICI(AV) algorithm. The signal z and its variance are inputs. The estimate $\hat{y}$ and its variance are outputs.

where

$$g_{h,\theta}(x) = w_{h,\theta}(x) / \sum_{x \in X} w_{h,\theta}(x), \tag{12.12}$$

and $w_{h,\theta}(x) > 0$ are the directional windows of the conical supports. The variance of these estimates is calculated as

$$\hat{\sigma}^2_{\hat{y}_{h,\theta_i}}(x) = \sum_s \sigma^2_{\hat{y}(X_s)} g^2_{h,\theta_i}(x - X_s) \tag{12.13}$$

and applied in the ICI.

The algorithm works as LPA-ICI(AV), assuming that only the variances for the ICI are updated while the estimates are calculated as linear ones with data-independent kernels $g_{h,\theta}$.

The flowchart of this recursive LPA-ICI(AV) is given in Fig. 12.1(a), where the contents of the steps are shown and the feedbacks indicate the links between the recursive steps. The block “noise model” is used to update the variance σ_z of the input signal z.

12.1.3.2 Steps of the recursive LPA-ICI(AV)

The recursive LPA-ICI(AV) filter is used in the recursive LPA-ICI(AV) algorithm. The steps of this algorithm differ by input data, calculation of variance for input data, and parameters of the used procedures.

- *Step 1: Initialization.* The input data are the observations z_s. The variances of the observations used in Eq. (12.13) are $\sigma^2_{\hat{y}(X_s)} = z_s$. This is the simplest “initial guess” for the observations’ variance.

  Experiments show that this choice works quite well. In particular, it is much better than a constant-variance assumption.

  The output data are the estimates $\hat{y}^{[1]}(x)$ calculated according to Eq. (12.11), where the superscript [1] shows the step number.
- *Step 2*: The input data are again the original observations z_s. The variances of the observations in Eq. (12.13) are defined according to the values of the estimates $\hat{y}^{[1]}(x)$ as $\sigma^2_{\hat{y}(X_s)} = \hat{y}^{[1]}(X_s)$. Thus, the initialization step is used only

to obtain the variance for the observations. These first two steps (steps 1 and 2) correspond to the LPA-ICI(AV) algorithm with one iteration for the variance update.

The outputs of step 2 are the estimates $\hat{y}^{[2]}(x)$ and the variance of these estimates $\sigma^2_{\hat{y}[2]}(x)$ calculated using the fused adaptive-scale directional estimates in Eqs. (8.62–8.63) and (8.65).

After this second step, the estimate $\hat{y}^{[2]}(x)$ is no longer Poissonian and the variance of this estimate is not equal to the mean value of the estimate.

- *Step 3*: The input data are the estimates $\hat{y}^{[2]}(x)$ instead of the original data z_s and the variances $\sigma^2_{\hat{y}[2]}(x)$ of the estimates used in Eq. (12.13) instead of $\sigma^2_{\hat{y}(X_s)}$ as the observation variances.

  The outputs of the algorithm are the estimates $\hat{y}^{[3]}(x)$and their variances $\sigma^2_{\hat{y}[3]}(x)$ calculated using the fused adaptive-scale directional estimates in Eqs. (8.62–8.63) and (8.65).

  Steps 4 and higher are identical to step 3 with input signal and variances obtained from the previous steps.

12.1.3.3 Optimization of the algorithm

Overall, the experimental work presented in this section is close to that produced for Gaussian image denoising in Sec. 8.3. Some work has been done to optimize the parameters of the algorithm. This optimized algorithm has been tested for multiple experiments, a part of which is demonstrated in the following.

We use narrow line-wise nonoverlapping sectorial kernels of type 10 (Sec. 7.3) with a shift-invariant window function w. The kernel support and scale are defined by the parameters $h \in H$:

$$H = \left\{ \begin{pmatrix} 1 \\ 1 \end{pmatrix}, \begin{pmatrix} 2 \\ 1 \end{pmatrix}, \begin{pmatrix} 3 \\ 1 \end{pmatrix}, \begin{pmatrix} 4 \\ 1 \end{pmatrix}, \begin{pmatrix} 6 \\ 1 \end{pmatrix}, \begin{pmatrix} 8 \\ 1 \end{pmatrix}, \begin{pmatrix} 10 \\ 1 \end{pmatrix}, \begin{pmatrix} 12 \\ 2 \end{pmatrix} \right\}. \tag{12.14}$$

The set H forms eight different pairs of h_1 and h_2 as in Eqs. (9.60–9.61). The ICI selects the best pair.

The estimates and the adaptive scales are calculated for eight directions $\theta_j = (j-1)\pi/4,\ j = 1, \ldots, 8$. These adaptive ICI estimates are fused using the multiwindow method with the final estimate calculated as the weighted mean in Eqs. (8.62–8.63).

The kernel $g_{h,\theta}$ is defined as a linear combination of the kernels of the zero and first orders:

$$g_{h,\theta} = \kappa g_{h,\theta}|_{m=(0,0)} + (1-\kappa) g_{h,\theta}|_{m=(1,0)}, \tag{12.15}$$

where the multiple index of m shows the orders of the kernels with respect to arguments x_1 and x_2, and κ is a parameter.

Experiments show that the convergence of the recursive LPA-ICI algorithm is very fast. A sufficient number of recursive steps is restricted by four.

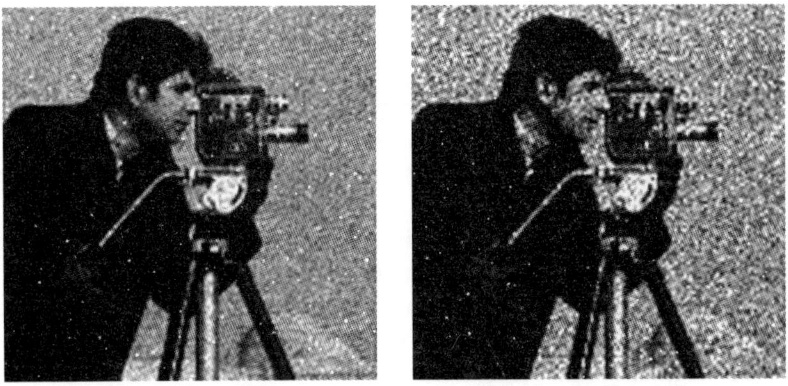

Noisy image with $\chi = 120$ Noisy image with $\chi = 30$

Figure 12.2 *Cameraman* fragment: Poisson noisy normalized images.

It was found that performance is better if the observations are not oversmoothed on a single iteration of the recursive algorithm because some smoothing is imbedded in the iterations themselves. Thus, a smaller value is selected for the threshold parameter $\Gamma = 0.7$. It is not a crucial point because the criteria values are more or less the same. However, the appearance of the estimates becomes better with this policy.

The parameter κ in the combined kernel in Eq. (12.15) is better taken with different values for different steps. We select $\kappa = 1$, that is, the zero-order model for the first step and $\kappa = 0.5$ for the rest.

12.1.3.4 Observations

In our simulations for the Poissonian observations, to achieve a different level of randomness (i.e., different SNRs) in the noisy observations, we multiply the true signal y (which has range [0, 1]) by a scaling factor:

$$z \sim \mathcal{P}(y \cdot \chi), \tag{12.16}$$

where $\chi > 0$ is a scaling factor. Scaling by χ allows us to get the random data z with a different level of random noise:

$$E\{z\} = y\chi, \; var\{z\} = y\chi. \tag{12.17}$$

The SNR in Eq. (12.3) is calculated as

$$\frac{E\{z\}}{\sqrt{var\{z\}}} = \sqrt{y\chi}. \tag{12.18}$$

For larger χ we have a smaller relative level of noise. In result imaging we always depict the signals and the estimates as being normalized by χ, that is, we show z/χ instead of z. This is done to preserve the mean value of the image intensity and make the results comparable for any χ.

The mean of the ratio z/χ is equal to y for any χ, the variance is calculated as

$$var\{z/\chi\} = \frac{var\{z\}}{\chi^2} = \frac{y\chi}{\chi^2} = \frac{y}{\chi}.$$

The last result agrees with Eq. (12.18) because a smaller χ means a larger level of randomness in the normalized signal.

Comparing the normalized *Cameraman* images in Fig. 12.2, we note that the observed level of noise for $\chi = 120$ is much lower than it is for $\chi = 30$.

In Fig. 12.3 we show the cross sections of this normalized *Cameraman* image for true and noisy images with $\chi = 30$ and $\chi = 120$. We note that the noise essentially disturbs the original image curve and the level of this clear random disturbance is obviously signal-dependent. We also recognize that for $\chi = 30$ the relative noise amplitude is significantly higher than that for $\chi = 120$.

We use this modeling of Poisson data because it allows us to compare our results with similar simulation scenarios appearing in a number of publications. In particular, we demonstrate good performance when comparing with the recently developed wavelet-based methods for Poisson data. Here we refer to the methods and modeling results in recent works [146], [167], [221], and [234].

12.1.4 Numerical experiments

Figures 12.4, 12.5, and 12.6 show the noisy and original images and the estimates obtained for the first initialization and the fourth step of the recursive LPA-ICI(AV)

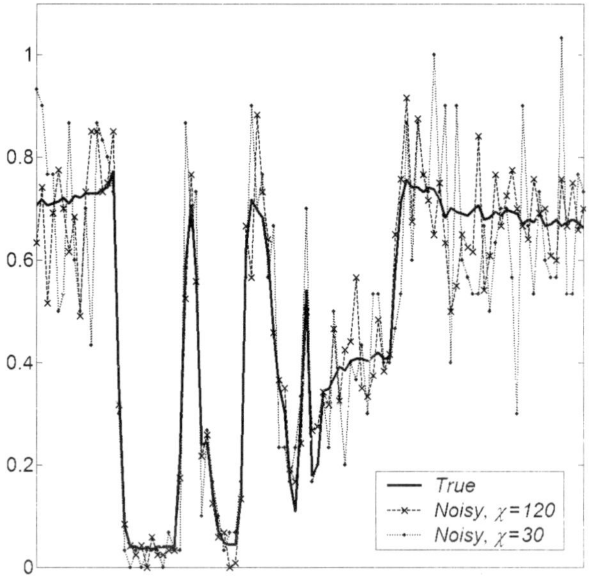

Figure 12.3 A cross section of the normalized *Cameraman* image: true image and Poisson noisy images with χ = 30 and χ = 120.

Figure 12.4 Filtering the *Cameraman* test image (fragment): data (first row, left) with a scaling factor equal to 60 (MSE = 23.9); first row, right—first step of the algorithm LPA-ICI(AV) (MSE = 7.90, IMSE = 4.81 dB); second row, left—fourth step of the algorithm LPA-ICI(AV) (MSE = 4.36, ISNR = 7.40 dB); second row, right—original image.

algorithm. The MSE values demonstrate a vast performance improvement in successive iterations. The quality of the final estimates is quite good visually and numerically. In particular, for the *Cameraman* image we achieve ISNR = 9.34 dB for $\chi = 30$, ISNR = 8.05 dB for $\chi = 60$, ISNR = 7.45 dB for $\chi = 90$, and ISNR = 6.82 dB for $\chi = 120$.

Some numerical results and comparisons with other methods for the *Cameraman* and *Lena* images are presented in Tables 12.1 and 12.2, respectively. The first row shows the values of the scaling factor χ. All other figures in the tables are the values of the MSE.

Starting from the second step, the recursive LPA-ICI(AV) algorithm shows the advance results, which are essentially better in comparison with the ones presented in [221] and [146]. Comparing the MSE values obtained for the successive steps to the recursive LPA-ICI(AV), we note that the main improvement is achieved in the first two to three steps.

Figure 12.5 Filtering *Lena* test image: top row, left—noisy data $\chi = 60$ (MSE = 29.0, SNR = 15.3 dB); top row, right—first step of the LPA-ICI(AV) algorithm (MSE = 7.07, ISNR = 6.13 dB); second row, left—fourth step (MSE = 2.62, ISNR = 10.4 dB); second row, right—original image.

The MSE values for the Spine image are shown in Table 12.3. The improvement obtained in the successive steps of the LPA-ICI (AV) algorithm is clearly seen.

We also show the results obtained using the Anscombe transform. For filtering the Anscombe transform data, the recursive (four-step) LPA-ICI algorithm is applied (Sec. 8.2) with the parameters optimized for Gaussian noise. The results are shown in the last row of Tables 12.1, 12.2, and 12.3.

We note that the obtained numerical values from the MSE are quite close to the best values achieved by the LPA-ICI(AV) algorithm when also using the four-step recursive filtering.

The Anscombe transform algorithm combined with the LPA-ICI developed for Gaussian noise filtering gives very good results. At least the numerical comparison with the results shown in [221] and [146] is in favor of the developed adaptive Anscombe transform algorithm. A visual comparison of the Anscombe transform algorithm versus the LPA-ICI(AV) algorithm favors the latter, but this advantage is not essential.

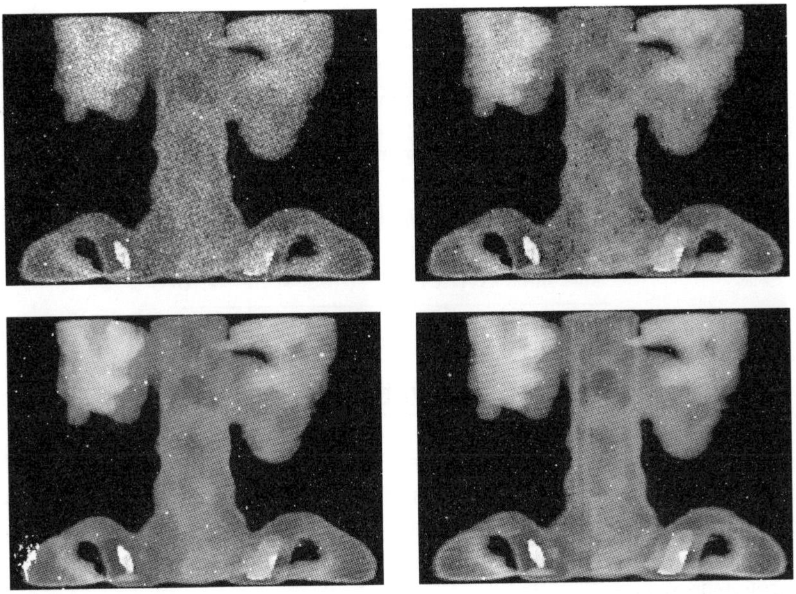

Figure 12.6 Filtering the *Spine* test image: data (first row, left) with scaling factor equal to 60 (MSE = 14.8, SNR = 15.1 dB); first row, right—first step of the algorithm LPA-ICI(AV) (MSE = 3.91, ISNR = 5.79 dB); second row, left—fourth step of the algorithm LPA-ICI(AV) (MSE = 1.56, ISNR = 9.77 dB); second row, right—original image.

Table 12.1 MSE comparison for *Cameraman* image.

χ-value	30	60	90	120
Noisy image	13.9	27.5	42.1	56.0
TN method [221]	2.76	7.73	14.11	21.59
Improved TN method [146]	2.13	5.37	9.22	13.59
LPA-ICI(AV), first step	3.75	8.04	13.4	18.8
LPA-ICI(AV), second step	2.28	5.30	8.96	13.2
LPA-ICI(AV), third step	1.79	4.50	7.79	11.8
LPA-ICI(AV), fourth step	**1.62**	**4.30**	**7.58**	**11.6**
Anscombe LPA-ICI	1.78	4.49	7.75	11.7

Table 12.2 MSE comparison for *Lena* image.

χ-value	30	60	90	120
Noisy image	14.5	29.0	43.6	58.0
TN method [221]	2.33	6.46	11.62	17.89
Improved TN method [146]	1.98	5.32	9.35	14.03
LPA-ICI(AV), first step	3.29	7.07	11.3	15.9
LPA-ICI(AV), second step	1.76	4.07	6.80	9.84
LPA-ICI(AV), third step	1.20	3.05	5.33	7.99
LPA-ICI(AV), fourth step	**0.99**	**2.62**	**4.72**	**7.19**
Anscombe LPA-ICI	1.23	3.13	5.44	8.16

Table 12.3 MSE for *Spine* image.

χ-value	30	60
Noisy image	7.45	14.8
LPA-ICI(AV), first iteration	1.91	3.91
LPA-ICI(AV), second iteration	1.08	2.37
LPA-ICI(AV), third iteration	0.77	1.79
LPA-ICI(AV), fourth iteration	**0.66**	**1.56**
Anscombe LPA-ICI	0.70	1.76

12.2 Indirect Poisson Observations

Nonlinear image deblurring procedures based on probabilistic considerations widely outperform conventional linear methods. Many nonlinear algorithms incorporate specific statistical models for the noise process in recorded data. In this section we consider these types of nonlinear probabilistic methods designed for the Poisson distribution data.

12.2.1 ML pure inverse

12.2.1.1 ML estimate

Since the random observation z has a Poisson distribution with mean $E\{z(x)\} = (y \circledast v)(x)$, the corresponding log-likelihood according to Eq. (11.59) has the form

$$L = \log l = \sum_s \{-(y \circledast v)(X_s) + z_s \log[(y \circledast v)(X_s)]\}, \tag{12.19}$$

where the term $\log(z_s!)$, which is independent on y, is omitted. The ML gives the estimate of y as

$$\hat{y} = \arg\max_y L. \tag{12.20}$$

Let us assume that the PSF is shift-invariant. Then the log likelihood Eq. (12.19) is of the form

$$L = \sum_s \left\{ -\sum_l v(s-l)y(l) + z_s \log\left[\sum_l v(s-l)y(l)\right]\right\}, \tag{12.21}$$

where s is the coordinate X_s.

The gradient and the Hessian matrix of L are, respectively, defined by the derivatives

$$\partial_{y(l)} L = -\sum_s \left[1 - \frac{z_s}{\sum_l v(s-l)y(l)}\right] v(s-l), \tag{12.22}$$

$$\partial_{y(l)}\partial_{y(l_1)} L = -\sum_s z_s \frac{v(s-l)v(s-l_1)}{\left[\sum_l v(s-l)y(l)\right]^2}. \tag{12.23}$$

The Hessian $\partial_{y(l)}\partial_{y(l_1)} L$ is a negative-definite matrix provided that the values of $v(l)$ are linearly independent. This means that the likelihood L is concave on $y(l)$ and the problem in Eq. (12.20) has a unique solution. Note that the linear independence of $v(l)$ means that $\sum_l v(s-l)b_l = 0$ only when all $b_l = 0$, that is, the convolution defined by $v(k)$ is invertible.

The ML solution in Eq. (12.20) is a nonlinear analogy of the naive (pure) inverse in Eq. (9.5), where we solve the linear convolution equation in a straightforward attempt to find the convolved values of the signal. This sort of approach does not assume any smoothness of y and results in a noisy and usually useless estimate. Nevertheless, the estimate in Eq. (12.20) is the ML solution. It is the best possible estimate taking into consideration both the systematic and random components of the Poisson observations simultaneously. A drawback of this solution is that it is pointwise. It estimates the values of $y(l)$ independently and does not assume any smoothness of the signal $y(l)$ as a function of the argument l.

We call this ML estimate a *Poisson naive inverse* or just *Poisson inverse*. To be practically efficient, the Poisson inverse requires some smoothing or filtering to attenuate the random components of this noisy estimate.

This smoothing can be introduced in a number of ways. The way that is most popular in the field employs regularization techniques that replace the likelihood function by a regularized likelihood with some penalty term inducing smoothing effects.

A second method involves post-processing (filtering) of the Poisson inverse or regularized Poisson inverse. The LPA-ICI adaptive filtering does a good job in this particular area.

Finally, note that the Poisson inverse is a typical ill-conditioned problem. The conditioning is defined by the PSF, but it is different from the one discussed in Sec. 9.1, because for the Poisson distribution it is an essentially nonlinear problem.

12.2.2 Richardson-Lucy method

The Poisson naive inverse can be obtained using the recursive *Richardson-Lucy* method introduced by Richardson and Lucy in image restoration and by Shepp and Vardi in emission tomography [14]. The necessary minimum conditions for L follow from

$$\partial_{y(l)} L = 0, \qquad l = 1, \ldots, n,$$

and according to Eq. (12.22) can be given in the form

$$1 = \frac{1}{A}\sum_s z_s \frac{v(s-l)}{\sum_l v(s-l)y(l)}, \qquad l = 1,\ldots,n. \tag{12.24}$$

where

$$A = \sum_s v(s-l).$$

For the normalized PSF we have $A = 1$. However, let us keep A in the equation because in the general case it may be not equal to one.

Multiplying Eq. (12.24) by $y(l)$, we arrive at an equation identical to Eq. (12.24) provided that $y(l) \neq 0$:

$$y(l) = \frac{1}{A}y(l)\sum_s z_s \frac{v(s-l)}{\sum_l v(s-l)y(l)}, \qquad l = 1,\ldots,n. \tag{12.25}$$

Gold's multiplicative recursive scheme ([93], pp. 114–116) applied to Eq. (12.25) has the form

$$y^{[k]}(l) = \frac{1}{A}y^{[k-1]}(l)\sum_s z_s \frac{v(s-l)}{\sum_l v(s-l)y^{[k-1]}(l)}, \\ l = 1,\ldots,n, \quad k = 1,2,\ldots, \tag{12.26}$$

where $^{[k]}$ are the numbers of iterations.

Using convolution notation we can represent Eq. (12.26) in a more compact form:

$$y^{[k]}(l) = \frac{1}{A}y^{[k-1]}(l)\sum_s z_s \frac{v(s-l)}{\hat{z}^{[k-1]}(s)} = y^{[k-1]}(l)\frac{1}{A}\left(\tilde{v} \circledast \frac{z}{\hat{z}^{[k-1]}}\right)(l),$$

$$\hat{z}^{[k-1]}(s) = \sum_l v(s-l)y^{[k-1]}(l) = (v \circledast y^{[k-1]})(l), \qquad l = 1,\ldots,n, \quad k = 1,2, \tag{12.27}$$

where $\tilde{v}$ stands for the conjugate PSF with kernel $\tilde{v}(l) = v(-l)$.

The recursion in Eq. (12.27) is known as the recursive *Richardson-Lucy* algorithm. It requires the calculation of two convolutions: $\hat{z}^{[k-1]}(s)$, which is the prediction for z_s, and the convolution of the ratio $z/\hat{z}^{[k-1]}$ with the conjugate kernel $\tilde{v}$. Because it is based on convolutions, the *Richardson-Lucy* recursion is a fast algorithm for each iteration.

Main properties of the algorithm:

- Each iteration is nonnegative with $y^{[k]} \geq 0$;
- The following identity holds:

$$\sum_s \hat{z}_s = \sum_s z_s,$$

that is, the total mean intensity of the observation image z_s coincides with the mean prediction intensity $\hat{z}_s$;

- The log-likelihood of $y^{[k]}(l)$ is a nondecreasing function of k,

$$L(y^{[k]}) \geq L(y^{[k-1]});$$

- Provided a positive initiation, $y^{(0)}(l) > 0$, $l = 1, \ldots, n$, the algorithm converges to the solution of Eq. (12.20);
- The convergence rate of the recursions of the Richardson-Lucy algorithm can be slow.

The computational efficiency of the estimated and achieved convergence rates can be improved using the Gauss-Newton style algorithms as in Eqs. (11.29–11.32) instead of the standard Richardson-Lucy algorithm [48].

12.2.3 EM pure inverse

The *Richardson-Lucy* algorithm can be derived using the *expectation-maximization* (*EM*) method. Let us present this general technique in the context of Poisson distribution data:

$$z(x) \sim \mathcal{P}\left(\mu(x)\right),$$

where $\mathcal{P}(\mu(x))$ denotes a Poisson distribution with the parameter μ.

The unknown intensities $\mu(x)$ are related to other unknown intensities $y(x)$ via a linear relation,

$$\mu(s) = \sum_{l=1}^{L} v(s,l)y(l), \qquad s = 1, \ldots, n, \tag{12.28}$$

where the nonnegative weights $v(s,l)$ are normalized to unity as

$$\sum_{s=1}^{n} v(s,l) = 1.$$

This model is more general than the shift-invariant convolution given in Eq. (12.6). Photons are emitted (from the emission space) according to the intensity $y(l)$. Those photons emitted from location l are detected (in detection space) at position s with transition probability $v(s,l)$.

It is assumed that the detected and counted photons have a Poisson distribution with the log-likelihood

$$L = \sum_{s}\left\{-\sum_{l} v(s,l)y(l) + z_s \log\left[\sum_{l} v(s,l)y(l)\right]\right\}. \tag{12.29}$$

Corresponding to the intensities y and μ we have *counts observed* (or *counts registered*) by sensors:

$$z_s \sim \mathcal{P}\left(\mu(s)\right), \tag{12.30}$$

and *unobserved counts* of photons emitted from the position l and registered at the position s:

$$\tilde{z}(s,l) \sim \mathcal{P}\left(v(s,l)y(l)\right). \tag{12.31}$$

If we would observe these *unobserved* counts $\tilde{z}(s,l)$ the really observed and recorded data z can be computed according to the balance equation as

$$z_s = \sum_{l=1}^{L} \tilde{z}(s,l). \tag{12.32}$$

Then the emitted and unobserved counts can be computed as

$$\tilde{z}(l) = \sum_{s=1}^{n} \tilde{z}(s,l). \tag{12.33}$$

These counts are the sums of the Poisson flows and have the Poisson distributions, respectively,

$$z_s \sim \mathcal{P}\left(\mu(s)\right), \qquad \tilde{z}(l) \sim \mathcal{P}\left(y(l)\right).$$

We differentiate between observed and unobserved photons by putting a tilde over the counts of the unobserved ones.

The unobserved counts $\tilde{z}(s,l)$ provide complete information about the state of the emission in the system and allow one to compute the amount of photons registered by sensors as well as those emitted by the sources.

The term "incomplete data" in the EM method implies the existence of two sample spaces: unobserved $\tilde{Z}$ (composed from $\tilde{z}(s,l)$) and observed Z (composed from z_s) and mapping from $\tilde{Z}$ to Z (defined by Eq. (12.32)). The corresponding $\tilde{z}(s,l)$ in the space $\tilde{Z}$ are not observed directly but only indirectly through z_s.

If $\tilde{z}(s,l)$ were known, we could avoid the inverse problem altogether and estimate the intensities $y(l)$ from the corresponding direct observations $\tilde{z}(l)$ in Eq. (12.33) (Sec. 12.1). The problem is to estimate $y(l)$ from the incomplete available observations Z.

The EM method uses the unobserved $\tilde{z}(s,l)$ as auxiliary parameters to simplify the calculation and maximize the likelihood. With the main goal of obtaining an ML estimate, the EM method proposes a recursive procedure based on updating the estimates of the unobserved variables.

Each iteration of the EM algorithm involves two steps, which are called the *expectation step* (E-step) and the *maximization step* (M-step). In the E-step the conditional expectation of the unobserved $\tilde{z}(s,l)$ is calculated provided that the z_s and $y(l)$ are known. In the M-step this conditional estimate of $\tilde{z}(s,l)$ is inserted into the log-likelihood and the ML gives the desirable estimates of $y(l)$.

If we have for the unobserved counts $\tilde{z}(s,l) \sim \mathcal{P}\left(v(s,l)y(l)\right)$ in Eq. (12.31) the corresponding log-likelihood will be of the form

$$L(y) = \sum_{s=1}^{n} \{-v(s,l)y(l) + \tilde{z}(s,l)\log[v(s,l)y(l)]\}, \tag{12.34}$$

where the term $\sum_{s=1}^{n} \log[\tilde{z}(s,l)!]$ is omitted as independent on y. The estimates of $y(l)$ provided given $\tilde{z}(s,l)$ follow from the maximization of this $L(y)$ on $y(l)$:

$$y(l) = \frac{\sum_{s=1}^{n} \tilde{z}(s,l)}{\sum_{s=1}^{n} v(s,l)}. \tag{12.35}$$

The precise definition of the EM *algorithm* for the model in Eq. (12.28) is as follows [36]. Suppose that $y^{[k-1]}$ denotes the current estimate of y after $(k-1)$ cycles of the algorithm.

The next cycle can be described in two steps, as follows:

E-step: Estimate the conditional expectation of $\tilde{z}(s,l)$ provided $y^{[k-1]}$ and z are given:

$$\tilde{z}^{[k-1]}(s,l) = E\{\tilde{z}(s,l)|y^{[k-1]}, z\} \tag{12.36}$$

M-step: Find $y^{[k]}$ from the optimization

$$y^{[k]} = \arg\max_{y} L(y), \tag{12.37}$$

where $L(y)$ is the log-likelihood of the unobserved counts with $\tilde{z}(s,l)$ replaced by the estimate $\tilde{z}^{[k-1]}(s,l)$.

The conditional expectation in Eq. (12.36) (E-step) is calculated as

$$\tilde{z}^{[k-1]}(s,l) = \frac{v(s,l)y^{[k-1]}(l)}{\sum_{l=1}^{L} v(s,l)y^{[k-1]}(l)} z_s, \tag{12.38}$$

and according to the M-step in Eq. (12.35)

$$y^{[k]}(l) = \frac{\sum_{s=1}^{n} \tilde{z}^{[k-1]}(s,l)}{\sum_{s=1}^{n} v(s,l)}. \tag{12.39}$$

Inserting Eq. (12.38) into Eq. (12.39) we obtain the following recursive algorithm:

$$y^{[k]}(l) = \frac{y^{[k-1]}(l)}{\sum_{s=1}^{n} v(s,l)} \sum_{s=1}^{n} z_s \frac{v(s,l)}{\sum_{l=1}^{L} v(s,l)y^{[k-1]}(l)}. \tag{12.40}$$

The last formula is the *Richardson-Lucy algorithm* derived for a general linear transform of y with weights $v(s,l)$.

Compare this result with the one derived earlier for the shift-invariant convolution model. Assuming that $v(s,l) \Rightarrow v(s-l)$ and that the summation is performed over unbounded areas, we obtain for Eq. (12.40)

$$y^{[k]}(l) = \frac{y^{[k-1]}(l)}{\sum_s v(s-l)} \sum_s z_s \frac{v(s-l)}{\sum_l v(s-l)y^{[k-1]}(l)}. \tag{12.41}$$

We may conclude that formulas (12.26) and (12.41) are identical.

Remember that in the derivation of Eq. (12.26) we started from the gradient calculation and used it in the multiplicative iterations. For the derivation of Eq. (12.40) we refer to the EM method, starting from the estimates of the unobserved counts. These two very different techniques arrive at identical results.

In the original papers by Richardson [193] and Lucy [147], no specific statistical model was used. The assumptions that z, y, and v are normalized probability densities and the Bayesian technique lead to the algorithm. An interesting study concerning this algorithm as well as a parallel between the linear and nonlinear deconvolution methods are given in [25]. The EM algorithm, theory, and application examples as well as a thorough review of the field can be found in [153].

12.2.4 Regularized inverse

The MLs defined by Eq. (12.27) or Eq. (12.40) are Poisson versions of the pure or naive inverse of the linear smoothing operators in Eqs. (12.7) and (12.28). If the operator v is ill-conditioned, the estimator is also ill-conditioned with all negative consequences, such as noise amplification in the estimate and low convergence rate of the iterations.A number of regularization techniques have been developed to improve the conditioning of the estimate and decrease its sensitivity to noise and computing errors.

The penalty term included in the likelihood is interpreted as a prior distribution of the estimated parameter (intensity y) in the Bayesian estimation or as a regularization term. The log-likelihood in Eq. (12.29) with the penalty term has the form

$$L_{reg}(y) = \sum_s \left\{ -\sum_l v(s,l)y(l) + z_s \log\left[\sum_l v(s,l)y(l)\right]\right\} - \alpha^2\psi(y), \tag{12.42}$$

where $\psi(y) \geq 0$ and α^2 is the regularization parameter. Then the estimate is

$$\hat{y} = \arg\max_y L_{reg}(y). \tag{12.43}$$

We use the negative penalty here because the ML assumes maximization of the criteria and the penalty works in the opposite direction, trying to decrease the value of the combined criterion.

The M-stage in Eq. (12.39) for Eq. (12.42) is routinely changed for [69]

$$y^{[k]}(l) = \frac{\sum_{s=1}^{n} \tilde{z}^{[k-1]}(s,l)}{\sum_{s=1}^{n} v(s,l) + \alpha^2 \partial_{y(l)}\psi(y^{[k-1]})}. \tag{12.44}$$

Then the algorithm in Eq. (12.40) takes the form

$$y^{[k]}(l) = \frac{y^{[k-1]}(l)}{\sum_{s=1}^{n} v(s,l) + \alpha^2 \partial_{y(l)}\psi(y^{[k-1]})} \sum_{s=1}^{n} z_s \frac{v(s,l)}{\sum_{l=1}^{L} v(s,l) y^{[k-1]}(l)}, \tag{12.45}$$

and the algorithm in Eq. (12.26) is a change for

$$y^{[k]}(l) = \frac{y^{[k-1]}(l)}{\sum_s v(s-l) + \alpha^2 \partial_{y(l)}\psi(y^{[k-1]})} \sum_s z_s \frac{v(s-l)}{\sum_l v(s-l) y^{[k-1]}(l)}. \tag{12.46}$$

Most proposals for the penalty term $\psi(y)$ in Eq. (12.42) define it as a function of the differences between the intensities in neighboring pixels [69], [132], [139].

In particular,

$$\psi(y) = \sum_{i,j} b_{i,j}\kappa(y_i - y_j),$$

where κ is a positive potential function and $b_{i,j}$ are some weights.

The main goal of this regularization is to impose a smoothness constraint on the reconstructed y. A larger regularization parameter α^2 in Eq. (12.42) naturally means a smoother $\hat{y}$. A number of proposals were produced to improve the regularization (e.g., [4], [169]).

Our approach relies on adaptive smoothing based on the ICI to suppress the noisy components in the estimates $y(x)$ and at the same time to preserve change-points, edges, and boundaries. These features are often of critical importance in diagnostics. Thus, we do not need strong smoothing from the regularizator. Because the penalty based on the differences $(y_i - y_j)$ may lead to oversmoothed estimates, we propose that the standard quadratic norm of the intensities can be used:

$$\psi(y) = \frac{1}{2}\sum_i y_i^2. \tag{12.47}$$

12.2.5 LPA-ICI filtering

The Poisson naive inverse or Poisson regularized inverse can be filtered using the quasi-likelihood LPA-ICI adaptive algorithms because the estimates have a signal-dependent variance. The Poisson inverse estimates $\hat{y}$ are defined as solutions of the problems in Eq. (12.20) or Eq. (12.43). The recursive *Richardson-Lucy* procedures in Eqs. (12.27) and (12.46) can be used to calculate the estimates. For application of the ICI algorithm we need the Poisson inverse estimates and their covariance matrices.

12.2.5.1 *Covariance of inverse Poisson estimates*

Univariate signal

According to the asymptotic properties of the ML in Eq. (11.7), the estimate in Eq. (12.20) is asymptotically unbiased with the covariance in the form

$$cov\{\hat{y}(l), \hat{y}(l_1)\} = (I^{-1})_{l,l_1}, \tag{12.48}$$

where I is the *Fisher information matrix*

$$\begin{aligned} I &= (I_{l,l_1})_{n\times n}, \\ I_{l,l_1} &= E\{\partial_{y(l)}L\partial_{y(l_1)}L\} = -E\{\partial_{y(l)}\partial_{y(l_1)}L\} \\ &= \sum_s y(s)\frac{v(s-l)v(s-l_1)}{\left[\sum_l v(s-l)y(l)\right]^2}. \end{aligned} \tag{12.49}$$

Here n is a size of the signal y and the derivatives $\partial_{y(l)}\partial_{y(l_1)}L$ are defined by Eq. (12.23).

It can be proved that the Hessian matrix $(E\{\partial_{y(l)}\partial_{y(l_1)}L\})_{n\times n}$ is negative definite provided that the values of $v(l)$ are linearly independent:

$$(E\{\partial_{y(l)}\partial_{y(l_1)}L\})_{n\times n} < 0.$$

Then the matrix I is invertible and the covariances of the Poisson inverse estimates can be calculated according to formula (12.48).

Multivariate signal

The Poisson inverse estimates in Secs. 12.2.1–12.2.4 are given in a form that is universally valid for signals of any dimensionality. To make the covariance formulas (12.48)–(12.49) applicable for a multivariate case, the signal y should be presented in vector form with the Fisher matrix in Eq. (12.49) composed from the expectations of the pairwise products of the elements of this vector of estimates.

Let $\mathbf{y}$ be this vector with elements $y_{(i)}$, $\mathbf{y} = (y_1, \ldots, y_i, \ldots, y_n)^T$. Assume that the ordering of y_i in the vector $\mathbf{y}$ is one-to-one linked with the pixel-wise original numbering of $y(l)$ on the 2D grid X. Then we can say that $l = l(i)$ is a function linking the position l of the pixel on the grid X with the position i of this element in the vector $\mathbf{y}$.

For the elements of this vector, the covariance matrix has a form similar to Eqs. (12.48–12.49), that is,

$$cov\{\hat{y}_i, \hat{y}_j\} = (I^{-1})_{i,j}, \tag{12.50}$$

where I is the *Fisher information matrix*

$$\begin{aligned} I &= (I_{i,j})_{n\times n}, \quad I_{i,j} = -E\{\partial_{y_{l(i)}}\partial_{y_{l(j)}}L\}, \\ I_{i,j} &= \sum_s y(s)\frac{v(s-l(i))v(s-l(j))}{\left[\sum_l v(s-l)y(l)\right]^2}. \end{aligned} \tag{12.51}$$

Thus, the items of the Fisher matrix I are calculated using the usual formula (12.49) with the positioning of these covariances in the matrix I defined by the pair (i, j).

When the covariances in Eq. (12.50) are calculated through the inverse matrix I^{-1} they can be defined as a function of the pixel numbers as

$$cov\{\hat{y}(l), \hat{y}(l_1)\} = cov\{\hat{y}_{i(l)}, \hat{y}_{i(l_1)}\}, \tag{12.52}$$

where $i = i(l)$ is a function inverse with respect to $l = l(i)$.

12.2.5.2 Covariance of regularized Poisson inverse

For the univariate signal, the Fisher matrix for the penalized likelihood in Eq. (12.42), with $\psi(y)$ in the form (12.47), is defined by

$$I_{l,l_1} = -E\{\partial_{y(l)}\partial_{y(l_1)}L_{reg}\} = \sum_s y(s)\frac{v(s,l)v(s,l_1)}{\left[\sum_l v(s,l)y(l)\right]^2} + \alpha^2\delta_{l,l_1}. \tag{12.53}$$

The covariance matrix of the regularized ML estimates is

$$cov\{\hat{y}(l), \hat{y}(l_1)\} = (I^{-1})_{l,l_1}. \tag{12.54}$$

This differs from the initial Fisher matrix in Eq. (12.49) by the diagonal term $\alpha^2\delta_{l,l_1}$. This term works as a standard Tichonov regularizator. We then have for the variances of the estimates

$$\sigma^2_{\hat{y}_l} = (I^{-1})_{l,l}. \tag{12.55}$$

For 2D signals, similar changes are produced for the elements in Eq. (12.51), which are replaced by

$$I_{i,j} + \alpha^2\delta_{l(i),l(j)}, \tag{12.56}$$

with further calculations leading to covariances in the form of Eq. (12.52).

12.2.5.3 LPA-ICI(AV) algorithm

In the LPA-ICI(AV) algorithm, only linear filters with shift-invariant kernels are used for estimation. These linear estimates are calculated according to

$$\hat{y}_{h,\theta}(x) = \sum_l g_{h,\theta}(x-l)\hat{y}(l) = (g_{h,\theta} \circledast \hat{y})(x), \tag{12.57}$$

where $\hat{y}$ is the Poisson inverse estimate.

The variance of these estimates is

$$\sigma^2_{\hat{y}_{h,\theta}}(x) = \sum_l \sum_{l_1} g_{h,\theta}(x-l)g_{h,\theta}(x-l_1)cov\{\hat{y}(l), \hat{y}(l_1)\}, \tag{12.58}$$

where $cov\{\hat{y}(l), \hat{y}(l_1)\}$ are the covariances of the inverse Poisson estimates calculated according to one of the groups of formulas (12.48)–(12.49), (12.51)–(12.52), or (12.53)–(12.56). The layout of the algorithm corresponds to the one shown in Fig. 11.2(a).

The estimates $\hat{y}_{h,\theta}(x)$ in Eq. (12.57) and their variances $\sigma^2_{\hat{y}_{h,\theta}}(x,h)$ in Eq. (12.58) are the inputs of the ICI block. The outputs are the adaptive-scale estimates $\hat{y}_{h^+,\theta}(x)$ and the variance $\sigma^2_{h^+,\theta}(x)$. For the multidirectional estimates, these adaptive-scale estimates and variances are calculated for all directions. The final block "fusing" aggregates the adaptive-scale estimates into the final one $\hat{y}(x)$. Its variance is $\sigma^2_{\hat{y}}$. The covariances $cov\{\hat{y}(l), \hat{y}(l_1)\}$ in Eq. (12.58) according to Eq. (12.51) depend on unknown signal values $y(l)$ and the final fused estimates are used for covariance updating.

12.2.5.4 Variance approximation

The above formulas require complex calculations because, for imaging, the matrix I is very large and its inversion can be a problem, for example, say, for the 256×256 image, $n = 256^2 = 65536$. One way to simplify the task is to take into consideration only diagonal and near-diagonal elements of the matrix I. In particular, if we use only diagonal terms of I, the inverse estimates are assumed to be uncorrelated with the variance of the estimates following from Eqs. (12.48–12.49) in the form

$$\sigma^2_{\hat{y}}(l) = \left\{ \sum_s y(s) \frac{v(s-l)v(s-l)}{\left[\sum_l v(s-l)y(l)\right]^2} \right\}^{-1}. \tag{12.59}$$

For the regularized inverse, this formula is modified as

$$\sigma^2_{\hat{y}}(l) = \left\{ \sum_s y(s) \frac{v(s-l)v(s-l)}{\left[\sum_l v(s-l)y(l)\right]^2} + \alpha^2 \right\}^{-1}. \tag{12.60}$$

Then, the formula for the variance of the LPA estimate in Eq. (12.58) is of the form

$$\sigma^2_{\hat{y}_{h,\theta}}(x) = \sum_l g^2_{h,\theta}(x-l)\sigma^2_{\hat{y}}(l) = (g^2_{h,\theta} \circledast \sigma^2_{\hat{y}})(x). \tag{12.61}$$

12.3 Local ML Poisson Inverse

In the previous section we used a two-stage procedure. The first stage is the Poisson nonlinear inverse of the linear blurring operator, and the second stage is a filtering of this inverse. Both operations are introduced independently as successive steps of some heuristic procedure.

This section is based on another idea. A different linear estimation procedure is derived from the local ML approach.

12.3.1 Local ML for indirect observations

We consider the indirect observations in Eq. (12.5) corresponding to the model where the blurring operator is given by the convolution. Similar to Eq. (11.9) introduce a local version of the log-likelihood as

$$L_h(x, C) = \sum_s \{-(y_h \circledast v)(X_s) + z_s \log[(y_h \circledast v)(X_s)]\} w_h(x - X_s). \tag{12.62}$$

Assuming further that the zero-order LPA is used for the convolution $(y_h \circledast v)(X_s)$, we arrive at

$$L_h(x, C) = \sum_s (-C + z_s \log C) w_h(x - X_s). \tag{12.63}$$

Then the ML estimate of C has the form in Eq. (11.61) as

$$\hat{C}_h(x) = (z \circledast g_h)(x),\ g_h(x) = \frac{w_h(x)}{\sum_x w_h(x)}, \tag{12.64}$$

where $\hat{C}_h(x)$ is the nonparametric estimate for $(y_h \circledast v)(x)$.

If $\hat{C}_h(x)$ is found, the estimate $\hat{y}_h$ of y_h is a solution of

$$(\hat{y}_h \circledast v)(x) = \hat{C}_h(x). \tag{12.65}$$

In the frequency domain, Eqs. (12.64) and (12.65) give

$$V(f)\hat{Y}_h(f) = G_h(f)Z(f). \tag{12.66}$$

Thus, we arrive at the linear inverse problem studied in Sec. 9.1. The unknown $\hat{Y}_h(f)$ is a solution of the ill-conditioned linear system in Eq. (12.66). Remember that the conditioning of this system is defined by the convolution kernel v.

The adaptive methods developed for the linear inverse are fully applicable to the linear Eq. (12.66). We use the linear inverse estimation of y from Eq. (12.66) jointly with the ICI rule. This rule gives the adaptive-scale h estimates for $\hat{y}_h$. Using a zero-order approximation in the smoothing kernel g_h means that the ICI defines maximum neighborhoods where y is nearly constant.

Thus, if both the y and the convolution $(y \circledast v)$ are approximately constant, then our passage to the linear estimate in Eq. (12.64) from the likelihood in Eq. (12.62) is completely legitimate. The ICI works to enable this hypothesis.

Note that in the likelihood in Eq. (12.62) we replace the unknown y by its smoothed (approximate) version $\hat{y}_h$, exactly as was done in Sec. 9.1, where the inversion in the linear observation model was considered.

Our implementation of the algorithm is based on the application of the ICI directly to the estimate $\hat{y}_h$. The experimental study shows the good performance of the algorithm. In addition, using the first-order kernels (instead of the order $m = 0$ kernels g_h in Eq. (12.64)) gives a better performance.

This demonstrates that the assumption that $(y \circledast v)$ is constant can be treated as quite approximate. The ICI scale selection allows us to use efficiently the linear filtering in Eq. (12.64) for the originally nonlinear ML Poisson distribution estimation problem.

12.3.2 Linear inverse plus LPA-ICI filtering

This algorithm uses the RI and RWI inversions developed for the linear Gaussian inverse in Sec. 9.2. Adaptation to the Poisson case requires some changes to the RWI inverse and the variance formulas exploited for the ICI rule.

The RI estimate for Eq. (12.66) is calculated according to formula (9.35) as

$$\hat{Y}_h^{RI}(f) = \frac{V^*(f)G_h(f)}{|V(f)|^2 + \alpha_1^2} Z(f). \tag{12.67}$$

The transfer function of this filter is

$$Q_{RI}(f, \alpha_1) = \frac{V^*(f)G_h(f)}{|V(f)|^2 + \alpha_1^2}$$

with the impulse response

$$q_{RI}(x) = \mathcal{F}^{-1}\left[Q_{RI}(f, \alpha_1)\right], \tag{12.68}$$

which is exactly as it is in Sec. 9.2.

We need the variance of this estimate. The observations are independent with the signal dependent variance equal to $(y \circledast v)(x)$. It follows that the variance of the linear estimate in Eq. (12.67) is

$$\sigma^2_{\hat{y}_h^{RI}}(x) = \sum_u q_{RI}^2(x-u)(y \circledast v)(u) = (q_{RI}^2 \circledast (y \circledast v))(x). \tag{12.69}$$

To calculate this variance we need some estimates of the signal y. The simplest ones are the observations themselves. Then the estimate of the variance of the RI is calculated as

$$\hat{\sigma}^2_{\hat{y}_h^{RI}}(x) = (q_{RI}^2 \circledast (z \circledast v))(x). \tag{12.70}$$

The ICI rule for the adaptive-scale selection and the fusing of the multi-directional estimates are the following steps of the RI procedure. These steps are completely similar to those used in Sec. 9.2. The final estimate is denoted as $\hat{y}^{RI}(x)$.

However, there is a serious difference with the methods developed for the linear inverse in Sec. 9.2. The estimate variance in Eqs. (12.69–12.70) is signal-dependent, and the estimates of the signal are used for calculations of this variance.

We call this algorithm the *regularized inverse with adaptive variance* (RI-AV). We use here the shortened name AV to emphasize the similarity between this algorithm and the LPA-ICI(AV) algorithm in Sec. 11.4.

12.3.2.1 Regularized variance-adaptive Wiener inverse (RWI-AV)

Let us consider Wiener filtering. The optimal transfer function for this filter is defined by the general formula in Eq. (9.30) as

$$Q_{RWI}(f) = \frac{V^*(f)G_h(f)}{|V(f)|^2 + n_1 n_2 S_\varepsilon(f)/|Y(f)|^2}, \qquad f \in F, \tag{12.71}$$

where $S_\varepsilon(f)$ is the spectrum of random noise.

In the Poissonian case in Eq. (12.6), $\varepsilon(x)$ is a zero-mean random variable with variance

$$\sigma_\varepsilon^2(x) = (y \circledast v)(x).$$

By the definition in Eq. (14.185), the spectrum $S_\varepsilon(f)$ is given as

$$S_\varepsilon(f) = \frac{1}{n_1 n_2} E\{|\mathcal{F}\{\varepsilon(x)\}|^2\}, \tag{12.72}$$

where, according to Eq. (3.50),

$$\mathcal{F}\{\varepsilon(x)\} = \sum_{k_1=0}^{n_1-1}\sum_{k_2=0}^{n_2-1} \varepsilon[k_1, k_2] W_{n_1}^{f_1 k_1} W_{n_2}^{f_2 k_2}.$$

Inserting it into Eq. (12.72) and calculating the expectation, we find

$$S_\varepsilon(f) = \frac{1}{n_1 n_2} \sum_{k_1=0}^{n_1-1}\sum_{k_2=0}^{n_2-1}\sum_{l_1=0}^{n_1-1}\sum_{l_2=0}^{n_2-1} W_{n_1}^{f_1 k_1} W_{n_2}^{f_2 k_2} W_{n_1}^{-f_1 l_1} W_{n_2}^{-f_2 l_2} E\{\varepsilon[k_1, k_2]\varepsilon[l_1, l_2]\}$$

$$= \frac{1}{n_1 n_2} \sum_{k_1=0}^{n_1-1}\sum_{k_2=0}^{n_2-1} E\{\varepsilon^2[k_1, k_2]\} = \frac{1}{n_1 n_2} \sum_{k_1=0}^{n_1-1}\sum_{k_2=0}^{n_2-1} (y \circledast v)(k_1, k_2).$$

Then the transfer function of the Wiener filter in Eq. (12.71) is

$$Q_{RWI}(f) = \frac{V^*(f) G_h(f)}{|V(f)|^2 + \Xi(y, v)/|Y(f)|^2}, \tag{12.73}$$

where

$$\Xi(y, v) = \sum_x (y \circledast v)(x)$$

is the total energy of the expectation of the observed signal.

We use the regularized version of the transfer function in Eq. (12.71) where the parameter α_2^2, similar to the one in Eq. (9.36), is exploited as

$$Q_{RWI}(f, \alpha_2) = \frac{V^*(f) G_h(f)}{|V(f)|^2 + \alpha_2^2 \Xi(y, v)/|Y(f)|^2}. \tag{12.74}$$

The RWI estimate is calculated according to formula (9.36) as

$$\hat{Y}_h^{RWI}(f) = Q_{RWI}(f, \alpha_2) Z(f), \tag{12.75}$$

with the impulse response of this filter given as

$$q_{RWI}(x) = \mathcal{F}^{-1}\{Q_{RWI}(f, \alpha_2)\}.$$

We also need to calculate the variance of the Wiener estimate. Similar to Eq. (12.69) we obtain

$$\sigma^2_{\hat{y}_h^{RWI}}(x) = \sum_u q^2_{RWI}(x - u)(y \circledast v)(u) = (q^2_{RWI} \circledast (y \circledast v))(x). \tag{12.76}$$

Some estimate of y is needed for the transfer function in Eq. (12.74) and for the variance formula (12.76).

It is natural to use the RI adaptive-scale estimate $\hat{y}^{RI}(x)$ as a pilot estimate of the signal y for the regularized Wiener estimator.
Then the RWI estimate and the variance of this estimate are

$$\hat{Y}_h^{RWI}(f) = \frac{V^*(f)G_h(f)}{|V(f)|^2 + \alpha_2^2 \Xi(\hat{y}^{RI}, v)/|\hat{Y}^{RI}(f)|^2} Z(f),$$
$$\Xi(\hat{y}^{RI}, v) = \sum_x (\hat{y}^{RI} \circledast v)(x), \tag{12.77}$$
$$\sigma^2_{\hat{y}_h^{RWI}}(x) = \left(q^2_{RWI} \circledast (\hat{y}^{RI} \circledast v)\right)(x),$$

where

$$q_{RWI}(x) = \mathcal{F}^{-1}\left\{\frac{V^*(f)G_h(f)}{|V(f)|^2 + \alpha_2^2 \Xi(\hat{y}^{RI}, v)/|\hat{Y}^{RI}(f)|^2}\right\}.$$

The remaining stages of the algorithm are the ICI rule for the adaptive-scale selection and the fusing of multidirectional estimates. These stages are completely similar to those used in Sec. 9.2. The final estimate is denoted as $\hat{y}^{RWI}(x)$.

Note that there is an essential difference between the introduced algorithm and the RWI methods developed for the linear inverse in Sec. 9.2. The estimate variance in Eq. (12.76) that was used in the ICI rule is signal-dependent. In Eq. (12.77), the RI estimate $\hat{y}^{RI}$ is used for this variance calculation.

We call this algorithm the *regularized Wiener inverse with adaptive variance* (RWI–AV). We again use the shortened name AV to emphasize the similarity of this algorithm to the LPA-ICI(AV) algorithm in Sec. 11.4.

12.3.2.2 Linear inverse versus Poisson inverse

The regularized inverse and the regularized Wiener inverse are linear filters that, to be accurate, are not relevant to the Poisson distribution problem with the signal-dependent observation's variance. In particular, the Wiener filter in Eq. (12.71) is an optimal linear solution for the stationary noise with a time-invariant spectrum $S_\varepsilon(f)$. This filter itself is not a good estimator for a nonstationary variance-varying signal. However, the RI and RWI filters for $h \in H$ generate a set of estimates that is sufficiently rich to select from, and the ICI efficiently does this selection work.

Simulation demonstrates an exceptionally good performance of the *linear inverse plus* LPA-ICI filtering. This algorithm is based mainly on convolution operations and is quite fast.

The alternative Poisson inverse (Sec. 12.2.1) is a nonlinear counterpart of the linear inverse. The Poisson inverse takes into consideration an accurate statistical model of the observations and potentially produces the best estimates. Thus, the Poisson inverse methods—starting from the nonlinear ML inverse operation—have a principal advantage versus the other methods based explicitly on linear filtering, even with the signal-dependent variances in the ICI rule. This complexity is a drawback of the Poisson-inverse-based algorithms.

12.3.3 Numerical experiments

12.3.3.1 *Deblurring Poisson* Cameraman *image*

The blur PSF is a uniform (box-car) 9×9 function. The level of the randomness of the data is controlled by the scale parameter $\chi > 0$ according to formula (12.16) as

$$z(x) \sim \mathcal{P}((y \circledast v)(x) \cdot \chi), \tag{12.78}$$

with

$$E\{z(x)\} = \chi \cdot (y \circledast v)(x), \qquad var\{z(x)\} = \chi \cdot (y \circledast v)(x).$$

We select $\chi = 17600$. Despite such a large value of χ, the nonuniformity of the noise variance is still an essential issue for the Poisson deblurring, as the following simulations will show. The actual pointwise values of the standard deviation σ_z are in the range of $[0, 0.0075]$ (assuming that the image is normalized to the range $[0, 1]$). Formula (9.41) gives for this noise BSNR = 32.5 dB.

We use the *linear inverse plus LPA-ICI algorithm* as it is defined by formulas (12.67) and (12.77). The parameters of the algorithm are as follows. The kernels and the scales h_1 and h_2 are as in Eqs. (12.14–12.15). The parameters for the regularized inverse and regularized Wiener inverse, respectively, are

$$\Gamma_1 = 1.5, \quad \Gamma_2 = 1.4, \qquad \alpha_1 = 0.03, \quad \alpha_2 = 0.28. \tag{12.79}$$

Figure 12.7 shows details of the blurred Poisson noisy observation and the reconstructed Cameraman image. The reconstruction is visually quite good, with most of the details properly restored and without significant distortions. The values of the ISNR and RMSE are given in the first row of Table 12.4.

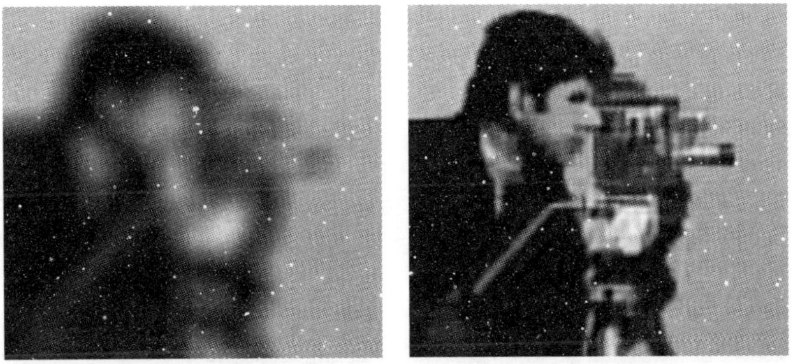

Figure 12.7 Deblurring the *Cameraman* image. Left is a fragment of noisy blurred observations. Right is the reconstructed image obtained by the proposed Poissonian adaptive deconvolution algorithm ISNR = 6.61 dB.

Table 12.4 RMSE and ISNR for *Cameraman* image for Poisson image reconstruction.

Algorithm \ Criterion	ISNR	RMSE
Linear inverse plus LPA-ICI filtering	**6.61**	**0.0428**
Optimized Gaussian LPA-ICI algorithm	6.03	0.0458
Gaussian LPA-ICI algorithm	5.38	0.0493

Let us compare the performance of this algorithm, which was designed specifically for the Poisson distribution, versus the algorithm designed for Gaussian data and exploited for Poisson data. To do this, we process the Poisson observations using the deconvolution algorithm developed for the Gaussian data in Sec. 9.2.3. The latter algorithm has the following parameters:

$$\Gamma_1 = 1.6, \quad \Gamma_2 = 1.3, \qquad \alpha_1 = 0.012, \quad \alpha_2 = 0.36. \tag{12.80}$$

The corresponding numerical results are given in the third row of Table 12.4. Further, we optimize the parameters of this Gaussian algorithm and find the optimal values for Γ_1 and Γ_2:

$$\Gamma_1 = 1.7, \qquad \Gamma_2 = 1.8. \tag{12.81}$$

Optimization improves the performance of the algorithm and gives the larger value for ISNR and the smaller value of RMSE shown in the second row of Table 12.4.

Images obtained using the Gaussian algorithms are shown in Fig. 12.8. We note an enhancement obtained by the parameter optimization. Comparison with the reconstructed image in Fig. 12.7 obtained by the algorithm developed for Poisson data demonstrates the obvious visual advantage of the latter. Figure 12.9 shows the adaptive scales selected by the ICI for the RI and RWI filters. It is remarkable how these scales reveal the features of the image across the corresponding direction.

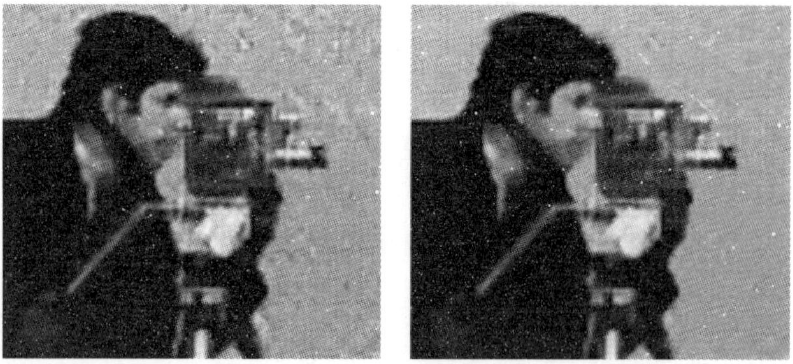

Figure 12.8 Filtering the Poisson data by the algorithm developed for the Gaussian ones. The standard selection of the algorithm parameters gives a poor estimate, ISNR = 5.38 dB (left). It is improved by optimization of some algorithm's parameters, ISNR = 6.03 dB (right).

RI-AV, $\theta=\pi/4$ RWI-AV, $\theta=0$

Figure 12.9 ICI adaptive scales, RI-AV and RWI-AV algorithms. Darker color corresponds to smaller scales.

12.4 Computerized Tomography

Tomography, from the Greek "tomos," meaning *section*, plus the international scientific "graphy," means a method of producing a 2D or 3D image of internal structures of an opaque object (human body, earth, engineering construction, etc.) by observations of differences in effects of propagation of energy waves impinging on those structures or emitted inside those structures. In computerized tomography a 3D image of an internal structure is reconstructed from a series of planar 2D projection images.

This finds use in such diverse fields as biomedical imaging (e.g., reviews for x-ray tomography [82]; positron emission tomography [170]; magnetic resonance imaging [237]; astronomy, radar techniques, seismology, and recently appearing internet tomography [32]).

The mathematical principles of computerized tomography first appeared in work by J. Radon as early as 1917. Today it is a wide research area with subjects specified by their applications as well as their solution methods. Basic approaches and techniques can be found in the following books: [14], [84], and [96].

12.4.1 Emission tomography

Emission computed tomography (ECT) is a diagnostic technique for medical imaging. There is a growing appreciation of the significance of the functional (as opposed to the anatomical) information that is provided by ECT and of its value for the purposes of medical diagnosis and monitoring the response to therapy. There are also significant advancements taking place, both in the instrumentation for data collection and in the computer methods for generating images from the measured data. These computer methods are designed to solve the inverse problem known as

Figure 12.10 An illustration for a typical emission tomography scanner. The label isotope atoms decay and emit positrons. The positron–electron annihilation creates two photons traveling in nearly opposite directions. The detectors in the ring around the body count emissions that occurred somewhere inside the imaginary straight tube connecting the two "firing" detectors D_1 and D_2.

"image reconstruction from projections." The major subfields of ECT are positron emission tomography (PET) and single-photon emission tomography (SPECT).

In this brief outline of the physics behind PET and SPECT and the data collection in these systems, we mainly adapt material from [131], [139], and [226].

The goal of the radiology specialty of nuclear medicine, which includes PET and SPECT, is to provide information about the distribution of a chosen molecule in space (and possibly also in time) inside the human body. For a molecule of biochemical or physiological importance, an image of its distribution within the body or a specific organ provides information about the functioning of the body or organ (including information about the locations of abnormalities such as tumors). The functional information obtained by PET and SPECT is complementary to the anatomical information that is available from other radiological imaging techniques, such as x-ray computed tomography and magnetic resonance imaging.

The techniques used in nuclear medicine involve labeling the chosen molecule with a radioactive atom and administering a dose of labeled molecules to the patient. The labeled molecules follow their specific biochemical pathways inside the body. The atoms used as labels are unstable isotopes and undergo radioactive decay at random, leading to the emission of gamma ray photons that can be detected outside the body by the detector systems of PET and SPECT scanners.

Figure 12.10 illustrates a typical PET scanner. The label isotope atoms decay and emit positrons. Within a few millimeters a positron interacts with an electron. The positron–electron annihilation creates two photons traveling in nearly opposite directions, as illustrated in Fig. 12.10(b). The emitted photons move along a line with a completely random orientation uniformly distributed in space. There is a ring of detectors surrounding the body and registering the arriving photons. The photons are counted in PET and SPECT scanners in different ways.

PET detectors count only the so-called coinciding photons. The situation is called a "true coincidence" when two photons from the same positron–electron annihilation travel without interaction to the detector and both are detected within a short time window. If one of the photons misses the detector pair for some reason, the time window contains only one detected photon and the event is discarded.

Thus, the detectors count emissions that occurr somewhere inside the imaginary straight tube connecting two "firing" detectors D_1 and D_2 in Fig. 12.10.

Let $\tilde{z}(s, l)$ be the count of the emissions that occur at a voxel l and registered by the sth tube. The voxel is a 3D pixel, volume element. A 3D body is built from a grid of voxels.

The data collected by the sth sensor pair (tube) are the counts

$$z_s = \sum_{l=1}^{L} \tilde{z}(s, l).$$

which are the total number of emissions registered in the sth tube. The sum

$$\tilde{z}(l) = \sum_{s=1}^{n} \tilde{z}(s, l)$$

defines the total number of annihilations that occur in the lth voxel.

The counts $\tilde{z}(s, l)$ and $\tilde{z}(l)$ are unobserved parameters of the system, and the counts z_s are the actual observed measurements. The links between the counts $\tilde{z}(s, l), \tilde{z}(l)$, and z_s are defined by the above formulas, which are identical to formulas (12.32) and (12.33).

The total tube counts in detectors are typically much smaller than the total number of emissions, because all of the photons traveling along the lines that do not cross the detector ring do not make "true coincidence" or are attenuated by the body's tissue would pass undetected.

The intensity of the emission is proportional to the concentration of unstable isotopes in the voxel. The problem is to estimate the distribution of this isotope in the body from the only available counts z_s.

The number of emissions in the voxel is random during the exposition and the Poisson distribution exploited to model the emission processes. Then the parameter of the Poisson distribution y characterizes the intensity of the emission and the concentration of the isotopes.

The balance model of the emission system has the form

$$\mu(s) = \sum_{l=1}^{L} v(s, l) y(l) + b(s), \qquad s = 1, \ldots, n, \tag{12.82}$$

where $y(l)$ is the intensity of the emission from the voxel l and $\mu(s)$ is the intensity of the observed counts z_s at the sensor s. These observed counts have a Poisson distribution

$$z_s \sim \mathcal{P}(\mu(s)), \qquad s = 1, \ldots, n. \tag{12.83}$$

The parameters $b(s) \geq 0$ in Eq. (12.82) stand for the background emission of the system (without the body). It is assumed that these parameters are known. For simplicity, we assume that $b(s) = 0$, while it can be easily taken into consideration.

In SPEC, the detectors count those photons that pass through the collimator and do not interact with the attenuating material on their pass through the body.

This does not require the "coincidence" effects of the pairwise count specific for the PET system.

In both PET and SPEC, the detectors accumulate counts for the lines (tube-shaped volumes) in space, where the counts are individual photons in SPEC and pairs of photons in PET. Apart from these effects, the total accumulated counts for each line indicates the total number of label atoms along the line that underwent radioactive decay, but it does not indicate where these atoms are located along the lines.

The model in Eq. (12.82) with the corresponding weights v is conventional for both SPEC and PET systems. The weights $v(s, l) \geq 0$ are treated as transition probabilities. This is a key element of modeling the emission system. If they are given, as we assume in what follows, the estimation of the intensities $y(l)$ from the observations z_s is a Poisson inverse problem.

The methods presented in the Sec. 12.2 are completely relevant and applicable to this problem.

We wish to emphasize that finding the weights $v(s, l)$ is a complex problem that is far from trivial. Simple geometrical ideas that could look reasonable for the simplified schemes in Fig. 12.10 are practically far from being satisfactory for the emission systems [49].

Overall, the purpose of the image reconstruction is to process these imperfect count data for large numbers of lines and detected photons to produce an image showing the distribution in space of the label atoms and molecules.

12.4.2 Transmission tomography

The development of x-ray computed tomography in the early 1970s revolutionized medical radiology. For the first time, physicians were able to obtain high-quality tomographic (cross-sectional) images of internal structures of the body. Over the years, the technical sophistication of x-ray systems has increased dramatically, and even today, it continues to mature, with new capabilities being researched and developed (e.g., [82], [96]).

Transmission tomographic images are reconstructed from a large number of measurements of x-ray transmissions through the body. These measurements are called *projections*. The source of radiation is outside the body, and it is significantly different than emission systems where the source is inside the body.

The resulting images are *tomographic maps* of the x-ray attenuation coefficients. The first practical instrument was developed in 1971 by Dr. G. N. Hounsfield in England and was used to image the brain. The projection data were acquired in approximately 5 minutes, and the tomographic image was reconstructed in approximately 20 minutes. Since then, x-ray computed tomography has developed dramatically and become the standard imaging procedure for virtually all parts of the body in thousands of facilities throughout the world.

A fundamental task of x-ray computed tomography is to make an extremely large number of highly accurate measurements of x-ray transmission through the body in a precisely controlled geometry. Projection data may be acquired in one of

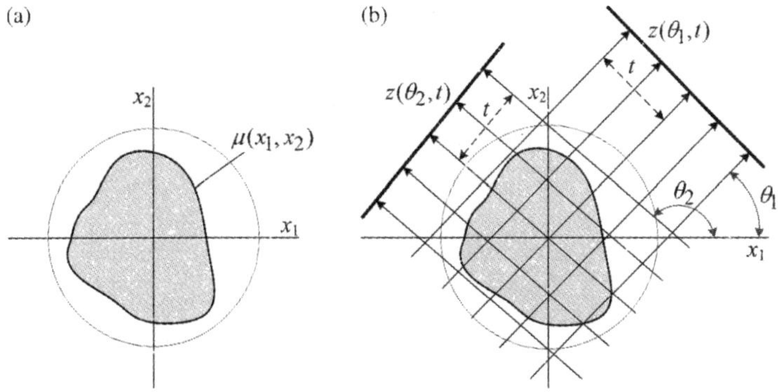

Figure 12.11 The angle θ defines the rotation of the beam and the coordinate t defines a location of the beam with the respect to the origin coordinate; $\mu(x_1, x_2)$ is a measured distribution density, and $z(\theta, t)$ sensor outputs.

several possible geometries based on the scanner configuration, scanning motions, and detector arrangement.

Parallel-beam geometry (Fig. 12.11) is the simplest and easiest geometry with which to understand the important transmission tomography principles. Multiple measurements of x-ray transmission are obtained using a single highly collimated x-ray pencil beam and detector.

The beam is translated in a linear motion across the patient to obtain a projection profile. The source and detector are then rotated around the patient's iso-center by approximately 1 deg and another projection profile is obtained. This translate-rotate scanning motion is repeated until the source and detector have been rotated by 180 degs. The highly collimated beam provides excellent rejection of radiation scattered in the patient; however, the complex scanning motion results in long scan time.

This geometry was used by Hounsfield in his original experiments. In Fig. 12.11, the angle θ defines the rotation of the beam and the coordinate t defines a location of the beam with the respect to the iso-center.

Rotated narrow fan beam (Fig. 12.12). Scan times were reduced significantly with the use of a narrow fan beam of x-rays and a linear detector array. A translate-rotate scanning motion is still employed. However, a large rotate increment was used, which resulted in shorter scan times. The reconstruction algorithms are slightly more complicated than those for the parallel-beam geometry.

Rotated fan beam and *detectors*. A fan beam of x-rays is rotated 360 degs around the iso-center. The beam is wide enough to completely contain the patient. A curved detector array consisting of several hundred independent detectors is mechanically coupled to the x-ray source, and both rotate together. No translation motion is used.

The most modern systems use *spiral or helical* 3D *scanners*. Multiple images are acquired while the body is translated through the geometry in a smooth continuous motion rather than stopping for each image. The reconstruction algorithms are quite sophisticated because they must accommodate the spiral or helical path

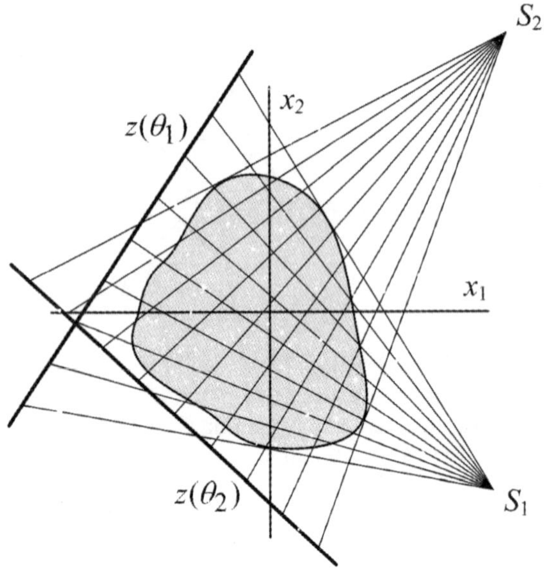

Figure 12.12 Rotated narrow-fan beam scanner.

traced by the x-ray sources around the body. Projection data can be acquired fast enough to image the beating heart without significant motion artifacts.

12.4.2.1 Measurement process

The primary goal in transmission tomography is to reconstruct 2D attenuation coefficients y of the radiation traveling through line-wise cross sections of the object.

Figure 12.13 shows a schematic version of the measurement process. In this configuration, a set of small voxels indexed by l forms narrow, straight tubes whereby $z_{0,s}$ photons are injected as a beam into the tube and z_s photons exit the tube and are registered by the detector. The remaining photons are absorbed into the body.

Let $y_{s,l}$ be the absorption coefficient of the voxel l in the tube s. The input flow of the photons is a Poisson random process with the parameter $E\{z_{0,s}\}$. The output number of photons is also Poissonian with the parameter

$$E\{z_s\} = E\{z_{0,s}\} \exp\left(- \sum_{l=1}^{l_s} y_{s,l} \right). \tag{12.84}$$

There is a large number of tubes at various orientations depending on the scanner geometry. The parallel-beam geometry (first generation of x-ray tomographs) is composed of parallel sets of tubes with different $E\{z_{0_s}\}$.

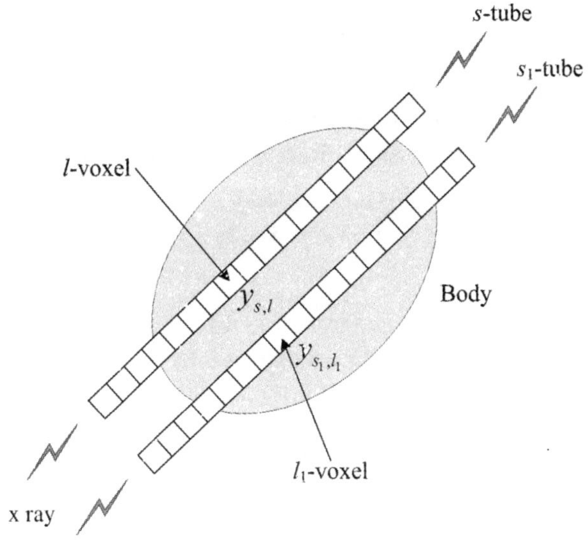

Figure 12.13 Measurement process for transmission x-ray tomography.

Equation (12.84) also can be represented in the form

$$E\{z_s\} = E\{z_{0_s}\} \exp\left(-\sum_{l \in L_s} y_{s,l} \right), \qquad (12.85)$$

where $\sum_{x \in L_s}$ means a summation over voxels belonging to the sth tube. Note that the summation of the absorption coefficients used in Eqs. (12.84) and (12.85) is valid provided that the x-rays are monochromatic.

Two types of models used for the transmission models are linear, which is based on the Radon transform, and nonlinear, which is basically exploited in the ML approach.

12.4.2.2 Linear model of transmission system

Radon transform

It follows from Eq. (12.85) that

$$\log \frac{E\{z_{0_s}\}}{E\{z_s\}} = \sum_{l \in L_s} y_{s,l}. \qquad (12.86)$$

If the output signal is defined as the logarithm of the ratio of intensities $E\{z_{0_s}\}$ and $E\{z_{0_s}\}$, it is equal to the sum of the absorption coefficients of the voxels along the transmission tube.

Let us exploit the concept of absorption density as a function of two variables with the notation $y(x_1, x_2)$. Thus, $y_{s,l} = y(x_1, x_2)\Delta_{s,l}$, where $\Delta_{s,l}$ is the size of the voxel and x_1, x_2 are the coordinates of the voxel.

The sum $\sum_{l \in L_s} y_{s,l}$ in Eq. (12.86) can be replaced by a line integral that can be presented in the form of the classical Radon transform [35], [82]:

$$\sum_{l \in L_s} y_{s,l} \simeq \int_{-\infty}^{\infty} \int_{-\infty}^{\infty} y(x_1, x_2)\delta(t - x_1 \cos\theta - x_2 \sin\theta)dx_1 dx_2. \tag{12.87}$$

The following coordinates are used to define the orientation of the line integration (tubes): The parameter t is the shortest distance from the origin of the coordinate system to the line, and θ is an angle defining the angular orientation of the line (Fig. 12.11(b)). All lines can be described by choosing

$$0 \leq \theta < \pi, \qquad t \geq 0. \tag{12.88}$$

If we assume that the observations in Eq. (12.86) are given in the Radon transform coordinates (θ, t), then Eq. (12.87) is a linear deterministic observation model,

$$\log \frac{E\{z_0(\theta, t)\}}{E\{z(\theta, t)\}} = \int_{-\infty}^{\infty} \int_{-\infty}^{\infty} y(x_1, x_2)\delta(t - x_1 \cos\theta - x_2 \sin\theta)dx_1 dx_2, \tag{12.89}$$

where $E\{z_0(\theta, t)\}$ is the intensity of the beam injected along the line with coordinates (θ, t), and $E\{z(\theta, t)\}$ is the intensity registered by the detector along the line with the same coordinates (θ, t). This expression is considered to be the equation with respect to an unknown distribution y provided that the observations are given.

Let us introduce and use the standard notation $Ry(\theta, t)$ for the Radon transform

$$Ry(\theta, t) = \int_{-\infty}^{\infty} \int_{-\infty}^{\infty} y(x_1, x_2)\delta(t - x_1 \cos\theta - x_2 \sin\theta)dx_1 dx_2. \tag{12.90}$$

Another equivalent representation of the Radon transform has the form of a 1D integral:

$$Ry(\theta, t) = \int_{-\infty}^{\infty} y(t\cos\theta - s\sin\theta,\ t\sin\theta + s\cos\theta)ds. \tag{12.91}$$

The integral in Eq. (12.91) can be interpreted as a projection of the density $y(x_1, x_2)$ on the line with the coordinates (θ, t).

The Radon transform is an interesting mathematical subject. Indeed, it maps the 2D function y in the projections $Ry(\theta, t)$. The first problem concerning the Radon transform is whether $y(x_1, x_2)$ can be restored from the projections given in the area of Eq. (12.88). The answer is yes—under the appropriate conditions on $y(x_1, x_2)$.

Let us give some illustrations of forward and inverse Radon transform mappings. First, assume that the distribution y is a delta function

$y = \delta(x_1 - x_1^*)\,\delta(x_2 - x_2^*)$. Inserting this delta function into Eq. (12.90), we have

$$Ry(\theta, t) = \int_{-\infty}^{\infty}\int_{-\infty}^{\infty} \delta(x_1 - x_1^*)\delta(x_2 - x_2^*)\delta(t - x_1\cos\theta - x_2\sin\theta)dx_1dx_2$$
$$= \delta(t - x_1^*\cos\theta - x_2^*\sin\theta).$$

Thus, the isolated bright point (delta function at the point x_1^*, x_2^*) in the spatial domain maps in the transform domain (θ, t) in the sinusoidal bright curve (delta function) of the form

$$t = x_1^*\cos\theta + x_2^*\sin\theta. \tag{12.92}$$

This t is a sinusoidal function of θ, and is why the (θ, t)-plot of the Radon transform is called the *sinogram* of y.

The Radon transforms of point-like objects are shown in Fig. 12.14(b). The signal y is equal to one in three isolated points and equal to zero for all other x (Fig. 12.14(a)). The Radon transform sinogram has the form of three harmonic functions. According to Eq. (12.92), each of these harmonics is generated by the nonzero points of y.

Figure 12.15(a) shows the "Shepp-Logan" phantom widely used in tomography to test the numerical accuracy of 2D reconstruction algorithms. It is a grayscale image that consists of one large ellipse (imitating the brain) containing several smaller ellipses (imitating features in the brain). We show in Fig. 12.15(b) the Radon transform of this phantom. It is calculated for $\theta = [0 : 179]$ using the codes given in the image processing toolbox of MATLAB.

The last image in Fig. 12.15(c) illustrates the inverse Radon transform returning the image of the phantom from its Radon transform. Numerically, a perfect reconstruction is not possible, and, as a result, this inverse transform is a disturbed, blurred version of the original phantom image.

The inverse Radon transform for the image in Fig. 12.14(b) is shown in Fig. 12.14(c). Because the inverse Radon transform is not perfect, it gives a blurred version of the original points from Fig. 12.14(a).

(a) (b) (c)

Figure 12.14 Radon transform illustration: (a) object (three isolated points), (b) Radon transform having a form of three harmonic functions, and (c) inverse Radon transform showing three slightly blurred points.

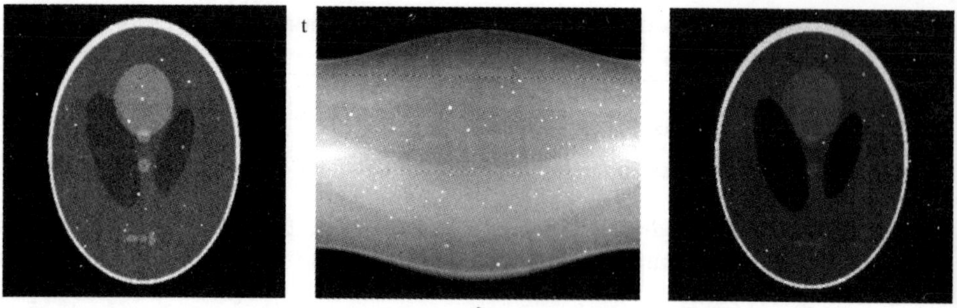

(a) (b) (c)

Figure 12.15 Radon transform illustration: (a) Shepp-Logan phantom, (b) Radon transform, (c) inverse Radon transform showing a blurred image of Shepp-Logan phantom.

Fourier transform Radon inverse

The integral Fourier transform of y is defined according to Eq. (4.40) as

$$\begin{aligned} Y(\lambda_1, \lambda_2) &= \mathcal{F}\{y(x_1, x_2)\} \\ &= \int_{-\infty}^{\infty}\int_{-\infty}^{\infty} y(x_1, x_2)\exp[-j(\lambda_1 x_1 + \lambda_2 x_2)]dx_1 dx_2, \end{aligned} \tag{12.93}$$

and the inverse as

$$\begin{aligned} y(x_1, x_2) = \mathcal{F}^{-1}\{Y(\lambda_1, \lambda_2)\} = \frac{1}{(2\pi)^2}\int_{-\infty}^{\infty}\int_{-\infty}^{\infty} Y(\lambda_1, \lambda_2) \\ \times \exp[j(\lambda_1 x_1 + \lambda_2 x_2)]d\lambda_1 d\lambda_2. \end{aligned} \tag{12.94}$$

Introduce the polar frequency variables ν and θ:

$$\begin{pmatrix} \lambda_1 \\ \lambda_2 \end{pmatrix} = \nu \cdot \begin{pmatrix} \cos\theta \\ \sin\theta \end{pmatrix}. \tag{12.95}$$

Inserting formula (12.95) into Eq. (12.93) gives

$$Y(\nu\cos\theta, \nu\sin\theta) = \int_{-\infty}^{\infty}\int_{-\infty}^{\infty} y(x_1, x_2)\exp[-j\nu(x_1\cos\theta + x_2\sin\theta)]dx_1 dx_2. \tag{12.96}$$

This formula allows one to link $Ry(\theta, t)$ with the integral Fourier transform. Let us represent the harmonic function in Eq. (12.96) in the following integral form:

$$\exp[-j\nu(x_1\cos\theta + x_2\sin\theta)] = \int_{-\infty}^{\infty}\exp(-j\nu t)\delta[t - (x_1\cos\theta + x_2\sin\theta)]dt.$$

Inserting the last representation in Eq. (12.96), we find

$$\begin{aligned}
&Y(\nu\cos\theta,\nu\sin\theta)\\
&=\int_{-\infty}^{\infty}\int_{-\infty}^{\infty} y(x_1,x_2)\int_{-\infty}^{\infty}\exp(-j\nu t)\delta[t-(x_1\cos\theta+x_2\sin\theta)]dt dx_1 dx_2\\
&=\int_{-\infty}^{\infty}\exp(-j\nu t)\left[\int_{-\infty}^{\infty}\int_{-\infty}^{\infty} y(x_1,x_2)\delta[t-(x_1\cos\theta+x_2\sin\theta)]dx_1 dx_2\right]dt\\
&=\int_{-\infty}^{\infty}\exp(-j\nu t)Ry(\theta,t)dt. \qquad (12.97)
\end{aligned}$$

These formulas are the contents of the famous *Fourier slice theorem* [82], [96]:

$$Y(\nu\cos\theta,\nu\sin\theta)=\int_{-\infty}^{\infty}\exp(-j\nu t)Ry(\theta,t)dt, \qquad (12.98)$$

which says that the 1D Fourier transform of the Radon transform $Ry(\theta, t)$ calculated with respect to the argument t,

$$\int_{-\infty}^{\infty}\exp(-j\nu t)Ry(\theta,t)dt, \qquad (12.99)$$

defines the 2D Fourier transform $Y(\nu\cos\theta, \nu\sin\theta)$ in Eq. (12.97) of the function $y(x_1, x_2)$.

The Fourier slice theorem makes it possible to invert the Radon transform by using a 1D Fourier transform of the Radon transform $Ry(\theta, t)$ in Eq. (12.99) and the 2D inverse Fourier transform of $Y(\nu\cos\theta, \nu\sin\theta)$ in Eq. (12.97).

However, a numerical implementation of this straightforward idea is far from being trivial. The main difficulties are linked with the interpolation of the Fourier transform from the polar frequency co-ordinates to the rectangular Cartesian ones for the frequencies λ_1 and λ_2. The interpolation errors influence the accuracy of the restoration of y from the Radon transform observations [82].

Inverse Radon transform

The numerical inverse Radon transform exists in a number of different forms [82], [148], [149]. In particular, the inverse formula can be reduced to the following integral:

$$\begin{aligned}
y(x) &= \frac{1}{2\pi}\int_0^{\pi}\int_{-\infty}^{\infty}\frac{1}{u_\theta(x)-t}\partial_t Ry(\theta,t)dt d\theta,\\
u_\theta(x) &= x_1\cos\theta + x_2\sin\theta.
\end{aligned} \qquad (12.100)$$

There is a singularity at $u_\theta(x) = t$ in the inner integral existing only in terms of the principal value. The integral in Eq. (12.100) can be defined in the following limit form [82]:

$$y_\alpha(x) = \frac{1}{2\pi}\int_0^\pi \int_{-\infty}^\infty Ry(\theta,t)H_\alpha(u_\theta(x)-t)dtd\theta,$$
$$y(x) = \lim_{\alpha\to 0} y_\alpha(x), \tag{12.101}$$

where

$$H_\alpha(v) = \begin{cases} \frac{1}{\alpha^2}, & \text{for } |v| \le \alpha, \\ -\frac{1}{v^2}, & \text{for } |v| > \alpha. \end{cases} \tag{12.102}$$

The convolution kernel H_α works in Eq. (12.101) as a filter with the parameter α. Note that the total area under $H_\alpha(v)$ is zero, $\int_{-\infty}^{\infty} H_\alpha(v)dv = 0$. For small α it approaches a derivative of the delta function, with the continuous frequency characteristic approximated by $|\lambda|$. Thus, it is a high-pass filter. The integral on θ in Eq. (12.101) smears the estimates obtained for each θ.

The Radon transform in Eq. (12.90) defines a forward projection of $y(x)$ on $Ry(\theta,t)$, while the back projection is a reconstruction of $y(x)$ from $Ry(\theta,t)$. Different filters can be used for H_α in Eq. (12.101). The methods of reconstruction based on the representation in Eq. (12.101) are called the *back projection* or *filtered back projection*. Some recent theoretical results concerning the integral back projection can be found in [148], [149].

Overall, the back projection is the inverse of the linear Radon operator. A difference with the inverse deconvolution considered in Sec. 9.1 is that Radon and inverse Radon operations are not local and not shift-invariant. A review of some of these topics can be seen in [82], where the back projection algorithms are discussed for the parallel beam and other configurations of the detectors, and where the numerical aspects of the Fourier transform for the inverse Radon transform are analyzed.

Conventional numerical back projection methods, such as Ram-Lak, Logan-Shepp, and Hann algorithms, are different mainly by the filter H_α in Eq. (12.102) [35], [91], [222]. The Ram-Lak filter is known as one of the best in terms of minimum image distortion. However, it is sensitive with respect to noise [82].

Noise in the linear model of a transmission system

The detector measurements depend on the Poisson counting z_s, z_{0_s}. In the linear model considered in Eq. (12.86) they are used in the form

$$\log\frac{z_{0_s}}{z_s} \simeq \sum_{l\in L_s} y_{s,l}. \tag{12.103}$$

These Poissonian observations transformed to $\log(z_{0_s}/z_s)$ are approximated by the Radon transform plus a random noise. For very large counts, the random components in Eq. (12.103) can be considered to be stationary white Gaussian noise [82]:

$$\log \frac{z_{0_s}}{z_s} = Ry(\theta_s, t_s) + \varepsilon(\theta_s, t_s), \tag{12.104}$$

where $\varepsilon(\theta, t) \sim N(0, \sigma_\varepsilon^2)$. Thus, we arrive at the linear indirect observation model with the Radon transform of the function of interest and the additive white Gaussian noise.

It is shown that the inverse Radon transform of $Ry(\theta_s, t_s) + \varepsilon(\theta_s, t_s)$ gives an estimate of y with an approximately stationary random error. The correlation function of this error is derived in the form [82]:

$$r_y(\rho) = \frac{\sigma_\varepsilon^2}{4\pi^2} \int_{-\infty}^{\infty} |H_\alpha(\lambda)|^2 J_0(2\pi\lambda\rho) d\lambda, \tag{12.105}$$

where $\rho = \sqrt{x_1^2 + x_2^2}$, $H_\alpha(\lambda)$ is the integral Fourier transform in Eq. (4.40) of the kernel H_α in Eq. (12.101), and J_0 is the first kind of zero-order Bessel function.

Taking a 1D Fourier transform of $r_y(\rho)$ gives the noise spectrum of $r_y(\rho)$. It is shown that it is proportional to $|H_\alpha(\lambda)|^2/|\lambda|$. For small α, this means that $|H_\alpha(\lambda)|^2/|\lambda| \sim |\lambda|$ [82]. The correlation $r_y(\rho)$ depends only on the distance ρ between the points in the 2D space.

12.4.2.3 Nonlinear ML model of transmission imaging

The probabilistic approach to transmission imaging assumes that the measurements are given by the Poisson counts z_s and the distribution of these random variables is taken into consideration. For the mean values in Eq. (12.84), we have a nonlinear link between the observations and the parameters to be estimated $y_{s,l}$:

$$E\{z_s\} = E\{z_{0_s}\} \exp\left(-\sum_{l \in L_s} y_{s,l}\right). \tag{12.106}$$

For the parallel beam system and data recorded according to the Radon transform, this gives

$$E\{z(\theta, t)\} = E\{z_0(\theta, t)\} \exp[-Ry(\theta, t)], \tag{12.107}$$

where $z(\theta, t)$ and $z_0(\theta, t)$ are the measurements produced for the direction θ and along the line (tube) with distance t from the origin. These θ and t define a set of voxels for L_s in Eq. (12.106).

It is convenient, as in the emission system model in Eq. (12.82), to consider a more general link between the parameters $y_{s,l}$ and the observations. Instead of the model in Eq. (12.106), we assume that

$$E\{z_s\} = E\{z_{0_s}\} \exp\left[-\sum_r a(s, r) y(r)\right], \tag{12.108}$$

where $\mathbf{y}$ is a vector of the parameters $y_{s,l}$, $y(r)$ is an rth item of this vector, and $a(s,r) \geq 0$ is a matrix of given weights, similar to $v(s,l)$ in Eq. (12.82), depending on the geometry of the system and the physics of the photon propagation in the body.

Let n be a length of the vector $\mathbf{y}$. As usual, the ML gives the universal formula for how to proceed. The log-likelihood function for the independent Poisson random z_s according to Eqs. (12.84) and (11.59) has the form

$$L = \sum_s [-E\{z_s\} + z_s \log E\{z_s\}].$$

Inserting Eq. (12.108), we find that

$$L = \sum_s \Bigg\{ -E\{z_{0_s}\} \exp\left[-\sum_r a(s,r)y(r)\right] + z_s \log E\{z_{0_s}\} - z_s \sum_r a(s,r)y(r) \Bigg\}. \quad (12.109)$$

Omitting the term $z_s \log E\{z_{0_s}\}$ independent on $\mathbf{y}$, we obtain

$$L = \sum_s b_s(\mathbf{y}),$$
$$b_s(\mathbf{y}) = -E\{z_{0_s}\} \exp\left[-\sum_r a(s,l_r)y(r)\right] - z_s \sum_r a(s,r)y(r). \quad (12.110)$$

The ML estimate of $\mathbf{y}$ is a solution of the problem

$$\hat{\mathbf{y}} = \arg\max_{\mathbf{y}} \sum_s b_s(\mathbf{y}). \quad (12.111)$$

We will need the gradient and the Hessian of the function $b_s(\mathbf{y})$ on $\mathbf{y}$. The corresponding derivatives are

$$\partial_{y(l)} b_s(\mathbf{y}) = \left\{ E\{z_{0_s}\} \exp\left[-\sum_r a(s,r)y(r)\right] - z_s \right\} a(s,l), \quad (12.112)$$

$$\partial_{y(l)} \partial_{y(k)} b_s(\mathbf{y}) = -E\{z_{0_s}\} \exp\left[-\sum_r a(s,r)y(r)\right] a(s,l)a(s,k), \quad (12.113)$$

with the vector gradient

$$\partial_{\mathbf{y}} b_s(\mathbf{y}) = [\partial_{y(1)} b_s(\mathbf{y}), \ldots, \partial_{y(n)} b_s(\mathbf{y})]^T,$$

and the Hessian matrix

$$\partial_{\mathbf{y}}\partial_{\mathbf{y}^T} b_s(\mathbf{y}) = [\partial_{y(l)}\partial_{y(k)} b_s(\mathbf{y})]_{n\times n},$$

$$\partial_{y(l)}\partial_{y(k)} b_s(\mathbf{y}) = -E\{z_{0_s}\}\exp\left[-\sum_r a(s,r)y(r)\right] a(s,l)a(s,k).$$

The Hessian is a definite negative matrix provided that the values $a(s,k)$ are linearly independent. Then, the likelihood L in Eq. (12.109) is concave on $\mathbf{y}$ and has a unique maximum, and the problem in Eq. (12.111) has a unique solution. The concavity of L means that the general recursive Gauss-Newton methods considered in Sec. 11.1.4 can be used to compute the estimates of $\mathbf{y}$.

Here we present a more traditional method developed for the Poisson inverse problem (Sec. 12.2). Following the derivation in Eqs. (12.24–12.27), we obtain from the minimum condition for L,

$$\partial_{y(l)} L = \sum_s \partial_{y(l)} b_s(\mathbf{y}) = 0,$$

that

$$\sum_s \partial_{y(l)} b_s(\mathbf{y}) = \sum_s \left\{ E\{z_{0_s}\}\exp\left[-\sum_r a(s,r)y(r)\right] - z_s \right\} a(s,l) = 0,$$
$$l = 1, \ldots,$$

and

$$1 = \frac{\sum_s E\{z_{0_s}\}a(s,l)\exp\left[-\sum_r a(s,r)y(r)\right]}{\sum_s a(s,l)z_s}.$$

Multiplying the last equation by $y(l)$, we have

$$y(l) = y(l)\frac{\sum_s E\{z_{0_s}\}a(s,l)\exp\left[-\sum_r a(s,r)y(r)\right]}{\sum_s a(s,l)z_s}.$$

Gold's multiplicative recursive scheme gives the recursive algorithm in the form

$$\hat{y}^{[k]}(l) = \hat{y}^{[k-1]}(l)\frac{\sum_s E\{z_{0_s}\}a(s,l)\exp\left[-\sum_r a(s,r)\hat{y}^{[k-1]}(r)\right]}{\sum_s a(s,l)z_s}, \tag{12.114}$$
$$l - 1, \ldots, n, \qquad k = 1, \ldots.$$

This recursive algorithm is known as the EM algorithm for transmission tomography. In this capacity it is derived from different statistical speculations than those in [131].

According to the general properties of the ML in Eq. (11.7), the estimates in Eq. (12.111) have a covariance in the form

$$cov\{\hat{y}(l), \hat{y}(l_1)\} = (I^{-1})_{l,l_1}, \qquad I_{l,l_1} = -E\{\partial_{y(l)}\partial_{y(l_1)}L\}$$
$$= \sum_s E\{z_{0_s}\}a(s,l)a(s,l_1)\exp\left[-\sum_r a(s,r)y(r)\right], \tag{12.115}$$

where I is the *Fisher information matrix.*

12.4.3 Adaptive LPA-ICI algorithms

12.4.3.1 Emission tomography

The emission tomography model in Eq. (12.83) coincides accurately with the model used for the Poisson inverse problem in Secs. 11.2 and 11.3. The algorithms described in these sections are completely applicable for emission tomography signal processing. The simulation results for these algorithms are given in Sec. 12.3.

12.4.3.2 Transition tomography

Nonlinear observation model

The recursive procedure in Eq. (12.114) gives estimates of the attenuation distribution y from the Poisson observations with the mean in Eq. (12.108). It solves the inverse problem of finding y from the convolutions $\sum_r a(s,r)y(r)$. This recursive procedure does not assume any links between $y(r)$ for different r and does not guarantee the smoothness of the attenuation distribution y. The estimates obtained in this way usually are highly noisy because of the ill-conditioning of the inverse problem. As a result of these recursive calculations, we have the estimates of the distribution y and their covariance in Eq. (12.115).

These estimates are similar to those considered in connection to the Poisson inverse of the emission tomography data. The algorithms LPA-ICI(AV) and LPA(AV)-ICI(AV) from Sec. 12.2.5 can be used for further filtering with the input data in Eqs. (12.114–12.115).

Linear observation model

The linear observation model in Eq. (12.104) despite the Gaussian noise, is quite different from the convolution model used in Chapter 9, where the PSF is a shift-invariant linear operator. The frequency characteristics of the PSF are exploited in the developed scale-adaptive RI and RWI algorithms. The Radon transform is a global (not local) linear operator, and its impulse response is shift-variant.

According to the observation model in Eq. (12.104), the data available for signal processing have an additive form,

$$q(\theta_s, t_s) = Ry(\theta_s, t_s) + \varepsilon(\theta_s, t_s), \qquad (12.116)$$

where $q(\theta_s, t_s) = \log(z_{0_s}/z_s)$ represents the observations and $\varepsilon(\theta_s, t_s) \sim \mathcal{N}(0, \sigma_\varepsilon^2)$ is
Gaussian white noise.

Let R_α^{-1} be the regularized inverse Radon transform defined in Eq. (12.101) by the kernels H_α, where α denotes a regularization parameter. Apply this operator to the data the Eq. (12.116) as

$$R_\alpha^{-1}(q(\theta, t)) = R_\alpha^{-1}(Ry(\theta, t)) + R_\alpha^{-1}(\varepsilon(\theta, t)). \qquad (12.117)$$

The first term in the right-hand side of Eq. (12.117) gives an approximation of the signal $y(x)$ depending on the parameter α,

$$R_\alpha^{-1}(Ry(\theta, t)) = \tilde{y}_\alpha(x) \rightarrow_{\alpha\rightarrow 0} y(x).$$

After the inverse Radon transform, we obtain in the original coordinates x

$$r(x) = \tilde{y}_\alpha(x) + \varepsilon_1(x), \qquad (12.118)$$

where $r(x) = R_\alpha^{-1}(q(\theta, t))$ are inverse Radon transformed observations and $\varepsilon_1(x) = R_\alpha^{-1}(\varepsilon(\theta, t))$ is the noise of these transformed data.

With reference to [82], we claim that the noise $R_\alpha^{-1}(\varepsilon(\theta, t))$ in Eq. (12.117) is stationary random with the correlation function given by Eq. (12.105).

The algorithm for the inversion of the Radon transform observations in Eq. (12.116) is composed from the following main steps:

1. The regularized inverse of the observations $q(\theta_s, t_s)$ is

$$r(x) = R_\alpha^{-1}(q(\theta, t)); \qquad (12.119)$$

2. The directional LPA-ICI filtering of $r(x)$ is

$$\hat{Y}_{\theta,h}(f) = G_{\theta,h}(f)R(f), \qquad (12.120)$$

where $R(f)$ is the DFT of $r(x)$ and $G_{\theta,h}(f)$ is the transfer function of the directional LPA filter, and the variances of these estimates are calculated as

$$\sigma^2_{\hat{y}_{\theta,h}}(h) = \frac{1}{n_1 n_2}\sum_f |G_{\theta,h}(f)|^2 S_{\varepsilon_1}(f), \qquad (12.121)$$

where $S_{\varepsilon_1}(f)$ is the discrete spectrum on the random noise ε_1 in Eq. (12.118) and n_1, n_2 define the size of the estimated $y(x)$;

3. The directional adaptive-scale LPA-ICI estimates $\hat{Y}_{\theta,h^+}(f)$ are fused into the final one.

The spectrum $S_{\varepsilon_1}(f)$ of the noise ε_1 could be calculated using formula (12.105). However, the digital inverse Radon transform is quite different from the integral one used in derivation of Eq. (12.105), and we prefer a more accurate and practical approach based on the estimation of the discrete version of the spectrum $S_{\varepsilon_1}(f)$.

Formulas (12.101) and (12.105) show that, with respect to noise, the inverse Radon transform can be considered to be a linear system with the continuous transfer function $H_\alpha(\lambda)$.

Assuming that a similar statement holds for the discrete inverse Radon transform, denote the corresponding discrete transfer function by $D(f)$, where f is a frequency of the DFT. Then the discrete spectrum $S_{\varepsilon_1}(f)$ of the noise ε_1 in Eq. (12.118) is calculated as

$$S_{\varepsilon_1}(f) = \sigma_\varepsilon^2 |D(f)|^2, \tag{12.122}$$

where σ_ε^2 is the variance of the discrete noise in the model in Eq. (12.116).

Provided that the algorithm of the discrete inverse Radon transform is given, the spectrum $S_{\varepsilon_1}(f)$ can be estimated using a straightforward *Monte-Carlo* modeling:

$$|\hat{D}(f)|^2 = \frac{1}{L} \sum_{j=1}^{L} |\mathcal{F}\{R_\alpha^{-1}(\varepsilon^{[j]})\}|^2, \tag{12.123}$$

where $\varepsilon^{[j]}(\theta, t)$ are independent Gaussian random variables with variance equal to one, $R_\alpha^{-1}(\varepsilon^{[j]})$ is the inverse discrete Radon transform of the random $\varepsilon^{[j]}$, and $\mathcal{F}\{R_\alpha^{-1}(\varepsilon^{[j]})\}$ is the DFT of this inverse. The index $^{[j]}$ numbers the L successive independent runs of the Monte-Carlo modeling.

Numerical experiments

A linear observation model is assumed for the transmission tomography system. As a test image we use the "Shepp-Logan" phantom shown in Fig. 12.15(a). We consider a reconstruction of this image from noisy Radon transform data in Eq. (12.116). The algorithm defined by formulas (12.119)–(12.123) is used.

In Fig. 12.16 we show the results as well as some images that are useful for comparison. In Figs. 12.16(a) and (b), one can observe the inverse Radon transform for the noiseless and noisy observations in Eq. (12.104) with $\sigma_\varepsilon = 0$ and $\sigma_\varepsilon = 4$, respectively. Degradation effects caused by the noise are obvious.

In Fig. 12.16(c), the image obtained by the proposed adaptive LPA-ICI algorithm is presented. This image is a result of the LPA-ICI filtering of the regularized Radon inverse image shown in Fig. 12.16(b). For comparison, the image obtained by the standard LPA-ICI algorithm assuming white additive noise with a constant variance in Eq. (12.118) is shown in Fig. 12.16(d).

The algorithms used to obtain images (c) and (d) in Fig. 12.16 are different only by calculation of the variance of the estimates. For the image shown in Fig. 12.16(c), we use the accurate formula (12.121), where the spectrum of the noise ε_1 is taken into consideration.

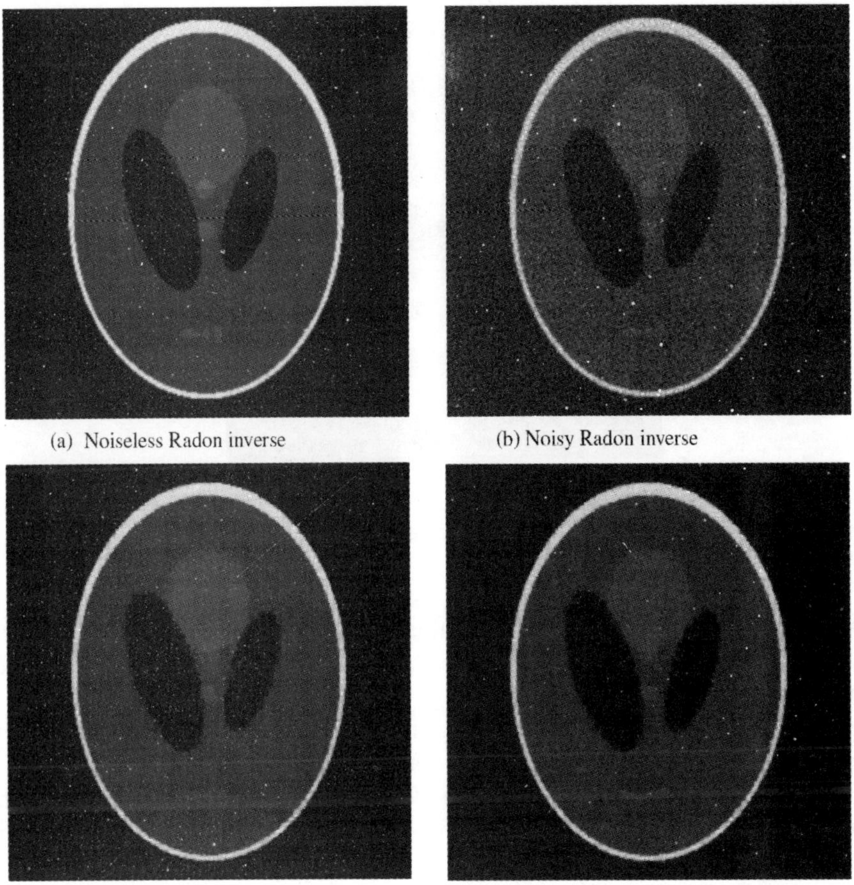

(a) Noiseless Radon inverse (b) Noisy Radon inverse

(c) LPA-ICI for Radon inverse (d) LPA-ICI usual

Figure 12.16 Inverse Radon transform filtering.

For the image shown in Fig. 12.16(d), we use for the variance the standard formula

$$\sigma^2_{\hat{y}_h}(h) = \frac{\sigma^2_\varepsilon}{n_1 n_2} \sum_f |G_h(f)|^2,$$

where the noise variance is constant.

Table 12.5 Phantom image: criterion values for the LPA-ICI algorithm developed for the Radon inverse and the usual LPA-ICI algorithm.

Algorithm \ Criteria	ISNR, dB	PSNR, dB	RMSE	MAE	MAXDIF	UIQI
LPA-ICI, Radon inv.	**9.66**	**24.13**	**15.85**	**7.78**	**221.11**	**0.1673**
LPA-ICI, usual	8.84	23.31	17.42	9.72	230.12	0.1643

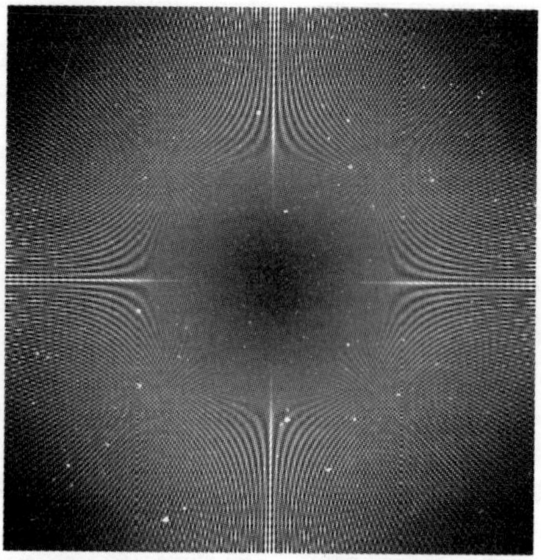

Figure 12.17 Imaging of the discrete spectrum $|D(f)|^2$ defined by formulas (12.123). The coordinates are discrete frequencies f_1, f_2 with zero values at the center of the image.

A comparison of these two LPA-ICI algorithms is given in Table 12.5. Numerically and visually, the algorithm described by formulas (12.119)–(12.123) gives better results.

Figure 12.17 shows the squared amplitude response $|D(f)|^2$ of the discrete inverse Radon transform computed according to formula (12.123).

The program *iradon.m* (with *Ram-Lak* filter and linear interpolation) from the image processing toolbox of MATLAB was used in these experiments. Thus, in Fig. 12.17 we can see the frequency characteristic of this particular implementation of the inverse Radon transform.

The LPA-ICI algorithm is used with the following parameters. The LPA degree $m = (0, 0)$. The threshold parameter $\Gamma = 1$. The window function w is uniform. The kernel supports are narrow sectorials defined by the scales

$$H = \left\{ \begin{pmatrix} 1 \\ 1 \end{pmatrix}, \begin{pmatrix} 2 \\ 1 \end{pmatrix}, \begin{pmatrix} 3 \\ 1 \end{pmatrix}, \begin{pmatrix} 4 \\ 1 \end{pmatrix}, \begin{pmatrix} 6 \\ 2 \end{pmatrix}, \begin{pmatrix} 9 \\ 3 \end{pmatrix} \right\}.$$

The number of directions used is four: $\theta_i = i \cdot \pi/4$, $i = 0, 1, 2, 3$.

Chapter 13

Multiresolution Analysis

The novel adaptivity introduced in this chapter can be viewed as the development of two independent ideas: wavelet multiresolution analysis and the adaptive-scale nonparametric regression estimation.

In nonparametric regression methods, the adaptive-scale selection is a key point. Lepski's approach, which was discussed above, and the ICI rule are the new statistics developed for adaptation. A common feature of these methods is that they all form a number of test estimates $\hat{y}_h(x)$ that are different by the scale $h \in H$ as well as by the data sets used for estimation, and statistical rules are exploited to select the best scale (window size) of the estimate. Thus, we yield the adaptive varying-scale pointwise adaptation.

The approach presented in this chapter results in a different type of scale-adaptive estimate. Mainly, instead of selecting the estimate with the unique best scale $h^+(x)$, we build a nonlinear estimate using all of the available estimates $\hat{y}_h(x)$, $h \in H$. This estimation is divided into two steps. The first step transforms the data into noisy spectrum coefficients (*multiresolution analysis*). In the second step, these noisy estimates of the spectrum are filtered by the thresholding procedure and used for estimation (*multiresolution synthesis*).

The LPA-based filters are exploited for the *nonparametric* (pointwise) *multiresolution spectrum analysis* and *synthesis*. In this way we introduce an extension of the conventional nonparametric local regression concepts and yield a wider class of adaptive-scale *multiresolution nonparametric regression* techniques with potentially better performance.

13.1 MR Analysis: Basic Concepts

"Multiresolution" assumes a special general approach/technology for data analysis and interpretation. In terms of imaging, it means that the image is considered to be a combination of some lower frequency, slowly varying, smooth background and high-frequency, rapidly varying sharp detail. The analysis starts from the background, and the details of different scales (sizes) are sequentially revealed and added to the background.

Multiresolution works as a sequential zoom system. First, we have a blurred image, when general contours (background) can be recognized only while details are smoothed away. Next, some of the largest details are revealed and added to the background. Further, step-by-step smaller details, contrast, sharpness appear in the image.

In general, multiresolution (MR) is a tool or a system of tools for decomposition of data in additive multiscale components. There is an enormous amount of problems in which MR analysis is the root of the signal processing.

Here we present the basic ideas of MR analysis, which are used for the development of its nonparametric kernel version. This is not a review of the field but an introduction of concepts given in the form that will be used later. A comprehensive presentation of MR analysis can be found in [33], [76], [150], [216], and [227]. In this section we mainly follow these works. For simplicity, we consider the restoration of continuous 1D functions while a nonparametric version of the approach in the following sections is given for multivariable discrete operators defined on regular grids.

Mathematically, the nonparametric version of MR analysis differs from the standard definitions that originated in the works on wavelet theory. However, this novel nonparametric technique shares with the standard one the basic idea of decomposing the signal into different scale components. Our use of the term MR emphasizes this common basic objective.

13.1.1 Orthogonal series

Assume that a continuous function of a continuous scalar argument $y(x)$, $x, y \in \mathbb{R}$ is square integrable, $||y||^2 = \int_{\mathbb{R}} y^2(x)dx < \infty$. The series

$$y_j(x) = \sum_{k=0}^{j} c_k \omega_k(x) \tag{13.1}$$

is used to approximate $y(x)$. If the coordinate functions $\omega_k(x)$ are orthonormal,

$$\langle \omega_k, \omega_l \rangle = \int_{\mathbb{R}} \omega_k(x)\omega_l(x)dx = \delta_{k,l},$$

then

$$c_k = \langle y, \omega_k \rangle = \int_{\mathbb{R}} y(x)\omega_k(x)dx, \tag{13.2}$$

and Eq. (13.1) gives

$$y_j(x) = \sum_{k=0}^{j} \langle y, \omega_k \rangle \omega_k(x). \tag{13.3}$$

We use this orthogonal series to explain the main elements of classical MR analysis.

Define the linear subspaces V_j generated by $\omega_k(x)$ as

$$V_j = \left\{ y(x) = \sum_{k=0}^{j} c_k \omega_k(x) : c_k \in \mathbb{R} \right\}, \qquad j = 0, 1, \ldots. \tag{13.4}$$

Then:

(A) The subspaces V_j are nested, $V_j \subset V_{j+1}$, and can be represented in the form

$$V_{j+1} = V_j \oplus W_{j+1}, \quad j = 0, 1, \ldots, \tag{13.5}$$

where $\oplus$ is the direct sum of the subspaces and the subspace W_{j+1} is an orthogonal complement of the subspace V_j defined by $\omega_{j+1}(x)$.

(B) The space

$$\bigcup_{j=0}^{\infty} V_j = V_0 \bigoplus_{j=1}^{\infty} W_j, \tag{13.6}$$

is dense in L_2.

Let us give some intuitive explanations of this mathematical material.

1. Assume that the basis functions $\omega_k(x)$ are ordered according to their smoothness or frequency contents, starting from the lower frequency term $\omega_0(x)$ and going to higher frequency terms as k grows. For example, let these functions be sines and cosines of the frequency $2\pi k$. The first $\omega_0(x)$ is a constant and all of the others are harmonics of the growing frequencies. The first summands in $\hat{y}_j(x)$ describe the background of the image while the higher frequency components add high-resolution details.
2. The model $y_j(x)$ includes $j+1$ summands, and the coefficients c_k, $k = 0, \ldots, j$, of these summands are calculated independently for each k according to Eq. (13.2). This independence is not compulsory but is a highly desirable property because it allows one to analyze and use the details separately.
3. The subspace V_j in Eq. (13.4) is a span of functions defined as a linear combination of ω_k, $k = 0, \ldots, j$. The subspace V_{j+1} includes all functions from V_j, that is, the subspaces V_j are nested (property (A)). The difference between the subspaces V_{j+1} and V_j is W_{j+1} defined by $\omega_{j+1}(x)$.
4. The union of all subspaces $\bigcup_{j=0}^{\infty} V_j$ includes all $\omega_k(x)$, $k \to \infty$. The statement that this union is dense in L_2 means that any function $y \in L_2$ can be represented in the form in Eq. (13.3), that is, there is a convergence $||y_j(x) - y(x)||_{L_2} \to 0$ as $j \to \infty$.

Expressions (13.3)–(13.5) include all formal elements of the MR. However, this is nothing more than the standard orthogonal series and the most important point concerning the construction of the MR basis ω_k is lacking here.

13.1.2 Wavelets

13.1.2.1 Localization

In MR analysis, a localization of the basis functions ω_k in space (time) and frequency domains is essential. Let us discuss this issue in more detail. The Fourier transform uses harmonic (sines and cosines) basis functions. The Fourier coefficients $\hat{c}_k$ display detail frequency spectrums of signals but destroy all space or time information. These coefficients do not tell anything about the location of nonstationarity or singularity in the signal because every coefficient is obtained as

$$\int_{\mathbb{R}} y(x)\cos(2\pi f_k x)dx, \int_{\mathbb{R}} y(x)\sin(2\pi f_k x)dx,$$

by integration over the whole signal interval.

Spatially local basis functions are required for the analysis of nonstationary signals. However, no basis function can give exact information on frequency and space localization at the same time.

Formalizing the notion of frequency and space/time uncertainty leads to lower bounds on the product of both. It can be given in the following standard form.

Let $\Theta(x)$ and $\Theta(\lambda)$, x, $\lambda \in \mathbb{R}$, be a univariate function and $\Theta(\lambda)$ be the integral Fourier transform of $\Theta(x)$:

$$\Theta(\lambda) = \mathcal{F}\{\Theta(x)\} = \int_{\mathbb{R}} \Theta(x)\exp(-j\lambda x)dx,$$

$$\Theta(x) = \mathcal{F}^{-1}\{\Theta(\lambda)\} = \frac{1}{2\pi}\int_{\mathbb{R}} \Theta(\lambda)\exp(j\lambda x)d\lambda.$$

In the space–frequency plane $(x,\ \lambda)$, the energy spread of Θ is symbolically represented by the *Heisenberg's uncertainty principle* ([216], Theorem 2.3) as

$$\bar{\sigma}_x\bar{\sigma}_\lambda \geq 1/2, \tag{13.7}$$

where

$$\bar{\sigma}_x^2 = \frac{\int_{\mathbb{R}} x^2|\Theta(x)|^2dx}{\int_{\mathbb{R}} |\Theta(x)|^2dx}, \qquad \bar{\sigma}_\lambda^2 = \frac{\int_{\mathbb{R}} \lambda^2|\Theta(\lambda)|^2d\lambda}{\int_{\mathbb{R}} |\Theta(\lambda)|^2d\lambda}.$$

Here $\bar{\sigma}_x^2$ and $\bar{\sigma}_\lambda^2$ are energy spreads of the signal and its Fourier transform with respect to the zero values of $\Theta(\lambda)$ and $\Theta(x)$. Small $\bar{\sigma}_x^2$ and $\bar{\sigma}_\lambda^2$ mean high concentrations of the corresponding energy in the signal or transform domains. The inequality in Eq. (13.7) shows that the product $\bar{\sigma}_x\bar{\sigma}_\lambda$ is at least $1/2$. The maximum value of this product, which is equal to $\frac{1}{2}$, is attained by the Gaussian function.

The inequality in Eq. (13.7) is used as a test of concentration of the basis function in space and frequency domains. If the product $\bar{\sigma}_x\bar{\sigma}_\omega$ is close to $\frac{1}{2}$, it means that the basis function has a concentration close to the optimal one. A good concentration in both space and frequency domains is a reasonable requirement for signal representation and analysis.

Let us calculate $\bar{\sigma}_x^2$ and $\bar{\sigma}_\lambda^2$ for the signal

$$\Theta_1(x) = \exp(-x^2/h^2). \tag{13.8}$$

The Fourier transform of $\Theta_1(x)$ is also the Gaussian function

$$\Theta_1(\lambda) = \sqrt{\pi}h \exp(-\lambda^2 h^2/4). \tag{13.9}$$

We wish to show that the same lower bound $\frac{1}{2}$ is achieved in Eq. (13.7) for all values of the scale parameter h in Eqs. (13.8–13.9).

Indeed, we have for $\bar{\sigma}_x^2$

$$\bar{\sigma}_x^2 = \frac{\int_{\mathbb{R}} x^2 \exp(-2x^2/h^2)dx}{\int_{\mathbb{R}} \exp(-2x^2/h^2)dx}.$$

Changing the integration variable $x/h \to u$, we obtain

$$\bar{\sigma}_x^2 = \frac{h^3 \int_{\mathbb{R}} u^2 \exp(-2u^2)du}{h \int_{\mathbb{R}} \exp(-2u^2)du} = h^2 A, \;\; A = \frac{\int_{\mathbb{R}} u^2 \exp(-2u^2)du}{\int_{\mathbb{R}} \exp(-2u^2)du} = \frac{1}{4}. \tag{13.10}$$

In a similar way we obtain for

$$\bar{\sigma}_\lambda^2 = \frac{\int_{\mathbb{R}} \lambda^2 \exp(-\lambda^2 h^2/2)d\lambda}{\int_{\mathbb{R}} \exp(-\lambda^2 h^2/2)d\lambda} = \frac{1}{h^2}B, \; B = \frac{\int_{\mathbb{R}} u^2 \exp(-u^2/2)du}{\int_{\mathbb{R}} \exp(-u^2/2)du} = 1. \tag{13.11}$$

The product

$$\bar{\sigma}_x^2 \bar{\sigma}_\omega^2 = AB = 1/4,$$

does not depend on h.

A smaller scale h results in a narrower peak of $\Theta_1(x)$ in Eq. (13.8) and a good energy concentration (small $\bar{\sigma}_x^2$ in Eq. (13.10)) in the signal domain. For same small h we have a smoother $\Theta_1(\lambda)$ in Eq. (13.9) with a poor energy concentration (larger $\bar{\sigma}_\lambda^2$ in Eq. (13.11)). For a larger h there is a narrow peak in the frequency domain and a smooth function $\Theta_1(x)$ in the space domain. This says that we can take some liberties when selecting the scale h, and Heisenberg's inequality cannot resolve this uncertainty.

13.1.2.2 Haar basis system

The Haar basis system is the simplest one that is well localized in the space domain and gives a uniform approximation of any continuous y. Here we consider the Haar basis on a real line and describe some of its properties, which are useful for the construction of general wavelet systems.

The Haar function is defined as an indicator of the segment (0, 1]:

$$\varphi(x) = \begin{cases} 1, & x \in (0, 1], \\ 0, & x \notin (0, 1]. \end{cases} \tag{13.12}$$

This function defines the subspace V_0 of the piecewise constant functions:

$$V_0 = \left\{ y(x) = \sum_{k \in Z} c_k \varphi(x-k) \,:\, c \in l_2 \right\}. \tag{13.13}$$

Clearly, if $y \in V_0$, then y is constant on $(k, k+1]$. In particular, for $x \in (0, 1]$, $y(x) = c_0\varphi(x)$.

Denote the translated (shifted) versions of φ as

$$\varphi_{0,k}(x) = \varphi(x-k), \qquad k \in Z.$$

The system $\{\varphi_{0,k}\}$ is an orthonormal basis in V_0,

$$\langle \varphi_{0,k}, \varphi_{0,l} \rangle = \int_{\mathbb{R}} \varphi(x-k)\varphi(x-l)dx = \delta_{k,l},$$

and $c_k = \langle \varphi_{0,k}, y \rangle$ in Eq. (13.13).

Now define a new linear subspace V_1 by

$$V_1 = \{y(x) = v(2x) \,:\, v \in V_0\}.$$

The space V_1 contains all functions that are constants on the intervals $(k/2, (k+1)/2]$, $k \in Z$. Obviously, $V_0 \subset V_1$ and the orthonormal basis of V_1 is given by the functions $\{\varphi_{1,k}\}$, where

$$\varphi_{1,k}(x) = \sqrt{2}\varphi(2x-k).$$

Here the scalar factor $\sqrt{2}$ is used for the normalization of the function as $\int_0^1 \varphi_{1,k}^2(x) = 2\int_0^{1/2} dx = 1$.

One can iterate this process and define a family of subspaces as

$$V_j = \{y = v(2^j x) \,:\, v \in V_0\}, \tag{13.14}$$

with the orthonormal bases

$$\varphi_{j,k}(x) = 2^{j/2}\varphi(2^j x - k), \qquad k \in \mathbb{Z}. \tag{13.15}$$

The functions $\varphi_{j,0}$ and $\varphi_{0,k}$ are shown in Fig. 13.1 for $j, k = 0, 1, 2, 3$. The *dilation* parameter j defines the scale (width) of the indicator function $\varphi_{j,k}(x)$ and k defines a *translation* of this function, its location on the x axis. The subset of the functions $\varphi_{j,k}(x)$ with the fixed j is orthonormal,

$$\langle \varphi_{j,k}(x), \varphi_{j,l}(x) \rangle = \delta_{k,l},$$

because these functions do not have overlapping supports.

The approximation of y on the subset V_j has the form

$$y_j(x) = \sum_{k \in \mathbb{Z}} c_{j,k}\varphi_{j,k}(x),\ c_{j,k} = \langle \varphi_{j,k}, y \rangle. \tag{13.16}$$

Figure 13.2(a) illustrates the approximating properties of the subset of $\varphi_{j,k}(x)$ with fixed j. This approximation is piecewise constant with the width of the segments

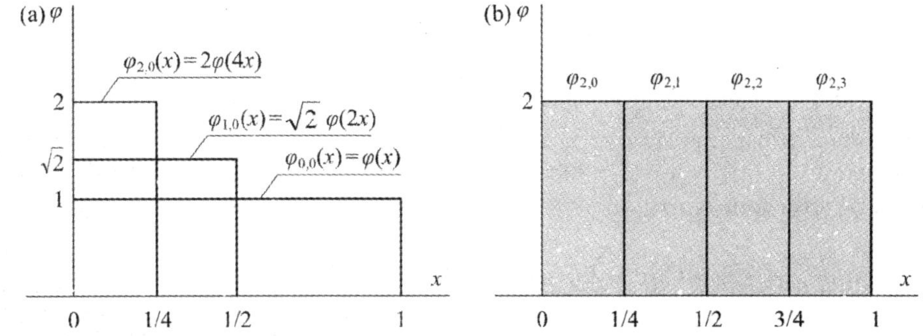

Figure 13.1 The Haar functions $\varphi_{j,0}$ are shown for $j = 0, 1, 2$ in (a) and the Haar functions $\varphi_{2,k}$ are shown for $k = 0, 1, 2, 3$ in (b). The *dilation* parameter j defines the scale (width) of the indicator function $\varphi_{j,k}(x)$ and k defines a *translation* of this function, its location on the x axis.

defined by j as 2^{-j}. A larger j gives a finer approximation with smaller segments of the piecewise y_j.

For comparison, we show in Fig. 13.2(b) the curves obtained by the sliding window smoothing using the integral kernel operator g_h in Eq. (4.17) obtained by the LPA with the window $\varphi(x)$ in Eq. (13.12) and order $m = 0$. This LPA kernel is

$$g_h(x) = \varphi(x/h)/h, \tag{13.17}$$

which corresponds to the dyadic dilation of the Haar basis $h_j = 2^{-j}$. With this scale, $g_{2^{-j}}(x) = 2^j \varphi(2^j x)$.

This kernel is different from the Haar basis $\varphi_{j,0}(x) = 2^{j/2}\varphi(2^j x)$ only by the normalization factor. In Fig. 13.2(b) we use the kernel $g_{2^{-j}}(x)$ with $j = 3$ and $j = 5$. The convolution integral operator with this kernel naturally gives a smooth (not piecewise constant) approximation of y.

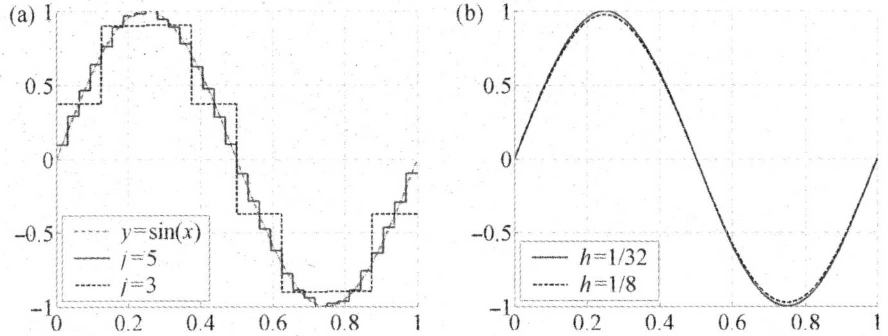

Figure 13.2 (a) Illustration of the approximation of the sine by the Haar functions of the scales $j = 3$ and $j = 5$. It is piecewise constant with the width of the segments defined by j as 2^{-j}. (b) Curves obtained by sliding window smoothing using the integral kernel operator g_h with the window $\varphi(2^j x)$ and $m = 0$. The kernel estimates of this sine are smoother than those in (a).

Thanks to Eq. (13.15), we have the inclusion of the subspaces V_j Eq. (13.14)

$$V_0 \subset V_1 \subset \ldots,$$

that is, these subspaces are nested.

Continuing this process infinitely, we approximate the whole space L_2. This means that the approximation

$$\hat{y}_J(x) = \sum_{k\in\mathbb{Z}} c_{J,k}\varphi_{J,k}(x),\ c_{J,k} = \langle\varphi_{J,k}, y\rangle, \tag{13.18}$$

converges to any $y \in L_2$ as $J \to \infty$, $||y - y_J||_{L_2} \to 0$.

The system $\varphi_{j,k}(x)$ is not a basis because $\varphi_{j,k}(x)$ are linearly dependent for different j.

How do we orthogonalize the system $\{\varphi_{j,k}\}_{j,k\in\mathbb{Z}}$? Here we follow the presentation of the material in [216].

Let us have the subspaces V_j defined above. Denote by W_1 the orthogonal complement of V_0 in V_1. In other words, the set $\{\varphi_{1,k}\}_{k\in\mathbb{Z}}$ represents V_1 as the sum of two subspaces:

$$V_1 = V_0 \oplus W_1.$$

This means that every $v_1 \in V_1$ can be represented as

$$v_1 = v_0 + w_1, \qquad v_0 \in V_0, \quad w_1 \in W_1,$$

where $< v_0, w_1 >= 0$.

Let us show that the subspace W_1 is a linear subspace spanned by a certain basis. Introduce a function

$$\psi(x) = \begin{cases} 1, & x \in (0, 1/2] \\ -1, & x \notin (1/2] \end{cases} \tag{13.19}$$

and the system

$$\psi_{1,k}(x) = \psi(x-k), \qquad k \in \mathbb{Z},$$

as a basis of W_1.

It can be checked that these functions are orthonormal

$$\langle\psi_{1,k}, \psi_{1,l}\rangle = \int_{\mathbb{R}} \psi(x-k)\psi(x-l)dx = \delta_{k,l},$$

and orthogonal to $\varphi_{0,k}$, that is,

$$\langle\varphi_{0,k}, \psi_{1,l}\rangle = 0.$$

Further, show that $\varphi_{1,k}(x)$, defining V_1, can be expressed through $\varphi_{0,k}$ and $\psi_{1,k}$ defining V_0 and W_1, respectively. The following transformations show it:

$$\varphi_{1,0}(x) = \sqrt{2}\varphi(2x) = \sqrt{2}[\varphi_{0,0}(x) + \psi_{1,0}(x)]/2 = [\varphi_{0,0}(x) + \psi_{1,0}(x)]/\sqrt{2},$$
$$\varphi_{1,1}(x) = \sqrt{2}\varphi(2x-1) = \sqrt{2}[\varphi_{0,0}(x) - \psi_{1,0}(x)]/2 = [\varphi_{0,0}(x) - \psi_{1,0}(x)]/\sqrt{2}.$$

Further, for any k,

$$\begin{aligned}\varphi_{1,k}(x) &= \sqrt{2}\varphi(2x-k) = \sqrt{2}[\varphi_{0,k}(x) + \psi_{1,k}(x)]/2 \\ &= [\varphi_{0,k}(x) + \psi_{1,k}(x)]/\sqrt{2}, \\ \varphi_{1,k+1}(x) &= \sqrt{2}\varphi(2x-k-1) = \sqrt{2}[\varphi_{0,k}(x) - \psi_{1,k}(x)]/2 \\ &= [\varphi_{0,k}(x) - \psi_{1,k}(x)]/\sqrt{2},\end{aligned}$$

Thus, we prove that

$$V_1 = V_0 \oplus W_1.$$

Continue this process and use translation and dilation to build the system of functions similar to $\varphi_{j,k}$ in Eq. (13.15) starting from ψ. Let

$$\psi_{j,k}(x) = 2^{j/2}\psi(2^j x - k),\ k \in \mathbb{Z}. \tag{13.20}$$

As it was for φ, the dilation parameter j defines the scale (width) and the translation parameter k defines a shift of this function, its location on the x axis. Examples of the functions $\psi_{j,k}$ are shown in Fig. 13.3.

For any scale j and translation k, the following link exists between the functions $\varphi_{j+1,k}(x)$, $\varphi_{j,k}(x)$ and $\psi_{j+1,k}(x)$:

$$\begin{aligned}\varphi_{j+1,2k}(x) &= [\varphi_{j,2k}(x) + \psi_{j+1,2k}(x)]/\sqrt{2}, \\ \varphi_{j+1,2k+1}(x) &= [\varphi_{j,2k}(x) - \psi_{j+1,2k}(x)]/\sqrt{2},\end{aligned}$$

and then

$$\psi_{j+1,2k}(x) = [\varphi_{j+1,2k}(x) - \varphi_{j+1,2k+1}(x)]/\sqrt{2}.$$

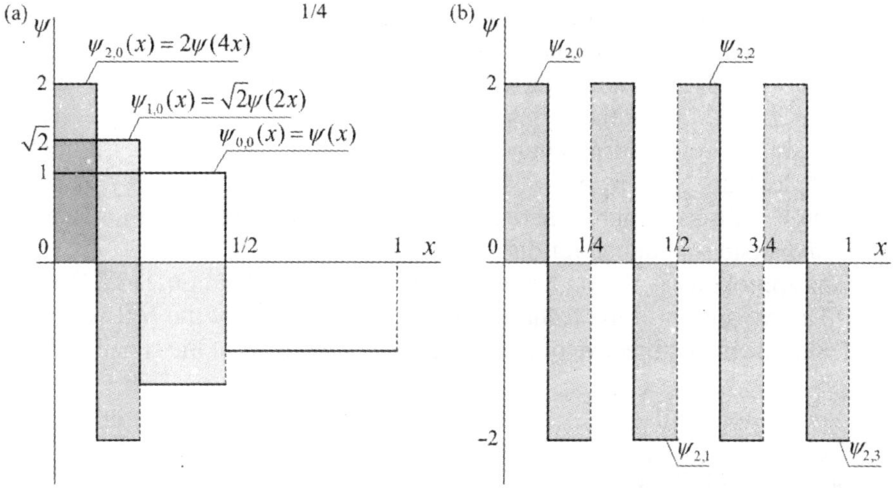

Figure 13.3 Examples of the wavelets $\psi_{j,0}$ and $\psi_{0,k}$ for j = 0 and k = 0, 1, 2 in Figure (a) and for j = 2 and k = 0, 1, 2, 3 in Figure (b).

It can be verified that $\psi_{j,k}(x)$ are orthonormal,

$$\langle \psi_{j,k}, \psi_{j,l} \rangle = \delta_{k,l}\delta_{j,i}, \qquad k, l \in \mathbb{Z},$$

and orthogonal to $\varphi_{i,k}(x)$, $i = 0, \ldots, j$:

$$\langle \psi_{j,k}, \varphi_{i,l} \rangle = 0, \qquad i = 0, \ldots, j, \quad k, l \in \mathbb{Z}.$$

Then

$$V_{j+1} = V_j \oplus W_{j+1},$$

and the subspace W_{j+1} is defined by the functions $\psi_{j+1,k}$:

$$W_{j+1} = \left\{ w_{j+1} : w_{j+1} = \sum_{k=-\infty}^{\infty} c_k \psi_{j+1,k} \right\}.$$

Every $y \in L_2(\mathbb{R})$ can be represented by the expansion

$$y(x) = \sum_{k \in \mathbb{Z}} \alpha_{0,k} \varphi_{0,k}(x) + \sum_{j=0}^{\infty} \sum_{k \in \mathbb{Z}} \beta_{j,k} \psi_{j+1,k}(x), \tag{13.21}$$

$$\alpha_{0,k} = \langle y, \varphi_{0,k} \rangle, \quad \beta_{j,k} = \langle y, \psi_{j+1,k} \rangle, \tag{13.22}$$

with the approximation of y in the subspace V_J as

$$\begin{aligned} \hat{y}_J(x) &= \sum_{k} \alpha_{0,k} \varphi_{0,k}(x) + \sum_{j=0}^{J-1} \sum_{k \in \mathbb{Z}} \beta_{j,k} \psi_{j+1,k}(x), \\ \alpha_{0,k} &= \langle y, \varphi_{0,k} \rangle, \quad \beta_{j,k} = \langle y, \psi_{j+1,k} \rangle, \end{aligned} \tag{13.23}$$

where $||y - \hat{y}_J||_{L_2} \to 0$ as $J \to \infty$.

For each J the series in Eqs. (13.18) and (13.23) define different form approximations of y on the same subspace V_J. The essential difference between these series is that Eq. (13.18) gives this approximation in terms of the scale functions with only the scale J used, while the series in Eq. (13.23) starts from the lower frequency term $\sum_k \alpha_0 \varphi_{0,k}(x)$ corresponding to the subspace V_0 and completes it, including the complements $\sum_{k \in \mathbb{Z}} \beta_{j,k} \psi_{j+1,k}(x)$ of the scales from $j = 0$ up to $j = J - 1$, to form the V_j subspace approximation. These complements are the differences of the signal expansions obtained for different scales.

The approximations $\hat{y}_J(x)$ given by formulas (13.18) and Eq. (13.23) are identical. The expansion in the form of Eq. (13.23) is a basis of the MR because it is used with the main objective of analyzing the components of the signal at different scales.

Let us make a few general remarks:

1. φ and ψ are called scale and wavelet functions, respectively. The terms "father" and "mother" wavelets also are used for these functions. The

following properties of φ and ψ:

$$\int_{\mathbb{R}} \varphi(x)dx = 1, \quad \int_{\mathbb{R}} \psi(x)dx = 0,$$

are obvious for the Haar case and can be generalized to any wavelet and scale function.

The functions $\varphi_{j,k}$, $\psi_{j,k}$ are *scaled* and *shifted* versions of the father and mother wavelets and

$$\int_{\mathbb{R}} \varphi_{j,k}(x)dx = 1, \quad \int_{\mathbb{R}} \psi_{j,k}(x)dx = 0.$$

The whole set of orthonormal functions is produced starting from two functions, φ and ψ only. The scale parameter h exploited in the previous chapters for LPA estimates is linked with the integer scale j used here as

$$h_j = 2^{-j}. \tag{13.24}$$

Thus, the dyadic scale parameter h_j is exploited in the wavelets and the large j corresponds to the fine scale with small h_j.

2. Localization in Haar's expansions is achieved in a straightforward manner. For $j = 0$, the scale function and wavelet have the segment $(0, 1]$ as their support. For larger scale j, the support becomes narrow and is equal to $(0, 2^{-j}]$. Thus, the supports of $\varphi_{j,k}$, $\psi_{j,k}$ are $(2^{-j}k, 2^{-j}(k+1)]$. It follows that

$$2^{j/2} \cdot \varphi_{j,0}(x) = 2^j \varphi(2^j x) \rightarrow_{j\rightarrow\infty} \delta(x),$$

that is, asymptotically, the scaled father wavelet $\varphi_{j,0}(x)$ multiplied by $2^{j/2}$ is the rectangular approximation of the Dirac delta function.

If the notation in Eq. (13.24) is used, we have

$$h_j^{-1/2}\varphi_{j,0}(x) = h_j^{-1}\varphi(x/h_j) \rightarrow_{h_j\rightarrow 0} \delta(x).$$

This behavior of the scalar function within the factor $h_j^{-1/2}$ is identical to the behavior of the smoothing kernel giving in the integral form $g_{h_j}(x) \rightarrow \delta(x)$ as $h_j \rightarrow 0$ (Sec. 4.3).

Asymptotically, the wavelet within the factor $2^{j/2}$ is an approximation of the derivative of the Dirac delta function,

$$2^{j/2} \cdot \psi_{j,0}(x) = 2^j \psi(2^j x) \rightarrow_{j\rightarrow\infty} \delta^{(1)}(x),$$

and we have the same for differentiating kernels given in the integral form $g_{h_j}^{(1)}(x) \rightarrow \delta^{(1)}(x)$ as $h_j \rightarrow 0$ (Sec. 4.3).

3. The series in Eqs. (13.18) and (13.23) which give identical results, are built using different basis functions: the scale functions φ in Eq. (13.18) and the

wavelets ψ in Eq. (13.23). The coefficients in Eq. (13.23) are

$$\alpha_{0,k} = \langle y, \varphi_{0,k} \rangle = \int_{2^{-j}k}^{2^{-j}(k+1)} y(x)\varphi_{0,k}(x)dx,$$

$$\beta_{j,k} = \langle y, \psi_{j+1,k} \rangle = \int_{2^{-j}k}^{2^{-j}(k+1)} y(x)\psi_{j+1,k}(x)dx.$$

The calculation of these coefficients assumes that a partition of the area of x on segments $(2^{-j}k, 2^{-j}(k+1)]$ is the support of the corresponding basis functions. These coefficients $\alpha_{0,k}, \beta_{j,k}$ are called the *wavelet's spectrum*. This spectrum gives the power distribution of the signal in space (variable k) and frequency (frequency is approximately proportional to j or inverse to the scale $h_j = 2^{-j}$).

The wavelet spectrum tells how much of the corresponding wavelet basis function "is present" in the total signal: A high absolute value coefficient means that at the given location and scale the singularity makes an important contribution.

This information is local in space and frequency: It tells where the singularity (jump in the signal) is and how far it ranges, that is, how large its scale is. This wavelet representation, that is, the set of details at different locations and different scales, is said to be closer to the way we look at the signal: At first sight, we see general features; on closer inspection, we discover the details.

The expressions in Eq. (13.22) are the analysis formulas while the representation in Eq. (13.21) is the synthesis. It is the perfect synthesis formula in Eq. (13.21) that gives an accurate reconstruction of y.

13.1.2.3 Wavelets

While Haar expansions exhibit all of the elements of MR analyses, the basic drawback is that the functions φ and ψ are not smooth, and, as a result, the series $\hat{y}_J$ are also not smooth and have a clearly seen piecewise constant character (Fig. 13.2a).

The *smooth scale* and *wavelet* functions satisfy the following vanishing moment conditions [150], [216]:

$$\int_{\mathbb{R}} x^i \varphi(x)dx = \delta_{i,0}, \qquad \int_{\mathbb{R}} x^i \psi(x)dx = 0, \qquad (13.25)$$
$$i = 0, \ldots, m-1.$$

Comparing this with Eqs. (4.30) and (4.31), we say that these φ and ψ have polynomial smoothness of order $m-1$ and can be treated like smoothing and differentiating LPA integral kernels, respectively. In particular, note that $\int_{\mathbb{R}} \varphi(x)dx = 1$ means that φ is a smoothing operator while $\int_{\mathbb{R}} \psi(x)dx = 0$ says that ψ is a differentiation operator.

We can also treat the kernels $g_h(x)$ and $g_h^{(r)}(x)$ as scale and wavelet functions. The multiscale bases $\varphi_{j,k}(x)$ and $\psi_{j,k}(x)$ for these smooth φ and ψ are built according to formulas (13.15) and (13.20).

These basis functions exhibit two types of orthogonalities: inter-scale and intra-scale. Interscale orthogonality means the orthogonality between the functions of the different scales:

$$\langle \psi_{j,k}(x), \psi_{i,k}(x) \rangle = \delta_{i,j}, \qquad \langle \varphi_{j,k}(x), \varphi_{i,k}(x) \rangle = \delta_{i,j}, \tag{13.26}$$

while intrascale orthogonality is the orthogonality of a function of the same scale but with shifted argument values:

$$\langle \psi_{j,k}(x), \psi_{j,l}(x) \rangle = \delta_{k,l}, \qquad \langle \varphi_{j,k}(x), \varphi_{j,l}(x) \rangle = \delta_{k,l}. \tag{13.27}$$

The properties in Eqs. (13.26–13.27) are far from being trivial, because they show that it is possible to find two such functions φ and ψ, and that their translation and dilation give the orthonormal basis. The very existence of this type of function is an interesting point. It is important to find the finite support φ and ψ. Such continuous scale and wavelet functions with supports of minimum size for any given order m in Eq. (13.25) were originally obtained by Daubechies [33].

The smooth scale functions $\varphi_{j,k}(x)$ and smooth wavelets $\psi_{j,k}(x)$ form an orthogonal MR as discussed above. This can be summarized in the following form. The standard MR wavelet expansion for univariate continuous data, $y(x)$, $x \in \mathbb{R}$, has the form of the parametric series in Eqs. (13.21–13.22), e.g., [76], [150], where $\varphi(x)$ and $\psi(x)$ are the scale function (father wavelet) and the wavelet (mother wavelet), respectively.

The intra- and interscale orthogonality of the functions $\varphi_{j,k}(x)$ and $\psi_{j,k}(x)$ is assumed. The double intra- and inter-scale orthogonality requirements make the design of the wavelet function a complex problem and seriously confine the classes of wavelet functions.

The father wavelet $\varphi(x)$ generates the following linear subspaces in L_2:

$$\begin{aligned} V_0 &= \{y(x) = \sum_s \varphi(x-s)c_s : \sum_s |c_s|^2 < \infty\}, \\ V_1 &= \{f(x) = y(2x) : y \in V_0\}, \\ &\ldots \\ V_j &= \{f(x) = y(2^j x) : y \in V_0\}, \end{aligned} \tag{13.28}$$

such that the subspaces V_j satisfy the properties in Eqs. (13.5–13.6).

The representation in Eq. (13.21) is composed of two summands. The first one is the basic lower frequency part given by the scale functions $\varphi_{0,k}(x)$. The second one is a high-frequency complement using the wavelets $\psi_{j,k}(x)$.

The orthogonal wavelets form a local space-frequency basis that can be interpreted as a tiling that covers the space-frequency space with wavelet space-frequency boxes. Figure 13.4(a) shows the space-frequency box of each $\psi_{j,k}$, which is translated by 2^j with the space and frequency widths scaled, respectively, by 2^j and 2^{-j}. In this figure, the j is interpreted as an integer frequency.

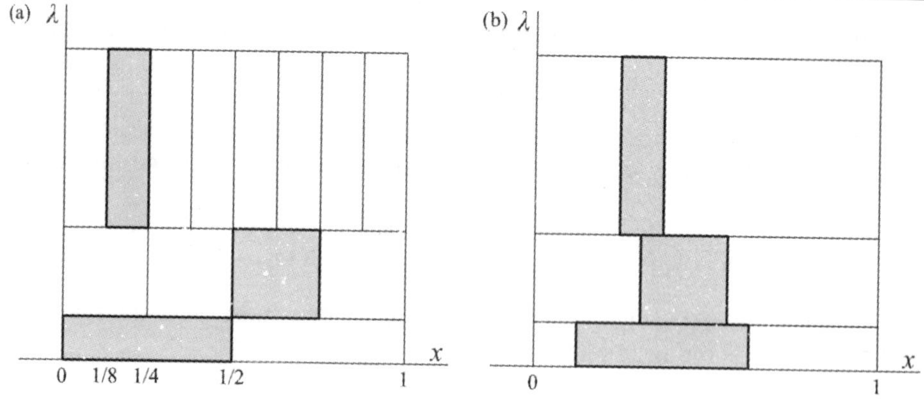

Figure 13.4 (a) Space-frequency boxes of the wavelets $\psi_{j,k}$ translated by 2^j with space and frequency widths scaled by 2^j and 2^{-j}, respectively. The tiling on x means a dyadic segmentation of the x axis specific for the wavelet transform. (b) Space-frequency boxes of kernel MR analysis. For comparison we use the same tiling on the frequency λ. The space-frequency boxes are sliding along the x axis. There is no segmentation on x for the kernel MR analysis.

13.2 Nonparametric LPA Spectrum

Consider the observation model in the form

$$z(x) = y(x) + \sigma\varepsilon(x), \tag{13.29}$$

where the multivariable $x \in \mathbb{Z}^d$ for simplicity is assumed to be an infinite regular integer, $z, y \in \mathbb{R}$ and ε is the standard Gaussian noise, $\varepsilon \sim \mathcal{N}(0, 1)$. The kernel estimate of $y(x)$ has the form of the d-dimensional convolution in Eq. (3.4):

$$\begin{aligned} \hat{y}_h(x) &= (g_h \circledast z)(x), \\ g_h(x) &= w_h(x)\phi^T(0)\Phi_h^{-1}\phi_h(x), \\ \Phi_h &= \sum_x w_h(x)\phi_h(x)\phi_h^T(x), \end{aligned} \tag{13.30}$$

where $h \in H$ is the scalar scale parameter of the estimate, and hereafter $\sum_x = \sum_{x \in \mathbb{Z}^d}$.

Recall that $g_h(x)$ defines a low-pass filter with the bandpass controlled by the value of h. Larger and smaller h give narrower- and wider-band low-pass filters, respectively.

Assume H is a finite or infinite set of positive scales and order its elements in descending order, starting from a large h_0 and going to a small h_J:

$$H = \{h_0 > h_1 > \cdots > h_J\}. \tag{13.31}$$

Note that this scale set is different from the set used in the ICI rule (e.g., Eq. (6.1)) where the scales in H form an ascent sequence.

We will assume that, for the smallest scale h_J in H, the LPA kernel $g_h(x)$ gives the identical operator

$$lim_{j\to J} g_{h_j}(x) = \delta_{x,0}. \tag{13.32}$$

This assumption is not restrictive and only defines the range of scales in Eq. (13.31) approaching sufficiently small h_J.

For instance, let $d = 1$, $m = 0$; the sampling interval is Δ and the window is Gaussian, $w_h(x) = \exp[-(x\Delta)^2/(2h^2)]$. Then, the kernel g_h is the Gaussian exponent

$$g_h(x) = B(h)\exp[-(x\Delta)^2/(2h^2)],$$

normalized by a constant $B(h)$. The maximum of this kernel is at $x = 0$ and the condition in Eq. (13.32) means that $g_h(1) \ll g_h(0)$, that is,

$$\exp[-\Delta^2/(2h_J^2)] \ll 1 \Longrightarrow h_J \ll \Delta.$$

If the window w_h and the kernel g_h have finite support,

$$S_h = supp\{w_h(x)\}, \tag{13.33}$$

we assume that the set H is such that the supports S_h are different and nested, that is,

$$S_{h_J} \subset S_{h_{J-1}} \subset \cdots \subset S_{h_0}, \tag{13.34}$$

with the largest support for h_0. These different supports guarantee that there are no identical kernels in the set of $g_h(x)$, $h \in H$.

13.2.1 Finite-difference LPA spectrum

Let us start from a simple decomposition of $\hat{y}_h(x)$ from Eq. (13.30) in a sum of differences of the estimates with different scales:

$$\begin{aligned} \hat{Y}_\beta(x) &= \hat{y}_{h_0}(x) + \sum_{j=1}^{J} \beta_j \cdot \Delta\hat{y}_j(x), \\ \Delta\hat{y}_j(x) &= (\Delta g_j \circledast z)(x), \\ \Delta g_j(x) &= g_{h_j}(x) - g_{h_{j-1}}(x), \qquad j = 1, 2, \ldots, J, \end{aligned} \tag{13.35}$$

where $\Delta\hat{y}_j(x) = \hat{y}_{h_j}(x) - \hat{y}_{h_{j-1}}(x)$ for $j = 1, 2, \ldots, J$, and $\beta = (\beta_1, \ldots, \beta_J)^T$ is a vector of coefficients.

Consider the items in the right-hand side of Eq. (13.35). The difference $\Delta\hat{y}_j(x)$ is a deviation of the estimate caused by a decrement of h from h_{j-1} to h_j. The largest $h = h_0$ means a coarser scale and strong smoothing with $\hat{y}_{h_0}(x) = (g_{h_0} \circledast z)(x)$ being the smoothest estimate giving the low-frequency picture of $y(x)$.

Smaller values of h correspond to a finer scale that detects higher-frequency details in the signal. In the sum in Eq. (13.35), the first term $\hat{y}_{h_0}(x)$ presents the smoothed background of the signal while the others add details about the different scales. Thus, Eq. (13.35) provides a decomposition of the signal in the different scale components. For $\beta_j = 1$, $1 \leq j \leq J$, we have a perfect reconstruction of the original data with no smoothing effects $\hat{Y}_\beta(x) = \hat{y}_{h_J}(x) = y(x)$.

Let us call $\Delta g_j(x)$ the nonparametric spectral (scale) analysis kernels and $\hat{Y}_\beta(x)$ the nonparametric regression spectral (scale) decomposition. The following properties are easy to verify:

- According to Eq. (3.15), the analysis kernels Δg_j, $j = 1, \ldots, J$, have vanishing moments up to order m:

$$(\Delta g_j \circledast x^k)(0) = 0, \qquad 0 \leq |k| \leq m. \tag{13.36}$$

  Thus, the analysis kernels are smooth with the polynomial smoothness defined by the power m of the LPA.
- The sum of the analysis kernels $\Delta g_j(x)$ assuming $\Delta g_0(x) = g_{h_0}(x)$ and Eq. (13.32) is the identical operator

$$\sum_{j=0}^{J} \Delta g_j(x) = g_{h_J}(x) = \delta_{x,0}. \tag{13.37}$$

- For any $h \in H$, the estimate $\hat{y}_h(x)$ in Eq. (13.30) can be represented in the form

$$\hat{y}_h(x) = \hat{Y}_{\beta(h)}(x), \tag{13.38}$$

  with the coefficients $\beta_j(h)$ in Eq. (13.35) defined as

$$\beta_j(h) = 1[h_j \geq h], \qquad h, h_j \in H, \tag{13.39}$$

  where the indicator function $1[x] = 1$ for $x \geq 0$ and $1[x] = 0$ for $x < 0$.

Equation (13.38) is verified by substituting β_j given by Eq. (13.39) into Eq. (13.35). A varying h in Eq. (13.39) controls the number of spectral items in the expansion in Eq. (13.35) and varies $\hat{Y}_{\beta(h)}(x)$ from the perfect reconstruction of the observed $z(x) = \hat{Y}_{\beta(h_J)}(x)$ to the most smoothed estimate $\hat{y}_{h_0}(x) = \hat{Y}_{\beta(h_0)}(x)$.

The problem of the adaptive-scale selection for the estimate $\hat{y}_h(x)$ can be reformulated as the selection of h in $\beta(h)$ from Eq. (13.39) for the estimate in Eq. (13.35). For h invariant on x, Eq. (13.39) gives the same scale selection for all x while h, which is dependent on x, results in the pointwise varying scale

$$\beta_j(h(x)) = 1[h_j \geq h(x)]. \tag{13.40}$$

13.2.2 Linear independence

Let us extend the supports of the kernels $g_h(x)$ to the maximum size support S_{h_0} by zero-padding $g_h(x)$:

$$g_{h_j}(x) = \begin{cases} g_{h_j}(x), & x \in S_{h_j} \subset \mathbb{Z}^d, \\ 0, & x \in S_{h_0}/S_{h_j} \subset \mathbb{Z}^d. \end{cases}$$

Now all of the kernels have the same supports and the inner product of the kernels can be defined as

$$\langle g_{h_i}, g_{h_j} \rangle = \sum_{x \in S_{h_0}} g_{h_i}(x) \cdot g_{h_j}(x), \tag{13.41}$$

with the quadratic norm

$$||g_{h_i}||^2 = \langle g_{h_i}, g_{h_i} \rangle = \sum_{x \in S_{h_0}} g_{h_i}^2(x).$$

Consider a linear combination of g_{h_i}

$$g = a_0 g_{h_0} + a_1 g_{h_1} + \cdots + a_J g_{h_J}.$$

If one or more kernels in this sum is a linear combination of the others, g may be generated without them. Then there is a redundancy in the set of kernels. This redundancy means that, in the equation

$$a_0 g_{h_0} + a_1 g_{h_1} + \cdots + a_J g_{h_J} = 0,$$

one or more of the a_i are nonzero and we say that the kernels are linearly dependent. For example, if the sum is zero and $a_k \neq 0$, then

$$g_{h_k} = -\frac{1}{a_k} \sum_{i=0, i \neq k}^{J} a_i g_{h_i}.$$

The *Gram matrix* can be used to check the linear independence of g_h, $h \in H$:

$$G = (G_{i,j})_{(J+1)\times(J+1)}, \quad G_{i,j} = \langle g_{h_i}, g_{h_j} \rangle.$$

The following is the test for linear independence. The kernels $g_h, h \in H$, are linearly independent iff the Gram determinant is nonzero:

$$\det G \neq 0. \tag{13.42}$$

In the following it is assumed that $g_{h_j}(x)$, $j = 0, \ldots, J$, are linearly independent.

The linear independence of $g_h, h \in H$, is mainly a restriction on the set of scales H. In particular, for the finite support w, the nested support condition in Eq. (13.34) usually suffices for linear independence. In any case, the linear independence of the kernels g_h, $h \in H$, is not a restrictive assumption. Note that if g_h, $h \in H$, are linearly independent, then $\Delta g_0 = g_{h_0}$, $\Delta g_j = g_{h_j} - g_{h_{j-1}}$ for $j = 1, 2, \ldots, J$, are also linearly independent.

13.2.3 Gram-Schmidt orthogonalization

The standard sequential Gram-Schmidt procedure can be used for orthogonalization of the linearly independent Δg_j or g_{h_j}, $j = 0, 1, \ldots, J$. This orthogonality of the kernel is used in the forthcoming spectrum analysis.

The procedure begins by initialization,

$$\mu_0 = \Delta g_0, \quad \omega_0 = \mu_0/||\mu_0||, \tag{13.43}$$

where the Euclidean norm $||\mu_0|| = \sqrt{\sum_{x \in S_{h_0}} \mu_0^2(x)}$. Construct ω_1 as follows:

$$\begin{aligned} \mu_1 &= \Delta g_1 - \langle \Delta g_1, \omega_0 \rangle \omega_0, \\ \omega_1 &= \frac{\mu_1}{||\mu_1||}. \end{aligned} \tag{13.44}$$

Assume that $\omega_1 \neq 0$. Then ω_1 and ω_0 are linearly independent and orthogonal. Continue this procedure, generating each new ω_i as follows:

$$\begin{aligned} \mu_i &= \Delta g_i - \sum_{k=0}^{i-1} \langle \Delta g_i, \omega_k \rangle \omega_k, \\ \omega_i &= \frac{\mu_i}{||\mu_i||}. \end{aligned} \tag{13.45}$$

This procedure maps Δg_j, $j = 0, 1, \ldots, J$, into the functions $(\omega_0, \ldots, \omega_J)$. Each μ_i is the difference between Δg_i and the sum of the projections of this Δg_i on all previous ω_k, $k = 1, \ldots, i-1$. If Eq. (13.42) holds, there will be $J+1$ nonzero μ_i, ω_i.

The norms $||\mu_i||$ and the inner products $\langle \Delta g_i, \omega_k \rangle$ are numbers in Eqs. (13.43–13.45). These equations can then be presented in the form

$$\begin{aligned} &\omega_0 ||\mu_0|| = \Delta g_0, \\ &\sum_{k=0}^{i-1} \langle \Delta g_i, \ \omega_k \rangle \omega_k + \omega_i ||\mu_i|| = \Delta g_i, \ i = 1, \ldots, J. \end{aligned}$$

This proves that the functions $(\omega_0, \ldots, \omega_J)$ satisfy the following linear equation:

$$Q\omega(x) = \Delta g(x), \tag{13.46}$$

where $\Delta g = (\Delta g_0, \ldots, \Delta g_J)^T$, $\omega = (\omega_0, \ldots, \omega_J)^T$, and Q is a nonsingular lower triangular matrix $(J+1) \times (J+1)$ with diagonal and nondiagonal elements, respectively:

$$Q_{ii} = ||\mu_i||, \quad Q_{ik} = \langle \Delta g_i, \omega_k \rangle, \quad k < i, \ \ i, k = 0, \ldots, J.$$

This matrix is nonsingular if Eq. (13.42) holds.

13.2.4 Orthogonal LPA spectrum analysis

Let us treat $\omega(x)$ as a $(J+1)$-vector kernel filter and apply it to the signal $y(x)$, $x \in \mathbb{R}^d$. Denote the output of this filtering as

$$\alpha(x) = (\omega \circledast y)(x), \alpha \in \mathbb{R}^{J+1}.$$

Call this output the *nonparametric spectrum* $\omega(x)$ of the input signal and the vector kernel $\omega(x)$ the *analysis filter.*

The following properties can be stated for these orthonormal kernels $\omega_j(x)$ and the corresponding outputs $\alpha_j(x)$, where $j = 0, \ldots, J$:

1. The outputs $\alpha_j(x)$ define a spectral analysis with components varying from a low-frequency (coarse scale) base signal $\alpha_0(x)$ to higher frequency (finer scale) signal increments $\alpha_j(x)$. A higher value j corresponds to a higher-frequency spectral component.
2. The analysis kernels have vanishing moments up to order m:

$$\begin{aligned} (\omega_0 \circledast x^k)(0) &= \delta_{|k|,0}/||\Delta g_0||, & 0 \le |k| \le m, \\ (\omega_j \circledast x^k)(0) &= 0,\ 1 \le |k| \le m, & j = 1, \ldots, J. \end{aligned} \tag{13.47}$$

3. The analysis kernels ω_j, $j = k+1, \ldots, J$, are orthogonal to g_{h_k}:

$$\langle \omega_j, g_{h_k} \rangle = 0,\ j = k+1, \ldots, J. \tag{13.48}$$

4. Let l_2 be a space of square-summable d-dimensional functions y. Introduce the accumulated kernels

$$\Omega_j(x) = \sum_{k=0}^{j} \omega_k(x), \quad j = 0, 1, \ldots, J$$

and define the linear subspaces W_j and V_j generated by the kernels ω_j and Ω_j, respectively, as

$$W_j = \{\Delta y_j = (\omega_j \circledast y) : y \in l_2\}, \quad j = 1, 2, \ldots, J, \tag{13.49}$$

$$V_j = \{y_j = (\Omega_j \circledast y) : y \in l_2\}, \quad j = 0, 1, \ldots, J. \tag{13.50}$$

The kernels ω_j are orthogonal on the design. However, this does not mean that the subspaces W_j are also orthogonal.

The subspaces V_{j+1} can be represented in the form

$$V_{j+1} = V_j + W_{j+1}, \qquad l = 0, 1, \ldots, J-1, \tag{13.51}$$

where the plus (+) means that for each element $y_{j+1} \in V_{j+1}$ there exist such $y_j \in V_j$ and $\Delta y_{j+1} \in W_{j+1}$ that $y_{j+1} = y_j + \Delta y_{j+1}$. The subspace W_{j+1} is a complement (nonorthogonal in general) of the subspace V_j. It follows that

$$V_{j+1} = V_0 + \left(\sum_{k=1}^{j+1} W_k \right), \qquad j = 0, \ldots, J-1.$$

One of the requirements on the sequence of subspaces V_j in MR analysis is completeness:

$$y_j(x) \to y(x) \qquad \text{as } j \to J.$$

The sequence V_j defined in Eq. (13.50) is complete because of Eq. (13.32).

13.2.5 Perfect reconstruction

The following is a *nonparametric kernel decomposition* of y based on the analysis kernels ω_j.

Proposition 16 [104] *Let* $y \in l_2(\mathbb{Z}^d)$*. Then*

$$y(x) = \sum_{j=0}^{J} \alpha_j(x) \cdot \omega_j(0), \tag{13.52}$$

$$\alpha_j(x) = (\omega_j \circledast y)(x). \tag{13.53}$$

Proof. Substituting Eq. (13.53) into Eq. (13.52), we obtain

$$y(x) = (K \circledast y)(x), \qquad K(x) = \sum_{j=0}^{J} \omega_j(x) \cdot \omega_j(0).$$

Prove that

$$K(x) = \delta_{x,0}. \tag{13.54}$$

The following two formulas follow from Eq. (13.46) and the orthonormality of $\omega(x)$:

$$K(x) = \omega^T(0)\omega(x) = \Delta g^T(0)(QQ^T)^{-1}\Delta g(x), \tag{13.55}$$

$$\sum_x \Delta g(x)\Delta g^T(x) = \sum_x Q\omega(x)\omega^T(x)Q^T = Q\sum_x \omega(x)\omega^T(x)Q^T = QQ^T. \tag{13.56}$$

Now we wish to calculate $K(x)$.

Let $\mathbf{1}$ be a vector of size $J+1$ where all elements are equal to 1. Then

$$\mathbf{1}^T \sum_x \Delta g(x) \cdot \Delta g^T(x) = \sum_x \left(\sum_{j=0}^{J} \Delta g_j(x) \right) \Delta g^T(x) = \sum_x \delta_{x,0} \Delta g^T(x)$$
$$= \Delta g^T(0). \tag{13.57}$$

Here we exploit that

$$\mathbf{1}^T \Delta g(x) = \sum_{j=0}^{J} \Delta g_j(x) = g_{h_J}(x) = \delta_{x,0}. \tag{13.58}$$

Multiplying Eq. (13.56) from the left-hand side by $\mathbf{1}^T$ and using Eq. (13.57), we find

$$\mathbf{1}^T \sum_x \Delta g(x) \Delta g^T(x) = \mathbf{1}^T Q Q^T \Longrightarrow g^T(0) = \mathbf{1}^T Q Q^T,$$

which means that

$$\Delta g^T(0)(Q Q^T)^{-1} = \mathbf{1}^T.$$

Inserting it in Eq. (13.55), we have

$$K(x) = \mathbf{1}^T \Delta g(x),$$

but, according to Eq. (13.58), $\mathbf{1}^T \Delta g(x) = \delta_{x,0}$. This proves Eq. (13.54). Then, $y(x) = (K \circledast y)(x) = y(x)$ completing the proof of the proposition. □

Equations (13.53) and (13.52) define, respectively, the analysis and synthesis steps of MR nonparametric analysis. Formula (13.52) gives the perfect reconstruction for any $y \in l_2(\mathbb{Z}^d)$.

We wish to observe that the expansion in Eq. (13.52) can be derived from different speculations using the local approximation concept. Indeed, let $y(x-u)$ be defined for $u \in \mathbb{Z}^d$ and $\omega_j(u)$, $j = 0, \ldots, J$ be orthonormal for $u \in \mathbb{Z}^d$. Looking for the approximation of $y(x-u)$ in the form

$$\hat{y}(x-u) = \sum_{j=0}^{J} c_j \omega_j(u), \tag{13.59}$$

and minimizing

$$\sum_u \left[y(x-u) - \sum_{j=0}^{J} c_j \omega_j(u) \right]^2,$$

we find that

$$c_j = \sum_u y(x-u)\omega_j(u).$$

According to the basic idea of the LPA, exploit the series Eq. (13.59) for $u = 0$ only, and note also that $c_j = \alpha_j(x)$ in Eq. (13.53). This immediately gives

$$\hat{y}(x) = \sum_{j=0}^{J} c_j \omega_j(0) = \sum_{j=0}^{J} \alpha_j(x)\omega_j(0)$$

and proves that the series in Eq. (13.59) leads to Eq. (13.52).

The synthesis formula in Eq. (13.52) is not unique. A very different version of the perfect reconstruction is given in the following proposition.

Proposition 17 [105] *Let* $y \in l_2(\mathbb{Z}^d)$ *and* $q_j = \sum_{l=j}^{J} Q_{l,j}$, *where* $Q_{l,j}$ *are elements of the matrix* Q *in Eq. (13.46). Then*

$$y(x) = \sum_{j=0}^{J} \alpha_j(x) \cdot q_j, \tag{13.60}$$

$$\alpha_j(x) = (\omega_j \circledast y)(x). \tag{13.61}$$

Proof. Substituting Eq. (13.61) into Eq. (13.60) transforms this representation into the convolution

$$y(x) = (K \circledast y)(x), \tag{13.62}$$

$$K(x) = \sum_{j=0}^{J} \omega_j(x) q_j. \tag{13.63}$$

Show that $K(x) = \delta_{x,0}$. Indeed, it follows from Eq. (13.46) that

$$\sum_{l=0}^{J} \Delta g_l(x) = \sum_{l=0}^{J} \sum_{j=0}^{J} Q_{l,j} \omega_j(x) = \sum_{j=0}^{J} \omega_j(x) \sum_{l=0}^{J} Q_{l,j}$$

$$= \sum_{j=0}^{J} \omega_j(x) \sum_{l=j}^{J} Q_{l,j} = \sum_{j=0}^{J} \omega_j(x) q_j = K(x).$$

However, according to Eq. (13.37), $\sum_{l=0}^{J} \Delta g_l(x) = g_{h_J}(x) = \delta_{x,0}$. This shows that $K(x) = \delta_{x,0}$ and completes the proof of this proposition. □

Formulas (13.61) and (13.60) yield the accurate MR spectral expansion of y that is valid for any $y \in l_2(\mathbb{Z}^d)$. Specifically, Eqs. (13.61) and (13.60) define, respectively, the analysis and synthesis steps of this MR analysis. The syntheses formula in Eq. (13.60) shows that the projections α_j of y on the MR analysis subspaces V_j taken with the weights q_j enables the perfect reconstruction for any $y \in l_2(\mathbb{Z}^d)$.

If $g_J(x) \neq \delta_{x,0}$, formula (13.60) can be used as an approximate synthesis formula that gives the reconstruction of y within the bias error of the kernel operator g_{h_J}. The accuracy of this operator was analyzed in Chapter 5. The following generalization of Eq. (13.60) is also valid:

$$\hat{y}_{h_i}(x) = \sum_{j=0}^{i} \alpha_j(x) q_j(i), \qquad q_j(i) = \sum_{l=j}^{i} Q_{l,j}, \tag{13.64}$$

where $\hat{y}_{h_i}$ is the LPA estimate of the scale h_i.

13.2.6 MR analysis for differentiation

For smoothing, an accurate identity operator is assumed for the scale J in the perfect reconstruction formulas (13.52) and (13.60). This sort of accurate discrete differentiation is not possible because there is no kernel for the discrete convolution in Eq. (3.7) that gives an accurate derivative for any $y \in l_2(\mathbb{Z}^d)$.

The approximate form of Eq. (13.60) is appropriate for the differentiation of nonparametric signals. Let us replace g_h in Eq. (13.46) by the differentiation kernels $g_h^{(r)}$, where r is the multi-index showing the order of the estimated derivative. Then, the Gram-Schmidt formulas define the orthogonal differentiation analysis kernels $\omega_j^{(r)}$ and the corresponding MR differentiation subspaces.

The analysis and syntheses formulas (13.60)–(13.61) are changed to the form:

$$\hat{y}^{(r)}(x) = \sum_{j=0}^{J} \alpha_j^{(r)}(x) q_j^{(r)}, \qquad \alpha_j^{(r)}(x) = (\omega_j^{(r)} \circledast y)(x). \tag{13.65}$$

Here

$$q_j^{(r)} = \sum_{l=j}^{J} Q_{l,j}^{(r)},$$

with $Q_{l,j}^{(r)}$ being the elements of the matrix $Q^{(r)}$ in the Gram-Shmidt equation

$$Q^{(r)} \omega^{(r)}(x) = \Delta g^{(r)}(x), \qquad x \in \mathbb{Z}^d, \tag{13.66}$$

where

$$\omega^{(r)} = [\omega_0^{(r)}(x), \omega_1^{(r)}(x), \ldots, \omega_J^{(r)}(x)]^T,$$
$$\Delta g^{(r)} = [\Delta g_0^{(r)}(x),\ \Delta g_1^{(r)}(x), \ldots, \Delta g_J^{(r)}(x)]^T, \qquad \Delta g_0^{(r)}(x) = g_0^{(r)}(x).$$

The $\omega_J^{(r)}$ is the finest scale of the differentiation analysis operator in the representation in Eq. (13.65). The interval of the scales from 0 to J used in Eq. (13.65) and specified in H defines the derivative estimates of different scales to select from or to use jointly in the combined MR estimation.

Because $\omega_J^{(r)}$ is not an accurate differentiating operator, formula (13.65) defines an approximate reconstruction of a derivative within an accuracy corresponding to the derivative estimate with the kernel $\omega_J^{(r)}$ having the finest scale in the set H, $h = h_J$. The accuracy of this approximation was analyzed in Chapter 5.

13.2.7 Examples of MR kernels

Examples of the MR analysis smoothing kernels ω_j are shown in Fig. 13.5. These kernels are obtained according to Eq. (13.46), where the g_h are derived from the LPA of the power $m = 2$ for the 2D Gaussian window $w = (1/2\pi)\exp(-||x||^2/2)$. The scale set $H = \{31, 21, 11, 5, 3, 1\}$ and the windows used in the kernel design are truncated to the squares $h \times h$. For $h = 1$, the zero-order kernel $m = 0$ is used

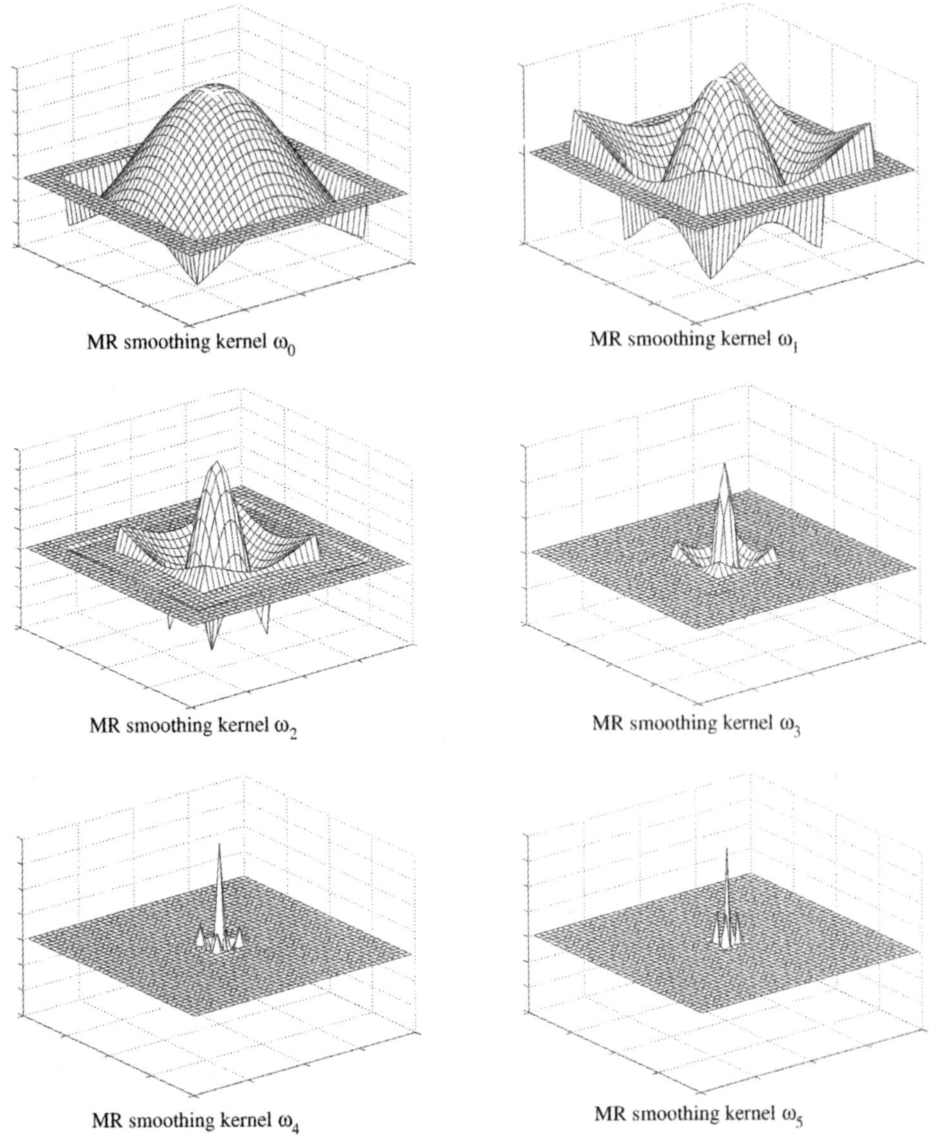

Figure 13.5 MR analysis smoothing kernels obtained using the LPA of power m = 2 and the 2D Gaussian window.

to obtain the accurate Kronecker delta function $\delta_{x,0}$. The first MR kernel ω_0 with $h_0 = 31$ defines a smoothing low-pass filter while other analysis kernels ω_j, $j = 1, \ldots, 5$, define band-pass filters.

The MR differentiating kernels $\omega_j^{(1,0)}(x)$ are shown in Fig. 13.6. These analysis differentiating kernels are obtained from $g_{h_j}^{(1,0)}(x)$ (differentiation on x_1) using the LPA of the power $m = 2$ and the 2D Gaussian window $w = (1/2\psi^2)$ $\exp[-||x||^2/(2\sigma^2)]$, $\sigma = 0.3$. The scale set $H = \{31, 21, 15, 11, 5, 3\}$ and the windows of the kernels are truncated to the squares $h \times h$. The first $\omega_0^{(1,0)}$ is the MR

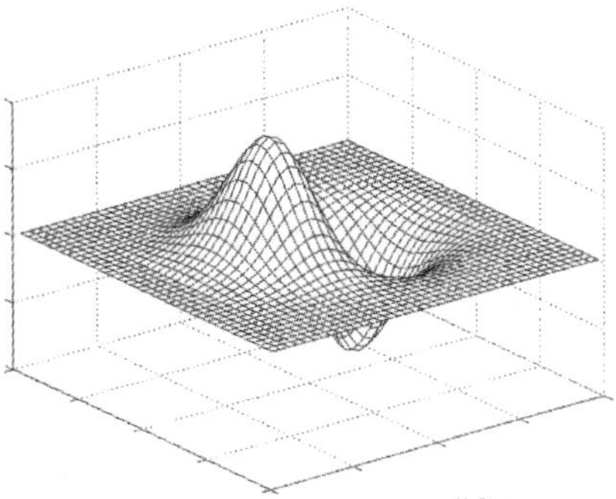

MR differentiating kernel $\omega_0^{(1,0)}(x)$

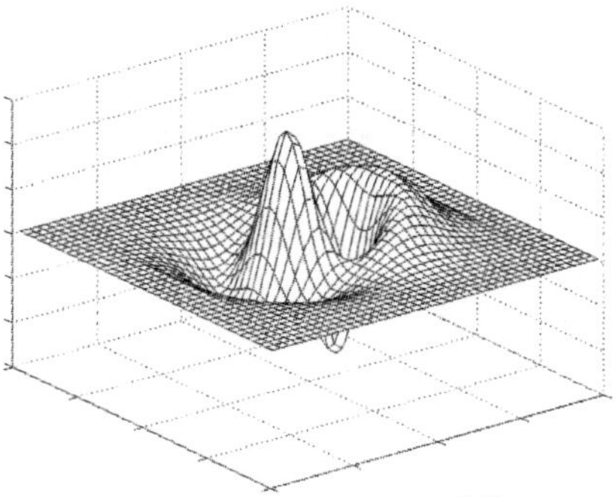

MR differentiating kernel $\omega_1^{(1,0)}(x)$

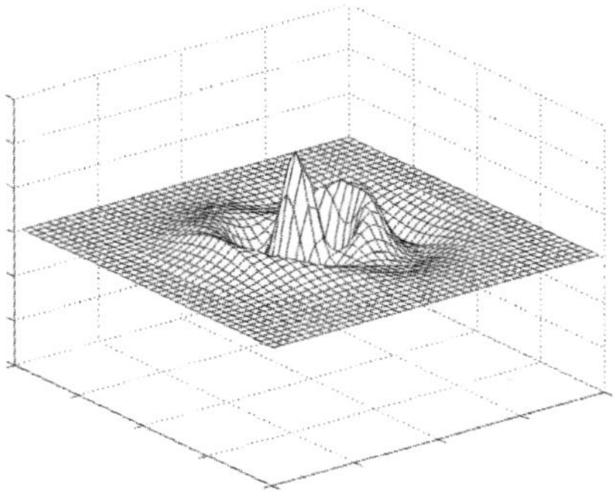

MR differentiating kernel $\omega_2^{(1,0)}(x)$

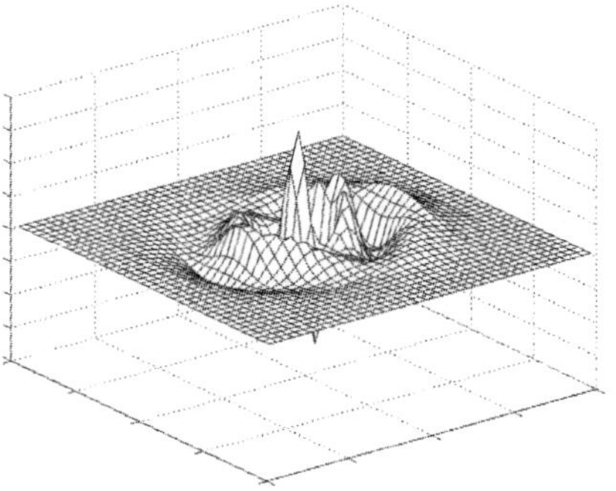

MR differentiating kernel $\omega_3^{(1,0)}(x)$

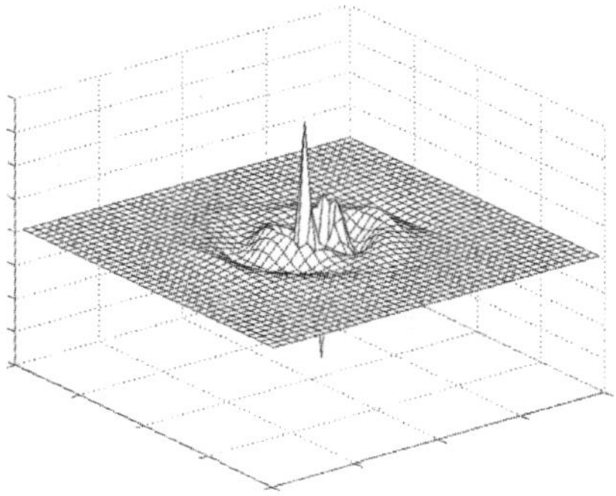

MR differentiating kernel $\omega_4^{(1,0)}(x)$

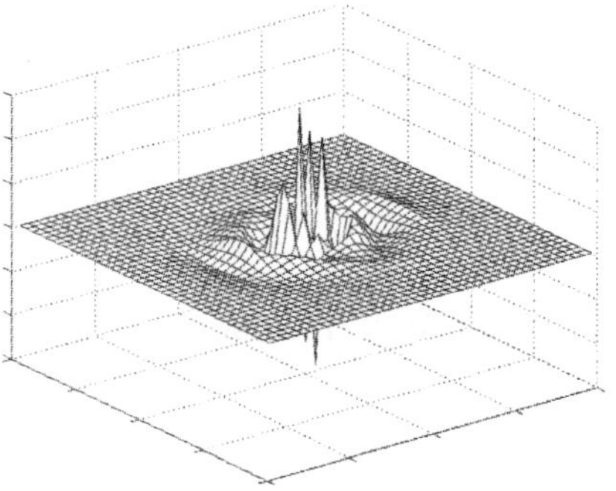

MR differentiating kernel $\omega_5^{(1,0)}(x)$

Figure 13.6 MR analysis differentiating kernels obtained using the LPA of power $m = 2$ and the 2D Gaussian window.

differentiation kernel of the largest scale $h_0 = 31$. The higher-scale kernels $\omega_j^{(1,0)}$, $j > 0$, become narrower.

The differentiation kernels have zero value for the central reference point $x = 0$ while the smoothing kernels $\omega_j(x)$ have maximum peaks at this point. The support of a square mask differentiating kernel should be larger than 1×1 for the LPA fitting with $m \geq 1$ to be admissible. For the minimum scale differentiating kernel, we use the mask 3×3, $h_J = 3$, which is the minimum h in the set H for differentiation. These examples are given for the symmetric square support kernels. The same technique can be used to design directional MR kernels.

13.3 Thresholding

13.3.1 Basic concepts

Let us apply the analysis kernels ω to the noisy observations in Eq. (13.29). The output

$$a(x) = (\omega \circledast z)(x),$$

can be presented as a sum of the deterministic $\alpha(x)$ and random $n(x)$ summands:

$$\begin{aligned} a(x) &= \alpha(x) + \sigma n(x), \\ \alpha(x) &= (\omega \circledast y)(x), \\ n(x) &= (\omega \circledast \varepsilon)(x). \end{aligned} \tag{13.67}$$

The elements of the vector $n(x)$ for a fixed x are independent of the identity covariance matrix:

$$\begin{aligned} E\{n(x)n^T(x)\} &= E\left\{(\omega \circledast \varepsilon)(x)(\omega^T \circledast \varepsilon)(x)\right\} \\ &= E\left\{\sum_k \omega(k)\varepsilon(x-k)\sum_l \omega^T(l)\varepsilon(x-l)\right\} \\ &= \sum_k\sum_l \omega(k)\omega^T(l)E\{\varepsilon(x-k)\varepsilon(x-l)\} \\ &= \sum_k\sum_l \omega(k)\omega^T(l)\delta_{k,l} \\ &= \sum_l \omega(l)\omega^T(l) = I. \end{aligned} \tag{13.68}$$

In scalar form, the output of the noisy signal analysis in Eq. (13.67) can be given as

$$a_j(x) = \alpha_j(x) + \sigma n_j(x), \qquad j = 0, \ldots, J, \tag{13.69}$$

where $n_j(x)$ is a standard Gaussian noise independent of j.

The synthesis formula (13.52) yields the estimate in the form

$$\hat{y}(x) = a_0(x)\cdot\omega_0(0) + \sum_{j=1}^{J} a_j(x)\cdot\omega_j(0), \tag{13.70}$$

$$a_j(x) = (\omega_j \circledast z)(x), \tag{13.71}$$

and this estimate is equal to the input signal, $\hat{y}(x) = z(x)$, because of the perfect reconstruction property of the expansion in Eq. (13.70). The kernels ω_j in Eq. (13.70) are local in space (via scaling h) and in frequency (provided smooth windows w in the LPA design).

This localization offers an advantage, because fewer scales j are usually needed to represent the signal y to a given level of approximation. The heuristic that we use is that the MR expansion of such a function in some basis is sparse, that is, only a few of the coefficients $\alpha_j(x)$ are big and the rest are small and thus negligible. Hence, to estimate the function, one has to estimate large coefficients and discard the rest.

This approach for wavelet expansions has proved to be useful and successful as shown in recent years by various authors (e.g., [150] and references herein).

Analysis of the spectral coefficients $\alpha_j(x)$ shows that they are redundant for many typical images. So, if a noisy estimate $\hat{\alpha}_j(x)$ of the coefficient $\alpha_j(x)$ is small, it is reasonable to regard it as mostly noise and to set it to zero; if it is big, it is reasonable to keep it. This is known as hard-thresholding. Soft-thresholding shrinks everything toward zero by a certain amount, thus reducing the variance of the estimation at the cost of a higher bias [76], [150].

Consider the "diagonal" estimator of $\alpha_j(x)$ using only $a_j(x)$ [150]. Assume that this estimator has a form

$$\hat{\alpha}_j(x) = \gamma_j a_j(x), \tag{13.72}$$

where $0 \leq \gamma_j \leq 1$ is an attenuation factor depending on $\hat{\alpha}_j(x)$. The estimate in Eq. (13.70) is changed to the form

$$\begin{aligned} \hat{y}(x) &= a_0(x) \cdot \omega_0(0) + \sum_{j=1}^{J} \gamma_j a_j(x) \cdot \omega_j(0), \\ a_j(x) &= (\omega_j \circledast z)(x), 0 \leq \gamma_j \leq 1. \end{aligned} \tag{13.73}$$

Thus, $\gamma_j a_j(x)$ is a nonlinear function of the observed $a_j(x)$. It is normal practice to not threshold the "coarse" component $a_0(x) \cdot \omega_0(0)$ even if it can be of a small magnitude, because it gives a lower frequency background of the signal nearly noiselessly.

The spectrum representation in Eq. (13.70) is clearly redundant. Indeed, if we have n observations, that is, x in Eq. (13.29) is defined on a grid with n nodes, then each spectrum coefficient $a_j(x)$ has the same size as a function of x, plus we have $J + 1$ of such coefficients. Thus, in total, instead of n original observations we obtain $n(J + 1)$ spectrum coefficients.

One of the objectives of spectrum analysis is to select from the redundant spectrum data a small number of coefficients that are sufficient for a good signal approximation. This sort of problem appears when we deal with a data compression problem, and a good basis enables one to use for the signal approximation a few terms of the series.

Thresholding of the noisy MR spectrum is a tool for signal denoising. Here we concern ourselves with this filtering problem only.

Random noise in the spectrum analysis in Eq. (13.71) is defined by $n_j(x) = (\omega_j \circledast \varepsilon)(x)$, where ε is the observation noise in Eq. (13.29). It follows from Eq. (13.68) that, for a fixed x, these random $n_j(x)$ are uncorrelated for different scales j:

$$E\{n_i(x)n_j(x)\} = \delta_{i,j},$$

because of the orthogonality of the analysis kernels ω_j.

However, for $x \neq x'$, the random $n_i(x)$ and $n_j(x')$ are correlated, in general, and uncorrelated provided that the distance between x and x' is so large that the corresponding estimates use uncorrelated observations z. Thus, at least for small h_j we have the spectrums $a_j(x)$, which are uncorrelated for nearly all x.

The representation of the spectrum in the form of Eq. (13.69), where the noisy spectrum $a_j(x)$ is an unbiased estimate of the true spectrum $\alpha_j(x)$ and the random errors are uncorrelated, is important for the development of theoretically justified thresholding rules.

The following are classical *thresholding algorithms* that define the attenuation factor γ_j in Eq. (13.73) as a nonlinear function of $a_j(x)$:

(a) Hard-thresholding:

$$\gamma(v) = 1(|v| \geq t); \tag{13.74}$$

(b) Soft-thresholding:

$$\gamma(v) = [1 - t/|v|]_+; \tag{13.75}$$

(c) Stein's thresholding:

$$\gamma(v) = [1 - t/|v|^2]_+; \tag{13.76}$$

where $t > 0$ is the threshold parameter.

The corresponding thresholding outputs $\gamma(v)v$ are shown in Figs. 13.7(a–c). The curve in Fig. 13.7(d) shows a more sophisticated thresholding that is continuous at $v = 0$.

The attenuation factors in Eqs. (13.72) and (13.73) are computed as

$$\gamma_j = \gamma(a_j/\sigma). \tag{13.77}$$

The thresholdings in Eqs. (13.74–13.76) imply clear heuristic ideas to discard lower-magnitude components of the spectrum.

To make it clear, assume for a moment that $\alpha_j = 0$; then $a_j = \sigma n_j$ and we have from Eqs. (13.74) and (13.77) that

$$\gamma_j = \gamma\left(\frac{a_j}{\sigma}\right) = \gamma(n_j) = 1(|n_j| \geq t).$$

Thus the term with $\alpha_j = 0$ is discarded from the estimate, as it should be, provided that $|n_j| \leq t$. For the Gaussian random n_j, the probability of this event is computed using the quantiles of the standard Gaussian distribution, which gives, in particular, $t = 2$ for the probability $p = 0.95$ and $t = 3$ for the probability $p = 0.997$. Thus, the threshold parameter is used for testing the hypothesis $\alpha_j = 0$ and controls the first kind of error of the thresholding rule. The parameter t has similar meanings for the rules in Eqs. (13.75) and (13.76).

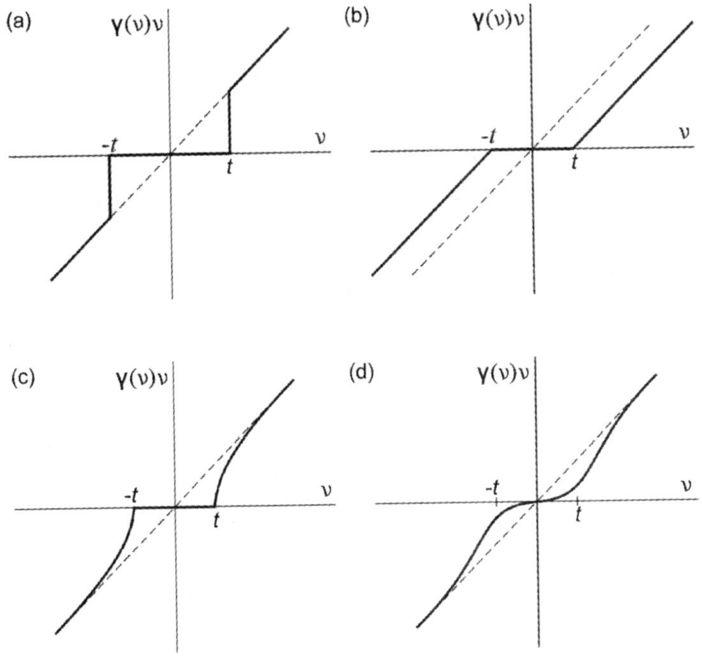

Figure 13.7 Examples of $\gamma(v)v$ used for thresholding of estimates.

The proper selection of t is very important for the efficiency of the thresholding. In particular, it is shown in [40] for wavelet transforms that the universal global threshold

$$t \sim \sqrt{2\log n} \tag{13.78}$$

gives a function reconstruction with a mean-square risk close to the minimax risk corresponding to the optimal but unknown thresholding rule.

The stochastic model in Eq. (13.69) yields a convenient tool for statistical arguments that can be used to derive optimal procedures as well as to justify heuristic ones.

Consider some results in this vein. Let the estimate of α_j in Eq. (13.69) be of the form of Eq. (13.72). Using the perfect reconstruction formula (13.52) for y and Eqs. (13.72) and (13.73) for the estimate, the MSE criterion gives

$$\begin{aligned} S_0(x) = E\{\|y(x) - \hat{y}(x)\|^2\} &= E\left\{\sum_{j=0}^{J}\{[\alpha_j(x) - \gamma_j a_j(x)]\omega_j(0)\}^2\right\} \\ &= E\left\{\sum_{j=0}^{J}[(1-\gamma_j)\alpha_j(x) + \sigma\gamma_j n_j(x)\omega_j(0)]^2\right\} \\ &= \sum_{j=0}^{J}\left[(1-\gamma_j)^2\alpha_j^2(x) + \sigma^2\gamma_j^2\right]\omega_j^2(0). \end{aligned} \tag{13.79}$$

Minimization of S_0 on γ_j results in

$$\gamma_j^* = \frac{\alpha_j^2(x)}{\alpha_j^2(x) + \sigma^2}, \tag{13.80}$$

which is the ideal attenuation factor, the so-called "oracle" solution. It depends on the unknown true spectrum $\alpha_j(x) = (\omega_j \circledast y)(x)$. This solution could be obtained if an *oracle* provides the information on unknown y. The corresponding ideal risk

$$\begin{aligned} S_0^*(x) &= E\{||y(x) - \hat{y}^*(x)||^2\} = \sum_{j=0}^{J} \frac{\sigma^2 \alpha_j^2}{\sigma^2 + \alpha_j^2} \omega_j^2(0), \\ \hat{y}^*(x) &= \sum_{j=0}^{J} \gamma_j^* \omega_j(0), \end{aligned} \tag{13.81}$$

can be used as a lower bound of the estimate accuracy.

Let us replace in Eq. (13.79) $\alpha_j^2(x)$ by $a_j^2(x) - \sigma^2$ and use instead of $S_0(x)$ the sample risk

$$S_1(x) = \sum_{j=0}^{J} [(1 - \gamma_j)^2 (a_j^2(x) - \sigma^2) + \sigma^2 \gamma_j^2] \omega_j^2(0),$$

which does not depend on the unknown true spectrum α_j.

It can be checked that

$$E\{S_1(x)\} = E\left\{ \sum_{j=0}^{J} \{(1 - \gamma_j)^2 [(\alpha_j(x) + \sigma n_j)^2 - \sigma^2] + \sigma^2 \gamma_j^2\} \, \omega_j^2(0) \right\} = S_0(x).$$

In terms of the expectation, the criteria $S_0(x)$ and $S_1(x)$ are equivalent.

Minimizing $S_1(x)$ on γ_j, provided $0 \le \gamma_j \le 1$, gives the adaptive attenuation factor as

$$\gamma_j(x) = \left[\frac{a_j^2(x) - \sigma^2}{a_j^2(x)} \right]_+ = \left[1 - \frac{\sigma^2}{a_j^2(x)} \right]_+ . \tag{13.82}$$

This is a pointwise-varying factor depending on x. It defines the so-called "Stein estimator" [212] and the Stein thresholding as in Eq. (13.76) with $t = 1$.

For random $a_j^2(x)$ with a large variance, this estimate is too noisy and cannot give a good estimate. However, if we smooth the criteria $S_1(x)$ on x, say, applying some normalized smoothing kernel $g_{\bar{h}} \ge 0$, $\sum_x g_{\bar{h}}(x) = 1$, we arrive at

$$\bar{S}_1(x) = (S_1 \circledast g_{\bar{h}})(x).$$

Similar to Eq. (13.82), minimization of $\bar{S}_1(x)$ on γ_j gives

$$\bar{\gamma}_j(x) = \left[1 - \frac{\sigma^2}{(a_j^2 \circledast g_{\bar{h}})(x)}\right]_+ . \tag{13.83}$$

Let us call this the *smoothed Stein's thresholding*, which can be more efficient then the original Stein's thresholding.

13.3.2 Control of error rate

Let us look at the thresholding problem from a statistical viewpoint. We start from the data in Eq. (13.69):

$$a_j(x) = \alpha_j(x) + \sigma n_j(x),$$

and restore the function y as

$$\hat{y}(x) = a_0(x) \cdot \omega_0(0) + \sum_{j=1}^{J} \gamma_j a_j(x) \cdot \omega_j(0), \tag{13.84}$$

$$\gamma_j = \gamma(a_j(x)/6), \tag{13.85}$$

where the rule in Eq. (13.74) is used.

The thresholding in Eq. (13.85) is closely related to the hypothesis testing, where each coefficient $a_j(x)$ is tested whether it is zero or not. The estimate building is based on the usage of the coefficients that are significantly different from zero.

The arguments given above for the thresholding parameter t are valid for a single test only. It has been said that the probability of the error is equal to $1 - p = 0.05$ or $1 - p = 0.01$, depending on $t = 2$ or 3 for a single test. However, for Eq. (13.85), this test should be repeated for each scale $j = 1, \ldots, J$ and each x. With an image of size 256×256, the number of fine scales $J = 5$ with error probability $1 - p = 0.05$ gives that $256 \cdot 256 \cdot 5 \cdot 0.05 = 16384$ coefficients $a_j(x)$ would be found significant (nonzero), even when all of the corresponding $\alpha_j(x)$ are zeros. Hence, we have a multiple test scenario, and a strong control of errors is needed.

It is well known from applications of wavelet thresholding that the data adaptive parameter t improves the performance of the algorithm in a very serious way. Many adaptive versions of thresholding are developed including: localized estimates of the unknown σ, the parameter t different for each scale, and t depending on correlation properties of the signal (some results and further references can be found in [76] and [150]).

Recently, new false discovery rate (FDR) criteria have been suggested as an alternative to the standard multiple hypotheses testing procedures [1], [13]. They were applied for wavelet estimators and proved to be quite efficient. These criteria

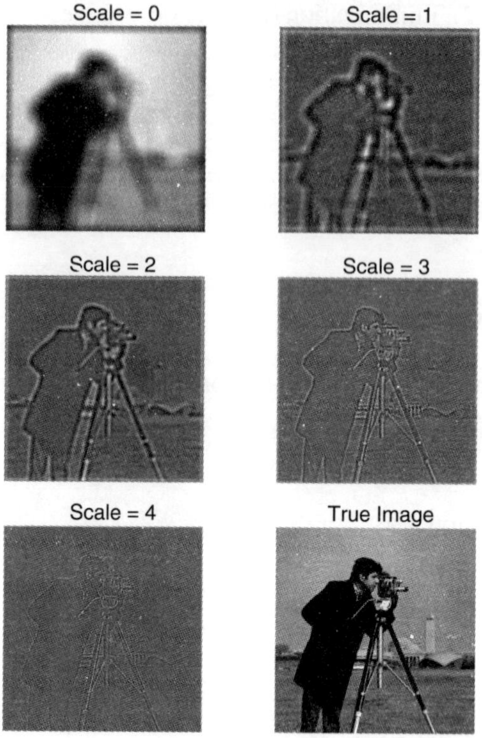

Figure 13.8 The LPA analysis spectrum of the noiseless *Cameraman* image. The spectrum images become sharper for larger values of the scale. The sum of all five expansion items shown in the last image is identical to the original *Cameraman* image.

control the expected proportion of incorrectly included coefficients among those chosen for the estimator. All of these techniques can be used for the considered nonparametric MR estimates.

13.3.3 Image processing

13.3.3.1 MR spectrum

The test signal is the 256×256 *Cameraman* image (8-bit grayscale) corrupted by an additive zero-mean Gaussian noise. The basic LPA kernels g_h are obtained for the uniform symmetric square window $w(x)$, $m = 1$, and for the finite set of scales $H = \{21, 11, 5, 3, 1\}$. The corresponding analysis kernels ω_j, $j = 0, 1, 2, 3, 4$, are computed according to Eq. (13.46).

Figure 13.8 illustrates the MR expansion in Eq. (13.52) for the noiseless image. It images the items of the MR expansion $\alpha_j(x)\omega_j(0)$ for the scales j. The first term of this expansion, $\alpha_0(x)\omega_0(0)$, presents a basic smooth lower-frequency component

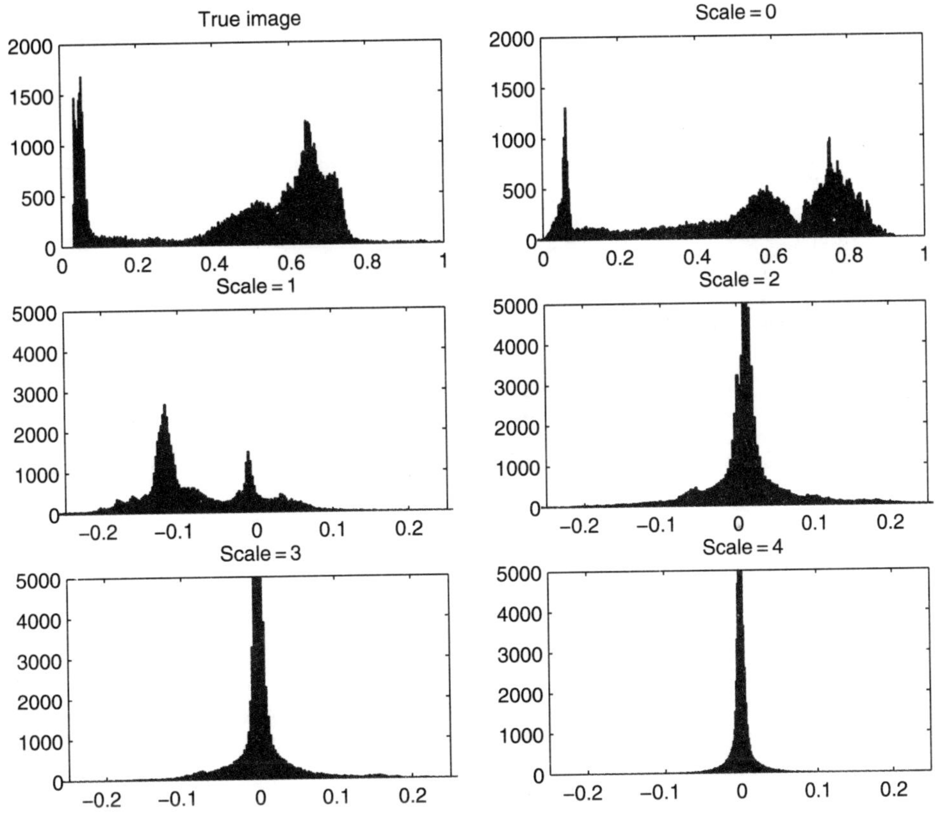

Figure 13.9 Histograms of the images shown in Fig. 13.8. The histogram for the scale equal to 0 is wide, covering nearly the whole segment [0, 1] and even a bit wider than the histogram of the true image. The histograms for the complement components of the MR expansion for the scales $j = 1, 2, 3, 4$ become narrower with a smaller number of nonzero items. The last scale-5 histogram is well-concentrated in a narrow neighborhood of zero.

of the image. The other terms with $j > 0$ serve as the complements of the basic one providing finer, higher-frequency details. The images of the expansion items become sharper for larger values of the scale j. The sum of all five MR expansion items shown in the last image of Fig. 13.8 gives a perfect reconstruction of the original image.

Histograms of the images from Fig. 13.8 are shown in Fig. 13.9. They illustrate the concept of the redundancy of the MR nonparametric regression expansion used in thresholding filtering. Indeed, the histogram for $j = 0$ covers a wide part of the image data segment [0, 1]. The histograms for the complementary components of the MR expansion are narrower and more pick-wise. The last scale, $j = 4$, has the smallest number of nonzero items, which are mainly concentrated in a narrow neighborhood of zero. Actually, this means that the space of the MR analysis for $j \geq 1$ is sparse, that is, only a few items of the MR nonparametric expansion are big and the rest are small and thus can be dropped.

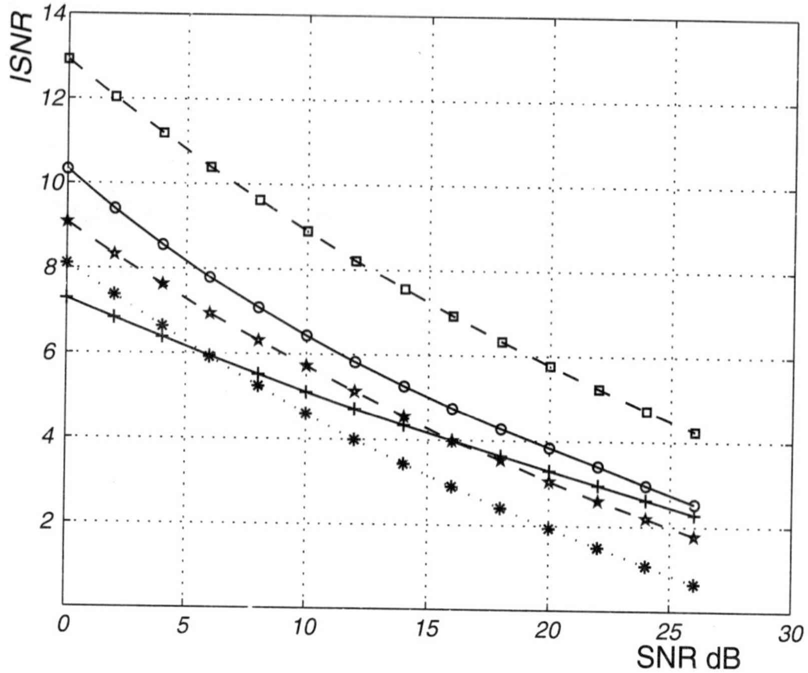

Figure 13.10 ISNR as a function of SNR of the noisy observations for the *Cameraman* image. The curves are given for the following four algorithms: soft-thresholding (star), Stein's (plus), smoothed Stein's (circle), ICI (asterisk). The threshold $t = 1.2$. The oracle estimation results are marked by squares. Overall, the advantage is in favor of the smoothed Stein algorithm, which outperforms the ICI algorithm by about 1 dB.

13.3.3.2 MR denoising

For image denoising we use the following thresholding procedures: hard, as in Eq. (13.74); soft, as in Eq. (13.75); Stein, as in Eq. (13.76); smoothed Stein, as in Eq. (13.83); oracle, as in Eq. (13.80). We wish to note that using the MR technique for the directional LPA kernels with fusing the obtained adaptive directional estimates allows one to dramatically improve the performance of the MR based denoising algorithms. The threshold t is a main design parameter of the hard, soft, and Stein algorithms. Multiple experiments for different images show that $t = 1.2$–1.5 is a sort of rule-of-thumb for the different scenarios.

Figure 13.10 shows the ISNR for these algorithms as a function of the SNR. The smoothed Stein algorithm demonstrates the best performance and outperforms the LPA-ICI algorithm approximately by 1dB. The soft-thresholding algorithm gives ISNRs that are about 1 dB worse than those for the ICI algorithm. The basic Stein algorithm shows figures better than the ICI algorithm only for SNR > 15 dB.

The oracle estimation naturally demonstrates the best performance with ISNR values about 2 dB higher than those for the smoothed Stein adaptive algorithm. The hard-thresholding demonstrates a much worse performance that the soft-thresholding algorithm. The results for the hard-thresholding are not shown in

Fig. 13.10. Overall, these results demonstrate that the MR-based algorithms achieve better results than the LPA-ICI algorithm.

13.3.3.3 MR differentiation

For noisy data, the thresholding incorporated in the differentiating expansion in Eq. (13.65) gives the derivative estimate in the form

$$\begin{aligned} \hat{y}^{(r)}(x) &= a_0^{(r)}(x) \cdot q_0^{(r)} + \sum_{j=1}^{J} \gamma_j a_j^{(r)}(x) \cdot q_j^{(r)}, \\ a_j^{(r)}(x) &= (\omega_j^{(r)} \circledast z)(x), \\ \gamma_j &= \gamma(a_j^{(r)}/\sigma), \end{aligned} \tag{13.86}$$

where the attenuation factors γ_j are defined for instance by Eqs. (13.74–13.76). Note that the standard deviation of the estimates $a_j^{(r)}$ is equal to σ because of the orthonormality of the analysis kernels $\omega_j^{(r)}$.

Let us illustrate an application of the MR algorithm in Eq. (13.86) for scale-adaptive differentiation. Two first derivatives of the *Cameraman* image intensity on x_1 (horizontal axis) and x_2 (vertical axis) are calculated. For the sake of simplicity of presentation and discussion, we apply a single window estimate with the symmetric window function w.

Figures 13.11 and 13.12 illustrate the formation of the derivatives for the noiseless image using the MR spectrum analysis in Eq. (13.65) produced by the kernels $\omega_j^{(r)}$. We use the LPA kernels of order $m = 1$ obtained for the uniform symmetric window with the scales $H = \{21, 15, 11, 5, 3\}$.

The last image in Figs. 13.11 and 13.12 is the sum of all five MR spectrum components corresponding to the expansion $\hat{y}^{(r)}(x) = \sum_{j=0}^{J} \alpha_j^{(r)}(x) \cdot q_j^{(r)}$. These are noiseless MR derivative spectrums. The influence of the scale on the differentiation operator is clearly seen. It varies from the smoothed derivative components given by the large-scale operators to the finer contour lines of the smaller-scale spectrums.

To produce a quantitative analysis of the differentiation of the noisy data, we need to compare the estimates with the accurate derivatives. Naturally, for the discrete *Cameraman* image, these accurate derivatives cannot be unknown.

Let us assume that the "accurate" numerical derivatives are defined as the estimates given by the corresponding differentiation kernels $g_h^{(1,0)}(x)$ and $g_h^{(0,1)}(x)$ for the noiseless *Cameraman* image provided that the scale $h \in H$ is equal to its minimum value $h = 3$. We take this estimate as the true derivative value.

Table 13.1 provides data illustrating an improvement that can be achieved by using the adaptive varying-scale estimators versus the estimators with a fixed invariant scale. The criteria are calculated as the means of the corresponding values obtained for the derivatives on x_1 and x_2.

Lines 1–5 of the table show the criteria values for the invariant scale estimators, and the last line corresponds to the MR soft-thresholding estimator ($t = 1.2$). It can be concluded that the best scale invariant estimator has the scale that is equal to

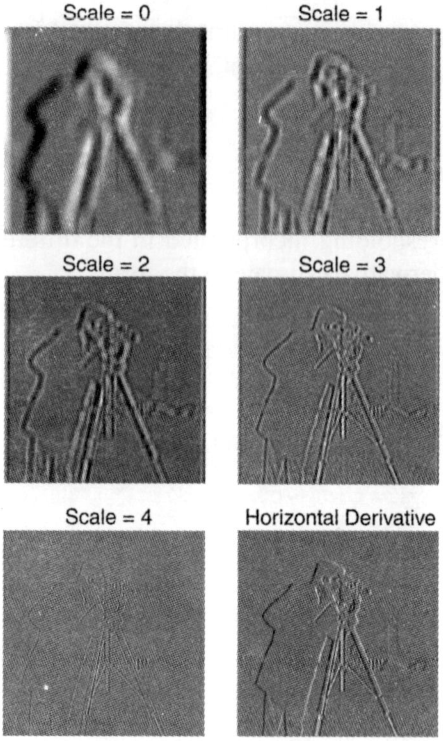

Figure 13.11 Spectrum expansion of the horizontal derivative using the noiseless *Cameraman* image. The zero scale corresponds to the first lower frequency (smooth) components of the derivative. The other items with larger scales serve as the complements of this basic item and provide finer sharp details. The sum of all five MR expansion items are shown in the last image.

the minimum value $h = 3$. It also can be noticed that the accuracy figures are very sensitive with respect to the scale value.

Further, comparing the criteria values of this best-scale invariant estimator versus corresponding values for the adaptive estimator, we can see clear improvements: about 30% for RMSE and about 40% for MAE. A smaller improvement can be seen in the value of MAXDIF, which is about 10%.

The visual effect of the adaptive-scale differentiation can be seen in Fig. 13.13. The images presented in this figure show the sum of absolute values of the estimates of the derivatives $\partial y/\partial x_1$ and $\partial y/\partial x_2$. The image on left-hand side is obtained by using the derivative estimator kernels $g_{h_5}^{(1,0)}$ and $g_{h_5}^{(0,1)}$ with the best-found invariant scale $h = 3$. The image on right-hand side is obtained by the MR soft-thresholding estimator.

Noisy components of the estimates clearly seen in the left-hand-side estimate with the fixed scale $h = 3$ are cleared out in the right-hand-side image while the fine details of the edges of the *Cameraman* are preserved. Visually, the kernel MR nonparametric differentiator demonstrates better results compared with the scale-invariant estimator with the best selection of scale values.

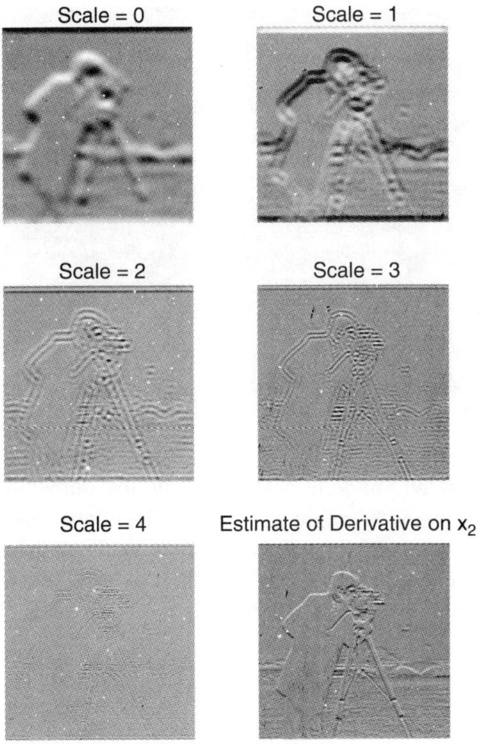

Figure 13.12 The spectrum expansion of the vertical derivative. The noiseless *Cameraman* image. The zero scale corresponds to the first lower frequency (smooth) components of the derivative. The other items with larger scales serve as the complements of this basic item and provide finer sharp details. The sum of all five MR expansion items are shown in the last image.

Table 13.1 Accuracy of the derivative estimation.

Invariant scale	RSME	MAE	MAXDIF
$h = 21$	0.0615	0.0275	0.4386
$h = 15$	0.0599	0.0271	0.4590
$h = 11$	0.0576	0.0263	0.4715
$h = 5$	0.0399	0.0216	0.4274
$h = 3$	0.0317	0.0252	0.1486
MR soft-thresholding	**0.0219**	**0.0145**	**0.1310**

13.4 Parallels with Wavelets

Let us summarize some similarities and differences between the introduced MR nonparametric regression and the corresponding multiresolution wavelet concepts as they were discussed in Sec. 13.1.2:

Figure 13.13 Images show the sums of the absolute values of the estimates of the derivatives $\partial_{x_1} y$ and $\partial_{x_2} y$. The left image is obtained using the derivative estimator kernels $g_h^{(1,0)}$ and $g_h^{(1,0)}$ with the best-found invariant scale $h = 3$. The right image is obtained by the MR soft-thresholding estimators with $t = 1.2$. Noisy components of the estimates clearly seen in the left-hand side image are cleared out in the right-hand side one while the fine edge details of the *Cameraman* image are well preserved.

1. The wavelet expansion in Eq. (13.21) is an orthogonal series with invariant coefficients $\{\alpha_{0,k}, \beta_{j,k}\}$ as in Eq. (13.22). As a function of x, this series is defined by the wavelets $\{\varphi_{0,k}(x), \psi_{j,k}(x)\}$, which means that Eq. (13.21) is a parametric expansion of y.

   In contrast, the expansion in Eq. (13.52) is nonparametric because its dependence on x cannot be parameterized and it goes through the coefficients $\alpha_j(x)$ of the expansion. There are no invariant coefficients in this expansion and the basis functions depend on x, which is typical for the standard parametric series.
2. Using the wavelets for a large set of data, say for imaging, requires segmentation of the data. The data are split into overlapping or nonoverlapping segments that are processed independently. This segmentation produces artifacts degradating the resulting signal. The shift-invariant wavelets are developed to repair these effects [150]. Signal processing based on kernel filters does not require the segmentation of data and does not produce this sort of artifact.
3. The kernels ω_0 and ω_j, $j = 1, \ldots, J$, in Eqs. (13.52) and (13.53) can be interpreted as father and mother wavelets, respectively, according to their role in the analysis and the vanishing moment conditions in Eq. (13.47). The kernel ω_0 defines a lower frequency background of the signal (father wavelet analysis) while the kernels ω_j, $j = 1, \ldots, J$, define higher-frequency complements to this background (mother wavelet analysis).
4. The subspaces W_j and V_j are different for the wavelet and the introduced MR analysis. The principal difference is that, in the wavelets, the W_{j+1} are orthogonal complements of V_j while there is no such orthogonality for the kernel MR analysis.

5. The dyadic scale $h_j = 2^{-i}$ is a crucial point of wavelet design and fast algorithm implementation. In the introduced nonparametric MR analysis, the scales defined by the set H are quite arbitrary. The most serious restriction concerns the linear independence of g_h for $h \in H$. The intra- and interscale orthogonality of wavelets defines the complexity of the design of these functions. In a nonparametric MR analysis, only the interscale orthogonality is used.
6. Analysis kernels ω_j similar to wavelets form a local space-frequency basis that can be interpreted as a tiling that covers the space-frequency space with space-frequency boxes. The scale parameter h is inversely proportional to the frequency. Comparing these boxes with the space-frequency boxes of the wavelets $\psi_{j,k}$ in Fig. 13.4(a) assumes that $h = 2^{-j}$. Then, the small j and large $h = 2^{-j}$ mean a wide estimation neighborhood, and large j and small $h = 2^{-j}$ correspond to a narrow estimation neighborhood.

   The kernel MR analysis space-frequency boxes slide along the x axis. This is not typical for the wavelet segmentation of the axis x (usually a multiple of 2^j). Thus, we arrive at the tilling shown in Fig. 13.4(b), which has the same space-frequency boxes as in Fig. 13.4(a) but with an arbitrary location along the x axis.
7. The wavelet design for multivariable functions is a serious problem that can be solved mainly by using the tensor product of the univariable wavelets. There are no principal difficulties with the design of LPA kernels of any dimensionality.
8. The introduced MR analysis is very redundant. The total size of this spectral presentation is equal to the data size multiplied by $(J + 1)$. The development of filter banks with downsampling of the MR spectrums is an interesting problem.

We conclude that the considered nonparametric approach is very different from the orthogonal series (in particular wavelet) techniques. It is very flexible and allows a lot of freedom for the design of filters (kernels) and signal processing.

Chapter 14

Appendix

14.1 Analytical Regular Grid Kernels

Formulas (3.10) and (3.11) define the smoothing and differentiating kernels on a regular grid in the form:

$$g_h[l] = w_h[l]\phi_h^T[l]\Phi_h^{-1}\phi[0] \tag{14.1}$$

and

$$g_h^{(r)}[l] = (-1)^{|r|} w_h[l]\phi_h^T[l]\Phi_h^{-1}\phi^{(r)}[0], \qquad l \in \mathbb{Z}^d, \tag{14.2}$$

with

$$\Phi_h = \sum_{l \in \mathbb{Z}^d} w_h[l]\phi_h[l]\phi_h^T[l]. \tag{14.3}$$

In this section we consider some particular cases when these formulas give analytical results. As a general idea, we use the polynomials $\phi_h[l]$, which are orthonormal with the weights $w_h[l]$. Then, Φ_h becomes the identical matrix and formulas (14.1) and (14.2) are simplified to

$$\begin{aligned} g_h[l] &= w_h[l]\phi_h^T[l]\phi[0], \\ g_h^{(r)}[l] &= (-1)^{|r|} w_h[l]\phi_h^T[l]\phi^{(r)}[0]. \end{aligned} \tag{14.4}$$

The orthonormal polynomials are different from those given in the standard form, $(-1)^{|k|}x^k/k!$, in Eq. (2.45). They are not monomials of the single power k but are composed from the monomials of different powers. Then, a few elements of the vectors $\phi(0)$ and $\phi^{(r)}(0)$ can be nonzero. It makes a difference because the monomials in Eqs. (2.22 or 2.45) have single nonzero elements in $\phi(0)$ and $\phi^{(r)}(0)$. For the sake of simplicity, we assume in the following that the sampling interval $\Delta = 1$.

14.1.1 1D kernels

Let us start from the 1D case, $l = l_1$, when formulas (14.1) and (14.2) are

$$g_h[l_1] = w_h[l_1]\phi_h^T[l_1]\Phi_h^{-1}\phi[0] \tag{14.5}$$

and

$$g_h^{(r_1)}[l_1] = (-1)^{r_1} w_h[l_1]\phi_h^T[l_1]\Phi_h^{-1}\phi^{(r_1)}[0], \tag{14.6}$$

$$\phi_h[l_1] = (\phi_{1,h}[l_1], \ldots, \phi_{M,h}[l_1])^T,$$

$$\Phi_h = \sum_{l_1=-\infty}^{\infty} w_h[l_1]\phi_h[l_1]\phi_h^T[l_1]. \tag{14.7}$$

In the scalar form, the orthonormality condition $\Phi_h = I_{M\times M}$ gives

$$\sum_{l_1} w_h[l_1]\phi_{r,h}[l_1]\phi_{s,h}[l_1] = \delta_{r,s}, \qquad r, s = 1, \ldots, M.$$

Then, the expressions in Eqs. (14.5) and (14.6) are simplified to

$$g_h[l_1] = w_h[l_1]\phi_h^T[l_1]\phi[0], \tag{14.8}$$

$$g_h^{(r_1)}[l_1] = (-1)^{r_1} w_h[l_1]\phi_h^T[l_1]\phi^{(r_1)}[0]. \tag{14.9}$$

Let us present some particular cases specified by the used window w. For simplicity, the index h is omitted in the orthonormal polynomials ϕ_h.

14.1.1.1 Rectangular symmetric window

Consider a rectangular symmetric window,

$$w_h[l_1] = \frac{1}{2h+1}, \quad |l_1| \le h, \tag{14.10}$$

where the parameter h is a nonnegative integer, $h \ge 0$. The discrete *Lagrange* polynomials of degree $m \le 2$, $|l_1| \le h$,

$$\phi_1[l_1] = 1, \qquad \phi_2[l_1] = \sqrt{\frac{3}{h(h+1)}}\, l_1, \tag{14.11}$$

$$\phi_3[l_1] = \sqrt{\frac{5}{h(h+1)(2h+3)(2h-1)}} \cdot [3l_1^2 - h(h+1)]$$

are orthonormal with weights as in Eq. (14.10). Orthonormality can be verified by calculating the matrix Φ_h in Eq. (14.7), which is a 3×3 identity matrix if the polynomials in Eq. (14.11) are orthonormal.

The elementary calculations according to Eqs. (14.8) and (14.9) give the following formulas for the kernels of the smoothing and differentiating operators [101], [102]:

(a) for $m = 0$, $h \ge 0$,

$$g_h[l_1] = \frac{1}{2h+1}, \quad |l_1| \le h; \tag{14.12}$$

(b) for $m = 1, h \geq 1, |l_1| \leq h$,

$$g_h[l_1] = \frac{1}{2h+1}, \tag{14.13}$$

$$g_h^{(1)}[l_1] = \frac{-3l_1}{h(h+1)(2h+1)}; \tag{14.14}$$

(c) for $m = 2, h \geq 1, |l_1| \leq h$,

$$g_h[l_1] = \left[1 + 5\frac{h(h+1) - 3l_1^2}{(2h+3)(2h-1)}\right]\frac{1}{2h+1}, \tag{14.15}$$

$$g_h^{(1)}[l_1] = \frac{-3l_1}{h(h+1)(2h+1)}, \tag{14.16}$$

$$g_h^{(2)}[l_1] = \frac{30[3l_1^2 - h(h+1)]}{h(h+1)(2h+3)(2h-1)(2h+1)}. \tag{14.17}$$

The vanishing moment Eqs. (3.15) are satisfied for the smoothing kernels $g_h[l_1]$: for $m = 0$ or $m = 1$,

$$\sum_{l_1=-h}^{h} g_h[l_1] = 1, \qquad \sum_{l_1=-h}^{h} l_1 \cdot g_h[l_1] = 0, \tag{14.18}$$

and for $m = 2$ or $m = 3$,

$$\sum_{l_1=-h}^{h} g_h[l_1] = 1, \qquad \sum_{l_1=-h}^{h} l_1^r \cdot g_h[l_1] = 0, \qquad r = 1, 2, 3; \tag{14.19}$$

For the differentiating kernel giving an estimate of the first derivative, the vanishing moment conditions in Eq. (3.16) are as follows for $g_h^{(1)}[l_1]$ and $m = 1$ or $m = 2$:

$$\sum_{l_1=-h}^{h} g_h^{(1)}[l_1] = 0, \qquad \sum_{l_1=-h}^{h} g_h^{(1)}[l_1] \cdot l_1 = -1, \qquad \sum_{l_1=-h}^{h} g_h^{(1)}[l_1] \cdot l_1^2 = 0. \tag{14.20}$$

For the differentiation kernels $g_h^{(2)}[l_1]$ giving an estimate of the second derivative, the conditions in Eq. (3.16) for $m = 2$ or $m = 3$ are of the form

$$
\begin{aligned}
\sum_{l_1=-h}^{h} g_h^{(2)}[l_1] &= 0, \\
\sum_{l_1=-h}^{h} g_h^{(2)}[l_1] \cdot l_1 &= 0, \\
\sum_{l_1=-h}^{h} g_h^{(2)}[l_1] \cdot \frac{l_1^2}{2} &= 1, \\
\sum_{l_1=-h}^{h} g_h^{(2)}[l_1] l_1^3 &= 0.
\end{aligned}
\tag{14.21}
$$

It is interesting to note that the kernels $g_h[l_1]$ are the same for $m = 0$ and $m = 1$ (formulas (14.12) and (14.13)), and the kernels $g_h^{(1)}[l_1]$ for the first derivative estimation are the same for $m = 1$ and $m = 2$ (formulas (14.14) and (14.16)). This property follows from the symmetry of the window $w_h[l_1]$ in Eq. (14.10). If the window is symmetric, we always have the same smoothing kernels $g_h[l_1]$ for $m = 2s$ and $m = 2s + 1$, $s = 0, 1, \ldots$, and the same differentiation kernels $g_h^{(r)}[l_1]$ for $m = r + 2s$ and $m = r + 2s + 1$.

The moment conditions in Eqs. (14.18–14.21) show the order of the corresponding kernels, that is, their ability to give accurate estimates for the polynomial functions of the corresponding powers. Thus, the kernel $g_h[l_1]$ shown in Eqs. (14.12) and (14.13) gives accurate estimates for the linear polynomials, while the kernel $g_h[l_1]$ shown in Eq. (14.15) gives accurate estimates for the polynomials of the third degree. For the design of the last kernel, we use the second-degree polynomial approximation only. The ability of this kernel to give the accurate estimate for the higher third-degree polynomial is increased because of the symmetry of the window, as was mentioned above. The differentiation kernel $g_h^{(1)}[l_1]$ shown in Eqs. (14.14) and (14.16) gives accurate estimates of the first derivative for the quadratic polynomials, and the differentiation kernel $g_h^{(2)}[l_1]$ shown in Eq. (14.17) gives accurate estimates of the second derivative for polynomials of the third degree.

Examples

The curves of the kernels $g_h[l_1]$ and $g_h^{(1)}[l_1]$ defined by formulas (14.15) and (14.16) are shown in Fig. 14.1. In this figure we also can see their amplitude frequency characteristics $|G_h|$ and $|G_h^{(1)}|$. The smoothing kernel $g_h[l_1]$ is a low-pass filter with the maximum amplitude frequency characteristic at $\bar{\omega} = 0$. Note that, according to Eq. (3.46), $G_h(0) = 1$ and, according to Eq. (3.47), $\partial_{\bar{\omega}} G_h(0) = 0$, $\partial_{\bar{\omega}}^2 G_h(0) = 0$. It can be seen in Fig. 14.1 that indeed $|G_h(0)| = 1$ and at least the first derivative $\partial_{\bar{\omega}} G_h(0)$ is equal to zero.

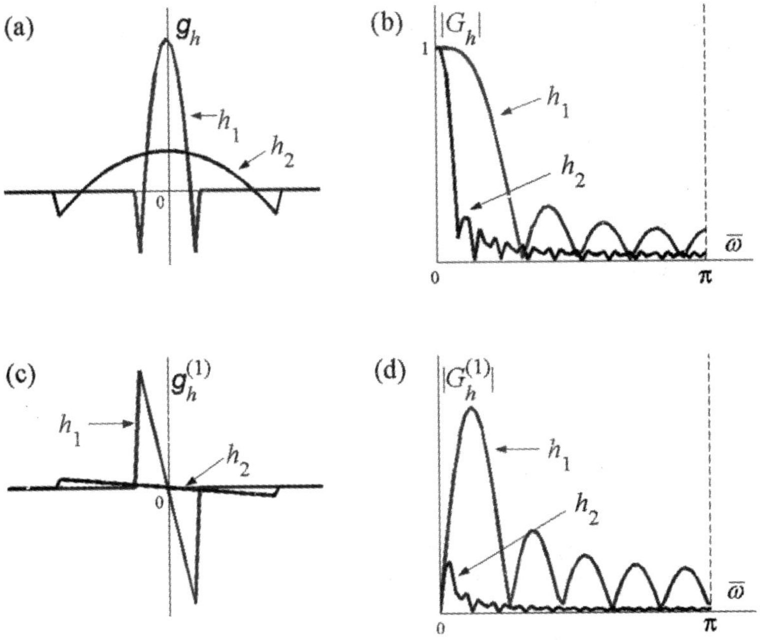

Figure 14.1 Kernels g_h and $g_h^{(1)}$ in Eqs. (14.15) and (14.16) defined by the Lagrange orthogonal polynomials of order $m = 2$ and their amplitude frequency characteristics $|G_h|$ and $|G_h^{(1)}|$.

The differentiating kernel $g_h^{(1)}[l_1]$ is a bandpass filter because its amplitude frequency characteristic $|G_h^{(1)}|$ is equal to zero at $\bar{\omega} = 0$. This confirms the vanishing moment condition in Eq. (3.48). The other conditions in Eq. (3.49) are $\partial_{\bar{\omega}} G_h^{(1)} = 1$ and $\partial_{\bar{\omega}}^2 G_h^{(1)} = 0$. We note that $|G_h^{(1)}|$ is a linear function at $\bar{\omega} = 0$. It can be seen from a comparison of all kernels for $h_2 > h_1$ that the kernels become wider for larger scales while their frequency responses become narrower.

14.1.1.2 Rectangular nonsymmetric window

Consider a rectangular nonsymmetric window,

$$w_h[l_1] = \frac{1}{h+1}, \quad 0 \leq l_1 \leq h, \tag{14.22}$$

where the parameter h is a nonnegative integer, $h \geq 0$. The corresponding discrete Lagrange polynomials of degree $m \leq 2$

$$\begin{aligned} \phi_1[l_1] &= 1, \qquad \phi_2[l_1] = \sqrt{\frac{3}{h(h+2)}}(2l_1 - h), \\ \phi_3[l_1] &= \sqrt{\frac{5}{h(h+2)(h+3)(h-1)}} \cdot [6l_1^2 - 6hl_1 + h(h-1)], \end{aligned} \tag{14.23}$$

are orthonormal with weights as in Eq. (14.22).

The elementary calculations according to Eqs. (14.8) and (14.9) give the formulas for the smoothing and differentiation kernels [101], [102]:

(a) for $m = 0$, $h \geq 0$,

$$g_h[l_1] = \frac{1}{h+1}, \quad 0 \leq l_1 \leq h; \tag{14.24}$$

(b) for $m = 1$, $h \geq 1$, $0 \leq l_1 \leq h$,

$$g_h[l_1] = \frac{2}{(h+2)(h+1)}(2h + 1 - 3l_1), \tag{14.25}$$

$$g_h^{(1)}[l_1] = \frac{-6(2l_1 - h)}{h(h+2)(h+1)}; \tag{14.26}$$

(c) for $m = 2$, $0 \leq l_1 \leq h$,

$$g_h[l_1] = \frac{3[3h^2 + 3h + 2 - 6(2h+1)l_1 + 10l_1^2]}{(h+2)(h+3)(h+1)}, \tag{14.27}$$

$$g_h^{(1)}[l_1] = \frac{18h(h-1)(2h+1) + 180h^2l^2 - 12l(16h^2 + 2h - 3)}{h(h+2)(h+3)(h^2-1)}, \tag{14.28}$$

$$g_h^{(2)}[l_1] = \frac{60[6l_1^2 - 6hl_1 + h(h-1)]}{h(h+2)(h+3)(h^2-1)}. \tag{14.29}$$

The vanishing moment conditions in Eqs. (3.15) and (3.16) can be verified for these kernels. Note that the orders of the zero moments in these conditions are exactly equal to the degree of the polynomials used in the kernel design. There are no order-increasing effects like the ones discussed above for the symmetric window kernels.

Let us illustrate this point. The following moment equations hold for $g_h[l_1]$ in Eq. (14.26):

$$\sum_{l_1=-h}^{h} g_h^{(1)}[l_1] = 0, \qquad \sum_{l_1=-h}^{h} g_h^{(1)}[l_1] \cdot l_1 = -1,$$

and these conditions coincide with the first two equations in Eq. (14.20). However, the third equation in Eq. (14.20), which follows from the symmetry of the window, does not have a place for the $g_h[l_1]$ in Eq. (14.26).

14.1.1.3 Exponential nonsymmetric window

Let an exponential nonsymmetric window be of the form

$$w[l_1] = (1-\gamma)\gamma^{l_1}, \qquad 0 \leq l_1 < \infty, \quad 0 < \gamma < 1, \tag{14.30}$$

where, in fact, γ is a scaling parameter. It is convenient to link it with the usual scale parameter h, which varies from 0 up to ∞:

$$\gamma = e^{-1/h}. \tag{14.31}$$

The corresponding polynomials orthonormal with the weight in Eq. (14.30) are

$$\phi_1[l_1] = 1, \ \phi_2[l_1] = \frac{1}{\sqrt{\gamma}}[l_1(1-\gamma) - \gamma], \tag{14.32}$$

$$\phi_3[l_1] = \frac{1}{2\gamma}[2\gamma^2 - (1+3\gamma)(1-\gamma)l_1 + (1-\gamma)^2 l_1^2].$$

Calculations give the following formulas for the kernels of the smoothing and differentiation operators in Eqs. (14.8) and (14.9) [101], [102]:

(a) for $m = 0$,

$$g_h[l_1] = w[l_1], \qquad 0 \le l_1 < \infty; \tag{14.33}$$

(b) for $m = 1, 0 \le l_1 < \infty$,

$$g_h[l_1] = [1 + \gamma - (1-\gamma)l_1]w[l_1],$$

$$g_h^{(1)}[l_1] = (1-\gamma)\left[1 - \frac{1}{\gamma}(1-\gamma)l_1\right]w[l_1]; \tag{14.34}$$

(c) for $m = 2, 0 \le l_1 < \infty$,

$$g_h[l_1] = \frac{1}{2}[2(1+\gamma+\gamma^2) - 3(1-\gamma^2)l_1 + (1-\gamma)^2 l_1^2]w[l_1],$$

$$g_h^{(1)}[l_1] = \frac{1}{4}(1-\gamma)\left\{6(1+\gamma) - (1/\gamma - 1)\left[4 + \frac{1}{\gamma}(1+3\gamma)^2\right]l_1 + (1/\gamma - 1)^2(1+3\gamma)l_1^2\right\}w[l_1], \tag{14.35}$$

$$g_h^{(2)}[l_1] = \frac{1}{2}(1/\gamma - 1)^2[2\gamma^2 - (1+3\gamma)(1-\gamma)l_1 + (1-\gamma)^2 l_1^2] \times w[l_1].$$

The vanishing moment conditions in Eqs. (3.15) and (3.16) hold for these kernels with orders of zero moments equal to the orders of the polynomials used in the design. There is no effect of the order increasing discussed above for the symmetric window kernels.

14.1.1.4 *Exponential symmetric window*

Consider an exponential symmetric window,

$$w[l_1] = \frac{1-\gamma}{1+\gamma}\gamma^{|l_1|}, \qquad -\infty < l_1 < \infty, \quad 0 < \gamma < 1,$$

where, in fact, γ is a scaling parameter that can be linked with the usual scale parameter h as in Eq. (14.31). The corresponding polynomials orthonormal with the weight $w[l_1]$ are

$$\phi_1[l_1] = 1, \qquad \phi_2[l_1] = \frac{1-\gamma}{\sqrt{2\gamma}} l_1,$$
$$\phi_3[l_1] = \frac{1}{\sqrt{2\gamma(1+8\gamma+\gamma^2)}}[-2\gamma^2 + (1-\gamma)^2 l_1^2]. \tag{14.36}$$

Calculations give the following formulas for the kernels of the smoothing and differentiation operators in Eqs. (14.8) and (14.9) [101], [102]:

(a) for $m = 0$,

$$g_h[l_1] = w[l_1];$$

(b) for $m = 1$,

$$g_h[l_1] = w[l_1],$$
$$g_h^{(1)}[l_1] = -\frac{(1-\gamma)^2}{2\gamma} l_1 w[l_1], \tag{14.37}$$

(c) for $m = 2$

$$g_h[l_1] = \left[1 + \frac{2\gamma - (1-\gamma)^2 l_1^2}{1+8\gamma+\gamma^2}\right] w[l_1],$$
$$g_h^{(1)}[l_1] = -\frac{(1-\gamma)^2}{2\gamma} l_1 w[l_1], \tag{14.38}$$
$$g_h^{(2)}[l_1] = \frac{(1-\gamma)^2}{\gamma(1+8\gamma+\gamma^2)}\left[-2\gamma + (1-\gamma)^2 l_1^2\right] w[l_1].$$

The vanishing moment Eqs. (14.18–14.21) hold for these symmetric window kernels.

Examples

The curves of kernels $g_h[l_1]$ and $g_h^{(1)}[l_1]$ defined by formulas (14.38) are shown in Fig. 14.2. In this figure we also can see the corresponding amplitude frequency characteristics $|G_h|$ and $|G_h^{(1)}|$. The smoothing kernel $g_h[l_1]$ is a low-pass filter with the maximum amplitude frequency characteristic at $\bar{\omega} = 0$.

The differentiating kernel $g_h^{(1)}[l_1]$ is a bandpass filter because its amplitude frequency characteristic $|G_h^{(1)}|$ is equal to zero at $\bar{\omega} = 0$. This confirms the vanishing moment condition in Eq. (3.48). Comparison with the curves in Fig. 14.1 shows that the exponential window defines the estimation kernels (smoothing and differentiation), which are smoother than those defined for the rectangular window.

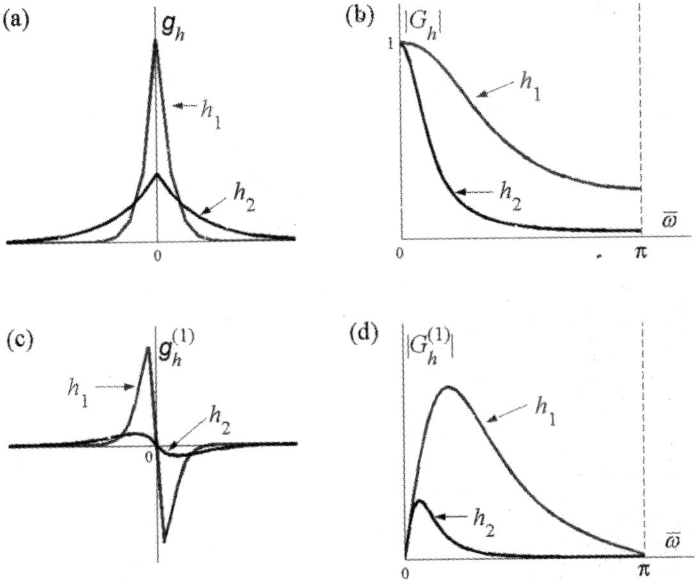

Figure 14.2 Kernels g_h and $g_h^{(1)}$ in Eq. (14.38) defined by the LPA with order $m = 2$ and with the exponential window and the amplitude frequency characteristics $|G_h|$ and $|G_h^{(1)}|$ of the kernels.

14.1.2 2D kernels

14.1.2.1 Separable 2D kernels

Let $g_h^{(r_1)}[l_1]$ and $g_h^{(r_2)}[l_2]$ be the 1D estimation kernels. Then 2D kernels can be derived as products of these kernels:

$$g_h^{(r)}[l_1, l_2] = g_h^{(r_1)}[l_1]g_h^{(r_2)}[l_2]. \tag{14.39}$$

Formula (14.39) defines the so-called separable kernel. The separable kernel can be smoothing or differentiating, depending on the values of r_1 and r_2.

In imaging, this type of kernel produces the corresponding operations independently on rows (x_1) and columns (x_2) of the image. Of course, the operators for rows and columns can be quite different, say by their orders, scales, and window functions used in the LPA. A specific feature of these operations is that the rows and columns are filtered independently.

In general, formulas (3.10) and (3.11) give inseparable kernels that cannot be represented in the form of the product in Eq. (14.39). These inseparable kernels allow one to produce filtering with row-column cross effects, when weights used for filtering of each row are column-dependent and the filtering of each column is row-dependent.

14.1.2.2 Inseparable 2D kernels

Consider a procedure that allows analytical results for the kernels that are, in general, inseparable. Let the 2D window w_h be a multiplicative function of integers l_1 and l_2:

$$w_h[l_1, l_2] = w_{h_1}[l_1]w_{h_2}[l_2], \tag{14.40}$$

where $w_{h_1}[l_1]$ and $w_{h_2}[l_2]$ are 1D windows with the scale parameters h_1 and h_2.

Assume that $\phi_{j,1}[l_1]$ and $\phi_{j,2}[l_2]$, $j = 0, 1, \ldots,$ are 1D polynomials of degree j orthonormal with the weights $w_{h_1}[l_1]$ and $w_{h_2}[l_2]$, respectively. The 2D polynomial $\phi[l_1, l_2]$ of degree k can be composed as a direct product of the sets of these 1D orthonormal polynomials as follows:

$$\begin{aligned} &\phi[l_1, l_2] = \phi_{j_1,1}[l_1]\phi_{j_2,2}[l_2], \\ &j_1 + j_2 = k, \end{aligned} \tag{14.41}$$

where the integers $j_1, j_2 \geq 0$ take values satisfying the equation $j_1 + j_2 = k$.

These 2D polynomials, built successively for $k = 0, 1, \ldots, m$, form a complete set of 2D polynomials of degree from 0 through m. It can be verified that these 2D polynomials are orthonormal with the 2D window $w_h[l_1, l_2]$ in Eq. (14.40).

These 2D orthonormal polynomials can be ordered according to their degrees similarly to Eq. (2.22):

$$\begin{aligned} &\phi_1[l_1, l_2] = \phi_{0,1}[l_1]\phi_{0,2}[l_2], \qquad \text{for } m = 0; \\ &\phi_2[l_1, l_2] = \phi_{1,1}[l_1]\phi_{0,2}[l_2], \quad \phi_3[l_1, l_2] = \phi_{0,1}[l_1]\phi_{1,2}[l_2], \qquad \text{for } m = 1; \\ &\phi_4[l_1, l_2] = \phi_{2,1}[l_1]\phi_{0,2}[l_2], \quad \phi_5 = \phi_{0,1}[l_1]\phi_{2,2}[l_2], \\ &\phi_6[l_1, l_2] = \phi_{1,1}[l_1]\phi_{1,2}[l_2], \qquad \text{for } m = 2; \text{ etc.,} \end{aligned} \tag{14.42}$$

with the total number of polynomials equal to $M = (m + 1)(m + 2)/2$.

In this way, we can design a complete set of 2D polynomials of degree less than or equal to m and orthonormal with the window $w_h[l], l = (l_1, l_2), h = (h_1, h_2)$. The estimator kernels can be obtained from the general multivariate formulas (14.4):

$$g_h[l] = w_h[l]\phi^T[l]\phi[0], \tag{14.43}$$

and

$$g_h^{(r)}[l] = (-1)^{|r|}w_h[l]\phi^T[l]\phi^{(r)}[0]. \tag{14.44}$$

All calculations are reduced to routine algebraic manipulations similar to the ones used for the 1D case. Let us present some of 2D kernels that are derived in this way.

14.1.2.3 Rectangular symmetric window

Consider a rectangular symmetric window,

$$w_{h_j}(l_j) = \frac{1}{2h_j + 1}, \qquad |l_j| \leq h_j, \quad j = 1, 2,$$

and

$$w_h[l] = w_{h_1}[l_1]w_{h_2}[l_2] = \frac{1}{(2h_1+1)(2h_2+1)}, \qquad |l_j| \le h_j, \quad j = 1, 2.$$

Using the orthonormal polynomials in Eq. (14.11) and the kernel formulas (14.43) and (14.44), we derive the kernels of the smoothing and differentiating operators of the different degrees m [101], [102]:

(a) for $m = 0$, $h_j \ge 0$,

$$g_h[l] = w_h[l]; \tag{14.45}$$

(b) for $m = 1$, $h_j \ge 1$,

$$g_h[l] = w_h[l], \tag{14.46}$$

$$g_h^{(1,0)}[l] = \frac{-3l_1}{h_1(h_1+1)} w_h[l], \tag{14.47}$$

$$g_h^{(0,1)}[l] = \frac{-3l_2}{h_2(h_2+1)} w_h[l]; \tag{14.48}$$

(c) for $m = 2$, $h_j \ge 1$,

$$g_h[l] = \left[1 + 5\sum_{j=1}^{2} \frac{h_j(h_j+1) - 3l_j^2}{(2h_j+3)(2h_j-1)}\right] \cdot w_h[l], \tag{14.49}$$

$$g_h^{(1,0)}[l] = \frac{-3l_1}{h_1(h_1+1)} w_h[l], \tag{14.50}$$

$$g_h^{(0,1)}[l] = \frac{-3l_2}{h_2(h_2+1)} w_h[l], \tag{14.51}$$

$$g_h^{(1,1)}[l] = \frac{9l_1l_2}{h_1h_2(h_1+1)(h_2+1)} w_h[l], \tag{14.52}$$

$$g_h^{(2,0)}[l] = \frac{30[3l_1^2 - h_1(h_1+1)]}{h_1(h_1+1)(2h_1+3)(2h_1-1)} w_h[l], \tag{14.53}$$

$$g_h^{(0,2)}[l] = \frac{30[3l_2^2 - h_2(h_2+1)]}{h_2(h_2+1)(2h_2+3)(2h_2-1)} w_h[l]. \tag{14.54}$$

Note that the smoothing operators in Eqs. (14.45) and (14.46) are identical to each other. A similar coincidence of Eq. (14.47) with Eq. (14.50) as well as Eq. (14.48) with Eq. (14.51) has a place for first-derivative estimation. This coincidence of the kernels corresponding to different degrees of the polynomial approximation is a result of the window symmetry property, $w_h[l] = w_h[-l]$, which was discussed above for the 1D case.

Also note that the kernel in Eq. (14.49) is inseparable, while all the others in Eqs. (14.46–14.48) and Eqs. (14.50–14.54) are separable. For instance, the kernel in

Eq. (14.48) can be represented in the separable form in Eq. (14.39) on the arguments l_1 and l_2 as

$$g_h^{(0,1)}[l] = w_{h_1}[l_1] \cdot w_{h_2}[l_2] \frac{-3l_2}{h_2(h_2+1)}.$$

Thus, the considered design procedure may give separable as well as inseparable kernels.

14.1.2.4 Rectangular nonsymmetric windows

Let rectangular nonsymmetric windows be of the form

$$w_{h_j}(l_j) = \frac{1}{h_j+1}, \qquad 0 \leq l_j \leq h_j, \quad j = 1, 2,$$

and

$$w_h[l] = w_{h_1}[l_1] w_{h_2}[l_2] = \frac{1}{h_1+1} \frac{1}{h_2+1}, \qquad 0 \leq l_j \leq h_j, \quad j = 1, 2.$$

Using the orthonormal polynomials in Eq. (14.23) and the kernel formulas (14.43) and (14.44), we derive the smoothing and differentiation kernels of the orders $m = 0, 1$:

(a) for $m = 0, h_j \geq 0$,

$$g_h[l] = w_h[l]$$

(b) for $m = 1, h_j \geq 1$,

$$g_h[l] = \left(1 + 3\sum_{s=1}^{2} \frac{h_j - 2l_j}{h_j + 2}\right) w_h[l],$$

$$g_h^{(1,0)}[l] = -\frac{6(2l_1 - h_1)}{h_1(h_1+2)} w_h[l],$$

$$g_h^{(0,1)}[l] = -\frac{6(2l_2 - h_2)}{h_2(h_2+2)} w_h[l].$$

14.1.2.5 Exponential nonsymmetric windows

Consider exponential nonsymmetric windows,

$$w_{h_j}(l_j) = (1 - \gamma_j)\gamma_j^{l_j}, \qquad 0 \leq l_j < \infty, \quad 0 < \gamma_j < 1, \quad j = 1, 2,$$

and

$$w_h[l] = w_{h_1}[l_1] w_{h_2}[l_2], \qquad 0 < \gamma_j < 1, \quad j = 1, 2,$$

where γ_j is a scaling parameter of the variable x_j, $\gamma_j = e^{-1/h_j}$, $j = 1, 2$. Using the corresponding orthonormal polynomials in Eq. (14.32), we derive the kernels of the smoothing and differentiation operators of orders $m = 0, 1$:

(a) for $m = 0$,

$$g_h[l] = w_h[l] \tag{14.55}$$

(b) and for $m = 1$,

$$\begin{aligned} g_h[l] &= \left\{1 + \sum_{j=1}^{2} \left[\gamma_j - (1-\gamma_j)l_j\right]\right\} \cdot w_h[l], \\ g_h^{(1,0)}[l] &= (1-\gamma_1)\left[1 - \frac{1}{\gamma_1}(1-\gamma_1)l_1\right] w_h[l], \\ g_h^{(0,1)}[l] &= (1-\gamma_2)\left[1 - \frac{1}{\gamma_2}(1-\gamma_2)l_2\right] w_h[l]. \end{aligned} \tag{14.56}$$

14.1.2.6 Exponential symmetric windows

Consider exponential symmetric windows,

$$w_{h_j}(l_j) = \frac{1-\gamma_j}{1+\gamma_j}\gamma_j^{|l_j|}, \qquad -\infty < l_j < \infty, \quad j = 1, 2,$$
$$w_h[l] = w_{h_1}[l_1]w_{h_2}[l_2].$$

Using the orthonormal polynomials in Eq. (14.36), we derive the smoothing and differentiating kernels for degrees $m = 0, 1, 2$:

(a) for $m = 0$,

$$g_h[l] = w_h[l], \qquad |l| < \infty; \tag{14.57}$$

(b) for $m = 1$, $|l| < \infty$, the kernels $g_h[l]$ coincide with those in Eq. (14.57) and

$$\begin{aligned} g_h^{(1,0)}[l] &= -\frac{(1-\gamma_1)^2}{2\gamma_1} l_1 w_h[l], \\ g_h^{(0,1)}[l] &= -\frac{(1-\gamma_2)^2}{2\gamma_2} l_2 w_h[l]; \end{aligned} \tag{14.58}$$

(c) for $m = 2$, the kernels $g_h^{(1,0)}[l]$ and $g_h^{(0,1)}[l]$ coincide with those in Eq. (14.58) and

$$g_h[l] = \left[1 + \sum_{j=1}^{2} \frac{2\gamma_j - (1-\gamma_j)^2 l_j^2}{1 + 8\gamma_j + \gamma_j^2}\right] w_h[l],$$
$$g_h^{(1,1)}[l] = \frac{l_1 l_2}{4\gamma_1\gamma_2}(1-\gamma_1)^2(1-\gamma_2)^2 w_h[l],$$
$$g_h^{(2,0)}[l] = \frac{(1-\gamma_1)^2}{\gamma_1(1+8\gamma_1+\gamma_1^2)}[-2\gamma_1 + (1-\gamma_1)^2 l_1^2] w_h[l], \tag{14.59}$$
$$g_h^{(0,2)}[l] = \frac{(1-\gamma_2)^2}{\gamma_2(1+8\gamma_2+\gamma_2^2)}[-2\gamma_2 + (1-\gamma_2)^2 l_2^2] w_h[l].$$

This type of analytical formula for the kernels can be given for any dimensionality d and for higher degrees m. These results are interesting because they can be exploited for recursive implementations of filtering.

However, in many cases, numerical computing of the kernel values is sufficient and even more convenient than the analytical formulas, which can be quite complex. All necessary numerical calculations can be produced based on the general formulas (3.10) and (3.11). Even the orthonormality of the polynomials is not very helpful because finally the kernels are given numerically by the corresponding arrays (masks) of their values.

14.2 LPA Accuracy

14.2.1 Proof of Proposition 4

In this section we present the proof of Proposition 4. For simplicity, we assume that $\Delta_1 = \cdots = \Delta_d = \Delta$.

We first derive Eq. (4.23) for Φ from the matrix Φ_h in Eq. (3.6). For small $\Delta \to 0$, we obtain

$$\Phi_h = \sum_l w_h(\Delta l)\phi_h(\Delta l)\phi_h^T(\Delta l)$$
$$= \frac{1}{\Delta^d}\sum_l w(\Delta l/h)\phi(\Delta l/h)\phi^T(\Delta l/h)\frac{\Delta^d}{h^d} \to \frac{1}{\Delta^d}\Phi, \tag{14.60}$$
$$\Phi = \int_{\mathbb{R}^d} w(u)\phi(u)\phi^T(u)du.$$

Here and in what follows, $\sum_l$ is a sum over an infinite regular grid $\sum_l = \sum_{l\in\mathbb{Z}^d}$.

In the limit replacement of the sum $\sum_l$ by the integral $\int_{\mathbb{R}^d}$, we assume for the continuous variable u that $\Delta l/h \to u$ and then $(\Delta/h)^d \to du$. Substituting

$$\Phi_h = \frac{1}{\Delta^d}\Phi \tag{14.61}$$

into $g_h(\Delta l)$ in Eq. (3.5), we have

$$g_h(\Delta l) = \frac{\Delta^d}{h^d} w(\Delta l/h)\phi^T(\Delta l/h)\Phi^{-1}\phi(0). \tag{14.62}$$

Further, inserting this formula into the signal estimate in Eq. (3.4), we obtain

$$\begin{aligned}\hat{y}_h(x) &= \sum_l g_h(\Delta l) z(x - \Delta l)\\ &= \sum_l w(\Delta l/h)\phi^T(\Delta l/h)\Phi^{-1}\phi(0) z\left(x - \frac{hl\Delta}{h}\right)\frac{\Delta^d}{h^d}\\ &\to \int_{\mathbb{R}^d} g(u) z(x - hu) du,\end{aligned}$$

where $g(x)$ is defined by Eq. (4.34). This proves Eq. (4.32).

In a similar way, substituting Eq. (14.61) into Eq. (3.7), we obtain for the kernel of the derivative estimate

$$g_h^{(r)}(l) = \left(\frac{-1}{h}\right)^{|r|} w\left(\frac{\Delta l}{h}\right)\phi^T\left(\frac{\Delta l}{h}\right)\Phi^{-1}\phi^{(r)}(0)\frac{\Delta^d}{h^d} \tag{14.63}$$

and for the estimate itself

$$\hat{y}_h^{(r)}(x) = \frac{1}{h^{|r|}}\int_{\mathbb{R}^d} g^{(r)}(u) z(x - hu) du,$$

where $g^{(r)}(x)$ is defined in Eq. (4.35). This proves Eq. (4.33) and completes the proof of Proposition 4.

Some extra conditions on the window $w(x)$ and the signal function $z(x)$ are required for the limit passages from the sums $\sum_l$ to the integrals $\int_{\mathbb{R}^d}$ to be valid. These requirements are not restrictive and can be given in different forms. In particular, it is sufficient to assume that the window w has a finite support and the signal z is continuous bounded. More details about these conditions can be found in the standard texts on mathematical analysis.

14.2.2 Proof of Proposition 7

Here we present the proof of Proposition 7.

Lemma 1 *Let $f(x)$, $x \in \mathbb{R}^d$, be a differentiable function from the class $\mathbb{C}^{Q+1}$ in Eq. (2.6). Then the following Taylor series representation is valid:*

$$f(x-u) = \sum_{t=0}^{Q} (-1)^t \sum_{|k|=t} \frac{1}{k!} u^k f^{(k)}(x) + (-1)^{Q+1} \sum_{|k|=Q+1} \frac{1}{k!} u^k f^{(k)}(x - \lambda u),$$
$$0 \le \lambda \le 1. \tag{14.64}$$

For $d = 2$, this Taylor series has the form

$$f(x_1 - u_1,\ x_2 - u_2)$$
$$= \sum_{t=0}^{Q} (-1)^t \sum_{k_1+k_2=t} \frac{1}{k_1! k_2!} u_1^{k_1} u_2^{k_2} f^{(k_1,k_2)}(x_1, x_2) + (-1)^{Q+1} \sum_{k_1+k_2=Q+1} \frac{1}{k_1! k_2!}$$
$$\times u_1^{k_1} u_2^{k_2} f^{(k_1,k_2)}(x_1 - \lambda_1 u_1\ x_2 - \lambda_1 u_2), 0 \le \lambda_1 \le 1,\ 0 \le \lambda_2 \le 1. \tag{14.65}$$

The first summand in Eq. (14.65) is a truncated Taylor series of degree Q. The second summand is the so-called residual term of the series in Lagrange form, where $\lambda = (\lambda_1, \lambda_2)$ is a 2D parameter. Note that Eq. (14.65) is an accurate equality.

In multi-index notation, this formula has the form of Eq. (14.64) and is valid for any d.

The assumption $f(x) \in \mathbb{C}^{Q+1}$ means that the derivatives in the residual term are bounded.

Lemma 2 *Let the hypotheses H1 and H2 of Proposition 7 hold for $y(x)$, $x \in \mathbb{R}^d$, $M = \min\{m+1, \alpha\}$; the observation model (Eq. (2.1)) be defined on the grid Δl, $l \in \mathbb{Z}^d$; and the errors ε be random zero-mean uncorrelated with variance σ^2. Then the following formulas are valid for the estimates in Eqs. (3.4–3.8) of the signal and the derivatives:*

(a) *for the bias of the estimation errors,*

$$m_{\hat{y}_h}(x,h) = (-1)^{M-1} \sum_{|k|=M} \frac{1}{k!} \sum_l g_h(\Delta l)(\Delta l)^k y^{(k)}(x - \lambda \Delta l), \tag{14.66}$$

$$m_{\hat{y}_h^{(r)}}(x,h) = (-1)^{M-1} \sum_{|k|=M} \frac{1}{k!} \sum_l g_h^{(r)}(\Delta l)(\Delta l)^k y^{(k)}(x - \lambda \Delta l), \quad |r| \le m; \tag{14.67}$$

(b) *for the variance of the estimation errors,*

$$\frac{1}{\sigma^2} \sigma^2_{\hat{y}_h}(h) = \sum_l [g_h(\Delta l)]^2, \tag{14.68}$$

$$\frac{1}{\sigma^2} \sigma^2_{\hat{y}_h^{(r)}}(h) = \sum_l [g_h^{(r)}(\Delta l)]^2. \tag{14.69}$$

Proof. (1) Consider the systematic error of estimation of the signal. First, assume that

$$m + 1 \leq \alpha. \tag{14.70}$$

Then $M = \min\{m+1, \alpha\} = m + 1$ and the Taylor series in Eq. (14.64) for $y(x - \Delta l)$ can be used with the $Q = m$ as

$$\begin{aligned} y(x - \Delta l) &= y_1(x, \Delta l) + y_2(x, \Delta l), \\ y_1(x, \Delta l) &= \sum_{t=0}^{m} (-1)^t \sum_{|k|=t} \frac{1}{k!} (\Delta l)^k y^{(k)}(x), \\ y_2(x, \Delta l) &= (-1)^{m+1} \sum_{|k|=m+1} \frac{1}{k!} (\Delta l)^k y^{(k)}(x - \lambda \Delta l), \qquad 0 \leq \lambda \leq 1. \end{aligned} \tag{14.71}$$

Since $y_1(x, \Delta l)$ is a polynomial of power m with respect to Δl, according to the moment conditions in Eq. (3.15), we obtain for the estimate in Eq. (3.4)

$$\sum_l g_h(\Delta l) y_1(x, \Delta l) = y(x). \tag{14.72}$$

Then, using Eq. (14.71),

$$\begin{aligned} E\{\hat{y}_h(x)\} &= (g_h \circledast y)(x) = \sum_l g_h(\Delta l) y_1(x, \Delta l) + \sum_l g_h(\Delta l) y_2(x, \Delta l) \\ &= y(x) + \sum_l g_h(\Delta l) y_2(x, \Delta l), \end{aligned}$$

and the estimation bias error has the form

$$\begin{aligned} m_{\hat{y}_h}(x, h) &= y(x) - E\{\hat{y}_h(x)\} = -\sum_l g_h(\Delta l) y_2(x, \Delta l) \\ &= (-1)^m \sum_l g_h(\Delta l) \sum_{|k|=m+1} \frac{1}{k!} (\Delta l)^k y^{(k)}(x - \lambda \Delta l). \end{aligned} \tag{14.73}$$

Now assume instead of Eq. (14.70) that

$$m + 1 > \alpha. \tag{14.74}$$

Then

$$M = \min\{m + 1, \alpha\} = \alpha$$

and the Taylor series Eq. (14.64) can be used with the degree $Q = \alpha - 1$. This gives the series in Eq. (14.71) as

$$
\begin{aligned}
y(x - \Delta l) &= y_1(x, \Delta l) + y_2(x, \Delta l), \\
y_1(x, \Delta l) &= \sum_{t=0}^{\alpha-1} (-1)^t \sum_{|k|=t} \frac{1}{k!} (\Delta l)^k y^{(k)}(x), \\
y_2(x, \Delta l) &= (-1)^{\alpha} \sum_{|k|=\alpha} \frac{1}{k!} (\Delta l)^k y^{(k)}(x - \lambda l \Delta), \qquad 0 \le \lambda \le 1.
\end{aligned}
\tag{14.75}
$$

Using Eq. (14.72), the error in Eq. (14.73) can be represented as

$$
\begin{aligned}
m_{\hat{y}_h}(x, h) &= y(x) - E\{\hat{y}_h(x)\} \\
&= y(x) - \left[\sum_l g_h(\Delta l) y_1(x, \Delta l) + \sum_l g_h(\Delta l) y_2(x, \Delta l) \right] \\
&= y(x) - \left[y(x) + \sum_l g_h(\Delta l) y_2(x, \Delta l) \right] \\
&= (-1)^{\alpha-1} \sum_l g_h(\Delta l) \sum_{|k|=\alpha} \frac{1}{k!} (\Delta l)^k y^{(k)}(x - \lambda l \Delta).
\end{aligned}
\tag{14.76}
$$

Equations (14.73) and (14.76) combined give Eq. (14.66) for the general case $M = \min\{m + 1, \alpha\}$.

For the derivative estimation we have the kernel $g_h^{(r)}(\Delta l)$ instead of the kernel $g_h(\Delta l)$. The proof of Eq. (14.67) is identical to that given above, except that, for the polynomial $y_1(x, \Delta l)$, according to the moment conditions in Eq. (3.16),

$$(g_h^{(r)} \circledast y_1)(x) = y^{(r)}(x).$$

(2) The formulas for the variance in Eqs. (14.68) and (14.69) immediately follow from Eqs. (3.4) and (3.7) provided that the noise is white with variance equal to σ^2. This completes the proof of Lemma 2. □

14.2.2.1 Proof of Proposition 7

Lemma 2 gives the general accuracy results for the discrete estimators valid for any fixed h and Δ. To prove Proposition 7, we consider the limit passages in formulas (14.66)–(14.69) as $\Delta \to 0$, $h \to 0$, and $\Delta/h \to 0$. The technique used here is similar to the one applied for the proof of Proposition 4, where the limit integral kernel estimates were derived from the discrete LPA estimates.

(1) Prove formula (5.11). According to Eq. (14.62), the discrete kernel g_h of the signal estimation is of the form

$$g_h(\Delta l) = \frac{\Delta^d}{h^d} g(\Delta l / h), \tag{14.77}$$

where g is the integral smoothing kernel in Eq. (4.34).

Inserting Eq. (14.77) into Eq. (14.66), we yield

$$
\begin{aligned}
(-1)^{M-1} m_{\hat{y}_h}(x, h) &= \sum_{|k|=M} \frac{1}{k!} \sum_{l} \frac{\Delta^d}{h^d} g\left(\frac{\Delta l}{h}\right) (\Delta l)^k y^{(k)}(x - \lambda \Delta l) \\
&= h^M \sum_{|k|=M} \frac{1}{k!} \sum_{l} \frac{\Delta^d}{h^d} g\left(\frac{\Delta l}{h}\right) \left(\frac{\Delta l}{h}\right)^k y^{(k)}(x - \lambda \Delta l).
\end{aligned}
\tag{14.78}
$$

Then

$$
\begin{aligned}
(-1)^{M-1} m_{\hat{y}_h}(x, h) = {} & h^M \sum_{|k|=M} \frac{1}{k!} \sum_{l} \frac{\Delta^d}{h^d} g\left(\frac{\Delta l}{h}\right) \left(\frac{\Delta l}{h}\right)^k y^{(k)}(x) + h^M \sum_{|k|=M} \frac{1}{k!} \\
& \times \sum_{l} \frac{\Delta^d}{h^d} g\left(\frac{\Delta l}{h}\right) \left(\frac{\Delta l}{h}\right)^k \left[y^{(k)}(x - \lambda \Delta l) - y^{(k)}(x)\right].
\end{aligned}
\tag{14.79}
$$

As $\Delta, h \to 0$, the first summand in Eq. (14.79) gives

$$
\begin{aligned}
& h^M \sum_{|k|=M} \frac{1}{k!} \sum_{l} \frac{\Delta^d}{h^d} g\left(\frac{\Delta l}{h}\right) \left(\frac{\Delta l}{h}\right)^k y^{(k)}(x) \\
& \to h^M \sum_{|k|=M} y^{(k)}(x) \frac{1}{k!} \int_{\mathbb{R}^d} g(u) u^k du + o(h^M),
\end{aligned}
\tag{14.80}
$$

where the integration variables u and du are used for $\Delta l / h$ and $(\Delta / h)^d$, and $o(h^M)$ shows the order of the asymptotic difference between the sum and the integral in Eq. (14.80).

Evaluate the second summand in Eq. (14.79). Using the Lipshitz condition in Eq. (5.8), we find

$$
|y^{(k)}(x - \lambda \Delta l) - y^{(k)}(x)| \leq K_L ||\lambda \Delta l||^{\gamma},
\tag{14.81}
$$

and then

$$
\begin{aligned}
& h^M \Big| \sum_{|k|=M} \frac{1}{k!} \sum_{l} \frac{\Delta^d}{h^d} g\left(\frac{\Delta l}{h}\right) \left(\frac{\Delta l}{h}\right)^k [y^{(k)}(x - \lambda \Delta l) - y^{(k)}(x)] \Big| \\
& \leq K_L h^M \sum_{|k|=M} \frac{1}{k!} \sum_{l} \frac{\Delta^d}{h^d} \Big| g\left(\frac{\Delta l}{h}\right) \left(\frac{\Delta l}{h}\right)^k \Big| \cdot ||\lambda \Delta l||^{\gamma} \\
& = K_L h^M \sum_{|k|=M} \frac{1}{k!} \sum_{l} \frac{\Delta^d}{h^d} \Big| g\left(\frac{\Delta l}{h}\right) \left(\frac{\Delta l}{h}\right)^k \Big| \cdot ||\lambda h \Delta l / h||^{\gamma} \\
& \to K_L h^M \sum_{|k|=M} \frac{1}{k!} \int_{\mathbb{R}^d} |g(u) u^k| \cdot ||\lambda h u||^{\gamma} du = o(h^M).
\end{aligned}
\tag{14.82}
$$

Note that in the multi-index notation used in Eq. (14.82)

$$||\lambda h u||^{\gamma} = \left[\sum_{i=1}^{d}(\lambda_i h_i u_i)^2\right]^{\gamma/2}.$$

Inserting Eqs. (14.80) and (14.82) into Eq. (14.79), we can see that

$$(-1)^{M-1} m_{\hat{y}_h}(x,h) = h^M \sum_{|k|=M} y^{(k)}(x)\frac{1}{k!}\int_{\mathbb{R}^d} g(u)u^k du + o(h^M).$$

This proves Eq. (5.11).

(2) Prove formula (5.12). The transformations are similar to the ones used in the above proof of formula (5.11). For the differentiating discrete kernel, according to Eq. (14.63), we have

$$g_h^{(r)}(\Delta l) = \frac{1}{h^{|r|}}\frac{\Delta^d}{h^d} g^{(r)}\left(\frac{\Delta l}{h}\right), \tag{14.83}$$

where $g^{(r)}(x)$ is an integral differentiating kernel as in Eq. (4.35).

Inserting Eq. (14.83) into Eq. (14.67), we yield

$$\begin{aligned}(-1)^{M-1} m_{\hat{y}_h^{(r)}}(x,h) &= \frac{1}{h^{|r|}}\sum_{|k|=M}\frac{1}{k!}\sum_{l}\frac{\Delta^d}{h^d} g^{(r)}\left(\frac{\Delta l}{h}\right)(\Delta l)^k y^{(k)}(x-\lambda\Delta l)\\ &= h^{M-|r|}\sum_{|k|=M}\frac{1}{k!}\sum_{l}\frac{\Delta^d}{h^d} g^{(r)}\left(\frac{\Delta l}{h}\right)\left(\frac{\Delta l}{h}\right)^k y^{(k)}(x-\lambda\Delta l)\end{aligned} \tag{14.84}$$

Then

$$\begin{aligned}&(-1)^{M-1} m_{\hat{y}_h^{(r)}}(x,h)\\ &= h^{M-|r|}\sum_{|k|=M}\frac{1}{k!}\sum_{l}\frac{\Delta^d}{h^d} g^{(r)}\left(\frac{\Delta l}{h}\right)\left(\frac{\Delta l}{h}\right)^k y^{(k)}(x) + h^{M-|r|}\sum_{|k|=M}\frac{1}{k!}\\ &\times \sum_{l}\frac{\Delta^d}{h^d} g^{(r)}\left(\frac{\Delta l}{h}\right)\left(\frac{\Delta l}{h}\right)^k [y^{(k)}(x-\lambda\Delta l) - y^{(k)}(x)].\end{aligned} \tag{14.85}$$

As $\Delta, h \to 0$, the fist summand in Eq. (14.85) gives

$$\begin{aligned}&h^{M-|r|}\sum_{|k|=M}\frac{1}{k!}\sum_{l}\frac{\Delta^d}{h^d} g^{(r)}(\Delta l/h)(\Delta l/h)^k y^{(k)}(x)\\ &= h^{M-|r|}\sum_{|k|=M} y^{(k)}(x)\frac{1}{k!}\int_{\mathbb{R}^d} g^{(r)}(u)u^k du + o(h^{M-|r|}).\end{aligned} \tag{14.86}$$

Evaluate the second summand in Eq. (14.85) using Eq. (14.81):

$$h^{M-|r|}\left|\sum_{|k|=M}\frac{1}{k!}\sum_{l}\frac{\Delta^d}{h^d}g^{(r)}\left(\frac{\Delta l}{h}\right)\left(\frac{\Delta l}{h}\right)^k[y^{(k)}(x-\lambda\Delta l)-y^{(k)}(x)]\right|$$
$$\leq K_L h^{M-|r|}\sum_{|k|=M}\frac{1}{k!}\sum_{l}\frac{\Delta^d}{h^d}\left|g^{(r)}\left(\frac{\Delta l}{h}\right)\left(\frac{\Delta l}{h}\right)^k\right|\cdot||\lambda\Delta l||^{\gamma}$$
$$=K_L h^{M-|r|}\sum_{|k|=M}\frac{1}{k!}\sum_{l}\frac{\Delta^d}{h^d}\left|g^{(r)}\left(\frac{\Delta l}{h}\right)\left(\frac{\Delta l}{h}\right)^k\right|\cdot||\lambda h\Delta l/h||^{\gamma}$$
$$=K_L h^{M-|r|}\sum_{|k|=M}\frac{1}{k!}\int_{\mathbb{R}^d}|g^{(r)}(u)u^k|\cdot||\lambda h u||^{\gamma}du=o(h^{M-|r|}). \tag{14.87}$$

Inserting Eqs. (14.86) and (14.87) into Eq. (14.85), we find that

$$(-1)^{M-1}m_{\hat{y}_h^{(r)}}(x,h)=h^{M-|r|}\sum_{|k|=M}y^{(k)}(x)\frac{1}{k!}\int_{\mathbb{R}^d}g^{(r)}(u)u^k du+o(h^{M-|r|}).$$

This proves Eq. (5.12).

(3) For the variance of the estimates we substitute Eqs. (14.77) and (14.83), respectively, into Eqs. (14.68) and (14.69). Calculations show that

$$\frac{1}{\sigma^2}\sigma^2_{\hat{y}_h}(h)=\sum_l[g_h(\Delta l)]^2=\frac{\Delta^d}{h^d}\sum_l\left[g\left(\frac{\Delta l}{h}\right)\right]^2\frac{\Delta^d}{h^d}$$
$$=\frac{\Delta^d}{h^d}\int_{\mathbb{R}^d}(g(u))^2du+o\left(\frac{\Delta^d}{h^d}\right),$$

and

$$\frac{1}{\sigma^2}\sigma^2_{\hat{y}_h^{(r)}}(h)=\sum_l\left[g_h^{(r)}(\Delta l)\right]^2=\frac{\Delta^d}{h^{d+2|r|}}\sum_l\left[g^{(r)}\left(\frac{\Delta l}{h}\right)\right]^2\frac{\Delta^d}{h^d}$$
$$=\frac{\Delta^d}{h^{d+2|r|}}\int_{\mathbb{R}^d}\left[g^{(r)}(u)\right]^2du+o\left(\frac{\Delta^d}{h^{d+2|r|}}\right).$$

This proves Eqs. (5.13) and (5.14) and completes the proof of Proposition 7.

14.3 ICI Rule

14.3.1 Proof of Proposition 10

In this section we present the proof of Proposition 10.

Lemma 3 *Let the estimates* $\hat{y}_{h^+}(x)$ *and* $\hat{y}^{(r)}_{h^+}(x)$ *be defined by the adaptive algorithms in Eqs. (6.13), (6.15), and (6.25), and let the random estimation errors* *satisfy the inequalities in Eqs. (6.5) and (6.23):*

$$\left|e_y^0(x,h)\right| \leq \chi_{1-\beta/2} \cdot \sigma_{\hat{y}_h}(x,h), \tag{14.88}$$

$$\left|e_{y^{(r)}}^0(x,h)\right| \leq \chi_{1-\beta/2} \cdot \sigma_{\hat{y}_h^{(r)}}(x,h). \tag{14.89}$$

Then the errors of the adaptive estimates have the following upper bounds:

$$|y(x) - \hat{y}_{h^+}(x)| \leq 3\Gamma\sigma_{\hat{y}_h}(x,h^*), \tag{14.90}$$

$$|y^{(r)}(x) - \hat{y}_{h^+}^{(r)}(x)| \leq 3\Gamma_{|r|}\sigma_{\hat{y}_h^{(r)}}(x,h_r^*), \tag{14.91}$$

where h^* and h_r^* are the ideal scales of the estimates given in Eqs. (5.23) and (5.30).

Proof. Consider the proof for the signal estimate. It was discussed in Sec. 6.1.1 that the ICI adaptive h^+ can be larger or smaller than the ideal scale h^*.

Let us start from the case $h^+ \geq h^*$. According to the definition of the adaptive-scale h^+, the confidence intervals $Q_j = Q(h_j)$ in Eq. (6.13) have a point in common for all $h \leq h^+$. As $h^+ \geq h^*$, this means that the intervals $Q_j = Q(h_j)$ also intersect with the ideal confidence interval $Q(h^*)$ corresponding to h^*. Combining these observations, we conclude that the distance between $y(x)$ and $\hat{y}_{h^+}(x)$ is at most a half-length of $Q(h^+)$, plus the length of $Q(h^*)$:

$$\left|y(x) - \hat{y}_{h^+}(x)\right| \leq \Gamma \cdot \sigma_{\hat{y}_h}(x,h^+) + 2\Gamma \cdot \sigma_{\hat{y}_h}(x,h^*). \tag{14.92}$$

Figure 14.3 illustrates this claim. The estimate $\hat{y}_{h^+}(x)$ is in the middle of the interval $Q(h^+)$ while the true $y(x)$ is somewhere in the interval $Q(h^*)$. Thus, the maximal distance between the estimate and the true signal is evaluated by the expression given in the right-hand side of the inequality in Eq. (14.92).

Because $h^+ \geq h^*$ in Eq. (6.8), we have for the corresponding standard deviations

$$\sigma_{\hat{y}_h}(x,h^+) \leq \sigma_{\hat{y}_h}(x,h^*).$$

Substituting it in Eq. (14.92) proves Eq. (14.90) for $h^+ \geq h^*$.

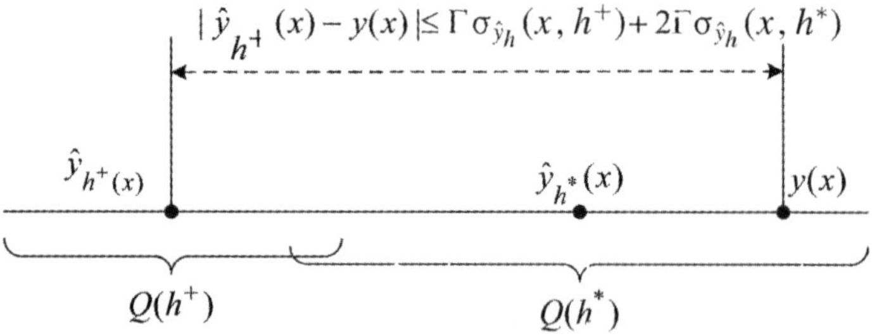

Figure 14.3 The estimate $\hat{y}_{h^+}(x)$ is in the middle of the interval $Q(h^+)$ while the true $y(x)$ is somewhere in the interval $Q(h^*)$. These intervals intersect. Then, the maximal distance between the estimate and the true value of the signal is equal to a half-length of $Q(h^+)$ plus the length of $Q(h^*)$.

For $h^+ < h^*$, the true $y(x)$ belongs to the adaptive confidence interval and

$$\left|y(x) - \hat{y}_{h^+}(x)\right| \leq 2\Gamma \cdot \sigma_{\hat{y}_h}(x, h^+). \tag{14.93}$$

The assumption of Proposition 10 on H,

$$h_1 = o(\log n/n), \qquad h_i - h_{i-1} = o(\log n/n), \tag{14.94}$$

enables H to include elements close to the ideal h^* and then $h^+ \simeq h^*$. It follows that $\sigma_{\hat{y}_h}(x, h^+) \simeq \sigma_{\hat{y}_h}(x, h^*)$ and the inequality in Eq. (14.90) follows from Eq. (14.93). This completes the proof of Eq. (14.90).

The proof of Eq. (14.91) is different only by the standard deviations used, $\sigma_{\hat{y}_h^{(r)}}(x, h)$, instead of $\sigma_{\hat{y}_h}(x, h)$, and the threshold $\Gamma_{|r|}$ instead of Γ. □

14.3.1.1 Proof of Proposition 10

The assumptions in Eqs. (14.88) and (14.89) mean that we do not have in our observations the unusually large random errors that occur with the small probability β. Of course, when $\left|e_y^0(x, h)\right| > \chi_{1-\beta/2} \cdot \sigma_{y_h}(x, h)$, the estimation error can be much larger than in the case when the random errors are bounded as in Eq. (14.88).

Let

$$I_{e_y^0} = 1\left[\left|e_y^0(x, h)\right| \leq \chi_{1-\beta/2} \cdot \sigma_{\hat{y}_h}(x, h)\right],$$

be the indicator of the random event in Eq. (14.88). This $I_{e_y^0}$ is equal to 1 if $\left|e_y^0(x, h)\right| \leq \chi_{1-\beta/2} \cdot \sigma_{\hat{y}_h}(x, h)$ and equal to zero if $\left|e_y^0(x, h)\right| > \chi_{1-\beta/2} \cdot \sigma_{\hat{y}_h}(x, h)$.

The average risk

$$\begin{aligned} E\{(y(x) - \hat{y}_{h^+}(x))^2\} &= E\{I_{e_y^0} \cdot [y(x) - \hat{y}_{h^+}(x)]^2\} \\ &\quad + E\{(1 - I_{e_y^0}) \cdot [y(x) - \hat{y}_{h^+}(x)]^2\} \end{aligned} \tag{14.95}$$

is the sum of two terms respectively corresponding to the usual small errors $\left|e_y^0(x, h)\right| \leq \chi_{1-\beta/2} \cdot \sigma_{\hat{y}_h}(x, h)$ and to unusual "pathologically large" errors ([66], [163]) when $\left|e_y^0(x, h)\right| > \chi_{1-\beta/2} \cdot \sigma_{\hat{y}_h}(x, h)$. According to Eq. (14.90) we have for the first summand in Eq. (14.95)

$$E\{I_{e_y^0} \cdot (y(x) - \hat{y}_{h^+}(x))^2\} \leq 9\Gamma^2 \sigma^2_{\hat{y}_h}(x, h^*). \tag{14.96}$$

Let $\chi_{1-\beta/2}$ be growing as $(\ln n)^{M/2M+d}$, where $n \sim 1/\Delta^d$ is the number of observations. Then, according to Eq. (6.10),

$$\Gamma_n = \gamma + \chi_{1-\alpha/2} = O\left((\ln n)^{\frac{M}{2M+d}}\right). \tag{14.97}$$

Consider the product $\Gamma^2 \sigma^2_{\hat{y}_h}(h^*)$ in Eq. (14.96), where $\Gamma = \Gamma_n$ is growing as above. Using formulas (5.13) and (5.23) for the variance of the estimate with the ideal scale h^*, we can verify that $h^* = O\left(\Delta^{\frac{d}{2M+d}}\right)$, $\sigma^2_{\hat{y}_h}(x, h^*) = O\left(\Delta^{\frac{2Md}{2M+d}}\right) =$

$O\left((1/n)^{\frac{2M}{2M+d}}\right)$. Then

$$\Gamma_n^2 \sigma_{\hat{y}_h}^2(x, h^*) = O\left(\left(\frac{\ln n}{n}\right)^{\frac{2M}{2M+d}}\right).$$

In this way we prove for the first summand in Eq. (14.95) that

$$E\{I_{e_y^0} \cdot [y(x) - \hat{y}_{h^+}(x)]^2\} = O\left(\left(\frac{\ln n}{n}\right)^{\frac{2M}{2M+d}}\right). \tag{14.98}$$

The following inequality holds for the probability of the random large error $\left|e_y^0(x, h)\right| / \sigma_{\hat{y}_h}(x, h) > \chi$:

$$P\{\left|e_y^0(x, h)\right| / \sigma_{\hat{y}_h}(x, h) > \chi\} \leq \exp(-c\chi^2) \tag{14.99}$$

with a certain constant $c > 0$.

We use here $\chi = \chi_{1-\beta/2} = O\left((\ln n)^{\frac{M}{2M+d}}\right)$. Then the probability in Eq. (14.99) of the large error is very small,

$$P\left\{\left|e_y^0(x, h)\right| / \sigma_{\hat{y}_h}(x, h) > (\ln n)^{\frac{M}{2M+d}}\right\} \leq \exp\left[-c(\ln n)^{\frac{2M}{2M+d}}\right]. \tag{14.100}$$

The inequality in Eq. (14.100) is exploited for the evaluation of the second summand in Eq. (14.95). The smallness of the probability of the larger error enables the smallness of $E\{(1 - I_{e_y^0}) \cdot [y(x) - \hat{y}_{h^+}(x)]^2\}$, which is intuitively clear.

It can be proven that the convergence rate of the risk in Eq. (14.95), including both the small and large terms, is of the order shown in Eq. (14.98). This proves formula (6.31) of Proposition 10.

The convergence rate in Eq. (6.32) for the derivative estimates can be obtained in a similar way.

The results formulated in the proposition and the technique used are similar to those presented in the cited [66] and [163], where the mathematical details can be seen.

In the paper [66] (Theorem 1, Corollary 1, Theorem 3), the univariable function and derivative adaptive estimation are considered. The analysis for a multivariable function estimation can be seen in [163] (Theorem 3.3.1).

14.4 Cross Validation

Cross validation (CV) is a technique that is used for parameter selection in parametric and nonparametric estimations (e.g., [214], [228]). The basic idea is to calculate an estimate based on part of the data and then choose the parameters of the estimate that give the best fit to the rest of the data.

Here we consider the so-called "leave-one-out" method. The estimate for $x = X_t$ is calculated without using the one pair of observations (z_t, X_t). The prediction error is defined as the difference between this estimate and the dropped observation z_t. The calculations are repeated in permutation manner for a set of left-out points and the quality-of-fit criterion is calculated based on the corresponding prediction errors. The estimate parameter is selected by minimizing this quality-of-fit criterion.

We show that this permutational procedure can be implemented with relatively minimal efforts and is applicable to various estimators. It is presented here for nonparametric estimates. This CV method can be used for selection of the scale h and the order m of LPA estimates.

14.4.1 Basic approach

Let us have n observations z_s, $s = 1, \ldots, n$, and let the estimates $\hat{y}_h(x)$ be defined by the general LPA formulas (2.55) and (2.56):

$$\hat{y}_h(x) = \sum_{s=1}^{n} g_h(x, X_s) z_s, \tag{14.101}$$

$$g_h(x, X_s) = w_h(x - X_s)\phi_h^T(x - X_s)\Phi_h^{-1}\phi(0),$$
$$\Phi_h = \sum_{s=1}^{n} w_h(x - X_s)\phi_h(x - X_s)\phi_h^T(x - X_s). \tag{14.102}$$

Introduce the estimates $\hat{y}_h^{[t]}(x)$ obtained provided that the observation at the point $x = X_t$ is omitted in the estimate. This means that

$$\hat{y}_h^{[t]}(x) = \sum_{s=1, s\neq t}^{n} g_h^{[t]}(x, X_s) z_s, \tag{14.103}$$

$$g_h^{[t]}(x, X_s) = w_h(x - X_s)\phi_h^T(x - X_s)\widetilde{\Phi}_h^{-1}\phi(0), \tag{14.104}$$

$$\widetilde{\Phi}_h = \sum_{s=1, s\neq t}^{n} w_h(x - X_s)\phi_h(x - X_s)\phi_h^T(x - X_s). \tag{14.105}$$

The quality-of-fit is evaluated according to the following CV quadratic criterion:

$$R_{CV}^2 = \frac{1}{n}\sum_{t=1}^{n}[r^{[t]}(X_t)]^2, \tag{14.106}$$

$$r_t^{[t]} = z_t - \hat{y}_h^{[t]}(X_t). \tag{14.107}$$

Here the residual of estimation $r_t^{[t]}$ is defined as the difference between the observation z_t and the estimate $\hat{y}_h^{[t]}(X_t)$ at the point $x = X_t$ omitted in the estimator in Eq. (14.103). This estimate $\hat{y}_h^{[t]}(X_t)$ works as an interpolator of the other observations for this omitted $x = X_t$. The residual can be composed as a sum of two different items:

$$\begin{aligned} r_t^{[t]} &= \tilde{e}^{[t]}(X_t) + \varepsilon_t, \\ \tilde{e}^{[t]}(X_t) &= y(X_t) - \hat{y}_h^{[t]}(X_t) \end{aligned} \tag{14.108}$$

where $\tilde{e}^{[t]}(X_t)$ is an interpolation error and the random noise is ε_t.

Note that the noise ε_t is independent with respect to the interpolation error $\tilde{e}^{[t]}(X_t)$, because the observation z_t contaminated by ε_t is dropped in the estimate $\hat{y}_h^{[t]}(X_t)$.

Using Eq. (14.108), we obtain for the expectation of Eq. (14.106)

$$E\{R_{CV}^2\} = E\left\{\frac{1}{n}\sum_{t=1}^{n}[\tilde{e}^{[t]}(X_t)]^2\right\} + \sigma^2, \tag{14.109}$$

where σ^2 is the variance of the random noise. Within the additive invariant term σ^2, the expectation of R_{CV}^2 is equal to the expectation of the criterion:

$$\widetilde{R}_{CV}^2 = \frac{1}{n}\sum_{t=1}^{n}[\tilde{e}^{[t]}(X_t)]^2. \tag{14.110}$$

The latter criterion gives the accuracy of the estimate because the error $\tilde{e}^{[t]}(X_t)$ is the difference between the true value of the signal and its estimate. This proves that for large n, when $\widetilde{R}_{CV}^2 \to E\{R_{CV}^2\}$, we can minimize R_{CV}^2 on h to find the optimal scale value minimizing the accuracy criterion $E\{\widetilde{R}_{CV}^2\}$.

Note that the standard accuracy criterion is defined as the squared sum of the estimation errors,

$$R^2 = \frac{1}{n}\sum_{t=1}^{n} e^2(X_t), \; e(X_t) = y(X_t) - \hat{y}_h(X_t). \tag{14.111}$$

The criteria in Eqs. (14.111) and (14.106) are different because, in the latter, the interpolation errors and interpolation properties of the "leave-one-out" estimator are evaluated.

14.4.2 Algorithm

A straightforward computational implementation of the CV according to formulas (14.103)–(14.106) is a demanding permutational procedure. Fortunately, there is an algorithm that reduces the complexity by using the basic estimates $\hat{y}_h(X_t)$ in Eq. (14.101) calculated only once for the complete set of observations. The following proposition gives the basic result.

Proposition 18 *Let the estimate* $\hat{y}_h(x)$ *be defined by Eqs. (14.101) and (14.102). Then the* CV *residual in Eq. (14.107) can be calculated as*

$$r_t^{[t]} = \frac{r_t}{1 - g_h(X_t, X_t)}, \qquad r_t = z_t - \hat{y}_h(X_t). \tag{14.112}$$

Proof. The estimates $\hat{y}_h(X_t)$ and $\hat{y}_h^{[t]}(X_t)$ are defined by

$$\hat{y}_h(X_t) = y(X_t, X_t) = \phi^T(0)\widehat{C},$$
$$\widehat{C} = \arg\min_C \sum_{s=1}^{n} w_h(X_t - X_s)[z_s - C^T \phi_h(X_t - X_s)]^2, \tag{14.113}$$

and

$$\hat{y}_h^{[t]}(X_t) = y^{[t]}(X_t, X_t) = \phi^T(0)\widehat{C}^{[t]},$$
$$\widehat{C}^{[t]} = \arg\min_C \sum_{s=1, s\neq t}^{n} w_h(X_t - X_s)[z_s - C^T \phi_h(X_t - X_s)]^2. \tag{14.114}$$

It is clear that the estimate $\widehat{C}$ of the vector C in Eq. (14.113) becomes identical to the estimate $\widehat{C}^{[t]}$ in Eq. (14.114) if the observation z_t satisfies the condition

$$z_t = C^T \phi(0). \tag{14.115}$$

In this case, the extra term in Eq. (14.113) corresponding to $s = t$ disappears.

For $C = \widehat{C}$ found from Eq. (14.113),

$$\phi^T(0)\widehat{C} = \hat{y}_h(X_t) \quad \text{and} \quad \hat{y}_h(X_t) = \sum_{s=1}^{n} g_h(X_t, X_s) z_s. \tag{14.116}$$

Then, Eq. (14.115) follows from

$$z_t = \sum_{s=1}^{n} g_h(X_t, X_s) z_s,$$

and it gives z_t as a function of other observations:

$$z_t = \frac{1}{1 - g_h(X_t, X_t)} \sum_{s=1, s\neq t}^{n} g_h(X_t, X_s) z_s. \tag{14.117}$$

Such z_t always exists provided that

$$g_h(X_t, X_t) \neq 1. \tag{14.118}$$

According to Eqs. (14.115) and (14.116),

$$\hat{y}_h(X_t) = \hat{y}_h^{[t]}(X_t) = z_t. \tag{14.119}$$

Using Eq. (14.117) we obtain

$$\hat{y}_h^{[t]}(X_t) = \frac{1}{1 - g_h(X_t, X_t)} \sum_{s=1, s\neq t}^{n} g_h(X_t, X_s) z_s.$$

This proves the following link between the kernel functions of the estimate with the complete and leave-one-out sets of observations:

$$g_h^{[t]}(X_t, X_s) = \frac{g_h(X_t, X_s)}{1 - g_h(X_t, X_t)}, \qquad s = 1, \ldots, n, s \neq t.$$

Now let us calculate the residuals $r^{[t]}(X_t)$

$$r_t^{[t]} = z_t - \hat{y}_h^{[t]}(X_t) = \frac{z_t[1 - g_h(X_t, X_t)] - \sum_{s=1,s\neq t}^{n} g_h(X_t, X_s)z_s}{1 - g_h(X_t, X_t)}$$

$$= \frac{z_t - \sum_{s=1}^{n} g_h(X_t, X_s)z_s}{1 - g_h(X_t, X_t)} = \frac{r_t^{[t]}}{1 - g_h(X_t, X_t)}.$$

This proves formula (14.112). □

Formula (14.112) shows that the weight $1/[1 - g_h(X_t, X_t)]$ enables one to transform the ordinary residual r_t, which can always be calculated easily, into the residual $r_t^{[t]}$ of the estimate with the omitted observation. Substituting Eq. (14.112) into formula (14.106) gives the CV criterion as

$$R_{CV}^2 = \frac{1}{n}\sum_{t=1}^{n} \frac{r_t^2}{[1 - g_h(X_t, X_t)]^2}. \tag{14.120}$$

This criterion can be modified to a more general form:

$$R_{CV} = \left\{\frac{1}{\#}\sum_{t\in I_{cont}} \frac{r_t^q}{[1 - g_h(X_t, X_t)]^q}\right\}^{1/q}, \qquad q > 0, \tag{14.121}$$

where the parameter q in Eq. (14.121) specifies the norm l_q used for the accuracy evaluation and # is the cardinality of I_{cont}. Here I_{cont} is a control set of observations. If $I_{cont} = \{s : s = 1, \ldots, n\}$, all of the observations are used in the CV criterion.

There are a number of ideas on how the control set should be formed. One of the usual ways is to form I_{cont} by excluding boundary points.

Let us comment on it for the 1D case. Assume that the observations are ordered as $x_1 < x_2 < \cdots < x_n$ and the symmetric window $w(x)$ is used for the LPA estimation. This means that, for x taking values in the middle of the interval $[x_1, x_n]$, the estimate $\hat{y}_h(x) = \sum_{s=1}^{n} g_h(x, X_s)z_s$ uses observations from both sides of x.

Another situation has a place for the boundary point. Let, say, $x = x_1$. Then the estimator $\hat{y}_h(x_1) = \sum_{s=1}^{n} g_h(x_1, X_s)z_s$ uses the observations with $X_s \geq x_1$ from one side of $x = x_1$. The existence of observations from both sides of x is important for interpolations tested by the criterion in Eq. (14.110). The ability to interpolate properly is much harder for the one-sided observation scenario than it is for the two-sided one. If we wish to apply our estimator for x taking values somewhere in the middle of the interval $[x_1, x_n]$, we need to imitate this scenario in our control set and exclude the boundary points from I_{cont}.

When the criterion function is defined, the estimate of the adaptive invariant scale is calculated as

$$\hat{h} = \arg\min_{h} R_{CV}^2.$$

The so-called generalized cross validation (GCV) is different from the one in Eq. (14.120) by the weight of the residuals that is used. It is of the form

$$R_{GCV}^2 = \frac{\sum_{t=1}^{n} r_t^2}{\sum_{t=1}^{n} [1 - g_h(X_t, X_t)]^2}, \qquad r_t = z_t - \hat{y}_h(X_t), \tag{14.122}$$

where the weights $[1 - g_h(X_t, X_t)]^2$ of the residuals used in Eq. (14.120) are replaced by their mean.

The CV methods can be applied for any multidimensional data with the above formulas given in the form that is universally valid for any d.

The version of the CV method that is considered here is the "leave-one-out" rule. This means that the available data are separated into two parts. One part is used for estimation and the other is used for the accuracy control. The accuracy control set is formed in a sliding mode when one observation is dropped from the estimate and is used for the control. However, this observation is used in the estimate when the control observation is changed.

14.4.3 Shift-invariant kernels

For the estimates with the shift-invariant kernels in the form of Eqs. (3.10) and (3.12) defined on the regular d-dimensional grid,

$$\begin{aligned} \hat{y}_h[t] &= (g_h \circledast z)[t], \\ g_h[l] &= w_h[l]\phi_h^T[l]\Phi_h^{-1}\phi(0), \quad \Phi_h = \sum_{l \in \mathbb{Z}^d} w_h[l]\phi_h[l]\phi_h^T[l], \end{aligned} \tag{14.123}$$

we have from Eq. (14.112) that

$$r_t^{[t]} = \frac{r_t}{1 - g_h[0]}, \qquad r_t = z[t] - \hat{y}_h[t], \quad r_t^{[t]} = z[t] - \hat{y}_h^{[t]}[t],$$

where $\hat{y}_h^{[t]}[t]$ is the estimate obtained in point t provided that the observation t is omitted from the set of the observations.

For the shift-invariant kernel, the weights $1/(1 - g_h[0])$ in the CV criterion are the same for all residuals, and the CV and GCV have the identical form:

$$R_{CV}^2 = R_{GCV}^2 = \frac{\sum_{t \in I_{cont}} (r_t^{[t]})^2}{\#(1 - g_h[0])^2}. \tag{14.124}$$

14.5 Directional LPA Accuracy

14.5.1 Proof of Proposition 12

In this section we present the proof of Proposition 12.

In the analysis, we follow the techniques in Secs. 14.2.1 and 14.2.2, with the only difference being that the multi-index power and the corresponding multi-index scale h are used here.

(1) Let us find the main term of the Taylor series defining the estimation error for the multi-index directional estimates. The systematic error of estimation is defined as

$$m_{\hat{y}_h}(x,h) = y(x) - E\{\hat{y}_h(x)\} = y(x) - \sum_l g_h(\Delta l) y(x-\Delta l). \qquad (14.125)$$

The Taylor series in Eqs. (8.22) and (8.23) for $y(x-\Delta l)$ with $Q = m$ has the form

$$y(x-\Delta l) = \sum_{t=0}^{m} (-1)^{|t|} \frac{(\Delta l)^t}{t!} \partial_x^t y(x) + R_m(\Delta l), \qquad (14.126)$$

with the residual term

$$R_m(\Delta l) = \sum_{t=0}^{|m|} (-1)^t \sum_{|k|=t} \frac{1}{k!} (\Delta l)^k \partial_x^k y(x) - \sum_{t=0}^{m} (-1)^{|t|} \frac{(\Delta l)^t}{t!} \partial_x^t y(x) + R_{|m|+1}(\Delta l), \qquad (14.127)$$

where

$$R_{|m|+1}(\Delta l) = (-1)^{|m|+1} \sum_{|k|=|m|+1} \frac{1}{k!} (\Delta l)^k \partial_x^k y(x-\lambda \Delta l). \qquad (14.128)$$

Inserting the series in Eqs. (14.126) into (14.125) and using the vanishing moment conditions in Eq. (8.19), we obtain for the signal estimation

$$\sum_l g_h(\Delta l) \sum_{t=0}^{m} (-1)^{|t|} \frac{(\Delta l)^t}{t!} \partial_x^t y(x) = y(x)$$

and then

$$\begin{aligned} m_{\hat{y}_h}(x,h) &= y(x) - \sum_l g_h(\Delta l) y(x-\Delta l) \\ &= y(x) - (y(x) + \sum_l g_h(\Delta l) R_m(\Delta l)) \\ &= -\sum_l g_h(\Delta l) R_m(\Delta l). \end{aligned} \qquad (14.129)$$

Remember that here, in multi-index notation, $\sum_{t=0}^{m} = \sum_{t_1=0}^{m_1} \cdots \sum_{t_d=0}^{m_d}$.

Then the Lagrange residual term $R_m(\Delta l)$ in Eq. (14.127) includes all of the items of Eq. (8.21) up to order $|m|$, excluding those given in the main term of Eq. (14.126), and includes the residual term $R_{|m|+1}(\Delta l)$ of order $|m|+1$:

$$\begin{aligned} - m_{\hat{y}_h}(x,h) &= \sum_l g_h(\Delta l)\left[\sum_{t=0}^{|m|}(-1)^t\sum_{|k|=t}\frac{1}{k!}(\Delta l)^k\partial_x^k y(x) - \sum_{t=0}^{m}(-1)^{|t|}\frac{(\Delta l)^t}{t!}\partial_x^t y(x)\right] \\ &+ (-1)^{|m|+1}\sum_l g_h(\Delta l)\sum_{|k|=|m|+1}\frac{1}{k!}(\Delta l)^k\partial_x^k y(x-\lambda\Delta l). \end{aligned} \tag{14.130}$$

The last summand is estimated for small Δl using the Lipshitz conditions in Eq. (8.32) as

$$\partial_x^k y(x-\lambda\Delta l) = \partial_x^k y(x) + O(||\Delta l||^\gamma), \qquad k = |m|+1. \tag{14.131}$$

Then Eq. (14.130) has the form

$$\begin{aligned} -m_{\hat{y}_h}(x,h) = \sum_l g_h(\Delta l)\Bigg[\sum_{t=0}^{|m|+1}(-1)^t\sum_{|k|=t}\frac{1}{k!}(\Delta l)^k\partial_x^k y(x) \\ - \sum_{t=0}^{m}(-1)^{|t|}\frac{(\Delta l)^t}{t!}\partial_x^t y(x)\Bigg] + (-1)^{|m|+1}\sum_l g_h(\Delta l) \\ \times \sum_{|k|=|m|+1}\frac{1}{k!}(\Delta l)^k O(||\Delta l||^\gamma). \end{aligned} \tag{14.132}$$

In these formulas, $(\Delta l)^k = \Pi_{i=1}^d(\Delta_i l_i)^{k_i}$.

It is not difficult to realize that, for each ith variable $\Delta_i l_i$ in Eq. (14.132), the lowest degree is equal to m_i+1 because all degrees up to m_i are withdrawn from $\sum_{t=0}^{|m|+1}(-1)^t\sum_{|k|=t}\frac{1}{k!}(\Delta l)^k\partial_x^k y(x)$ by the main term of the Taylor series $\sum_{t=0}^{m}(-1)^{|t|}[(\Delta l)^t/t!]\partial_x^t y(x)$. It follows that the lowest degrees (main terms) of the Taylor series in Eq. (14.126) appearing in the bias in Eq. (14.125) are $(\Delta_i l_i)^{m_i+1}$ for each of the variables $i = 1,\ldots,d$.

(2) Derive the limit integral formulas for the discrete kernels $g_h(x)$ and $g_h^{(r)}(x)$. First note that, as $\Delta \to 0$, we have for Φ_h in Eq. (8.28):

$$\begin{aligned} \Phi_h &= \sum_k w_h(k\Delta)\phi_h(k\Delta)\phi_h^T(k\Delta) \\ &= \frac{1}{\Pi_{i=1}^d\Delta_i}\sum_k w\left(\frac{k\Delta}{h}\right)\phi\left(\frac{k\Delta}{h}\right)\phi^T\left(\frac{k\Delta}{h}\right)\cdot\Pi_{i=1}^d\frac{\Delta_i}{h_i} \\ &\to \frac{1}{\prod_{i=1}^d\Delta_i}\Phi, \end{aligned} \tag{14.133}$$

$$\Phi = \int_{R^d} w(x)\phi(x)\phi^T(x)dx.$$

Using this formula for Φ_h we obtain

$$
\begin{aligned}
g_h(x) &= w_h(x)\phi^T(0)\Phi_h^{-1}\phi_h(x) \\
&= \frac{1}{\prod_{i=1}^d h_i} w(x/h)\phi^T(0)\Phi_h^{-1}\phi\left(\frac{x}{h}\right) \\
&\to \frac{\prod_{i=1}^d \Delta_i}{\prod_{i=1}^d h_i} w(x/h)\phi^T(0)\Phi^{-1}\phi\left(\frac{x}{h}\right) = \frac{\prod_{i=1}^d \Delta_i}{\prod_{i=1}^d h_i} g\left(\frac{x}{h}\right), \qquad (14.134) \\
g(x) &= w(x)\phi^T(0)\Phi^{-1}\phi(x).
\end{aligned}
$$

For the differentiating kernel in Eq. (8.29) we obtain in a similar way using Eq. (14.133)

$$
\begin{aligned}
g_h^{(r)}(x) &= \frac{(-1)^{|r|}}{h^r} w_h(x)\phi_h^T(x)\Phi_h^{-1}\phi^{(r)}(0) \\
&= \frac{(-1)^{|r|}}{h^r}\frac{1}{\prod_{i=1}^d h_i} w(x/h)\phi^T\left(\frac{x}{h}\right)\Phi_h^{-1}\phi^{(r)}(0) \\
&\to \frac{(-1)^{|r|}}{h^r}\frac{\prod_{i=1}^d \Delta_i}{\prod_{i=1}^d h_i} w(x/h)\phi^T\left(\frac{x}{h}\right)\Phi^{-1}\phi^{(r)}(0) \\
&= \frac{1}{h^r}\frac{\prod_{i=1}^d \Delta_i}{\prod_{i=1}^d h_i} g^{(r)}\left(\frac{x}{h}\right), \qquad (14.135) \\
g^{(r)}(x) &= (-1)^{|r|} w(x)\phi^T(x)\Phi^{-1}\phi^{(r)}(0).
\end{aligned}
$$

(3) Now we are ready to calculate the asymptotic formula for the bias error in Eq. (14.125). Let us insert the smoothing kernel in Eqs. (14.134) into (14.132) and calculate the limit values for the corresponding sums:

$$
\begin{aligned}
&- m_{\hat{y}_h}(x,h) \\
&= \frac{\prod_{i=1}^d \Delta_i}{\prod_{i=1}^d h_i}\sum_l g\left(\frac{\Delta l}{h}\right) \\
&\times \left[\sum_{t=0}^{|m|+1}(-1)^t \sum_{|k|=t}\frac{h^k}{k!}\left(\frac{\Delta l}{h}\right)^k \partial_x^k y(x) - \sum_{t=0}^{m}(-1)^{|t|}\frac{h^t\left(\frac{\Delta l}{h}\right)^t}{t!}\partial_x^t y(x)\right] \\
&+ (-1)^{|m|+1}\frac{\prod_{i=1}^d \Delta_i}{\prod_{i=1}^d h_i}\sum_l g\left(\frac{\Delta l}{h}\right)\sum_{|k|=|m|+1}\frac{h^k}{k!}\left(\frac{\Delta l}{h}\right)^k O(||\Delta l||^{\gamma}). \qquad (14.136)
\end{aligned}
$$

Here we exploit the limit passage

$$
\frac{\prod_{i=1}^d \Delta_i}{\prod_{i=1}^d h_i}\sum_l g(\Delta l/h)\frac{1}{k!}(\Delta l/h)^k \to \frac{1}{k!}\int g(u)u^k du, \qquad (14.137)
$$

where $\Delta l/h$ serves as a discrete analog of the integration variable u and $\prod_{i=1}^d \Delta_i/\prod_{i=1}^d h_i$ is a discrete analog of $du = \prod_{i=1}^d du_i$.

For the last term in Eq. (14.136) we have

$$\frac{\Pi_{i=1}^{d}\Delta_i}{\Pi_{i=1}^{d}h_i}\sum_{l} g\left(\frac{\Delta l}{h}\right)\sum_{|k|=|m|+1}\frac{h^k}{k!}\left(\frac{\Delta l}{h}\right)^k O(||\Delta l||^{\gamma})$$

$$= \sum_{|k|=|m|+1}\frac{h^k}{k!}\sum_{l} g\left(\frac{\Delta l}{h}\right)\left(\frac{\Delta l}{h}\right)^k O\left(\left\|h\left(\frac{\Delta l}{h}\right)\right\|^{\gamma}\right)\frac{\Pi_{i=1}^{d}\Delta_i}{\Pi_{i=1}^{d}h_i}$$

$$\to \sum_{|k|=|m|+1}\frac{h^k}{k!}\int g(u)u^k O(||hu||^{\gamma})du = \sum_{|k|=|m|+1} o(h^k) = o(h^{|m|+1}). \quad (14.138)$$

Using Eqs. (14 137) and (14.138), we find for Eq. (14.136) that

$$-m_{\hat{y}_h}(x,h) = \sum_{t=0}^{|m|+1}(-1)^t\sum_{|k|=t} h^k A_g^{(k)}\partial_x^k y(x)$$
$$-\sum_{t=0}^{m}(-1)^{|t|}h^t A_g^{(t)}\partial_x^t y(x) + o(h^{|m|+1}) \quad (14.139)$$

where

$$A_g^{(k)} = \frac{1}{k!}\int_{R^d} g(u)u^k du.$$

In these formulas, $h^k = \Pi_{i=1}^{d} h_i^{k_i}$.

The degrees of polynomial items in Eq. (14.132) for $m_{\hat{y}_h}(x,h)$ correspond to the degrees of h_i in Eq. (14.139). Thus, for each ith scale h_i in Eq. (14.139), the lowest power is equal to $m_i + 1$ because all powers up to m_i are withdrawn by the main term of the Taylor series $\sum_{t=0}^{m}(-1)^{|t|}\partial_x^t y(x)h^t A_g^{(t)}$ and the next power $m_i + 1$ exists in the term $\sum_{t=0}^{|m|+1}(-1)^t\sum_{|k|=t}\partial_x^k y(x)h^k A_g^{(k)}$.

It follows that the main term of the estimation bias can be given in the form

$$-m_{\hat{y}_h}(x,h) = \sum_{i=1}^{d}(-1)^{m_i+1}h_i^{m_i+1}y^{(m_i+1)}(x)A_{g,i} + o\left(\sum_{i=1}^{d}h_i^{m_i+1}\right),$$
$$A_{g,i} = \frac{1}{(m_i+1)!}\int_{R^d} g(u)u_i^{m_i+1}du. \quad (14.140)$$

This proves Eq. (8.33) for $M_i = m_i + 1$, $i = 1,\ldots,d$. For the more general case when $M_i = \min(m_i + 1, \alpha_i)$, the result can be verified by following the technique used in the proof of Lemma 2.

(4) To calculate the bias for the differentiating estimator, we substitute the Taylor series in Eq. (14.126) into

$$m_{\hat{y}_h^{(r)}}(x,h) = y^{(r)}(x) - E\{\hat{y}_h^{(r)}(x)\} = y^{(r)}(x) - \sum_{l} g_h^{(r)}(\Delta l)y(x-\Delta l).$$
$$(14.141)$$

Using the vanishing moment conditions in Eq. (8.20), we obtain for the derivative estimation

$$\sum_l g_h^{(r)}(\Delta l)\sum_{t=0}^{m}(-1)^{|t|}\frac{(\Delta l)^t}{t!}\partial_x^t y(x) = y^{(r)}(x)$$

and then

$$\begin{aligned} m_{\hat{y}_h^{(r)}}(x,h) &= y^{(r)}(x) - \sum_l g_h^{(r)}(\Delta l) y(x-\Delta l) \\ &= y^{(r)}(x) - \left[y^{(r)}(x) + \sum_l g_h^{(r)}(\Delta l) R_m(\Delta l)\right] \\ &= -\sum_l g_h^{(r)}(\Delta l) R_m(\Delta l). \end{aligned}$$

Inserting the differentiating kernel in Eq. (14.135) into the last expression and using a transformation similar to the one leading to Eq. (14.132), we find that

$$\begin{aligned} &- m_{\hat{y}_h^{(r)}}(x,h) \\ &= \frac{1}{h^r}\frac{\prod_{i=1}^d \Delta_i}{\prod_{i=1}^d h_i}\sum_l g^{(r)}\left(\frac{\Delta l}{h}\right) \\ &\times \left[\sum_{t=0}^{|m|+1}(-1)^t\sum_{|k|=t}\frac{h^k}{k!}\left(\frac{\Delta l}{h}\right)^k \partial_x^k y(x) - \sum_{t=0}^{m}(-1)^{|t|}\frac{h^t\left(\frac{\Delta l}{h}\right)^t}{t!}\partial_x^t y(x)\right] \\ &+ (-1)^{|m|+1}\frac{1}{h^r}\frac{\prod_{i=1}^d \Delta_i}{\prod_{i=1}^d h_i}\sum_l g^{(r)}\left(\frac{\Delta l}{h}\right)\sum_{|k|=|m|+1}\frac{h^k}{k!}\left(\frac{\Delta l}{h}\right)^k o\left(\left\|\frac{h\Delta l}{h}\right\|^{\gamma}\right). \end{aligned} \tag{14.142}$$

Calculating the limits of the sums in Eq. (14.142) we arrive at

$$\begin{aligned} -m_{\hat{y}_h^{(r)}}(x,h) = &\frac{1}{h^r}\sum_{t=0}^{|m|+1}(-1)^t\sum_{|k|=t}h^k A_{g^{(r)}}^{(k)}\partial_x^k y(x) \\ &- \frac{1}{h^r}\sum_{t=0}^{m}(-1)^{|t|}h^t A_{g^{(r)}}^{(t)}\partial_x^t y(x) + \sum_{|k|=|m|+1}h^{k-r}\cdot O(||h||^{\gamma}), \end{aligned} \tag{14.143}$$

where

$$A_{g^{(r)}}^{(k)} = \frac{1}{k!}\int_{R^d} g^{(r)}(u)u^k du.$$

After that, the speculations concerning the analysis of the degrees in Eq. (14.143) with respect to h_i give (similar to Eq. (14.140))

$$-m_{\hat{y}_h^{(r)}}(x,h) = \sum_{i=1}^{d}(-1)^{m_i+1}h^{m_i+1-r_i}y^{(m_i+1)}(x)A_{g^{(r)},i} + o\left(\sum_{i=1}^{d} h_i^{m_i+1-r_i}\right),$$
$$A_{g^{(r)},i} = \frac{1}{(m_i+1)!}\int_{R^d} g^{(r)}(u)u_i^{m_i+1}du \tag{14.144}$$

This proves Eq. (8.34) for $M_i = m_i + 1$. For the general case when $M_i = \min(m_i + 1, \alpha_i)$, as was mentioned for signal estimation, the result can be verified by following the technique used in the proof of Lemma 2.

(5) For the variance of the estimation errors we have

$$\frac{1}{\sigma^2}\sigma^2_{\hat{y}_h}(x,h) = \sum_l [g_h(\Delta l)]^2, \tag{14.145}$$

$$\frac{1}{\sigma^2}\sigma^2_{\hat{y}_h^{(r)}}(x,h) = \sum_l [g_h^{(r)}(\Delta l)]^2. \tag{14.146}$$

The integral formulas are obtained by the usual limit passages.

Using Eq. (14.134) we have

$$\sum_l (g_h(\Delta l))^2 = \frac{\Pi_{i=1}^d \Delta_i}{\Pi_{i=1}^d h_i}\sum_l g^2\left(\frac{\Delta l}{h}\right)\frac{\Pi_{i=1}^d \Delta_i}{\Pi_{i=1}^d h_i}$$
$$\rightarrow \frac{\Pi_{i=1}^d \Delta_i}{\Pi_{i=1}^d h_i}\int_{R^d} g^2(u)du + o\left(\frac{\Pi_{i=1}^d \Delta_i}{\Pi_{i=1}^d h_i}\right).$$

This proves Eq. (8.35).

In a similar way we have from Eq. (14.135) that

$$\sum_l [g_h^{(r)}(\Delta l)]^2 = \frac{1}{h^{2r}}\frac{\Pi_{i=1}^d \Delta_i}{\Pi_{i=1}^d h_i}\sum_l \left[g^{(r)}\left(\frac{\Delta l}{h}\right)\right]^2\frac{\Pi_{i=1}^d \Delta_i}{\Pi_{i=1}^d h_i}$$
$$\rightarrow \frac{1}{h^{2r}}\frac{\Pi_{i=1}^d \Delta_i}{\Pi_{i=1}^d h_i}\int_{R^d}[g^{(r)}(u)]^2du + o\left(\frac{1}{h^{2r}}\frac{\Pi_{i=1}^d \Delta_i}{\Pi_{i=1}^d h_i}\right)$$
$$= \frac{\Pi_{i=1}^d \Delta_i}{\Pi_{i=1}^d h_i^{1+2r_i}}\int_{R^d}[g^{(r)}(u)]^2du + o\left(\frac{\Pi_{i=1}^d \Delta_i}{\Pi_{i=1}^d h_i^{1+2r_i}}\right).$$

This proves Eq. (8.36) and completes the proof of Proposition 12.

14.5.2 Convergence rate for signal estimation

Proposition 19 *Under the assumptions of Proposition 12 and assuming that $M_i = m_i + 1$, $i = 1, \ldots, d$, we obtain the following:*

(a) The ideal scales h^* are defined by

$$h_1^* = \left[\frac{\sigma^2 \Pi_{i=1}^d \Delta_i}{P_1(x)} B_g\right]^{1/\alpha}, \tag{14.147}$$

$$h_k^* = (h_1^*)^{(m_1+1)/(m_k+1)} G_k(x), \tag{14.148}$$

$$\alpha = (m_1 + 1)\left[2 + \sum_{k=1}^{d} \frac{1}{(m_k + 1)}\right], \tag{14.149}$$

where

$$G_k(x) = \left[\frac{L_{m_1+1}(x)(m_1 + 1)A_{g,1}}{L_{m_k+1}(x)(m_k + 1)A_{g,k}}\right]^{1/(m_k+1)},$$

$$P_1(x) = 2\alpha_1[L_{m_1+1}(x)A_{g,1}]^2(m_1 + 1)\Pi_{i=1}^d G_i(x),$$

$$\alpha_1 = (m_1 + 1)\sum_{k=1}^{d} \frac{1}{(m_k + 1)}.$$

(b) The convergence rate for the ideal scales and $\Pi_{i=1}^d \Delta_i \sim 1/n$, where n is a large number of observations, is

$$\bar{l}_{\hat{y}_h}(x, h^*) = O\left(\left(\frac{\sigma^2}{n}\right)^{\beta_1}\right), \tag{14.150}$$

$$\beta_1 = \frac{1}{1 + (1/2)\sum_{k=1}^d [1/(m_k + 1)]}. \tag{14.151}$$

Proof. The derivative $\partial_{h_k} \bar{l}_{\hat{y}_h}(x, h)$ for $\bar{l}_{\hat{y}_h}(x, h)$ defined in Eq. (8.37) gives

$$\partial_{h_k} \bar{l}_{\hat{y}_h}(x, h) = 0$$

for the ideal scales in the form

$$\partial_{h_k} \bar{l}_{\hat{y}_h}(x, h) = 2(m_k + 1)h_k^{m_k} L_{m_k+1}(x)A_{g,k}\left[\sum_{i=1}^{d} h_i^{m_i+1} L_{m_i+1}(x)A_{g,i}\right] - \sigma^2 \frac{\Delta^d}{h_k \Pi_{i=1}^d h_i} B_g = 0, \qquad k = 1, \ldots, d. \tag{14.152}$$

Here and in what follows we omit the superscript * for the ideal h to make the notation simpler.

These equations can be rewritten as

$$2(m_k+1)h_k^{m_k+1}L_{m_k+1}(x)A_{g,k}\left[\sum_{i=1}^{d}h_i^{m_i+1}L_{m_i+1}(x)A_{g,i}\right]\Pi_{i=1}^{d}h_i \tag{14.153}$$
$$=\sigma^2\Delta^d B_g,\quad k=1,\ldots,d.$$

The right-hand sides of the equations are the same for all k. It then follows that

$$(m_k+1)h_k^{m_k+1}L_{m_k+1}(x)A_{g,k}=(m_l+1)h_l^{m_l+1}L_{m_l+1}(x)A_{g,l},\qquad k,\ l=1,\ldots,d.$$

These equations define the ratios of the ideal scale parameters h_k and h_l. They allow one to express all of the scales through h_1:

$$h_k=\left[\frac{(m_1+1)h_1^{m_1+1}L_{m_1+1}(x)A_{g,1}}{(m_k+1)L_{m_k+1}(x)A_{g,k}}\right]^{1/(m_k+1)}. \tag{14.154}$$

This proves formula (14.148).

Using Eq. (14.154), we can check by direct calculations that

$$\Pi_{i=1}^{d}h_i=h_1^{\alpha_1}\Pi_{i=1}^{d}\left[\frac{(m_1+1)L_{m_1+1}(x)A_{g,1}}{(m_i+1)L_{m_i+1}(x)A_{g,i}}\right]^{1/(m_i+1)}, \tag{14.155}$$

where

$$\alpha_1=\sum_{k=1}^{d}\frac{m_1+1}{m_k+1},$$

and

$$\sum_{i=1}^{d}h_i^{m_i+1}L_{m_i+1}(x)A_{g,i}=h_1^{m_1+1}\alpha_1 L_{m_1+1}(x)A_{g,1}. \tag{14.156}$$

Let us consider Eq. (14.153) for $k=1$ and substitute h_i defined according to Eq. (14.154). Then we find the formula for the ideal scale h_1 as

$$h_1^{m_1+1}=\frac{\sigma^2\Delta^d B_g}{2\left[\sum_{i=1}^{d}h_i^{m_i+1}L_{m_i+1}(x)A_{g,i}\right](m_1+1)L_{m_1+1}(x)A_{g,1}\Pi_{i=1}^{d}h_i}$$
$$=\frac{\sigma^2\Delta^d B_g}{2h_1^{m_1+1}\alpha_1(L_{m_1+1}(x)A_{g,1})^2(m_1+1)h_1^{\alpha_1}\Pi_{i=1}^{d}\left[\frac{(m_1+1)L_{m_1+1}(x)A_{g,1}}{(m_i+1)L_{m_i+1}(x)A_{g,i}}\right]^{1/(m_i+1)}}.$$

In this transformation, we use Eqs. (14.155) and (14.156).

Then, finally we obtain for h_1

$$h_1^{\alpha} = \frac{\sigma^2 \Delta^d B_g}{2\alpha_1 [L_{m_1+1}(x) A_{g,1}]^2 (m_1+1) \Pi_{i=1}^d \left[\frac{(m_1+1) L_{m_1+1}(x) A_{g,1}}{(m_i+1) L_{m_i+1}(x) A_{g,i}} \right]^{1/(m_i+1)}}, \tag{14.157}$$

$$\alpha = \alpha_1 + 2(m_1+1) = \sum_{k=1}^{d} \frac{m_1+1}{m_k+1} + 2(m_1+1). \tag{14.158}$$

This proves formula (14.147).

The minimal value of the criterion $\bar{l}^*_{\hat{y}_h}(x,h)$ in Eq. (8.37) with ideal h defined according to Eq. (14.154) can be expressed as

$$\bar{l}^*_{\hat{y}_h}(x,h) = \sigma^2 \frac{\Delta^d}{\Pi_{i=1}^d h_i} B_g (1 + r_1), \tag{14.159}$$

$$r_1 = \frac{[h_1^{m_1+1} \alpha_1 L_{m_1+1}(x) A_{g,1}]^2}{\Delta^d \sigma^2 B_g} \Pi_{i=1}^d h_i. \tag{14.160}$$

Using Eqs. (14.155) and (14.156), we have

$$\begin{aligned} r_1 &= \frac{[h_1^{m_1+1} \alpha_1 L_{m_1+1}(x) A_{g,1}]^2}{\Delta^d \sigma^2 B_g} \Pi_{i=1}^d h_i \\ &= \frac{[h_1^{m_1+1} \alpha_1 L_{m_1+1}(x) A_{g,1}]^2}{\Delta^d \sigma^2 B_g} h_1^{\alpha_1} \Pi_{i=1}^d \left[\frac{(m_1+1) L_{m_1+1}(x) A_{g,1}}{(m_i+1) L_{m_i+1}(x) A_{g,i}} \right]^{1/(m_i+1)} \\ &= h_1^{(m_1+1)2+\alpha_1} \frac{[\alpha_1 L_{m_1+1}(x) A_{g,1}]^2}{\Delta^d \sigma^2 B_g} \Pi_{i=1}^d \left[\frac{(m_1+1) L_{m_1+1}(x) A_{g,1}}{(m_i+1) L_{m_i+1}(x) A_{g,i}} \right]^{1/(m_i+1)} \\ &= h_1^{\alpha} \frac{2\alpha_1 [L_{m_1+1}(x) A_{g,1}]^2 (m_1+1)}{\Delta^d \sigma^2 B_g} \Pi_{i=1}^d \left[\frac{(m_1+1) L_{m_1+1}(x) A_{g,1}}{(m_i+1) L_{m_i+1}(x) A_{g,i}} \right]^{1/(m_i+1)} \\ &\quad \times \frac{\alpha_1}{2(m_1+1)}. \end{aligned} \tag{14.161}$$

It follows from Eq. (14.157) that

$$h_1^{\alpha} \frac{2\alpha_1 [L_{m_1+1}(x) A_{g,1}]^2 (m_1+1)}{\Delta^d \sigma^2 B_g} \Pi_{i=1}^d \left[\frac{(m_1+1) L_{m_1+1}(x) A_{g,1}}{(m_i+1) L_{m_i+1}(x) A_{g,i}} \right]^{1/(m_i+1)} = 1.$$

Then we find for r_1 in Eq. (14.161) that

$$r_1 = \frac{\alpha_1}{2(m_1+1)} = \frac{1}{2(m_1+1)} \sum_{i=1}^{d} \frac{m_1+1}{m_i+1} = \frac{1}{2} \sum_{i=1}^{d} \frac{1}{m_i+1}.$$

Thus,

$$\bar{l}^*_{\hat{y}_h}(x,h) = \sigma^2 \frac{\Delta^d}{\Pi_{i=1}^d h_i} B_g(1+\gamma^2), \tag{14.162}$$

$$\gamma = \sqrt{\frac{1}{2} \sum_{k=1}^{d} \frac{1}{m_k+1}}. \tag{14.163}$$

In this way we prove that the ratio of the squared bias to the variance for the ideal h is a constant γ^2 depending only on the orders of the polynomials and the dimension d. Note also that, for $m_1 = \cdots = m_d = m$, formulas (14.163) and (5.26) are identical provided that in the last one $M = m+1$.

It follows from Eq. (14.157) that

$$h_1 = O((\sigma^2 \Delta^d)^{1/\alpha})$$

and from Eq. (14.155) that

$$\Pi_{i=1}^d h_i = O(h_1^{\alpha_1}) = O((\sigma^2 \Delta^d)^{\alpha_1/\alpha}).$$

Then we find from Eq. (14.162) that

$$\bar{l}^*_{\hat{y}_h}(x,h) = O\left(\sigma^2 \frac{\Delta^d}{(\sigma^2 \Delta^d)^{\alpha_1/\alpha}}\right) = O(\sigma^2 \Delta^d)^{1-\alpha_1/\alpha} = O(\sigma^2 \Delta^d)^{1-\alpha_1/\alpha}.$$

Note that

$$1 - \alpha_1/\alpha = \frac{\alpha - \alpha_1}{\alpha} = \frac{2(m_1+1)}{2(m_1+1) + \sum_{k=1}^d \dfrac{m_1+1}{m_k+1}} = \frac{1}{1 + \dfrac{1}{2} \sum_{k=1}^d \dfrac{1}{m_k+1}}.$$

This gives

$$\bar{l}^*_{\hat{y}_h}(x,h) = O(\sigma^2 \Delta^d)^{\beta_1}, \beta_1 = \frac{1}{1 + \dfrac{1}{2} \sum_{k=1}^d \dfrac{1}{m_k+1}}.$$

For $\Delta^d \sim 1/n$, this proves Eq. (14.150) and completes the proof of Proposition 19. □

14.5.3 Convergence rate for derivative estimation

Proposition 20 *Under the assumptions of Proposition 12 and assuming that* $M_i = m_i + 1$, $i = 1, \ldots, d$, *we obtain the following:*

(a) The ideal scales h^* *are defined by*

$$h_1^* = \left[\frac{\Delta^d \sigma^2}{P_1^{(r)}(x)} B_{g(r)}\right]^{1/\alpha_3}, \tag{14.164}$$

$$h_k^* = (h_1^*)^{(m_1+1-r_1)/(m_k+1-r_k)} G_k^{(r)}(x), \tag{14.165}$$

$$\alpha_3 = 2(m_1 + 1 - r_1)\left(1 + \frac{1}{2}\sum_{k=1}^{d} \frac{1 + 2r_k}{m_k + 1 - r_k}\right), \tag{14.166}$$

where

$$G_k^{(r)}(x) = \left[\frac{(1 + 2r_k)(m_1 + 1 - r_1)L_{m_1+1}(x)A_{g(r),1}}{(1 + 2r_1)(m_k + 1 - r_k)L_{m_k+1}(x)A_{g(r),k}}\right]^{1/(m_k+1-r_k)},$$

$$P_1^{(r)}(x) = 2\alpha_2[L_{m_1+1}(x)A_{g(r),1}]^2 \frac{(m_1 + 1 - r_1)}{(1 + 2r_1)^2} \Pi_{i=1}^d [G_i^{(r)}(x)]^{(1+2r_i)}, \tag{14.167}$$

$$\alpha_2 = \sum_{k=1}^{d} \frac{m_1 + 1 - r_1}{m_k + 1 - r_k}(1 + 2r_k).$$

(b) The convergence rate for the ideal scales and $\Pi_{i=1}^d \Delta_i \sim 1/n$, *where* n *is a large number of observations, is*

$$\bar{l}_{\hat{y}_h^{(r)}}(x, h^*) = O\left(\left(\frac{\sigma^2}{n}\right)^{\beta_2}\right), \tag{14.168}$$

$$\beta_2 = \frac{1}{1 + \frac{1}{2}\sum_{k=1}^d (1 + 2r_k)/(m_k + 1 - r_k)}. \tag{14.169}$$

Proof. The proof is similar to the proof given for Proposition 19. The derivative $\partial_{h_k} \bar{l}_{\hat{y}_h^{(r)}}(x, h)$ for $\bar{l}_{\hat{y}_h^{(r)}}(x, h)$ defined in Eq. (8.38) gives

$$\begin{aligned}
&\partial_{h_k} \bar{l}_{\hat{y}_h^{(r)}}(x, h) \\
&= 2\left[\sum_{i=1}^{d} h_i^{m_i+1-r_i} L_{m_i+1}(x)A_{g(r),i}\right](m_k + 1 - r_k)h_k^{m_k-r_k} L_{m_k+1}(x)A_{g(r),k} \\
&\quad - \sigma^2 \frac{(1 + 2r_k)\Delta^d}{h_k \Pi_{i=1}^d h_i^{1+2r_i}} B_{g(r)} = 0, k = 1, \ldots, d.
\end{aligned}$$

Here and in what follows we omit the superscript * for the ideal h to make the notation simpler.

These equations can be rewritten as

$$\frac{1}{1+2r_k}\left[\sum_{i=1}^{d} h_i^{m_i+1-r_i} L_{m_i+1}(x) A_{g(r),i}\right](m_k+1-r_k)h_k^{m_k+1-r_k} L_{m_k+1}(x) A_{g(r),k}$$

$$\times \Pi_{i=1}^{d} h_i^{1+2r_i} = \tfrac{1}{2}\sigma^2 \Delta^d B_{g(r)}, \quad k=1,\ldots,d. \tag{14.170}$$

Taking a pair of these equations for different k, and calculating the ratio of the corresponding left- and right-hand parts of the equations, we obtain

$$\frac{1}{1+2r_k}(m_k+1-r_k)h_k^{m_k+1-r_k} L_{m_k+1}(x) A_{g(r),k}$$

$$= \frac{1}{1+2r_l}(m_l+1-r_l)h_l^{m_l+1-r_l} L_{m_l+1}(x) A_{g(r),l}, \quad k,l=1,\ldots,d.$$

These equations define the rations of the ideal scale parameters h_k and h_l. They allow one to express all of the scales through h_1:

$$h_k = \left[\frac{(1+2r_k)(m_1+1-r_1)h_1^{m_1+1-r_1} L_{m_1+1}(x) A_{g(r),1}}{(1+2r_1)(m_k+1-r_k) L_{m_k+1}(x) A_{g(r),k}}\right]^{1/(m_k+1-r_k)}. \tag{14.171}$$

This proves formula (14.165).

Using Eq. (14.171), we can check by direct calculations that

$$\Pi_{i=1}^{d} h_i^{(1+2r_i)} = h_1^{\alpha_2} \Pi_{i=1}^{d}\left[\frac{(1+2r_i)(m_1+1-r_1) L_{m_1+1}(x) A_{g(r),1}}{(1+2r_1)(m_i+1-r_i) L_{m_i+1}(x) A_{g(r),i}}\right]^{(1+2r_i)/(m_i+1-r_i)}, \tag{14.172}$$

where

$$\alpha_2 = \sum_{i=1}^{d} \frac{m_1+1-r_1}{m_i+1-r_i}(1+2r_i),$$

and

$$\sum_{i=1}^{d} h_i^{m_i+1-r_i} L_{m_i+1}(x) A_{g(r),i} = \frac{1}{(1+2r_1)} h_1^{m_1+1-r_1} \alpha_2 L_{m_1+1}(x) A_{g(r),1}. \tag{14.173}$$

Let us consider one of the equations in Eq. (14.170) for $k = 1$ and insert in it h_i, which is defined according to Eq. (14.171). Then we find the formula for the ideal scale h_1 as

$$\begin{aligned} h_1^{m_1+1-r_1} &= \frac{\sigma^2 \Delta^d B_{g(r)}(1+2r_1)}{2\left[\sum_{i=1}^d h_i^{m_i+1-r_i} L_{m_i+1}(x) A_{g(r),i}\right](m_1+1-r_1) L_{m_1+1}(x) A_{g(r),1} \Pi_{i=1}^d h_i^{1+2r_i}} \\ &= \frac{\sigma^2 \Delta^d B_{g(r)}(1+2r_1)^2}{2h_1^{m_1+1-r_1} \alpha_2 [L_{m_1+1}(x) A_{g(r),1}]^2 (m_1+1-r_1) h_1^{\alpha_2} W}, \end{aligned} \tag{14.174}$$

where

$$W = \Pi_{i=1}^d \left[\frac{(1+2r_i)(m_1+1-r_1) L_{m_1+1}(x) A_{g(r),1}}{(1+2r_1)(m_i+1-r_i) L_{m_i+1}(x) A_{g(r),i}} \right]^{(1+2r_i)/(m_i+1-r_i)}.$$

In this transformation we use Eqs. (14.172) and (14.173).

Then, finally we obtain for h_1

$$h_1^{\alpha_3} = \frac{\sigma^2 \Delta^d B_{g(r)}(1+2r_1)^2}{2\alpha_2 [L_{m_1+1}(x) A_{g(r),1}]^2 (m_1+1-r_1) W}, \tag{14.175}$$

$$\begin{aligned} \alpha_3 &= \alpha_2 + 2(m_1+1-r_1) \\ &= 2(m_1+1-r_1) + \sum_{i=1}^d \frac{m_1+1-r_1}{m_i+1-r_i}(1+2r_i). \end{aligned} \tag{14.176}$$

This proves formula (14.164).

The minimal criterion value $\bar{l}_{\hat{y}_h^{(r)}}(x, h)$ in Eq. (8.38) with ideal h defined according to Eq. (14.171) can be expressed as

$$\bar{l}_{\hat{y}_h^{(r)}}(x, h) = \sigma^2 \frac{\Delta^d}{\Pi_{i=1}^d h_i^{1+2r_i}} B_{g(r)}(1+r_2), \tag{14.177}$$

$$r_2 = \frac{\left[\sum_{i=1}^d h_i^{m_i+1-r_i} L_{m_i+1}(x) A_{g(r),i}\right]^2}{\Delta^d \sigma^2 B_{g(r)}} \Pi_{i=1}^d h_i^{1+2r_i}. \tag{14.178}$$

Using Eqs. (14.172) and (14.173), we have

$$r_2 = \frac{\left[\sum_{i=1}^{d} h_i^{m_i+1-r_i} L_{m_i+1}(x) A_{g(r),i}\right]^2}{\Delta^d \sigma^2 B_{g(r)}} \Pi_{i=1}^{d} h_i^{1+2r_i}$$

$$= \frac{[h_1^{m_1+1-r_1} \alpha_2 L_{m_1+1}(x) A_{g(r),1}]^2}{(1+2r_1)^2 \Delta^d \sigma^2 B_{g(r)}} h_1^{\alpha_2} W$$

$$= \frac{h_1^{\alpha_3}}{\Delta^d} \frac{[\alpha_2 L_{m_1+1}(x) A_{g(r),1}]^2}{(1+2r_1)^2 \sigma^2 B_{g(r)}} W$$

$$= \frac{\alpha_2}{2(m_1+1-r_1)} = \frac{1}{2} \sum_{k=1}^{d} \frac{1+2r_k}{m_k+1-r_k}. \tag{14.179}$$

Thus,

$$\bar{l}_{\hat{y}_h^{(r)}}(x,h) = \sigma^2 \frac{\Delta^d}{\Pi_{i=1}^{d} h_i^{1+2r_i}} B_{g(r)} (1+\gamma_r^2), \tag{14.180}$$

where

$$\gamma_r = \sqrt{\frac{1}{2} \sum_{i=1}^{d} \frac{1+2r_k}{m_i+1-r_i}}. \tag{14.181}$$

In this way we prove that the ratio of the squared bias to the variance for an ideal h is a constant γ^2 depending only on the orders of the polynomials and the dimension d.

Note that it follows from Eq. (14.175) that

$$h_1 = O((\sigma^2 \Delta^d)^{1/\alpha_3})$$

and

$$\Pi_{i=1}^{d} h_i = 0(h_1^{\alpha_2}) = O((\sigma^2 \Delta^d)^{\alpha_2/\alpha_3}).$$

Then

$$\bar{l}_{\hat{y}_h^{(r)}}(x,h) = O\left(\sigma^2 \frac{\Delta^d}{(\sigma^2 \Delta^d)^{\alpha_2/\alpha_3}}\right) = O(\sigma^2 \Delta^d)^{1-\alpha_2/a_3} = O(\sigma^2 \Delta^d)^{1-\alpha_2/a_3},$$

because

$$1 - \frac{\alpha_2}{\alpha_3} = \frac{\alpha_3 - \alpha_2}{\alpha_3} = \frac{2(m_1+1-r_1)}{2(m_1+1-r_1) + \sum_{i=1}^{d} \frac{m_1+1-r_1}{m_i+1-r_i}(1+2r_i)}$$

$$= \frac{1}{1 + \frac{1}{2} \sum_{i=1}^{d} \frac{1+2r_i}{m_i+1-r_i}},$$

which gives

$$\bar{l}_{\hat{y}_h^{(r)}}(x,h) = O(\sigma^2\Delta^d)^{\beta_2},\ \beta_2 = \frac{1}{1+\frac{1}{2}\sum_{i=1}^{d}\frac{1+2r_i}{m_i+1-r_i}}.$$

For $\Delta^d = 1/n$, this proves Eq. (14.168) and completes the proof of the proposition. □

14.6 Random Processes

14.6.1 Spectrum of random process

When working in the frequency domain it is convenient to calculate the variance of random variables using frequency domain transfer functions. Here we derive some formulas useful for these calculations (e.g., [117], [189]).

Let y, z, and g be a 2D input signal, output signal, and impulse response function of a linear discrete system,

$$z(x) = (y \circledast g)(x).$$

In the frequency domain, this gives

$$Z(f) = Y(f)G(f), \tag{14.182}$$

where $Z(f)$, $Y(f)$, and $G(f)$ are DFTs of the signals and the transfer function of the system, respectively.

Assume that the input y is a zero-mean stationary random process. It allows a spectral representation defined in Eq. (3.51)

$$y(x_1,x_2) = \frac{1}{n_1 n_2}\sum_{f_1=0}^{n_1-1}\sum_{f_2=0}^{n_2-1} Y(f_1,f_2)W_{n_1}^{-f_1x_1}W_{n_2}^{-f_2x_2}, \tag{14.183}$$

where the Fourier transform coefficients $Y(f_1,f_2)$ are zero-mean random independent with variance $E\{|Y(f_1,f_2)|^2\}$. Remember that $W_{n_1} = e^{-j2\pi/n_1}$, $W_{n_2} = e^{-j2\pi/n_2}$ in Eq. (14.183). The independence of $Y(f)$, $f = (f_1,f_2)$, means that

$$E\{Y(f)Y^*(f')\} = 0 \qquad \text{if } f \neq f'.$$

Calculate the expectation from both sides of the Parseval formula Eq. (3.57) rewritten for $y(x_1,x_2)$ and $Y(f_1,f_2)$. This gives

$$\sum_{x_1=0}^{n_1-1}\sum_{x_2=0}^{n_2-1} E\{y^2(x_1,x_2)\} = \frac{1}{n_1 n_2}\sum_{f_1=0}^{n_1-1}\sum_{f_2=0}^{n_2-1} E\{|Y(f_1,f_2)|^2\}. \tag{14.184}$$

Thus, the total energy of the signal y calculated in the spatial domain (left-hand side of Eq. (14.184)) is equal to the total energy calculated over the frequency domain (right-hand side of Eq. (14.184)).

Let the signal energy spectrum be defined as

$$S_y(f_1, f_2) = \frac{1}{n_1 n_2} E\{|Y(f_1, f_2)|^2\}. \tag{14.185}$$

Then

$$\frac{1}{n_1 n_2} \sum_{f_1=0}^{n_1-1} \sum_{f_2=0}^{n_2-1} E\{|Y(f_1, f_2)|^2\} = \sum_{f_1=0}^{n_1-1} \sum_{f_2=0}^{n_2-1} S_y(f_1, f_2).$$

The signal spectrum in Eq. (14.185) gives the distribution of the energy in Eq. (14.184) in the frequency domain. Let

$$\sigma_y^2 = E\{y^2(x_1, x_2)\}$$

be an invariant variance (power) of the random y.

It follows from Eq. (14.184) that

$$\sigma_y^2 = \frac{1}{n_1 n_2} \sum_{f_1=0}^{n_1-1} \sum_{f_2=0}^{n_2-1} S_y(f_1, f_2) = \frac{1}{n_1 n_2} \sum_{f \in F} S_y(f_1, f_2), \tag{14.186}$$

which expresses the variance σ_y^2 through the energy spectrum $S_y(f_1, f_2)$ of the random process y. It shows that the variance σ_y^2 is equal to the mean of the energy spectrum $S_y(f_1, f_2)$ calculated over the frequencies of the DFT. The spectrum $S_y(f_1, f_2)$ gives a distribution of the energy of the signal y in the frequency domain.

For a white-noise signal with independent components of equal power, we have from Eq. (14.186) that

$$\sigma_y^2 = S_y(f_1, f_2),$$

that is, all frequency components of the signal y have the same energy equal to σ_y^2.

Derive the formulas for the spectrum transform. The spectrum of z is defined according to Eq. (14.185) as

$$S_z(f_1, f_2) = \frac{1}{n_1 n_2} E\{|Z(f_1, f_2)|^2\}. \tag{14.187}$$

Using Eq. (14.182), we find

$$\begin{aligned} S_z(f_1, f_2) &= \frac{1}{n_1 n_2} E\{|G(f_1, f_2)|^2 |Y(f_1, f_2)|^2\} \\ &= |G(f_1, f_2)|^2 \frac{1}{n_1 n_2} E\{|Y(f_1, f_2)|^2\} \\ &= |G(f_1, f_2)|^2 S_y(f_1, f_2). \end{aligned}$$

Thus, we arrive at the well-known formula for the spectrum transformation

$$S_z(f_1, f_2) = |G(f_1, f_2)|^2 S_y(f_1, f_2). \tag{14.188}$$

This gives for variance

$$\sigma_z^2 = \frac{1}{n_1 n_2} \sum_{f_1=0}^{n_1-1} \sum_{f_2=0}^{n_2-1} |G(f_1, f_2)|^2 S_y(f_1, f_2). \tag{14.189}$$

In conclusion, we remind the reader again that the DFT spectrums impose periodicity assumptions on the random signals and spectrums.

14.6.2 Minimization in frequency domain

We are looking for the best estimate of y_h to minimize the quadratic criterion in Eq. (9.17). In the frequency domain, this means minimization on $Y_h(f)$:

$$\min_{Y_h(f)} R_\psi^2. \tag{14.190}$$

The necessary minimum conditions can be derived by the following standard procedure. Represent $Y_h(f)$ in the form

$$Y_h(f, \xi) = Y_h^{opt}(f) + \xi \delta Y_h(f), \tag{14.191}$$

where $Y_h^{opt}(f)$ is an optimal solution of Eq. (14.190), $\delta Y_h(f)$ is an arbitrary variation of this optimal solution, and ξ is an arbitrary parameter.

Inserting Eq. (14.191) into Eq. (14.190) and considering R_ψ^2 as function of ξ, $R_\psi^2(\xi)$, we note that, by definition, $R_\psi^2(\xi)$ achieves the minimum value at $\xi = 0$, which means that

$$\partial_\xi R_\psi^2(\xi)|_{\xi=0} = 0.$$

Introduce Eq. (9.18) in the form

$$R_\psi^2(\xi) = \frac{1}{n_1 n_2} \sum_{f \in F} r_\psi^2(f, \xi), \tag{14.192}$$

$$r_\psi^2(f, \xi) = [Z_h(f) - V(f) Y_h(f, \xi)][Z_h^*(f) - V^*(f) Y_h^*(f, \xi)] \\ + \alpha^2 Y_h(f, \xi) |D(f)|^2 Y_h^*(f, \xi)$$

and calculate the derivative ∂_ξ. Because ξ is a parameter of $Y_h(f, \xi)$ and $Y_h^*(f, \xi)$, we find that

$$\partial_\xi R_\psi^2(\xi) = \frac{1}{n_1 n_2} \sum_{f \in F} \partial_\xi r_\psi^2(f, \xi)$$
$$= \frac{1}{n_1 n_2} \sum_{f \in F} [\partial_{Y_h(f,\xi)} r_\psi^2(f, \xi) \cdot \partial_\xi Y_h(f, \xi) + \partial_{Y_h^*(f,\xi)} r_\psi^2(f, \xi) \cdot \partial_\xi Y_h^*(f, \xi)].$$

It follows from Eq. (14.191) that

$$\partial_\xi Y_h(f, \xi) = \delta Y_h(f),$$

and then

$$\partial_\xi R_\psi^2(\xi) = \frac{1}{n_1 n_2} \sum_{f \in F} \partial_\xi r_\psi^2(f, \xi)$$
$$= \frac{1}{n_1 n_2} \sum_{f \in F} [\partial_{Y_h(f,\xi)} r_\psi^2(f, \xi) \cdot \delta Y_h(f) + \partial_{Y_h^*(f,\xi)} r_\psi^2(f, \xi) \cdot \delta Y_h^*(f)].$$

This derivative is equal to zero at $\xi = 0$:

$$\frac{1}{n_1 n_2} \sum_{f \in F} [\partial_{Y_h(f,\xi)} r_\psi^2(f, \xi) \cdot \delta Y_h(f) + \partial_{Y_h^*(f,\xi)} r_\psi^2(f, \xi) \cdot \delta Y_h^*(f)]|_{\xi=0} = 0. \quad (14.193)$$

The variation $\delta Y_h(f)$ is an arbitrary function of f. Note also that $\delta Y_h(f)$ is a complex-valued function consisting of real and imaginary parts that are arbitrary and independent. This means that we may assume that $\delta Y_h(f)$ and $\delta Y_h^*(f)$ are independent functions. It then follows from Eq. (14.193) that

$$\partial_{Y_h(f,\xi)} r_\psi^2(f, \xi)|_{\xi=0} = 0, \qquad f \in F, \quad (14.194)$$

and

$$\partial_{Y_h^*(f,\xi)} r_\psi^2(f, \xi)|_{\xi=0} = 0, \qquad f \in F. \quad (14.195)$$

Equation (14.195) is a complex conjugate version of Eq. (14.194). Thus, these equations are equivalent.

The function $r_\psi^2(f, \xi)$ is linear with respect to $Y_h(f, \xi)$ and $Y_h^*(f, \xi)$. Thus, the derivative $\partial_{Y_h(f,\xi)}$ is linear on $Y_h^*(f, \xi)$ and the derivative on $\partial_{Y_h^*(f,\xi)}$ is linear on $Y_h(f, \xi)$. To obtain the equation with respect to $Y_h(f, \xi)$, we prefer to use the necessary minimum condition in the form of Eq. (14.195).

For $\xi = 0$, Eq. (14.195) can be given in more simple notation as

$$\partial_{Y_h^*(f)} r_\psi^2(f) = 0, \qquad f \in F, \quad (14.196)$$

where $r_\psi^2(f)$ is defined in Eq. (9.19), or as

$$\partial_{Y_h^*(f)} R_\psi^2 = 0, \qquad f \in F. \quad (14.197)$$

Thus, the derivation of the equation for the optimal solution is reduced to the calculation of the derivatives in Eqs. (14.196) and (14.197).

14.6.3 Vector random processes

14.6.3.1 Vector and matrix differentiation

Consider the vectors $\mathbf{a} \in \mathbb{R}^n$, $\mathbf{b} \in \mathbb{R}^n$, $\mathbf{c} \in \mathbb{R}^m$ and the $n \times m$ matrix $\mathbf{A} = (A_{ij})_{n \times m}$. Let the derivative $\partial_\mathbf{a}$ be a vector of the derivatives $\partial_\mathbf{a} = (\partial_{a_1}, \ldots, \partial_{an})^T$ or the vector derivative, and the derivative $\partial_\mathbf{A}$ be a matrix of the derivatives $\partial_\mathbf{A} = (\partial_{A_{ij}})_{n \times m}$ or the matrix derivative.

The following formulas hold for the vector derivatives:

$$\partial_{\mathbf{a}}(\mathbf{a}^T\mathbf{b}) = \mathbf{b}, \tag{14.198}$$

$$\partial_{\mathbf{b}}(\mathbf{a}^T\mathbf{b}) = \mathbf{a}, \tag{14.199}$$

$$\partial_{\mathbf{a}}(\mathbf{a}^T\mathbf{A}\mathbf{c}) = \mathbf{A}\mathbf{c}, \tag{14.200}$$

$$\partial_{\mathbf{c}}(\mathbf{a}^T\mathbf{A}\mathbf{c}) = \mathbf{A}^T\mathbf{a}. \tag{14.201}$$

Indeed, for the derivative $\partial_{\mathbf{a}}(\mathbf{a}^T\mathbf{b})$, we have

$$\partial_{\mathbf{a}}(\mathbf{a}^T\mathbf{b}) = \left(\partial_{a_i}\sum_{k=1}^{n} a_k b_k\right)_{i=1,\dots,n} = (b_i)_{i=1,\dots,n} = \mathbf{b}.$$

This proves Eq. (14.198).

Since $\mathbf{a}^T\mathbf{A}\mathbf{c} = \sum_{k=1}^{n} a_k \sum_{l=1}^{m} A_{kl}c_l$, we yield for the derivative $\partial_{\mathbf{a}}(\mathbf{a}^T\mathbf{A}\mathbf{c})$

$$\partial_{\mathbf{a}}(\mathbf{a}^T\mathbf{A}\mathbf{c}) = \left(\partial_{a_i}\sum_{k=1}^{n} a_k \sum_{l=1}^{m} A_{kl}c_l\right)_{i=1,\dots,n} = \left(\sum_{l=1}^{m} A_{il}c_l\right)_{i=1,\dots,n} = \mathbf{A}\mathbf{c}.$$

This proves Eq. (14.200).

Formulas (14.199) and (14.201) can be proven in a similar manner. The following formulas hold for the matrix derivatives

$$\partial_{\mathbf{A}}(\mathbf{a}^T\mathbf{A}\mathbf{c}) = \mathbf{a}\mathbf{c}^T, \tag{14.202}$$

$$\partial_{\mathbf{A}^T}(\mathbf{a}^T\mathbf{A}\mathbf{c}) = \mathbf{a}^T\mathbf{c}, \tag{14.203}$$

and

$$\partial_{\mathbf{A}}(\mathbf{a}^T\mathbf{Q}\mathbf{A}\mathbf{S}\mathbf{c}) = \mathbf{Q}^T\mathbf{a}\mathbf{c}^T\mathbf{S}^T, \tag{14.204}$$

where the matrices $\mathbf{Q}$ and $\mathbf{S}$ are of relevant sizes.

It is a good exercise to check these formulas. In particular, for the derivative $\partial_{\mathbf{A}}(a^TAc)$, we calculate the derivative with respect to the A_{ij} element of the matrix A. This gives

$$\partial A_{ij}(\mathbf{a}^T\mathbf{A}\mathbf{c}) = \partial A_{ij}\sum_{k=1}^{n} a_k \sum_{l=1}^{m} A_{kl}c_l = a_i c_j,$$

and then we have for that matrix $\partial_{\mathbf{A}}(a^TAc)$

$$[\partial A_{ij}(\mathbf{a}^T\mathbf{A}\mathbf{c})]_{n\times m} = (a_i c_j)_{\substack{i=1,\dots,n\\ j=1,\dots,m}} = \mathbf{a}\mathbf{c}^T.$$

This proves Eq. (14.202).

Formula (14.203) can be verified in a similar way. The quadratic form $\mathbf{a}^T\mathbf{Q}\mathbf{A}\mathbf{S}\mathbf{c} = \mathbf{u}^T\mathbf{A}\mathbf{v}$, where $\mathbf{u}^T = \mathbf{a}^T\mathbf{Q}$ and $\mathbf{v} = \mathbf{S}\mathbf{c}$. Using Eq. (14.202), we have $\partial_{\mathbf{A}}(\mathbf{u}^T\mathbf{A}\mathbf{v}) = \mathbf{u}\mathbf{v}^T$, and inserting formulas for $\mathbf{u}$ and $\mathbf{v}$ we find $\mathbf{u}\mathbf{v}^T = \mathbf{Q}^T\mathbf{a}\mathbf{c}^T\mathbf{S}^T$. This proves Eq. (14.204).

14.6.3.2 Spectrum of random processes

Here we present a vector version of the spectral representation of the random processes. Let $\mathbf{y} \in \mathbb{R}^L$, $\mathbf{z} \in \mathbb{R}^N$ be input and output vector signals and $\mathbf{g}$ be an impulse response $N \times L$ matrix of a linear system

$$\mathbf{z}(x) = (\mathbf{g} \circledast \mathbf{y})(x), \quad z_i(x) = \sum_{j=1}^{L} (g_{ij} \circledast y_j)(x), \qquad i = 1, \ldots, N.$$

In the frequency domain this gives

$$\mathbf{Z}(f) = \mathbf{G}(f)\mathbf{Y}(f), \qquad f \in F, \tag{14.205}$$

where the capital letters $\mathbf{Z}(f)$, $\mathbf{Y}(f)$, and $\mathbf{G}(f)$ stand for the DFT of the corresponding vector signals and the transfer matrix of the system.

Assume that the input $\mathbf{y}(x)$ is a zero-mean stationary random process. It allows a spectral representation, which is

$$\mathbf{y}(x_1, x_2) = \frac{1}{n_1 n_2} \sum_{f_1=0}^{n_1-1} \sum_{f_2=0}^{n_2-1} \mathbf{Y}(f_1, f_2) W_{n_1}^{-f_1 x_1} W_{n_2}^{-f_2 x_2}, \tag{14.206}$$

where, according to Eq. (3.52), $W_{n_1} = e^{-j2\pi/n_1}$ and the Fourier transform vector coefficients $\mathbf{Y}(f_1, f_2)$ are zero-mean random independent with the covariance matrix $E\{\mathbf{Y}(f_1, f_2)\mathbf{Y}^H(f_1, f_2)\}$. The index H stands for the Hermitian transpose denoting the transpose and complex conjugate operations, $\mathbf{Y}^H = (\mathbf{Y}^*)^T$.

The independence of $\mathbf{Y}(f)$ means that

$$E\{\mathbf{Y}(f)\mathbf{Y}^H(f')\} = 0 \qquad \text{if } f \neq f'.$$

Using Eq. (3.56), we can write

$$\sum_{x_1=0}^{n_1-1} \sum_{x_2=0}^{n_2-1} \mathbf{y}(x_1, x_2)\mathbf{y}^T(x_1, x_2) = \frac{1}{n_1 n_2} \sum_{f_1=0}^{n_1-1} \sum_{f_2=0}^{n_2-1} \mathbf{Y}(f_1, f_2)\mathbf{Y}^H(f_1, f_2). \tag{14.207}$$

Calculate the expectation from both sides of this formula:

$$\sum_{x_1=0}^{n_1-1} \sum_{x_2=0}^{n_2-1} E\{\mathbf{y}(x_1, x_2)\mathbf{y}^T(x_1, x_2)\} = \frac{1}{n_1 n_2} \sum_{f_1=0}^{n_1-1} \sum_{f_2=0}^{n_2-1} E\{\mathbf{Y}(f_1, f_2)\mathbf{Y}^H(f_1, f_2)\}. \tag{14.208}$$

The covariance matrix of the random vector $\mathbf{y}$ is defined as

$$\mathbf{r}_{\mathbf{y}}(x_1, x_2) = E\{\mathbf{y}(x_1, x_2)\mathbf{y}^T(x_1, x_2)\}.$$

The energy spectrum matrix of this vector is defined as

$$\mathbf{S}_{\mathbf{y}}(f_1, f_2) = \frac{1}{n_1 n_2} E\{\mathbf{Y}(f_1, f_2)\mathbf{Y}^H(f_1, f_2)\}. \tag{14.209}$$

With this notation, formula (14.208) has the form

$$\sum_{x_1=0}^{n_1-1}\sum_{x_2=0}^{n_2-1} r_{\mathbf{y}}(x_1, x_2) = \sum_{f_1=0}^{n_1-1}\sum_{f_2=0}^{n_2-1} S_{\mathbf{y}}(f_1, f_2). \tag{14.210}$$

The diagonal items $[\mathbf{r}_{\mathbf{y}}(x_1, x_2)]_{k,k}$ of the matrix $\mathbf{r}_{\mathbf{y}}(x_1, x_2)$ are variances of the corresponding kth elements of the vector $\mathbf{y}$:

$$\sigma_{y_k}^2 = E\{|y_k(x)|^2\} = [r_{\mathbf{y}}(x_1, x_2)]_{k,k}.$$

If these diagonals $[\mathbf{r}_{\mathbf{y}}(x_1, x_2)]_{k,k}$ are constant, we obtain from Eq. (14.210) for the variances of $y_k(x)$

$$\sigma_{y_k}^2 = \frac{1}{n_1 n_2}\sum_{f_1=0}^{n_1-1}\sum_{f_2=0}^{n_2-1}[\mathbf{S}_{\mathbf{y}}(f_1, f_2)]_{k,k}, \tag{14.211}$$

where $[\mathbf{S}_{\mathbf{y}}(f_1, f_2)]_{k,k}$ is the kth diagonal element of the energy spectrum matrix $\mathbf{S}_{\mathbf{y}}(f_1, f_2)$.

We will use this formula for the calculation of the estimate variances.

Let us derive the formulas for the spectrum transform. The spectrum of $\mathbf{z}(x)$ is defined according to Eq. (14.209) as

$$\mathbf{S}_{\mathbf{z}}(f_1, f_2) = \frac{1}{n_1 n_2} E\{\mathbf{Z}(f)\mathbf{Z}^H(f)\}. \tag{14.212}$$

Using Eq. (14.205), we find

$$\begin{aligned}\mathbf{S}_{\mathbf{z}}(f_1, f_2) &= \frac{1}{n_1 n_2} E\{\mathbf{G}(f)\mathbf{Y}(f)[\mathbf{G}(f)\mathbf{Y}(f)]^H\}\\ &= \frac{1}{n_1 n_2}\mathbf{G}(f) E\{\mathbf{Y}(f)\mathbf{Y}^H(f)\}\mathbf{G}^H(f)\\ &= \mathbf{G}(f)\mathbf{S}_{\mathbf{y}}(f_1, f_2)\mathbf{G}^H(f).\end{aligned}$$

Thus, we arrive at the well-known formula for the spectrum transformation,

$$\mathbf{S}_{\mathbf{z}}(f_1, f_2) = \mathbf{G}(f)\mathbf{S}_{\mathbf{y}}(f_1, f_2)\mathbf{G}^H(f). \tag{14.213}$$

This gives for variance in Eq. (14.211)

$$\sigma_{z_k}^2 = \frac{1}{n_1 n_2}\sum_{f_1=0}^{n_1-1}\sum_{f_2=0}^{n_2-1}[\mathbf{S}_{\mathbf{z}}(f_1, f_2)]_{k,k}. \tag{14.214}$$

14.7 3D Inverse

Formulas for the regularized inverse and the Wiener regularized inverse filters are derived in this section.

14.7.1 Regularized inverse

According to the basic idea of the regularized inverse, the loss function R_f in Eq. (9.71) given in the frequency domain is replaced by the criterion with the penalty term

$$R_{f,\psi}^2 = R_f^2 + \psi_{\alpha_1}(f), \tag{14.215}$$

$$R_f^2 = \sum_{i=1}^{n_1} |E_i(f)|^2 = \mathbf{E}^H(f)\mathbf{E}(f), \qquad \mathbf{E}(f) = \mathbf{Z}(f) - \mathbf{V}(f)\mathbf{Y}(f),$$

$$\psi_{\alpha_1}(f) = \alpha_1^2 \mathbf{Y}^H(f)\mathbf{Y}(f),$$

where α_1^2 is the regularization parameter. The minimum condition for $R_\psi^2(f)$ has a form similar to Eq. (9.73):

$$\partial_{\mathbf{Y}^H(f)} R_{f,\psi} = 0. \tag{14.216}$$

This results in the equation

$$[\mathbf{V}^H(f)\mathbf{V}(f) + \alpha_1^2 I_{l\times l}]\widehat{\mathbf{Y}}^{RI}(f) = \mathbf{V}^H(f)\mathbf{Z}(f) \tag{14.217}$$

and gives the regularized inverse solution in the form

$$\widehat{\mathbf{Y}}^{RI}(f) = [\mathbf{V}^H(f)\mathbf{V}(f) + \alpha_1^2 I_{l\times l}]^{-1}\mathbf{V}^H(f)\mathbf{Z}(f). \tag{14.218}$$

It can be written as

$$\widehat{\mathbf{Y}}^{RI}(f) = \mathbf{Q}_{RI}\mathbf{Z}(f),$$

where

$$\mathbf{Q}_{RI}(f, \alpha_1) = [\mathbf{V}^H(f)\mathbf{V}(f) + \alpha_1^2 I_{l\times l}]^{-1}\mathbf{V}^H(f) \tag{14.219}$$

is the transfer matrix of the regularized inverse operator.

For $\alpha_1 = 0$, this gives the transfer matrix of the naive inverse operator:

$$\mathbf{Q}_{RI}(f) = [\mathbf{V}^H(f)\mathbf{V}(f)]^{-1}\mathbf{V}^H(f) \tag{14.220}$$

The spectrum of the random component of the solution in Eq. (14.218) is calculated as

$$\mathbf{S}_{\hat{\mathbf{Y}}^{RI}}(f) = \mathbf{Q}_{RI}(f, \alpha_1)\sigma^2\mathbf{Q}_{RI}^H(f, \alpha_1). \tag{14.221}$$

Then, using Eq. (14.214), we obtain for the variance of $\hat{y}_k^{RI}(x)$

$$\sigma_{\hat{y}_k^{RI}}^2 = \frac{1}{n_1 n_2}\sum_{f\in F}(\mathbf{S}_{\hat{\mathbf{Y}}^{RI}}(f))_{k,k} = \frac{1}{n_1 n_2}\sum_{f\in F}[\mathbf{Q}_{RI}(f, \alpha_1)\sigma^2\mathbf{Q}_{RI}^H(f, \alpha_1)]_{k,k}. \tag{14.222}$$

14.7.2 LPA regularized inverse

The approach that we developed to the solution of the equation set in Eq. (9.70) is based on the LPA-ICI adaptive filtering. This approach is a direct generalization for the multivariate case of the technique developed in Sec. 9.1.

We estimate the smoothed image intensity $\mathbf{y}_h$ instead of the original one $\mathbf{y}$ with the link between them defined by the smoothing scalar kernel g_h:

$$\mathbf{y}_h(x) = (g_h \circledast \mathbf{y})(x). \tag{14.223}$$

The convolution of the scalar kernel g_h with the vector signal y means a vector composed from the convolutions of the elements of this vector with the kernel g_h. In the frequency domain, this means that

$$\mathbf{Y}_h(f) = \mathbf{Y}(f)G_h(f), \tag{14.224}$$

where $\mathbf{Y}_h(f) = \mathcal{F}\,\{\mathbf{y}_h(\tilde{x})\}$.

Apply the operator g_h to both sides of Eq. (9.70). Then, in the frequency domain, we have

$$\mathbf{Z}_h(f) = \mathbf{V}(f)\mathbf{Y}_h(f) + \boldsymbol{\varepsilon}_h(f), \tag{14.225}$$

where $\mathbf{Z}_h(f) = \mathbf{Z}(f)G_h(f)$, $\boldsymbol{\varepsilon}_h(f) = \boldsymbol{\varepsilon}(f)G_h(f)$. In this way we replace the original observation equation by the equation for the smoothed signal. To solve Eq. (14.225) using the regularization technique, we introduce a regularized loss function similar to Eq. (14.215) in the form

$$R^2_{\psi,h}(f) = R^2_h(f) + \psi_{\alpha_1}(f),$$

$$R^2_h(f) = \sum_{i=1}^{n_1} |E_{h,i}(f)|^2 = \mathbf{E}^H_h(f)\mathbf{E}_h(f),\ \mathbf{E}_h(f) = \mathbf{Z}_h(f) - \mathbf{V}(f)\mathbf{Y}_h(f)$$

$$\psi_{\alpha_1}(f) = \alpha_1^2\mathbf{Y}^H_h(f)\mathbf{Y}_h(f).$$

The loss function $R^2_{\psi,h}(f)$ is minimized on $\mathbf{Y}_h(f)$. Comparing the loss function $R^2_{\psi,h}(f)$ with $R^2_{f,\psi}(f)$ in Eq. (14.215), we note that they are different only by $\mathbf{Z}_h(f)$ used in $R^2_{\psi,h}(f)$ instead of $\mathbf{Z}(f)$ in $R^2_{f,\psi}(f)$. Then we can use the solution of Eq. (14.218) and all formulas following from this solution by just replacing $\mathbf{Z}(f)$ by $\mathbf{Z}_h(f)$. In this way we obtain the regularized inverse solution $\widehat{\mathbf{Y}}^{RI}_h(f)$ in the form

$$\widehat{\mathbf{Y}}^{RI}_h(f) = [\mathbf{V}^H(f)\mathbf{V}(f) + \alpha_1^2 I_{l\times l}]^{-1}\mathbf{V}^H(f)\mathbf{Z}_h(f) = G_h(f)\widehat{\mathbf{Y}}^{RI}(f), \tag{14.226}$$

where $\widehat{\mathbf{Y}}^{RI}(f)$ is defined by Eq. (14.218).

Denote the regularized transfer matrix as

$$\mathbf{Q}_{RI}(f,\alpha_1) = G_h(f)[\mathbf{V}^H(f)\mathbf{V}(f) + \alpha_1^2 I_{l\times l}]^{-1}\mathbf{V}^H(f). \tag{14.227}$$

Then the estimate in Eq. (14.226) has the form

$$\widehat{\mathbf{Y}}^{RI}_h(f) = \mathbf{Q}_{RI}(f,\alpha_1)\mathbf{Z}(f). \tag{14.228}$$

The spectrum matrix of variance of $\widehat{\mathbf{Y}}_h(f)$ according to Eq. (14.213) is

$$\mathbf{S}_{\widehat{\mathbf{Y}}^{RI}_h}(f) = \mathbf{Q}_{RI}(f,\alpha_1)\sigma^2\mathbf{Q}^H_{RI}(f,\alpha_1).$$

Then the variance of the regularized inverse estimate $\hat{y}_{h,k}^{RI}(x) = \mathcal{F}^{-1}\{\widehat{Y}_{h,k}^{RI}(f)\}$ is calculated using Eq. (14.214) as

$$\sigma^2_{\hat{y}_{h,k}^{RI}} = var\{\hat{y}_{h,k}^{RI}(x)\} = \frac{1}{n_1 n_2}\sum_{f\in F}[\mathbf{Q}_{RI}(f,\alpha_1)\sigma^2\mathbf{Q}_{RI}^H(f,\alpha_1)]_{k,k}. \tag{14.229}$$

14.7.3 LPA-regularized Wiener inverse

We are looking for the optimal linear estimate of the smoothed vector signal $\mathbf{y}_h(x)$ in the form of the linear matrix shift-invariant operator

$$\hat{\mathbf{y}}(x) = (\mathbf{q}_{RWI} \circledast \mathbf{z})(x), \tag{14.230}$$

with the matrix kernel $\mathbf{q}_W$ to be found. The vector observation signal $\mathbf{z}$ is defined by the observation model in Eq. (9.69) and the corresponding quadratic risk is

$$R^2 = E\left\{\sum_x [\mathbf{y}_h(x) - \hat{\mathbf{y}}(x)]^T[\mathbf{y}_h(x) - \hat{\mathbf{y}}(x)]\right\}. \tag{14.231}$$

It is assumed that the signal $\mathbf{y}_h$ is deterministic while the noise $\boldsymbol{\varepsilon}$ is random. Thus, the expectation in R^2 is calculated with respect to the random noise only.

According to the Parseval theorem in Eq. (3.57),

$$\sum_x [\mathbf{y}_h(x) - \hat{\mathbf{y}}(x)]^T[\mathbf{y}_h(x) - \hat{\mathbf{y}}(x)]$$
$$= \frac{1}{n_1 n_2}\sum_{f\in F}[\mathbf{Y}_h(f) - \widehat{\mathbf{Y}}(f)]^H[\mathbf{Y}_h(f) - \widehat{\mathbf{Y}}(f)]. \tag{14.232}$$

In the frequency domain we have for Eq. (14.230)

$$\widehat{\mathbf{Y}}(f) = \mathbf{Q}_{RWI}(f)\mathbf{Z}(f), \tag{14.233}$$

where $\mathbf{Q}_{RWI}(f) = \mathcal{F}\{\mathbf{q}_{RWI}(x)\}$.

Substituting Eq. (14.233) in Eq. (14.232), we have

$$R^2 = E\left\{\frac{1}{n_1 n_2}\sum_{f\in F}[\mathbf{Y}_h(f) - \mathbf{Q}_{RWI}(f)\mathbf{Z}(f)]^H[\mathbf{Y}_h(f) - \mathbf{Q}_{RWI}(f)\mathbf{Z}(f)]\right\},$$

where, according to Eq. (9.70),

$$\mathbf{Z}(f) = \mathbf{V}(f)\mathbf{Y}(f) + \boldsymbol{\varepsilon}(f).$$

The necessary minimum conditions similar to Eq. (9.72) have the form of the partial matrix derivative:

$$\partial_{\mathbf{Q}^H_{RWI}(f)} R^2 = 0,\ f \in F. \tag{14.234}$$

To compute this derivative, we use formula Eq. (14.202), which gives

$$\begin{aligned}
&\partial_{\mathbf{Q}^H_{RWI}(f)}[\mathbf{Y}_h(f) - \mathbf{Q}_{RWI}(f)\mathbf{Z}(f)]^H[\mathbf{Y}_h(f) - \mathbf{Q}_{RWI}(f)\mathbf{Z}(f)] \\
&= -\partial_{\mathbf{Q}^H_{RWI}(f)}\mathbf{Z}^H(f)\mathbf{Q}^H_{RWI}(f)[\mathbf{Y}_h(f) - \mathbf{Q}_{RWI}(f)\mathbf{Z}(f)] \\
&= -[\mathbf{Z}^H(f)]^T[\mathbf{Y}_h(f) - \mathbf{Q}_{RWI}(f)\mathbf{Z}(f)]^T.
\end{aligned}$$

Inserting this derivative into Eq. (14.234) and transposing the result, we arrive at the set of equations

$$E\{[\mathbf{Y}_h(f) - \mathbf{Q}_{RWI}(f)\mathbf{Z}(f)] \cdot \mathbf{Z}^H(f)\} = 0, \qquad f \in F,$$

which follows by

$$\mathbf{Q}_{RWI}(f)E\{\mathbf{Z}(f)\mathbf{Z}^H(f)\} = E\{(\mathbf{Y}_h(f)\mathbf{Z}^H(f)\}, \qquad f \in F,$$

This gives the transfer matrix for the Wiener filter in the form

$$\mathbf{Q}_{RWI}(f) = E\{\mathbf{Y}_h(f)\mathbf{Z}^H(f)\}(E\{\mathbf{Z}(f)\mathbf{Z}^H(f)\})^{-1}, \qquad f \in F. \qquad (14.235)$$

Let us use a more general model for the random noise than is assumed in Eq. (9.69). The random $\boldsymbol{\varepsilon}$ is zero-mean stationary and random with the spectrum $\mathbf{S}_\varepsilon(f)$, that is, $E\{\boldsymbol{\varepsilon}(f)\} = E\{\boldsymbol{\varepsilon}^*(f)\} = 0$ and, according to Eq. (14.209), $E\{\boldsymbol{\varepsilon}(f)\boldsymbol{\varepsilon}^H(f)\} = n_1 n_2 \mathbf{S}_\varepsilon(f)$. For the white noise in Eq. (9.69), this means that $\mathbf{S}_\varepsilon(f) = \boldsymbol{\sigma}^2$, where the diagonal matrix $\boldsymbol{\sigma}^2$ is given by Eq. (9.78).

The expectation over this random $\boldsymbol{\varepsilon}$ gives for Eq. (14.235)

$$\begin{aligned}
E\{\mathbf{Y}_h(f)\mathbf{Z}^H(f)\} &= E\{\mathbf{Y}_h(f)[\mathbf{V}(f)\mathbf{Y}(f) + \boldsymbol{\varepsilon}(f)]^H\} \\
&= E\{\mathbf{Y}_h(f)\mathbf{Y}^H(f)\mathbf{V}^H(f)\} + E\{\mathbf{Y}_h(f)\boldsymbol{\varepsilon}^H(f)\} \\
&= \mathbf{Y}_h(f)\mathbf{Y}^H(f)\mathbf{V}^H(f)
\end{aligned}$$

and

$$\begin{aligned}
E\{\mathbf{Z}(f)\mathbf{Z}^H(f)\} &= E\{[\mathbf{V}(f)\mathbf{Y}(f) + \boldsymbol{\varepsilon}(f)][\mathbf{V}(f)\mathbf{Y}(f) + \boldsymbol{\varepsilon}(f)]^H\} \\
&= \mathbf{V}(f)\mathbf{Y}(f)\mathbf{Y}^H(f)\mathbf{V}^H(f) + E\{\boldsymbol{\varepsilon}(f)\boldsymbol{\varepsilon}(f)^H\} \\
&= \mathbf{V}(f)\mathbf{Y}(f)\mathbf{Y}^H(f)\mathbf{V}^H(f) + n_1 n_2 \mathbf{S}_\varepsilon(f).
\end{aligned}$$

Then the transfer matrix of the Wiener filter in Eq. (14.235) is

$$\begin{aligned}
\mathbf{Q}_{RWI}(f) = \mathbf{Y}_h(f)\mathbf{Y}^H(f)\mathbf{V}^H(f) \cdot [&\mathbf{V}(f)\mathbf{Y}(f)\mathbf{Y}^H(f)\mathbf{V}^H(f) \\
&+ n_1 n_2 \mathbf{S}_\varepsilon(f)]^{-1},\ f \in F.
\end{aligned}$$

Using Eq. (14.224), it can also be written as

$$\begin{aligned}
\mathbf{Q}_{RWI}(f) = G_h\mathbf{Y}(f)\mathbf{Y}^H(f)\mathbf{V}^H(f) \cdot [&\mathbf{V}(f)\mathbf{Y}(f)\mathbf{Y}^H(f)\mathbf{V}^H(f) \\
&+ n_1 n_2 \mathbf{S}_\varepsilon(f)]^{-1},\ f \in F. \qquad (14.236)
\end{aligned}$$

The optimal estimate of y_h is

$$\widehat{\mathbf{Y}}_h^{RWI}(f) = \mathbf{Q}_{RWI}(f)\mathbf{Z}(f). \tag{14.237}$$

It is the optimal linear inverse operator. Inserting the regularization parameter α_2 into the transfer matrix $Q_{RWI}(f)$, we arrive at the regularized Wiener inverse matrix filter with a transfer matrix:

$$\begin{aligned}\mathbf{Q}_{RWI}(f,\alpha_2) &= G_h(f)\mathbf{Y}(f)\mathbf{Y}^H(f)\mathbf{V}^H(f)\\ &\cdot [\mathbf{V}(f)\mathbf{Y}(f)\mathbf{Y}^H(f)\mathbf{V}^H(f) + n_1n_2\alpha_2^2\mathbf{S}_\varepsilon(f)]^{-1},\ f \in F.\end{aligned} \tag{14.238}$$

The regularized Wiener inverse estimate similar to Eq. (14.237) has the form

$$\widehat{\mathbf{Y}}_h^{RWI}(f) = \mathbf{Q}_{RWI}(f,\alpha_2)\mathbf{Z}(f). \tag{14.239}$$

The variance of this estimate is calculated according to Eq. (14.214) as

$$\sigma^2_{\hat{y}_{h,k}^{RWI}}(x) = var\{\hat{y}_{h,k}^{RWI}(x)\} = \frac{1}{n_1n_2}\sum_{f\in F}[\mathbf{Q}_{RWI}(f,\alpha_2)\mathbf{S}_\varepsilon(f)\mathbf{Q}^H_{RWI}(f,\alpha_2)]_{k,k}. \tag{14.240}$$

14.8 Nonlinear Methods

14.8.1 Proof of Proposition 13

In this section we prove Proposition 13. It is assumed in the asymptotic that h and Δ are small as $\Delta \to 0$, $h \to 0$, $h/\Delta \to \infty$. These assumptions are used to have small estimation errors:

$$e_y(x,h) = y(x) - \hat{y}_h(x) \to 0.$$

Let $X_s = \Delta s$ in Eq. (10.49). Changing the summation variables, $X_s = x - \Delta s$, the necessary minimum condition in Eq. (10.49) can be given in the form

$$\begin{aligned}&\sum_s w_h(\Delta s)\psi^{(1)}(r_s) = 0,\\ &r_s = y(x-\Delta s) + \varepsilon_s - \hat{y}_h(x),\end{aligned} \tag{14.241}$$

where r_s is the residual of estimation, $\hat{y}_h(x) = \widehat{C}$, and $\varepsilon_s = \varepsilon(x - \Delta s)$.

The polynomial expansion of $y(x - \Delta s)$ on small Δs gives for the residuals

$$\begin{aligned}r_s &\simeq y(x) - y^{(1)}(x)\Delta s + y^{(2)}(x)(\Delta s)^2/2 + \varepsilon_s - \hat{y}_h(x)\\ &= e_y(x,h) - y^{(1)}(x)\Delta s + y^{(2)}(x)(\Delta s)^2/2 + \varepsilon_s.\end{aligned}$$

Assuming that $[e_y(x,h) - y^{(1)}(x)\Delta s + y^{(2)}(x)(\Delta s)^2/2]$ is small, the first two terms of the Taylor series for $\psi^{(1)}(r_s)$ give

$$\psi^{(1)}(r_s) \simeq \psi^{(1)}(\varepsilon_s) + \psi^{(2)}(\varepsilon_s)[e_y(x,h) - y^{(1)}(x)\Delta s + y^{(2)}(x)(\Delta s)^2/2],$$

which results for Eq. (14.241) in

$$\sum_s w_h(\Delta s)\{\psi^{(1)}(\varepsilon_s) + \psi^{(2)}(\varepsilon_s)[e_y(x,h) - y^{(1)}(x)\Delta s + y^{(2)}(x)(\Delta s)^2/2]\} \simeq 0. \tag{14.242}$$

The random $\Delta \sum_s w_h(\Delta s)\psi^{(2)}(\varepsilon_s)$ converges in probability to the asymptotic expectation:

$$\Delta \sum_s w_h(\Delta s)\psi^{(2)}(\varepsilon_s) \xrightarrow{P} E\{\psi^{(2)}\} > 0. \tag{14.243}$$

Note that here

$$\Delta \sum_s w_h(\Delta s) = \sum_s w(\Delta s/h)\frac{\Delta}{h} \to \int_{\mathbb{R}} w(u)du = 1, \tag{14.244}$$

and the mathematical expectation $E\{\psi^{(2)}\}$ is calculated on the random ε.

To prove Eq. (14.243), we calculate the mean-square error between $\Delta \sum_s w_h(\Delta s)\psi^{(2)}(\varepsilon_s)$ and the expectation $E\{\psi^{(2)}\}$. This error can be represented as

$$\Delta \sum_s w_h(\Delta s)\psi^{(2)}(\varepsilon_s) - E\{\psi^{(2)}\} = \Delta \sum_s w_h(\Delta s)(\psi^{(2)}(\varepsilon_s) - E\{\psi^{(2)}\}) + \left[\Delta \sum_s w_h(\Delta s) - 1\right] E\{\psi^{(2)}\}.$$

Since, according to Eq. (14.244), $[\Delta \sum_s w_h(\Delta s) - 1]E\{\psi^{(2)}\} \to 0$, we have

$$\Delta \sum_s w_h(\Delta s)\psi^{(2)}(\varepsilon_s) - E\{\psi^{(2)}\} \to \Delta \sum_s w_h(\Delta s)[\psi^{(2)}(\varepsilon_s) - E\{\psi^{(2)}\}].$$

Then we obtain for the mean-square error

$$\begin{aligned}
&E\left\{\left[\Delta \sum_s w_h(\Delta s)\psi^{(2)}(\varepsilon_s) - E\{\psi^{(2)}\}\right]^2\right\} \\
&\to E\left\{\left[\Delta \sum_s w_h(\Delta s)(\psi^{(2)}(\varepsilon_s) - E\{\psi^{(2)}\})\right]^2\right\} \\
&= \frac{\Delta^2}{h^2} \sum_s w^2(\Delta s/h) E\{[\psi^{(2)}(\varepsilon_s) - E\{\psi^{(2)}\}]^2\} \\
&= E\{[\psi^{(2)}(\varepsilon_s) - E\{\psi^{(2)}\}]^2\} \frac{\Delta^2}{h^2} \sum_s w^2(\Delta s/h) \\
&\to E\{[\psi^{(2)}(\varepsilon_s) - E\{\psi^{(2)}\}]^2\} \frac{\Delta}{h} \int_{\mathbb{R}} w^2(u)du \to 0.
\end{aligned}$$

This proves the mean-square convergence of $\Delta \sum_s w_h(\Delta s)\psi^{(2)}(\varepsilon_s)$ to $E\{\psi^{(2)}\}$ and the convergence in probability Eq. (14.243) follows from the mean-square convergence.

In a similar way, we obtain the mean-square and in-probability convergence:

$$\begin{aligned}
\sum_s w_h(\Delta s)\Delta s\psi^{(2)}(\varepsilon_s) &= \frac{1}{h}\sum_s w(\Delta s/h)\Delta s\psi^{(2)}(\varepsilon_s) \\
&\xrightarrow{P} \frac{h}{\Delta}\int_{\mathbb{R}} w(u)udu \cdot E\{\psi^{(2)}\}, \qquad (14.245) \\
\sum_s w_h(\Delta s)(\Delta s)^2\psi^{(2)}(\varepsilon_s) &= \frac{1}{h}\sum_s w(\Delta s/h)(\Delta s)^2\psi^{(2)}(\varepsilon_s) \\
&\xrightarrow{P} \frac{h^2}{\Delta}\int_{\mathbb{R}} w(u)u^2du \cdot E\{\psi^{(2)}\}.
\end{aligned}$$

Substituting Eqs. (14.243) and (14.245) into Eq. (14.242), we obtain with respect to the estimation error $e_y(x, h)$

$$\begin{aligned}
&\sum_s w_h(\Delta s)\psi^{(1)}(\varepsilon_s) + e_y(x,h)\frac{1}{\Delta}E\{\psi^{(2)}\} - y^{(1)}(x)\frac{h}{\Delta}\int_{\mathbb{R}} w(u)udu \cdot E\{\psi^{(2)}\} \\
&\quad + y^{(2)}(x)\frac{h^2}{2\Delta}\int_{\mathbb{R}} w(u)u^2du \cdot E\{\psi^{(2)}\} = 0,
\end{aligned}$$

and

$$\begin{aligned}
e_y(x,h) &= y^{(1)}(x)h\int_{\mathbb{R}} w(u)udu - y^{(2)}(x)\frac{h^2}{2}\int_{\mathbb{R}} w(u)u^2du \\
&\quad - \frac{\Delta\sum_s w_h(\Delta s)\psi^{(1)}(\varepsilon_s)}{E\{\psi^{(2)}\}}.
\end{aligned}$$

Then

$$\begin{aligned}
e_y(x,h) &= m_{\hat{y}_h}(x,h) + e_y^0(x,h), \\
m_{\hat{y}_h}(x,h) &= y^{(1)}(x)h\int_{\mathbb{R}} w(u)udu - y^{(2)}(x)\frac{h^2}{2}\int_{\mathbb{R}} w(u)u^2du, \qquad (14.246) \\
e_y^0(x,h) &= -\frac{\Delta\sum_s w_h(\Delta s)\psi^{(1)}(\varepsilon_s)}{E\{\psi^{(2)}\}},
\end{aligned}$$

where $m_{\hat{y}_h}(x, h)$ and $e_y^0(x, h)$ are the bias and random components of the estimation error.

Because, according to Eq. (10.53), $E\{\psi^{(1)}(\varepsilon_s)\} = 0$, we have $E\{e_y^0(x, h)\} = 0$ and the estimation bias is

$$m_{\hat{y}_h}(x,h) = y^{(1)}(x)h\int_{\mathbb{R}} w(u)udu - y^{(2)}(x)\frac{h^2}{2}\int_{\mathbb{R}} w(u)u^2du.$$

This proves Eq. (10.57).

Small values of the orders $o(h)$, $o(h^2)$ in Eq. (10.57) correspond to the errors of using integrals instead of the corresponding sums. The variance of the error is

$$\begin{aligned} var\{e_y(x,h)\} &= E[e_y^0(x,h)]^2 \\ &= \frac{\Delta^2 \sum_s w_h^2(\Delta s) E\{(\psi^{(1)})^2\}}{[E\{\psi^{(2)}\}]^2} = \frac{E\{(\psi^{(1)})^2\}\Delta^2 \sum_s w^2(\Delta s/h)}{[E\{\psi^{(2)}\}]^2 h^2} \\ &\to \frac{\Delta}{h}\frac{E\{(\psi^{(1)})^2\}}{(E\{\psi^{(2)}\})^2}\int_{\mathbb{R}} w^2(u)du + o\left(\frac{\Delta}{h}\right) \\ &= V\frac{\Delta}{h}\int_{\mathbb{R}} w^2(u)du + o\left(\frac{\Delta}{h}\right). \end{aligned}$$

This proves Eq. (10.56) and completes the proof of the proposition.

14.8.2 Proof of Proposition 14

In this section we prove Proposition 14.

(1) Let $X_s = \Delta s$ then the necessary minimum condition for Eq. (10.4) has the form

$$\begin{aligned} &\sum_s w_h(x - \Delta s)\psi^{(1)}(r_s)\phi_h(x - \Delta s) = 0, \\ &r_s = y(\Delta s) + \varepsilon_s - C^T\phi_h(x - \Delta s), \end{aligned} \tag{14.247}$$

where r_s is the residual of estimation and $\varepsilon_s = \varepsilon(\Delta s)$. Note that $s \in \mathbb{Z}^d$ is a d-dimensional multi-index and the sampling period Δ is scalar.

Change the summation variables in Eq. (14.247) to $x - \Delta s \Longrightarrow \Delta s$. Then these equations take a form that is more convenient for accuracy analysis,

$$\begin{aligned} &\sum_s w_h(\Delta s)\psi^{(1)}(r_s)\phi_h(\Delta s) = 0, \\ &r_s = y(x - \Delta s) + \varepsilon_s - C^T\phi_h(\Delta s),\ \varepsilon_s = \varepsilon(x - \Delta s). \end{aligned} \tag{14.248}$$

The polynomial expansion of $y(x - \Delta s)$ on Δs gives, according to Eq. (14.64) (with $Q = m$ in this formula),

$$\begin{aligned} y(x - s\Delta) &= y_1(x, s\Delta) + y_2(x, s\Delta), \\ y_1(x, s\Delta) &= \sum_{t=0}^{m}(-1)^t\sum_{|k|=t}\frac{1}{k!}(s\Delta)^k y^{(k)}(x), \end{aligned} \tag{14.249}$$

$$y_2(x, s\Delta) = (-1)^{m+1}\sum_{|k|=m+1}\frac{1}{k!}(s\Delta)^k y^{(k)}(x - \lambda s\Delta), \tag{14.250}$$

where $y_1(x, s\Delta)$ is the main polynomial part of the series and $y_2(x, s\Delta)$ is a residual term of the series in the Lagrange form.

The main polynomial part of the series

$$y_1(x, s\Delta) = \sum_{t=0}^{m} \sum_{|k|=t} \frac{1}{k!}(-s\Delta)^k y^{(k)}(x)$$

can be written as

$$y_1(x, s\Delta) = \phi^T(\Delta s)C^*(x), \tag{14.251}$$

where $C^*(x)$ is a vector composed of the elements $\partial_x^k y(x)$, which are the true values of the signal and its derivatives ordered in $C^*(x)$ according to the order of the multivariate polynomials in Eq. (2.45).

The vector $\phi(\Delta s)$ in Eq. (14.251) is identical to the vector ϕ of the polynomials $(-s\Delta)^k/k!$ used in the linear multivariable LPA estimate in Eq. (2.43).

Using Eq. (14.251), we have for Eq. (14.249)

$$y(x - s\Delta) = C^{*T}(x)\phi(\Delta s) + y_2(x, s\Delta). \tag{14.252}$$

Let us represent the first item in Eq. (14.252) in a form corresponding to the truncated Taylor series exploited in the LPA, where the polynomials ϕ are scaled by h. Using the scaled polynomials ϕ_h instead of ϕ, we obtain

$$C^{*T}(x)\phi(\Delta s) = \widetilde{C}^{*T}(x)\phi_h(\Delta s), \tag{14.253}$$

that is, $\widetilde{C}_k^*$ is equal to C_k^* multiplied by h^l with $l \geq 0$ equal to the power of the corresponding polynomial in $\phi_k(\Delta s)$. Then Eq. (14.252) is of the form

$$y(x - s\Delta) = \widetilde{C}^{*T}(x)\phi_h(\Delta s) + y_2(x, s\Delta). \tag{14.254}$$

According to the definition of the LPA estimates in Eq. (2.32), the C_k gives the estimates of $\widetilde{C}_k^*$ and the difference $\widetilde{C}^* - C$ is a vector error of the LPA coefficients.

(2) Let us now derive the equation linearized with respect to the estimation errors. Inserting Eq. (14.254) into Eq. (14.248) and using the introduced vector $\widetilde{C}^*$, we have for the residuals r_s

$$\begin{aligned} r_s &= y(x - \Delta s) - C^T\phi_h(\Delta s) + \varepsilon_s \\ &= \widetilde{C}^{*T}\phi_h(\Delta s) - C^T\phi_h(\Delta s) + y_2(x, s\Delta) + \varepsilon_s \\ &= \delta C^T(x)\phi_h(\Delta s) + y_2(x, s\Delta) + \varepsilon_s, \end{aligned}$$

where

$$\delta C = \widetilde{C}^* - C. \tag{14.255}$$

This gives for Eq. (14.248)

$$\sum_s w_h(\Delta s)\psi^{(1)}(\delta C^T\phi_h(\Delta s) + y_2(x, s\Delta) + \varepsilon_s)\phi_h(\Delta s) = 0. \tag{14.256}$$

Assuming that $\delta C^T\phi_h(\Delta s) + y_2(x, s\Delta)$ is small, the first two terms of the Taylor series for $\psi^{(1)}$ give

$$\psi^{(1)}(r_s) \simeq \psi^{(1)}(\varepsilon_s) + \psi^{(2)}(\varepsilon_s)[\delta C^T\phi_h(\Delta s) + y_2(x, s\Delta)].$$

Inserting this into Eq. (14.256), we have

$$\sum_s w_h(\Delta s)\{\psi^{(1)}(\varepsilon_s)\phi_h(\Delta s) + \psi^{(2)}(\varepsilon_s)[\phi_h(\Delta s)\phi_h^T(\Delta s)\delta C + y_2(x, s\Delta)\phi_h(\Delta s)]\} = 0. \tag{14.257}$$

It can be shown, as was done for Eq. (14.243), that the random matrix

$$\Delta^d \sum_s w_h(\Delta s)\psi^{(2)}(\varepsilon_s)\phi_h(\Delta s)\phi_h^T(\Delta s)$$

converges in probability to the asymptotic expectation, that is,

$$\Delta^d \sum_s w_h(\Delta s)\psi^{(2)}(\varepsilon_s)\phi_h(\Delta s)\phi_h^T(\Delta s) \xrightarrow{P} E\{\psi^{(2)}\}\Phi,$$

where, according to Eq. (14.60), $\Delta^d \sum_s w_h(\Delta s)\phi_h(\Delta s)\phi_h^T(\Delta s) \to \Phi = \int_{\mathbb{R}^d} w(u) \times \phi(u)\phi^T(u)du$. Substituting this limit into Eq. (14.257), we obtain

$$\sum_s w_h(\Delta s)\dot{\psi}^{(1)}(\varepsilon_s)\phi_h(\Delta s) + E\{\psi^{(2)}\}\Phi\delta C\frac{1}{\Delta^d} + \sum_s w_h(\Delta s)\psi^{(2)}(\varepsilon_s)y_2(x, s\Delta)\phi_h(\Delta s) = 0, \tag{14.258}$$

which can be solved with respect to δC:

$$\delta C = -\frac{\Delta^d}{E\{\psi^{(2)}\}}\Phi^{-1}\left\{\sum_s w_h(\Delta s)[\psi^{(1)}(\varepsilon_s)\phi_h(\Delta s) + y_2(x, s\Delta)\phi_h(\Delta s)\psi^{(2)}(\varepsilon_s)]\right\}. \tag{14.259}$$

(3) Let us derive the formulas for the estimation bias. The expectation of the vector error in Eq. (14.259) is

$$E\{\delta C\} = -\Phi^{-1}\Delta^d \sum_s w_h(\Delta s)y_2(x, s\Delta)\phi_h(\Delta s). \tag{14.260}$$

Using Eq. (14.250) and the standard limit passages from sums to integrals as $h,\ \Delta \to 0$ and $h/\Delta \to \infty$, we obtain

$$\Delta^d \sum_s w_h(\Delta s) y_2(x, s\Delta)\phi_h(\Delta s)$$
$$= (-1)^{m+1}\Delta^d \sum_s w_h(\Delta s) \sum_{|k|=m+1} \frac{1}{k!}(s\Delta)^k y^{(k)}(x - \lambda s\Delta)\phi_h(\Delta s)$$
$$= (-1)^{m+1}h^{m+1} \sum_s \frac{\Delta^d}{h^d} w(\Delta s/h) \sum_{|k|=m+1} \frac{1}{k!}(s\Delta/h)^k y^{(k)}(x - \lambda hs \cdot \Delta/h)$$
$$\times \phi(\Delta s/h)$$
$$\to (-1)^{m+1}h^{m+1} \sum_{|k|=m+1} \frac{1}{k!} \int_{\mathbb{R}^d} \{y^{(k)}(x) + [y^{(k)}(x - \lambda hu) - y^{(k)}(x)]\}$$
$$\times u^k w(u)\phi(u)du$$
$$\to (-1)^{m+1}h^{m+1} \sum_{|k|=m+1} \frac{1}{k!} y^{(k)}(x) \int_{\mathbb{R}^d} u^k w(u)\phi(u)du + o(h^{m+1}).$$

Here we use the Lipshitz conditions in Eq. (5.8), which give

$$(-1)^{m+1}h^{m+1} \sum_{|k|=m+1} \frac{1}{k!} \int_{\mathbb{R}^d} [y^{(k)}(x - \lambda hu) - y^{(k)}(x)] \cdot u^k w(u)\phi(u)du = o(h^{m+1}).$$

Then we have from Eq. (14.260)

$$E\{\delta C\} \to (-1)^m h^{m+1} \sum_{|k|=m+1} \frac{1}{k!} y^{(k)}(x)\Phi^{-1} \int_{\mathbb{R}^d} u^k w(u)\phi(u)du + o(h^{m+1}).$$

To obtain the estimation error for the signal and the kth derivatives from δC^T, according to the link between the estimate of C and the estimates of the signal and derivatives in Eq. (2.43), it should be multiplied by $(-1/h)^{|r|}\phi^{(r)}(0)$. Thus,

$$m_{\hat{y}_h^{(r)}}(x, h) = (-\frac{1}{h})^{|r|} E\{(\delta C)^T\}\phi^{(r)}(0)$$
$$= (-1)^m h^{m+1}(-\frac{1}{h})^{|r|} \sum_{|k|=m+1} \frac{1}{k!} y^{(k)}(x) \int_{\mathbb{R}^d} u^k w(u)\phi^T(u)\Phi^{-1}\phi^{(r)}(0)du$$
$$+ o(h^{m+1-|r|})$$
$$= (-1)^m h^{m+1-|r|} \sum_{|k|=m+1} y^{(k)}(x)\frac{1}{k!} \int_{\mathbb{R}^d} u^k g^{(r)}(u)du + o(h^{m+1-|r|}), \tag{14.261}$$

where

$$g^{(r)}(x) = (-1)^{|r|}\phi^T(u)\Phi^{-1}\phi^{(r)}(0).$$

This proves Eq. (10.86).

(4) Let us derive the formulas for the variance of estimation. For the random component of δC we have from Eqs. (14.259) and (14.260)

$$\delta C^0 \triangleq \delta C - E\{\delta C\} = -\frac{\Delta^d}{E\{\psi^{(2)}\}}\Phi^{-1}\sum_s w_h(\Delta s)\{\psi^{(1)}(\varepsilon_s)\phi_h(\Delta s) + y_2(x, s\Delta)\phi_h(\Delta s)[\psi^{(2)}(\varepsilon_s) - E\{\psi^{(2)}\}]\}.$$

Because $|y_2(x, s\Delta)| = o(h^{m+1})$, the corresponding term can be omitted in the calculation of the variance and we can use δC^0 in the form

$$\delta C^0 \simeq -\frac{\Delta^d}{E\{\psi^{(2)}\}}\Phi^{-1}\sum_s w_h(\Delta s)\psi^{(1)}(\varepsilon_s)\phi_h(\Delta s).$$

For the estimation errors of the signal and derivatives, this gives

$$\begin{aligned} e^0_{y^{(r)}}(x,h) &= (\delta C^0)^T\phi^{(r)}(0)\left(-\frac{1}{h}\right)^{|r|} \\ &= -\frac{\Delta^d}{E\{\psi^{(2)}\}}\left(-\frac{1}{h}\right)^{|r|}\sum_s w_h(\Delta s)\phi_h^T(\Delta s)\Phi^{-1}\phi^{(r)}(0)\psi^{(1)}(\varepsilon_s) \\ &= -\frac{\Delta^d}{E\{\psi^{(2)}\}}\sum_s g_h^{(r)}(\Delta s)\psi^{(1)}(\varepsilon_s), \end{aligned}$$

where

$$g_h^{(r)}(\Delta s) = \frac{(-1)^{|r|}}{h^{|r|}} w_h(\Delta s)\phi_h^T(\Delta s)\Phi^{-1}\phi^{(r)}(0).$$

Then the variance of $e^0_{y^{(r)}}(x,h)$ is calculated as

$$\begin{aligned} var\{e^0_{y^{(r)}}(x,h)\} &= E\{(e^0_{y^{(r)}}(x,h))^2\} = V\Delta^{2d}\sum_s [g_h^{(r)}(\Delta s)]^2 \\ &= V\frac{\Delta^{2d}}{h^{2|r|}}\sum_s\left[\frac{1}{h^d}w(\Delta s/h)\phi^T(\Delta s/h)\Phi^{-1}\phi^{(r)}(0)\right]^2 \\ &= V\frac{\Delta^d}{h^{d+2|r|}}\sum_s [w(\Delta s/h)\phi^T(\Delta s/h)\Phi^{-1}\phi^{(r)}(0)]^2\frac{\Delta^d}{h^d} \\ &\to V\frac{\Delta^d}{h^{d+2|r|}}\int_{\mathbb{R}^d}[w(u)\phi^T(u)\Phi^{-1}\phi^{(r)}(0)]^2 du + o\left(\frac{\Delta^d}{h^{d+2|r|}}\right) \\ &= V\frac{\Delta^d}{h^{d+2|r|}}\int_{\mathbb{R}^d}[g^{(r)}(u)]^2 du + o\left(\frac{\Delta^d}{h^{d+2|r|}}\right). \end{aligned}$$

This proves Eq. (10.87).

14.8.3 Proof of Proposition 15

In this section we prove Proposition 15.

(1) The signal values y_s corresponding to observations z_s can be written as

$$y_s = y(X_s) = y(x - (x - X_s)).$$

Using the Taylor series in Eq. (14.64) with respect to $(x - X_s)$, we obtain

$$y_s = \sum_{t=0}^{m} (-1)^t \sum_{|k|=t} \frac{1}{k!}(x - X_s)^k y^{(k)}(x) + (-1)^{m+1} \sum_{|k|=m+1} \frac{1}{k!} \times (x - X_s)^k y^{(k)}(x - \lambda(x - X_s)), \quad 0 \leq \lambda \leq 1. \tag{14.262}$$

or

$$y_s = y_1(x, X_s) + y_2(x, X_s), \tag{14.263}$$

where

$$y_1(x, X_s) = \sum_{t=0}^{m} (-1)^t \sum_{|k|=t} \frac{1}{k!}(x - X_s)^k y^{(k)}(x)$$

is the main polynomial part of the series and

$$y_2(x, X_s) = (-1)^{m+1} \sum_{|k|=m+1} \frac{1}{k!}(x - X_s)^k y^{(k)}(x - \lambda(x - X_s))$$

is a residual.

With the notation in Eqs. (14.251) and (14.253), the main polynomial term $y_1(x, X_s)$ can be represented in the form corresponding to the polynomials used in the LPA as

$$y_1(x, X_s) = \phi^T(x - X_s)C^* = \phi_h^T(x - X_s)\widetilde{C}^*.$$

Here $C^*(x)$ is a vector composed of the elements $y^{(k)}(x)$, where the true values of the signal and its derivatives are ordered in $C^*(x)$ according to the order of the polynomials in ϕ and $\widetilde{C}_k^*$ is equal to C_k^* multiplied by h^l with $l \geq 0$ equal to the power of the corresponding polynomial. Thus, Eq. (14.263) has the form

$$y_s = \phi_h^T(x - X_s)\widetilde{C}^* + y_2(x, X_s). \tag{14.264}$$

(2) Let us derive the error equations in the variations from the true signal values. The necessary minimum conditions in Eq. (11.11) can be rewritten as

$$\sum_s \{\partial_y \ln f(z_s, y_s - y_s + y_h(x, X_s))\} w_h(x - X_s)\phi_h(x - X_s) = 0,$$

where $y_h(x, X_s) = \phi_h^T(x - X_s)C$ is the LPA of the signal.

Inserting in this equation y_s from Eq. (14.264) and replacing $y_h(x, X_s)$ by $\phi_h^T(x - X_s)C$, we find

$$\sum_s \{\partial_y \ln f(z_s, y_s - \phi_h^T(x - X_s)\delta C - y_2(x, X_s))\} w_h(x - X_s)\phi_h(x - X_s) = 0, \tag{14.265}$$

where δC is defined by Eq. (14.255). Assuming that δC and $y_2(x, X_s)$ are small, the linear terms of the Taylor series give for Eq. (14.265)

$$\sum_s [\partial_y \ln f(z_s, y_s)] w_h(x - X_s)\phi_h(x - X_s) - \sum_s [\partial_y^2 \ln f(z_s, y_s)] \times w_h(x - X_s)\phi_h(x - X_s)[\phi_h^T(x - X_s)\delta C + y_2(x, X_s)] \simeq 0. \tag{14.266}$$

Under appropriate conditions, as $h \to 0$ and the observation number becomes larger, $h/\Delta \to \infty$, similar to the proof of Eq. (14.243), the following convergence in probability can be shown:

$$\sum_s [\partial_y^2 \ln f(z_s, y_s)] w_h(x - X_s)\phi_h(x - X_s)[\phi_h^T(x - X_s)\delta C + y_2(x, X_s)] \xrightarrow{P} \sum_s E\{\partial_y^2 \ln f(z_s, y_s)\} w_h(x - X_s)\phi_h(x - X_s) \times [\phi_h^T(x - X_s)\delta C + y_2(x, X_s)]\}. \tag{14.267}$$

In the second part of this formula, the sums should be replaced by the corresponding integrals; however, these calculations are produced latter.

Using Eq. (14.267) in Eq. (14.266) gives

$$\sum_s [\partial_y \ln f(z_s, y_s)] w_h(x - X_s)\phi_h(x - X_s) - \sum_s E\{\partial_y^2 \ln f(z_s, y_s)\} w_h(x - X_s)\phi_h(x - X_s) \times [\phi_h^T(x - X_s)\delta C + y_2(x, X_s)] = 0. \tag{14.268}$$

Remember that, according to Eq. (11.21),

$$E\{\partial_y^2 \ln f(z_s, y_s)\} = -I_y(X_s),$$

where $I_y(X_s)$ is the Fisher information calculated at the point X_s. Then Eq. (14.268) can be rewritten as

$$\sum_s [\partial_y \ln f(z_s, y_s)] w_h(x - X_s)\phi_h(x - X_s) + \sum_s I_y(X_s) w_h(x - X_s)\phi_h(x - X_s) \times [\phi_h^T(x - X_s)\delta C + y_2(x, X_s)] = 0. \tag{14.269}$$

We solve this equation with respect to δC:

$$\delta C = -\Phi_h^{-1}(I_y)\sum_s \partial_y \ln f(z_s, y_s)\cdot w_h(x-X_s)\phi_h(x-X_s) \tag{14.270}$$

$$-\,\Phi_h^{-1}(I_y)\sum_s I_y(X_s)w_h(x-X_s)\phi_h(x-X_s)y_2(x,X_s), \tag{14.271}$$

where it is denoted

$$\Phi_h(I_y) = \sum_s I_y(X_s)w_h(x-X_s)\phi_h(x-X_s)\phi_h^T(x-X_s).$$

The linearized Eq. (14.270) is a basis of the accuracy analysis.

(3) Let us calculate the bias and variance of the estimate. Because of Eq. (11.19), $E\{\partial_y \ln f(z,y)\} = 0$, the first summand in Eq. (14.270) is the zero-mean random variable and

$$E\{\delta C\} = -\Phi_h^{-1}(I_y)\sum_s I_y(X_s)w_h(x-X_s)\phi_h(x-X_s)y_2(x,X_s). \tag{14.272}$$

Subtracting Eq. (14.272) from Eq. (14.270), we arrive at

$$\delta C - E\{\delta C\} = -\Phi_h^{-1}(I_y)\sum_s \partial_y \ln f(z_s, y_s)\cdot w_h(x-X_s)\phi_h(x-X_s). \tag{14.273}$$

Calculate the covariance of this vector:

$$\begin{aligned} cov\{\delta C\} &= E\{(\delta C - E\{\delta C\})(\delta C - E\{\delta C\})^T\} \\ &= \Phi_h^{-1}(I_y)W_h(I_y)\Phi_h^{-1}(I_y), \qquad (14.274) \\ W_h(I_y) &= \sum_s I_y(X_s)w_h^2(x-X_s)\phi_h(x-X_s)\phi_h^T(x-X_s). \end{aligned}$$

It is used in this calculation that according to Eq. (11.20),

$$E\{(\partial_y \ln f(z_s, y_s))^2\} = I_y(X_s).$$

(4) Passage to the limit in the formulas for the bias. We have for the bias Eq. (14.272) and Eq. (14.274) for the covariance. The limit passages are defined by the standard transformations using $X_s = s\Delta$, $h \to 0$, $\Delta \to 0$, and $\Delta/h \to \infty$.

We first consider the matrix $\Phi_h(I_y)$. The usual successive steps of this transformations are

$$\begin{aligned} \Phi_h(I_y) &= \frac{1}{h^d}\sum_s I_y(s\Delta)w((x-s\Delta)/h)\phi((x-s\Delta)/h)\phi^T((x-s\Delta)/h) \\ &\to \frac{1}{\Delta^d h^d}\int_{\mathbb{R}^d} I_y(v)w((x-v)/h)\phi((x-v)/h)\phi^T((x-v)/h)dv. \end{aligned}$$

Changing the variable of integration $(x - v)/h = u$, we obtain

$$\Phi_h(I_y) = \frac{1}{\Delta^d}\int_{\mathbb{R}^d} I_y(x - hu)w(u)\phi(u)\phi^T(u)du$$
$$\rightarrow \frac{I_y(x)}{\Delta^d}\int_{\mathbb{R}^d} w(u)\phi(u)\phi^T(u)du = \frac{I_y(x)}{\Delta^d}\Phi, \tag{14.275}$$

where Φ is given in Eq. (14.60).

It is assumed in the limit passage in Eq. (14.275) that the Fisher information $I_y(x)$ is smooth enough with respect to x. In particular, this requirement may have a form of the Lipshitz condition in Eq. (5.8) as

$$|I_y(x) - I_y(x')| \le L_I||x - x'||^{\gamma}, \qquad 0 < \gamma \le 1. \tag{14.276}$$

Calculations for the bias in Eq. (14.272) give

$$\sum_s I_y(s\Delta)w_h(x - \Delta s)\phi_h(x - \Delta s)y_1(x, \Delta s)$$
$$= \frac{(-1)^{m+1}}{h^d}\sum_s I_y(s\Delta)w((x - \Delta s)/h)\phi((x - \Delta s)/h)$$
$$\times \sum_{|k|=m+1}\frac{1}{k!}(x - \Delta s)^k y^{(k)}(x - \lambda(x - \Delta s))$$
$$= \frac{(-1)^{m+1}}{\Delta^d h^d}\sum_{|k|=m+1}\frac{1}{k!}\sum_s I_y(s\Delta)w((x - \Delta s)/h)\phi((x - \Delta s)/h)$$
$$\times (x - \Delta s)^k y^{(k)}(x - \lambda(x - \Delta s))\Delta^d$$
$$\rightarrow \frac{(-1)^{m+1}}{\Delta^d h^d}\sum_{|k|=m+1}\frac{1}{k!}\int_{\mathbb{R}^d} I_y(v)w((x - v)/h)\phi((x - v)/h)$$
$$\times (x - v)^k y^{(k)}(x - \lambda(x - v))dv$$
$$= \frac{(-1)^{m+1}h^{m+1}}{\Delta^d}\sum_{|k|=m+1}\frac{1}{k!}\int_{\mathbb{R}^d} I_y(x - hu)w(u)\phi(u)u^k y^{(k)}(x - \lambda hu)du$$
$$\rightarrow \frac{(-1)^{m+1}h^{m+1}}{\Delta^d}I_y(x)\sum_{|k|=m+1} y^{(k)}(x)\frac{1}{k!}\int_{\mathbb{R}^d} w(u)u^k\phi(u)du + o\left(\frac{h^{m+1}}{\Delta^d}\right).$$

In this last transformation, we use the fact that the Lipshitz conditions in Eq. (14.276) are fulfilled for the Fisher information $I_y(x)$ and for $y^{(k)}$.

Substituting the last formula and Eq. (14.275) into Eq. (14.272) gives

$$E\{\delta C\} = -(-1)^{m+1}h^{m+1}\sum_{|k|=m+1} y^{(k)}(x)\frac{1}{k!}\int_{\mathbb{R}^d} w(u)u^k\Phi^{-1}\phi(u)du + o(h^{m+1}). \tag{14.277}$$

According to Eq. (11.10), the bias of the estimate of the signal as well as of the derivatives can be given in the form

$$
\begin{aligned}
&E\{y_h^{(r)}(x) - \hat{y}_h^{(r)}(x)\} \\
&\quad = \left(-\frac{1}{h}\right)^{|r|} E\{\delta C^T\}\phi^{(r)}(0) \\
&\quad = -(-h)^{m+1-|r|} \sum_{|k|=m+1} y^{(k)}(x)\frac{1}{k!}\int_{\mathbb{R}^d} w(u)u^k\phi^T(u)\Phi^{-1}\phi^{(r)}(0)du + o(h^{m+1-|r|}) \\
&\quad = (-1)^m h^{m+1-|r|} \sum_{|k|=m+1} \frac{1}{k!}y^{(k)}(x)\int_{\mathbb{R}^d} g^{(r)}(u)u^k du + o(h^{m+1-|r|}),
\end{aligned}
$$

where $g^{(r)}(x) = (-1)^{|r|}w(x)\phi^T(x)\Phi^{-1}\phi^{(r)}(0)$. This proves Eq. (11.22).

(5) Passage to the limit in the formulas for the variance. Consider the variance of estimation starting from formula (14.274). We have for the matrix $W_h(I_y)$

$$
\begin{aligned}
W_h(I_y) &= \frac{1}{h^{2d}}\sum_s I_y(\Delta s)w^2((x-\Delta s)/h)\phi((x-\Delta s)/h)\phi^T((x-\Delta s)/h) \\
&= \frac{1}{\Delta^d h^{2d}}\sum_s I_y(\Delta s)w^2((x-\Delta s)/h)\phi((x-\Delta s)/h)\phi^T((x-\Delta s)/h)\Delta^d \\
&\to \frac{1}{\Delta^d h^{2d}}\int_{\mathbb{R}^d} I_y(v)w^2((x-v)/h)\phi((x-v)/h)\phi^T((x-v)/h)dv \\
&= \frac{1}{\Delta^d h^d}\int_{\mathbb{R}^d} I_y(x-hu)w^2(u)\phi(u)\phi^T(u)du \\
&\to \frac{1}{\Delta^d h^d} I_y(x)\int_{\mathbb{R}^d} w^2(u)\phi(u)\phi^T(u)du.
\end{aligned}
$$

Substituting this $W_h(I_y)$ and $\Phi_h(I_y)$ from Eq. (14.275) into Eq. (14.274), we find

$$
\begin{aligned}
cov\{\delta C\} &= E\{(\delta C - E\{\delta C\})(\delta C - E\{\delta C\})^T\} \\
&= I_y^{-1}(x)\frac{\Delta^d}{h^d}\Phi^{-1}\int_{\mathbb{R}^d} w^2(u)\phi(u)\phi^T(u)du\Phi^{-1}. \qquad (14.278)
\end{aligned}
$$

Recalculate the variance of the coefficients into the variances of the estimates using Eq. (11.10) to obtain

$$var\{\hat{y}_h(x)\} = E\{(y^{(r)}(x) - \hat{y}_h^{(r)}(x))^2\} = \frac{1}{h^{2|r|}}[\phi^{(r)}(0)]^T cov\{\delta C\}\phi^{(r)}(0)$$

$$= I_y^{-1}(x)\frac{\Delta^d}{h^{d+2|r|}}\int_{\mathbb{R}^d}\{w(u)[\phi^{(r)}(0)]^T\Phi^{-1}\phi(u)\}[w(u)\phi^T(u)\Phi^{-1}\phi^{(r)}(0)]$$

$$= I_y^{-1}(x)\frac{\Delta^d}{h^{d+2|r|}}\int_{\mathbb{R}^d}[g^{(r)}(u)]^2du. \tag{14.279}$$

This proves Eq. (11.23).

References

1. Abramovich, F. and Y. Benjamini, "Adaptive thresholding of wavelet coefficients," *Computational Statistics and Data Analysis*, vol. 22, pp. 351–361, 1996.
2. Abramovich, F. and B. Silverman, "Wavelet decomposition approaches to statistical inverse problems," *Biometrika*, vol. 85, pp. 115–129, 1998.
3. Akaike, H., "A new look at the statistical model identification," *IEEE Trans. Automatic Control*, vol. 19, pp. 716–723, 1974.
4. Alenius, S. and U. Ruotsalainen, "Generalization of median root prior reconstruction," *IEEE Trans. Medical Imaging*, vol. 21, no. 11, pp. 1413–1420, 2002.
5. Analoui, M. and J.P. Allebach, "Model-based halftoning by direct binary search," *Proc. SPIE/IS&T Symp. Electronic Imaging Science and Technology*, vol. 1666, pp. 96–108, 1992.
6. Andrews, D.F., P.J. Bickel, F.R. Hampel, P.J. Huber, W.H. Rogers, and J.W. Tukey. *Robust estimation of location: Survey and advances*. Princeton Univ. Press, Princeton, NJ, 1972.
7. Anscombe, F.J., "The transformation of Poisson, binomial and negative binomial data," *Biometrika*, vol. 35, pp. 246–254, 1948.
8. Arce, G.R. "A general weighted median filter structure admitting negative weights," *IEEE Trans. Signal Processing*, vol. 46, no. 12, pp. 3195–3205, 1998.
9. Astola, J. and P. Kuosmanen. *Fundamentals of nonlinear digital filtering*. CRC Press, Boca Raton, FL, 1997.
10. Banham, M.R. and A.K. Katsaggelos, "Digital image restoration," *IEEE Signal Processing Magazine*, vol. 14, no. 2, pp. 24–41, 1997.
11. Bates, D.M. and D.G. Watts. *Nonlinear regression analysis and its applications*. John Wiley and Sons, New York, 1988.
12. Baxes, G. *Digital image processing: Principles and applications*. John Wiley and Sons, New York, 1994.
13. Benjamini, Y. and W. Liu, "A step-down multiple hypotheses testing procedure that controls the false discovery rate under independence," *J. Statist. Planning Inference*, vol. 82, no. 1–2, pp. 163–170, 1999.
14. Bertero, M. and P. Boccacci. *Introduction to inverse problems in imaging*. Institute of Physics, Bristol and Philadelphia, 1998.
15. Blake, A. and A. Zisserman. *Visual reconstruction*. MIT Press, Cambridge, MA, 1987.
16. Boccacci, P. and M. Bertero, "Image-restoration methods: basics and algorithms," in *Confocal and two-photon microscopy: Foundations, applications, and advances*, Diaspro, A. (ed.), Wiley-Liss, New York, pp. 253–269, 2002.
17. Box, G.E.P. and D.R. Cox, "An analysis of transformations," *J. Royal Statist. Soc.*, Series B 26, pp. 211–243, 1964.

18. Bracewell, R.N. *The Fourier transform and its applications*. McGraw-Hill, New York, NY, 1986.
19. Brown, R. *Smoothing, forecasting and prediction of discrete time series*. Prentice-Hall, Englewood Cliffs, NJ, 1963.
20. Burden, R.L. and J.D. Faires. *Numerical analysis*. PWS Publishing, Boston, 1993.
21. Burt, P.J. and E.H. Adelson, "The Laplacian pyramid as a compact image code," *IEEE Trans. Commun.*, vol. 31, no. 4, pp. 532–540, 1983.
22. Candès, E.J. and D.L. Donoho, "Recovering edges in ill-posed inverse problems: optimality of curvelet frames," *Ann. Stat.*, vol. 30, no. 3, pp. 784–842, 2002.
23. Cavalier, L. and A. Tsybakov, "Sharp adaptation for inverse problems with random noise," *Prob. Theory Related Fields*, vol. 123, pp. 323–354, 2002.
24. Carasso, A.S., "Direct blind deconvolution," *SIAM J. Appl. Math.*, vol. 61, no. 6, pp. 1980–2007, 2001.
25. Carasso, A.S., "Linear and nonlinear image deblurring: a document study, " *SIAM J. Numeric. Anal.*, vol. 36, pp. 1659–1689, 1999.
26. Carr, J.C., A.H. Gee, R.W. Prager, and K.J. Dalton, "Quantitative visualization of surfaces from volumetric data," *Proc. WSGC'98—Sixth Inter. Conf. Central Europe on Computer Graphics and Visualization*, Plzen, Czech Republic, 1998.
27. Chang, P.C., C.S. Yu, and T.H. Lee, "Hybrid LMS-MMSE inverse halftoning technique," *IEEE Trans. Image Processing*, vol. 10, no. 1, pp. 95–103, 2001.
28. Chen, S.S., Donoho D., and M.A. Saunders, "Atomic decomposition by basis pursuit," *SIAM Rev.*, vol. 43, no. 1, pp. 129–159, 2001.
29. Cleveland, W.S., "Robust locally weighted regression and smoothing scatterplots," *J. Amer. Statist. Assoc.*, vol. 74, pp. 829–836, 1979.
30. Cleveland, W.S. and S.J. Devlin, "Locally weighted regression: an approach to regression analysis by local fitting," *J. Amer. Statist. Assoc.*, vol. 83, pp. 596–610, 1988.
31. Cleveland, W.S. and C. Loader, "Smoothing by local regression: principles and methods," in *Statistical theory and computational aspects of smoothing*, Hardel, W., and Schimek, M. (eds.), Heidelberg Physica-Verlag, pp. 10–49, 1996.
32. Coates, M., Hero A.O., Nowak R., and B. Yu, "Internet tomography," *IEEE Signal Processing Mag.*, vol. 19, no. 3, pp. 47–65, 2002.
33. Daubechies, I. *Ten lectures on wavelets*. SIAM, Philadelphia, 1992.
34. Davis, P.J. *Interpolation and approximation*. New York, Blaisdell Publishing, 1963.
35. Deans, S.R. *The Radon transform and some of its applications*. John Wiley and Sons, New York, 1983.
36. Dempster, A.P., Laird, N.M., and D.B. Rubin, "Maximum likelihood from incomplete data via the *EM* algorithm," *J. Royal Statist. Soc.*, Series B, vol. 39, no. 1, pp. 1–38, 1977.
37. Dias, J.M.B., "Fast GEM wavelet-based image deconvolution algorithm," *Proc. IEEE Int. Conf. on Image Processing, ICIP*, vol. 2., 2003.
38. Djurović, I. and L.J. Stanković, "Nonparametric algorithm for local frequency estimation in multidimensional signals," *IEEE Trans. Image Processing*, vol. 13, no. 4, pp. 467–474, 2004.
39. Donoho, D.L., "Nonlinear wavelet methods for recovery of signals, densities, and spectra from indirect and noisy data," in *Proc. Symposia in applied mathematics: Different perspectives on wavelets*, Daubechies, I. (ed.), vol. 47, pp. 173–205, American Mathematical Society, Providence, RI, 1993.

40. Donoho, D.L. and I.M. Johnstone, "Ideal spatial adaptation by wavelet shrinkage," *Biometrika*, vol. 81, pp. 425–455, 1994.
41. Donoho, D.L. and I.M. Johnstone, "Adapting to unknown smoothness via wavelet shrinkage," *J. Amer. Statist. Assoc.*, vol. 90, no. 432, pp. 1200–1224, 1995.
42. Donoho, D.L.,"Wedgelets: nearly minimax estimation of edges," *Ann. Stat.*, vol. 27, pp. 859–897, 1999.
43. Donoho, D.L., "Non-linear solution of linear inverse problems by wavelet-vaguelette decomposition," *Appl. Comput. Harm. Anal.*, vol. 2, pp. 101–126, 1995.
44. Engl, H.W., M. Hanke, and A. Neubauer. *Regularization of inverse problems*. Kluwer Academic Publishers, Dordrecht, The Netherlands, 1996.
45. Fan, J., "On the optimal rates of convergence for nonparametric deconvolution problems," *Ann. Stat.*, vol. 21, no. 4, pp. 1900–1925, 1993.
46. Fan, J. and I. Gijbels. *Local polynomial modelling and its application*. Chapman and Hall, London, 1996.
47. Fan, J., M. Farmen, and I. Gijbels, "Local maximum likelihood estimation and inference," *J. Royal Statist. Soc.*, Ser. B, vol. 60, pp. 591–608, 1998.
48. Fessler, J.A., "Penalized maximum-likelihood image reconstruction using space-alternative generalized *EM* algorithms," *IEEE Trans. Image Processing*, vol. 4, no. 10, pp. 1417–1429, 1995.
49. Fessler, J.A., "Aspire 3.0 user's guide: a sparse iterative reconstruction library," Communication and Signal Processing Lab., Department of Electrical and Computer Engineering, University of Michigan, Ann Arbor. Online: http//:www.eecs.umich.edu/~fessler.
50. Figueiredo, M.A.T. and R.D. Nowak, "An EM algorithm for wavelet based image reconstruction," *IEEE Trans. Image Processing*, vol. 12, no. 8, pp. 906–916, 2003.
51. Fletcher, R., *Practical methods of optimization*. 2nd ed., John Wiley and Sons, New York, 1987.
52. Floyd, R.W. and L. Steinberg, "An adaptive algorithm for spatial gray scale," *Int. Symp. Digest Tech. Papers, Soci. Inform. Displays*, pp. 36–37, 1975.
53. Foi, A., "Anisotropic nonparametric image processing: Theory, algorithms and applications." Tesi, Dipartimento di Matematica, Politecnico di Milano, Italy, 2005.
54. Foi, A., V. Katkovnik, K. Egiazarian, and J. Astola, "A novel anisotropic local polynomial estimator based on directional multiscale optimizations," *Proc. Sixth Int. Conf. Mathematics in Signal Processing*, Cirencester, UK, pp. 79–82, 2004.
55. Foi, A., V. Katkovnik, K. Egiazarian, and J. Astola, "Inverse halftoning based on the anisotropic *LPA-ICI* deconvolution," in *Proc. 2004 Int. TICSP Workshop on Spectral Methods and Multirate Signal Processing, SMMSP 2004*, Vienna, Austria, Astola, J. et al. (eds), TTY-PAINO, Tampere, Finland, pp. 49–56, 2004.
56. Foi, A., R. Bilcu, V. Katkovnik, and K. Egiazarian, "Anisotropic local approximations for pointwise adaptive signal-dependent noise removal," *XIII European Signal Proc. Conf., EU-SIPCO 2005*, Sept. 4, pp 2005.
57. Foi, A., S. Alenius, M. Trimeche, V. Katkovnik, and K. Egiazarian, "A spatially adaptive Poisson image deblurring," *IEEE 2005 Int. Conf. Image Processing, ICIP 2005*, Sept. 4, pp. 925–928, 2005.
58. Freeman, W.T. and E.H. Adelson, "The design and use of steerable filters," *IEEE Trans. Pattern Analysis and Machine Intelligence*, vol. 13, no. 9, pp. 891–906, 1991.

59. Friedlander, F.G. *Introduction to the theory of distributions*. 2nd ed., Cambridge University Press, Cambridge, UK, 1999.
60. Galatsanos, N.P. and A.K. Katsaggelos, "Methods for choosing the regularization parameter and estimation the noise variance in image restoration and their relation," *IEEE Trans Image Processing*, vol. 1, no. 3, pp. 322–336, 1992.
61. Ganesan, L. and P. Bhattacharyya, "Edge detection in utnextured and textured images—a common computational framework," *IEEE Trans. Systems, Man, and Cybernetics, Part B: Cybernetics*, vol. 27, no 5, pp. 823–834, 1997.
62. Gel'fand, I.M. and G.E. Shilov. *Generalized functions, vol. 1: Properties and operations*. Harcourt Brace, New York, 1977.
63. Gel'fand, I.M. and G.E. Shilov. *Generalized functions, vol. 2: Spaces of fundamental and generalized functions*. Harcourt Brace, New York, 1977.
64. Gilbo, E.P. and I.B. Chelpanov. *Signal processing on the basis of sequential selection*. Sovetskoe Radio, Moscow, 1975 (in Russian).
65. Glasbey, C.A. and G.W. Morgan. *Image analysis for the biological sciences*. New York, John Wiley and Sons, New York, 1995.
66. Goldenshluger, A. and A. Nemirovski, "On spatial adaptive estimation of nonparametric regression," *Math. Meth. Statistics*, vol. 6, pp. 135–170, 1997.
67. Goldenshluger, A., "On pointwise adaptive nonparametric deconvolution," *Bernoulli*, vol. 5, pp. 907–925, 1999.
68. Gonzalez, R.C. and R.E. Woods. *Digital image processing*. 2nd ed., Prentice Hall, Englewood Cliffs, NJ, 2002.
69. Green, P., "Bayesian reconstruction from emission tomography data using modified *EM* algorithm," *IEEE Trans. Medical Imaging*, vol. 9, no. 1, pp. 84–93, 1990.
70. Hamza, Ben A., P. Luque, J. Martinez, and R. Roman, "Removing noise and preserving details with relaxed median filters," *J. Math. Imaging and Vision*, vol. 11, no. 2, pp. 161–177, 1999.
71. Hansen, P.C., "Deconvolution and regularization with Toeplitz matrices," *Numeric. Algorithms*, vol. 29, pp. 323–378, 2002.
72. Haralick, R.M., "Edge and region analysis for digital image data," *Comput. Graphics Image Processing*, vol. 12, pp. 60–73, 1980.
73. Haralick, R.M., "Digital step edges from zero crossing of second directional derivatives," *IEEE Trans. Pattern Analysis and Machine Intelligence*, vol. 16, no. 1, pp. 58–68, 1984.
74. Hardi, R.C. and K.E. Barner, "Rank conditioned rank selection filters for signal restoration," *IEEE Trans. Image Processing*, vol. 3, no. 2, pp. 192–206, 1994.
75. Härdle, W. *Applied nonparametric regression*. Cambridge University Press, Cambridge, UK, 1990.
76. Härdle, W., G. Kerkyacharian, D. Picard, and A. Tsybakov. *Wavelets, approximation and statistical applications*. Lecture Notes in Statistics, vol. 129, Springer-Verlag, New York, 1998.
77. Hastie, T.J. and C. Loader, "Local regression: automatic kernel carpentry" (with discussion), *Statist. Sci.*, vol. 8, no. 2, pp. 120–143, 1993.
78. Hastie, T.J. and R.J. Tibshirani. *Generalized linear models*. Chapman and Hall, London, 1990.
79. Hein, S. and A. Zakhor, "Halftone to continuous-tone conversion of error-diffusion coded images," *IEEE Trans. Image Processing*, vol. 4, no. 2, pp. 208–216, 1995.

80. Heyde, C.C., *Quasi-likelihood and its applications.* Springer-Verlag, New York, 1997.
81. Hillery, A.D. and R.T. Chin, "Iterative Wiener filters for image restoration," *IEEE Trans. Signal Processing*, vol. 39, pp. 1892–1899, 1991.
82. Hiriyannaiah, H., "X-ray computed tomography for medical imaging, " *IEEE Signal Processing Mag.*, vol. 14, no. 2, pp. 42–59, 1997.
83. Horn, B. and B. Schunck, "Determining optical flow," *Artif. Intell.*, vol. 17, pp. 185–203, 1981.
84. Hsieh, J. *Computed tomography*. SPIE Press, Bellingham, WA, 2003.
85. Huang, J.S. and D.H. Tseng, "Statistical theory of edge detection," *Graphics and Image Processing*, 43, pp. 337–346, 1988.
86. Huber, P.J. *Robust statistics*. John Wiley and Sons, New York, 1981.
87. Hurvich, C.M., J.S. Simonoff, and C.-L. Tsai, "Smoothing parameter selection in nonparametric regression using an improved Akaike information criterion," *J. Royal Statist. Soci.*, Ser. B, vol. 60, pp. 271–293, 1998.
88. Hwang, H. and R.A. Haddad, "Adaptive median filters: new algorithms and results," *IEEE Trans. Image Processing*, vol. 4, no. 4, pp. 499–502, 1995.
89. Hwang, H. and R.A. Haddad, "Multilevel nonlinear filters for edge detection and noise suppression," *IEEE Trans. Signal Processing*, vol. 42, no. 2, pp. 249–258, 1994.
90. Hwang, H. and R.A. Haddad, "Adaptive median filters: new algorithms and results," *IEEE Trans. Image Processing*, vol. 44, pp. 499–502, 1995.
91. Jain, A.K. *Fundamentals of digital image processing*. Prentice Hall, London, 1989.
92. Jähne, B. *Practical handbook on image processing for scientific and technical applications*. CRC Press, Boca Raton, FL, 2004.
93. Jansson, P.A. (ed.). *Deconvolution of images and spectra*. Academic Press, New York, 1997.
94. Jarvis, J.F., C.N. Judice, and W.H. Ninke, "A survey of techniques for the display of continuous tone pictures on bi-level displays," *Computer Graphics Image Processing*, vol. 5, pp. 13–40, 1976.
95. Joshi, J. and M.I. Miller, "Maximum a posteriori estimation with good's roughness for three-dimensional optical-sectioning microscopy," *J. Opt. Soc. Am. Ser. A.*, vol. 10, no. 5, pp. 1078–1085, 1993.
96. Kak, A.C. and M. Slaney. *Principles of tomographic imaging*. IEEE Press, New York, 1987.
97. Kalifa, J., S. Mallat, and B. Rouge, "Deconvolution by thresholding in mirror wavelet bases," *IEEE Trans. Image Processing*, vol. 12, no. 4, pp. 446–457, 2003.
98. Katkovnik, V., "Problem of approximating functions of many variables," *Autom. Remote Control*, vol. 32, no. 2, Part 2, pp. 336–341, 1971.
99. Katkovnik, V., "Homogeneous integral averaging operators obtained by the method of least squares," *Autom. Remote Control*, vol. 32, no. 11, Part 1, pp. 1767–1775, 1971.
100. Katkovnik, V., "Linear and nonlinear methods of nonparametric regression analysis," *Soviet J. Autom. and Inform. Sci.*, vol. 5, pp. 25–34, 1979.
101. Katkovnik, V. *Linear estimation and stochastic optimization problems.* Nauka, Moscow, 1976 (in Russian).
102. Katkovnik, V. *Nonparametric identification and smoothing of data* (*local approximation methods*). Nauka, Moscow, 1985 (in Russian).

103. Katkovnik, V., "A new method for varying adaptive bandwidth selection," *IEEE Trans. Signal Processing*, vol. 47, no. 9, pp. 2567–2571, 1999.
104. Katkovnik, V. "A multiresolution nonparametric regression for spatially adaptive image de-noising." *IEEE Signal Processing Letts.*, vol. 11, no. 10, pp. 798–801, 2004.
105. Katkovnik, V, "Multiresolution kernel nonparametric regression: a new approach to pointwise spatial adaptation," *Discrete Signal Processing*, vol. 15, no. 1, pp. 73–116, 2005.
106. Katkovnik, V., K. Egiazarian, and J. Astola, "Adaptive window size image de-noising based on intersection of confidence intervals (*ICI*) rule," *J. Math. Imaging and Vision,* vol. 16, pp. 223–235, 2002.
107. Katkovnik, V., K. Egiazarian, and J. Astola, "Application of the *ICI* principle to window size adaptive median filtering," *Signal Processing*, vol. 83, no. 2, pp. 251–257, 2003.
108. Katkovnik, V., K. Egiazarian, and J. Astola. *Adaptive varying scale methods in image processing*, TICSP Series 19, TTY, Monistamo, Tampere, 2003.
109. Katkovnik, V. and I. Shmulevich, "Kernel density estimation with adaptive varying window size," *Pattern Recognition Letts.*, vol. 23, no. 12, pp. 1641–1648, 2002.
110. Katkovnik, V., K. Egiazarian, and J. Astola, "A spatially adaptive nonparametric regression image deblurring," *IEEE Trans. Image Processing*, vol. 14, no. 10, pp. 1469–1478, 2005.
111. Katkovnik, V., A. Foi, K. Egiazarian, and J. Astola, "Directional varying scale approximations for anisotropic signal processing," *Proc. EUSIPCO 2004*, pp. 101–104, 2004.
112. Katkovnik, V., A. Gershman, and L.J. Stankovic, "Sensor array signal tracking using a data-driven window approach," *Signal Processing*, vol. 80, no. 12, pp. 1507–2515, 2000.
113. Katkovnik, V. and L.J. Stankovic, "Periodogram with varying and data-driven window length," *Signal Processing*, vol. 67, no. 3, pp. 345–358, 1998.
114. Kassam, S.A. and H.V. Poor, "Robust techniques for signal processing: a survey," *Proc. IEEE*, vol. 73, no. 3, pp. 433–481, 1985.
115. Kaufman, A. *Volume visualization.* IEEE Computer Society Press, Los Alamos, CA, 1991.
116. Kaufman, A., D. Cohen, and R. Yagel. "Volumetric shading techniques" in *Volume visualization*, Kaufman, A. (ed.), IEEE Computer Society Press, Los Alamos, CA, 1991, Chap. 4, pp. 169–173.
117. Kay, S.M. *Modern spectral estimation, theory and applications.* Prentice Hall, Englewood Cliffs, NJ, 1988.
118. Kerkyacharian, G., O. Lepski, and D. Picard, "Nonlinear estimation in anisotropic multi-index denoising," *Probability Theory and Related Fields*, vol. 121, no. 2, pp. 137–170, 2001.
119. Kerkyacharian, G., O. Lepski, and D. Picard, "Nonlinear estimation in anisotropic multi-index denoising II," Technical Report N810, pp. 1–21, Department of Statistics, University Paris VI, March, 2003.
120. Kim, H.Y. and R. de Queiroz, "Inverse halftoning by decision tree learning," *Proc. IEEE Int. Conf. on Image Processing, ICIP*, Barcelona, Spain, 2003.

121. Kingsbury, N., "Complex wavelets for shift invariant analysis and filtering of signals," *Appl. Computat. Harmonic Anal.*, vol. 10, no. 3, pp. 234–253, 2001.
122. Kite, T.D., B.L. Evans, T.L. Sculley, and A.C. Bovik, "Digital image halftoning as 2-D delta-sigma modulation," *Proc. IEEE Int. Conf. Image Processing*, vol. I, pp. 799–802, 1997.
123. Kite, T.D., B.L. Evans, A.C. Bovik, and T.L. Sculley, "Modeling and quality assessment of halftoning by error diffusion," *IEEE Trans. Image Proc.*, vol. 9, pp. 909–922, 2000.
124. Kite, T.D., N. Damera-Venkata, B.L. Evans, and A.C. Bovik, "A fast, high-quality inverse halftoning algorithm for error diffused halftones," *IEEE Trans. Image Processing*, vol. 9, no. 9, pp. 1583–1592, 2000.
125. Kite, T.D., B.L. Evans, and A.C. Bovik, "Fast rehalftoning and interpolated halftoning algorithms with flat low-frequency response," *Proc. Inter. Conf. Image Processing, 1999*, vol. 3, pp. 602–603, Kobe, Japan, 1999.
126. Klemelä, J. and A. Tsybakov, "Sharp adaptive estimation of linear functionals," *Ann. Statistics*, vol. 29, no. 6, pp. 1537–1566, 2001.
127. Korostelev, A. and A. Tsybakov. *Minimax theory of image reconstruction.* Springer-Verlag, New York–Heidelberg–Berlin, 1993.
128. Kuan, D.T., A.A. Sawchuk, T.C. Strand, and P. Chavel, "Adaptive noise smoothing filter for images with signal dependent noise," *IEEE Trans. Pattern Anal. Mach. Intell.*, vol. 7, pp. 165–177, 1985.
129. Lagendijk, R.L. and J. Biemond. *Iterative identification and restoration of images.* Kluwer, Boston, MA, 1991.
130. Lancaster, P. and K. Saulcauskas, "Surfaces generated by moving least squares methods," *Math. Computation*, vol. 37, no. 155, pp. 141–158, 1981.
131. Lange, K. and R. Carson, "*EM* reconstruction algorithms for emission and transmission tomography," *J. Computer Assisted Tomography*, vol. 8, no. 2, pp. 306–316, 1984.
132. Lange, K., "Convergence of *EM* image reconstruction algorithm with Gibbs smoothing," *IEEE Trans. Medical Imaging*, vol. 9, no. 12, pp. 439–446, 1990.
133. Lee, J.S., "Speckle analysis and smoothing of synthetic aperture radar images," *Comput. Graphics Image Process.*, vol. 17, pp. 24–32, 1981.
134. Lee, J.S., "Digital image enhancement and noise filtering by use of local statistics," *IEEE Trans. Pattern Anal. Mach. Intell.*, vol. 2, pp. 165–168, 1980.
135. Lepskii, O., "On one problem of adaptive estimation in Gaussian white noise," *Theory Probab. Appl.*, vol. 35, no. 3, pp. 454–466, 1990.
136. Lepskii, O., "Asymptotically minimax adaptive estimation I: Upper bounds. Optimally adaptive estimates," *Theory Probab. Appl.*, vol. 36, pp. 682–697, 1991.
137. Lepski, O. and V. Spokoiny, "Optimal pointwise adaptive methods in nonparametric estimation," *Ann. Statistics*, vol. 25, no. 6, pp. 2512–2546, 1997.
138. Lepski, O., E. Mammen, and V. Spokoiny, "Ideal spatial adaptation to inhomogeneous smoothness: an approach based on kernel estimates with variable bandwidth selection," *Ann. Statistics*, vol. 25, no. 3, pp. 929–947, 1997.
139. Lewitt, R.M. and S. Matej, "Overview of methods for image reconstruction from projections in emission computed tomography," *Proc. IEEE*, vol. 91, no. 10, pp. 1588–1611, 2003.

140. Li, S. and W.K. Liu, "Reproducing kernel hierarchical partition of unity, Part I—Formulation and theory," *Int. J. Numerical Methods Engineering*, vol. 45, pp. 289–317, 1999.
141. Li, S. and W.K. Liu, "Reproducing kernel hierarchical partition of unity, Part II—Applications," *Int. J. Numerical Methods Engineering*, vol. 45, pp. 251–288, 1999.
142. Lin, H.M. and A.N. Wilson, Jr., "Median filters with adaptive length," *IEEE Trans. Circuits Syst.*, vol. 35, no. 6, pp. 675–690, 1988.
143. Lin, H.M. and A.N. Wilson, Jr., "Median filters with adaptive length," *IEEE Trans. Circuits and Systems*, vol. 35, no. 6, pp. 675–690, 1988.
144. Liu, J. and P. Moulin, "Complexity-regularized image restoration," *Proc. IEEE Int. Conf. Imag. Proc.—ICIP'1998,* Chicago, IL, vol. 1, pp. 555–559, 1998.
145. Loader, C. *Local regression and likelihood.* Series in Statistics and Computing, Springer-Verlag, New York, 1999.
146. Lu, H., Y. Kim, and J.M.M. Anderson, "Improved Poisson intensity estimation: denoising application using Poisson data," *IEEE Trans. Image Processing,* vol. 13, no. 8, pp. 1128–1135, 2004.
147. Lucy, L.B., "An iterative technique for the rectification of observed distributions," *Astron. J.* vol. 79, pp. 745–754, 1974.
148. Madych, W.R., "Radon's inverse formulas," *Trans. Ameri. Math. Soc.*, January, pp. 1–17, 2004.
149. Madych, W.R., "Tomography, approximate reconstruction, and continuous wavelet transform," *Appl. Computat. Harmonic Analysis*, vol. 7, pp. 54–100, 1999.
150. Mallat, S., *A wavelet tour of signal processing*, Academic Press, New York, 1999.
151. Mallows, C.L. "Some comments on C_p," *Technometrics*, vol. 15, pp. 661–675, 1973.
152. McCullagh, P. and J.A. Nelder. *Generalized linear models.* 2nd ed., Chapman and Hall, London, 1999.
153. McLachlan, G.J. and T. Krishnana, *The EM algorithm and its extensions.* John Wiley and Sons, New York, 1997.
154. McNally, J., T. Karpova, J. Cooper, and J. Conchello, "Three-dimensional imaging by deconvolution microscopy," *Methods*, vol. 19, pp. 373–385, 1999.
155. Mercurio, D. and V. Spokoiny, "Estimation of time-dependent volatility via local change point analysis," WIAS, preprint 904, 2004.
156. Meşe, M. and P.P. Vaidyanathan, "Look up table (LUT) method for inverse halftoning," *IEEE Trans. Image Processing*, vol. 10, no. 10, pp. 1566–1578, 2000.
157. Meşe, M. and P.P. Vaidyanathan, "Tree-structured method for LUT inverse halftoning and for image halftoning," *IEEE Trans. Image Processing*, vol. 11, no. 6, pp. 644–655, 2002.
158. Nadaraya, E.A., "On estimating regression," *Theory Prob. Appl.*, vol. 9, pp. 141–142, 1964.
159. Nayar, F.E., K. Ikeuchi, and T. Kanade. "Surface reflection: physical and geometrical perspectives," *IEEE Trans. Pattern Anal. Mach. Intell.*, vol. 13, no. 7, pp. 611–634, 1991.
160. Neelamani, R., R. Nowak, and R. Baraniuk, "Model based inverse halftoning with wavelet vaguelette deconvolution," *Proc. IEEE Int. Conf. Image Processing, ICIP'00*, Vancouver, Canada, vol. 3, pp. 973–976, 2000.
161. Neelamani, R., H. Choi, and R.G. Baraniuk, "ForWaRD: Fourier-wavelet regularized deconvolution for ill-conditioned systems," *IEEE Trans. Image Processing*, vol. 52, no. 2, pp. 418–433, 2004.

162. Neelamani, R., R. Nowak, and R. Baraniuk, "WInHD: Wavelet-based inverse halftoning via deconvolution," Preprint, Department of ECE, Rice University, Houston, TX, USA, pp. 1–22, 2003.
163. Nemirovski, A., *Topics in non-parametric statistics*, Lecture notes in mathematics, 1738, Springer-Verlag, New York, pp. 85–277, 2000.
164. Neter, J., M.H. Kutner, C.J. Nachtsheim, and W. Wasserman, *Applied linear regression models*, 4th ed., McGraw-Hill, Irwin, 2003.
165. Ng, M.K. and N.K. Bose, "Mathematical analysis of super-resolution methodology," *IEEE Signal Processing Mag.*, vol. 20, no. 3, pp. 62–74, 2003.
166. Nikias, C.L. and M. Shao, *Signal processing with alpha-stable distributions and applications*. John Wiley and Sons, New York, 1995.
167. Nowak, R.D. and R.G. Baraniuk, "Wavelet-domain filtering for photon imaging systems," *IEEE Trans. Image Processing,* vol. 8, no. 5, pp. 666–678, 1999.
168. Nowak, R.D. and E.D. Kolaszyk, "A statistical multiscale framework for Poisson inverse problem," *IEEE Trans. Inform. Theory,* vol. 46, no. 5, pp. 1811–1825, 2000.
169. Nuyts, J. and J. Fessler, "A penalized -likelihood image reconstruction method for emission tomography, compared to post-smoothed maximum-likelihood with matched spatial resolution," *IEEE Trans. Medical Imaging*, vol. 22, no. 9, pp. 1042–1052, 2003.
170. Ollinger, J.M. and J.A. Fessler, "Positron-emission tomography," *IEEE Signal Processing Mag.*, vol. 14, no. 1, pp. 43–55, 1997.
171. Oppenheim, A.V., R.W. Schafer, and J.R. Buck. *Discrete-time signal processing*. 2nd ed., Prentice Hall, Englewood Cliffs. NJ, 1999.
172. O'Sullivan, F., "A statistical perspective on ill-posed inverse problems," *Statist. Sci.*, vol. 1, pp. 502–527, 1986.
173. Papoulis, A. *The Fourier integral and its applications.* McGraw-Hill, New York, 1987.
174. Papoulis, A. *Probability and statistics*. Prentice-Hall, Englewood Cliffs, NJ, 1990.
175. Parzen, E., "On estimation of a probability density function and mode," *Ann. Statist.*, vol. 81, pp. 215–222, 1962.
176. Pawley, J.B. (ed.). *Handbook of biological confocal microscopy*. Plenum, New York, 1995.
177. Park, Suog Cheol, Min Kyu Park, and Moon Gi Kang, "Super-resolution image reconstruction: a technical overview," *IEEE Signal Processing Mag.*, vol. 20, no. 3, pp. 21-36, 2003.
178. Perona, P., "Deformable kernels for early vision," *IEEE Trans. Pattern Anal. Machine Intell.*, vol. 17, no. 5, pp. 488–499, 1995.
179. Petersen, I.F., "Statistical optimization by smoothing," *Engineer. Cybernet.*, N2, 1969 (in Russian).
180. Polzehl, J. and V. Spokoiny, "Adaptive weights smoothing with applications to image restoration," *J. Royal Statist. Soc.*, Series B, vol. 62, pp. 335–354, 2000.
181. Polzehl, J. and V. Spokoiny. "Local likelihood modeling by adaptive weights smoothing," WIAS, preprint 787, 2002.
182. Polzehl, J. and V. Spokoiny, "Image denoising: pointwise adaptive approach," *Ann. Statistics*, vol. 31, no. 1, pp. 30–57, 2003.
183. Polzehl, J. and V. Spokoiny. "Varying coefficient regression modeling," WIAS, preprint 818, 2003.
184. Polyak, B.T. *Introduction to optimization*. Optimization Software, Inc., New York: Publications Division, 1987.

185. Polyak, B.T. and J.Z. Tsypkin, "Robust identification," *Automatica*, vol. 16, pp. 53–63, 1980.
186. Pomalaza-Raez, C.A. and C.D. McGillem, "An adaptive, nonlinear edge preserving filter," *IEEE Trans. Acoust. Speech Signal Processing*, vol. 32, pp. 571–576, 1984.
187. Pratt, W.K., *Digital image processing*. John Wiley and Sons, New York, 1978.
188. Preza, C. and J. Conchello, "Depth-variant maximum likelihood restoration for three-dimensional fluorescence microscopy," *J. Opti. Soc. Amer.*, A, vol. 21, no. 9, pp. 1593–1601, 2004.
189. Priestley, M.B., *Spectral analysis and time series*. Academic Press, London, 1989.
190. Qiu, P., "The local piecewise linear kernel smoothing procedure for fitting jump regression surfaces," *Technometrics*, vol. 46, no. 1, pp. 87–98, 2004.
191. Ripley, B., *Statistical inference for spatial processes.* Cambridge University Press, Cambridge, 1988.
192. Rice, J., "Bandwidth selection for nonparametric regression," *Ann. Statistics*, vol. 12, pp. 1215–1230, 1984.
193. Richardson, W.H., "Bayesian-based iterative method of image restoration," *J. Opti. Soci. Amer.*, vol. 62, pp. 55–59, 1972.
194. Rousseeuw, P.J. and A.M. Leroy. *Robust regression and outlier detection*. John Wiley and Sons, New York, 1987.
195. Ruppert, D., "Empirical-bias bandwidths for local polynomial nonparametric regression and density estimation," *J. Amer. Statist. Associ.*, vol. 92, no. 439, pp. 1049–1062, 1997.
196. Savitzky, A. and M.J.E. Golay, "Smoothing and differentiation of data by simplified least squares procedures," *Analyt. Chem.*, vol. 36, pp. 1627–1639, 1964.
197. Seber, G.A. and C.J. Wild. *Nonlinear regression.* Wiley, New York, 1989.
198. Schucany, W.R., "Adaptive bandwidth choice for kernel regression," *J. Ameri. Statist. Assoc.*, vol. 90, pp. 535–540, 1995.
199. Simonoff, J.S. *Smoothing methods in statistics.* Springer-Verlag, New York, 1998.
200. Sicuranza, G. and S. Mitra (eds.). *Nonlinear image processing*. Academic Press, New York, 2000.
201. Simoncelli, E.P. and H. Farid, "Steerable wedge filters for local analysis," *IEEE Trans. Image Processing,* vol. 5, pp. 1377–1382, 1996.
202. Simoncelli, E.P. and W.T. Freeman, "The steerable pyramid: a flexible architecture for multi-scale derivative computation," *Second IEEE Int. Conf. Image Processing*, Washington, DC, 1995.
203. Shin, B.S and G.S. Shin, "Fast normal estimation using surface characteristics," *Proc. 6th IEEE Visualization Conf. (Visualization '95)*, 1995.
204. Shevlyakov, G.L. and N.O. Vilchevski. *Robustness in data analysis: criteria and methods.* Modern Probability and Statistics Series 1, Brill Academic Publishers, Leiden, Boston, Tokyo, 2006.
205. Shmulevich, I. and G.R. Arce, "Spectral design of weighted median filters admitting negative weights," *IEEE Trans. Signal Proc. Lett.*, vol. 8, no. 12, pp. 313–316, 2001.
206. Sorenson, J.A. and M.E. Phelps. *Physics in nuclear medicine*. Grune & Stratton, New York, 1980.
207. Snyder, D.L., A.M. Hammound, and R.L. White, "Image recovery from data acquired with a charge-couple-device camera," *J. Optical Soci. Amer.*, A., vol. 10, no. 5, pp. 1014–1023, 1993.

208. Spokoiny, V., "Estimation of a function with discontinuities via local polynomial fit with an adaptive window choice," *Ann. Statistics*, vol. 26, no. 4, pp. 1356–1378, 1998.
209. Stanković, L., "Performance analysis of the adaptive algorithm for bias-to-variance trade-off. " *IEEE Trans. Signal Processing*, vol. 52, no. 5, pp. 1228–1234, 2004.
210. Starck, J.L., E.J. Candes, and D.L. Donoho, "The curvelet transform for image denoising," *IEEE Trans. Image Processing*, vol. 11, no. 6, pp. 670–684, 2002.
211. Starck, J.L. and E. Pantin, "Deconvolution in astronomy: a review," *Publ. Astronom. Soc. Pacific*, vol. 114, pp. 1051–1062, 2002.
212. Stein, C.M., "Estimation of the mean of a multivariate normal distribution," *Ann. Statist.*, vol. 9, pp. 1135–1151, 1981.
213. Stiller, C. and J. Konrad, "Estimating motion in image sequences," *IEEE Signal Processing Mag.*, vol. 16, pp. 70–91, 1999.
214. Stone, C.J., "Cross-validatiry choice and assessment of statistical prediction (with discussion)," *J. Royal Statist. Soc.*, Ser. B, vol. 36, pp. 11–147, 1974.
215. Stone, C.J., "Consistent nonparametric regression (with discussion)," *Ann. Statistics*, vol. 5, no 4, pp. 595–645, 1977.
216. Strang, G. and T. Nguyen. *Wavelets and filter banks*. Wellesley Cambridge Press, Boston, 1996.
217. Stewart, C.V., "Robust parameter estimation in computer vision," *SIAM Rev.*, vol. 41, no. 3, pp. 513–537, 1999.
218. Suoranta, R. and K. Estola, "Robust median filter with adaptive window length," *IEEE Inter. Symp. Circuits and Systems*, vol. 1, pp. 108–111, 1991.
219. Tibshirani, R.J. and T.J. Hastie, "Local likelihood estimation," *J. Amer. Statist. Assoc.*, vol. 82, pp. 559–567, 1987.
220. Tikhonov, A.N. and V.Y. Arsenin. *Solution of ill-posed problems*. John Wiley and Sons, New York, 1977.
221. Timmermann, K.E. and R. Nowak, "Multiscale modeling and estimation of Poisson processes with application to photon-limited imaging," *IEEE Trans. Inform. Theory*, vol. 45, no. 3, pp. 846–862, 1999.
222. Toft, P. "The Radon transform. Theory and applications." Ph.D. Thesis, Department of Mathematical Modelling, Technical University of Denmark, 1996.
223. Tsybakov, A., "Robust reconstruction of functions by the local-approximation methods," *Problems Informat. Transmission*, vol. 22, pp. 133–146, 1986.
224. Ulichney, R. *Digital halftoning*. MIT Press, Cambridge, MA, 1987.
225. Unal, G.B. and A.E. Cetin, "Restoration of error-diffused images using projection onto convex sets," *IEEE Trans. Image Processing*, vol. 10, no. 12, pp. 1836–1841, 2001.
226. Vardi, Y., A. Shepp, and L. Kauffman, "A statistical model for positron emission tomography," *J. Amer. Statist. Assoc.*, vol. 80, no. 389, pp. 8–20, 1985.
227. Vetterli, M. and J. Kovačević, *Wavelets and subband coding*. Prentice Hall, Englewood Cliffs, NJ, 1995.
228. Wahba, G., *Spline models for observational data*. CBMS-NSF Regional Conference Series in Applied Mathematics 59, SIAM, Philadelphia, 1990.
229. Walker, J.S. *Fourier analysis*. Oxford University Press, New York, 1988.
230. Watson, G.S., "Smooth regression analysis," *Sankhya*, Ser. A, vol. 26, pp. 359–372, 1964.

231. Wang, Z. and A.C. Bovik, "A universal image quality index," *IEEE Signal Processing Lett.*, vol. 9, no. N3, pp. 81–84, 2002.
232. Wedderburn, R.W.M., "Quasilikelihood functions, generalized linear models and the Gauss-Newton method," *Biometrica*, vol. 61, pp. 439–447, 1974.
233. Wendt, P.D., E.J. Coyle, and N.C. Gallagher, "Stack filters," *IEEE Trans. Acoust., Speech, Signal Processing*, vol. ASSP-34, pp. 898–919, 1986.
234. Willett, R.M. and R.D. Nowak, "Platelets: a multiscale approach for recovering edges and surfaces in photon-limited medical imaging," *IEEE Trans. Medical Imaging*, vol. 22, no. 3, pp. 332–350, 2003.
235. Wong, P.W., "Inverse halftoning and kernel estimation for error diffusion," *IEEE Trans. Image Processing*, vol. 4, no. 4, pp. 486–498, 1995.
236. Wong, P.W. and N.D. Memon, "Image processing for halftones," *IEEE Signal Processing Mag.*, vol. 20, no. 4, pp. 59–70, 2003.
237. Wright, G.A., "Magnetic resonance imaging," *IEEE Signal Processing Mag.*, vol. 14, no. 1, pp. 56–66, 1997.
238. Xiong, Z., M.T. Orchard, and K. Ramchandran, "Inverse halftoning using wavelets," *IEEE Trans. Signal Processing*, vol. 8, pp. 1479–1482, 1999.
239. Yagel, R., D. Cohen, and A. Kaufman, "Normal estimation in 3D discrete space," *The Visual Computer*, vol. 8, no. 5–6, pp. 278–291, 1992.
240. Yang, R., L. Yin, M. Gabbouj, J. Astola, and Y. Neuvo, "Optimal weighted median filters under structural constraints," *IEEE Trans. Signal Processing*, vol. 43, pp. 591–604, 1995.

Index

Authors' Biographies

Vladimir Katkovnik earned his M.Sc., Ph.D. and Doctor of Sc. degrees from the Leningrad Polytechnic Institute in Russia in 1960, 1964, and 1974, respectively. He was formerly at the Department of Mechanics and Control Processes of this Institute from 1964–1991, at the Statistics Department of the University of South Africa, Pretoria, from 1991–1999, and in the Mechatronics Department of Kwangju Institute of Science and Technology in South Korea from 2001–2003. Since 2003 he has been a Research Professor of the Signal Processing Institute at Tampere University of Technology in Finland. His research interests include stochastic signal processing, linear and nonlinear filtering, nonparametric estimation, imaging, nonstationary systems, and time-frequency analysis. He has published seven books and about 200 papers.

Karen Egiazarian is a Senior Member of IEEE. He earned his M.Sc. from Yerevan State University in Armenia in 1981, his Ph.D. from Moscow Lomonosov's State University, Russia, in 1986, and his D. Tech. from Tampere University of Technology, Finland, in 1994. He has formerly been at the Signal Processing Laboratory, Institute of Problem of Informatics and Automation at the National Academy of Sciences of Armenia from 1981–1993, and with the Signal Processing Laboratory, Tampere University of Technology, from 1993. Since 1999, he has been a Professor of the Institute of Signal Processing at Tampere University of Technology. Prof. Egiazarian's research interests include signal and image processing, applied mathematics, and digital logic. He has published one book, three book chapters and about 350 articles.

Jaakko Astola is a Fellow of IEEE. He earned his B.Sc., M.Sc., Licentiate, and Ph.D. from Turku University in Finland in 1972, 1973, 1975, and 1978, respectively. He was formerly at the Institute for Mathematical Sciences of Kyoto University in Japan from 1976–1977, at the Department of Information Technology at the Lappeenranta University of Technology in Finland from 1979–1987, at the Eindhoven University of Technology of The Netherlands in 1984, and at the Department of Applied Mathematics at Tampere University, Finland, from 1987–1992. Since 1993 he has been a Professor of Signal Processing and the Director of the Tampere International Center for Signal Processing of Tampere University of Technology. His research interests include signal processing, coding theory, spectral techniques, and statistics. He has published five books and about 500 papers.